F.M. Faranda • L. Guglielmo • G. Spezie (Eds.)

Mediterranean Ecosystems: Structures and Processes

Springer
Milano
Berlin
Heidelberg
New York
Barcelona
Hong Kong
London
Paris
Singapore
Tokyo

F.M. Faranda • L. Guglielmo • G. Spezie (Eds.)

Mediterranean Ecosystems

Structures and Processes

Professor Dr. Francesco Maria Faranda
Università di Genova
Istituto di Scienze Ambientali Marine
Viale Benedetto XV, 5
I-16132 Genova
Italy

Professor Dr. Letterio Guglielmo
Università di Messina
Dipartimento di Biologia ed Ecologia Marina
Salita Sperone, 31
I-98166 Messina-S. Agata
Italy

Professor Dr. Giancarlo Spezie
Istituto Universitario Navale
Istituto di Meteorologia e Oceanografia
Via Amm. Acton, 38
I-80133 Napoli
Italy

Cover figure:
Mediterranean Basin: Sea Surface Temperature & Vegetation with ATSR-2

European Space Agency – Esa
Produced from ESA remote sensing data
Image processed by: ESA/ERSIN/EURISY, Frascati
®ESA/CCLRC/RAL/NERC/BNSC 1997

Springer-Verlag Italia
a member of BertelsmannSpringer Science+Business Media GmbH

ISBN 88-470-0114-5

Library of Congress Cataloging-in-Publication Data: applied for

Typesetting: Photo Life, Vimodrone (Milano)
Printing and binding: Staroffset, Cernusco S/N (Milano)
Cover design: Simona Colombo

Printed in Italy

SPIN: 10781852

Acknowledgements

The papers accepted for publication in this volume were reviewed by an international board of eminent scientists. The Editors wish to thank the reviewers for their assistance and cooperation, which has greatly facilitated the relatively prompt publication of the volume. Many thanks are due to all the participants of the Ischia Workshop and to the invited speakers Profs. Battaglia, Hopkins and Ricci. The Editors are also grateful to Dr. Antonio Capone for secretarial assistance and to Mr. Vincenzo Bonanzinga, Dr. Giovanni Brancato and Mr. Nicola Donato for their assistance in preparing the graphics and in elaborating the texts and indexes. This volume was financially supported by Consorzio Nazionale Interuniversitario per le Scienze del Mare (CoNISMa).

Foreword

This new book, as described in the Preface, presents a collection of contributions by several participants of the First National Congress of Marine Sciences organized by the CoNISMa.

The 62 selected papers, grouped into seven sections, deal with a number of significant and current topics of marine research conducted at Mediterranean stations mostly situated along the Italian coast. The focus of each section is clearly expressed in the headings.

The research sites may be conceived as "key" or focal points in the study of marine ecosystems. Indeed, the information collected transcends regional interests by promoting better understanding of the structures and processes characterizing a wide area of the Mediterranean basin.

The reader may easily note that in the entire book special attention is devoted to the observed changes and the ensuing diversity in time and space. Temporal diversity ranges from paleontological scales to shorter time scales. The latter is the case, for instance, with chemical species, circulation patterns, energy fluxes linking particulate matter with microbial communities, structure of food webs, bacterial, plant and animal species, etc.

As to the disciplinary aspects, the majority of papers reflect a modern approach, in both conceptual and methodological terms, and a marked tendency to overcome the boundaries of specific disciplines. This applies particularly to the evaluation of intraspecific biodiversity, where methods borrowed from molecular biology are largely utilized.

A multidisciplinary approach, with a large contribution from physiology and biochemistry, is also adopted in those studies aiming to evaluate environmental quality and to search for causes and mechanisms responsible for deleterious biological effects. This kind of approach may find practical applications, for instance, in the rational management of coastal wetlands.

Apart from the valuable scientific results presented at the meeting, a novelty to be stressed is the role played by the consortium. Far from being confined to providing space to scientific debates, other important goals of the CoNISMa are the identification of new themes, relevant also from the viewpoint of possible applications; the organization and coordination of multidisciplinary research programs; and the appreciable task of circulating the product in the national and international scientific community.

Prof. Bruno Battaglia
Biology Department
University of Padova
Padova (Italy)

Preface

The papers published in this volume reflect the present knowledge of the Mediterranean gained by the Italian scientific community. These contributions, even though quite representative, do not completely exhaust the range of activities conducted in the framework of Italian marine research. However, they show an extremely broad range of these activities, which practically encompasses all fields of present interest of the marine sciences. In addition, these works show how these different aspects find their synthesis in research characterized by strong multidisciplinarity.

These papers were selected from the communications presented at the First National Congress of Marine Sciences, held on the Island of Ischia, off Naples, 11-14 November 1998, organized by CoNISMa in cooperation with the Associazione Italiana di Oceanografia e Limnologia (AIOL), the Società Italiana di Biologia Marina (SIBM) and the Società Italiana di Ecologia (SItE). The Italian National Marine Science Consortium (CoNISMa), which groups together 24 Italian universities, was founded in 1994 to promote and coordinate research activities in the various disciplines of marine science. The sea, if understood as an element of the national territory, obviously calls for choices in terms of management and government, which need to be supported by constantly updated scientific knowledge.

One of the tasks of CoNISMa, according to its statute, is to hold a National Congress every 2 years aimed at providing the scientific community with an opportunity to spread and update its knowledge in the marine sciences. The works collected in this volume, after being refereed by senior scientists from various fields of marine science, have been grouped together according to cultural areas.

We decided to follow this criterion, rather than a geographical one, or one which favoured the environmental characterization of the studied areas, to be consistent with the title of the volume, thus reflecting the approach that we believe most effective as a guide for the reader.

However, it is worth emphasizing that the multidisciplinary character and content of these papers leads to different "transversal" reading and clustering possibilities, which the informed reader will be free to design and undertake autonomously.

The editors and CoNISMa hope that this volume will be useful to the marine research community, and that its circulation will increase and strengthen the interest for the scientific, technical and cultural activities aimed at a better understanding of the Mediterranean Sea.

The Editors

Contents

Physical and Chemical Variability of Marine Coastal Environments

Environmental Monitoring

Plankton Ecology

Fish Recruitment and Feeding Ecology

Benthos Ecology

Change and Biodiversity

Sediment Processes and Material Recycling

Contributors

Marine Geology and Morphobathymetry in the Bay of Naples (South-Eastern Tyrrhenian Sea, Italy)

G. Aiello, V. Budillon, G. Cristofalo, B. D'Argenio, G. de Alteriis, M. De Lauro, L. Ferraro, E. Marsella, N. Pelosi, M. Sacchi, and R. Tonielli

ABSTRACT

Recent multibeam bathymetry (Elac, Bottomchart MK2) and high resolution seismics (Subbottom Chirp and 1-4 kJ Sparker source), acquired in the frame of an on-going programme of sea-floor mapping of Naples and Salerno Bays (south-eastern Tyrrhenian margin, Italy) and financed by the National Geological Survey of Italy, allows to put new insights into the recent evolution of the bay. The morphology and stratigraphy of the continental shelf and slope appear strongly controlled by the interplay of volcanism and canyoning that acted along the Magnaghi and Dohrn axes. Detailed bathymetry reveals the complexity of the drainage pattern which consists of a previously unknown, dense network of minor tributary channels. At places, the Dohrn and Magnaghi canyon walls are intensively affected by slope instability, as evidenced by numerous submarine slides and scars involving large volumes of sediments. Previously unreported mound-shaped morphological highs ("Bacarozzi" Facies), Holocene reworked sediments and sea-bottom creep appear on acoustic Chirp profiles in the inner sectors of the bay and seem to be related to volcano-sedimentary processes. On the contrary, sedimentation over the shelf at the southern edge of the bay (Sorrento-Capri) seems less influenced by volcanic activity and seabed features include Late Pleistocene regressive sand bodies and Holocene patch reefs, coastal dunes and depositional terraces.

Introduction

An on-going programme of sea-floor mapping in Naples and Salerno Bays (south-eastern Tyrrhenian margin), is among the main research projects of the Geomare Sud Institute, CNR Naples, Italy. Financed by the National Geological Survey of Italy (CARG project), this research aims at producing, during the years 1998-2000, geological maps of selected marine coastal zones at the 1: 50.000 scale. Recent criteria for geological mapping of Italian marine areas (joint CNR-SGN Committee, see Catalano et al. 1996) have been followed.

For a review of recent geological and geophysical studies over the Naples offshore area, see Latmiral et al. (1971), Finetti and Morelli (1974), Pescatore et al. (1984), Fusi et al. (1991), Milia (1996), Marsella et al. (1996), Aiello et al. (1997a, b), Cinque et al. (1997), De Lauro et al. (1997).

Materials and Methods

Most of the data presented here have been acquired in the frame of three oceanographic cruises organised by the Geomare Sud and by other institutions during the 1996-1998 time span. The first cruise (R/V Urania, GMS97-1; October-November, 1996) was dedicated to the acquisition of a high-resolution multibeam survey of the Bay of Naples (below the -50 m isobath) and of a narrow belt south of the Sorrento Peninsula (Figs. 1, 2). An ELAC Nautik Bottomchart MK2 multibeam system was adopted with a bathymetric range of -50 / -3000 m. The ELAC system is equipped with a transducer of 256 beams arranged to generate a 15 to 150° fan width; the vertical accuracy is about 0.25% of water depth whereas the spatial resolution (the size of the "footprint") ranges from 1 to 5 m. The maximum horizontal coverage is 7.5 times the

Istituto di Ricerca Geomare Sud, CNR, Via A. Vespucci 9, 80142 Napoli, Italy

F.M. Faranda, L. Guglielmo, G. Spezie (eds)
Mediterranean Ecosystems: Structures and Processes

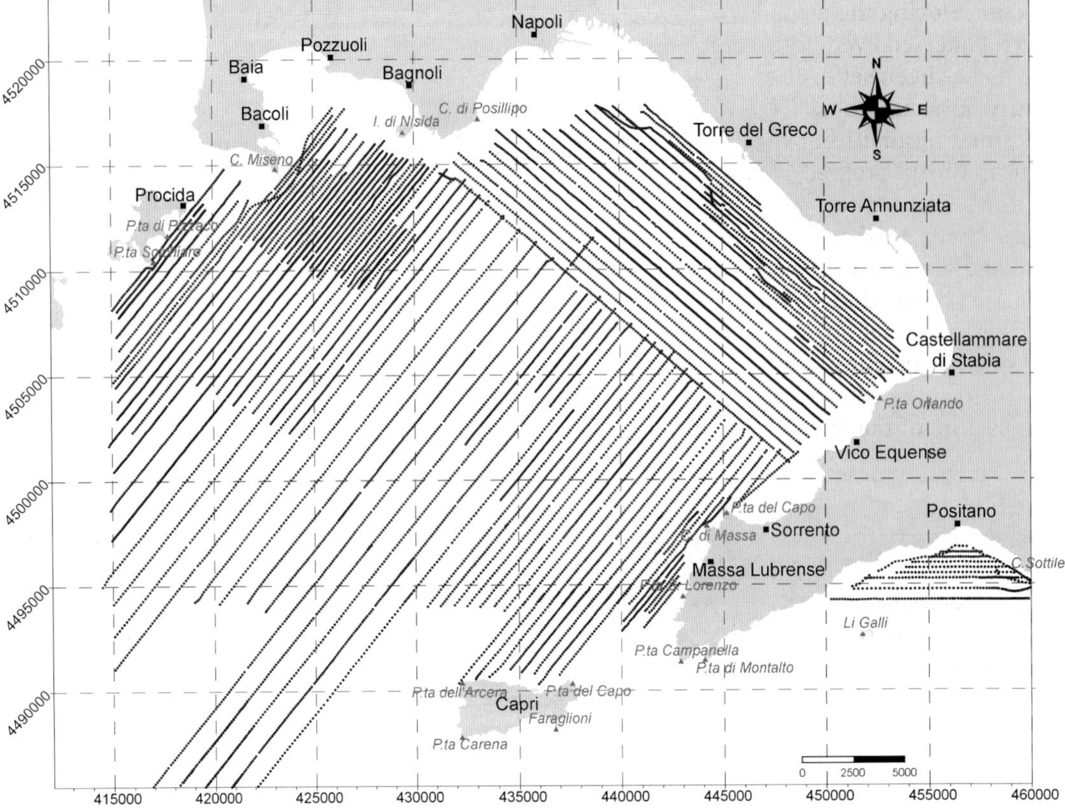

Fig. 1. Navigation lines of Multibeam and Subbottom Chirp survey in the Bay of Naples

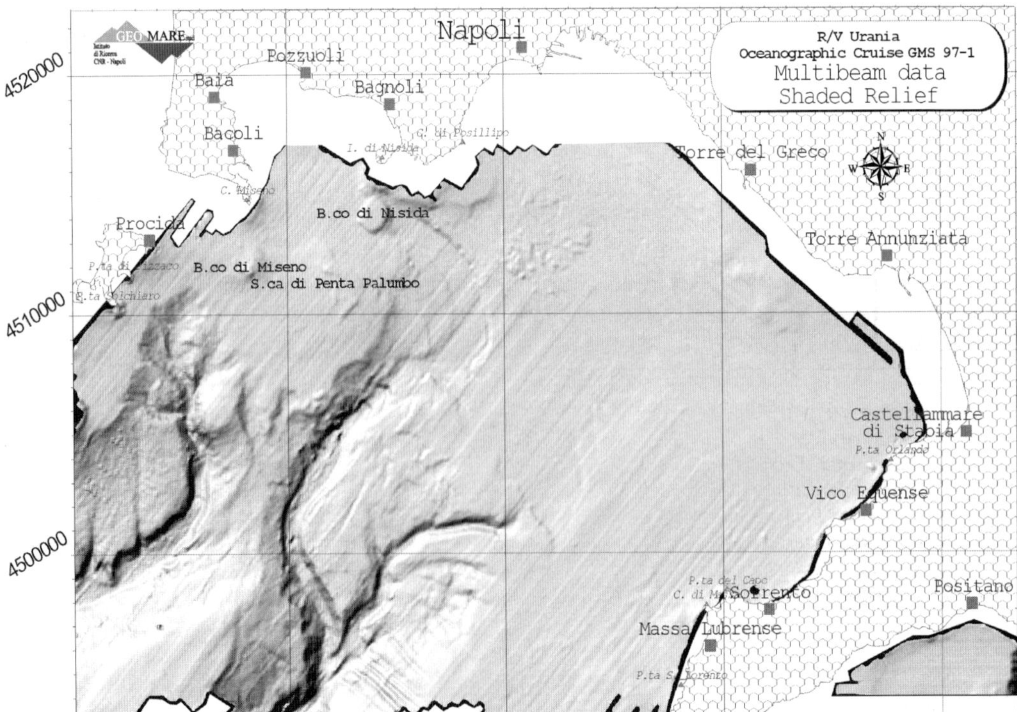

Fig. 2. Shaded relief map of the Bay of Naples based on ELAC Nautik Bottomchart MK2 Multibeam data

water depth. During the acquisition CTD profiles have been collected every 6 hours. The line spacing (Fig. 1) has been set to obtain a variable overlap (from 30%, up to 100%) between swaths. During the acquisition about 1500 km of Subbottom Chirp profiles have been recorded. During the second cruise, (GMS98-1, May 1998), 110 cores (for a total of more than 450 m), 18 dredges and 41 box-cores have been collected over the area coupled to a grid of more than 500 km of single-channel Sparker 1 kJ and 4 kJ and magnetometric profiles. The third survey[1] was chiefly aimed at the detailed sea-floor exploration of some sectors of the area through the deep tow TOBI sonar (Southampton Oceanography Centre, UK) which can provide detailed acoustic imagery of the sea-floor (Chiocci et al. 1998). The analysis of the sampled sediments and of the acoustical/geophysical data acquired during these last two cruises is still in progress.

Results

The continental shelf in the bay has a variable width ranging from 2.5 km (off the western side of Capri island) to about 10-15 km (off the Sorrento coast); offshore the Vesuvian area the

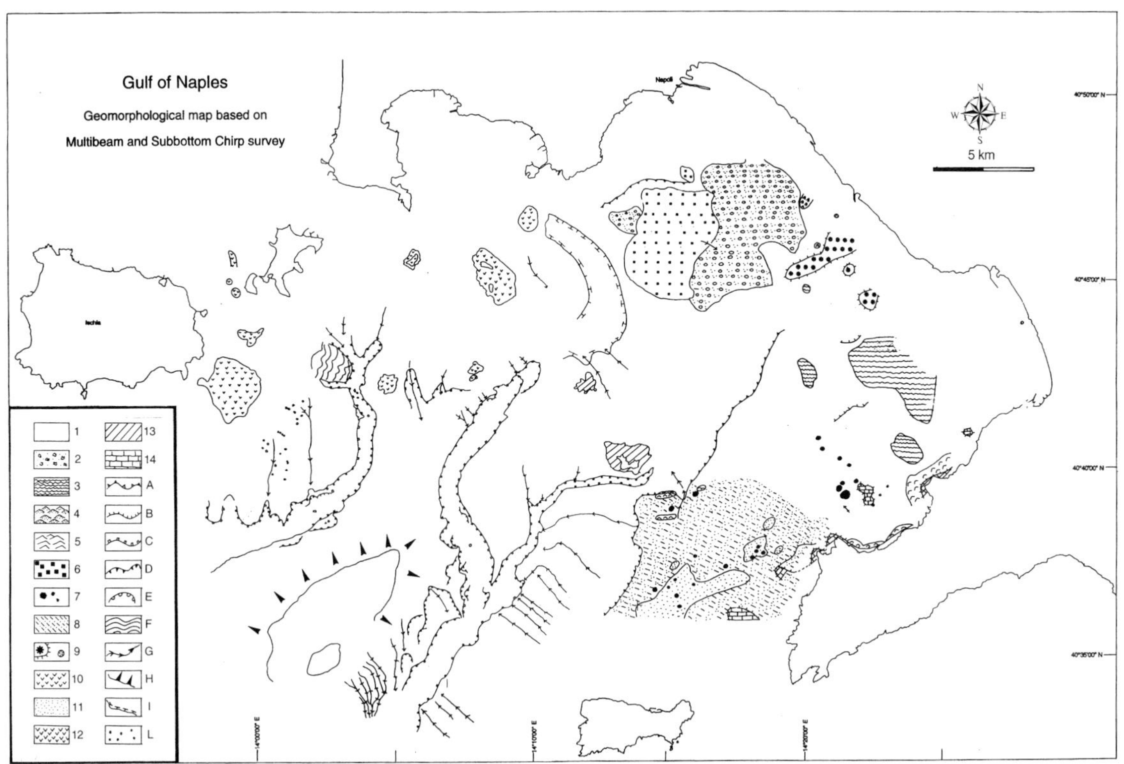

Fig. 3. Geomorphological map of the Bay of Naples. Deposits: *Late Pleistocene-Holocene*: *1* post-glacial drape; *2* reworked sediments (mud flows, lahars); *3* sediments deformed by creep; *4* sand lobes; *5* coastal dunes (buried by post-glacial drape); *6* enigmatic mounds ("Bacarozzi" Facies); *7* algal patch reefs (post-glacial); *8* coastal regressive sands (buried by post-glacial drape); *9* submerged or buried parasitic vents or maars (Somma Vesuvius); *10* volcanic edifices (tuff cones, domes) and their remnants. *Middle-Late Pleistocene*: *11* coastal regressive sand (last glacial); *12* volcanic edifices (tuff cones, domes) and their remnants; *13* remnants of the Middle-Late Pleistocene shelf; *Cretaceous*: *14* outcrops of carbonatic acoustic basement. Landform: *A* shelf break; *B* terrace rim; *C* canyon wall; *D* slide and scar along canyon walls; *E* slump scar; *F* gravity slump; *G* drainage pattern; *H* "Banco di Fuori" slope; *I* "Ammontatura" channel rim; *L* debris flow outrunner blocks

[1] TIVoLI cruise (Tobi use in Italian Volcanic Islands, settember-october 1998) Chief Scientist: Prof. F.L. Chiocci; Centre of Study for the Quaternary and the Environmental Evolution, CNR, Rome - Geomare Sud, CNR, Naples - Istituto Ciencias del Mar, Barcellona, Spain - Southampton Oceanography Centre, UK.

morphological break, which normally separates the continental shelf from the upper slope, is not clearly identified, while offshore the Posillipo hill a morphological break is present near the 90 m isobath. Such a complex physiography is accounted for by the interference of subaerial and submarine volcanism coupled to tectonism and linear erosion operated by the Dohrn and Magnaghi canyons and their tributary channels.

The whole continental shelf is covered by recent (< 10 k years) drape and most of the seabed features have been recognised through seismic interpretation in Holocene sediments (Fig. 3); where the resolution of profiles does not allow the identification of the Holocene drape, the analysis of bottom samples (box-cores) has confirmed the wide occurrence of a thin veneer of recent sediments.

Multibeam data show the occurrence of two non volcanic bathymetric highs located next to the heads of the Dohrn canyon at the centre of the bay. Seismic stratigraphy reveals that they consist of deltaic sediments arranged with clinoform patterns and related to the distal part of a fossil prograding wedge fed by the paleo-Sarno river. These remnants have an elevation with respect to the surrounding average depth that is in the order of 20-30 m. This anomaly can be accounted for by the differential erosion that has occurred during the shelf subaerial exposure related to eustatic lowstand. Seismic evidence allows to infer a pre- "Campanian Ignimbrite" age (e.g. 35 ka) of this erosional phase.

In the Torre del Greco offshore area submerged or buried parasitic vents, genetically related to the Somma-Vesuvius volcano, appear concentrated along a NE-SW trend. A tectonic lineament, showing evidence of normal faulting, has been previously suggested by Finetti and Morelli (1974) in the Somma Vesuvius area, based on the interpretation of multichannel seismic profiles. In the offshore region between Torre del Greco and Napoli the occurrence of a Holocene unit, rapidly thinning seawards and characterised by chaotic reworked sediments, suggests the emplacement of either *lahar*-type deposits due to eruptive episodes or mud flows (Fig. 4). This unit is bordered to the west by mound-shaped highs ("Bacarozzi" Facies), with an acoustically transparent seismic facies (Fig. 5), cropping out at the sea bottom or partially buried

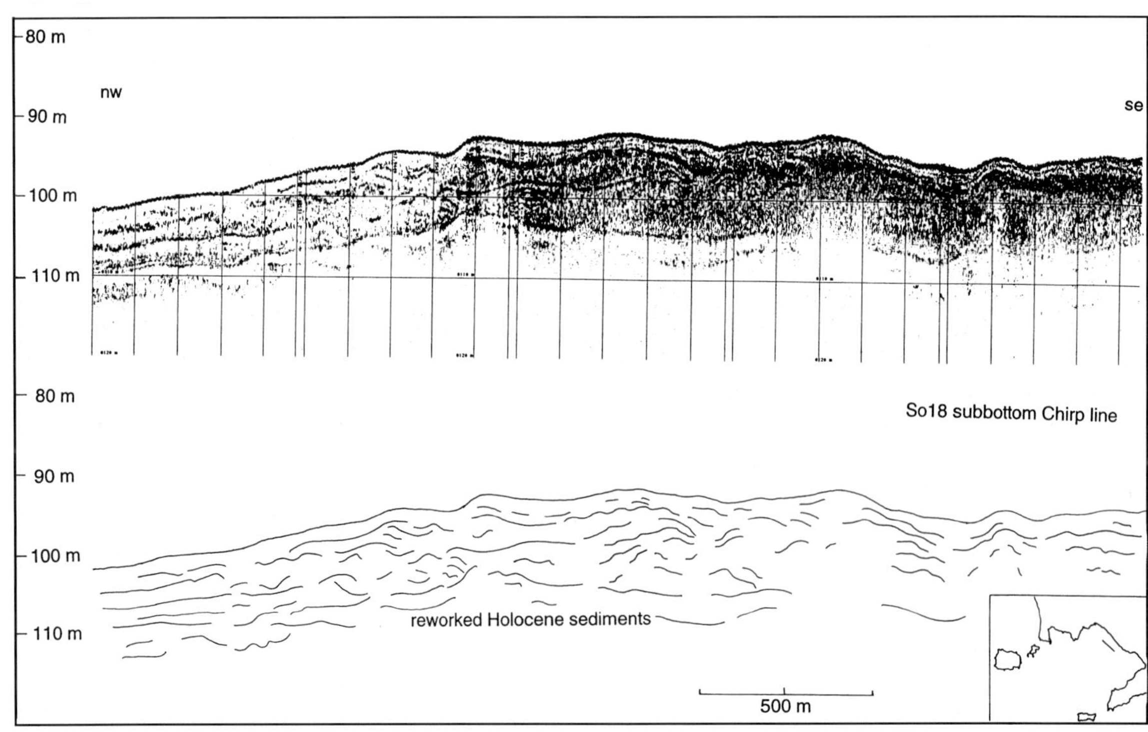

Fig. 4. Subbottom chirp profile showing the partially chaotic reflections interpreted as Holocene reworked sediments, genetically linked to the activity of the Somma Vesuvius strato-volcano (mud flows and *lahars*)

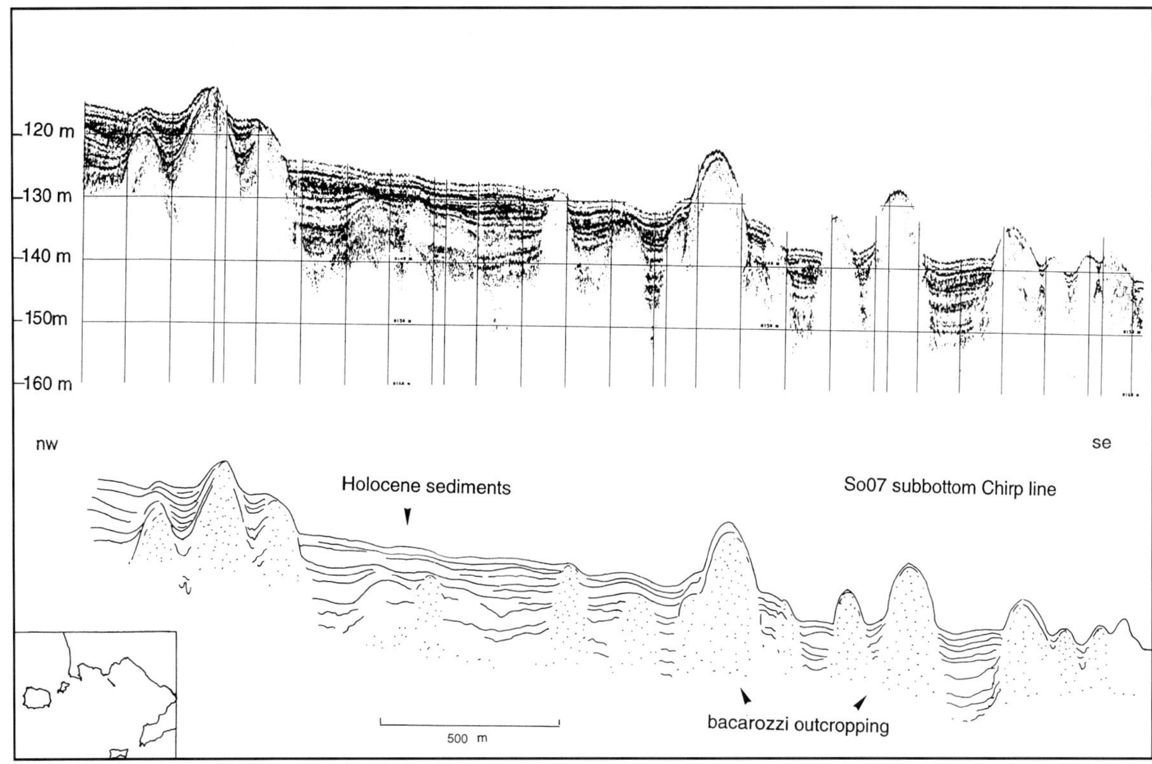

Fig. 5. Subbottom chirp profile showing the enigmatic "Bacarozzi" facies: their lithology consists of shales with intercalated mid to coarse volcanogenic sands and pumice horizons

by sediments. These morphologic highs, never previously mentioned and now under investigation, have been sampled by coring and consist of shaley mounds, with poorly sorted mid to coarse volcanogenic sand and thick pumice levels.

Offshore the Sarno river mouth, at the southeastern corner of the bay, a typical estuarine sedimentation is present, producing a thick Holocene wedge; creep movements at the sea bottom, probably controlled by horizons with high contents of organic matter, involve Holocene sediments above a sharp surface of separation, which may correspond to the last *maximum flooding surface* (Capotondi et al. 1994), formed 5000-6000 years ago (Fig. 6). In the area offshore of Sorrento – Punta Campanella the Holocene wedge tapers towards the SW and the Subbottom Chirp profiles show the occurrence of coastal regressive sands covered by a thin Holocene drape, sometimes cropping out at the sea bottom as relic sandy ridges.

The continental shelf south of the Punta Campanella-Capri alignment (in the Salerno Gulf) has not been affected by significant vertical movements from the 5e isotopic sub-stage (Cinque and Romano 1990) and glacio-eustatic sea level fluctuations represent the main factors controlling the sea-bottom morpho-evolution. On the contrary, subsidence occurred north of this alignment, at the shelf margin, since evidence of subaerial erosion is found at depths of 140-150 m. Algal patch reefs, coastal dunes as well as submerged depositional terraces (Budillon et al. in press), often characterise this area and seem to be related to the last glacio-eustatic sea level rise.

The erosion and the transport of sediments in the bay of Naples are chiefly driven by the Dohrn and Magnaghi canyons, which engrave the continental slope and shelf and expose hundred of metres of a Middle-Late Pleistocene prograding wedge made of clastic and volcanic deposits (Fig. 7). A major morphostructural high ("Banco di Bocca Grande", also called "Banco di Fuori") separates the two canyons and presumably is formed by a Mesozoic carbonate block which resulted from the regional uplift and tilting of the acoustic basement. The width of the

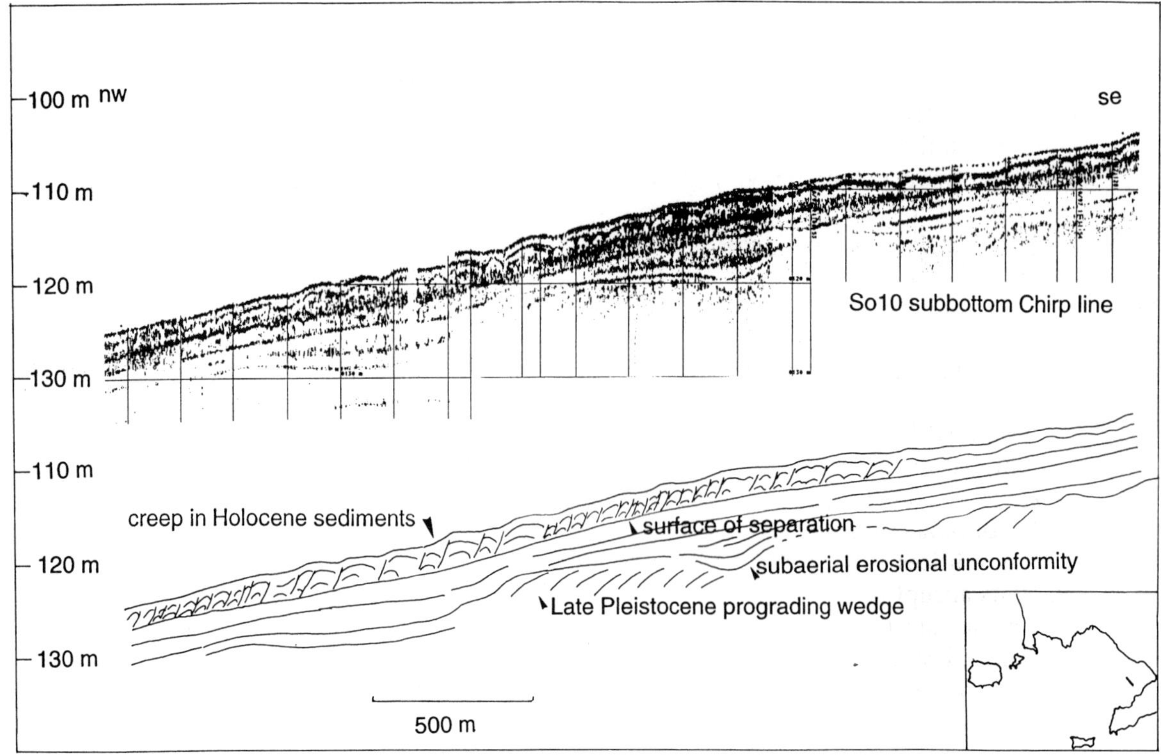

Fig. 6. Subbottom chirp profile showing creep in Holocene sediments off the Sarno river mouth. Creep movements involve the sediments above a surface of separation, interpreted as a maximum flooding surface

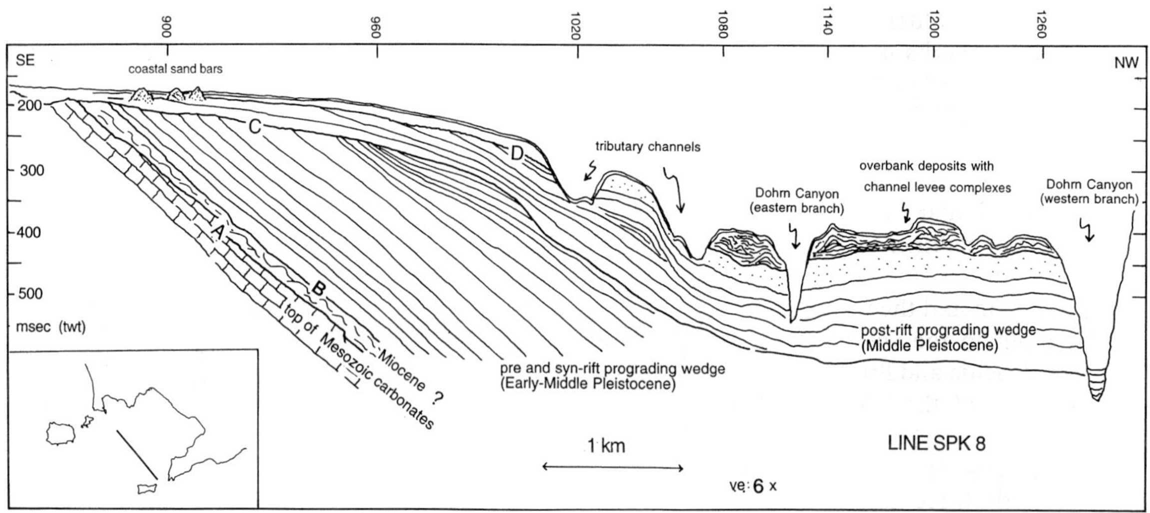

Fig. 7. Line drawing of the Sparker line SP8 showing the canyoning engraving down to the Middle-Late Pleistocene prograding units. Note the occurrence of a peculiar seismic unit interpreted as overbank deposits with channel-levée complexes. *A* top of Mesozoic carbonatic acoustic basement; *B* top of Miocene flysch deposits; *C* erosional truncation involving the Early-Middle Pleistocene prograding wedge; *D* erosional truncation involving the Middle Pleistocene prograding wedge

canyons ranges from a few hundred metres to more than 1 km. Their depth ranges from 250 m at the shelf edge to some 1300 m at the junction with the bathyal plain, the dip of their walls attains some 35° in the steepest sectors.

The Dohrn Canyon starts on two major curved branches. The northern branch merges into the shelf through a 1.5 km wide and a 20-40 m deep channel ("Ammontatura") characterised by a flat bottom and asymmetrical levees and by curved shape in plain view (Figs. 2 and 3). The origin of the "Ammontatura" shallow channel can tentatively be correlated to the hydrodynamic regime driven by the northern axis of Dohrn canyon. Its emplacement seems to post-date the latest stages of erosion and transport in the canyon and to pre-date the formation of the most recent volcanic edifices in the bay of Pozzuoli, as suggested by its abrupt termination towards the Nisida submarine volcanic edifice.

The southern branch is laterally fed by several tributary channels, whose pattern seems partly fault controlled. The location of this branch and several lines of evidence based on seismic stratigraphy suggest a genetic link between the activity of the Dohrn canyon and the Sarno paleo-drainage system during sea level lowstands. A seismic sequence, 30-40 m thick, overlying older undisturbed reflectors and including overbank deposits with channel-levee complexes (Fig. 7), has been located on both sides of the Dohrn southern branch. These deposits do not seem to be related to the main branches, which result deeply incised, but to the presence of tributary channels, controlling the overflowing of sediment fluxes in the surrounding areas.

The erosion and the transport of the volcanoclastic input in the western sector of the bay (offshore of Ischia and Procida islands) has acted along the axis of the Magnaghi Canyon which appears unrelated to present or past fluvial drainage system on land. Its head is typically trilobate with three main tributary channels joining basinwards into the main axis. Multibeam survey shows the occurrence of erosional phases presently acting on the continental slope located south-east of Procida Island and of debris "outrunner" blocks (Prior et al. 1984) a few km south-east of the Banco di Ischia volcanic edifice (Figs. 2 and 3).

A relative abundance of U-shaped morphologies with respect to the V-shaped, typically erosional, profiles has been observed, suggesting recent phases of canyon filling. Nevertheless several submarine slides and scars are still evident on the canyon's walls especially along the western flanks and on the continental slope as well. They involve large volumes of sediments (sometimes up to 10^6 m^3), in the bathymetric range from -250 m to -800/900 m. Factors known to control submarine slides include loading and underconsolidation of sediments due to rapid sedimentation, oversteepening of slopes consequent to phases of deep linear erosion, earthquakes and sea-level changes (Saxov and Nieuwenhuis 1982; O' Leary and Dobson 1992). This appears to be one of the most significant problems to be investigated further, due to the risk of anomalous waves over the coastal zones, originated by large submarine instability.

Acknowledgements. The authors thank the scientific party, the officers and the crew of cruises GMS97-1 and GMS98-1. The paper has been supported by the National Research Council grants developed to Geomare sud Institute and by the National Geological Survey of Italy (CARG project). The authors wish to thank two anonymous reviewers for improving the comprehension of the text.

References

Aiello G, Budillon F, de Alteriis G, Di Razza O, De Lauro M, Marsella E, Pelosi N, Pepe F, Sacchi M, Tonielli R (1997a) Seismic exploration of the perityrrhenian basins in the Latium-Campania offshore. Abstr Int Congr "ILP Task Force: Origin of Sedimentary Basins", Torre Normanna (Palermo Italy) June 1997

Aiello G, Budillon F, de Alteriis G, Ferranti L, Marsella E, Pappone G, Sacchi M (1997b) Late Neogene tectonics and basin evolution of the Southern Italy Tyrrhenian margin. Abstr Int Congr "ILP Task Force: Origin of Sedimentary Basins", Torre Normanna (Palermo Italy) June 1997

Budillon F, Cristofalo GC, Tonielli R (in press.) Segnalazione di terrazzi deposizionali sommersi in Penisola Sorrentina: Atlante dei terrazzi deposizionali sommersi lungo le coste italiane. Mem descrittive della carta geol d'Italia, vol 54

Capotondi L, Asioli A, Borsetti AM (1994) Ecozonazione a Foraminiferi plantonici e paleoceanografia dell'ultima deglaciazione nel Mediterraneo centrale. Atti 10° Congr AIOL Alassio 4-6 Nov 1992: 107-115

Catalano R, Bartolini C, Fabbri A, Lembo P, Marani M, Marsella E, Roveri M, Ulzega A (1996) Linee guida al rilevamento geologico nelle aree marine da sottoporre al Servizio Geologico Nazionale. Commissione di studio del CNR per la cartografia geol marina, Rapporto finale

Chiocci FL, Aiello G, Baraza J, De Lauro M, Ercilla G, Estrada F, Tonielli R, Farran M, Martorelli E, Senatore MR, Tommasi P, Bosman A, Cristofalo G, Ortolani U, Sarli R (1998) Ischia debris avalanche: first evidence of a lateral collapse of the Ischia southern flank from TOBI images. Presented at Conference on Hard Soils and Soft Rocks, Napoli, october 1998

Cinque A, Romano P (1990) Segnalazione di nuove evidenze di antiche linee di riva in Penisola Sorrentina (Campania). Geogr Fis Din Quat 13 (1): 23-36

Cinque A, Aucelli PPC, Brancaccio L, Mele R, Milia A, Robustelli G, Romano P, Russo F, Russo M, Santangelo N, Sgambati D (1997) Volcanism, tectonics and recent geomorphological change in the Bay of Napoli. Suppl Geogr Fis Din Quat III, T.2: 123-141

De Lauro M, Sacchi M, D'Argenio B (1997) Rilievi batimetrici con ecoscandaglio multifascio (multibeam) nel Golfo di Napoli. Abst 16° Congr Naz Gruppo Nazionale Geofisica della Terra Solida. Roma, Italy, pp 19

Finetti I, Morelli C (1974) Esplorazione sismica a riflessione dei Golfi di Napoli e Pozzuoli. Boll Geofys Teor Appl 16(62-63): 175-222

Fusi N, Mirabile L, Camerlenghi A, Ranieri G (1991) Marine geophysical survey of the Gulf of Naples (Italy): relationship between submarine volcanic activity and sedimentation. Mem Soc Geol Ital 47: 95-114

Latmiral G, Segre AG, Bernabini M, Mirabile L (1971) Prospezioni sismiche per riflessione nei Golfi di Napoli e Pozzuoli ed alcuni risultati geologici. Boll Soc Geol Ital 90: 163-172

Marsella E, de Alteriis G, Pepe F, Pelosi N, Aiello G, Catalano R, D'Argenio B, De Lauro M, Sacchi M, Sulli A (1996) Prospezione sismica multicanale di alta risoluzione per lo studio dei bacini peritirrenici nei settori Campano e Siciliano. Tech Rep n.5, Istituto di ricerca Geomare sud, CNR, Napoli (cruise GMS96-01)

Milia A (1996) Evoluzione tettono-stratigrafica di un bacino peritirrenico: il Golfo di Napoli. PhD thesis. Univ Napoli, Federico II, pp 184

O' Leary DW, Dobson MR (1992) Southeastern New England continental rise: origin and history of slide complexes. In: C.W. Poag CW, de Graciansky PC (eds) Geologic evolution of Atlantic continental rises. Van Nostrand Reinhold, New York, pp 101-121

Pescatore T, Diplomatico G, Senatore MR, Tramutoli M, Mirabile L (1984) Contributi allo studio del Golfo di Pozzuoli: aspetti stratigrafici e strutturali. Mem Soc Geol Ital 27: 133-149

Prior DB, Bornhold BD, Johns MW (1984) Depositional characteristics of a submarine debris flow. J Geol 92: 707-727

Saxov S, Nieuwenhuis JK (1982) Marine slides and other mass movements. Plenum Press, New York

CHAPTER 2

Physical and Geochemical Transports Along the Ancona Coastal Area

G. Budillon[1], M. Grotti[2], P. Rivaro[2], G. Spezie[1], and S. Tucci[3]

ABSTRACT

In the last two years a quasi-synoptic research programme has been carried out in the Adriatic Sea. The project, named PRISMA-2 (Programma di RIcerca e Sperimentazione per il Mare Adriatico - 2nd phase) is a multidisciplinary study focused on the physical and biogeochemical processes which determine the fate and the distribution of the chemical species discharged in the Adriatic Sea.

During the summer and winter legs (July-September 1996 and February-April 1997 respectively) a lot of collected data allowed the study and the comparison of the physical and geochemical parameters of the coastal waters offshore in the Ancona area where the cyclonic circulation of the Adriatic Sea is strongly intensified. In this region the coastal flow is characterized by the presence of fresh waters coming from the river discharges and then it is very rich in pollutants and nutrients.

Estimated transports of fresh water coming from the northern Adriatic Sea, defined by our hydrological and current measurements, reveal a strong temporal variability that is consistent with the change of the river discharges from the summer to the winter season. Also the distribution of nutrients within the water column exhibits a high temporal variability related to the fast evolution of biological activities and water masses modification. Summer transports of suspended matter from north are clearer than in the winter season. In particular, during the winter considerable differences in the quantities collected are noted.

Introduction

The Adriatic Sea is an elongated basin northwest-southeast oriented, connected with the Mediterranean Sea only in its southern part.

On the basis of its hydrodynamical and morphological characteristics, the Adriatic Sea is traditionally separated in three sub-basins. The northern sector is a shallow shelf area with a gently sloping and an average bottom depth of about 35 m. The middle area is 140 m deep on average and it is characterized by two depressions reaching 260 m. The southern sector connects the Adriatic Sea to the Mediterranean Sea through the Otranto Strait.

The circulation and the hydrology of the Adriatic Sea have been recently analysed by Artegiani et al. (1997a, b) and by Zavatarelli et al. (1998) on the basis of historical data distributed in the periods 1911-14 and 1947-83. The baroclinic general circulation has been classified into several currents and gyres. The development and the intensity of these structures reveal a strong seasonal variability especially in the northern part (Artegiani et al. 1999).

The circulation of the North Adriatic Sea (N.A.S.) is primarily driven by the thermohaline forcing (Franco et al. 1982), which is responsible for the flushing time of the system (Hopkins et al. 1999), and by the infrequent but strong forcing that is constituted by the wind stress especially in the northern part (Malanotte Rizzoli and Bergamasco 1983). The N.A.S. is also an unusual environment, since the extremely differ-

[1] Istituto di Meteorologia e Oceanografia, Istituto Universitario Navale di Napoli, Via Acton 38, 80127 Napoli, Italy
[2] Dipartimento di Chimica e Chimica Industriale, Sezione di Chimica Analitica ed Ambientale, Università di Genova, Italy
[3] Dipartimento per lo Studio del Territorio e delle sue Risorse, Università di Genova, Italy

F.M. Faranda, L. Guglielmo, G. Spezie (eds)
Mediterranean Ecosystems: Structures and Processes

ent biochemical water characteristics, related to the nutrient-rich river runoff, create a spatially complex system with a high temporal variability.

The geostrophic response to the forcing caused by both the accumulation of less dense water (river runoff), producing a higher sea level in the northern part, and the forcing due to the internal accumulation of dense waters (both forcings sustain a southward pressure force), create a surface and a bottom current system in the western boundary (along the Italian coast) with both vertical and horizontal shear (Hopkins et al. 1997).

An important difference between these two southward currents concerns the time variability in their forcing: the former is governed by the local change of the river runoff, while the latter follows the winter dense-water formation.

Both these coastal currents are responsible for the net southward export of mass and of geochemical properties.

The main goal of this paper is to highlight the most relevant relationship between the physical and geochemical fields and to evaluate the coastal transports of water, suspended and dissolved matter in the area offshore of Ancona as derived by the measurements carried out during the summer 1996 and the winter 1997 in the frame of the PRISMA 2 project.

Materials and Methods

Field activity

The cruises took place from August to September 1996 and February-March 1997 offshore in the Ancona area. During the summer cruises five legs were performed measuring only instrumental data (e.g. CTD data) and two legs collecting also geochemical samples. During February-March cruises, thirteen legs were made and only during one of them geochemical samples were collected.

The spatial and temporal distributions of the sampling locations during the physical-geochemical cruises are shown in Fig. 1. All cruises were performed as hydrological transects from the coast to the open sea in order to be perpendicular to the expected flows. The length of these transects was established in order to sample both the coastal (fresh) waters and the open sea (saltier) waters. The average distance between the stations along the transects was also established on the basis of the local value of the internal Rossby radius of deformations in order to resolve the mesoscale structures; this radius is 6.5 km during the summer cruises while in the winter season the temperature profiles are practically uniform with depth and Rossby radius is not easily estimated.

Physical parameters were measured by a Seabird 9/11+ CTD and water samples were collected with 10 litres Niskin bottles. The subsam-

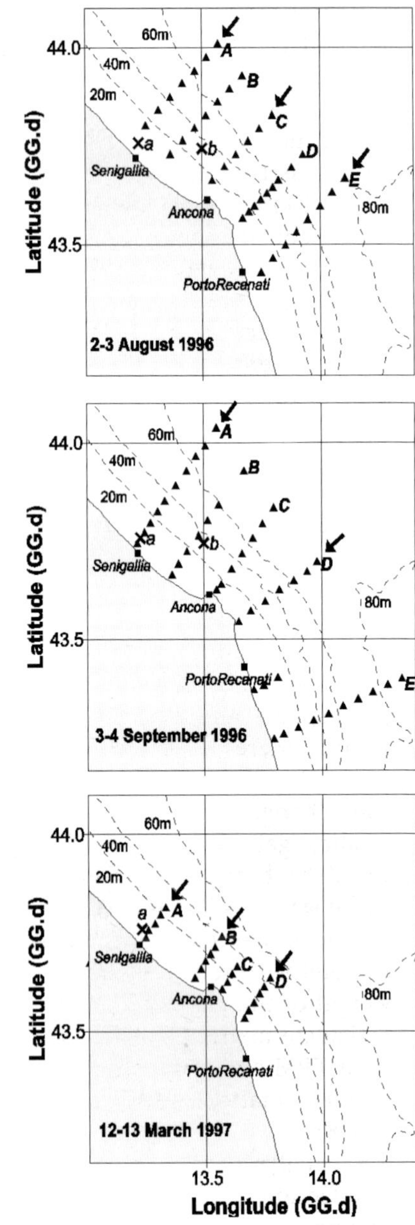

Fig. 1. The study area, station (*triangle*) and mooring (*cross*) locations. The *arrows* indicate the sections where physical and geochemical data have been collected simultaneously

ples for the determination of nutrients (silicate, phosphate, nitrate, nitrite and ammonium) were collected directly from the Niskin bottle, filtered throughout 0.7 μm GFF filter and stored at -30 °C in 100 ml LDPE (low density poly-ethylene) containers, until analysis. Water samples for the determination of suspended matter were filtered through prewashed 47 mm diameter Millipore (cellulose acetate-nitrate) filters having a nominal diameter of 0.45 μm and stored in petri dishes at -30°C.

Continuous current measurements were acquired using three moored instruments located in two different positions (A and B, Fig. 1). The coastal mooring (12 m depth) was equipped with one current meter at the nominal depth of 5 m; the offshore mooring was located at 40 m depth with two instruments at the nominal depth of 5 m and 18 m. All instruments recorded hourly average data during the summer period while only the hourly averaged time series of the coastal mooring is available for the winter season. High frequencies (tidal and inertial oscillations) were filtered applying a low-pass numerical filter to the hourly averaged current data.

Samples Analysis

Measurements of nutrients were performed by an Autoanalyzer Techicon II, according to the Hansen and Grasshoff methods (1983). Estimated detection limits for nitrate, nitrite, ammonium, phosphate and silicate were 0.05, 0.01, 0.05, 0.05, 0.1 μM, respectively.

Total particulate matter (TPM) was determined by weight change of the sample filter after drying over silica gel. To correct for weight changes induced by humidity reference filters from the same lot were used. Combustible organic matter was determined by ashing the filter at 1000°C in preweighted porcelain crucibles; the increase in weight of the crucible is reported as inorganic particulate matter (IPM) and the difference between TPM and IPM was reported as organic particulate matter (OPM) (Strickland and Parsons 1968).

Dimensional analysis was performed immediately after collection with a Coulter Counter Multisizer II using a 140 μm tube (Krank 1980). Particle sizes were separated in 256 channels in a range between 0.6-65 μm.

Transport Estimations

Using the hydrological and low-pass filtered current meter data the longshore transports were calculated.

Transport values were obtained on the basis of the classical geostrophic method imposing the zero reference level, on first step, at the depth of 5 m. In such a shallow area, it is well known that the geostrophic method has obvious limitation because the friction with the bottom can not be considered negligible. This evident restriction is not dramatic when the barotropic component of the motion is prevailing, with respect to the baroclinic one, and it is available from direct measurements. For this purpose the baroclinic fields were adjusted using the low-pass filtered current meter time series recorded close to the coast and more offshore (Fig. 1), both at the depth of 5 m. Daily averaged data were used to calculate the current filed with the geostrophic method.

The summer evaluations were made using the data recorded by the two surface current meters and linearly interpolated to get the reference speed for each couple of stations. Because of the lack of current data, the winter evaluations were obtained using only the information coming from the coastal current meter close to Senigallia. This approximation did not affect our estimation since in winter the coastal flow, as discussed in the next section, is confined in a smaller area and therefore it was well monitored by the coastal instrument.

In order to evaluate the fresh water concentration this simple relationship was applied:

$$C = \frac{\bar{S} - S}{\bar{S}}$$

where $\bar{S}$ is the salinity of the local MLIW ($\bar{S}$ = 38.4 in our computations) and S the salinity of the in situ data.

The calculated fresh water, the velocity data and the measured geochemical concentrations were interpolated on a regular grid by using the kriging method with 1 km horizontal and 2 m vertical resolution. In order to normalize the data an anisotropic ratio of 500 was imposed during the gridding procedures. The transport matrix was obtained by multiplying two by two the corresponding elements of concentration and corresponding velocity matrices. By integrating over all the section we got the total transport for each transect.

Results and Discussion

The Vertical Structure

In the summer period, the solar heating affects the surface layer forming a well developed thermocline (Fig. 2). Below it two water masses may be identified mainly on the basis of their salinity values. Maximum salinity values highlight the influence of the Modified Levantine Intermediate Water (MLIW) in the eastern (right) flank of the sections, while close to the bottom, in the western continental slope, relatively low salinity and temperature values are mostly indicative of waters originated in the Northern Adriatic Sea during the winter season (Artegiani et al. 1989). Moreover, close to the surface, the vertical structure of the water column exhibits the presence of a coastal low salinity sector originated by the river outflow in the N.A.S.. This flow meanders close to the coast extending its influence between 15-30 km offshore.

Chemical parameters showed a strong temporal variability related to the fast evolution of biological activities and water masses modification. In Fig. 2, the nitrate and silicate distributions found along section A in August sampling are reported. As it can be seen, their distribution along the water column showed the typical trend for summer season, with low levels in the surface layer due to post bloom depletion and higher concentrations near the bottom, because of remineralization processes. This feature is also justified by the stable vertical stratification that prevents mixing of the sub-pycnocline layer. Similar trends were found in the remaining sections of August sampling. In particular, nitrate concentration ranged from values below the detection limit to 1.5 μM, silicate from about 1 mM to 8 μM, phosphate from 0.11 to 0.30 μM (Frache et al. in press a, b). Chemical parameters did not show any significant changes in their distribution in September, except for nitrate, which showed higher concentrations in the surface layer of the nearshore stations close to Ancona. Concerning suspended matter, in August low concentrations were found, with an average value of 2.4 mg/l and a percentage of inorganic fraction of about 26%; dimensional analysis using number of particles per volume units showed a range 35.000 to 50.000 particle/cm^3 with an average value of 43.800 particle/cm^3. The higher values in concentration and

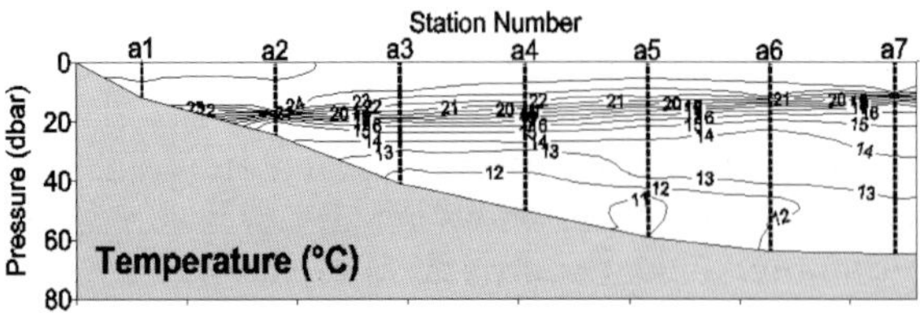

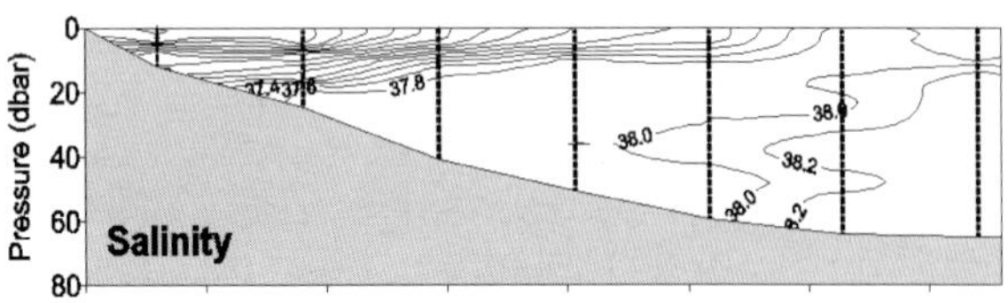

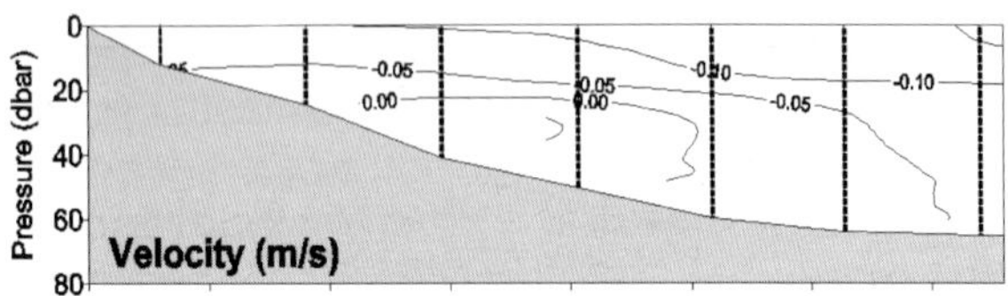

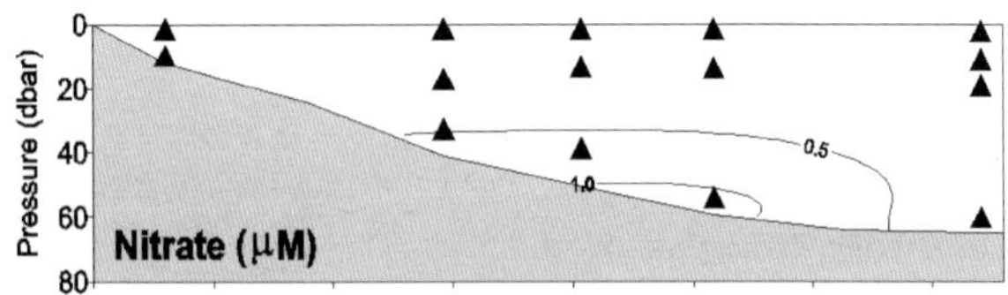

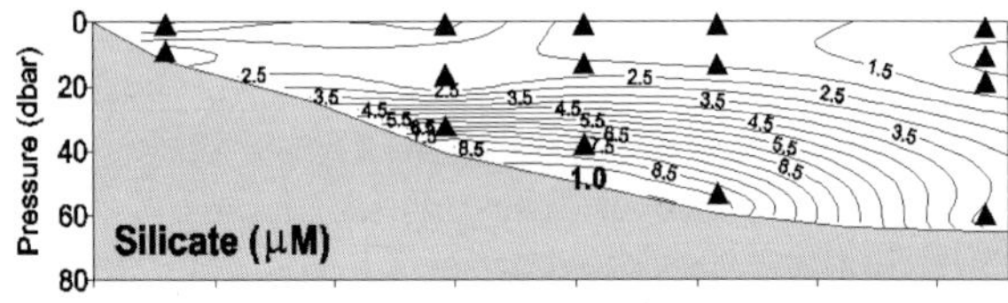

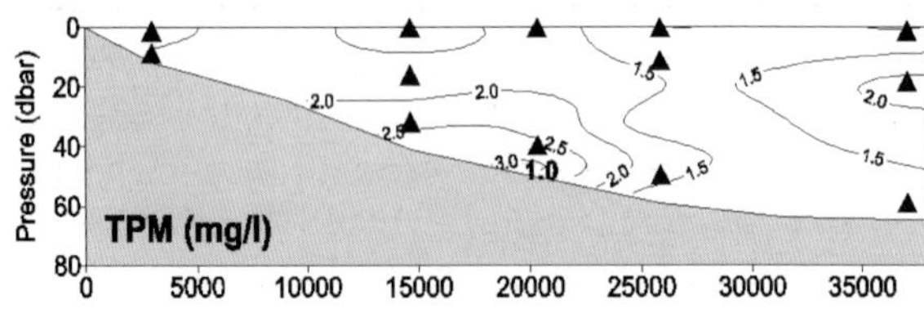

Fig. 2. The vertical structure: 2-3 August 1996 – Section A. *Triangles* indicate the depth of the geochemical samples

number of particles were found near the bottom and this is clearly shown for TPM in Fig. 2. Spectral analysis on the samples showed a main mode centred on 3 μm (92% of the 36 samples) and a secondary mode centred on 20-50 μm. The period before the sampling was characterized by good weather conditions and negligible local river out flow; therefore this condition has permitted the sedimentation for the particles of high dimension permitting the suspension in the water column only for the smaller particles (average dimension: 4.48 μm); the low concentrations measured in his period are the direct consequence of this situation. In September concentration of suspended matter was higher than in August with an average value of 4.2 mg/l and with the highest values close to 15.0 mg/l. The percentage of inorganic matter is about 43% with an increase of 17% compared with the samples collected in August. The increase of the inorganic fraction is caused by the input of land origin; this aspect, together with the increase of the particulate average size (5.83 μm), has contributed to double the TPM concentration in the water column. As for the nitrate concentrations this increase was mainly found in the coastal stations located near Ancona. The determination of particle numbers showed values similar to those found in August with an average value of 44.500 particles/cm^3. Also the spectral characteristics in suspended matter confirmed a main mode centred on 3 μm (75% of the 24 samples) and a secondary mode between 20 and 40 μm.

In winter the thermohaline field gave a different picture: the absence of the thermocline and the presence of colder and fresher waters close to the coast, modified the density field that exhibits a strong barotropic characteristic (Fig. 3). Moreover the area involved by the coastal low salinity flow was smaller than during the summer period, as shown in Fig. 3, being the typical extension of the coastal flow about 8-10 km from the coast. In the same picture nitrate showed both a strong gradient, moving from the nearshore to the offshore stations, the former being more influenced by the presence of fresh water coming from the river discharges and a general decrease with depth.

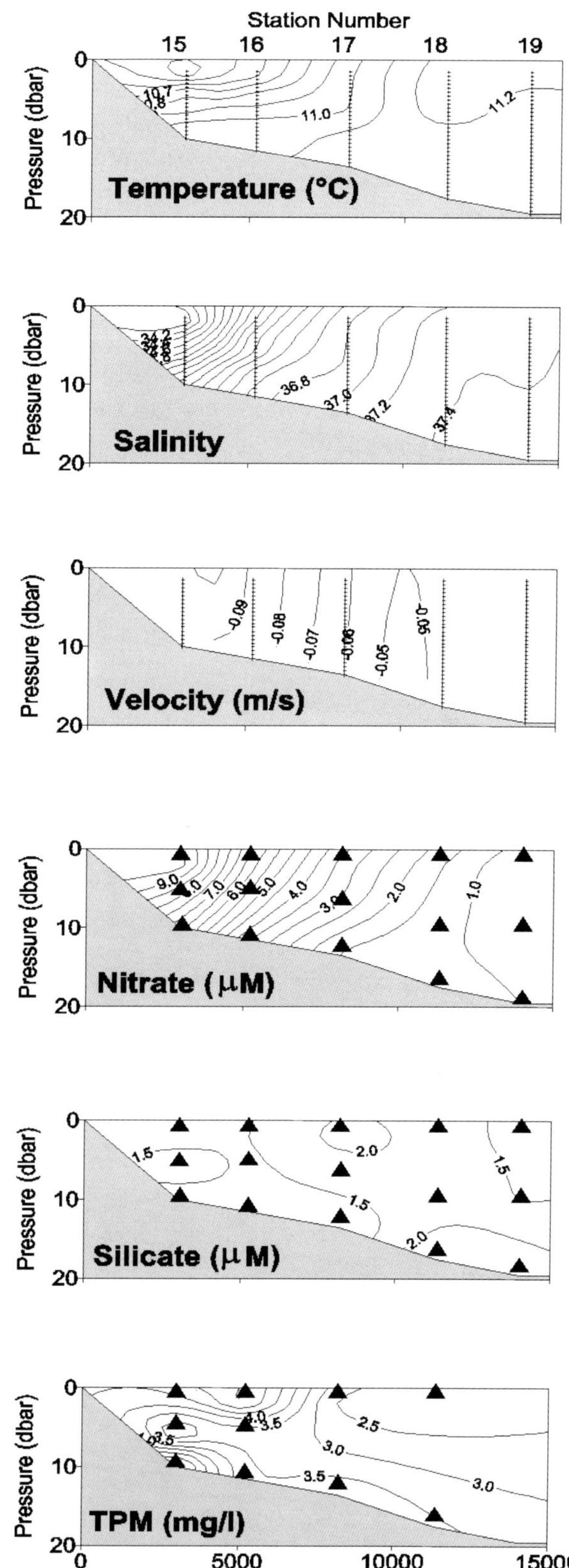

Fig. 3. The vertical structure: 12-13 March 1997 – Section A. *Triangles* indicate the depth of the geochemical samples

Nitrate concentrations ranged from 15 to 0.42 μM. The same distribution did not occur for phosphate and silicate; in fact phosphate ranged from detection limit to 0.22 μM, while silicate were less than 4 μM, without showing any west-east gradient (Frache et al. in press a, b). Concentrations of suspended matter were higher than those found in summer and the highest values were around 20 mg/l with an average value of 6 mg/l and the inorganic fraction was about 44%. Particle numbers for volume units were half of those found in summer; values were in the range 15.000 to 25.000 particle/cm^3 with an average value of 19.000 particle/cm^3. Spectral analysis confirmed also in winter the importance of the cluster centred on 3 μm (45% of the 38 samples). Also in those samples are frequent bimodal spectra with a variable mode included from 10 to 40 μm. This sampling has been performed a few days after a storm, consequently the water column contains elevated quantities of suspended matter with high dimension. With respect to the summer values, the mean size is higher with a value of 7.5 μm and a particle number of about half. The TPM distribution along section A (Fig. 3) showed the presence of two maxima: the first, in the surface layer, related to the coastal flow and the second, near the bottom, due to resuspension phenomena (Puskaric et al. 1992).

Current Data

Hourly current measurements were filtered in order to eliminate the high frequencies like tidal and inertial oscillations that did not give any significant contribution to the long term transport. The longshore filtered component revealed an averaged current of 3 cm/s in summer and 5 cm/s in winter in the southeastern direction but

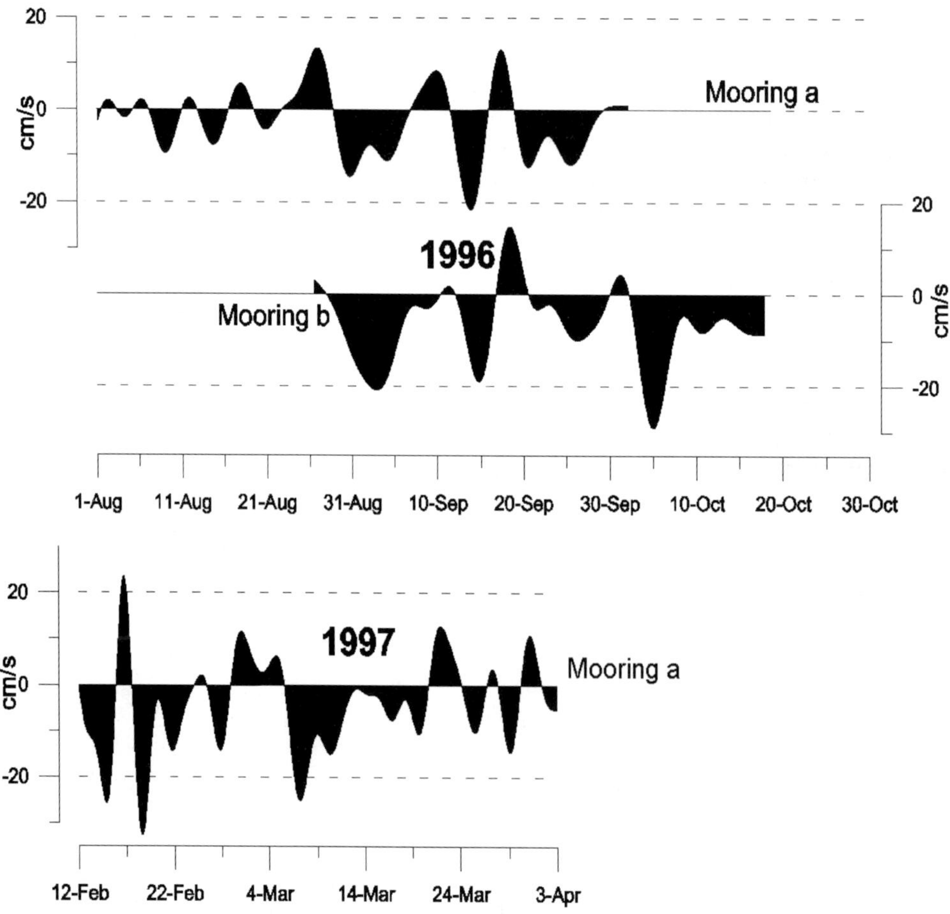

Fig. 4. The near-surface current time series, hourly low-pass filtered values. Positive (negative) velocities indicate a longshore flow with a northwest (southeast) direction

characterized by some inversions in direction (Fig. 4). More detail about the current measurements and variability in this period can be found in Budillon et al. (2000a, b).

Fresh Water Transport

The fresh water transport calculated through all the sections is reported in Fig. 5. In the same figure, the trend of Po River discharge (A. Bergamasco, personal communication) during the sampling period is shown, revealing a good agreement with the estimated fresh water transport off Ancona coast.

In particular the mean fresh water transport was calculated to be about 1200 m^3/s for the summer cruises and 1080 m^3/s for the winter ones. These values were consistent with the total fresh inflow of all rivers discharged in the N.A.S. that is about twice that of the Po River contribution (1120 m^3/s and 1260 m^3/s during the summer and winter cruises respectively). The cross-correlation analysis performed between the fresh water estimations and the Po River discharge revealed different time lag for the summer and winter data; more accurate information can be obtained investigating the complete PRISMA-2 data set that includes also spring and autumn cruises.

We wish to outline that because the speed of the coastal flow is higher in winter than in summer but the fresh water input similar, the area affected by the fresh water is expected to be smaller in winter. This remark is in good agreement both with our measurements and with the geostatistical analysis performed by Budillon and Ruta (1998).

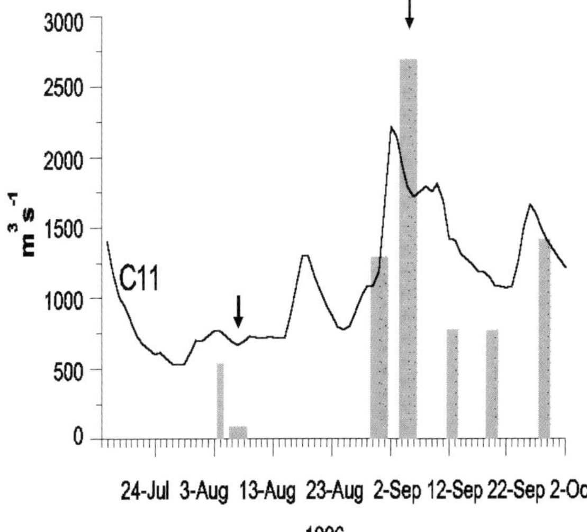

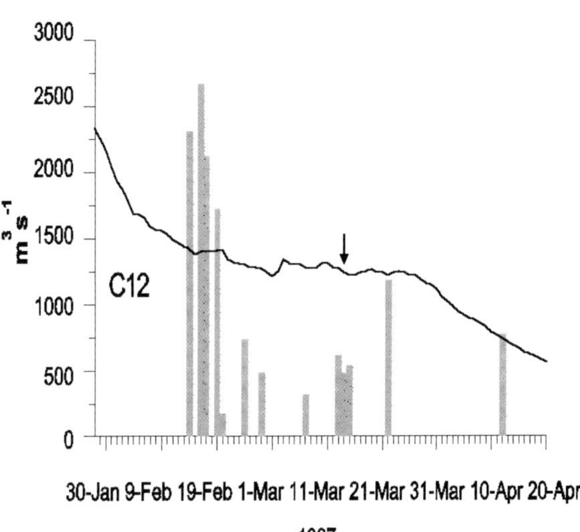

Fig. 5. Fresh water southward flow (*histogram*) and Po discharge (*line*). The histogram width is representative of the temporal length of the hydrological sampling. The *arrows* indicate the geochemical sampling

Geochemical Transport

Geochemical transport was computed by considering all sections marked by the arrows in Fig. 1. Transport mean values and relative 95% confidence intervals of nitrate, phosphate, silicate, TPM, OPM and particle numbers, estimated for August, September and March surveys are reported in Fig. 6. Southward transport of all considered parameters are higher in September than in August. Since the differences between the concentrations of parameters during the summer sampling were not so considerable, the current variability overrode in the transport calculations.

Geochemical transport values referred to September sampling were higher than those calculated for March even if the summer concentrations of parameters as well as the current speed were lower. In fact, as previously discussed, the area involved in the southern transport proved more extended in summer.

Only the nitrate transport value for March did not prove significantly different from that of September since the concentration in winter was much higher than in summer.

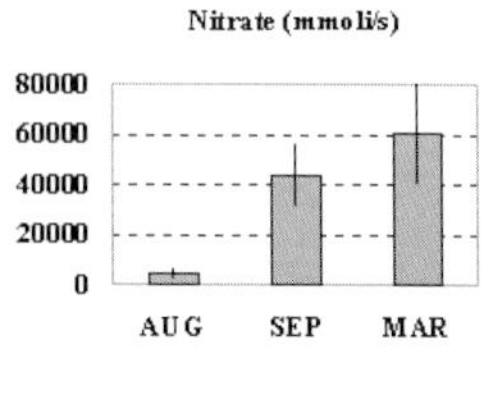

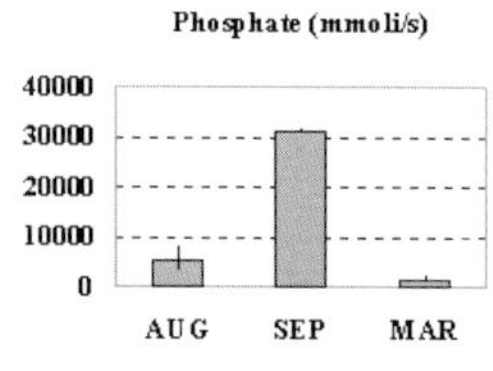

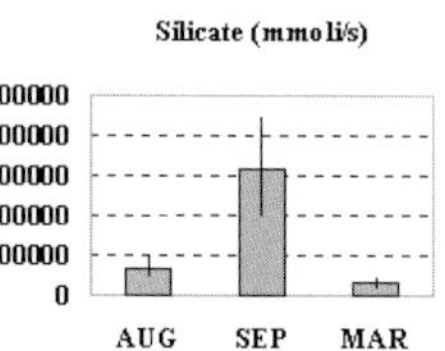

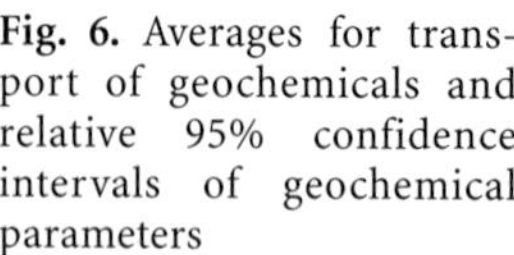
Fig. 6. Averages for transport of geochemicals and relative 95% confidence intervals of geochemical parameters

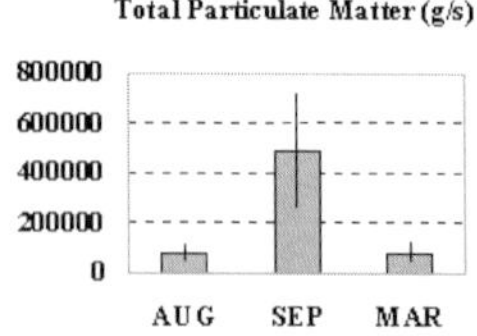

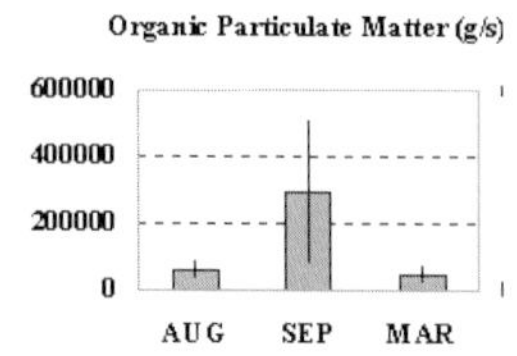

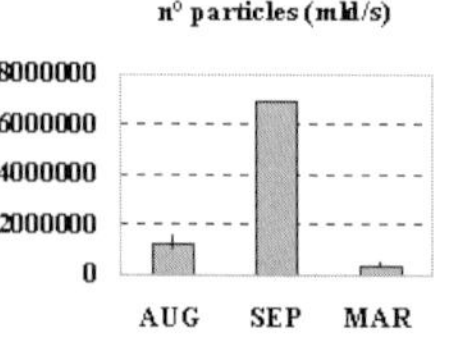

These results refer to the transport by the coastal flow and cannot be compared with those obtained by Gacic et al. (1997), whose computation was applied to the whole section from the Italian to the Croatian coast, including the offshore transport of geochemical material.

Conclusions

One of the major tasks of the PRISMA-2 project in the Ancona area was to evaluate the mass and geochemical transports coming from the Northern Adriatic Sea. As a first approach data collected during three surveys carried out in two different seasons were considered.

Fresh water transport along the Ancona coast proved similar during the summer and winter cruises. These estimations are consistent with the total input due to all the rivers in N.A.S.. The coastal near-surface current measured in this area was lower in the summer (mean about 3 cm/s) than in the winter (mean about 5 cm/s) while the area occupied by the fresh flow was smaller in the summer preserving similar fresh transport.

The high variability in trasport of all the geochemical parameters due to the coastal flow seems to be more dependent on dynamical forcing than on changing concentrations, except for Nitrate. The southward transport in September proved 10 times higher for nitrates and about 4-6 times higher for the other parameters than in August. Due to the smaller area involved by the southward flow in winter, the transport of geochemicals in March was not significantly different from that found in August, except for nitrate, because of their much higher winter concentrations.

Acknowledgements. This study was performed as part of the Italian National "Programma di Ricerca e Sperimentazione per il Mare Adriatico" 2nd phase (PRISMA 2).

References

Artegiani A, Azzolini R, Salusti E (1989) On the dense water in the Adriatic Sea. Oceanol Acta 12(2): 151-160

Artegiani A, Bregant D, Paschini E, Pinardi N, Raicich F, Russo A (1997a) The Adriatic Sea general circulation. I. Air-sea interactions and water mass structure. J Phys Oceanogr 27: 1492-1514

Artegiani A, Bregant D, Paschini E, Pinardi N, Raicich F, Russo A (1997b) The Adriatic Sea general circulation. II. Baroclinic circulation structure. J Phys Oceanogr 27: 1515-1532

Artegiani A, et al. (1999) Condizioni oceanografiche della fascia costiera fra Senigallia e Porto Recanati nel periodo 3 agosto – 27 settembre 1996. Progetto Prisma II fase. Sottoprogetto 1. Oceanogr Fis Chim e Biol Tech Rep (in press)

Budillon G, Ruta G (1998) Rappresentazione di dati idrologici e analisi geostatistica in ambiente GIS (IDRISI) nell'area costiera anconetana. I Conv Naz Sci Mare, Ischia (NA) 11-14 November

Budillon G, Ortona A, Paschini E (2000a) Analisi dei trasporti di fresh water nel sistema costiero Anconetano. Proc AIOL, Vol. XIII(II): 23-32

Budillon G, Paschini E, Simioli A, Zambianchi E (2000b) Surface dynamics of a coastal area off Ancona (Adriatic Sea). This volume

Frache R, Abelmoschi ML, Bottinelli C, Grotti M, Rivaro P, Soggia F (2000a) Distribuzione di nutrienti, ossigeno e metalli in tracce in presenza di strutture frontali nel Mare Adriatico: Progetto Prisma II fase: Sottoprogetto 1. Oceanograf Fis Chim Biol Tech Rep 1 (in press)

Frache R, Abelmoschi ML, Bottinelli C, Grotti M, Guarini R, Rivaro P, Soggia F (2000b) Distribuzione di nutrienti, ossigeno e metalli in tracce in presenza di strutture frontali nel Mare Adriatico: Progetto Prisma II fase: Sottoprogetto 1. Oceanograf Fis Chim Biol Tech Rep 2 (in press)

Franco P, Jeftic L, Malanotte Rizzoli P, Michelato A, Orlic M (1982)

Descriptive model of the Northern Adriatic. Oceanol Acta 5: 379-389

Gacic M, Civitarese G, Ursella L (1997) Spatial and seasonal variability of water and biogeochemical fluxes in the Adriatic Sea. Cont Shelf Research, pp 1-23

Hansen HP, Grasshoff K (1983) Automated chemical analysis. In: Grasshoff K, Ehrhardt M, Kremling K (eds) Methods of sea water analysis (2nd ed). Verlag Chemie, Weinheim D, pp 347-374

Hopkins TS, Artegiani A, Bignami F, Russo A, (1999) Water masses modification in the Northern Adriatic, a preliminary assesment from the Elna data set. Proceeding of the ELNA Meeting. Portonovo 23-27 April 1996. Ecosystem research report N° 32, The Adriatic Sea, pp 3-29

Krank K (1980) Variability of particulate matter in a small coastal inlet. Can J Fish Aquat Sci: 1209-1215

Malanotte Rizzoli P, Bergamasco A (1983) The dynamics of the coastal region of the Adriatic Sea. J Phys Oceanogr 13: 353-373

Puskaric S, Fowler SW, Miquel JC (1992) Temporal changes in particulate flux in the Northern Adriatic Sea. Estuar Coast Shelf Sci 35: 267-287

Strickland JDH, Parsons TR (1968) A practical handbook of seawater analysis. Bull Fish Res Board Can 167: 1-311

Zavatarelli M, Raicich F, Bregant D, Artegiani A (1998) Climatological biogeochemical characteristics of the Adriatic sea. J Marine Sys, pp 1-37

CHAPTER 3

Surface Dynamics of a Coastal Area off Ancona (Adriatic Sea)

G. Budillon[1], E. Paschini[2], A. Simioli[1], and E. Zambianchi[1]

ABSTRACT

In the framework of the PRISMA 2 programme, an HF Coastal Dynamics Applications Radar (CoDAR) has been utilized starting in July 1997, in order to identify surface frontal structures characterizing the Italian coast of the Adriatic Sea and to study the surface circulation, which plays a fundamental role in heat and momentum transfer processes between coastal and offshore areas.

Near-surface currents in the coastal area between Ancona and Senigallia derived from the shore-based HF radar have been mapped at weekly intervals at a resolution of 1.5 km over an area of approximately 30 x 30 km.

Preliminary results show a southeastward (i.e. alongshore) coastal jet that extends its influence as far as 10 km from the coast; farther offshore a cyclonic eddy, with a diameter of 10-20 km typically occurs. This circulation can occasionally be reversed showing an anticyclonic eddy that forces a northward coastal current with a far less energetic signature.

Introduction

Recently the Italian coastal areas of the Adriatic Sea have been intensively investigated in the framework of the Italian national PRISMA-2 research programme, a multidisciplinary study focussed on the physical and biochemical processes that determine the fate and the distribution of pollutants discharged in the Adriatic Sea.

The general circulation of the Adriatic Sea, as derived by several experimental and modelling studies, shows a seasonal and annual cyclonic characteristic. Typically the surface circulation in the southern and middle Adriatic Sea consists of a smooth flow encircling cyclonically the entire basin, while during the winter the general circulation consists of several cyclonic gyres (Artegiani et al. 1997; Malanotte Rizzoli and Bergamasco 1983; Franco et al. 1982; Paschini et al. 1993).

The cyclonic circulation of the Adriatic Sea is intensified in the central coastal region around 43.5 N, where the local bathymetry shows a strong slope over a wide longshore area (Artegiani et al. 1980)

The flow along the Italian coastline is also characterized by the presence of fresh waters of river runoff origin. In particular, coastal waters off Ancona show a strong seasonal variability of the physical and bio-chemical parameters depending on river discharges. The boundary between this coastal freshwater strip and the open sea is often characterized by the development and growth of mesoscale instabilities (see, e.g. Bracalari et al. 1989, and references therein).

In the framework of the PRISMA 2 activities, a special emphasis has been devoted to the study of the coastal area off Ancona, both for its geographical setting and for the proximity of one of the major cities on the Adriatic coast. In addition to the traditional measurements extensively carried out over the whole basin, the objective of characterizing the surface circulation in the region immediately offshore (and upstream, with respect to the prevailing coastal current) from the city was pursued by means of a coastal radar which was set up and has been working since summer 1997. The data presented and discussed here, in particular, refer to the months of August, September and October 1997.

[1] Istituto di Meteorologia e Oceanografia, Istituto Universitario Navale, Via Acton 38, 80133 Napoli, Italy
[2] Istituto Ricerca Pesca Marina, CNR, Ancona, Italy

F.M. Faranda, L. Guglielmo, G. Spezie (eds)
Mediterranean Ecosystems: Structures and Processes

The use of HF radar signal to characterize the surface circulation of a sea area was first explored by Crombie (1955) (for a review see Lipa and Barrick 1983; Prandle 1991; Fernandez 1993), who observed that those signals were backscattered by ocean waves of precisely half the radar wavelength. The general working principle of these instruments is based on the Doppler shift of the Bragg-scattering waves propagating radially towards or away from the radar. This yields, obviously, the necessity of utilizing at least two radar installations in order to get a characterization of the two-dimensional surface velocity field.

Over the last years two systems have been developed and subsequently utilized based on the above principles: the Coastal Ocean Dynamics Applications Radar (CODAR, see Barrick et al. 1977) and the Ocean Surface Current Radar (OSCR, see Prandle 1987; Shay et al. 1995). This latter utilizes an 85 m long, 16 or 32 element phased-array antenna, emitting a narrow beam electronically guided over the studied area. The CODAR antennae, on the contrary, consist of two crossed loops and a monopole, and are comparatively a very compact system. Radial velocities are retrieved from the sea-echo Doppler spectra by a least squares method (Lipa and Barrick 1983).

Drawbacks and advantages of the two systems are an obvious consequence of the above: the OSCR implies a relatively simple signal processing technique; on the other hand, it is a voluminous system, and its installation is quite demanding. The CODAR is a much more compact, easy to set up device, but requires heavier data postprocessing. While until the recent past most published results refer to OSCR measurements (Prandle and Ryder 1985; Prandle 1991; Prandle et al. 1993; Shay et al. 1995), the refinement of data analysis techniques has lately led to the assessment of the accuracy of CODAR systems which have been successfully employed off the coast of British Columbia (Masson 1996) and in Monterey Bay (Paduan and Rosenfeld 1996).

For the above reasons, a CODAR system was selected to map surface currents in the area off Ancona.

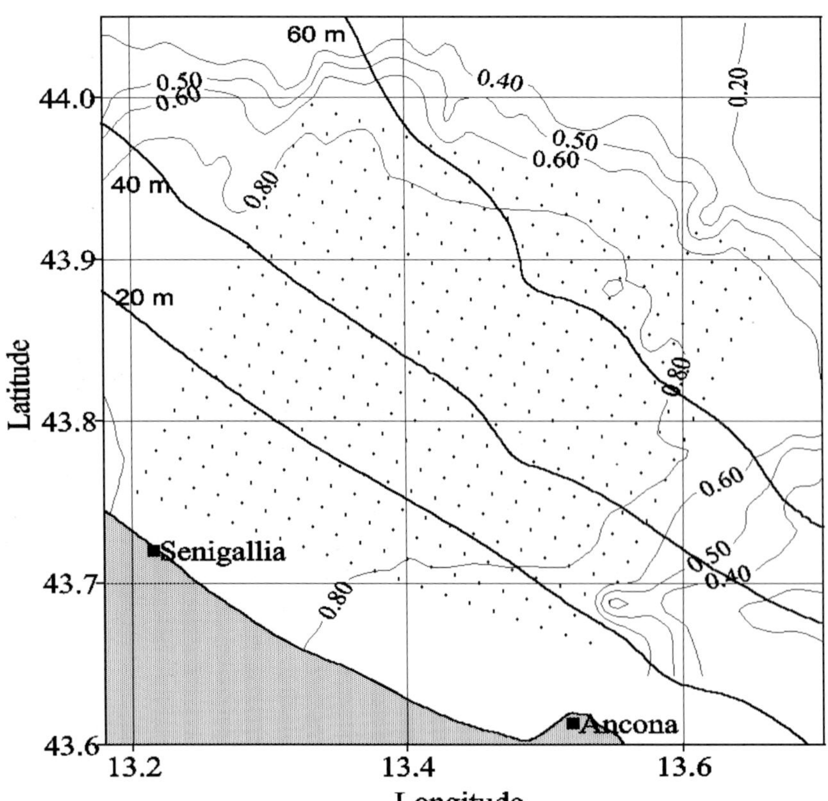

Fig. 1. Map of the studied area. The *dots* represent the 30 x 30 km radar measurement locations considered in the present work. *Contours* refer to the normalized data density over the three month period August through October 1997

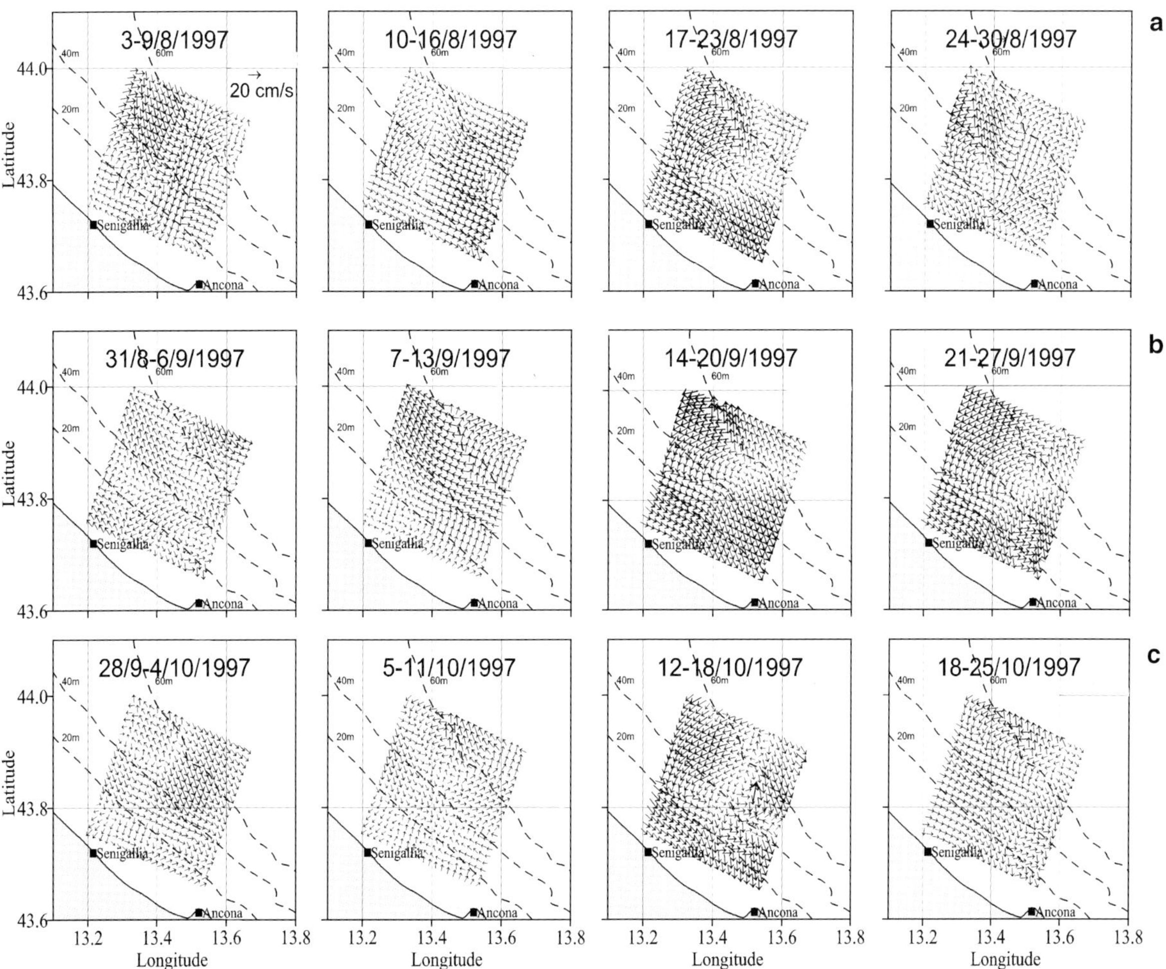

Fig. 2a-c. **a** Weekly averaged surface currents (August 1997). **b** Weekly averaged surface currents (September 1997). **c** Weekly averaged surface currents (October 1997)

Materials and Methods

The instrument utilized for this study is a Seasonde CODAR Ocean Sensor, which operates at 47-50 MHz. The two antennae were located at Senigallia and Ancona on the western coast of the Adriatic Sea, at a distance from each other of approximately 30 km. The complete array of current measurement locations is shown in Fig. 1. At increasing distance of the observation point from the antennae the measurement reliability decreases for distortion errors; additional errors can be due to a too narrow view angle. Lipa and Barrick (1983) suggest, in consequence of the above, to take into consideration only data relative to a matrix of 20 x 20 km for a setting analogous to ours. After a first analysis aimed at assessing the consistency of data outside this core area with those inside it, we have decided to take into account measurement spanning over a 30 x 30 km area, as discussed in Budillon et al. (2000a). Besides distortion, another problem which may arise in radar data is the presence of spatial and/or temporal gaps; this can be due to obvious hardware reasons but also to problems with the direction finding algorithms or with the signal maximum range fluctuations (Paduan and Rosenfeld 1996). The contours in Fig. 1 show the data density normalized over the total nominal amount of data for the measurement period considered in this study. As can be seen, the 30 x 30 km square in which data do not suffer from major distortion approximately coincides with the locations characterized by a normalized den-

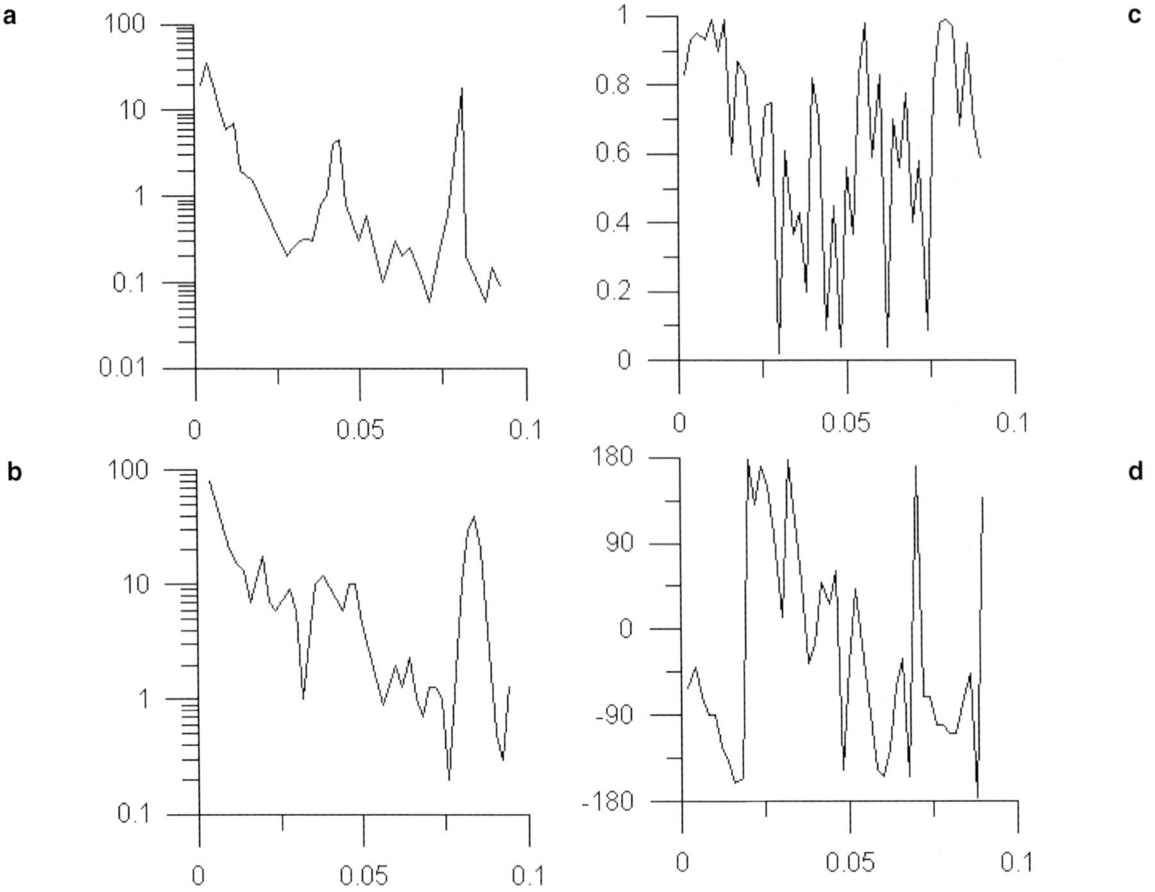

Fig. 3a-d. Comparison between alongshore velocities measured by the moored currentmeter off Senigallia (power spectrum in **a**) and at one selected location close to the mooring location (power spectrum in **b**). **c** and **d** show coherence and phase of the signals, respectively

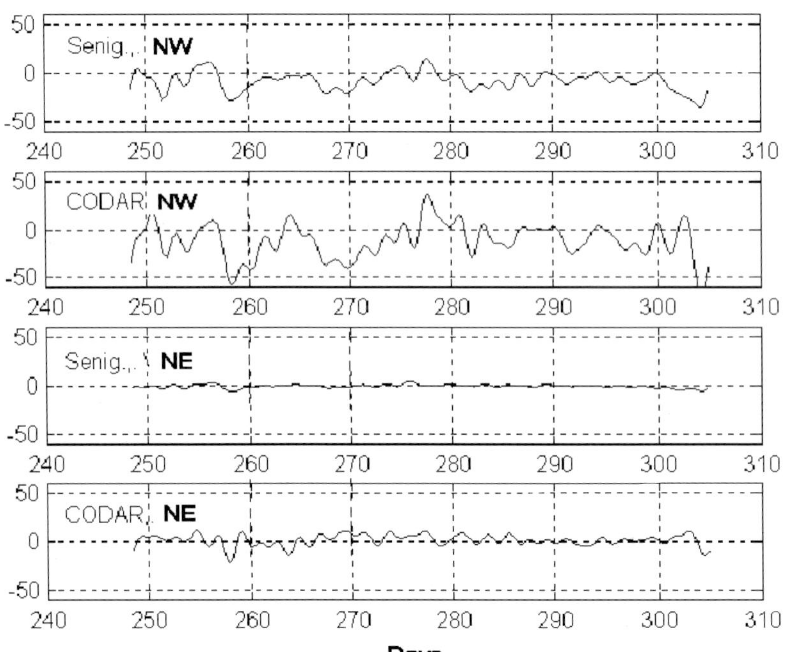

Fig. 4. Comparison between velocities measured by the CODAR and by the moored currentmeter off Senigallia for the cross-shore component (*upper half* of the figure) and for the alongshore component (*lower half*)

sity higher than 0.8 over the whole considered period, the area to which the present analysis was therefore limited. Monthly coverage was also checked, and the three months considered resulted to be quite evenly represented.

In the adopted configuration, the system provides data averaged over 1 h periods. In order to filter out semi-diurnal and diurnal tides as well as inertial oscillations, data were low-pass filtered with a Chebishev filter with a 40 h cutoff and averaged over weekly periods. Spatial/temporal gaps were filled by means of linear interpolation with neighbouring (in space and time) data.

Results and Concluding Remarks

In Fig. 2 we present weekly surface current maps in the studied area relative to the months of August, September and October 1997.

The surface circulation is dominated by a southward coastal jet and by a single cyclonic eddy farther offshore. This dynamical feature is very frequent in our data set and it is often detectable also in the hourly data; its typical dimension is of the order of 2 times the Rossby radius of deformation, which was estimated in the same area and period at around 6.5 km (Budillon et al. 2000b).

The coastal current shows typical velocities of the order of 10-20 cm/s and extends its influence for 10-15 km from the coast. The cyclonic circulation shows a symmetric field approximately located in the centre of the studied area. We expect the topography to be important in setting the location and the persistence of the eddy in this region. A major forcing setting up the eddy field can obviously be related to the local field of wind stress.

This quasi-permanent circulation pattern does not occur in two situations detected at the beginning of August and at the beginning of September, when the coastal jet direction is almost completely reversed and the investigated area is dominated by an anticyclonic local circulation, most likely due to meteorological conditions.

The accuracy of the surface current measurements was assessed by a comparison with data gathered by a current meter located at 5 m depth and moored off Senigallia, i.e. in the western portion of the domain spanned by the radar signal. The outcome of the comparison is shown in Figs. 3 and 4: power spectra of the alongshore currentmeter data and of the hourly radar ones (collected at one selected location close to the mooring site) are shown in Fig. 3a and b, respectively. Figure 3c and d display coherence and phase between them; the coherence results quite high both in the high frequency peaks associated with the tides, obviously, and in the low frequencies of the residual circulation. Figure 4 shows the time series, after low-pass filtering, of the current records simultaneously measured by the two systems. As can be seen, the agreement between the two data sets is quite good, especially in the alongshore component. Velocities measured by the currentmeter are systematically lower than the radar ones; this can explained by the fact that radar data refer to the immediate surface, whereas the currentmeter was moored at 5 m depth, where we expect the dynamics to be weaker.

The CODAR measurements in this area were also compared with velocities drawn by surface drifters launched off Senigallia in August 1997. Even though comparing Eulerian and Lagrangian data is never straightforward the agreement is quite satisfactory, at least in qualitative terms (see Mazzoldi et al. 1999, for a description of the validation experiment and a thorough discussion of the subtle points of the data analysis).

Acknowledgements. This work was carried out in the framework of the activities of Task 1 of PRISMA-2 Project, funded by the Italian Ministry for Scientific Research. The authors wish to thank M. Moretti and A. Ortona for helpful discussions.

References

Artegiani A, Azzolini R, Moretti M, Vultaggio M (1980) A first survey on the Ancona coastal area currents. Ann Ist Univ Navale 49-50: 151-160

Artegiani A, Bregant D, Paschini E, Pinardi N, Raicich F, Russo A (1997) The Adriatic Sea general circulation. II. Baroclinic circulation structure. J Phys Oceanogr 27: 1515-1532

Barrick DE, Evans MW, Weber BL (1977) Ocean surface currents mapped by radar. Science 198: 138-144

Bracalari M, Salusti E, Sparnocchia S, Zambianchi E (1989) On the role of the froude number in geostrophic coastal currents around Italy. Oceanol Acta 12:161-166

Budillon G, Paschini E, Russo A, Simioli A (2000a) Primi risultati codar sulle correnti marine superficiali nell'area anconetana. Proc AIOL, Vol XIII(II): 23-32

Budillon G, Grotti M, Rivaro P, Spezie G, Tucci S (2000b) Physical and geochemical transports along the Ancona coastal area (PRISMA-2 Project). This volume

Crombie DD (1955) Doppler Spectrum of Sea Echo at 13.56 Mc/s. Nature 175: 681-682

Fernandez DM (1993) High-frequency radar measurements of coastal ocean surface currents. PhD diss Stanford Univ

Franco P, Jeftic L, Malanotte Rizzoli P, Michelato A, Orlic M (1982) Descriptive model of the Northern Adriatic. Oceanol Acta 5: 379-389

Lipa BJ, Barrick DE (1983) Least-squares method for the extraction of surface currents from CODAR crossed-loop data: application at ARSLOE. IEEE J Oceanic Eng OE8: 226-253

Malanotte Rizzoli P, Bergamasco A (1983) The dynamics of the coastal region of the Adriatic Sea. J Phys Oceanogr 13: 353-373

Masson D (1996) A case-study of wave-current interaction in a strong tidal current. J Phys Oceanogr 26: 359-372

Mazzoldi A, Dalla Porta G, Gasparri A (1999) Misure di correnti costiere superficiali mediante radar costiero HF. Proceedings XIV Congr. AIOL XIII(1): 1-10

Paduan JD, Rosenfeld LK (1996) Remotely sensed surface currents in Monterey Bay derived from shore-based HF radar. J Geophys Res 101 C9: 20669-20686

Paschini E, Artegiani A, Pinardi N (1993).The mesoscale eddy field of the Middle Adriatic Sea. Deep-Sea Res 40: 1365-1377

Prandle D (1987) The fine-structure of nearshore tidal and residual circulations revealed by HF radar surface current measurements. J Phys Oceanogr 17: 231-254

Prandle D (1991) A new view of near-shore dynamics based on observations from HF radar. Prog Oceanogr 27: 403-438

Prandle D, Ryder DK (1985) Measurements of surface currents in liverpool bay by high-frequency radar. Nature 315: 128-131

Prandle D, Loch SG, Player R (1993) Tidal flow through the Straits of Dover. J Phys Oceanogr 23: 23-37

Shay LK, Graber HC, Ross DB, Chapman RD (1995) Mesoscale ocean surface current structure detected by high-frequency radar. J Atmos Oceanic Technol 12: 881-900

Hydrological Characteristics and Dynamics of the Northern Adriatic during Late Summer and Autumn 1997

V. Cardin and L. Ursella

ABSTRACT

Horizontal and vertical distributions of the hydrological data monitored during the autumn campaign of the PRISMA 2 project in the Northern Adriatic in 1997 show the presence of various mesoscale cyclonic and anticyclonic structures, as well as a strong coastal front separating water of fluvial origin from that of the open sea. The surface layer, in the southern part of the studied area, presents higher density and salinity than in the northern part due to the cyclonic nature of the basin-scale Adriatic general circulation. On the contrary, the presence of a low density core in the bottom layer probably indicates an anticyclonic circulation. The application of a statistical method, *cluster analysis*, to temperature and salinity data have permitted the individuation of three groups with distinct characteristics, corresponding to three different water masses. Their spatial extension is in concordance with the oceanographic conditions already observed in the Northern Adriatic. The dynamic characteristics were obtained from ship-mounted ADCP current measurements to which the "detiding" method of Candela has been applied. The steady part, in agreement with the general circulation of the Adriatic Sea, indicates currents flowing southwards along the Italian coast; additionally there are indications of a sub-basin cyclonic circulation to the north of the Po river mouth. On the contrary, the low frequency residual current field shows a variable spatial structure, with currents alternating northward and southward, that broadly agrees with the horizontal and vertical distribution of the thermohaline parameters. Finally, it is observed that where water of fluvial origin is present, the residual current field indicates southward flow. The M2 tidal ellipses, as expected, are polarised in the centre of the basin along the NW-SE direction, while along the boundaries, they follow the coastline orientation.

Introduction

The Adriatic Sea is an elongated basin located between the Italian peninsula and the Balkans. Its northern end is very shallow and gently sloping, and can be divided into a shallower region where our study was carried out, and a deeper region, separated by the 40-m isobath. This part of the Adriatic, as also the rest of the basin, is subject to strong meteorological forcing producing a clear seasonal variability in circulation (Artegiani et al. 1997a). The river runoff is particularly strong in the northern area, affecting the circulation through buoyancy input. Due to this particular characteristic, the entire Adriatic Sea is considered as a dilution basin with an average annual fresh water gain of about 1metre, since precipitation and evaporation almost compensate each other (Raicich 1996). In the baroclinic circulation of the Adriatic Sea currents and gyres can be identified, whose strength and occurrence vary seasonally mainly in the northern and middle basins. During autumn, the western and eastern coastal areas are characterised by an intensification of the alongshore currents and generally the circulation is cyclonic presenting three main cyclonic gyres in the northern, middle and southern basins (Artegiani et al. 1997b). During spring and summer, the water column is strongly stratified due to a vertical temperature gradient, and fresh water inputs mainly affect the surface layer. In autumn and

Istituto Nazionale di Oceanografia e di Geofisica Sperimentale (OGS), Borgo Grotta Gigante 42/c, 34010 Sgonico (TS), Italy

F.M. Faranda, L. Guglielmo, G. Spezie (eds)
Mediterranean Ecosystems: Structures and Processes

winter, the water column is homogeneous, and the fresh water remains in the coastal zone, being separated from the off-shore waters by a frontal system which runs parallel to the coast (Franco 1989) associated with a southward surface flux. Furthermore, the Northern Adriatic has been recognised as a site of convection and water mass formation on the continental shelf (Malanotte-Rizzoli 1991). Strong surface cooling and evaporation due to the outbreaks of cold dry air coming mainly from the north-east (Bora wind) are responsible for the generation of dense water whose density varies between 29.4 kg/m^3 and 29.8 kg/m^3 (Artegiani and Salusti 1987).

Data and Methods

Temperature and conductivity were measured using a high resolution CTD during late summer and autumn in the northern Adriatic as part of the autumn campaign within the framework of the PRISMA 2 Project. The campaign was carried out from September 24 untill November 6 1997, with a total of 118 stations from which only 59 were used for the hydrological analysis (Fig. 1). The derived parameters, such as salinity and sigma-t s(S,t,0) are obtained by application of algorithms for the computation of fundamental seawater properties (UNESCO 1983).

The horizontal maps of temperature, salinity and sigma-t of the surface layer herein analysed were obtained from the application of the kriging method and correspond to the means of the first 4 dbar for the surface layer and the last 5 dbar for the bottom layer.

Temperature and salinity data were organised in a matrix and, after standardisation of the data, the *cluster analysis* statistical procedure (Legendre and Legendre 1983; Murtagh and Heck 1987; Cardin and Celio 1997) with a complete linkage method based on the similarity ratio matrix has been applied. The application of this method is justified by the nature of the data and from the need to highlight the cohesion of the groups maximising either the similarity within the groups or the differences among them.

Current data obtained from a ship-mounted

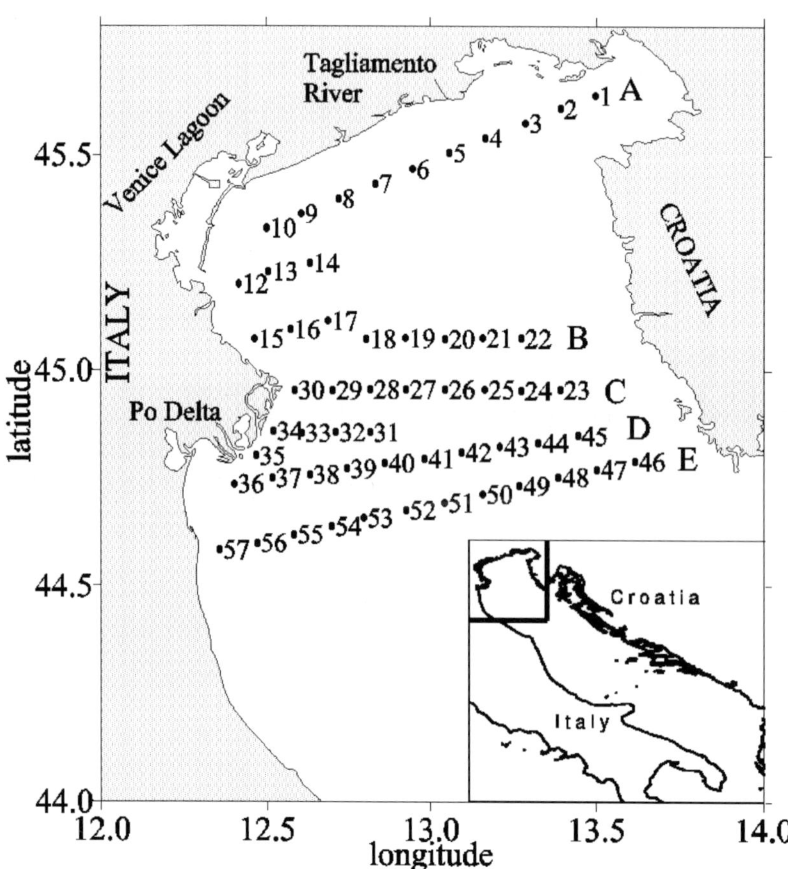

Fig. 1. Location of CTD stations monitored from 24 to 28 September 1997, during the PRISMA 2 campaign

ADCP were collected over nearly all of the autumn campaign along the ship track, with repetition of different transects, even when no CTD sampling was done. The lines along which current data are available are depicted in Fig. 6. The vertical resolution is of 2.5 m, while the horizontal one is of about 10-20 m, depending on the ship speed. The instrument is not able to make measurements in the first 8 m near the surface and in the layer near the bottom, corresponding to 15% of the whole depth. The surface and bottom layer currents have been extrapolated for each component, considering in the surface layer a constant value equal to the first available datum, and in the bottom layer, a logarithmic decay from the deepest datum to the zero bottom value. The accuracy in the current measurement is of the order of one cm/s, due also to the fact that the bottom tracking is always available and the speed of the ship is, in this way, measured with the same accuracy as the water velocities. The possible misalignment of the transducers with respect to the ship's stern and eventual errors in the gyrocompass are corrected in the first part of the processing. The data have then been averaged horizontally every 500 m along each transect. Thereafter, the method developed by Candela et al. (1992) was applied to this data set in order to obtain the circulation of the Northern Adriatic. It permits to separate three components in the signal: the barotropic part of the tide, the "steady" and the "residual" current. The steady component represents the time-independent part of the current over the period considered, while the residual one contains the remaining spatio-temporal variability. The method uses the least square procedure to fit horizontally the data to a function sum of a non-depending time term and of a linear combination of functions like sine and cosine of the tidal frequency, whose coefficients are functions of the spatial co-ordinates of the point considered. As suggested by the author of the method and already tested in a previous work (Gacic et al. 1998), polynomials were the most appropriate as the basis for the steady term and for the coefficients of the tides. The residual term presents a time variability on the order of a day or greater, and thus may contain signals related to wind effects or to discharge pulses of the rivers, among which the Po plays the main role. In fact, the river discharge is seen in the residual field only in concomitance with a sudden variation in the mean flow.

Results and Discussion

From its elongated shape, the T-S diagram indicates a high variability mainly in temperature, but also in salinity; in fact these parameters present values varying from 13.5 to 23°C and 30 to 38. Corresponding sigma-t variations are from 21 kg/m^3 to 28.5 kg/m^3. The presence of three main different water bodies can be observed: the light waters, located to the left of the upper part of the diagram are warmer (20-23°C) and less salty (30-35). The dense waters instead, are located at the bottom and present lower values in temperature but higher in salinity (around 14°C and 37-38). The intermediate waters could be recognised in the central part of the diagram, being characterised mainly by two close branches. The application of *cluster analysis* provides more information than the classical T-S diagram analysis, because it gives not only the characteristics of the possible water bodies presented in the area, but also their spatial distribution. The data have been separated into three clusters, each one being associated with a water mass possessing distinct characteristics (Fig. 2). The first cluster groups together water probably of a fluvial origin with a temperature around 20.91±0.63°C, and a relatively low salinity of 34.77±0.88. It is confined principally to the surface layer along the Italian coast in correspondence to river mouths and lagunal zones, and tends to flow along the coast varying in width and position as a function of the river flow inputs and the mesoscale variability. Due to the low fresh water discharge in the area (Po river discharge rate ~940 m^3/s) during the period when data were collected, this water is restricted to an area extending ~20 km out from the coast. The second cluster comprises water with a temperature similar to that of the preceding group (21.02±0.89°C), but with a higher mean salinity: 36.34±0.63. It is found in the intermediate layer of the coastal zone and reaches the surface out to sea. Finally, the third cluster is associated with water from the bottom layer below the pycnocline having a relatively low temperature (15.15±1.3°C) and high salinity (37.42±024).

The horizontal distributions of the thermohaline properties at the surface (Fig. 3a) show the presence of various mesoscale cyclonic and anticyclonic structures, as also a strong coastal front separating water of fluvial origin from that of the open sea. In the southern part of the studied

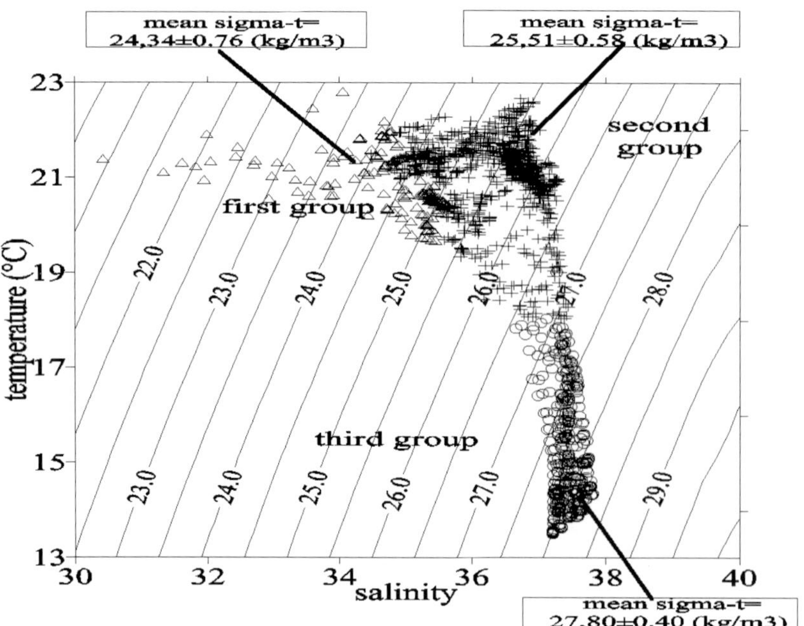

Fig. 2. T-S diagram showing the water masses obtained from the cluster analysis

area, a central zone of higher density and salinity due to the cyclonic nature of the basin-scale Adriatic general circulation is observed.

A different spatial pattern is seen for the bottom layer (Fig. 3b), where the distribution of the variables is defined as the averages in the 5 m layer above the seafloor. The density field is distinguished by a front that separates the denser waters situated in the central part of the basin from those less dense located off the western coastal zone. This front is determined, apart from the lower salinity of the coastal area, by the higher temperature due to the shallower bottom near the shore that permits a penetration of the seasonal heating down to the bottom. In the centre of the southern part of the studied area, a low density core (~27 kg/m^3) generated by the elevated temperature (~17.8°C) is observed; this

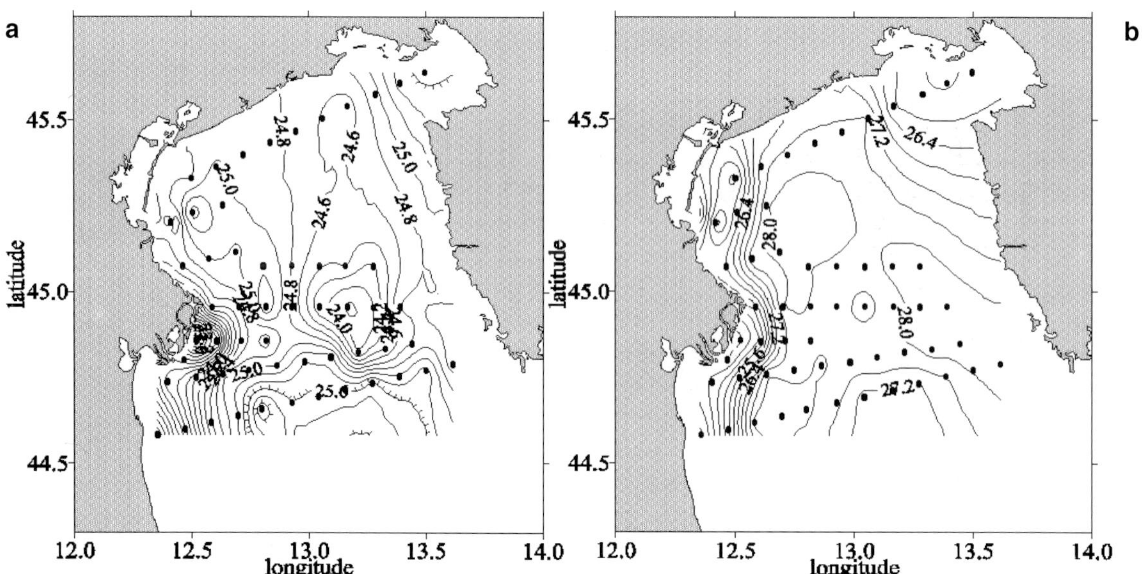

Fig. 3a,b. Horizontal distribution of sigma-t for: a surface layer; b bottom layer

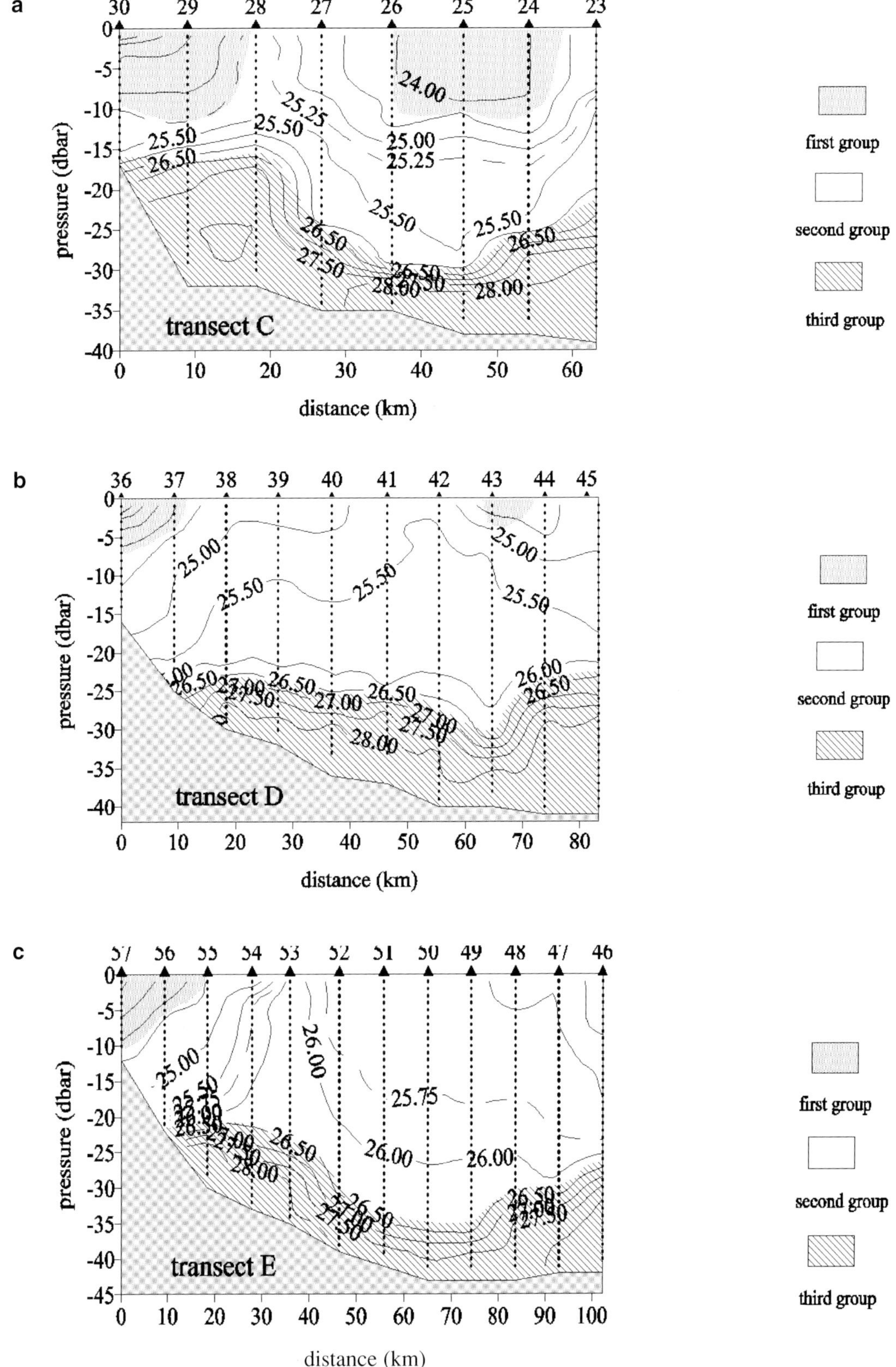

Fig. 4a-c. Vertical distribution of sigma-t (kg·m^{-3}) for: **a** transect *C*; **b** transect *D*; **c** transect *E*

may indicate the presence of an anticyclonic circulation opposite to that observed in the surface layer.

The vertical sections of the studied parameters across the Adriatic Sea during this period show the deepening of the thermocline from the north to south. At the northern end of the basin, the thermocline is positioned at about 8-10 m depth, while in the southern part (transect E), it almost reaches the bottom (5 m from the bottom). In Fig. 4, three vertical sections are presented corresponding to transect C, located off the northern mouth of the Po river, transect D off the southern mouth and transect E situated in the southern part of the studied area. Generally, the salinity conditions the sigma-t distribution above the thermocline, while below it, the temperature is the more important controlling factor. The doming of the isopycnals in the intermediate layer tends to divide the surface layer into two parts indicating the presence of various mesoscale cyclonic and anticyclonic structures.

For what concerns the analysis of the ADCP data, the overall length of the data record permits to calculate only the M2 tidal component. The transport ellipses obtained for M2 are polarised in the centre of the basin along the NW-SE direction (coincident with the Adriatic longitudinal axis), while near the Italian coast they change their orientation, almost following the shoreline. The maximum tidal amplitude is found in the central part of the Adriatic Sea where the major semi-axals shows values of around 6 m^2/s corresponding to a velocity of 10 cm/s; the minimum is found immediately north of the Po river delta, where the major semi-axis decreases to 0.5 m^2/s with a corresponding velocity of about 1 cm/s (Fig. 5). In the central part of the basin the behaviour of the tidal transport ellipses is coherent with the results obtained from numerical models (Cavallini 1985; Malacic et al. 1998). On the contrary, in the northernmost part, just outside the Gulf of Trieste, the ellipses in general have smaller semi-axes and they are rotated by about 90 degrees with respect to the ones coming from the models; this is probably due to insufficient data coverage in these zones and to the lack of imposition of boundary conditions.

The structure obtained for the steady part of the current field shows a southward current along the Italian coast, as well as the tendency to maintain a sub-basin cyclonic circulation in the northernmost area to the north of the Po River delta (Fig. 6). The existence of this cyclone has already been revealed by Borzelli et al. (1992) with a surface drifter experiment. The structure

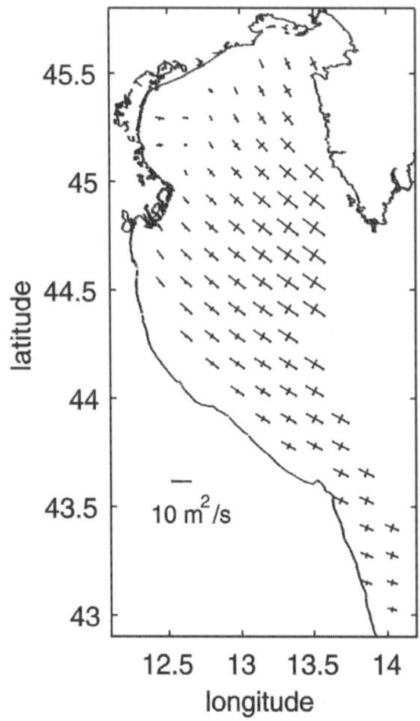

Fig. 5. Tidal transport ellipses for the M2 tidal component

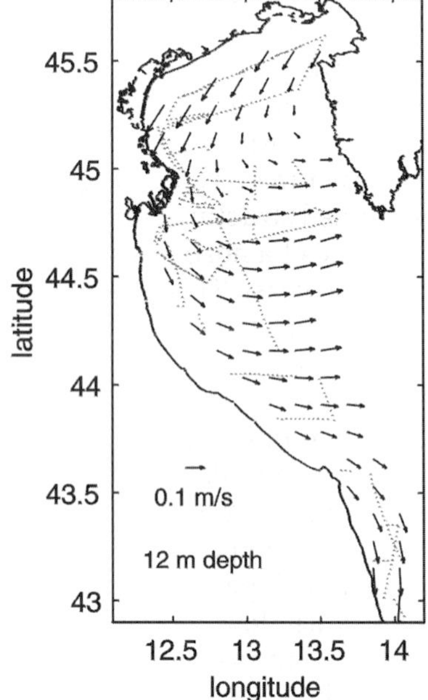

Fig. 6. Horizontal distribution of steady currents at 12 m depth. Data locations are indicated

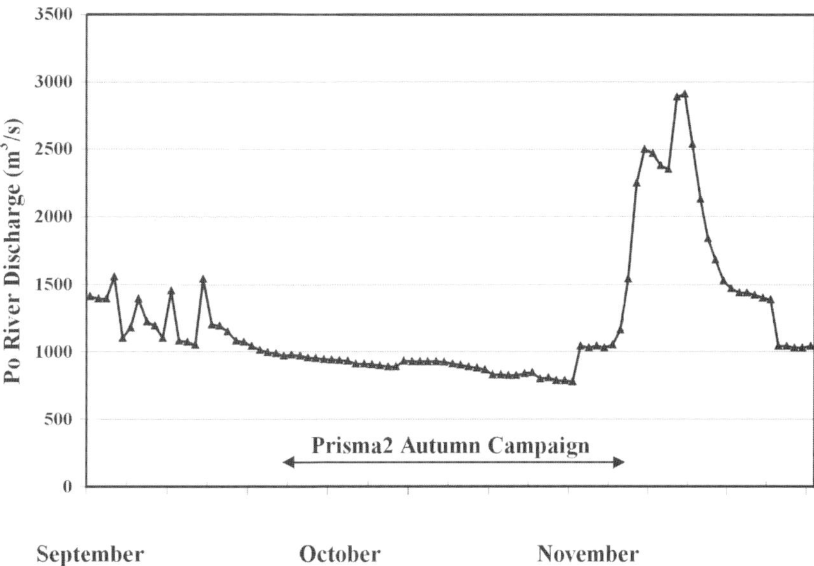

Fig. 7. Po river discharge for October-November 1997

found in the steady field is maintained almost throughout the water column and its behaviour seems to be barotropic, at least concerning the main features. It should be remembered that these velocities are representative of the general behaviour of the circulation over the whole campaign considered; therefore the barotropic behaviour is probably due to the fact that, for the most part of the period, the water column is homogenised over the major part of the studied area. Finally it must be pointed out that the steady field accounts for the basin-wide general circulation, but not for the small scale structures.

On the contrary, the residual pattern changes from transect to transect, reflecting the variability associated with river discharges and mesoscale structures. Therefore, it presents a more complicated structure and shows occasionally the presence of a northward flowing current along the Italian coast. This feature is unusual if seen in terms of the Adriatic general circulation, but it must be remembered that the residual field presents a variability on the order of several days, and to obtain the measured current at one point, the steady and tidal terms must be added to this one. From the Po river discharge (Fig. 7), it is evident that no important pulses are present during the studied period, so that no strong southward residual flow is expected at any of the transects situated in front or to the south of the Po river delta. This means that no direct response is expected and any eventual current has to be explained in terms of wind effects or mesoscale structures detached from the coastal steady current. The vertical distribution of the residual current components perpendicular to transects C (data collected on 27/09/97), D (data collected on 27/09/98) and E (data collected on 28/09/97), is depicted in Figs. 8, 9 and 10, respectively. It shows clearly that effectively no southward flow is present at transects D and E. The only exception is transect C, for which the velocity pattern can be explained in terms of a mesoscale eddy detached from the mean southward current. A strong southward flow is also present at some transects in the northernmost part of the studied area. All of this zone is characterised by an along-shore narrow current due to the discharge of rivers, that flows along the Italian coast with the coast on the right. Unfortunately, no data about the discharge of these rivers are available. Looking at these sections in detail, it can be seen that transect A presents a southward velocity in correspondence to the Tagliamento river mouth, while the rest of the section is occupied by the northward current. Perhaps this is due to the fact that the transect is distant from the coast and it does not reach the coastal boundary layer; also, the ADCP data do not cover completely the hydrological section. Proceeding with the sections toward the south, the southern flow on the western side becomes stronger, reaching the maximum just south of the southernmost mouth of the Venice lagoon. As the dimensions of the current reversals are of the order of 10 km, the observed structures could be associated with

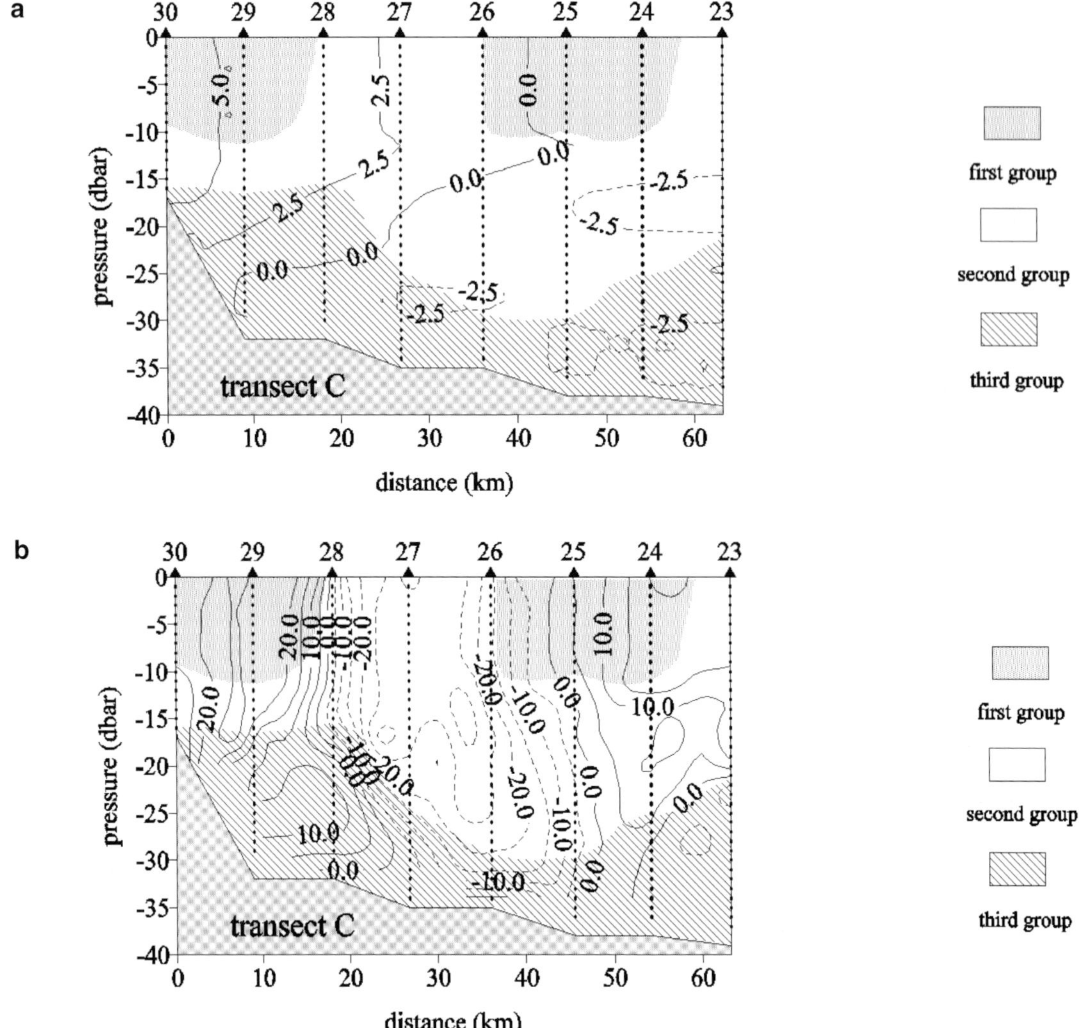

Fig. 8a,b. **a** Steady and **b** residual across section current field at transect *C*. *Positive* values indicate southward flow, while *negative* ones mean northward flow. Water masses individuated by the cluster analysis are superimposed

mesoscale eddies. The difference in the coastal current intensities at various transects is probably due to the different distance from the coast of each section.

In the surface layer, where water of fluvial origin is present, the residual field seems to fit the water mass pattern obtained from the cluster analysis better than the steady one. On the contrary, in the bottom layer, the water mass pattern is similar to the steady field. As the residual field presents a variability on the order of a few days, the correlation found at the surface may suggest a strong temporal variability of the field of mass as well. Also, as the residual currents reflect very closely the mass distribution, their behaviour seems to be geostrophic in a first approximation. Comparing at the surface, for the period during which the CTD data were collected, the distribution of the residual field of different transects with the spatial pattern obtained from the cluster analysis, it can be established that in most cases where water of fluvial origin is present, the residual current field indicates southward flow. The spatial correspondence of these structures is only qualitative and sometimes is not perfect; this is probably because the resolution of the residual field is half a kilometre, while the one of the field of mass is about 10 km and does not permit to resolve smaller structures. The similarity has been observed almost over the entire area. In general, the alternance of the inflows and outflows in the residual current field broadly agrees

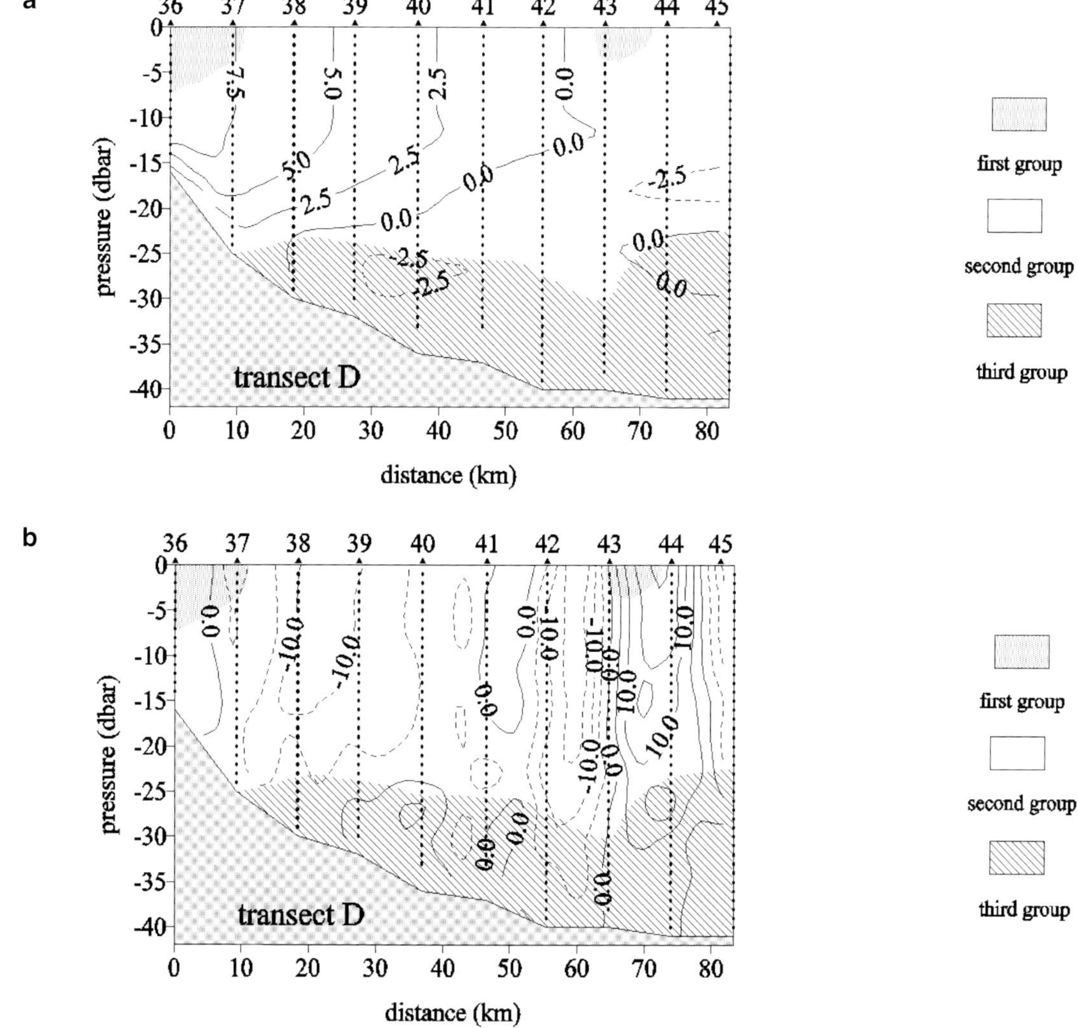

Fig. 9a,b. **a** Steady; **b** residual across section current field at transect *D*. *Positive* values indicate southward flow, while *negative* ones mean northward flow. Water masses individuated by the cluster analysis are superimposed

with the horizontal and vertical distributions of the thermohaline parameters. Keeping this in mind and looking at the three transects analysed (C, D and E) in more detail, the characteristics of the hydrological and current fields can be related. The two cores of water of fluvial origin present at transect C (indicated by the first group in the cluster analysis) are evident in both the horizontal and vertical density fields, and may have a unique origin, the Po river. A series of pulses in the Po river discharge, during the first part of September may be the origin of a core of fresh water, observed as two structures split by the northward flow in Figs. 8b, 9b and 10b. The two mesoscale structures present a different behaviour: the first one tends to flow southward along the Italian coast, and it is seen at the south-western part of the basin; the second one seems to join the anticyclonic gyre positioned along transect C, seen partially at transect D (small core of first group water and southward residual flow), and absent at transect E. Between these two structures, there is a doming of the isopycnals in correspondence to station 28, which is the centre of a cyclonic gyre in the residual field pattern. The same thing happens at section E around station 53, where a core of high density water is present; the residual field, as expected, presents a strong (30 cm/s) northward current east of this station, which decreases west of it, but does not reverse the sign. This is probably due to the fact that the horizontal gradient on the west side of

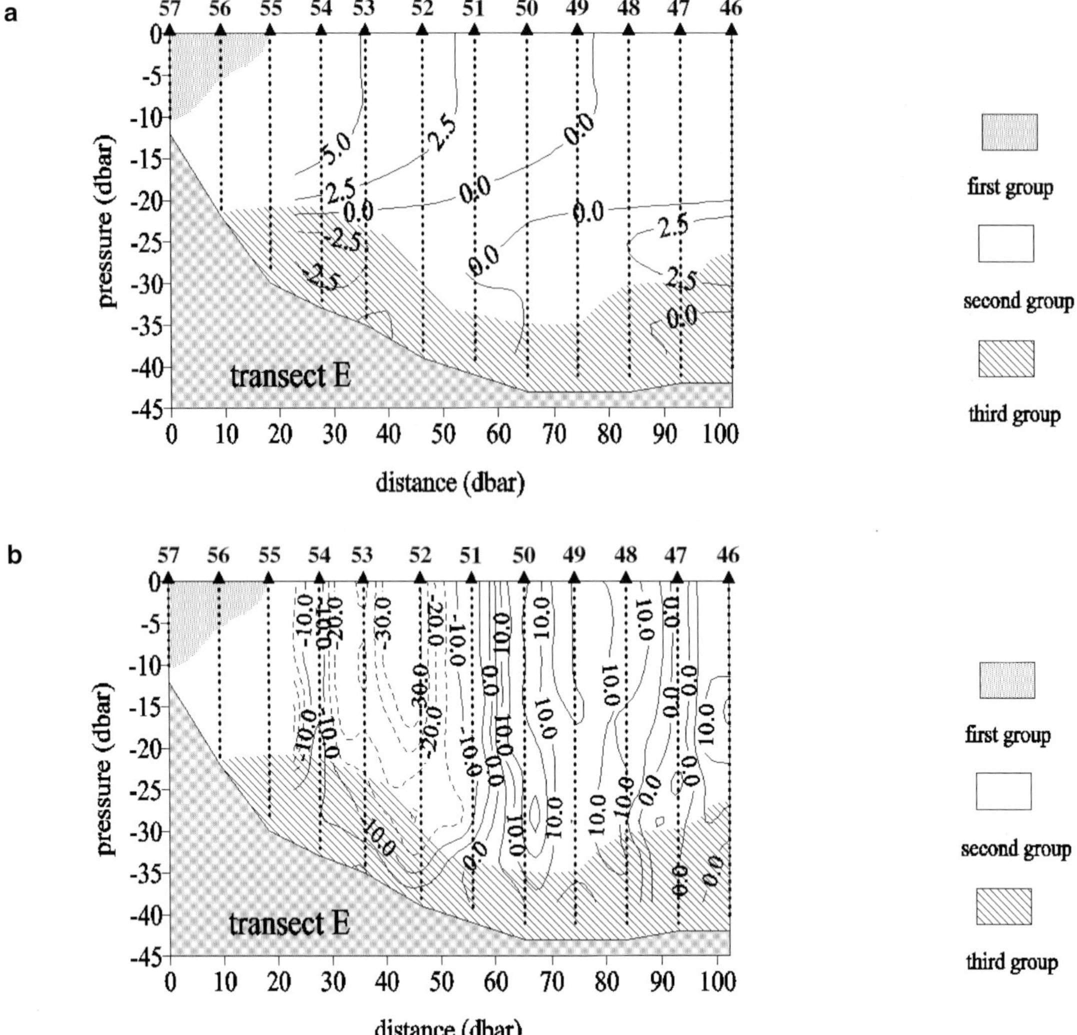

Fig. 10a,b. a Steady and **b** residual across section current field at transect *E*. *Positive* values indicate southward flow, while *negative* ones mean northward flow. Water masses individuated by the cluster analysis are superimposed

the dome is much weaker than on the other side, and moreover the ADCP data do not cover completely the part of the transect west of the dome.

Conclusions

The application of the cluster analysis to temperature and salinity data shows a close correlation with the oceanographic conditions present in the northern Adriatic, and permits to identify three main groups that could be associated with water masses.

The horizontal distributions of thermohaline properties throughout the water column indicate very clearly the presence of various cyclonic and anticyclonic structures corroborating the importance of the mesoscale dynamics in the Adriatic Sea. A strong coastal front of fluvial origin running southwards separates the denser waters, localised in the central part of the basin, from those less dense located off the western coastal zone.

The method of Candela applied to the ADCP data allowed to determinate the M2 tidal component as well as the steady and residual part of the current field; results that seems to be coherent and in concordance with the oceanographic conditions of the Northern Adriatic. The transport ellipses obtained for M2 are polarised in the centre of the basin along the NW-SE direction, while near the Italian coast they run almost parallel to it. The steady field presents a barotropic behaviour, in which it is possible to distinguish a southward current along the Italian coast and a cyclonic circulation in the northernmost part of the basin. Finally, the high variability observed in the residual field should be associated with river discharges and mesoscale structures. Therefore, it can be concluded that the steady field describes the basin-wide general circulation, while small scale structures are mainly associated with the residual field.

The comparison of the results obtained from the cluster analysis and the current field shows a better correlation between the mass pattern and the residual component; in particular it can be established that where water of fluvial origin (first group) is located, a southward flow is present. The high correlation between the field of mass and the time-dependent component of the current field which varies on a time scale of several days, suggests that geostrophy represents a rather good approximation for this part of the flow.

Acknowledgements. This research has been undertaken in the framework of the Programma di Ricerca e Sperimentazione del Mare Adriatico Project (PRISMA 2). We thank Dr. Miroslav Gacic for helpful comments on the manuscript.

References

Artegiani A, Salusti E (1987) Field observations of the flow of dense water on the bottom of the Adriatic Sea during the winter of 1981. Oceanol Acta 10 (4): 387-391

Artegiani A, Bregant D, Paschini E, Pinardi N, Raicich F, Russo A (1997a) The Adriatic sea general circulation. I. Air-sea interactions and water mass structure. J Phys Oceanogr 27(8): 1492-1514

Artegiani A, Bregant D, Paschini E, Pinardi N, Raicich F, Russo A (1997b) The Adriatic sea general circulation. II. Baroclinic circulation structure. J Phys Oceanogr 27(8): 1515-1532

Borzelli G, Ligi R, Ferulano E (1992) Surface circulation in the Northern Adriatic Sea as revealed by a drifter experiment. Nuovo Cimento 15(3): 265-274

Candela J, Beardsly RC, Limeburner R (1992) Separation of tidal and subtidal currents in ship-mounted Acoustic Doppler Current Profiles observations. J Geophys Res 97: 769-788

Cardin V, Celio M (1997) Cluster analysis as a statistical method for identification of the water bodies present in the Gulf of Trieste (Northern Adriatic Sea). Boll Teor Geofis Appl 38(1-2): 119-135

Cavallini F (1985) A three-dimensional numerical model of tidal circulation in the northern Adriatic Sea. Boll Oceanol Teor Appl 3: 205-218

Franco P (1989) Osservazioni sull'oceanografia fisica e chimica dell'Adriatico. In: Atti del Convegno, Lo stato di salute dell'Adriatico: problemi e prospettive. Urbino 23-24 Maggio 1989: 21-29

Gacic M, Civitarese G, Ursella L (1998) Spatial and seasonal variability of water and biogeochemical fluxes in the Adriatic Sea. In: Malanotte-Rizzoli P, Eremeev VN (eds) The eastern Mediterranean as a laboratory basin for the assessment of contrasting ecosystems. Kluwer Academic Publishers, Amsterdam, pp 335-357

Legendre L, Legendre P (1984b) La structure des donnèes écologie. Ecologie numérique 2nd ed. 2. Masson, Paris et les Presses de l'Université du Québec, 363 pp

Malacic V, Viezzoli D (1998) Tidal dynamics in the Gulf of Trieste - Northern Adriatic. In: 35th CIESM Congr Proc, vol 35(1), pp 172-173

Malanotte-Rizzoli P (1991) The Northern Adriatic Sea as a prototype of convection and water mass formation on the continental shelf. In: Chu PC, Gascard JC (eds) Deep convection and deep water formation in the oceans (Elsevier oceanography series, 57) Elsevier, Amsterdam, pp 229-239

Murtagh F, Heck A (1987) Multivariate data analysis. Astrophysics and Space Science Library N° 131. Kluwer Academic Publishers, Dordrecht, 205 pp

Raicich F (1996) On the fresh water balance of the Adriatic Sea. J Mar Syst 9: 305-319

CHAPTER 5

Biological Utilisation of the Major Nutrients in the Western Coastal Waters of the North Adriatic Sea

S. Cozzi and G. Catalano

ABSTRACT

The availability and distribution of the dissolved inorganic and organic fractions of nitrogen (DIN and DON) and phosphorus (DIP and DOP) are extremely variable in the north-western Adriatic Sea. Freshwater inputs, mainly from the Po River, and biological processes significantly affect the nutrient behaviour in this region. The purpose of this contribution is to describe these processes by the comparison of the dissolved nitrogen and phosphorus distributions to computed estimates of their biological production or consumption in the marine environment. The variations of dissolved nitrogen were mainly due to freshwater inputs of DIN, especially in winter (up to 18 μmol-N•dm^{-3}), while a high marine release of DON was observed in summer (up to +22 μmol-N•dm^{-3} compared to the contribution of advection). The DIP concentrations were generally lower than 0.06 μmol-P•dm^{-3}, while DOP values ranged from 0.01 to 0.1 μmol-P•dm^{-3}. The evident inverse relationship between DIP and DOP behaviours indicated that the phosphorus availability in the coastal zone is mainly regulated by recycling processes rather than by terrestrial inputs.

Introduction

The knowledge of the distribution of inorganic and organic nutrients and their biological pathways are fundamental to understanding a marine ecosystem. This is particularly important in a region like the northern Adriatic Sea, where nutrient availability is strongly influenced by the concurrent effects of heavy terrestrial inputs, transport due to water circulation and biological processes (Degobbis and Gilmartin 1990; Zoppini et al. 1995; Artegiani et al. 1997). Their fate has been related to undesirable phenomena that have occurred in this area: unusually intense algal blooms, near anoxia or anoxia in bottom layers and more frequent mucilage formation events than in other Adriatic regions (Vollenweider et al. 1992; Degobbis et al. 1995; Obernosterer and Herndl 1995; Tommasino 1996; Malone et al. 1999).

The aim of this work is to describe the availability of dissolved nitrogen and phosphorus, and the effects of biological phenomena of removal and release on their inorganic and organic fractions, in the western waters of the northern Adriatic Sea, which are heavily influenced by the Po River plume. These processes are discussed by comparing the relevant distributions of concentration to computed estimates of their conservative (driven by mixing processes) and non-conservative (basically attributed to biological assimilation or regeneration) components in the marine environment, obtained assuming the existence of a net nutrient balance during the dilution of the Po River waters in the basin.

Materials and Methods

The presented data were collected during two cruises of the PRISMA 2 "Biogeochemical Cycles" Research Project (June 1996 and February 1997). In both cruises, we chose transects (EF and EF-A, respectively), which are representative of the mixing between seawater and the Po River waters (Fig. 1), considered to be the main allocthonous source of nutrients in the northern Adriatic Sea.

CNR, Istituto Talassografico di Trieste, Viale Romolo Gessi 2, 34123 Trieste, Italy

F.M. Faranda, L. Guglielmo, G. Spezie (eds)
Mediterranean Ecosystems: Structures and Processes

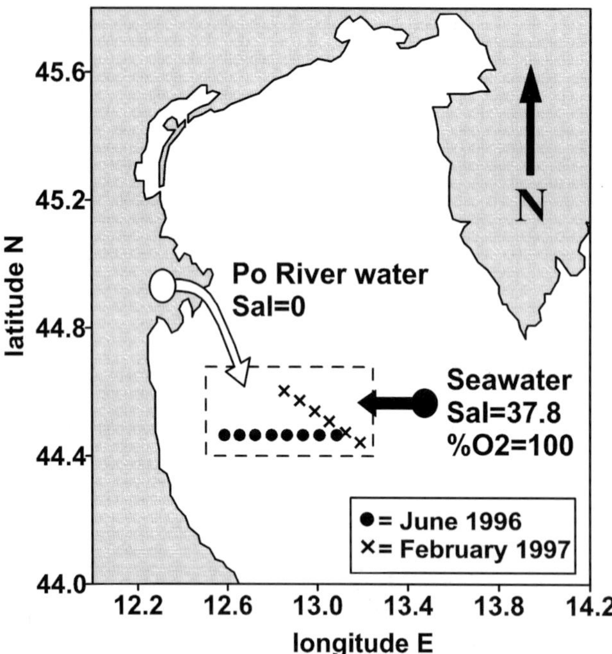

Fig. 1. Sampling stations of the PRISMA 2 "Biogeochemical Cycles" Research Project: June 1996 transect EF (•) and February 1997 transect EF-A (×)

Both transects cut across the frontal system delimiting the zone directly influenced by the river plume, and the offshore zone influenced by the diluted river inputs. This has permitted the estimate of inorganic and organic nutrient balances in both these environments, in late winter and early summer, during different phases of the evolution of the coastal front.

At every station, salinity data were computed from the downcast, and samples for the determination of the inorganic nutrients and dissolved organic nitrogen (DON) and phosphorus (DOP) were filtered through Whatman GF/F glass microfibre filters. Analyses of dissolved inorganic nitrogen (DIN = nitrates + nitrites + ammonium) and dissolved inorganic phosphorus (DIP = phosphates) were performed by standard spectrophotometric methods (Grasshoff 1983). The DON and DOP concentrations were determined by the photo-oxidation (UV + hydrogen peroxide) method of Walsh (1989).

The DIN, DIP, DON and DOP concentrations biologically consumed or produced [$\Delta-(i)$] were estimated by subtracting the values ascribed to freshwater dilution from the measured concentrations:

$$\Delta-(i) = C(i) - (C(\mathrm{riv})i \cdot W(\mathrm{riv})) - (C(\mathrm{sea})i \cdot W(\mathrm{sea}))$$

where $C(i)$ is the concentration ($\mu\mathrm{mol}\cdot\mathrm{dm}^{-3}$) of the nutrient ($i$) measured in the sample and $C(\mathrm{riv})i$ and $C(\mathrm{sea})i$ are, respectively, the nutrient (i) concentrations in Po River waters (monthly average, n=4, at Po Pontelagoscuro and Po Serravalle stations, ARPA Regione Emilia Romagna, Sede Provinciale of Ferrara, personal communication) and in the sea. The freshwater and seawater fractions in the sample [$W(\mathrm{riv})$ and $W(\mathrm{sea})$ respectively] were calculated by assigning a salinity zero to the Po River waters and 37.8 for seawater. This value of salinity and every C(sea)i concentration are an average of the relevant data from the stations farthest from the Italian Coast, at depths not significantly influenced by terrestrial inputs, where nutrient assimilation and regeneration rates were assumed approximately to be balanced (oxygen saturation =100%).

By assuming that physico-chemical processes of sedimentation have a negligible influence on the considered parameters, $\Delta-(i)<0$ means that the concentration of the nutrient (i), resulting from the mixing between fresh and seawater was further reduced by phytoplankton assimilation. Thus $\Delta-(i)>0$ indicates that the regeneration prevails over assimilation processes with a consequent increase in nutrient concentrations. If $\Delta-(i)$ is close to zero, assimilation and regeneration should be nearly balanced so that a conservative behaviour of nutrient concentrations would be expected.

Results and Discussion

The following figures (Figs. 2, 5) show both distributions of concentration and Δ-(*i*) balances for each considered nutrient along the transects of Fig. 1. Black lines, visible in the figures, indicate the position of the halocline which separates, in both seasons, an upper layer, more influenced by the freshwater inputs (range of salinity from 29.0 up to 35.0), from a bottom layer and/or offshore zone (salinity higher than 35.0) less influenced by continental inputs.

Dissolved Nitrogen in Late Winter

In this period, the distribution of the DIN concentrations (Fig. 2a) shows a decreasing trend from the upper layer (18 μmol-N·dm^{-3}) to the bottom (1 μmol-N·dm^{-3}), inversely related to the salinity, which indicates the conspicuous input of inorganic nitrogen associated with the advected Po River waters. The DON concentrations show (Fig. 2b) a more homogeneous distribution than the DIN fraction (from 6 up to 14 μmol-N·dm^{-3}) which represents the most abundant pool of the dissolved nitrogen.

In spite of the high DIN concentrations, Δ-(DIN) (Fig. 2c) is negative in the upper layer because of the biological uptake of this inorganic nutrient (values from −6 up to −18 μmol-N·dm^{-3}). On the contrary, the lower DIN concentrations under the halocline are related to a *quasi*-conservative [Δ-(DIN) close to zero] behaviour of the inorganic nitrogen.

The Δ-(DON) values (Fig. 2d) are positive in the upper layer (up to +7 μmol-N·dm^{-3}), indicating a biological release of DON subsequent to the phytoplankton activity, while they are close to zero in the deeper part of the water column indicating, also for DON, a conservative behaviour in the bottom layer.

Dissolved Phosphorus in Late Winter

The distribution of DIP (Fig. 3a) does not show the advection of phosphate due to the freshwater input in the upper layer but, on the contrary, its highest values of concentration are observed in the deeper offshore zone (up to 0.05 μmol-P·dm^{-3}). The DOP concentrations (Fig. 3b) decrease from the inshore (0.07 μmol-P·dm^{-3}) to the offshore (0.01 μmol-P·dm^{-3}) zone, inversely

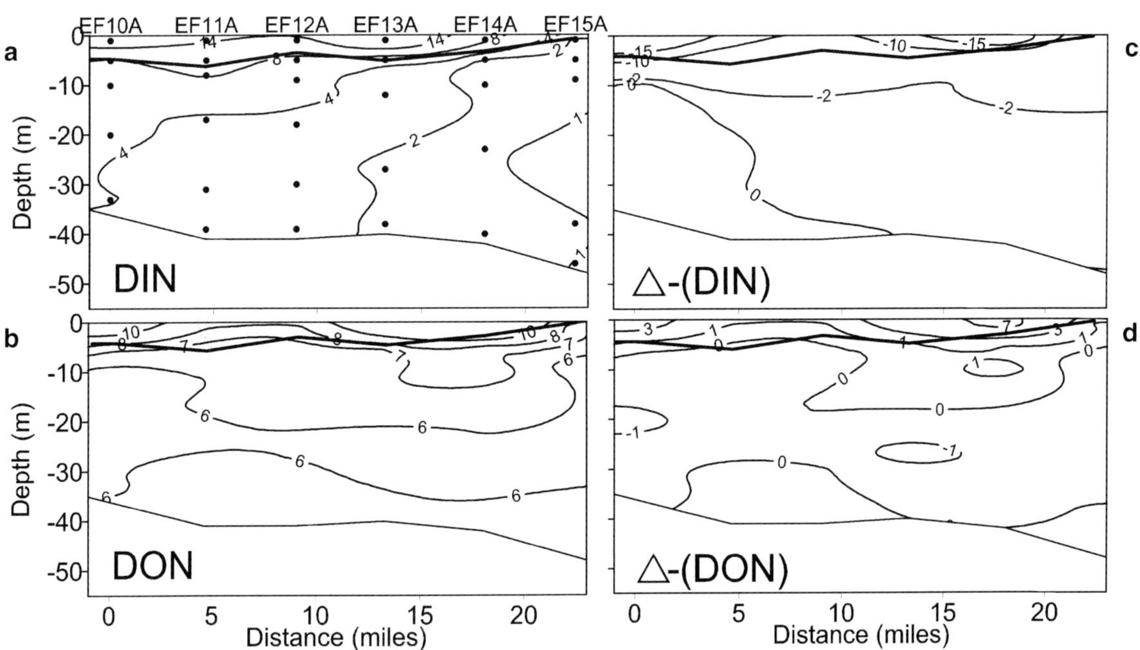

Fig. 2. Concentrations and Δ-(i) values (μmol-N·dm^{-3}) of dissolved inorganic and organic nitrogen (*DIN* and *DON*) in February 1997 (transect EF-A)

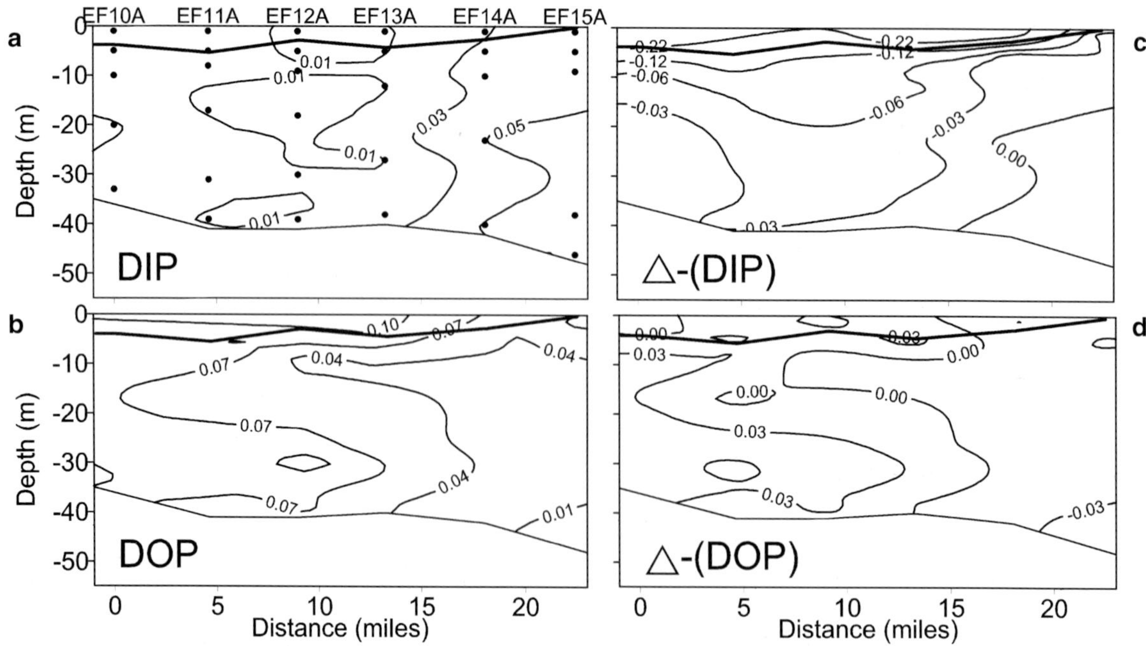

Fig. 3. Concentrations and Δ-(*i*) values (μmol-P·dm^{-3}) of dissolved inorganic and organic phosphorus (*DIP* and *DOP*) in February 1997 (transect EF-A)

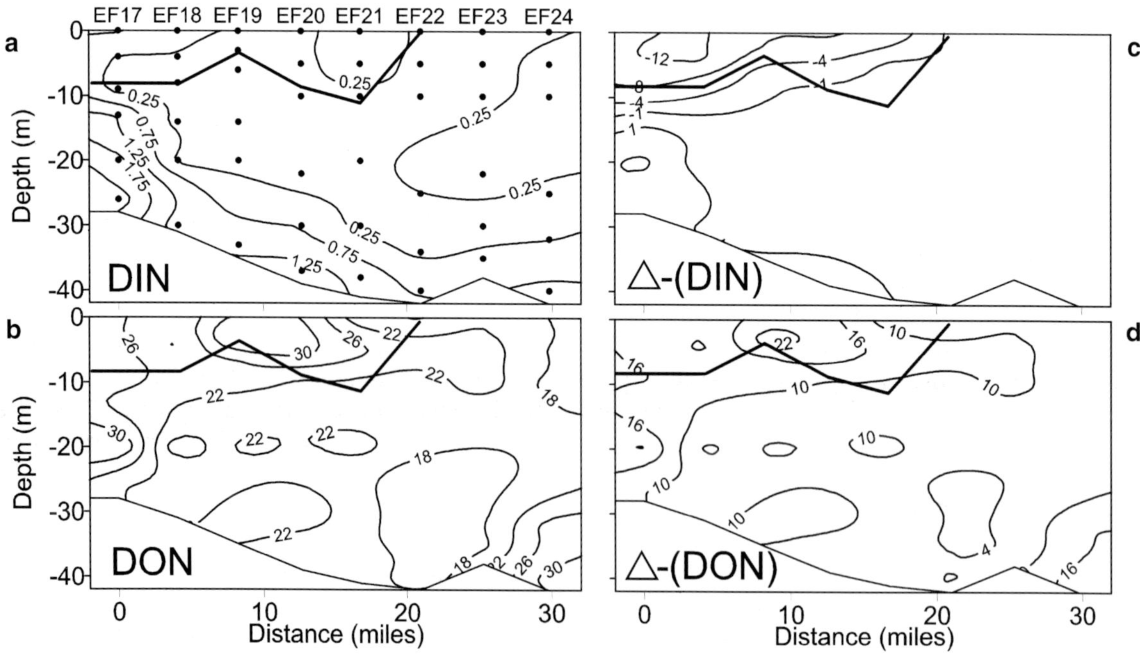

Fig. 4. Concentrations and Δ-(*i*) values (μmol-N·dm^{-3}) of dissolved inorganic and organic nitrogen (*DIN* and *DON*) in June 1996 (transect EF)

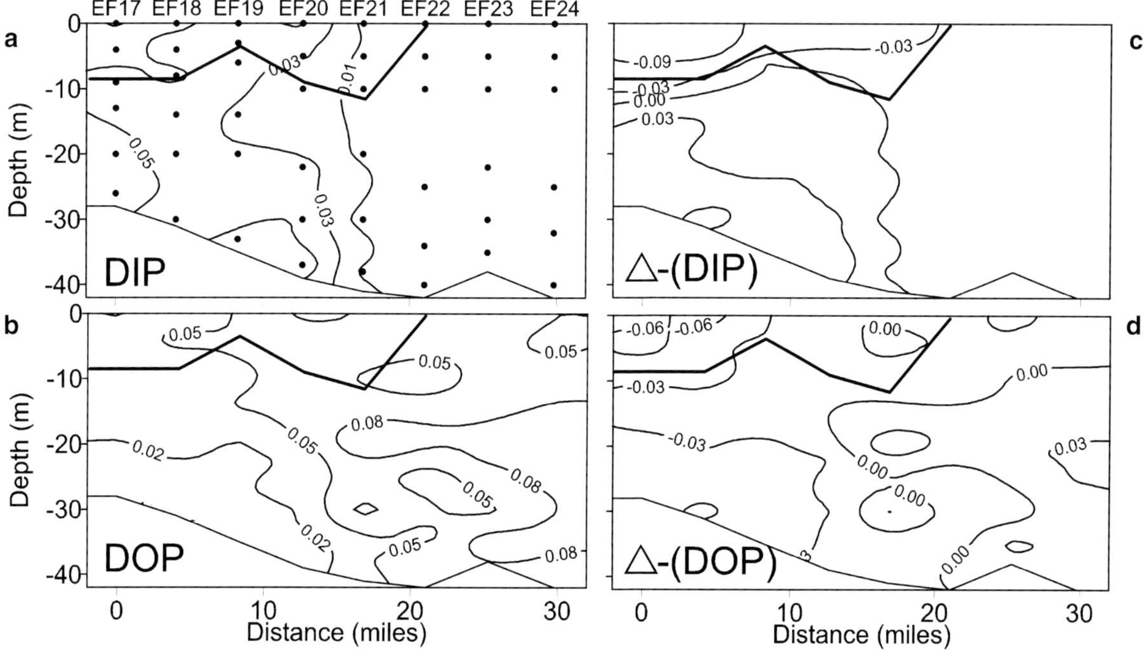

Fig. 5. Concentrations and Δ-(*i*) values (μmol-P·dm^{-3}) of dissolved inorganic and organic phosphorus (*DIP* and *DOP*) in June 1996 (transect EF)

to the DIP trend. Both these patterns indicate the prevalence of the biological processes rather than terrestrial inputs in regulating phosphorus availability.

The Δ-(DIP) distribution (Fig. 3c) shows a trend similar to that of Δ-(DIN): the most negative values are located in the upper layer (from -0.13 up to -0.35 μmol-P·dm^{-3}) because of the phytoplankton uptake of the nutrient. Considering the dissolved organic fraction, the Δ-(DOP) distribution (Fig. 3d) indicates a positive balance in the coastal zone (+0.03 μmol-P·dm^{-3}) and a negative balance (up to -0.03 μmol-P·dm^{-3}) in the deeper offshore zone.

Dissolved Nitrogen in Summer

During summer, the low DIN concentrations (Fig. 4a) in the upper layer (less than 0.25 μmol-N·dm^{-3}) indicate that biological assimilation has depleted the continental input of the inorganic nitrogen in the layer from the surface down to 10 metres depth. In this situation, the main source of DIN (up to 2 μmol-N·dm^{-3}) are the remineralisation processes of nitrogen in the bottom layer. The DON fraction is very high, with concentrations (Fig. 4b) from 18 up to 35 μmol-N·dm^{-3}, representing almost the entire quantity of the available dissolved nitrogen in seawater.

The assimilation and regeneration processes of DIN are confirmed by the Δ-(DIN) distribution (Fig. 4c) which shows negative values in the inshore upper layer (up to -12 μmol-N·dm^{-3}) and positive values (up to +3 μmol-N·dm^{-3}) in the inshore deeper layer. The biological release of DON in the marine environment (Fig. 4d) is visible in all stations of the transect [Δ-(DON) from +4 up to +22 μmol-N·dm^{-3}] but it is particularly high where the continental inputs of DIN have been strongly consumed (the inshore upper layer).

Dissolved Phosphorus in Summer

The DIP distribution (Fig. 5a) clearly shows a decreasing trend from the zone near the Italian coast (0.05 μmol-P·dm^{-3}) to the offshore zone (less than 0.01 μmol-P·dm^{-3}) but without a significant relationship to the position of the halocline. As in winter, the DOP trend (Fig. 5b) is opposite to the DIP trend, indicating a coupling between the inorganic and organic fractions of phosphorus in the marine environment.

In this case, the Δ–(DIP) and Δ–(DOP) distributions (Fig. 5c and d) indicate that the DIP availability in the inshore zone is determined by both continental input and assimilation/remineralisation processes. In the upper layer, inorganic phosphorus is partially consumed [Δ–(DIP) values up to –0.09 μmol–P·dm^{-3} in Fig. 5c], while it is regenerated in the bottom layer by remineralisation [Δ–(DIP) equal to +0.03 μmol–P·dm^{-3}]. On the contrary, the offshore zone shows a depletion of DIP in the entire water column related to close to zero Δ–(DIP) values. Regarding the organic fraction (Fig. 5d), the main features are the negative Δ–(DOP) values in the coastal zone (–0.03 μmol–P·dm^{-3}) linked to phosphorus remineralisation and the quasi-conservative behaviour of DOP passing to the offshore zone.

Conclusions

The distributions of concentration and the presented nutrient balance in the northern Adriatic Sea highlight the roles of the continental inputs and biological (assimilation and regeneration) processes as sources or sinks of nitrogen and phosphorus. In particular, the necessity of the inclusion of their dissolved organic fractions in order to assess the total eutrophic state of the basin and to understand better the seasonal cycling of these elements is emphasised. In fact, during some periods, DON and DOP can reach almost up to 100% of the total dissolved nutrient.

The Po River discharge is an important source for both DIN and DIP, but, as far as phosphorus is concerned, the river supply is consumed within the coastal zone also in winter, while this does not occur for DIN. The inverse relationships between the inorganic and organic fractions are more evident for phosphorus than for nitrogen. These results also strongly support the hypothesis that, in the northern Adriatic Sea, phosphorus plays a greater role as a limiting element than nitrogen.

With respect to the nutrient balance, the Δ–(i) distributions point out the greater importance of biological cycling in controlling nitrogen and phosphorus availability (inorganic and organic) when compared to advection and mixing processes, also within a few miles from the coast. The photosynthetic extracellular release by phytoplankton (Fogg 1983; Lancelot 1983; Fogg 1996), the dissolution of particulate organic matter by attached bacteria (Azam et al. 1993), the activity of grazers (Ducklow and Carlson 1992; Strom et al. 1997) and the virus-induced bacteria lysis (Bratbak et al. 1990) are all important sources of dissolved organic matter. Further research is needed to establish the relative importance of these mechanisms in the northern Adriatic Sea.

Acknowledgements. The authors wish to thank the ARPA, Regione Emilia Romagna, Sede Provinciale of Ferrara for the Po River nutrient data.

References

Artegiani A, Bregant D, Paschini E, Pinardi N, Raicich F, Russo A (1997) The Adriatic Sea general circulation. I. Air-sea interactions and water mass structure. J Phys Oceanogr 27: 1492-1514

Azam F, Smith DC, Steward GF, Hangström Å (1993) Bacteria-organic matter coupling and its significance for oceanic carbon cycling. Microb Ecol 28: 167-179

Bratbak G, Hedal M, Norland S, Thingstad TF (1990) Viruses as partners in spring bloom microbial trophodynamics. Appl Environ Microbiol 56: 1400-1405

Degobbis D, Gilmartin M (1990) Nitrogen, phosphorus, and biogenic silicon budgets for the northern Adriatic Sea. Oceanol Acta 13 1: 31-45

Degobbis D, Fonda Umani S, Franco P, Malej A, Precali R, Smodlaka N (1995) Changes in the northern Adriatic ecosystem and the hypertrophic appearance of gelatinous aggregates. Sci Total Environ 165: 43-58

Ducklow HW, Carlson CA (1992) Oceanic bacterial production. In: Marshall KC (ed) (Advances in microbial ecology, 12) Plenum Press, New York, pp 113-181

Fogg GE (1983) The ecological significance of extracellular products of phytoplankton photosynthesis. Bot Mar 24: 3-14

Fogg GE (1996) The extracellular products of algae. Oceanogr Mar Biol Annu Rev 4: 195-212

Grasshoff K (1983) Methods of seawater analysis. In: Grasshoff K, Ehrhardt M, Kremling K (eds). Verlag Chemie, Weinheim, pp 125-215

Lancelot C (1983) Factors affecting phytoplankton extracellular release in the Southern Bight of the North Sea. Mar Ecol Prog Ser 12: 115-121

Malone TC, Malej A, Harding LW (1999) Ecosystem at the land-sea margin. In: Malone TC, Malej A (eds) Coastal Estuarine Studies 55, AGU, Washington, DC, pp 381

Obernosterer I , Herndl GJ (1995) Phytoplankton extracellular release and bacterial growth: dependence on the inorganic N:P ratio. Mar Ecol Prog Ser 116: 247-257

Strom SL, Benner R, Ziegler S, Dagg MJ (1997) Planktonic grazers are potentially important source of marine dissolved organic carbon. Limnol Oceanogr 42(6): 1364-1374

Tommasino MG (1996) It is feasible to predict "slime blooms" or "mucilage" in the northern Adriatic Sea? Ecol Model 84: 189-198

Vollenweider RA, Marchetti R, Viviani R (1992) Marine Coastal eutrophication. In: Vollenweider RA, Marchetti R, Viviani R (eds) Elsevier, Amsterdam, pp 21-581

Walsh TW (1989) Total dissolved nitrogen in sea water: a new-high-temperature combustion method and a comparison with photo-oxidation. Mar Chem 26: 295-311

Zoppini A, Pettine M, Totti C, Puddu A, Artegiani A, Pagnotta R (1995) Nutrients, standing crop and primary production in western coastal waters of the Adriatic Sea. Estuarine, Coastal Shelf Sci 41: 493-513

CHAPTER 6

Seasonal Variability of the Hydrography in the Adriatic Sea: Water Mass Properties and Circulation

B. Manca[1], P. Franco[2], and E. Paschini[3]

ABSTRACT

Recent quasi-synoptic seasonal observations (May 1995-February 1996) carried out within the framework of the national programme PRISMA (Programma di Ricerca e Sperimentazione del Mare Adriatico) have significantly contributed to a better understanding of longitudinal water exchanges across four sections of the Adriatic Sea. The observations show the complex variability due to forcing mechanisms (i.e. the atmospheric forcing, the fresh-water buoyancy input and the inflow of warm and highly saline water originating in the Eastern Mediterranean Sea) and the natural seasonal variability, as an inherent component of the system. During winter, a homogeneous cold and dense water mass is formed in the northern shelf area ($\sigma_\theta \cong 29.40$ kg·m^{-3}), while open-ocean deep convective movements lead to dense-water formation in the southern cyclonic gyre ($\sigma_\theta \cong 29.18$ kg·m^{-3}). The Northern Adriatic dense water (NADW) flows southward along the western continental shelf. A part of NADW flows over the Pelagosa sill with a potential density of about 29.14 kg/m^3 however, without filling the deepest part of the Southern Adriatic basin. The latter fraction reaches the Otranto Strait at the shelf break and intrudes into the Northern Ionian Sea. The Adriatic deep water (ADW), formed through open-ocean convection in the Southern Adriatic Sea, flows over the sill of Otranto to form the main component of the Eastern Mediterranean deep water. The Adriatic Sea is defined on the yearly scale as a dilution basin; thus, an inflow of water from the Ionian Sea, saltier and warmer than the Adriatic waters, i.e. the Levantine intermediate water (LIW), occurs. The LIW enters the Otranto Strait and penetrates, diluted, towards the eastern portion of the Southern Adriatic Sea where it is entrained cyclonically in the gyre. A branch of modified LIW extends over the Pelagosa sill and intrudes into the northern shelf regions. The maximum northward flow of the LIW was observed during the autumn, probably compensating the fresh water buoyancy input in the north that had occurred mainly in spring and early autumn.

Introduction

The Adriatic Sea is a semi-enclosed quasi-rectangular basin (≈ 800 km by latitude and 200 km by longitude) of the Eastern Mediterranean Sea, communicating with the Ionian Sea through the narrow (≈75 km) Otranto Strait. Topographically (Fig. 1) the Adriatic Sea presents a wide northern continental shelf region sloping from 20 m to a maximum deep trench of about 270 m in the middle of the basin. To the south, between the shallower Pelagosa sill (≈130 m) and the deeper Otranto sill (≈800 m), the basin is characterised by a large topographic depression reaching down to 1200 m. Here, the 200 m isobath marks the shelf break which extends ≈30-50 km off the western coastline. The shelf is steeper in proximity to the eastern shoreline. Thus, the physical processes occurring in the northern shelf region are strongly influenced by its geographic location and topography. On the other hand, the physical processes that occur in the southern basin are potentially comparable to those of the ocean, mainly the open-ocean deep convection.

[1] Istituto Nazionale di Oceanografia e di Geofisica Sperimentale, OGS, Borgo Grotta Gigante 42/c, 34010 Sgonico (TS), Italy
[2] CNR, Istituto di Biologia del Mare, Venezia, Italy
[3] CNR, Istituto di Ricerca sulla Pesca Marittima, Ancona, Italy

F.M. Faranda, L. Guglielmo, G. Spezie (eds)
Mediterranean Ecosystems: Structures and Processes

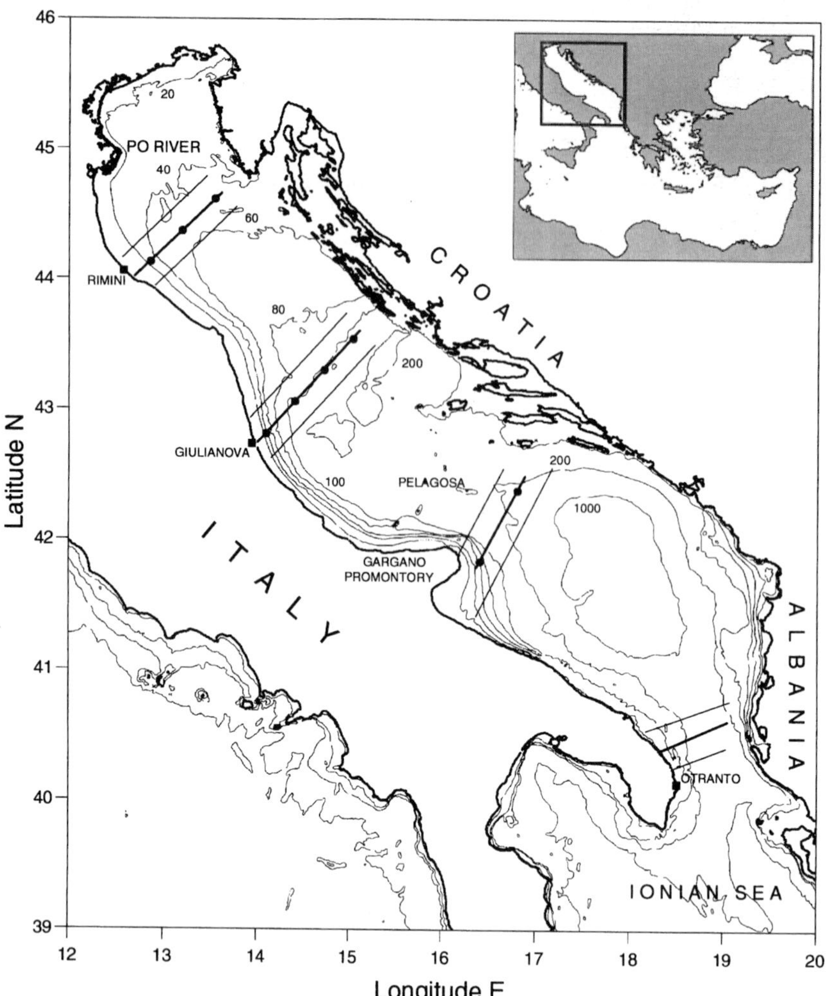

Fig. 1. The Adriatic Sea bottom topography and its geographic position in the Eastern Mediterranean. The *thick* and *thin lines* indicate the hydrographic sections occupied during the PRISMA programme. The position of the current measurements are marked by *dots*. Depths are given in metres

However, a strong interaction between the shelf and the open-sea waters may be expected, due to the combination of topographic and baroclinic constraints.

The main factors influencing the Adriatic Sea's hydrography are: the fresh water buoyancy input, coming mainly from river discharges in the northern shelf region and leading to a southward flow of the western coastal current which transports the diluted Adriatic surface water (ASW); the saline water input from the Ionian Sea, which leads to a northward flow along the eastern coast transporting Ionian surface water (ISW) and Levantine intermediate water (LIW); the atmospheric forcing, which leads to a two-layer baroclinic circulation during summer and to dense-water formation during winter. In fact, the Adriatic Sea has been noted as an area of convective movements and dense-water formation, namely the North Adriatic dense water (NADW), over the northern shelf by air-sea heat fluxes that induce the complete overturning of the water column during winter (Franco 1972; Malanotte Rizzoli 1991). Moreover, open-ocean deep convection processes occur in the Southern Adriatic Sea (Ovchinnikov et al. 1985; Manca and Bregant 1998), due to a topographically controlled cyclonic circulation, producing the Adriatic deep water (ADW).

In the past, many authors from in-situ data and modelling efforts described the main aspects of the baroclinic and the wind-driven circulation in the Adriatic Sea. Papers by Buljan and Zore-Armanda (1976), Franco et al. (1982), Orlic et al. (1992), Malanotte Rizzoli and Bergamasco (1983), Bergamasco and Gacic (1996), Artegiani et al. (1997), and related literature, offer wide syntheses of the basic processes which drive the Adriatic Sea's hydrography and dynamics. The insufficient data set, especially in the southern basin, makes a coherent and satisfactory description of the northern and southern basins quite difficult, as does the study of the shallow northern region synoptically with the deep southern one. Some basic processes, such as the dense-water formation in the southern Adriatic and the possible contribution of the NADW to the ADW have been very poorly documented. Knowledge of the role of fresh water input in the northern shelf area, as well as the considerable amount of saline water received from the Ionian Sea through the Otranto Strait, which drives the entire Adriatic baroclinic circulation, is still lacking.

The main objective of the sub-project "longitudinal fluxes", within the PRISMA programme (**P**rogramma di **RI**cerca e **S**perimentazione del **M**are **A**driatico), was the study on seasonal timescales of the water and the biogeochemical fluxes, across four significant transversal sections of

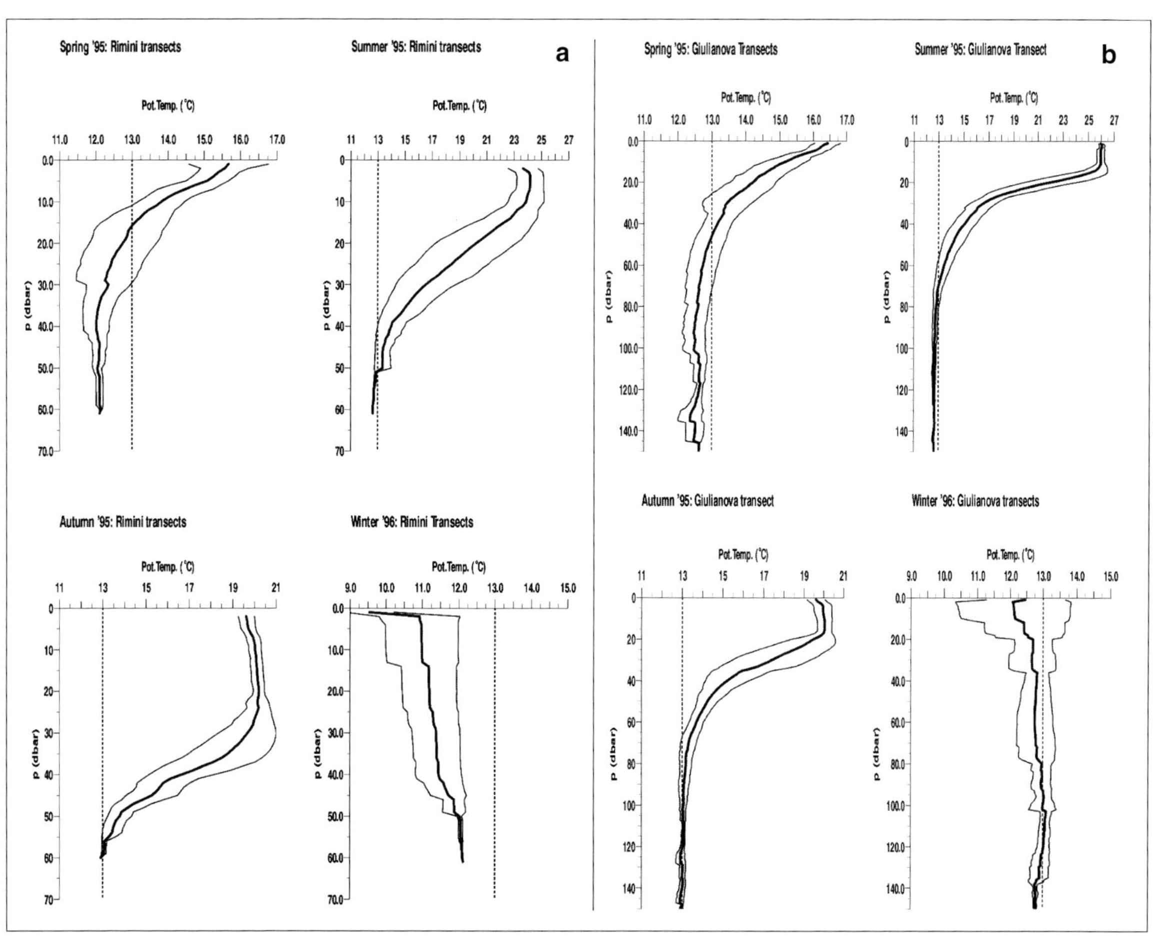

Fig. 2a,b. Spatially averaged vertical profiles of temperature during the seasonal cruises in the northern transects **a** and middle transects **b**. The thin lines denote one σ standard deviation

the Adriatic Sea. This paper refers mainly to the overall seasonal variability of the dynamics of the Adriatic Sea, focussing on the influence of the fresh water input and dense water formation on the general circulation patterns. The coastal currents and interior jets connecting the sub-basin gyres, and the water exchange regime between the northern, middle and southern basins are shown, and compared using direct current measurements.

Measurements

Four multidisciplinary surveys were made every three months (May, August, November 1995, and February 1996) at stations located in four areas (Fig. 1), each of them traversed by three parallel transects: in the northern shallow zone (the Rimini transects), in the middle Adriatic (the Giulianova transects), over the steep continental slope of the southern basin (the Gargano transects), and at the southern opening to the Ionian Sea (the Otranto transects). Temperature and salinity profiles were collected with a Seabird SBE-911 *plus* CTD system coupled with a 24-bottle General Oceanics rosette sampler. Water samples were analysed and SIS digital reversing thermometer readings were taken to provide data for the calibration of salinity and temperature; samples were also collected for biogeochemical determinations. Continuous Eulerian current measurements were conducted by means of current meter moorings deployed along the same sections.

Results and Discussion

Seasonal Thermal, Salinity and Density Cycles

The seasonal thermal cycle in the northern and middle Adriatic Sea is shown by the space-averaged vertical profiles of temperature computed separately for the Rimini and the Giulianova transects (Fig. 2). In winter, the most distinguishing feature is the noticeable decrease of temperature at the surface due to the maximum heat loss of the seasonally cold waters. The cooling effect involves the entire water column in the northern shelf area and reaches a depth of 80-100 m in the central basin. A vertically homogeneous cold and dense water mass, namely the NADW, is formed in the northern shelf area ($\sigma_\theta > 29.40$ kg·m^{-3}) by turbulent convective movements that reduce the vertical density gradients. A further, progressive decrease of temperature in the bottom layer of the central basin occurs in spring. This is due to the advection of the cold NADW which, formed during the previous winter, reaches the Giulianova sections later on.

In spring and summer, the upper thermocline becomes more pronounced, reaching a maximum depth of 50 m. The NADW is topped by the warm surface layer and is entrained southwards by the western coastal current. The NADW has been identified by Manca and Giorgetti (1998) as a sub-surface cold core water (< 13 °C) flowing close to the western shelf in the southern basin throughout the year.

In the southern Adriatic Sea, the layer below the seasonal thermocline is subjected to a thermal variability (not shown) mostly related to the lateral advection of the NADW and to the inflow of the Mediterranean water. However, the relatively fresh water from the northern basin flows along the Italian coast leaving the open-sea waters exposed to buoyancy losses during winter. An overturning of the water column occurred during winter 1996 down to 600 m. This horizon was marked by the isopycnal 29.18 kg·m^{-3} (Manca and Bregant 1998).

The seasonal salinity and density distributions along the Rimini transect are shown in Figs. 3 and 4, respectively. The salinity field consistently shows two-layer haline stratification with important variations typified by the freshwater river discharges occurring throughout the year. The major Po river discharge with repetitive impulses of about 4000 m^3·s^{-1} occurs from 15 April to 15 July 1995, affecting the corresponding spring and summer surveys. The salinity front separating the coastal water from the open-sea water extends from the sea surface to the western shelf and further, in spring (Fig. 3a). The fresh water flows southwards in a narrow coastal boundary region of about 30-40 km up to the southernmost section at the Otranto Strait. The return flow is constituted by the saline core water at the interior of the basin that presents a vertical homogeneity right down to the bottom. In the northern basin, preliminary evidence seems to indicate that the presence of less saline water further to the east may very likely be associated with meandering current

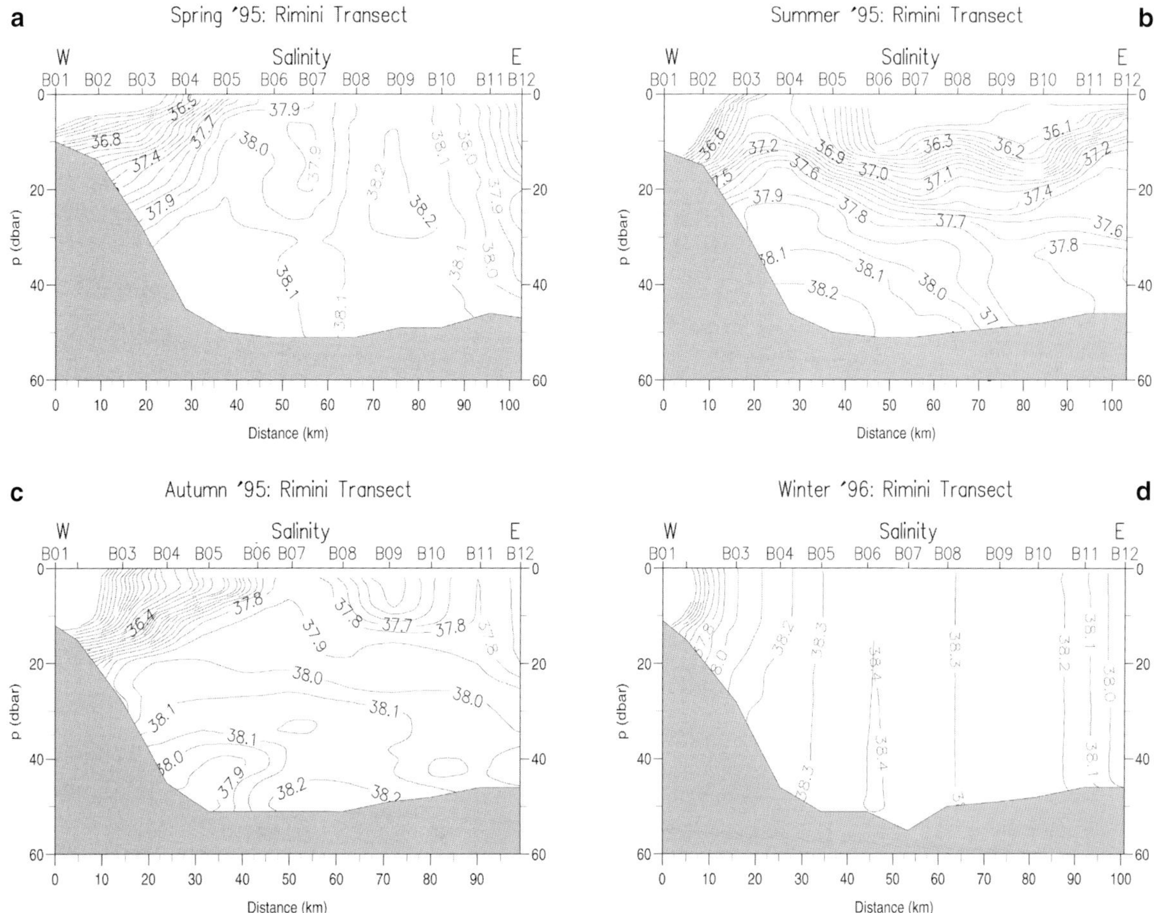

Fig. 3a-d. Salinity distributions along the northern Rimini transect in: **a** spring '95; **b** summer '95; **c** autumn '95; **d** winter '96. The identifications at the *top of the panels* denote the positions of the stations

entraining diluted water into an eastward recirculation, between two adjacent cyclonic structures. The first is the cyclonic gyre which encompasses the northward branch of the general cyclonic circulation as shown by the doming isopycnals in Fig. 4a and further typified in the subsequent Fig. 8b; the second is a cyclonic eddy situated in the shallow north-western part, north of the Po river (P. Franco 1970; Franco and Michelato 1992). The presence of the eddy, during the stratified conditions of 1995, results from data other than those of the present study also (P. Franco, personal communication). A coherent anticyclone may appear in between. It is described as a non permanent feature by Franco (1970), Zore-Armanda and Mladinic (1976), Franco et al. (1982), and Brana and Krajcar (1995), but can not be substantiated in this study due to the absence of data in the north-eastern region. In summer, due to the surface heating and to the warmer riverine outflow, the fresh water spreads to the interior of the basin, resulting in a noticeable layered structure of the water column (Fig. 3b). However, an erosion of the pycnocline due to meso-scale eddy motions was noticed (Fig. 4b). In autumn, a further intense riverine impulse ($\approx 3500\ m^3/s$) occurs. A salinity front again marks a near coastal strip of fresh water that, similarly as in spring, reaches the southern Otranto sections. It seems that the release of the relatively fresher water pool, stored in the northern part, occurs close to the bottom layer along the base of the western shelf (Fig. 3c), in an intermittent manner. In winter, as a consequence of the maximum heat loss, the northern shelf region is characterised by a vertically mixed water column reaching down to the bottom (Fig. 3d). As opposed to the summer situation, very strong horizontal density gradients favour the southward flow of the recently

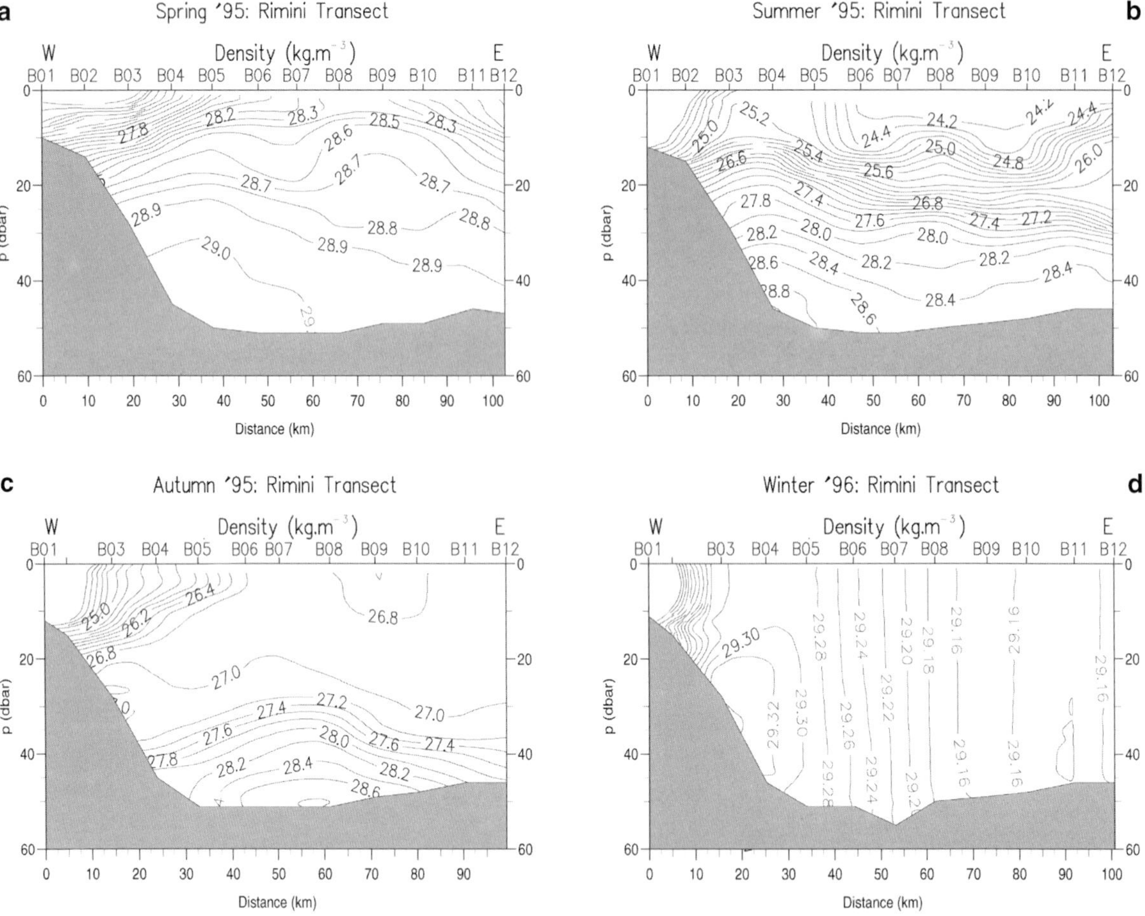

Fig. 4a-d. The same as Fig. 3, but for density

formed NADW (σ_θ > 29.30 kg·m^{-3}) along the western continental slope (Fig. 4d) and the northward flow of more saline water in the interior of the basin.

A very remarkable seasonal variability is evident comparing the vertical salinity and density distributions along the Gargano transect, shown in Figs. 5 and 6, respectively. The western continental shelf/slope is occupied by the relatively fresh but denser water than that of the open-sea, identified by its core values of θ < 13 °C, $S \cong 38.35$ and $\sigma_\theta \cong 29.14$ kg·m^{-3}. This represents the branch of the NADW, which survives the deepening in the trench of the Middle Adriatic and flows over the Pelagosa sill. The maximum core density of $\sigma_\theta \cong 29.16$ kg·m^{-3} has been observed in summer. This may give an indication of the flow rate of the core associated with the NADW. The most prominent feature is the seasonal variability of the lateral extension of LIW ($S > 38.60$) as seen in the bottom layer. This highlights the temporal variability of the flow rate regime of the LIW within this region. The extension area occupied by LIW shows a maximum in summer and autumn.

The seasonal vertical distributions of salinity at the Otranto transect (Fig. 7) are mostly indicative of the complex variability of the water exchange pattern between the Adriatic and Ionian Seas. The most prominent feature is a three-layer structure. The upper layer is mainly characterised by the diluted water of Adriatic origin, the intermediate layer is subjected to a variable extension of the highly saline LIW tongue ($S > 38.70$), and the bottom layer is permanently occupied by the densest and less saline ADW ($S < 38.65$). The NADW ($S < 38.60$) contributes to the Adriatic water exiting along the western shelf. In spring and winter, the NADW reaches the southern section more abundantly,

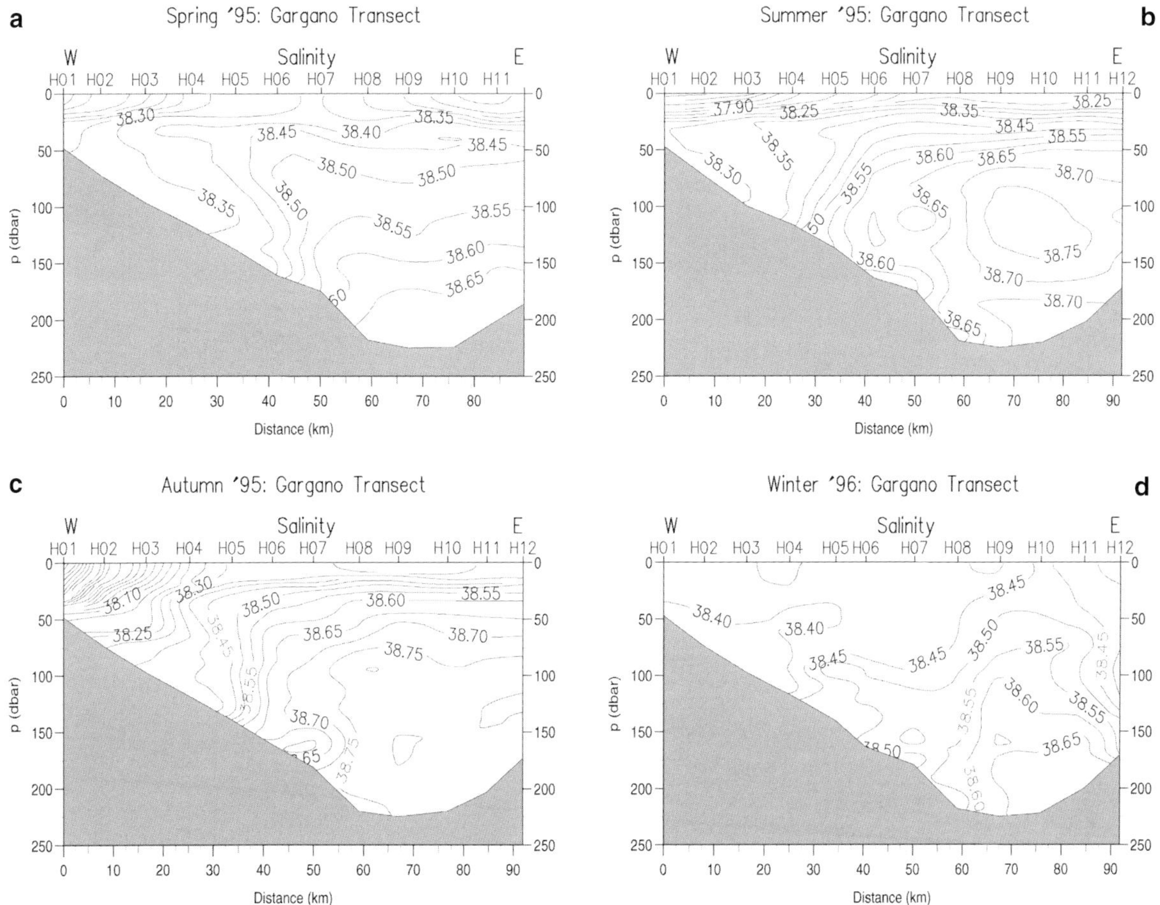

Fig. 5a-d. Salinity distributions along the Gargano transect in: **a** spring '95; **b** summer '95; **c** autumn '95; **d** winter '96. The identifications at the top of the panels denote the positions of the stations

detaches from the shelf and flows in the intermediate layer along the western continental slope. A salinity front leaning toward the Italian slope separates the NADW from the LIW. The LIW lies prevalently against the eastern flank between 100 and 700 m. The maximum LIW inflow occurs in summer and autumn accompanied also by the maximum of salinity in its core. It is important to notice that the NADW and LIW extents are complementary in the strait interior. The notable enhancement of salinity in the LIW core, as found at the Otranto ($S > 38.85$) and Gargano ($S > 38.75$) sections, seems to be due to important anomalies of the intermediate layer in the Ionian Sea. This is reflected in the maximum of salinity observed in the northern transect too ($S > 38.40$), during the following winter 1996 (Fig. 3d), giving an indication of the time that the Ionian water needs to reach the northern sections.

Dynamics and Circulation

Northern and Middle Adriatic Sea

The variability of the circulation at sub-basin scale in the Northern and Middle Adriatic Sea is discussed from the data collected in the spring and autumn of 1995. These data constitute the transition from two extreme situations, the winter one that is essentially characterised by a vertical homogeneity of the water column and a summer one when the water column is highly stratified. Both situations have been widely investigated using modelling exercises and making syntheses from observations (see Malanotte-Rizzoli and Bergamasco 1983; Franco and Michelato 1992). Here, the basin's response to the fresh water buoyancy input is studied when such a forcing event occurs, essentially in spring and autumn, as in 1995.

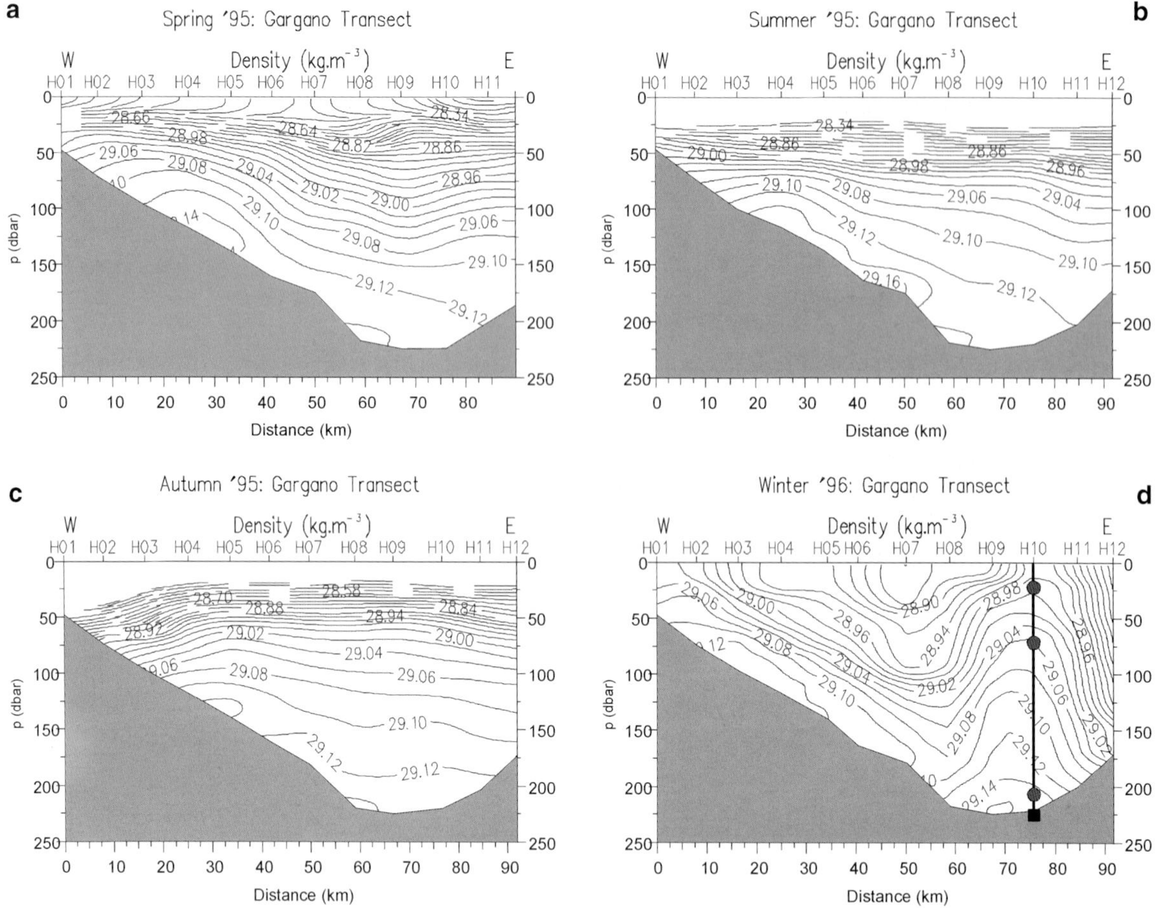

Fig. 6a-d. The same as Fig. 5, but for density. The mooring array of currentmeters at station H10 is shown

Close connections exist between the property distributions, exemplified by the salinity, and the circulation of the Adriatic Sea. The salinity distribution and the geopotential anomaly computed at the sea surface are used to describe the upper level, while the distribution of salinity at the isopycnal surface, chosen below the pycnocline, and the topography of this isopycnal seems adequately to represent the baroclinic structure of the Northern Adriatic circulation. The spring and autumn 1995 situations are depicted in Fig. 8 and 9, respectively. The analysis at the surface further highlights the sub-basin scale features, which are superimposed on the general well-known density-driven flow pattern. Figures 8a and 9a quite clearly show the near-coastal jet transporting southwards diluted water throughout the basin. This will appear clearly in the southern basin too. During spring, the saline water of southern origin intrudes into the northern basin along a mid-longitudinal axis. In autumn, the northward flow occurs mostly along the eastern flank. The latter situation is mostly dictated by the amount of riverine input, which strengthens the northern flow pattern in the eastern flank of the basin. A very interesting result emerges from the comparison between Figs. 8b and 9b. The double cyclonic circulation structure, which marks the sub-basin scale dynamics in the spring, weakens in the northern basin in autumn. The cyclonic vorticity does not change significantly in the middle of the basin, indicating that the gyre is more likely of a permanent nature, as also emerges from historical data analyses (Artegiani et al. 1997). Furthermore, the structures observed at the surface change significantly with depth (Figs. 8d and 9d). Here, the dynamics are represented by the topography of the isopycnal 28.80 $kg \cdot m^{-3}$ in spring and 28.40 $kg \cdot m^{-3}$ in autumn. In spring, the cyclonic vorticity, clearly evident at the surface, changes to an anticyclonic one below the pycnocline, both in the Northern and Middle Adriatic Sea. In autumn, a barotropic homogeneous motion prevails. The examination of Figs. 8c and 9c reveals the pool of saline water on the eastern side and the penetration of its filaments into the northern basin. Comparing them with Figs. 8d and 9d gives an indication of the path of advection due to the existing circulation.

The Southern Adriatic Sea

The topography of the southern basin exercises a strong control over the circulation. The Pelagosa sill acts as a barrier limiting the penetration of the Ionian waters into the northern basin. The geopotential anomaly and sea-surface salinity analyses of the data collected during the seasonal surveys in the southern basin are depicted in Figs. 10 and 11, respectively. The best coverage obtained during the spring survey results from a co-operative investigation in the southern basin, when three ships were contemporarily at sea, within the framework of the projects "OTRANTO MAST/MTP" and "OTRANTO GAP" of NATO's SACLANTCEN in La Spezia.

The maps show a consistent cyclonic circulation and the rim current parallel to the western coast transporting fresh water to the southern Otranto sections. The sea-surface salinity patterns generally follow the geopotential anomaly, showing a maximum salinity in the centre of the gyre. The sea-surface temperature (not shown) indicates the absence of any permanent structure throughout the year with the exception of winter, when the gyre intensifies, bringing cold and saline water from the deep layer in contrast to the warm surface water of Ionian origin.

The northern-bound rim current at the shelf break bifurcates. While one branch turns southwest to close the cyclonic gyre, the other basical-

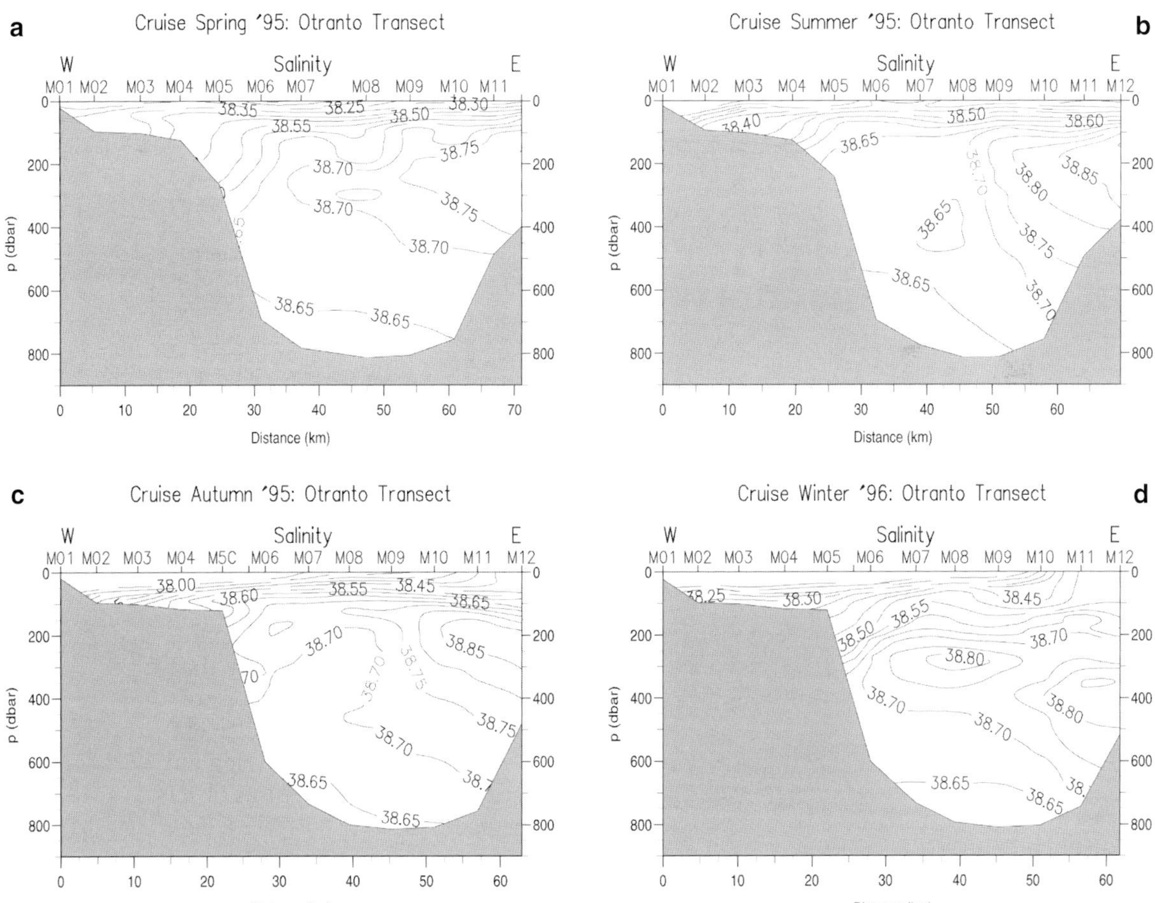

Fig. 7a-d. Salinity distributions along the Otranto transect in: **a** spring '95; **b** summer '95; **c** autumn '95; **d** winter '96. The identifications at the *top of the panels* denote the positions of the stations

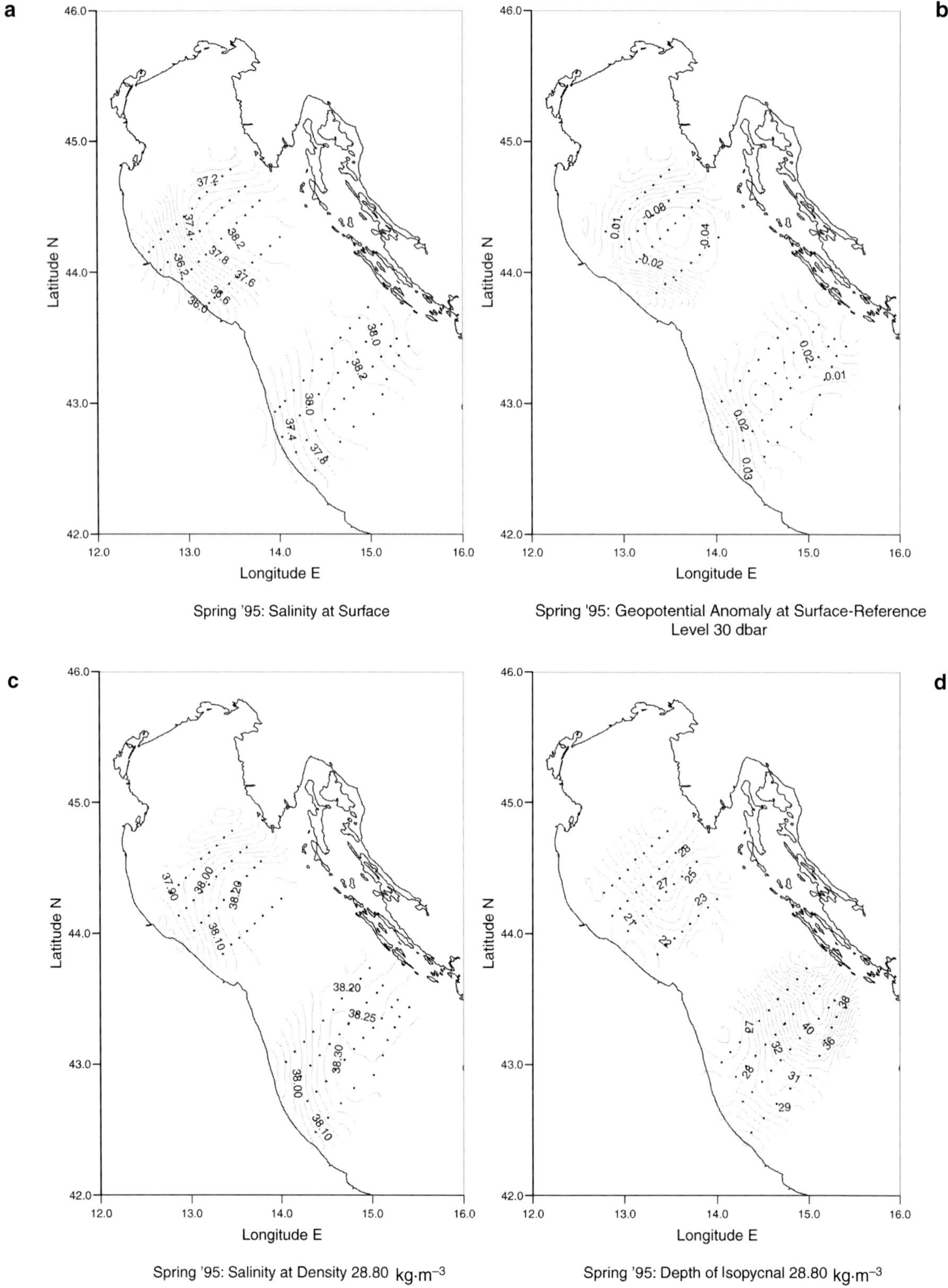

Fig. 8a-d. Surface salinity distribution and geopotential anomaly (m^2/s^2) referenced to 30 dbar (*upper panels*); salinity on the isopycnal 28.80 kg·m^{-3} and depth (m) of isopycnal (*lower panels*), Spring 1995. The *dots* denote the positions of the stations

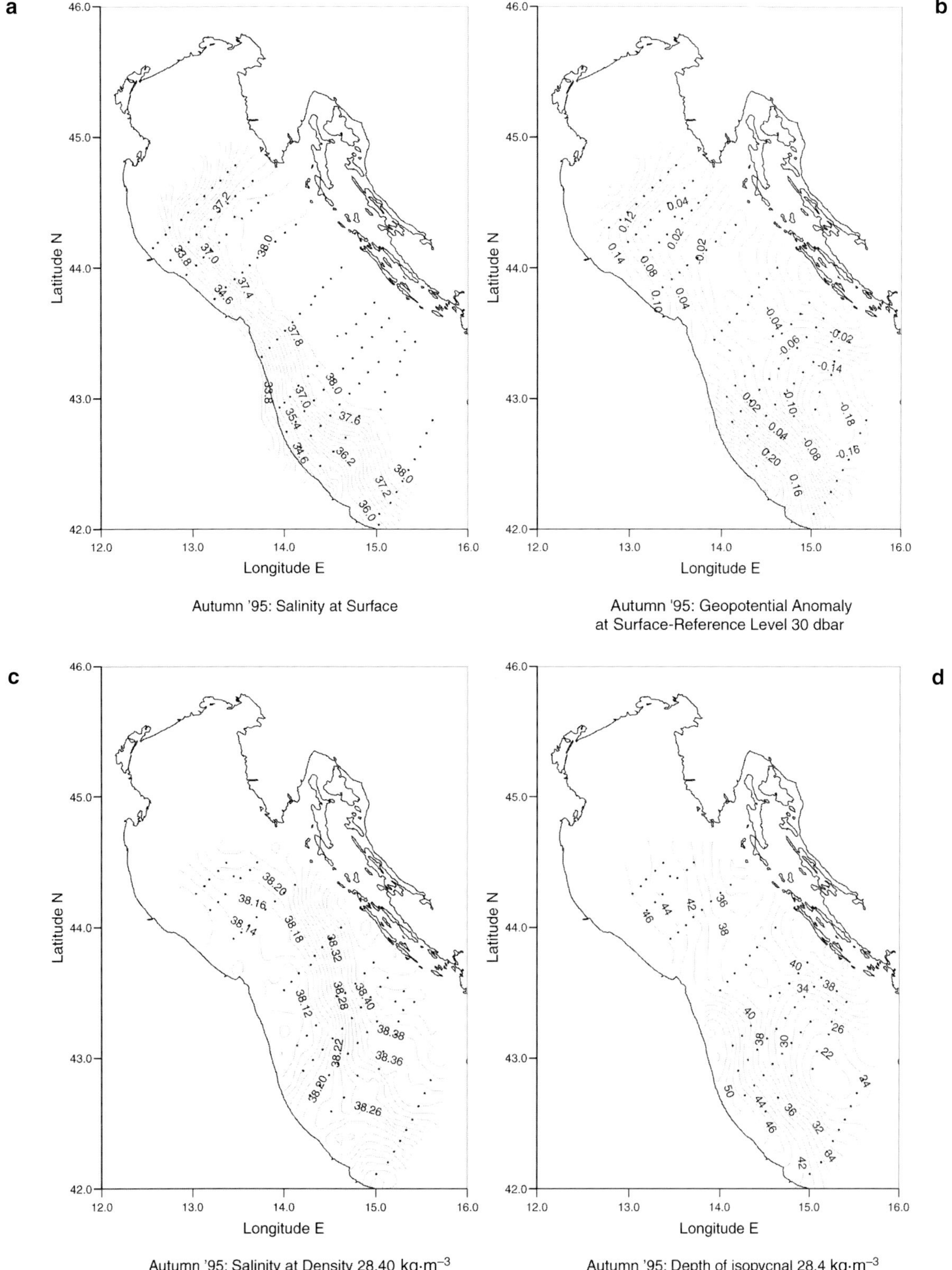

Fig. 9a-d. The same as Fig. 8, but for isopycnal 28.40 kg·m^{-3}, in autumn 1995

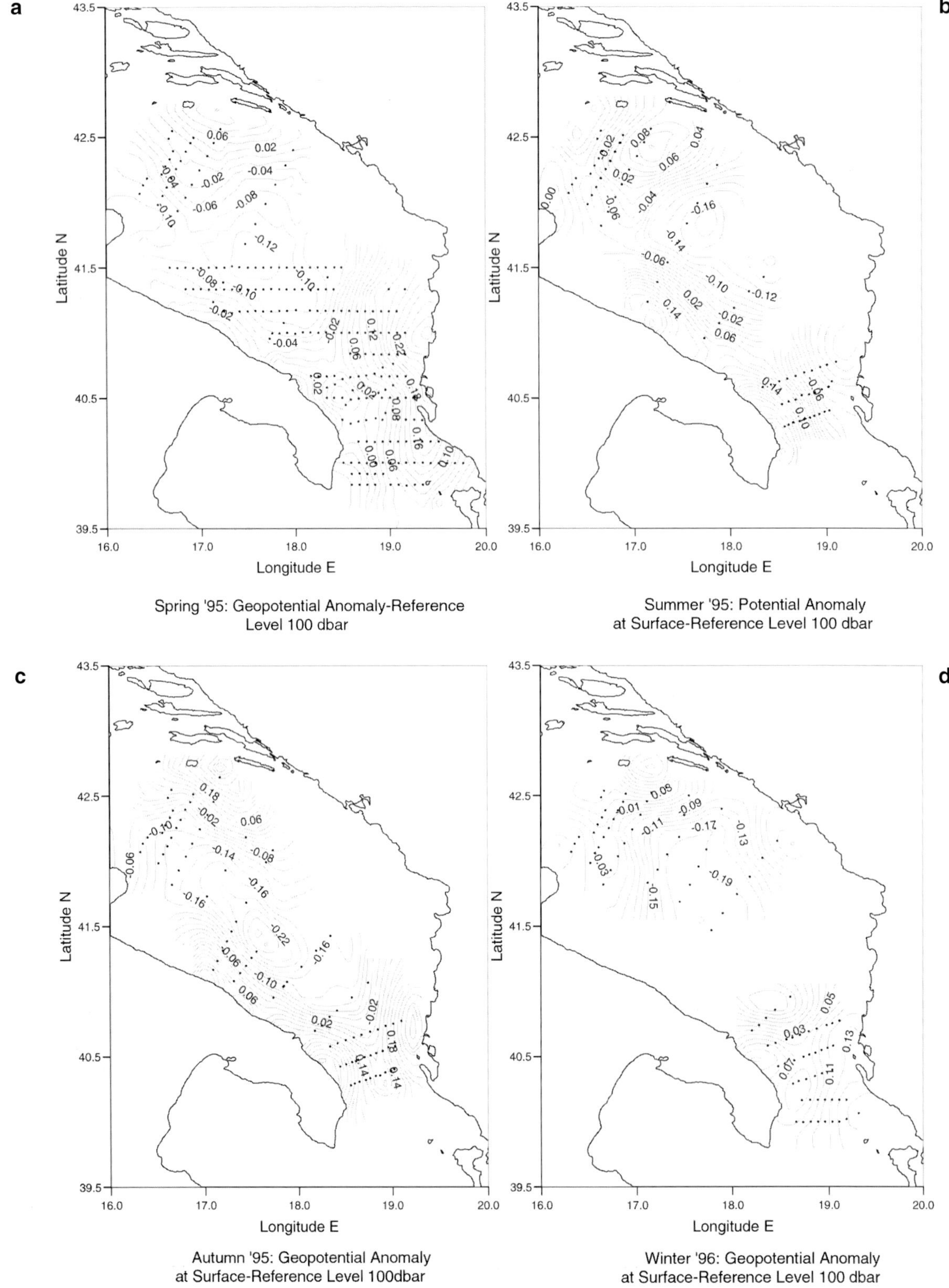

Fig. 10a-d. Geopotential anomalies (m^2/s^2) at the surface referenced to 100 dbar (the *dots* denote the station used for the map) in: **a** spring '95, **b** summer '95, **c** autumn '95, and **d** winter '96

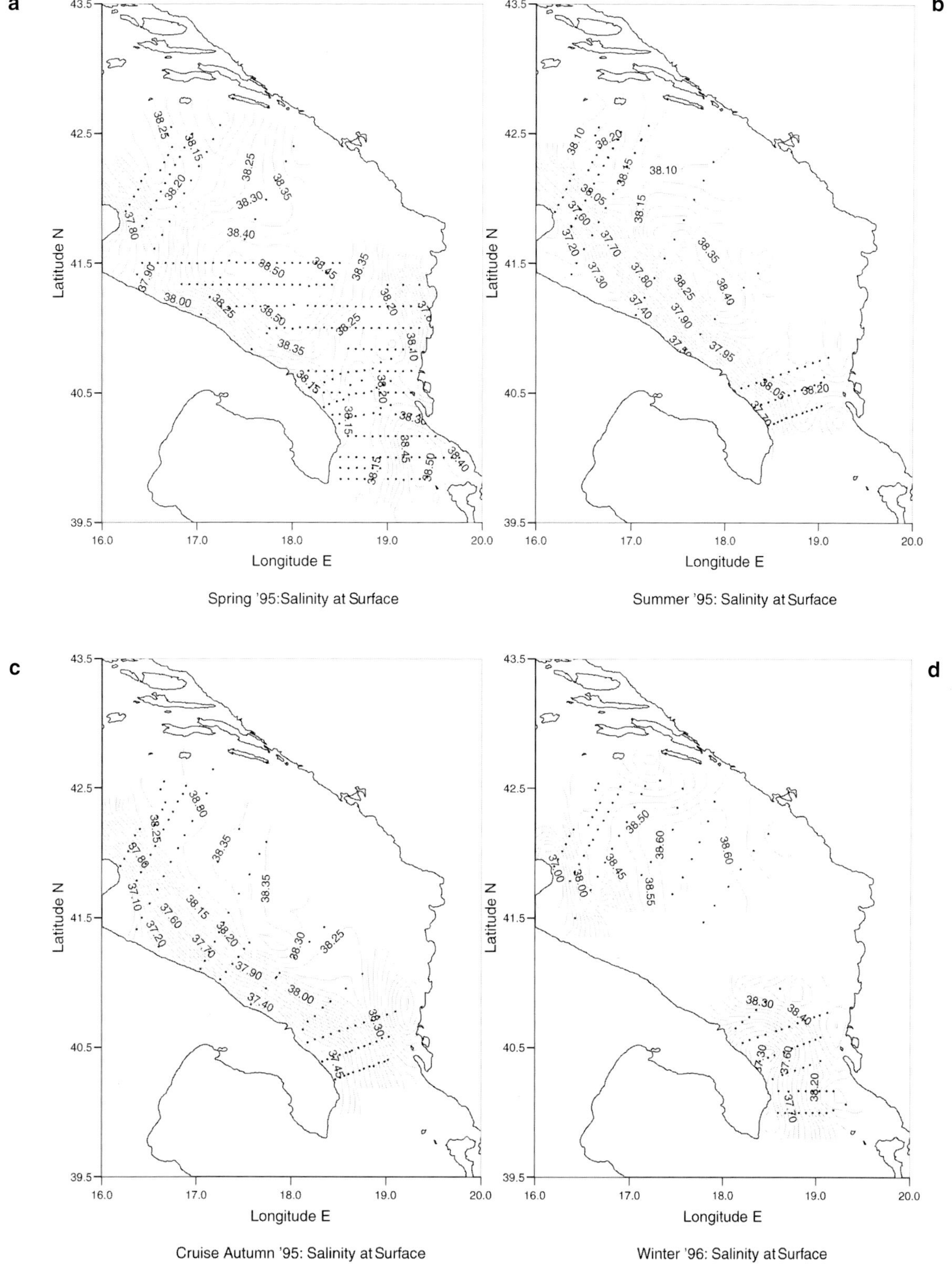

Fig. 11a-d. The same as Fig. 10, but for the horizontal distribution of salinity at the surface ly continues to flow northwards, driven by a coherent current generated by an anticyclone near the eastern coast. Only the portion of saline water which lies in the upper 120 m layer reaches the Northern Adriatic generating the pressure gradients responsible for the general Adriatic cyclonic circulation. The anticyclonic eddy has been consistently observed during all the surveys. The general features of the circulation at the surface do not change significantly with depth, indicating the strong barotropic character of the southern gyre.

One of the major questions concerning the hydrography of the Adriatic Sea is the definition of the water exchange at the Pelagosa sill, which ideally separates two basins. The available time series of current measurements at station H10 (for its position, see Fig. 6) show that southward flowing currents prevail in the surface and in the intermediate layers (Fig. 12). Quite energetic impulses are superimposed indicating a strong meso-scale activity in the surface layer with some reversals of the flow regime. Hydrographic measurements show that the reversals may be related to the lateral oscillations of the anticyclonic eddy situated close to the eastern coast. In winter, the meso-scale activity intensifies and extends mostly down to the bottom thus affecting the whole water column. The currents flowing in the bottom layer (Fig. 12, lower panel) do not differ substantially from those in the layers above, intensities apart. A south-westward flow dominates clearly indicating that the station in question is located in the northern rim current of the cyclonic gyre. Episodic events, such as those that occurred at the end of the measured period, evidence a northward transportation of the LIW. The resulting currents have an average intensity of about 20 ± 12 $cm\cdot s^{-1}$ at 22 m, where frequent episodes of current higher than 35 $cm\cdot s^{-1}$ are found. The currents weaken progressively towards the bottom, with average intensities of $\approx 8\pm7$ $cm\cdot s^{-1}$ at 70 m and $\approx 5\pm6$ $cm\cdot s^{-1}$ close to the bottom.

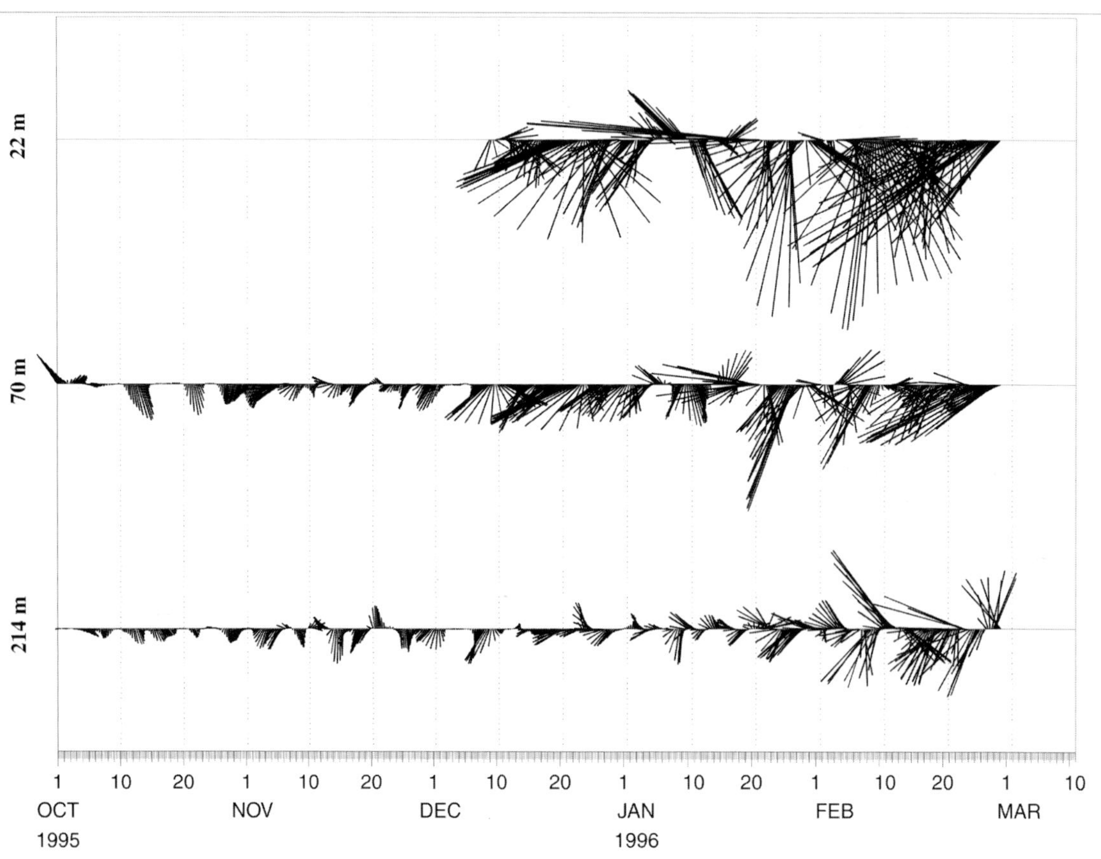

Fig. 12. Current velocity vectors of low-pass filtered (24 hours) time series, sub-sampled every 6 hours, from the three levels of the mooring at station H10 along the Gargano transect, October 1995-February 1996

Conclusions

The observations obtained within the framework of the PRISMA programme concern the whole of the Adriatic basin and, for the first time, cover an entire annual seasonal cycle. This study examines the complex seasonal and spatial variability of the existing dynamics, thereby forming a basis for further studies on transports and biogeochemical flux estimates.

In particular, we have investigated the general circulation and the seasonal variability controlled by its own internal system. The heat exchange with the atmosphere and with the Eastern Mediterranean through the Otranto Strait, as well as the fresh-water buoyancy input from river discharges, create enough pressure gradients in the basin interior generating intense large scale meandering currents. The water column in the northern shallow part almost completely overturns in winter. In the southern basin, moreover, open-ocean deep convection generates dense-water down to 600 m in the centre of the topographically controlled cyclonic gyre. The southern gyre was present throughout the year and was identified at the surface in the centre by higher salinity values than those at the border, due to the upwelling of saline water from intermediate layers. Previous studies, based on averaged seasonal sea-surface temperature from almost a decade of AHVRR data sets (Gacic et al. 1997), show on the contrary the gyre to be a recurrent feature.

Seasonal changes in the hydrographic structure are shown in cross-sections located in the northern and the southern basin. The cold and dense water formed in the northern shelf region, i.e. the NADW, spreads southwards and in the following spring period contributes to a decrease of temperature in the deep layer of the middle basin. On the way towards the southern basin, the NADW has been very clearly defined and characterised both at the Gargano and Otranto transects. This has occurred without its sinking in the deepest part of the southern basin owing to the low core density ($\sigma_\theta \cong 29.14$ kg·m^{-3}) in 1995.

The return path is constituted by the saline Ionian water that intrudes into the Adriatic Sea, through the Otranto Strait, along the eastern coast. Thus, the LIW intrudes into the Adriatic Sea between 100 and 700 m. The salinity in the LIW core, situated at 200 m, ranges from 38.75 to 38.95. The considerable increase of the core salinity is mostly related to a long-term variability of the thermohaline properties and pathways into the Ionian. However, on a seasonal time scale, a larger volume of LIW intrudes into the Adriatic Sea in summer and in autumn, whereas a much weaker influence was documented in winter and spring. The LIW is to a large extent entrapped in the southern gyre and, due to the topographic constraints at the Gargano sections, turns southwards along the western flank. The Pelagosa sill acts as a physical barrier, limiting the penetration of the LIW into the northern basin. The current time series at the shelf break show that the flow is very energetic with significant low-frequency oscillations that are coherent from the surface down to the bottom. The energy appears to increase remarkably in winter.

This exploratory study shows the need to complement the results obtained from analyses of the thermohaline field with further analyses of current time series, where available, with the primary objectives of: (1) obtaining definitive and comprehensive water-transport estimates; (2) quantifying the dynamics of the Adriatic Sea, providing information on the variability and intensities of the general circulation.

Acknowledgements. This work was supported by contribution n. 94.04550.PG03 from the Consiglio Nazionale delle Ricerche in Rome. We wish to thank the masters and crews of the R.V. Urania for their technical assistance on board. We thank the SACLANTCEN in La Spezia, Italy, for providing the hydrographic data collected by R/V Alliance and Magnaghi during the OTRANTO GAP experiment. The help of Luciano Perini and Laura Ursella in currentmeter data analysis is gratefully acknowledged. Finally, we acknowledge two anonymous reviewers for their constructive criticisms on the previous version of the manuscript.

References

Artegiani A, Bregant D, Paschini E, Pinardi N, Raicich F, Russo A (1997) The Adriatic Sea general circulation. I. Air-sea interaction and water mass structure. II. Baroclinic circulation structure. J Phys Oceanogr 27: 1492-1532

Bergamasco A, Gacic M (1996) Baroclinic response of the Adriatic Sea to an episode of bora wind. J Phys Oceanogr 26: 1354-1369

Brana JH, Krajcar V (1995) General circulation of the Northern Adriatic Sea: results of long term measurements. Estuarine Coastal Shelf Sci 40: 421-434

Buljan M, Zore-Armanda M (1976) Oceanographical properties of the Adriatic Sea. Oceanogr Mar Biol Annu Rev 14: 11-98

Franco P (1970) Oceanography of Northern Adriatic Sea. 1. Hydrologic features: cruises July-August and October-November 1965. Arch Oceanol Limnol 16 Suppl: 1-93

Franco P (1972) Oceanography of Northern Adriatic Sea. 2. Hydrologic features: cruises January-February and April-May 1966. Arch Oceanol Limnol 17 Suppl: 1-97

Franco P, Jeftic L, Malanotte-Rizzoli P, Michelato A, Orlic M (1982) Descriptive model of the Northern Adriatic. Oceanol Acta 5 (3): 379-389

Franco P, Michelato A (1992) Northern Adriatic sea: oceanography of the basin proper and of the western coastal zone. Elsevier Sci Publ, Sci Total Environ Suppl 1992, pp 35-62

Gacic M, Marullo S, Santoleri R, Bergamasco A (1997) Analysis of the seasonal and interannual variability of the sea surface temperature field in the Adriatic Sea from AVHRR data (1984-1992). J Geoph Res 102: 22937-22946

Malanotte Rizzoli P (1991) The Northern Adriatic Sea as a prototype of convection and water mass formation on the continental shelf. In: Chu PC, Gascard JC (eds) Deep convection and deep water formation in the oceans. (Elsevier Ocean Series 57) Elsevier, Amsterdam, pp 229-239

Malanotte Rizzoli P, Bergamasco A (1983) The dynamics of the coastal region of the Northern Adriatic Sea. J Phys Oceanogr 13: 1105-1130

Manca B, Bregant D (1998) Dense water formation and circulation in the Southern Adriatic Sea during winter 1996. Rapp Commun Int Mer Mediterr 35: 176-177

Manca B, Giorgetti A (1998) Thermohaline properties and circulation patterns in the Southern Adriatic Sea from May 1995 to February 1996. In: Piccazzo M (ed) Atti 12° AIOL, Isola di Vulcano, 1996, vol II, pp 399-414

Orlic M, Gacic M, La Violette PE (1992) The currents and circulation of the Adriatic Sea. Oceanol Acta 15 (2): 109-124

Ovchinnikov IM, Zats VI, Krvosheya VG, Udodov AI (1985) Formation of deep Eastern Mediterranean waters in the Adriatic Sea. Oceanology 25 (6): 704-707

Zore-Armanda M, Mladinic G (1976) Some results of direct current measurements in the north-west part of the Adriatic Sea (in Croatian). In: Hidrografski Godišnjak 1974, Hydrographic Institute of the Republic of Croatia (ed), Split, pp 61-67

CHAPTER 7

Polysaccharide Production by Microalgae from the Adriatic Sea

L. Boni, M. Cangini, A. Grifoni, F. Guerrini, and R. Pistocchi

ABSTRACT

Polysaccharide production by different phytoplankton organisms from the Adriatic Sea was measured under optimal and stress conditions. The highest sugar amount was produced and released by diatoms and, in all the algae tested, total and extracellular polysaccharide content increased during growth. The amount of extracellular polysaccharides was found to increase under different stress conditions such as a high N/P ratio, presence of bacteria associated with P limitation, presence of metals.

Introduction

A well known aspect of phytoplankton physiology is represented by the release of organic compounds which usually consist of photosynthetic products that are temporarily in excess and are discarded rather than accumulated as reserve material (Fogg 1983). In diatoms the production of extracellular polymeric substances represents a relevant and physiological aspect of their growth. Polysaccharides are the primary component of these substances and are known to have several functions such as to promote motility, permit sessile adhesion or colony formation, avoid dessiccation, stabilize sediments, favour association with bacteria, protect against physical or chemical changes of the environment and others (Hoagland et al. 1993). Polysaccharide production and release is known to increase under some stress conditions such as the presence of a high N/P ratio (Myklestad 1977). We applied this and other environmental stresses in the attempt to understand when polysaccharide extrusion was enhanced and, in particular, we studied this aspect in microalgae isolated from the Northern Adriatic Sea, an area periodically affected by mucilages.

Material and Methods

The algae were cultured in f/2 medium (Guillard and Ryther 1962), with an unchanged or decreased phosphorus content; sugars were measured according to Dubois et al. (1956) and the work scheme of Pistocchi et al. (1997).

Results

Polysaccharide Production

Polysaccharide production was measured in the dinoflagellates *Gonyaulax polyedra* (Gon), *Gymnodinium* sp. (Gymn), *Prorocentrum micans* (Pror), and in the diatoms *Thalassiosira* sp. (Thal), *Asterionellopsis glacialis* (Ast), *Achnanthes brevipes* (Ach), *Cylindrotheca closterium* (C. clost), *Cylindrotheca fusiformis* (C. fus), during different periods of growth (days 7, 12, 16, 21). Both total and extracellular polysaccharides accumulated mostly in the stationary phase of growth and the highest amount (expressed as µg/ml culture) was produced and released extracellularly by the algae belonging to the diatom group. In particular, the species

Università di Bologna, Scienze Ambientali, Via Tombesi dall'Ova 55, 48100 Ravenna, Italy

F.M. Faranda, L. Guglielmo, G. Spezie (eds)
Mediterranean Ecosystems: Structures and Processes

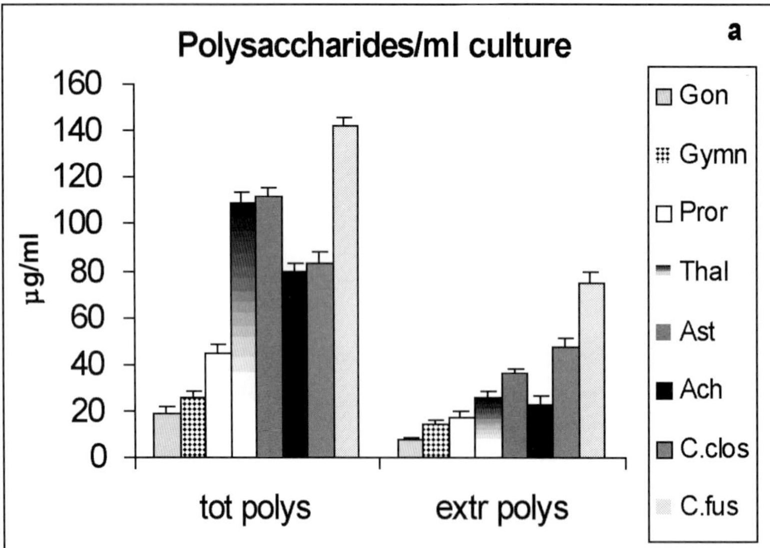

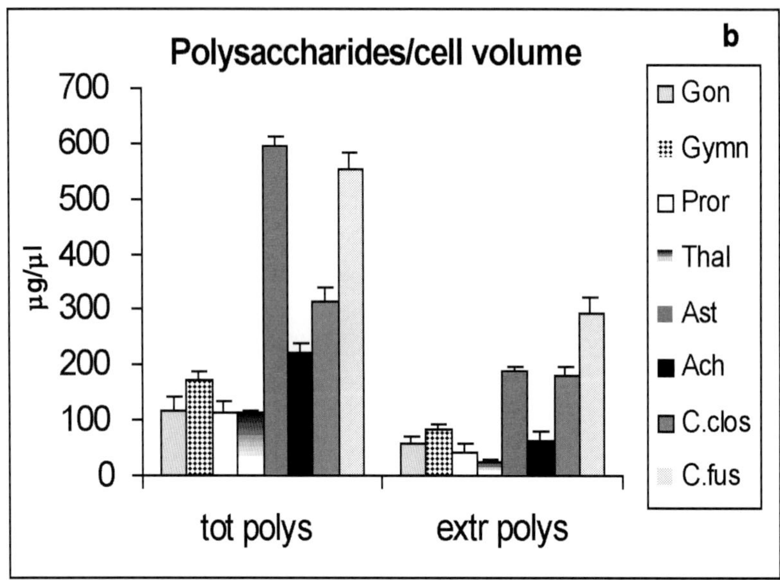

Fig. 1a,b. Total (*tot*) and extracellular polysaccharides (*extr polys*) produced by different phytoplankton species at day 21 of growth, **a** as µg/ml culture or **b** as µg/µl cell volume. Data are the means ± SD of three different experiments

C. closterium and *C. fusiformis* released in the external medium the highest percentage of polysaccharides with respect to the total amount produced. However, as the different species varied much in size, we also considered the amount of polysaccharides produced on a cell volume basis. In Fig. 1 are compared the amount of total and extracellular polysaccharides produced by the different species at day 21, both as µg/ml culture (Fig. 1a) or as mg/ml cell volume (Fig. 1b). It is possible to observe that the dinoflagellates produced the lowest amount in both cases and, when the content was expressed on a cell volume basis, the greater variation occurred for *A. glacialis*.

Effect of N/P Ratio

In experiments performed with the diatom *C. fusiformis* the value of the N/P ratio (24 in f/2 medium) was increased by reducing the addition of inorganic phosphate to 1/3 or 1/6 of the usual amount. The media obtained had N/P ratio values of 73 and 146, respectively. In these media the algae grew less, so that final cell concentrations in 1/3 P and 1/6 P cultures were 39 and 73% respectively of that present in f/2 control culture (Guerrini et al. 1998). The cells under phosphate limitation showed, however, a higher total and extracellular carbohydrate production which increased with increasing limitation (Fig. 2). A similar pattern was observed with *A. brevipes*, when the P limitation was increased by reducing the P content to 1/20 (N/P ratio = 485) of that present in the control medium. Also in this case the amount of extracellular polysaccharides excreted by *A. brevipes* grown in 1/20 P medium, was 1.2 to 2-fold higher than that released by control cells, on a per cell basis (Cangini et al. 1996).

Effect of the Presence of Bacteria

C. fusiformis was grown both in axenic cultures and in the presence of bacteria isolated from marine sediments. The two conditions were adopted to grow the algae under three different nutrient regimes: unchanged f/2 medium and the same medium modified by adding a phosphorus content equal to 1/3 and 1/6 of the control medium. In all the different media there was an increase in carbohydrate content (both total and extracellular) when bacteria were present; the effect became more marked with increasing phosphorus depletion (Guerrini et al. 1998). In Fig. 2 are reported the values of extracellular polysaccharide production by *C. fusiformis* grown alone or in the presence of bacteria, under optimal or P-limited conditions.

Effect of Metals

Gonyaulax polyedra, Gymnodinium sp., *P. micans, C. closterium, C. fusiformis, A. brevipes, A. glacialis, Thalassiosira* sp. were grown in the presence of two different copper concentrations and *C. fusiformis, A. brevipes, P. micans* in the presence of two different cadmium concentrations. The two concentrations were chosen in order to avoid complete inhibition of algal growth, so that metabolic activities could be measured (Pistocchi et al. 1997). The algae showed a different sensitivity to the same metal and usually the diatoms were more resistant than dinoflagellates, with the exception of *A. glacialis* and *Thalassiosira* sp. Although diatoms and dinoflagellates differed very much in the amount of polysaccharides extruded, almost all species showed an increased release of polysaccharides in the external medium when the metal was present. This effect was observed especially when less inhibitory concentrations of the two metals were adopted, while the highest metal concentrations were effective in stimulating polysaccharide production only when algal growth was not too severely affected. In Fig. 3 the results obtained with copper are reported. A similar pat-

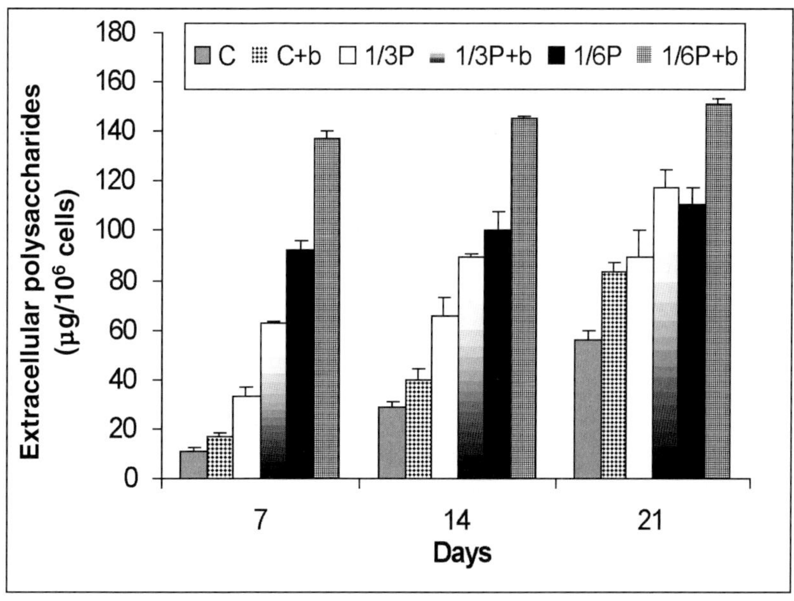

Fig. 2. Extracellular polysaccharides produced by axenic *C. fusiformis* cultures grown in control medium (*C*) or exposed to two different P limitations (*1/3P, 1/6P*) and by cultures grown in the same conditions but in the presence of bacteria (+ *b*). Data are the means ± SD of three different experiments

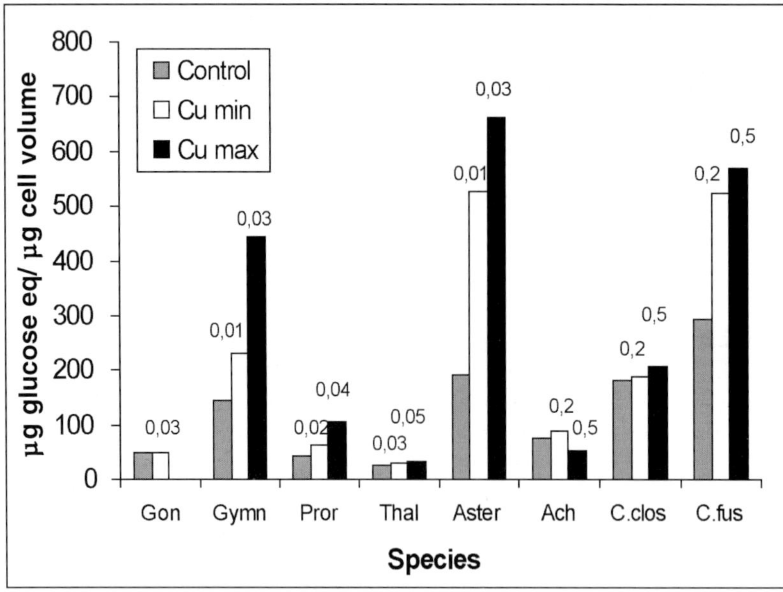

Fig. 3. Extracellular polysaccharides produced by the algae grown in the presence of two different copper concentrations, on a cell volume basis. The different concentrations (mg/l) to which the algae were exposed are reported on the graph. Data are the means three different experiments; SD are not given for graphical purposes but did not exceed 10%

tern was also observed in the presence of cadmium (data not shown).

Conclusions

All the different results showed, not unexpectedly, that an increase in polysaccharide extrusion occurred when various stress conditions were present. We can, however, attribute a different significance to this increase. The nutrient stress, for example, causes a metabolic switch from protein synthesis to carbohydrate synthesis in that the photosynthetic activity proceeds to a normal extent and only carbonic hydrates can be produced in high amounts. The fact that the presence of bacteria can increase this effect is ecologically relevant. In fact bacteria, which usually act as remineralizers, under nutrient depletion become competitors with microalgae for inorganic nutrients so that the nutrient stress becomes even higher and the algae are more stimulated to produce polysaccharides. The higher polysaccharide production in the presence of the metals can be attributed to a mechanism of detoxification in that the negative charges usually present on these molecules can interact with the positively charged metals. In fact we found that dinoflagellates, which produced the lowest carbohydrate amount, accumulated a higher metal concentration intracellularly (Pistocchi et al. 1997), thus the concentration to which they could be exposed was very low. Studies are in progress to understand if differences in polysaccharide composition can account for the different sensitivities observed between the various species.

References

Cangini M, Guerrini F, Trost P, Pistocchi R, Scagliarini S, Boni L (1996) Effect of phosphorus limitation on enzymes of carbon metabolism in diatoms. In: 1st Eur Phycol Congr, Aug 11-18 1996, Cologne, pp 47

Dubois M, Gilles KA, Hamilton JK, Rebers PA, Smith F (1956) Colorimetric method for determination of sugar and related substances. Anal Chem 28: 350-356

Fogg GE (1983) The ecological significance of extracellular products of phytoplankton photosynthesis. Bot Mar 26: 3-14

Guerrini F, Mazzotti A, Boni L, Pistocchi R (1998) Bacterial-algal interactions in polysaccharide production. Aquat Microb Ecol 15: 247-253

Guillard RRL, Ryther JH (1962) Studies on marine planktonic diatoms. I. *Cyclotella nana* Hustedt and Detonula confervacea (Cleve) Gran. Can J Microbiol. 8: 229-39

Hoagland KD, Rosowski JR, Gretz MR, Roemer SC (1993) Diatom extracellular polymeric substances: function, fine structure, chemistry, and physiology. J Phycol 29: 537-566

Myklestad S (1977) Production of carbohydrates by marine planktonic diatoms. II. Influence of the N/P ratio in the growth medium on the assimilation ratio, growth rate, and production of cellular and extracellular carbohydrates by *Chaetoceros affinis* var. *Willei* (Gran) Hustedt and *Skeletonema costatum* (Grev.) Cleve. J Exp Mar Biol Ecol 29: 161-179

Pistocchi R, Guerrini F, Balboni V, Boni L (1997) Copper toxicity and carbohydrate production in the microalgae *Cylindrotheca fusiformis* and *Gymnodinium* sp. Eur J Phycol 32: 125-132

Evaluation of Environmental Quality for the Management of Brackish Wetlands: Use of Bioindicators and Biomarkers in the Pontine Lakes

S. Casini, S. Aurigi, M.C. Fossi, F. Monaci, and S. Focardi

ABSTRACT

The aim of this study was to investigate the quality of the waters of the Pontine Lakes by evaluating the chemical stress status of several edible species. A series of chemical and biochemical tools was used to evaluate environmental impact of contaminants on several species of fish. The following fish were sampled in the lakes of Fogliano and Caprolace in the period 1994-1997: *Chelon labrosus, Sparus aurata, Liza saliens, Liza aurata, Liza ramada, Mugil cephalus, Gobius niger and Solea vulgaris.* Common contaminants (trace elements, chlorinated hydrocarbons) were assayed in the samples and a series of biomarkers (mixed function mono-oxygenases, esterases and porphyrins) were analysed. The somatic liver index was determined. The results show that trace element concentrations are below EEC statuary levels in edible parts. With regard to chlorinated hydrocarbons, some degree of contamination by PCBs was found. The MFO system activities showed differences between the two lakes and between different stations in a lake, confirming that this enzyme system is a good biomarker for evaluating chemical stress status of organisms in relation to environmental quality.

Introduction

The idea of using biomarkers (McCarthy and Shugart 1990) for ecotoxicological studies is first found in the pioneering studies of Bayne et al. (1976) and Payne et al. (1987) in marine environments. In the last 20 years biomarkers have been used in a broad range of environmental situations (Bayne et al. 1985; McCarthy and Shugart 1990; Depledge and Fossi 1994, Depledge 1994). In fact, the concept of biomarkers for evaluating risk in marine, terrestrial and fresh water environments has captured the attention of control agencies and is currently being assessed by several research commissions. Various definitions of biomarkers have been proposed in the last decade, but the most appropriate is probably the following given by the National Academy of Science (NRC 1987): "A biomarker is a xenobiotically induced variation in cellular or biochemical components or processes, structure, or functions that is measurable in a biological system or samples". In most ecotoxicological studies, a single biomarker is not able to give enough information on the health status of the organism, and there is a need for the application of a suite of biomarkers.

This study was carried out at Caprolace and Fogliano coastal lakes, two of the Pontine Lakes, shallow water saline ecosystems located in the Circeo National Park, south of Rome, Italy.

The aim of this study was to investigate water quality in the Pontine Lakes and the chemical stress in commercial fish species, by residue analysis (trace elements, organochlorines) and testing of a suite of biomarkers such as mixed function oxidases (MFO), B esterases, and porphyrins. The MFO activities are induced by liposoluble contaminants (PAHs, TCDDs, PCBs), B esterase activity is inhibited by organophosphate and carbamate insecticides (Ludke et al. 1975), porphyrins accumulation is induced by liposoluble contaminants, some insecticides and herbicides and some heavy metals.

Materials and Methods

In June 1994 specimens of *Chelon labrosus, Sparus aurata, Liza saliens, Liza aurata, Liza*

Dipartimento di Biologia, Università di Siena, Via delle Cerchia 3, 53100 Siena, Italy

F.M. Faranda, L. Guglielmo, G. Spezie (eds)
Mediterranean Ecosystems: Structures and Processes

ramada, Mugil cephalus, Gobius niger, and *Solea vulgaris* were sampled in Fogliano lake. In July 1995 specimens of *Liza aurata* were sampled in Fogliano and Caprolace lakes (midle of the lakes). In June 1997 specimens of *Liza aurata* were sampled in two sites in Fogliano lake: A (by a sluice that divides the lakes waters from a polluted shipway, Rio Martino) and B (midle of the lake); and one in Caprolace lake (midle of the lake, B1). All specimens were dissected immediately after sacrifice the livers and brains were kept in liquid nitrogen and muscle at –20°C until analysis.

After separation of the microsomal fraction, the following MFO activities were evaluated: ethoxyresorufin-o-deethylase (EROD) and benzyloxyresorufin-o-deethylase (BROD) by the method of Lubet et al. (1985); benzopyrene-mono-oxygenase (BPMO) by the method of Kurelec et al. (1977). Acetylcholinesterase activity was measured in brain homogenate by the method of Ellman (1961).

Porphyrins (copro, uro and protoporphyrins) were measured in the liver by the method of Grandchamp et al. (1980). Somatic Liver Index [S.L.I. = (liver weight/body weight) x 100] was calculated for each specimen.

Qualitative and quantitative evaluation of the levels of hexachlorobenzene, total DDTs and total PCBs was performed in muscle by gas chromatography, using a Perkin-Elmer Autosystem model equipped with Ni63 electron capture detector and type SBP-5 (Supelco) bonded-phase fused silica capillary columns.

Trace elements (Cd, Pb, Se, Zn, Cu, Cr, Ni), were detected by plasma emission spectrometry according to the method of Broekaert (1994), mercury by flow injection mercury system by the method of Stoeppler and Backhaus (1978).

Results

The results of biomarker analysis of the 1994 samples are shown in Table 1. The highest BPMO, EROD and BROD activities were measured in the three species of *Liza* and in *Mugil cephalus*. *Gobius niger* had the lowest EROD and BROD activities. The lowest BPMO activity was measured in *Solea vulgaris*. AChE activity did not vary substantially between the different species.

Trace metal concentrations found in liver are shown in Table 2. Mercury concentrations were less than 0.2 µg/g fresh weight (f.w.) in all species. The highest concentrations were found in *Gobius niger* (0.19 µg/g f.w.). Cadmium concentrations were extremely low. Lead concentrations

Table 1. Biomarkers (mean values and standard deviations) in several fish species sampled in 1994 at Fogliano lake (*min*, minutes; *prot*, protein)

Species (*n*)	AChE activity (µmol/min/g brain)	BPMO activity (A.U./mg prot./min)	EROD activity (pmol/min/mg prot.)	BROD activity (pmol/min/mg prot.)
Solea vulgaris (7)	38.65 (18.6)	2.1 (2.7)	72.8 (56.8)	2.32 (1.18)
Liza aurata (4)	34.34 (5.1)	23.4 (16.7)	148.2 (234.9)	3.15 (2.55)
Liza saliens (3)	34.01 (7.7)	25.9 (21.7)	254.4 (251.5)	3.40 (1.28)
Liza ramada (3)	33.3 (5.2)	24.1 (11.1)	288.2 (177.0)	3.78 (3.69)
Mugil cephalus (5)	45.1 (26.4)	16.3 (12.2)	261.5 (231.1)	11.91 (8.63)
Gobius niger (2)	41.5 (16.7)	12.3 (16.1)	20.0 (23.7)	1.17 (1.20)

Table 2. Mean values (µg/g fresh weight) of trace elements in pooled livers of several fish species sampled in 1994 at Fogliano lake

Species (*n*)	Hg	Cd	Pb	Se	Fe	Zn	Cu	Cr	Ni
Chelon labrosus (1)	0.08	0.09	1.10	0.92	176	46	175	0.43	1.54
Sparus aurata (1)	0.09	< 0.01	< 0.01	3.93	193	55	211	< 0.01	1.82
Liza aurata (3)	0.14	0.20	0.85	9.40	456	54	511	1.82	9.89
Liza saliens (2)	0.10	0.16	0.69	1.53	413	74	479	2.32	6.45
Liza ramada (6)	0.12	0.14	< 0.01	7.48	568	59	639	1.51	6.92
Mugil cephalus (4)	0.13	0.21	0.70	6.09	605	112	656	0.16	2.81
Gobius niger (1)	0.19	0.34	1.30	7.46	67	45	82	3.45	1.55

Table 3. Range of concentrations (ng/g fresh weight) of organochlorines in muscle of several fish species sampled in 1994 at Fogliano lake

Species (*n*)	Length (cm)	PCBs	pp'DDE	DDT	HCB	HCH
Solea vulgaris (5)	23.0-23.5	3-674	0.3-156	0.5-177	0.01-1.00	0.005-2.000
Liza aurata (4)	22.0-23.0	12-23	0.5-2.5	0.6-2.8	0.02-0.05	0.005-0.030
Liza saliens (3)	17.0-23.0	60-150	1.0-1.5	2.0-4.0	0.02-0.03	0.010-0.080
Liza ramada (3)	16.4-19.8	11-50	0.6-4.9	0.8-4.8	0.05-0.07	0.010-0.020
Mugil cephalus (5)	38.5-44.0	7-31	0.3-1.2	0.4-1.4	0.01-0.04	0.003-0.020
Gobius niger (2)	13.8-14.0	80-240	0.2-5.6	0.4-9.7	0.10-1.20	0.030-0.400

were higher in *Gobius niger* (1.3 µg/g), presumably due to the benthic habitat of this species.

Organochlorine analysis in muscle (Table 3) revealed hexachlorobenzene (HCB), lindane, pp'DDT and its metabolites and several PCBs congeners. The HCB, DDT and lindane concentrations were low. Interestingly, a specimen of *Solea vulgaris* had a concentration of 674 ng/g f.w., similar to the levels typical of heavily contaminated areas by PCBs. Similar high levels were also found in a specimen of *Liza saliens* and one of *Gobius niger.*

The results of biomarker studies in specimens of *Liza aurata* sampled in 1995 are shown in Fig. 1. EROD and particularly BROD activities were statistically higher in fish from Caprolace lake ($P < 0,05$). AChE activities were lower (though not significantly) in specimens from Caprolace. The results of porphyrin analysis seem to confirm higher chemical stress in fish from Caprolace lake. Copro, proto and total porphyrins were higher in livers of specimens from this lake. Organochlorine concentrations are shown in Table 4. The HCB levels were higher in muscle of specimens of *Liza aurata* from Caprolace lake than in the same species from

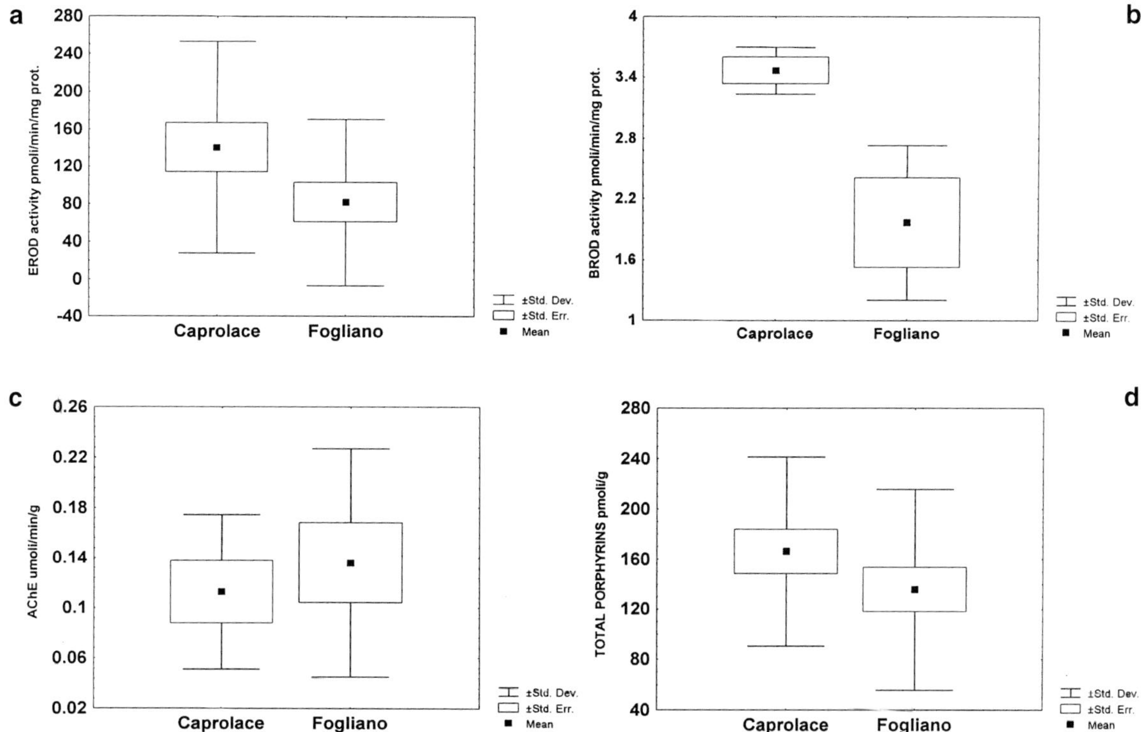

Fig. 1a-d. Mean values, standard errors and standard derivations of 4 different biomarkers measured in liza aurata sampled in 1995 at Fogliano ($n = 16$) and Caprolace ($n = 15$) Lakes

Table 4. Range of concentrations (ng/g fresh weight) of organochlorines in muscle of *Liza aurata* sampled in 1995 at Fogliano and Caprolace lake

Lake	(*n*)	HCB	ΣDDT	ΣPCB
Fogliano	(19)	0.005-0.35	2.0-32.7	10-154
Caprolace	(20)	0.01-1.97	1.5-31.5	3-150

Fogliano lake. The DDTs concentration, were higher than in 1994. The PCB levels were higher in fish from Fogliano than Caprolace lake.

Biomarkers in *Liza aurata* captured in 1997 are shown in Fig. 2. The BPMO and BROD activities were higher in fish from sites B1 (Caprolace) and A (Fogliano, by a sluice that divides the lake waters from a polluted shipway, Rio Martino) with respect to B (Fogliano). The difference was statistically significant ($P < 0{,}05$) for BROD activity. Highest EROD activity was recorded at site A, as was the Somatic Liver Index value ($P < 0{,}05$).

These data confirm the 1995 results, and reveal an "anomaly" at site A, close to a potential source of liposoluble contaminants and trace metals (the shipway). Esterases did not show differences among the different sampling sites. Total porphyrins were higher in fish from site A ($P < 0{,}05$).

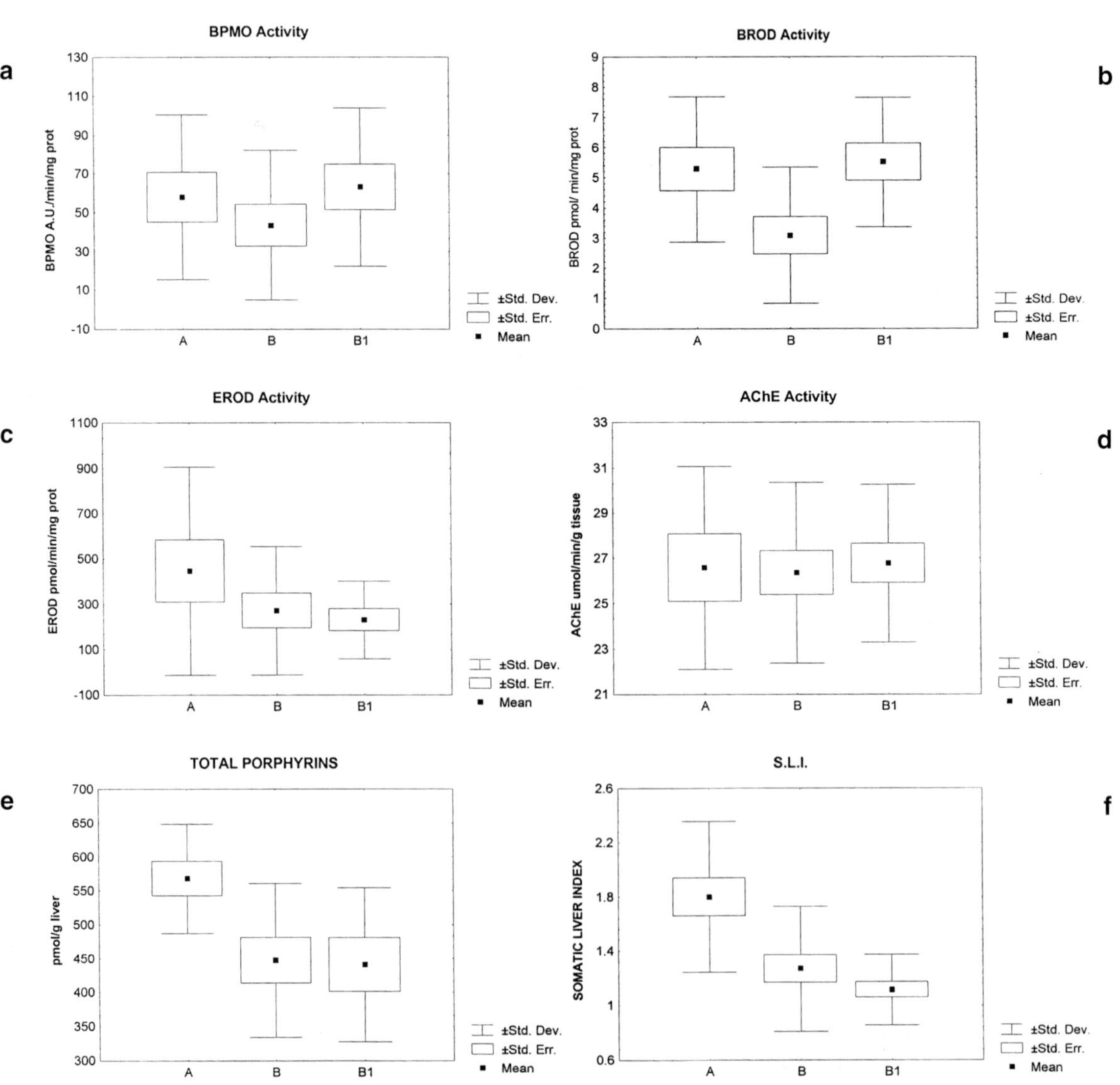

Fig. 2a-f. Mean values, standard errors and standard derivations of 6 different biomarkers measured in Liza aurata sampled in 1997 at Fogliano and Caprolace Lakes. *A* = Fogliano – Rio Martino (n = 10); *B* = Fogliano (n = 8), *B1* = Caprolace (n = 12)

Table 5. Range of concentrations (ng/g fresh weight pooled muscle) and total Hg (µg/g dry weight pooled livers) in *L. aurata* sampled in 1997

Site	(*n*)	HCB	∑DDT	∑PCB	Hg
Fogliano B	(5)	0.009-0.4	1.8-29.8	6.1-49.5	0.055-0.111
Caprolace B1	(5)	0.008-0.35	0.9-17.3	1.5-28.3	0.061-0.116

Total mercury concentrations were analysed in pools of liver from specimens sampled at stations B and B1 (Table 5). Relatively low values, less than those of the first sampling, were found in both stations.

Organochlorine levels were also measured in pooled muscle of specimens from sites B and B1 (Table 5). The HCB levels were lower in Caprolace with respect to 1995, DDT levels were similar to 1995, PCB concentrations lower (almost half those found in 1995 on average).

Discussion and Conclusions

The results of this study show that fish in the two lakes contain traces of anthropogenic contaminants. In the first stage, the levels of trace elements were not found to be anomalous from an ecotoxicological point of view. Concentrations of Hg in liver were below the EEC limit of 0.5 µg/g f.w. for edible parts (liver normally has levels 2-5 times greater than muscle). The results for chlorinated hydrocarbons indicate the highest levels of PCBs. Analysis of DDT metabolites showed highest concentrations of pp'-DDE in all species. This confirms the continuous decrease in DDT residues since its use was limited, as indicated by results in Tyrrhenian fish (Focardi et al. 1983).

In the second stage only *Liza aurata* was used as a bioindicator. The results obtained for contaminants show that Lake Fogliano is slightly more contaminated than Caprolace, however PCB levels were of the same order of magnitude as in fish of the southern Tyrrhenian sea (Corsolini et al. 1995). With regard to MFO system activities induced by fat soluble contaminants, interesting results were obtained in the comparison between lakes and between stations in the same lake. The MFO levels were about 80% higher in Caprolace in both years of the study. In sampling for 1997 higher MFO activities and porphyrin concentrations were also found at station A which is near the sludge of Rio Martino, a potential source of fat soluble contaminants. Changes in biomarker values might be due to contaminants other than organochlorines and heavy metals.

In summary, contaminants (trace metals, organochlorines) analysed throughout the study show relatively low levels, with some exception, and below the limits fixed by "low" ; this confirms the suitability of these waters for aquaculture. The extremely sensitive responses of biomarkers have underlined some differences in the specimens, when comparing the two lakes and sites inside each lake (e.g. site A). The biomarkers measured therefore turned out to be very useful in detecting small variations in chemical stress.

References

Bayne BL, Brown DA, Burns K, Dixon DR, Ivanovici A, Livingstone DR, Lowe DM, Moore MN, Stebbing ARD, Widdows J (1985) The effects of stress and pollution on marine animals. Prager Sci, pp 384

Bayne BL, Livingstone DR, Moore MN, Widdows J (1976) A cytochemical and biochemical index of stress in *Mytilus edulis*. Mar Pollut Bull 7: 221-224

Broekaert JAC (1994) Plasma optical emission and mass spectrometry. I. Zeev B. In: Alfassi: determination of trace elements, VHC, pp 192-243

Corsolini S, Focardi S, Kannan K, Tanabe S, Borrell A, Tatsukawa R (1995). Congener profile and toxicity assessment of polychlorinated biphenyls in dolphins, sharks and tuna fish from Italian coastal waters. Mar Environ Res 40: 33-53

Depledge MH (1994) The rational basis for the use of biomarkers as ecotoxicological tools. In: Fossi MC, Leonzio C (eds) Nondestructive biomarkers in vertebrates. Lewis Publishers, Boca Raton, pp 271-296

Depledge MH, Fossi MC (1994) The role of biomarkers in environmental assessment. 2. Invertebrate. Ecotoxicology 3:161-172

Ellman L, Courtey KD, Andreas VJr., Featherstone RM (1961) A new rapid colorimetric determination of cholinesterase activity. Biochem Pharmacol 7: 88-98

Focardi S, Bacci E, Leonzio C, Cristetig C (1983) Chlorinated idrocarbons in marine animals from the Northern Tyrrhenian Sea (N.W. Mediterranean). Thalassia Jugosl 20: 37-43

Grandchamp B, Deybach JC, Grelier M, Deverneuil H, Nordmann Y (1980) Studies of porphyrin synthesis in fibroblasts of patients with congenital erythropoietic porphyria and one patient with homozygous coproporphyria. Biochem Biophys Acta 629: 577-586

Kurelec B, Britvic S, Rijavec M, Muller WEG, Zahn RK (1977) Benzo(a)pyrene mono-oxygenase induction in marine fish - Molecular response to oil pollution. Mar Biol 44: 211-216

Lubet RA, Nims RW, Mayer RT, Cameron JW, Schechtman LM (1985) Measurement of cytochrome P450 dependent dealkylation of alkoxyphenoxazones in hepatic S9s and hepatocyte homogenates: effects of dicumarol. Mutat Res 142: 127-131

Ludke JL, Hill EF, Dieter MP (1975) Cholinesterase (ChE) response and related mortality amog bird fed Che inhibitors. Arch Environ Contam Toxicol 1: 1-21

McCarthy F, Shugart LR (1990) Biomarkers of environmental contamination. Lewis Publishers, USA

NRC (National Research Council) (1987) Committee on biological markers. Environ Health Perspect 74: 3-9

Payne JF, Rahimtula AD, Porter EL (1987) Review and perspective on the use of mixed-function oxygenase enzymes in biological monitoring. Comp Biochem Physiol 86C: 233-245

Stoeppler M, Backhaus F (1978) Pre-treatment studies with biological and environmental materials, I. System of pressurized multisample decomposition. Fresenius Z Anal Chem 291: 116-120

Toxicological Evaluation of Organochlorine Levels on some Fish Specimens from Adriatic Sea

I. Corsi, S. Aurigi, and S. Focardi

ABSTRACT

The aim of this study is to assess the organochlorine levels and their toxic impact in some fish specimens of commercial importance in terms of utilisation for human diet. Fish samples of *Merluccius merluccius, Sprattus sprattus, Sardina pilchardus, Scomber japonicus, Boops boops* and *Mullus barbatus* were collected in different sites along the Adriatic Sea during winter 1996 and 1997. Polychlorinated biphenyls including mono-, and di-*ortho* congeners, dichlorodiphenyltrichloro-ethane (DDT) including its derivates and hexachlorobenzene (HCB) were analysed to evaluate their bioaccumulation in the food chain. The 2,3,7,8-TCDD equivalents (TEQs) were assessed based on congener concentrations and toxic equivalency factors (TEFs). The PCB, DDT and HCB levels detected were higher in *Merluccius merluccius* followed by *Sardina pilchardus* and on the same lower range that of *Boops boops, Mullus barbatus, Scomber japonicus, Sprattus sprattus.* In terms of toxicological evaluation, the TEQs approach calculated for some PCB congeners, revealed the highest values for liver of *Merluccius merluccius* followed by that of *Sardina pilchardus* muscle fillets

Introduction

Being 800 km in length and 200 km in width and having unique oceanographical and environmental characteristics, the Adriatic Sea is one of the main fishing areas of the Mediterranean. Shallow waters, lagoons and deltas, varying salinity and temperatures as well as aspects linked to human impact such as continental water inflow are features that make the Adriatic basin suitable for studying contamination and pollution (Giordani et al. 1997). Moreover, the presence of two circular streams between the North and the Centre not only separate these two areas but also accounts for the differences in the processes and aspects that characterise this basin.

In the present study, polychlorinated biphenyls (PCBs), dichlorodiphenyltrichloroethane (DDT) and hexachlorobenzene (HCB) were biomonitored in two different sites along the Italian coast assuming that the concentrations of chemicals in biological samples are likely to reflect the present levels in the marine environment (Porte and Albaiges 1993).

Since *dioxin-like compounds,* such as the organochlorines have been proved to bioaccumulate in fish, this finding encourages the development of research concerning the behaviour of these chemicals in aquatic species (Hellou et al. 1993; Lech and Bend 1980).

This study is also focused on the comparison of the levels of organochlorines between liver and muscle tissues as the liver is rich in lipids and bioaccumulates mostly lipophilic substances and because humans assume muscle tissue as food.

To determine target tissue and find which species accumulate most contaminants, we took concentrations detected for each class of contaminants and compared them among the same species from different sampling sites as well as comparing those between the different species fished in the North and the Central Adriatic.

Finally, to evaluate potential toxicological risk for the aquatic organism itself and for their predators including man, we applied the toxic equivalents (TEQs) concept which has been introduced to simplify risk assessment and regu-

Dipartimento di Scienze Ambientali, Università di Siena, Via delle Cerchia 3, 53100 Siena, Italy

F.M. Faranda, L. Guglielmo, G. Spezie (eds)
Mediterranean Ecosystems: Structures and Processes

latory control for *dioxin-like compounds* such as PCBs (Safe 1990, 1993; Ahlborg et al. 1994). It was recognised that as the recommended toxic equivalency factors (TEFs) have been developed for exposure scenarios then the corresponding values may be appropriate for body burden consideration and ecotoxicity purposes (Ahlborg et al. 1994).

Materials and Methods

Sampling was carried out in 2 stages: the first in the Northern Adriatic off the coast of Venetum and the second in the Central Adriatic close to the Apulean shores (Fig. 1). Specimens of *Merluccius merluccius*, *Sprattus sprattus*, *Sardina pilchardus*, *Scomber japonicus*, *Boops boops* and *Mullus barbatus* were captured during the prespawning period with common fishing procedure, wrapped in aluminium foil, stored on ice and frozen at −20°C on reaching the laboratory. Livers and muscle fillets were removed and grouped in several pools to minimise length and weight ranges. Fresh samples were used for determining organochlorines concentration according to the method of Wakimoto (1971). Instrumental analysis was performed with a Perkin-Elmer Autosystem model gas chromatography GC–ECD equipped with Ni^{63} electron capture detector and type SPB-5 bonded-phase, fused silica capillary column. A mixture of specific isomers was used for calibration, recovery evaluation and quantification such as Aroclor 1260, DDT isomers and HCB supplied by Supelco Inc.; specific PCB congeners were supplied by Dr. Ehrenstorfer (GmbH).

Results and Discussion

Data were evaluated by comparing organochlorine levels in liver and muscle among fish species according to the student's *t*-test for statistical significance ($P<0.05$) and by correlating these levels in each matrix for each class of contaminants using Pearson Product Moment correlation.

The highest organochlorine levels were recorded in the Northern Adriatic according to the observed highest degree of organochlorine contamination of this area (Table 1). Comparing the organochlorine levels among the species from this area, the liver of *Merluccius merluccius* seems to accumulate significantly higher levels

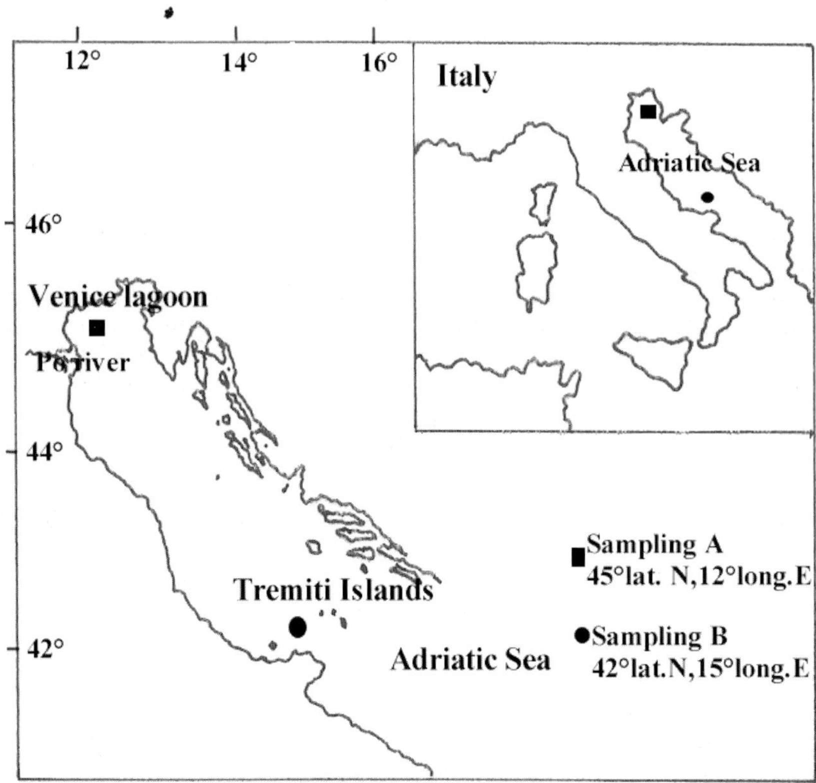

Fig. 1. Map of the sampling areas along the Adriatic Sea

Table 1. Summary of parameters and results of residues analysis for all specimens analysed

			Liver				Muscle		
	N° of samples	Weight (g)	Length (cm)	PCBs μgKg^{-1}	*p,p'* DDE μgKg^{-1}	HCB μgKg^{-1}	PCBs μgKg^{-1}	*p,p'* DDE μgKg^{-1}	HCB μgKg^{-1}
Northern Adriatic									
Merluccius merluccius	14	78.5 ± 43.1	21.94 ± 4.3	253.81 ± 64.9[a]	96.6 ± 16.4[a]	0.54 ± 0.4[a]	23.6 ± 2[a]	3.33 ± 3.1[a]	0.09 ± 0.1[a]
Sprattus sprattus	18	21.8 ± 3.3	14.3 ± 0.9	14.02 ± 5.4[a]	7.65 ± 2.4	5.46 ± 3.9	60.83 ± 4.6[a]	9.91 ± 1.8	0.36 ± 0.2
Sardina pilchardus	7	37.39 ± 7.2	16 ± 1.1	118.9 ± 96.9	37.11 ± 22.7	4.28 ± 0.6	104.1 ± 20.3	21.47 ± 14	1.62 ± 1.0
Central Adriatic									
Merluccius merluccius	19	66.74 ± 89.1	17.63 ± 6	165.19 ± 71.5[a]	58.5 ± 28.1[a]	0.81 ± 1.1	48.84 ± 94.6[a]	13 ± 18.9[a]	0.20 ± 0.3
Sardina pilchardus	28	32.2 ± 6.5	15.39 ± 1.1	84.72 ± 16.5[a]	9.20 ± 2.1	4.74 ± 11.4	49.2 ± 22.4[a]	5.94 ± 1.8	0.21 ± 0.3
Boops boops	19	48.64 ± 19.9	15.44 ± 1.6	71.49 ± 31.7[a]	3.11 ± 2.2[a]	0.7 ± 0.5[a]	9.02 ± 3.4[a]	0.55 ± 0.2[a]	0.09 ± 0.1[a]
Scomber japonicus	30	54.13 ± 21.6	17 ± 1.9	27.54 ± 9.1[a]	4.59 ± 1.6	0.15 ± 0.1	6.92 ± 3[a]	2.46 ± 3.2	0.2 ± 0.1
Mullus barbatus	20	37 ± 17.1	12.6 ± 2.1	38.77 ± 24.1	4.43 ± 2.5	0.32 ± 0.3	17.3 ± 11.2	2.03 ± 1.1	0.10 ± 0.1

Organochlorines expressed as wet weight basis

[a] Statistically significant between liver and muscle levels ($p<0.05$; t-test)

(253.8 μgKg^{-1}) than those recorded in *Sardina* (118.9 μgKg^{-1}) and *Sprattus* (14 μg Kg^{-1}). Samples of *Merluccius merluccius* coming from the Central Adriatic exhibited also the highest levels of organochlorines (165.2 μgKg^{-1}), although lower compared to those detected in the North.

Among other fish specimens monitored, *Sardina pilchardus* registered levels of organochlorines comparable with those of *Merluccius merluccius* both in the North and in the Centre. Several studies have shown statistically significant correlation between organochlorines concentration in sediment and in liver tissue of a variety of benthic species such as *Mullus barbatus* (Georgakopoulos-Gregoriades et al. 1991; Varanasi et al. 1993) and referring to our study the levels detected in *Mullus barbatus,* both in muscle and in liver, were significantly lower than those of *Sardina* and *Merluccius.* Although these results are referred to a relatively closed area such as the Adriatic basin, it is likely that no specific organochlorines bioaccumulation pattern was observed among the benthic and pelagic species we monitored, but rather was widespread among all our species with some high values among the pelagic ones. Regarding the levels of total DDT detected, as a consequence of the application of a strong extraction technique with 1NKOH in ethanol, these results were referred to the sum of 4,4'DDE and 4,4'DDT transformed in 4,4'DDE and did not included the 4,4'DDD which disappeared. The 4,4'DDE levels showed an identical distribution of PCBs in all the specimens analysed as confirmed by the high number of correlation found between PCB and DDT levels in liver tissue for most of the species (Figs. 2-4). Surprisingly, the same trend of correlation was found also in muscle fillets in all the fish species analysed as reported in Table 2. Regarding the levels of HCB, the liver of *Merluccius merluccius* only from the North Adriatic and of *Boops boops* from the Centre were shown to accumulate significantly higher levels. Moreover our results seems to confirm the capability of liver tissue to accumulate organic contaminants to a greater extent but, on the other hand, their presence in muscle tissue should be taken into account in relation to the risk posed by the wide presence of these species in human diet. The highest levels of PCBs recorded in most of the species may be evaluated also

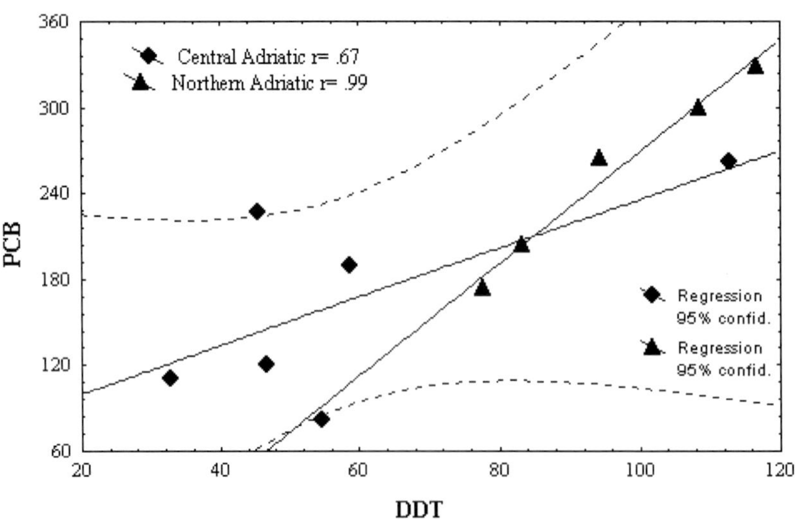

Fig. 2. Correlation between PCB and DDT levels in liver of *Merluccius merluccius* sampled in the Northern and in the Central Adriatic ($P < 0.05$)

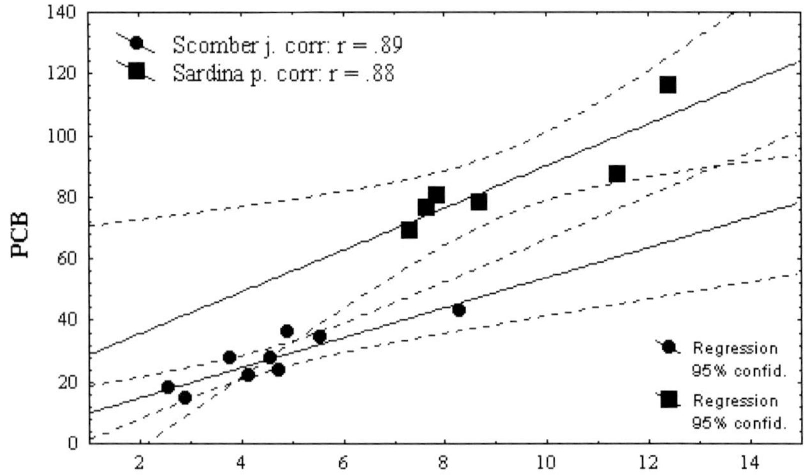

Fig. 3. Correlation between PCB and DDT levels in liver tissue of *Scomber japonicus* and *Sardina pilchardus* from the Central Adriatic ($P < 0.05$)

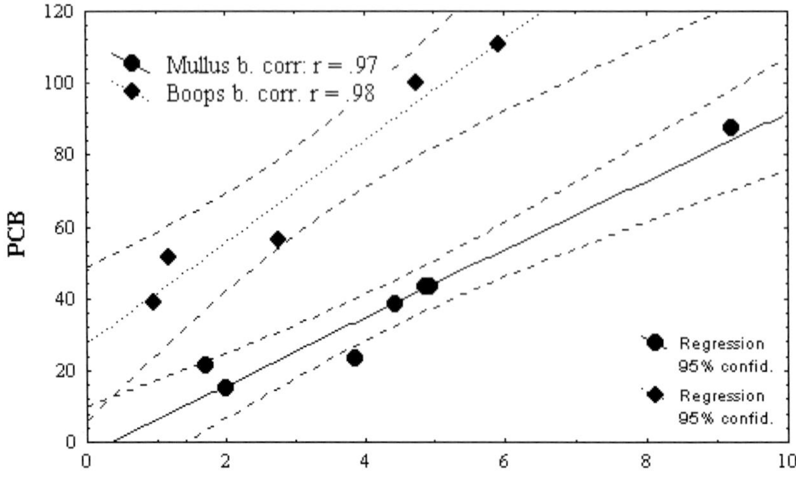

Fig. 4. Correlation between PCB and DDT levels in liver tissue of *Mullus barbatus* and *Boops boops* from the Central Adriatic ($P < 0.05$)

Table 2. Correlation between PCB and DDT levels in muscle fillets

Northern Adriatic	*r* values	Central Adriatic	*r* values
Sprattus sprattus[a]	0.72	*Sardina pilchardus*	0.77
Sardina pilchardus		*Merluccius merluccius*	0.93
Merluccius merluccius	0.98	*Boops boops*	0.65
		Scomber japonicus	0.86
		Sardina pilchardus	0.77
		Mullus barbatus	0.98

[a] Liver tissue correlation $r = 0.99$

from a toxicological point of view in the light of the new concept of toxic equivalents (TEQs), recently introduced to simplify risk assessment and regulatory control for dioxin-like compounds including PCBs (Ahlborg et al. 1994). The interim described by WHO/IPCS indicated some PCB congeners (mono-*ortho* and di-*ortho*) for calculating toxic equivalency factors (TEF) for human intake such as IUPAC number PCB-118, PCB-156 and PCB-189. By using this calculation for the TEFs, we reported the TEQ values for all the species studied with reference to liver and muscle levels (Table 3). The highest values were reported for mono- and di-*ortho* in liver of *Merluccius merluccius* (TEQ values: 1.02-3.89 and 3.99-0.9) and for muscle of *Sardina pilchardus* (TEQ values 0.58-0.34 and 0.52-0.32). The values reported by Ahlborg et al. (1994)

Table 3. Toxic equivalents (TEQs) according to the WHO/IPCS TEFs approach cited by Ahlborg et al. (1994)

Northern Adriatic	TEQ mono-ortho				TEQ di-ortho		
	PCB 118	PCB156	PCB189	Total	PCB180	PCB170	Total
Liver							
Merluccius merluccius	2.12	1.66	0.12	3.89	0.53	3.46	3.99
Sardina pilchardus	0.26	0.14	0.03	0.43	0.07	0.46	0.53
Sprattus sprattus	0.41	0.307	0.003	0.72	0.06	0.43	0.49
Muscle							
Merluccius merluccius	0.022	0.05	0.005	0.08	0.017	0.11	0.13
Sardina pilchardus	0.31	0.23	0.036	0.58	0.073	0.443	0.52
Sprattus sprattus	0.28	0.14	0.008	0.43	0.04	0.26	0.30
Central Adriatic	**TEQ mono-ortho**				**TEQ di-ortho**		
	PCB 118	**PCB156**	**PCB189**	**Total**	**PCB180**	**PCB170**	**Total**
Liver							
Merluccius merluccius	0.62	0.37	0.028	1.02	0.13	0.77	0.9
Sardina pilchardus	0.33	0.18	0.02	0.53	0.076	0.437	0.51
Scomber japonicus	0.09	0.07	0.01	0.17	0.02	0.14	0.16
Mullus barbatus	0.13	0.096	0.013	0.24	0.03	0.2	0.23
Boops boops	0.17	0.22	0.01	0.4	0.06	0.4	0.46
Muscle							
Merluccius merluccius	0.26	0.11	0.009	0.38	0.04	0.25	0.29
Sardina pilchardus	0.22	0.11	0.006	0.34	0.05	0.28	0.32
Scomber japonicus	0.02	0.02	0.002	0.04	0.005	0.038	0.04
Mullus barbatus	0.086	0.04	0.003	0.13	0.014	0.09	0.1
Boops boops	0.027	0.025	0.001	0.05	0.007	0.053	0.06

referred to salmon coming from the Great Lakes, a widely polluted area, were 46.8 for mono-*ortho* and 8.3 for di-*ortho* which highly exceed ours.

Moreover, in comparing our values with those reported for *Thunnus thynnus thynnus* from the Tyrrhenian Sea (Corsolini et al. 1995) which are 128.55-69.4 for mono-*ortho* and 8.33-4.63 for di-*ortho* in muscle and liver respectively, our data are much lower. Thus the Adriatic species studied by us do not indicate a potential risk in terms of human fish consumption according to this finding.

In this respect, the approach reported by the JMG guidelines (1992) indicates a threshold of the "upper" range of 5000 ng g^{-1} for polychlorobiphenyl levels in fish tissue that is extremely high compared to our data. Therefore muscle fillets are presumably not a dangerous source of organochlorines for the fish consumer in the Adriatic basin, at least not the muscle fillets of the species monitored in this study.

Although our results do not allow us to define the Adriatic Sea as a sink of pollution, the presence of detectable levels of these contaminants evidence the necessity to continue monitoring of them until they are completely eliminated.

Acknowledgements. This research was funded by the Italian National Research Council (CNR) as a part of a research programme called PRISMA 2 for the safeguard of the Adriatic Sea.

References

Ahlborg UG, Becking GC, Birnbaum LS, Brouwer A, Derks HJGM, Feeley M, Golor G, Hanberg A, Larsen JC, Liem AKD, Safe SH, Schlatter C, Waern F, Younes M, Yrjanheikki E (1994) Toxic equivalency factors for dioxin-like PCBs: report on a WHO-ECEH and IPCS consultation, December 1993. Chemosphere 28: 1049-1067

Corsolini S, Focardi S, Kurunthachalam K, Tanabe S, Borrell A, Tatsukawa R (1995) Congener profile and toxicity assessment of polychlorinated biphenyls in dolphins, sharks and tuna collected from Italian coastal waters. Mar Environ Res 40(1): 33-53

Gergakopoulos-Gregoriades E, Vassilopoulou V, Stergiou KI (1991) Multivariate analysis of organochlorines in red mullet from Greek waters. Mar Pollut Bull 22(5): 237-241

Giordani P, Miserocchi S, Balboni V, Malaguti A, Lorenzelli R, Honsell G, Poniz P (1997) Factors controlling trophic conditions in the North-West Adriatic basin: seasonal variability. Mar Chem 58: 351-360

Hellou J, Warren WG, Payne JF (1993) Organochlorines including polychlorinated biphenyls in muscle, liver and ovaries of cod, *Gadus morhua*. Arch Environ Contam Toxicol 25: 497-505

Joint Monitoring Group (JMG) of the Oslo and Paris Commission (1992) A compilation of standards and guidance values for contaminants in fish crustaceans and mollusc for the assessment of possible hazards to human health. JMG 17/3/10E

Lech JJ, Bend JR (1980) Relationship between biotransformation and the toxicity and fate of xenobiotic chemicals in fish. Environ Health Perspect 34: 115-131

Porte C, Albaiges J (1993) Bioaccumulation patterns of hydrocarbons and polychlorinated byphenyls in bivalves, crustaceans and fishes. Arch Environ Contam Toxicol 26: 273-281

Safe S (1990) Polychlorinated biphenyls (PCBs), dibenzo-*p*-dioxins (PCDDs), dibenzofurans (PCDFs) and related compounds: environmental and mechanistic considerations that support the development of toxic equivalency factors. Crit Rev Toxicol 21: 51-88

Safe S (1993) Development of bioassay and approaches for the risk assessment of 2,3,7,8-tetrachlorodibenzo-p-dioxin and related compounds. Environ Health Perspect Suppl 101: 317-325

Varanasi U, Stein JE, Reichert WL, Tilbury KL, Krhan MM, Chan SL (1993) Chlorinated and aromatic hydrocarbons in bottom sediments, fish and marine mammals in US coastal waters: Laboratory and field studies of metabolism and accumulation. In: Walker CH, Livingstone DR (eds) Persistent pollutants in marine ecosystems. Pergamon Press, Oxford, pp 83-115

Wakimoto T, Tatsukawa R, Ogawa T (1971) Analytical methods of PCBs (in Japanese). J Enviro Pollut Contam 7: 517-522

Biomarkers in the Teleost Fish *Diplodus puntazzo*: a Study on Animals from an Unpolluted Environment (Brackish Water Pond Acquatina-Lecce, Italy)

M.E. Giordano[1], M.G. Lionetto[1], S. Vilella[2], and T. Schettino[1]

ABSTRACT

In the present work metallothioneins, EROD and acetylcholinesterase, biomarkers specific for different classes of pollutants (heavy metals, organic pollutants and organophosphate and carbamate pesticides respectively) were studied in the teleost fish, *Diplodus puntazzo,* an important species for aquaculture and fishery in the juvenile stage held in the brackish water pond of Acquatina (Lecce - Italy).

The EROD activity was determined on liver by using the method recommended by UNEP (1997), acetylcholinesterase activity was measured on muscle by the Ellman's method (1961), while metallothioneins were measured on liver, intestine, gills and kidney by using the method introduced by Viarengo et al. (1997) on mussels. This last method, slightly modified to be applied to the fish tissue samples, allows us to clearly detect a significant increase of the metallothionein content in the liver, intestine, gills and kidney of *Diplodus puntazzo* exposed to sublethal concentrations of cadmium (220 µg/l for a week).

Chemical analysis excluded the presence of heavy metals and pesticides either in the Acquatina pond (in water and sediments) or in the tissues of the fish. This allowed us to attribute the monthly variations found in the biomarkers to physiological and pollutant independent variations. We studied the relation between each of the three biomarkers and the water temperature and somatic indices (body wet weight and body length) with the aim to provide basal information for a correct evaluation of the status of health of fish in coastal sea areas exposed to a higher anthropic impact and in areas used for aquaculture.

Introduction

"A biomarker is a xenobiotically induced variation in cellular or biochemical components or process structures or functions that is measurable in a biological system or sample" (NRC 1987). Biomarkers can provide sensitive and specific measure of the exposure of living organisms to xenobiotics, using samples obtained from the field. No single biomarker can provide all the information necessary to evaluate exposure or its significance, but the response of one biomarker can provide information that improves interpretation of the others. Therefore, the integrated use of several biomarkers, each specific for a different class of pollutant, is very useful in marine biomonitoring programmes for evaluating the impact of environmental pollutants on marine organisms and for earlier safeguard actions. It must be taken into account that the responses of biomarkers can be influenced by natural factors such as age, seasonal variations, sexual cycle etc, which must be considered for an accurate identification of responses induced by toxic chemicals.

In the present work we studied natural variations in three different biomarkers, metallothioneins (Mt), etoxiresorufin-o-deethylase (EROD) and acetylcholinesterase (AChE), specific for different classes of pollutants, in the teleost fish *Diplodus puntazzo*. The fish species chosen is easily available, relatively abundant and widely distributed along the Apulian coasts; it is very important for aquaculture and fishery, but to date not studied in the field of biomark-

[1] Laboratorio di Fisiologia Generale e Ambientale e [2] Laboratorio di Fisiologia Comparata, Dipartimento di Biologia, Università di Lecce, 73100 Lecce, Italy

F.M. Faranda, L. Guglielmo, G. Spezie (eds)
Mediterranean Ecosystems: Structures and Processes

ers. Fishes were held for a year in Acquatina (Lecce-ITALY), a brackish water pond in connection with the Adriatic Sea. Chemical analysis excluded the presence of contaminants (heavy metal and pesticides) in this pond (either in the water or in the sediments) (Table 1) and in the fish tissues (not shown), allowing us to chose Acquatina as uncontaminated environment for studying natural variations in the biomarker levels. In fact, the aim of this monthly biomonitoring was to obtain useful reference knowledge for a correct evaluation of the status of health of fish in coastal sea areas exposed to a higher anthropic impact and in areas used for aquaculture.

The three biomarkers studied are specific for different classes of pollutants and their integrated use can provide information necessary to evaluate the exposure of marine organisms to the most common pollutants along our coasts.

Metallothionein are low molecular weight (6-7000 D), cysteine-rich (20-30%), metal-binding proteins whose neosynthesis represents a specific response of the organisms to pollution by heavy metals such as Cu, Zn, Cd and Hg (Engel and Roesijadi 1987; Viarengo 1989). Levels of metallothioneins are utilized as an early indicator of the biological effects of heavy metals, which are an important source of pollution in coastal areas of industrialized countries (Engel and Roesijadi 1987; Roesijadi 1992).

Table 1. Chemical analysis on water and sediments of the brackish water pond Acquatina (Lecce-Italy) performed in October 1997, just before starting the biomonitoring study. Chemicals analyses were repeated in April 1998, during biomonitoring, obtaining analogue results. Heavy metals were measured by atomic absorption spectrophotometry, pesticides (organophosphate, organochloride, organonitrogen, which are the most common in agriculture of this area) by mass spectrometry. Levels in the sediment were reported as w.w.

	Water (ppm)	Sediments (ppm)
Pesticides (organophosphate, organochloride organonitrogen)	Absent	Absent
Cadmium	<0.0040	<0.004
Lead	0.155	0.314
Zinc	0.198	<0.004
Copper	<0.015	<0.015
Mercury	<0.001	0.179

Etoxiresorufin-O-deethylase (EROD) is an expression of the activity of cytocrome P-450I A1, whose increase in fish liver is strongly correlated to the exposure of the organism to organic xenobiotic pollutants (Haasch et al. 1989).

Acetylcholinesterase (AChE) activity is inhibited by two major classes of pesticides, the organophosphates and carbamates. This means that AChE measurements are probably the best specific index of pesticide poisoning in most marine species (Galgani and Bocquene 1988).

Materials and Methods

Materials

An homogeneous stock of the teleost fish *Diplodus puntazzo* (9 months old) was purchased from a commercial source (Ittica Ugento, Lecce Italy) and held in the brackish water pond of Acquatina, Lecce Italy. Fishes were regularly fed during the monitoring period and were monthly sampled from November 1997 to October 1998.

All chemicals, reagent grade, were purchased from Sigma (St. Louis, MO).

Methods

Metallothionein (Mt) content in liver, intestine, gills and kidney was determined by using the spectrophotometric method introduced by Viarengo et al. (1997), slightly modified (the composition of the homogenization buffer was changed as follow: sucrose 0.15 M, Tris HCl 20 mM, pH 8.6 with HCl) to be applied to fish tissues. This method allows us to obtain a final fraction enriched in Mt with a minimal negligible contamination of other SH-proteins, as determined by electrophoresis.

The EROD activity was measured in the microsomal fraction of liver by using the sprectrofluorimetric method recommended by UNEP (1997).

Acetylcholinesterase activity was determined in muscle by the spectrophotometric method introduced by Ellman et al. (1961).

The statistical significance of the difference between control and treated groups in the toxicity experiments was analysed by the Student's *t*-test, One-Way ANOVA and Newman-Kules post

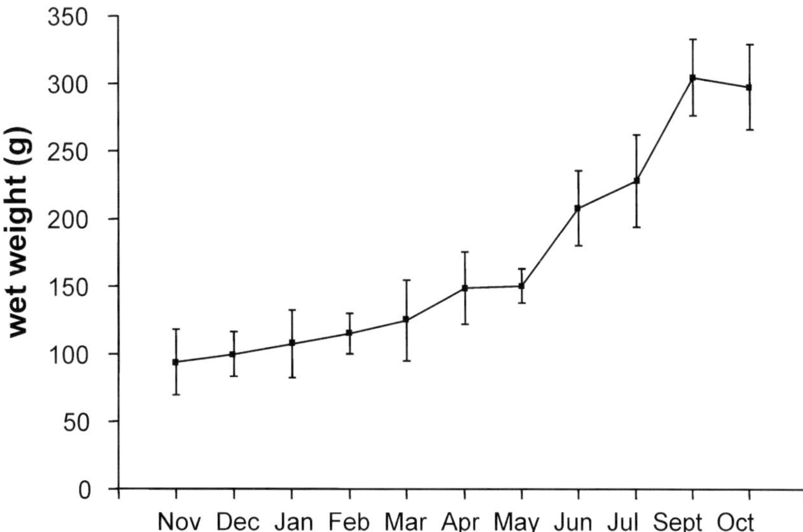

Fig. 1. Monthly measurements of *Diplodus puntazzo* wet weight. Data are expressed as mean ± SD

test, where indicated. Statistical correlations were performed by the GraphPad Prism Programme (2.1 version).

Results

In the year of monitoring the average monthly temperature of the water (measured in the area where fishes were held) ranged between 11.4 and 27.2°C with the minimum value in February and the maximum value in July; the average wet weight and length of the animals grew from the initial average value of 93.8 g and 16.3 cm respectively to 306.6 g and 23.2 cm. In Fig. 1 the time-course of the animal wet weight during the monitoring year is reported. This time-course shows two distinct phases: the first one, from November to May, characterised by a slow growth and the second one from June to September characterised by a faster growth. The growth stopped in October. Gonads started to appear in some specimens from September, but at the end of monitoring (October 1998) a clear distinction between male and female was still not possible, so the complete sexual maturity was still not reached. Chemical analysis excluded the presence of contaminants (heavy metal and pesticides) in this pond (either in the water or in the sediments) (Table 1) and in the fish tissues (not shown).

As concerns metallothionein monthly determinations, they were performed in different organs which are very sensitive to heavy metal pollution: gills and intestine, which are the first interfaces of the animals with the environment, and liver and kidney which are the main sites of accumulation and detoxification of heavy metals (Cinier et al. 1997).

For the metallothionein content evaluation we utilized the spectrophotometric method introduced by Viarengo et al. (1997) for mussels, slightly modified to be applied to the fish tissue samples (see Methods). Therefore, the field monitoring was preceded by toxicity experiments in the laboratory to test the sensitivity of the spectrophotometric method utilised to detect metal exposure in *Diplodus puntazzo*. As reported in Fig. 2, when the animals were exposed for a week to 220 µg/l of Cd added to the water, the metallothionein content in intestine, gills, kidney and liver significantly increased ($P<0.01$) with respect to the control.

The monthly determination of the metallothionein content in intestine, liver, gills and kidney (Fig. 3) revealed differences in these four organs. In fact, in the intestine a slight increase of these proteins was observed from December to April. Statistical analysis performed by One-Way ANOVA and Newman-Kules post test revealed a significant ($P<0.05$) difference between February (where the minimum value of temperature was registered) and all the other months (except March), which, on the other hand, did not reveal any significanct differences between each other. In the gills the increase in the coldest months (January and February) appeared more marked. In fact, the gill metallothionein content measured in January and February resulted highly significantly different ($P<0.001$) with respect to all the other months, while there was not any significant difference in the other months. On the contrary, in the liver a general upwards trend was observed in spring and summer months with the maximum value in June. Regarding the kidney, each monthly value was obtained from a pool of seven tissues, because of the small amount of kidney samples obtained; therefore, unlike the other organs, these values lack in standard deviation. Kidney metallothionein content data showed a high seasonal variation with increased value in autumn and winter months.

The ANOVA analysis and post test applied to all the months (reported above) or to three month groups (a season) (not reported) could not represent the best way to evidence a seasonal trend and could cancel slight differences between one month and the other. Therefore, for a better understanding of the variability observed in these data, the metallothionein content in the four organs studied was correlated with an environmental seasonal factor (temperature of the water) and somatic indices (body wet weight and body length). Intestinal and gill metallothioneins showed a significant negative correlation only with temperature; liver revealed a significant positive correlation with all the para-

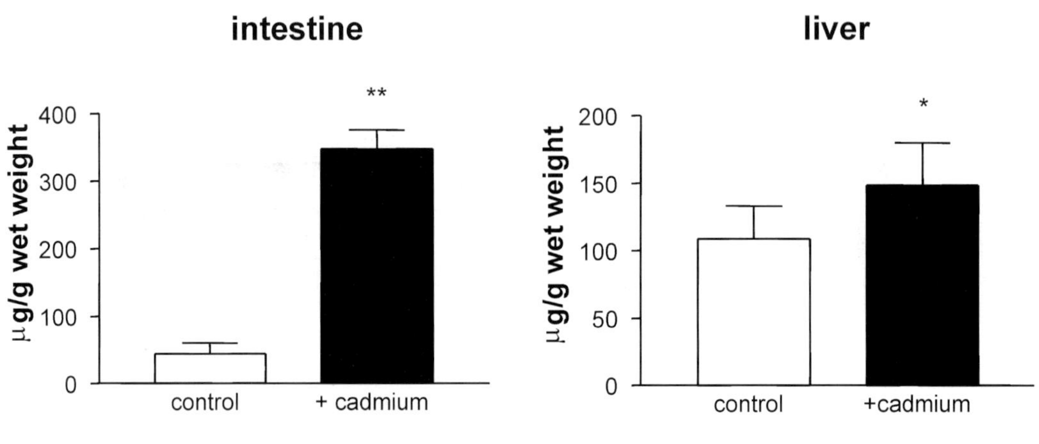

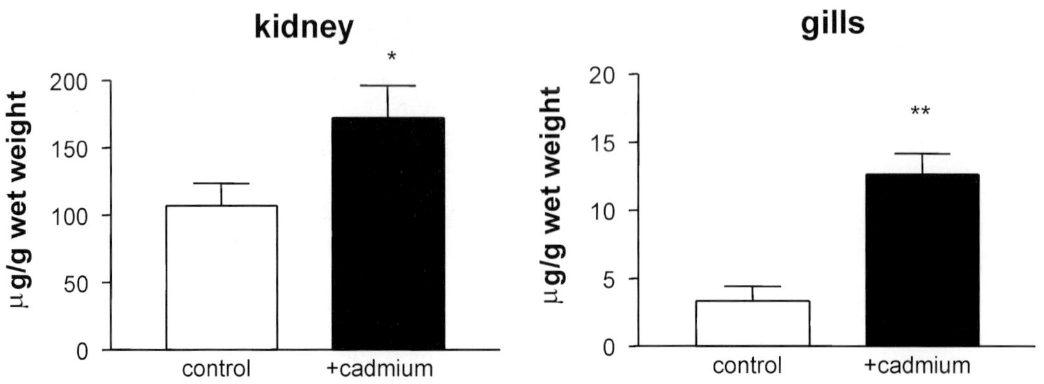

Fig. 2. Effect of the exposure of *Diplodus puntazzo* to Cd 220 mg/ml for a week on the metallothionein contents in the intestine, gills, liver and kidney. Data are expressed as mean ± SD. The statistical significance of the difference between control (3 animals) and treated (3 animals) groups was analysed by the Student's t test

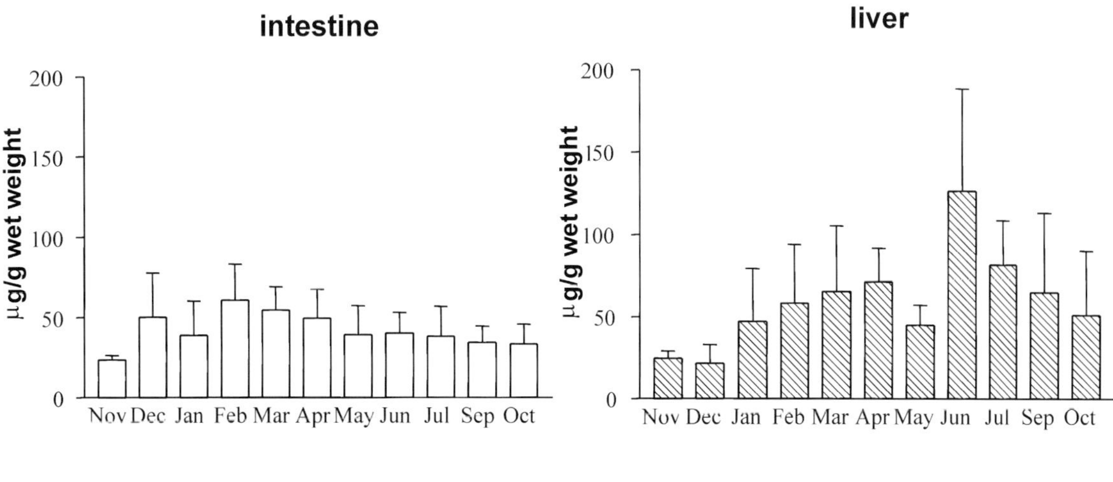

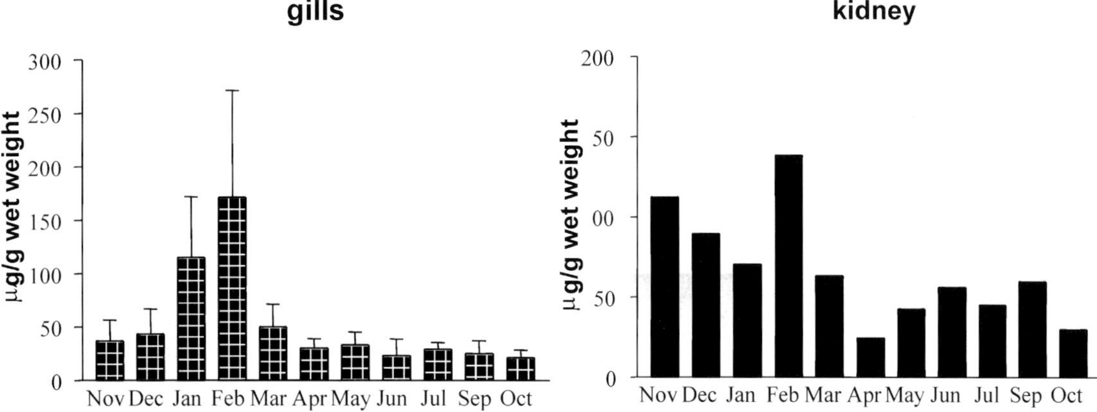

Fig. 3. Monthly determination of metallothionein content in intestine, liver, gills and kidney of *Diplodus puntazzo*. Determinations were performed on 7 animals separately. Data are expresssed as mean ± SD. Kidney values were obtained from a pool of 7 tissues

Table 2. Correlations between metallothionein content in different tissues and somatic indices (body wet weight and body length) and monthly average temperatures

Metallothionein	Body wet weight	Body length	Water temperature
Intestine	$r = -0.2146$ $P = 0.0887$ $n = 64$	$r = -0.2093$ $P = 0.0969$ $n = 64$	$r = -0.2463$ $P = 0.0498$ $n = 64$
Liver	$r = 0.3481$ $P = 0.0052$ $n = 63$	$r = 0.3535$ $P = 0.0045$ $n = 63$	$r = 0.3606$ $P = 0.0037$ $n = 63$
Gills	$r = -0.4647$ $P = 0.1498$ $n = 67$	$r = -0.4826$ $P = 0.1327$ $n = 67$	$r = -0.4103$ $P = 0.0006$ $n = 67$
Kidney	$r = -0.5331$ $P = 0.0913$ $n = 11$	$r = -0.6075$ $P = 0.0474$ $n = 11$	$r = -0.4379$ $P = 0.1780$ $n = 11$

meters studied (Table 2). For the kidney only body length was significantly correlated with the metallothionein values.

As reported in Fig. 4, we monthly determined EROD activity in liver of *Diplodus puntazzo* and the total hepatic induction index (THI), which represents the ability of the organism to metabolise xenobiotics via the MFO (mixed function oxidase) system, calculated as follows: THI=enzyme activity/body wet weight (Boersma et al. 1984). It is possible to observe a general downwards trend either in EROD activity or in THI; the reduction appeared more evident in July, September and October. Although One-Way ANOVA did not reveal any significant differences between months, the correlation between the reduction in EROD and THI and either water temperature and somatic indices was significant (Table 3).

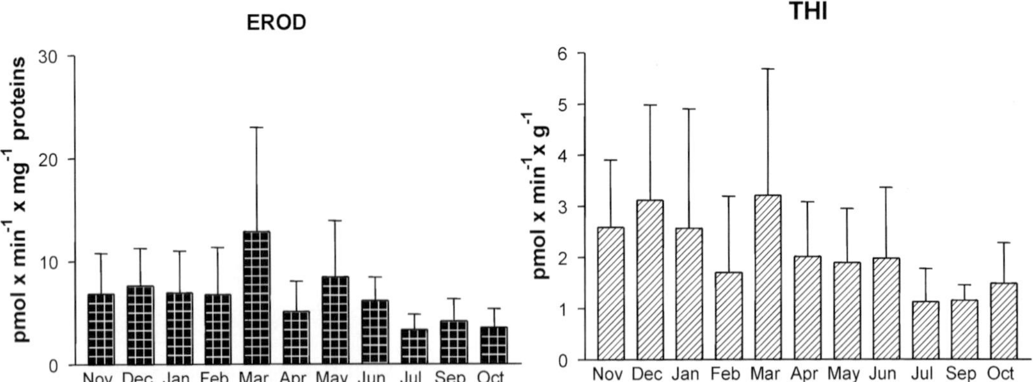

Fig. 4. Monthly determination of hepatic EROD activity and THI in *Diplodus puntazzo*. Determinations were performed on 7 animals separately. Data are expressed as mean ± SD

As reported in Fig. 5, the monthly determination of acetylcholinesterase activity did not show marked seasonal variations; this is confirmed by the fact that no correlation with temperature, body weight and body length was observed (Table 3).

Discussion

Due to increased industrial and other anthropogenic activities during the last decades, there is an increasing demand for sensitive and relatively inexpensive diagnostic biomarkers, applicable for biomonitoring programmes to assess the human impact on the environment. It must be taken into account that the responses of biomarkers can be influenced by the physiological status of the organisms in relation to their life cycle and by natural stress, including seasonal variations in temperature, salinity, etc.. The knowledge of seasonal fluctuations in biomarkers is of crucial importance when these responses have to be used to detect pollution induced stress in marine organisms. Therefore, in this work metallothionein, EROD and acetylcholinesterase, biomarkers specific for different classes of pollutants, have been studied in specimens of *Diplodus puntazzo*, held in the brackish water pond of Acquatina (Lecce - Italy). Chemical analysis excluded the presence of heavy metals and pesticides either in the Acquatina pond (in water and sediments) (Table 1) or in the tissues of the fish (not shown). The study has been carried out monthly on an homogeneous stock of specimens, 9 months old at the beginning of the study to 21 months old at the end. During this period the fishes grew in the body mass until gonads started to appear. Therefore, biomarkers were studied during the juvenile stage of the fish, scarcely investigated in the literature, excluding the variability due to the reproductive cycle (which will be investigated next). Moreover, in order to avoid variability in the biochemical parameters tested due to the nutritional state, fishes were regularly fed during the monitoring year. Data obtained were related to an environmental seasonal factor, the temperature of the pond water, which showed seasonal variations.

As regards metallothionein determination, the method utilised proved to be sensitive to variations in metallothionein content in tissues of *Diplodus puntazzo*. In fact, it allowed us to clearly detect a significant increase of this pro-

Table 3. Correlations between EROD activity, THI, acetylcholinesterase and somatic indices (body wet weight and body length) and monthly average temperatures

	Body wet weight	Body length	Water temperature
EROD	r = -0.2608 P = 0.0317 n = 68	r = -0.2771 P = 0.0221 n = 68	r = -0.2429 P = 0.0459 n = 68
THI	r = -0.3355 P = 0.0052 n = 68	r = -0.3463 P = 0.0038 n = 68	r = -0.2450 P = 0.0441 n = 68
Acetyl-cholinesterase	r = -0.06685 P = 0.6118 n = 60	r = -0.2141 P = 0.1004 n = 60	r = -0.04354 P = 0.7264 n = 67

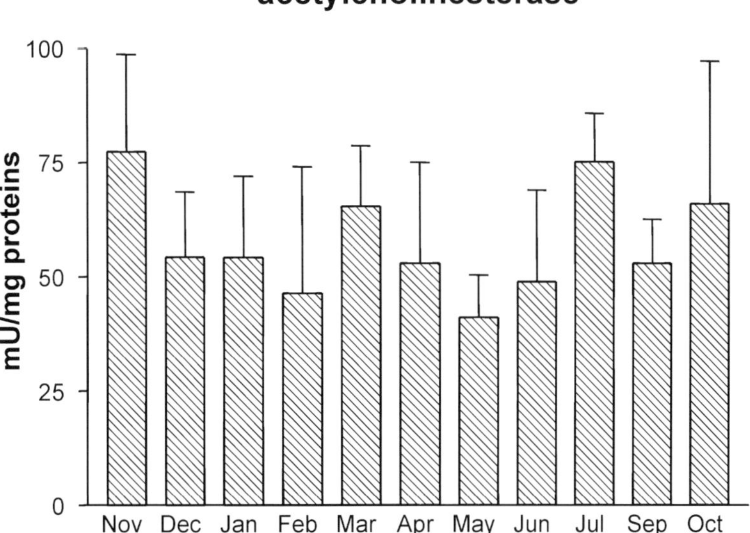

Fig. 5. Monthly determination of acetylcholinesterase activity in *Diplodus puntazzo* muscle. Determinations were performed on 7 animals separately. Data are expressed as mean ± SD

tein content in the liver, intestine, gills and kidney (Fig. 2) of fishes exposed to sublethal concentrations of cadmium (220 µg/l for a week), suggesting that all the organs studied can be utilised for heavy metal exposure monitoring. However, differences were observed in the monthly determinations of the metallothionein content in the different organs. In fact, intestine and gill metallothionein content appeared not to be influenced by the growth of the animals during the year, but were shown to be strongly influenced by low temperature, as indicated by the highest values reached in the coldest months and by the significant negative correlation with temperature. Liver metallothionein content increased significantly with water temperature according to data reported in goldfish by Carpené et al. (1992) and revealed a significant dependence from the growth of the animals. In the kidney the dependence from the parameters studied was not clear, probably because of the small amount of data available. A possible explanation of the differences observed in the metallothionein content in the different tissues could be the different role played by metallothioneins in different organs in the homeostatic balance of essential heavy metals; alternatively it is possible to postulate the presence of different metallothionein isoforms differentially regulated in the organs studied. In fact, as reported by Roesijadi (1992), multiple forms of metallothioneins were found in several species of marine organisms, but their differential role in cell function has yet to be studied further. Moreover, it is interesting that the metallothionein content of intestine and gills, which represent the first interfaces between the organism and the environment, showed the same behaviour with respect to environmental temperature. Moreover, in spite of the scarce use of intestinal and gill Mt levels in biomonitoring measurements, this study suggests their usefulness as heavy metal pollution indicators, since they are metal inducible (Fig. 2), but, unlike liver, insensitive to the growth of the animal during the juvenile state. However, their increase in the coldest months must be considered.

Regarding EROD and the related THI, water temperature and age of animals exhibited a significant influence on the variability of the observed data, as indicated by the negative correlations (Table 3), according to previous studies on the liver of the dab *Limanda limanda* (Lange et al. 1998; Sleiderink and Boon 1996).

Unlike EROD, acetylcholinesterase activity did not show any significant correlation with temperature and age of the animals (Table 3), according to data reported for other animals like birds (Michell and White 1982). Therefore, the variation observed in the monthly distribution could be attributable just to an interindividual variability.

This study, which will be continued in the future and followed by in-laboratory experiments, will give useful reference knowledge and basal information on biomarker variations in the juvenile state of fish on which scarce information is available in literature. These data contribute to a correct evaluation of the physiological status of fish in coastal sea areas in view of the safeguard of fish resources, fundamental for important human activities such as aquaculture and fishery.

Acknowledgements. This work was supported by a grant from "Accordo di Programma tra Provincia di Lecce e Università di Lecce".

References

Boersma DC, Brownlee LJ, Hollebone BR (1984) A total hepatic induction index of metabolism for lipophilic xenobiotics. J Appl Toxicol 4: 187-193

Carpene E, Camatti A, Isani G, Cattani O, Cortesi P (1992) Cadmium-metallothionein in liver and kidney of goldfish (*Crassius auratus*): effects of temperature and salinity. Ital J Biochem 41: 273-282

Cinier CD, Petitramel M, Faure R, Garin D (1997) Cadmium bioaccumulation in carp (*Cyprinus carpio*) tissues during long-term high exposure: analysis by inductively coupled plasma-mass spectrometry. Ecotoxicol Environ Safety 38: 137-143

Ellman GL, Courtney KD, Andres VJ, Featherstone RM (1961) A new and rapid colorimetric determination of acetylcholinesterase activity. Biochem Pharmacol 7: 88-95

Engel DW, Roesijadi G (1987) Metallothioneins: a monitoring tool. In: Vernberg WB, Calabrese A, Thurberg FP, Vernberg FJ (eds) Pollution physiology of estuarine organisms. Univ South Carolina Press, Columbia, pp 421-438

Galgani F, Bocquene G (1988) A method for routine detection of organophosphates and carbamates in sea-water. Environ Tecnol Lett 10: 311-322

Haash MN, Wejksnora PJ, Stegeman JJ, Lech JJ (1989) Cloned rainbow trout liver $P_{(1)}450$ complementary DNA as a potential environmental monitor. Toxicol Appl Pharmacol 98: 362-368

Lange U, Saborowski R, Siebers D, Buchholz F, Karbe L (1998) Temperature as a key factor determining the regional variability of the xenobiotic-inducible ethoxyresorufin-O-deethylase activity in the liver of dab (*Limanda limanda*). Can J Fish Aquat Sci 55: 328-338

Mitchell CA, White DH (1982) Seasonal brain acetylcholinesterase activity in three species of shore birds over wintering in Texas. Bull Environ Contam Toxicol 29: 360-365

NRC (National Research Council) (1987) Committee on biological markers. Environ Health Perspect, 74: 3-9

Roesijadi G (1992) Metallothioneins in metal regulation and toxicity in aquatic animals. Aquat Toxicol 22: 81-113

Sleidernk HM, Boon JP (1996) Temporal induction pattern of hepatic cytocrome P450 1A in thermally acclimated dab (*Limanda limanda*) treated with 3,3',4,4'-tetrachlorobiphenyl (CB77). Chemosphere 32: 2335-2344

UNEP (United Nations Environment Programme) (1997) Meeting of experts to review the Mediterranean Action Plan Biomonitoring Programme. Malta 29 September-1 October, 1997

Viarengo A (1989) Heavy metals in marine invertebrates: mechanisms of regulation and toxicity at the cellular level. CRC Crit Rev Aquat Sci 1: 295-317

Viarengo A, Ponzano E, Dondero F, Fabbri R (1997) A simple spectrophotometric method for metallothionein evaluation in marine organisms: an application to Mediterranean and Antarctic molluscs. Mar Environ Res 44: 69-84

CHAPTER 11

Metallothionein and Glutathione as Stress Indicators in Bivalves in the Lagoon of Venice

P. Irato, A. Cassini, G. Santovito, F. Cattalini, and V. Albergoni

ABSTRACT

Bivalves are often used as indicators of metal pollution, as their digestive glands and gills are important target organs for metal accumulation. With the aim of studying the effects of stress conditions (heavy metal pollution) using metallothionein (MT) and glutathione (GSH) as markers, we studied three species of Bivalvia: *Mytilus galloprovincialis*, which feeds on suspended matter filtered from sea water by cilia on the gill surfaces, and *Tapes philippinarum* and *Scapharca inequivalvis*, which live on sediments. These species were sampled in spring 1998 in two stations with different pollution levels, at Chioggia and Marghera in the Lagoon of Venice. Digestive gland MT contents were significantly different in the same station in the three species: the specimens with the greatest MT contents were *T. philippinarum*. The same species at Marghera was also the one with the greatest amounts of Mn, Co and Fe in the digestive gland. GSH digestive gland contents and Mn, Co and Fe concentrations were statistically greater in *M. galloprovincialis* from Chioggia with respect to *M. galloprovincialis* from Marghera. GSH digestive gland contents of *S. inequivalvis* were also higher in the species from Chioggia. Comparing digestive gland results of the three species at the same two stations, *M. galloprovincialis* and *S. inequivalvis* had significantly greater amounts of GSH than *T. philippinarum*. As regards gills, *M. galloprovincialis* from Chioggia had significantly greater amounts of GSH than *M. galloprovincialis* from Marghera; in the latter station *S. inequivalvis* had significantly greater amounts of GSH and metals than *M. galloprovincialis*. We also looked for relationships between different parameters: length or weight with MT or GSH.

Introduction

Marine bivalve molluscs are often used as indicators of metal pollution. Since the digestive gland (a site for storage) and gills (the site of uptake) are important target organs for metal accumulation, these tissues were chosen to investigate biological responses to metals such as levels of metallothionein (MT) and glutathione. In mussels, metals are primarily complexed by MTs (Viarengo 1989), low-molecular-weight, cysteine-rich proteins, which are capable of sequestering and detoxifying excess intra-cellular metal. Consequently, MTs are regarded as potentially specific indicators of metal pollution. Glutathione (GSH in its reduced form) is considered one of the most important hydrophilic antioxidant agents involved in the protection of cell membranes from lipid peroxidation by scavenging oxygen radicals (Meister 1989). Mussels exposed to environmentally high levels of metals undergo an increase in lipid peroxidation, as demonstrated by the enhanced accumulation of lipofuscin within tertiary lysosomes of specimens found in metal-polluted environments (Regoli 1992). With the aim of studying the effects of stress conditions (heavy metal pollution) using MT and GSH as markers, we studied three species of Bivalvia: *Mytilus galloprovincialis*, which feeds on suspended matter filtered from sea water by cilia on the gill surfaces, and *Tapes philippinarum* and *Scapharca inequivalvis*, which live on organic matter in sediments. Marine sediments act as sinks for a variety of pollutants which may be concentrated up to 1000 times higher than in background water levels. Ingested sediment particles are carried into the

Dipartimento di Biologia, Università di Padova, Via U. Bassi 58/B, 35121 Padova, Italy

F.M. Faranda, L. Guglielmo, G. Spezie (eds)
Mediterranean Ecosystems: Structures and Processes

alimentary tract, where they are desorbed from the particles and absorbed by the animal during digestion (Burbidge et al. 1994).

Materials and Methods

In May 1998, specimens of *M. galloprovincialis*, *T. philippinarum* and *S. inequivalvis* were sampled in two stations with different pollution levels at Chioggia and Marghera in the Lagoon of Venice. The shell length of the animals was measured. Soft tissues were separated from the shell, and digestive glands and gills were removed, weighed and frozen at −80°C. Subsequently, glutathione determination and MT assays were performed on the samples, dividing the organs into two parts. For total glutathione determination, the samples were homogenized (1:5 w/v ratio) in 5% sulphosalicilic acid and centrifuged at 15,000 rpm for 30 min at 4°C. Total GSH contents were measured in the resulting supernatant by Anderson's (1985) method and referred to wet weights. For MT assays, the samples were homogenized (1:4 w/v ratio) in 20 mM Tris HCl pH 7.5, added with 0.006 mM leupeptine, 0.5 mM PMSF (phenylmethylsulphonylfluoride) as antiproteolitic agents, and 0.01% β-mercaptoethanol as reducing agent, and centrifuged at 30,000 g for 60 min at 4°C. The MT contents were measured in the resulting supernatant by the silver saturation method (Scheuhammer and Cherian 1991) and the values referred to total protein determined by the method of Lowry et al. (1951). The digestive glands and gills of other individual animals were homogenized with the same buffer as the MT assays and digested with HNO_3 AristaR in PFA vessels in a CEM MDS-2000 microwave furnace. Metal concentrations (Cr, Mn, Fe, Co, Ni, Cu, Zn, Cd) were determined with an inductively coupled plasma-atomic emission spectrometry (ICP-AES) Spectroflame Modula sequential and referred to dry weights (identical amounts of homogenate used for digestion were dried in an oven at 130°C).

Values are shown as means ± standard error. Statistical analysis was performed with the PRIMER statistical program: Student's t-test for comparison of pairs or ANOVA for different groups followed by the Student-Newman-Keuls test for determination of significant differences ($P = 0.05$). Linear regression analyses were also performed.

Results and Discussion

Sediment metal analyses (conducted by Casellato S. and Degetto S. in 1999; unpublished data) highlighted the important fact that Marghera has three times more Cr than Chioggia and that Cd is only found in Marghera. Results show that MT contents, in both digestive glands and gills of all three species, were not significantly different between the two stations with different levels of pollution. Instead, digestive gland MT contents were significantly different in the same station among the three species: *T. philippinarum* had the greatest MT contents, followed by *S. inequivalvis* and *M. galloprovincialis* (Fig. 1a). The greatest amount of MT in the digestive gland of *T. philippinarum* is probably due to a higher concentration of Cu (Table). Moreover, it is known that Cu has higher affinity for MT with respect to Zn, whereas Zn is a good inductor. We therefore hypothesize that, in *T. philippinarum*, Cu is linked to MT induced by Zn, so that Zn can then induce other MT. *T. philippinarum* from Marghera was the species with the greatest amounts of Mn, Co and Fe (Table). Gill MT contents were not significantly different in the same station among the three species (Fig. 1b).

GSH digestive gland contents were statistically greater in *M. galloprovincialis* from Chioggia than in the same species from Marghera ($P = 0.001$). In proportion, some metals such as Mn, Co and Fe were also greater in *M. galloprovincialis* from Chioggia than in specimens from Marghera (Table). GSH digestive gland contents of *S. inequivalvis* were also greater in the specimens from Chioggia ($P = 0.006$).

Comparing GSH contents in the digestive gland of the three species in the same two stations, *T. philippinarum* had significant lower amounts of GSH than *M. galloprovincialis* and *S. inequivalvis* (Fig. 2a). This situation may be due to the very high induction of MT in *Tapes*.

Instead, considering gills, the only statistically significant differences were: (1) *M. galloprovincialis* from Chioggia had a significantly greater amount of GSH than *M. galloprovincialis* from Marghera ($P = 0.009$); (2) at Marghera, *S. inequivalvis* had a significantly greater amount of GSH than *M. galloprovincialis* (Fig. 2b). The latter datum is in keeping with the metal contents: the gills of *S. inequivalvis* had more metal concentrations (except Cu) than the other two species.

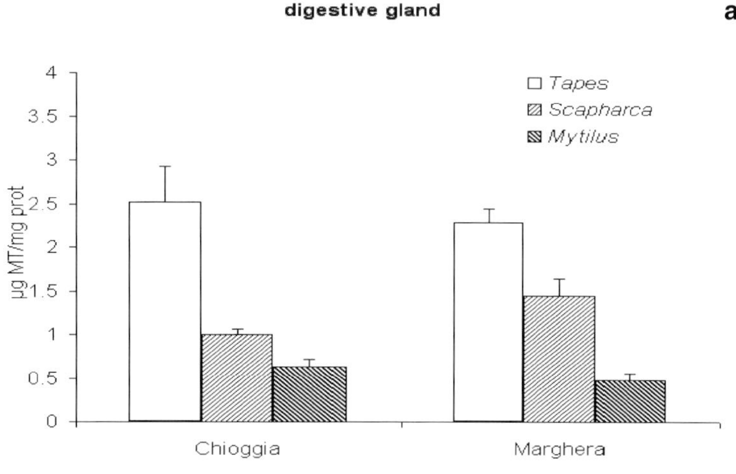

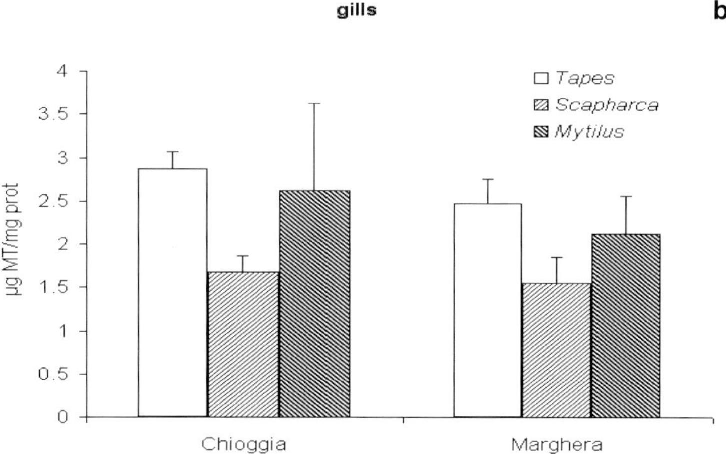

Fig. 1a,b. MT concentrations in *M. galloprovincialis*, *T. philippinarum* and *S. inequivalvis* collected from two sites (n = 7 specimens per site). **a** Digestive glands: significant differences: Chioggia: *Mytilus versus Tapes*, *Mytilus versus Scapharca*, *Tapes versus Scapharca*; Marghera: *Mytilus versus Tapes*, *Mytilus versus Scapharca*, *Tapes versus Scapharca*. **b** Gills; no significant differences

Table. Tissue metal concentrations (µg of metal/g of dry weight) determined in organisms sampled in two areas of Lagoon of Venice

Marghera	Cr	Mn	Fe	Co	Ni	Cu	Zn	Cd
Tapes								
Digestive gland	21.2±8.1	9.8±1.7	268.2±36.9	19.3±4.3	15.8±5.2	41.6±3.7	116.5±16.3	0.7±0.2
Gills	8.2±3.0	11.0±1.3	292.9±49.3	16.8±3.3	6.3±1.6	10.9±0.7	101.2±12.3	1.4±0.3
Scapharca								
Digestive gland	20.4±12.4	2.4±0.5	178.1±17.8	2.6±1.0	14.0±9.0	19.8±1.1	61.5±2.7	0.8±0.1
Gills	46.5±12.4	29.0±5.2	514.0±32.6	50.7±23.9	40.4±10.4	8.9±2.9	208.4±48.1	8.9±2.5
Mytilus								
Digestive gland	5.1±1.7	4.2±1.3	101.7±15.4	6.2±1.5	6.8±1.3	19.2±1.2	115.2±13.4	1.6±0.1
Gills	4.6±2.0	1.7±0.3	26.8±5.6	14.9±3.8	9.8±1.1	19.4±3.1	115.5±12.7	0.6±0.2
Chioggia	**Cr**	**Mn**	**Fe**	**Co**	**Ni**	**Cu**	**Zn**	**Cd**
Tapes								
Digestive gland	8.9±3.5	12.0±2.3	276.1±56.4	54.0±15.1	8.5±2.6	25.2±2.7	83.7±12.3	0.3±0.0
Gills	18.8±6.3	37.2±12.0	425.4±70.4	55.1±13.2	15.0±4.2	12.5±2.0	108.6±15.7	0.6±0.1
Mytilus								
Digestive gland	6.5±1.2	19.6±1.1	347.7±63.6	22.5±7.9	6.8±0.8	16.2±0.9	72.1±12.4	0.9±0.1
Gills	4.4±2.0	6.7±0.5	138.5±17.4	4.3±1.1	7.5±0.8	15.7±6.0	71.4±11.2	0.5±0.1

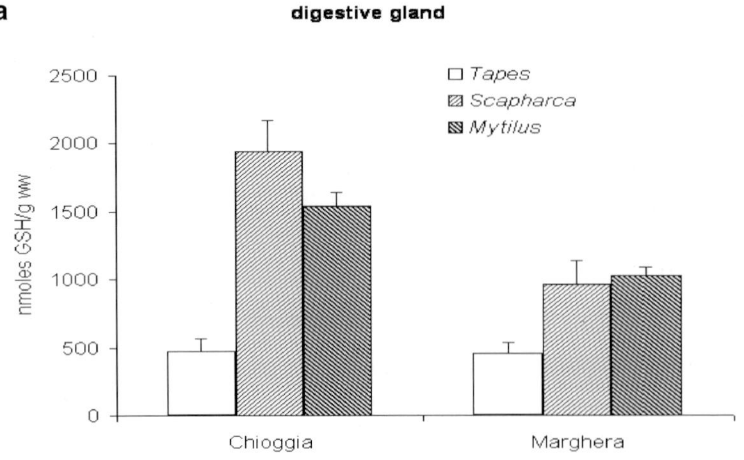

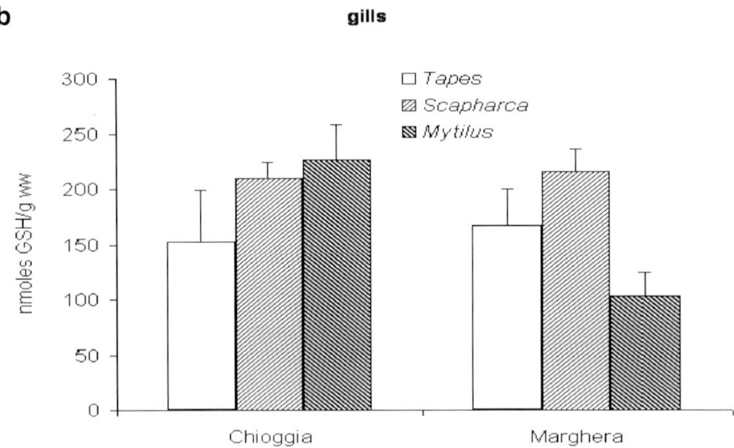

Fig. 2a,b. GSH concentrations in *M. galloprovincialis*, *T. philippinarum* and *S. inequivalvis* collected from two sites ($n = 7$ specimens per site). **a** Digestive glands: significant differences: Chioggia and Marghera: *Mytilus versus Tapes*, *Tapes* versus *Scapharca*. **b** Gills: significant differences: Marghera: *Mytilus versus Scapharca*

MT concentrations were similar in the digestive glands and gills of *T. philippinarum* and *S. inequivalvis*, but greater in the gills than in the digestive glands of *M. galloprovincialis*. Conversely, GSH concentrations were higher in digestive glands than in gills (from 3 times in *T. philippinarum* to 10 times in *M. galloprovincialis* from Marghera), in accordance with results reported by Viarengo et al. (1995) in a study on the Antarctic scallop *Adamussium colbecki* and the Mediterranean scallop *Pecten jacobaeus*.

In aquatic organisms such as molluscs, tissue metal concentrations may be greatly influenced by environmental conditions and the physiological status of individual specimens (Wright et al. 1985). One endogenous factor which must be considered is specimen weight (Cossa et al. 1980). Moreover, metallothionein-like proteins (MTL) and metal concentrations are generally correlated positively. So weight also appears as an important factor in explaining variations in MTL levels. In a study on *Crassostrea gigas*, an inverse relationship was generally observed between weight and metals or MTL concentrations in whole soft tissues (Mouneyrac et al. 1998). Seasonal variability sometimes interferes with the influence of weight. In a study on the marine bivalve *Macoma balthica*, Bordin et al. (1997) reported that MTL induction is greater during cold rather than warm seasons. As their study, as far as we know, is the first in which the seasonal impact of MTL induction has been shown to be important, we looked for relationships between different parameters: length or weight with MT or GSH. We found only one inverse relationship between specimen length and GSH contents in *T. philippinarum* from Marghera ($r = -0.962$; $P = 0.000$). In other cases, no relationship could be detected. We intend to collect the three species of bivalvia during the forthcoming year, in order to evaluate seasonal differences.

In conclusion, our results led us to conclude that MT and GSH contents in mussels may be considered as biomarkers of heavy metal exposure, and that their use should be promoted also in field monitoring. MT and GSH biomarkers indicate an anomalous situation, which may also be caused by several other factors. Metal monitoring is necessary for knowledge on which metals occur in and are accumulated by animals. The above two monitoring methods are not alternatives, but are complementary. The two techniques used together may provide a considerable improvement over current monitoring approaches which are based on contaminant analysis alone.

Acknowledgements. Grant: Cofinanziamento MURST, project "Conservazione della biodiversità e gestione sostenibile dei bioti salmastri delle coste italiane".

References

Anderson ME (1985) Determination of glutathione and glutathione disulfide in biological samples. Methods Enzymol 113: 548-555

Bordin G, McCourt J, Cordiero Raposo F, Rodriguez AR (1997) Metallothionein-like metalloproteins in the Baltic clam *Macoma balthica*: seasonal variations and induction upon metal exposure. Mar Biol 129: 453-463

Burbidge FJ, Macey DJ, Webb J, Talbot V (1994) A comparison between particulate (elemental) zinc and soluble zinc ($ZnCl_2$) uptake and effects in the mussel, *Mytilus edulis*. Arch Environ Contam Toxicol 26: 466-472

Cossa D, Bourget E, Pouliot D, Piuze J, Chanut JP (1980) Geographical and seasonal variations in the relationship between trace metal content and body weight in *Mytilus edulis*. Mar Biol 58: 7-14

Lowry OH, Randall RJ, Rosebrough NJ, Farr AL (1951) Protein measurement with the Folin-phenol reagent. J Biol Chem 193: 265-275

Meister A (1989) On the biochemistry of glutathione. In: Taniguchi N, Higashi T, Sakamoto S, Meister A (eds) Glutathione centennial: molecular perspectives and clinical implications. Academic Press, San Diego, pp 3-22

Mouneyrac C, Amiard JC, Amiard-Triquet C (1998) Effects of natural factors (salinity and body weight) on cadmium, copper, zinc and metallothionein-like protein levels in resident populations of oysters *Crassostrea gigas* from a polluted estuary. Mar Ecol Progr Ser 162: 125-135

Regoli F (1992) Lysosomal responses as a sensitive stress index in biomonitoring heavy metal pollution. Mar Ecol Progr Ser 84: 63-69

Scheuhammer AMV, Cherian MG (1991) Quantification of metallothionein by silver saturation. Methods Enzymol 205: 75-83

Viarengo A (1989) Heavy metals in marine invertebrates: mechanisms of regulation and toxicity at the cellular level. Rev Aquat Sci 1: 295-317

Viarengo A, Canesi L, Garcia Martinez P, Peters LD, Livingstone DR (1995) Pro-oxidant processes and antioxidant defense systems in the tissues of the Antarctic scallop (*Adamussium colbecki*) compared with the Mediterranean scallop (*Pecten jacobaeus*). Comp Biochem Physiol 111B: 119-126

Wright DA, Mihursky JA, Phelps HL (1985) Trace metals in Chesapeake Bay oysters: intra-sample variability and its implications for biomonitoring. Mar Environ Res 16: 181-197

Evaluation of Biological Stress Indices in *Tapes philippinarum* from the Lagoon of Venice through Monitoring of Natural Populations and Transplantation Experiments

M.G. Marin[1], N. Nesto[2], and L. Da Ros[2]

ABSTRACT

Monitoring of natural populations of the clam *Tapes philippinarum* from two sites in the Lagoon of Venice was carried out together with a transplantation experiment. The sites were chosen for their different hydrological conditions and pollution levels. Several physiological and histochemical stress-related parameters have been measured. Clearance rate and scope for growth are significantly affected by higher anthropogenic contamination levels acting on natural clam beds, though they are not responsive in transplanted animals after a six-week exposure. On the contrary "survival in air" response and lysosomal latency time are confirmed as early warning indices, particularly suitable for both monitoring approaches. Clam *T. philippinarum* is proposed as an indicator species in order to obtain an integrated biological evaluation of coastal seawater and sediment quality.

Introduction

Bivalve molluscs are commonly used as indicator organisms in evaluating the environmental quality of marine coastal ecosystems, mostly for their sessile and filter-feeding habits (Goldberg et al. 1978). Two different approaches have been followed, namely: (1) measuring the bioaccumulation of organic and inorganic chemical contaminants in target organs or in whole tissues (evaluation of the contaminant body burden); (2) quantifying biological responses to environmental stress (evaluation of biomarkers of exposure). Monitoring programmes can be carried out on both native and transplanted populations, the latter technique being based on the comparison of chemical and/or biological characteristics in samples collected from one population and exposed, after translocation, to different environmental conditions. The advantages of experimental exposure through translocation experiments are based on the possibility to expose organisms with common life history and genetic make up at sites to be monitored at investigators' choice (de Kock and Kramer 1994).

In the present studies, several physiological and histochemical stress-related parameters have been monitored in two natural populations of the clam *Tapes philippinarum* and after a transplantation experiment at the same locations in the Lagoon of Venice.

The sampling sites were chosen on the basis of their organic and trace metal contaminants level, reflecting different anthropogenic impact (Martin et al. 1994; Fattore et al. 1997a, 1997b; Frignani et al. 1997).

Materials and Methods

Two natural populations of *T. philippinarum*, one from a polluted industrial area, Porto Marghera, and the second one from a relatively clean site though featuring highly variable hydrologic parameters, like temperature, salinity and dissolved oxygen, Palude della Rosa (Bianchi et al. 1997), were monitored twice a year. Moreover, a transplantation experiment was carried out on organisms taken from a reference commercial growing area (Val Dogà) and translocated for six weeks at the same monitoring sites. Samplings on natural populations were performed in April and July, and the transplantation experiment in April, 1998. While sampling the clams, the water

[1] Dipartimento di Biologia, Università di Padova, Via G. Colombo 3, 35121 Padova, Italy
[2] Istituto di Biologia Marina, CNR, Castello 1364/A, 30122 Venezia, Italy

F.M. Faranda, L. Guglielmo, G. Spezie (eds)
Mediterranean Ecosystems: Structures and Processes

temperature, salinity and suspended particulate organic matter values were recorded.

The physiological responses (clearance rate and oxygen uptake) were measured in static systems on sixteen individual clams of standard body size (4 cm length) from each site, according to the methodology described by Smaal and Widdows (1994). Each physiological rate was then converted to energy equivalent (Jg^{-1} h^{-1}) in order to obtain the scope for growth, which represents the difference between the energy gained from the food and the energy lost via metabolic energy expenditure (Widdows 1993). Moreover, the scope for growth was calculated by using a standardized ration level of 0.4 mg POM^{-1} and an absorption efficiency of 0.45 in order to show underlying pollution-induced stress, when all other natural and potential environmental stressors were held constant (Widdows et al. 1997).

"Survival in air" was tested on thirty clams from each monitoring site: the animals were subjected to anoxia by air exposure at 18°C in humidified chambers (Eertman et al. 1993). Survival was assessed daily until a 100% mortality rate was reached.

Lysosomal membrane stability based on the latency of N-acetyl-β-hexosaminidase was estimated on digestive gland tissue of ten clams, frozen immediately after dissection, preserved in supercooled hexane (-70°C) and then processed as described by Moore (1976).

Analysis of survival was performed according to the method of Kaplan and Meier (1958); the significance of differences between groups was tested using the Gehan and Wilcoxon test (Gehan 1965). The ANOVA test was used for all other statistical comparisons.

Results and Discussion

Monitoring of the physiological responses (i.e. clearance rate, oxygen uptake and scope for growth) in the natural populations of *T. philippinarum* revealed a general worsening of conditions in clams from the most polluted area (P. Marghera) in July (Table 1). In particular, the clearance rate is 40% lower than in clams from Palude della Rosa (significant for $P<0.01$). Oxygen uptake is never significantly different in clams from the two monitored sites, showing a fluctuating behaviour with lower values at Palude della Rosa in April, and at P. Marghera in July. The scope for growth values slow down to 0.63 J h^{-1} g^{-1} in July (5.37 J h^{-1} g^{-1} at Palude della Rosa, significantly different for $P<0.05$).

On the contrary, for the transplantation experiment physiological measurements do not reveal significant differences between transplanted clams and the reference population (Table 2).

The decrease in the clearance rate observed for the clam population of P. Marghera confirms this physiological measurement as the most responsive to pollutants among the components of the energy budget (Widdows and Donkin

Table 1. Monitoring of natural population of *T. philippinarum*: physiological, biochemical and environmental parameters (mean ± 95% C.I.). Statistical comparison between sites in the same month (*n.s.* not significant)

	n	April		July	
		Porto Marghera	Palude della Rosa	Porto Marghera	Palude della Rosa
Clearance rate (l $h^{-1}g^{-1}$)	16	2.182 ± 0.543	2.654 ± 0.193	2.143 ± 0.391	3.539 ± 0.958
		n.s.		**	
O_2 uptake(μmol h^{-1} g^{-1})	16	17.260 ± 3.377	13.090 ± 0.193	18.074 ± 6.743	20.363 ± 6.423
		n.s.		n.s.	
Scope for growth (J $h^{-1}g^{-1}$)	16	1.165 ± 2.116	5.020 ± 0.193	0.630 ±3.808	5.365 ± 3.943
		n.s.		*	
Latency (min)	10	8.5 ± 2.5	26.5 ± 1.5	9.0 ± 3.7	22.0 ± 3.9
		***		***	
POM (mg l^{-1})	4	1.595 ± 0.057	1.269 ± 0.275	0.847 ± 0.175	1.781 ± 0.108
		**		***	
Temperature (°C)	-	17.9	14.0	25.0	26.0
Salinity (‰)	-	26.73	31.56	28.93	32.90

* $P<0.05$; ** $P<0.01$; *** $P<0.001$

Table 2. Transplantation experiment of *T. philippinarum*: physiological, biochemical and environmental parameters (mean ± 95% C.I.). Statistical comparison between transplanted and reference clams and, in brackets, between transplanted clams (*n.s.* not significant)

	n	Val Dogà	Porto Marghera	Palude della Rosa
Clearance rate (l $h^{-1}g^{-1}$)	16	2.198 ± 0.305	2.250 ± 0.462 n.s.	2.240 ± 0.735 n.s.(n.s.)
O_2 uptake (μmol $h^{-1}g^{-1}$)	16	16.621 ± 2.245	15.911 ± 1.384 n.s.	15.750 ± 3.896 n.s.(n.s.)
Scope for growth (J $h^{-1}g^{-1}$)	16	1.520 ± 1.498	2.060 ± 1.901 n.s.	2.091 ± 2.801 n.s.(n.s.)
Latency (min)	10	24 ± 4.801	17 ± 2.613 **	22 ± 2.613 n.s.(n.s.)
POM (mg l^{-1})	4	1.118 ± 0.048	0.789 ± 0.504	3.232 ± 0.302
Temperature (°C)	-	18.0	17.5	18.0
Salinity (‰)	-	30.04	29.29	26.65

* *P*<0.05; ** *P*<0.01; *** *P*<0.001

1992). Nevertheless, constant values of all physiological parameters registered after the transplantation experiment may indicate the need for a longer exposure period in order to significantly affect the responses of clams.

Curves of survival in air (Fig. 1a,b) show a better condition for clams of Palude della Rosa in April (*P*< 0.001), though they are quite similar in July. As reported by Eertman et al. (1993) for the blue mussel *Mytilus edulis*, this biologi-

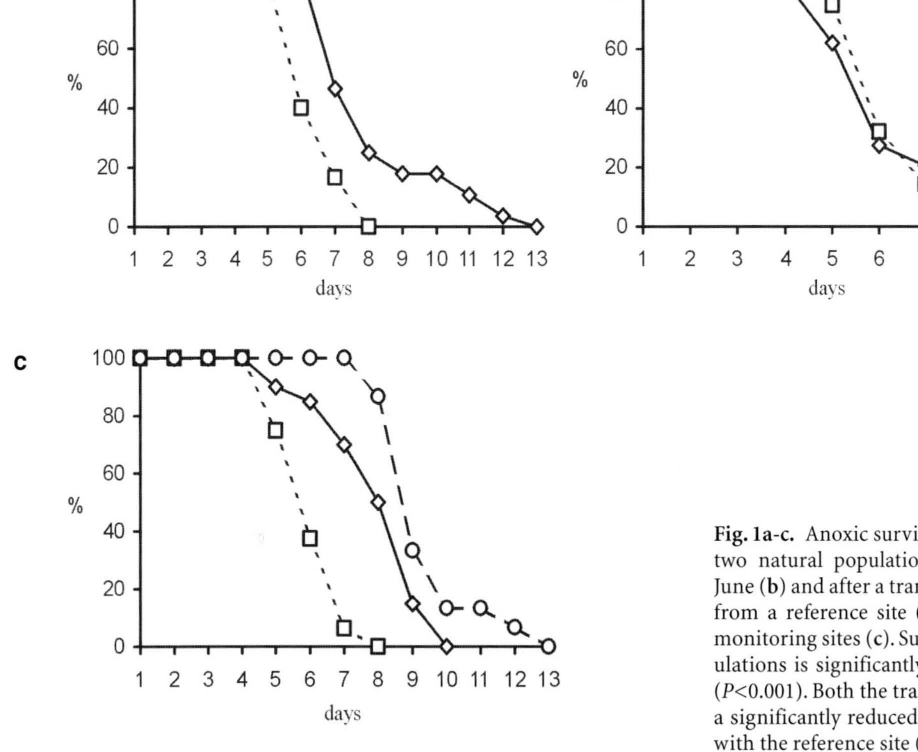

Fig. 1a-c. Anoxic survival *curves* of clams from two natural populations in April (**a**) and in June (**b**) and after a transplantation experiment from a reference site (Val Dogà) to the same monitoring sites (**c**). Survival of the native populations is significantly different only in April (*P*<0.001). Both the transplanted samples show a significantly reduced survival in comparison with the reference site (P. Marghera, *P*<0.001; P. Rosa, *P*<0.01), P. Rosa being significantly different from P. Marghera (*P*<0.001)

cal response exhibits seasonal influences, mainly in relation to the spawning period, when mean anoxic survival time reaches the lowest value. A phase difference between the two clam populations during spawning period, which in Venice lagoon occurs from June to October (Breber 1996), could have acted by lowering the sensitivity of the "survival in air" test. Furthermore, more restrictive environmental conditions (mostly temperature and oxygen availability) can exert their detrimental effect particularly on clams from Palude della Rosa, where these natural environmental stressors come to play a predominant role. As for the transplanted clams, these have significantly reduced survival times ($P<0.01$ for Palude della Rosa and $P<0.001$ for P. Marghera) in comparison with the reference animals of Val Dogà (Fig. 1c); moreover, a gradient in clam conditions is revealed, as the survival time of *T. philippinarum* at P. Marghera is lower than those at Palude della Rosa ($P<0.01$).

The lysosomal latency test exhibits a significant decrease ($P<0.001$) of lysosomal membrane stability in digestive gland cells of clams from the most polluted site, on both sampling seasons (Table 1: mean latency time 68% and 59% lower in April and July, respectively). Moreover, a significant ($P<0.01$), though lower reduction of the latency time is observed on clams transplanted at P. Marghera (Table 2), where the histochemical index shows the same slowing down trend reported for "survival in air" response in the translocation experiment.

As a whole the results indicate different levels of responsiveness for the biological indices measured, mostly depending on season and exposure time at the monitoring sites. Because of their rapid response to changing environmental conditions, resistance to aerial exposure and lysosomal latency time appear to be particularly suitable in defining early biological warning systems. Finally, the use of *T. philippinarum* as an indicator species in estuarine environments is suggested as a potential tool in order to obtain an integrated biological response evaluating both water column and sediment quality. In conclusion, the overall results obtained in this study confirm that the bivalve transplantation method is a useful approach in marine coastal biomonitoring. In particular, it allows to reach valuable and quicker results in comparison with seasonal traditional monitoring, being a technique less time consuming, even though similarly sensitive.

Acknowledgements. This work was funded by the MURST project "Conservazione della biodiversità e gestione sostenibile dei biotopi salmastri delle coste italiane" and by grants from AMAV, Venezia and E.U. in the frame of E.U. WATERS Project: LIFE 96 ENV/IT/00103.

References

Bianchi F, Acri F, Alberighi L, Bastianini M, Boldrin A, Cavalloni B, Cioce F, Comaschi A, Rabitti S, Socal G, Turchetto M (1997) The lagoon of Venice: a biological variability study. Unesco, Paris

Breber P (1996) Allevamento della vongola verace in Italia. Cleup, Padova

de Kock WC, Kramer KJM (1994) Active biomonitoring (ABM) by translocation of bivalve molluscs. In: Kramer KJM (ed) Biomonitoring of coastal waters and estuaries. CRC Press, Boca Raton, Florida, pp 51-84

Eertman RHM, Wagenvoort AJ, Hummel H, Smaal AC (1993) "Survival in air" of the blue mussel *Mytilus edulis* L. as a sensitive response to pollution-induced environmental stress. J Exp Mar Biol Ecol 170: 179-175

Fattore E, Benfenati E, Mariani G, Cools E, Vezzoli G, Fanelli R (1997a) Analysis of organic micropollutants in sediment samples of the Venice Lagoon, Italy. Wat Air Soil Pollut 99: 237-244

Fattore E, Benfenati E, Mariani G, Fanelli R, Evers EHG (1997b) Patterns and sources of polychlorinated dibenzo-p-dioxins and dibenzofurans in sediments from the Venice Lagoon, Italy. Environ Sci Technol 31: 1777-1784

Frignani M, Bellucci LG, Langone L, Muntau H (1997) Metal fluxes to the sediments of the northern Venice Lagoon. Mar Chem 58: 275-292

Gehan EA (1965) A generalized Wilcoxon test for comparing arbitrarily singly censored samples. Biometrika 52: 203-223

Goldberg ED, Bowen VT, Farrington JH, Harvey G, Martin JH, Parker PL, Riseborough RW, Robertson W, Schneider E, Gamble E (1978) The mussel watch. Environ Conserv 5: 1-25.

Kaplan EL, Meier P (1958) Non parametric estimation from incomplete observations. J Am Stat Assoc 53: 457- 481

Martin JM, Huang WW, Yoon YY (1994) Level and fate of trace metals in the Lagoon of Venice. Mar Chem 46: 371-386

Moore MN (1976) Cytochemical demonstration of latency of lysosomal hydrolases in digestive cells of the common mussel, *Mytilus edulis,* and changes induced by thermal stress. Cell Tissue Res 175: 279-287

Smaal AC, Widdows J (1994) The scope for growth of bivalves as an integrated response parameter in biological monitoring. In: Kramer KJM (ed) Biomonitoring of coastal waters and estuaries. CRC Press, Boca Raton, Florida, pp 247-267

Widdows J (1993) Marine and estuarine invertebrate toxicity tests. In: Calow P (ed) Handbook of Ecotoxicology, vol I. Blackwell Scientific, Oxford, pp 145-166

Widdows J, Donkin P (1992) Mussels and environmental contaminants: bioaccumulation and physiological aspects. In: Gosling E (ed.) The mussel *Mytilus.* Elsevier Press, Amsterdam, pp 383-424

Widdows J, Nasci C, FossatoVU (1997) Effects of pollution on the scope for growth of mussels (*Mytilus galloprovincialis*) from the Venice lagoon, Italy. Mar Environ Res 43: 69-79

Concentrations of Trace Metals (Cd, Cu, Fe, Pb) in *Posidonia oceanica* Seagrass of Liscia Bay, Sardinia (Italy)

M. Baroli[1], A. Cristini[2], A. Cossu[3], G. De Falco[1], V. Gazale[4], C. Pergent-Martini[5], and G. Pergent[5]

ABSTRACT

Heavy metal concentrations were measured in *Posidonia oceanica* seagrass from a Mediterranean site (Liscia bay, N Sardinia, Italy) in order to assess the degree of metal pollution in the area. Metals Cd, Cu, Fe and Pb were measured in sheaths and rhizomes dated and dissected by lepidochronology procedures, which may offer a timeseries of metal concentrations in the environment (Pergent 1990; Romeo et al. 1995).

Comparison between metal concentrations according to lepidochronological years in sheaths and rhizomes produced significant differences: Cd and Cu concentrations were similar in the two tissues, while Fe and Pb showed higher values in sheaths, with differences up to 2 orders of magnitude. Rhizome concentrations showed no trend related to lepidochronological years, whereas Cu, Fe and Pb tended to decrease over time in sheaths.

Values detected in sheaths and rhizomes of *P. oceanica* from Liscia bay were generally low, compared with other Mediterranean sites, with the exception of Fe: this confirms that the site is a clean area without anthropogenic enrichment and with metal concentrations representing background values.

Introduction

Posidonia oceanica is a long-lived benthic species widespread in the Mediterranean. It is the only marine phanerogam capable of forming durable structures ("matte"), it may absorb metals directly from the water column or from interstitial water in sediments, and has a high capacity for concentrating chemical pollutants and to accumulate trace metals occurring in the environment during its life cycle (Calmet et al. 1988; Malea and Haritonidis 1989). In fact, in the last 10 years *P. oceanica* has been used as a biological "recorder" of marine environmental quality (Augier 1985; Maserti et al. 1988). The capability of *P. oceanica* to concentrate a range of pollutants such as trace metals has been clearly established (Augier et al. 1977; Maserti et al. 1988; Malea and Haritonidis 1989).

When *P. oceanica* leaves die only the blade falls away and the sheathing base remains fixed to the rhizome (Giraud 1977; Pergent 1987). Sheaths and rhizomes decay extremely slowly and may persist within a "matte" for millennia (Boudouresque et al. 1983). Sheath thickness has been shown to vary cyclically according to position along the rhizome (Crouzet 1981; Pergent 1990). These cyclical variations have a chronological significance with each cycle, defined by two thickness minima, corresponding to a one-year period of growth (Pergent 1990). Recent works show that *P. oceanica* can memorise trace metal contents over several decades (Romeo et al. 1995; Pergent-Martini 1998).

The aim of the present work was to measure heavy metal concentrations in *P. oceanica* seagrass from a Mediterranean site (Liscia bay, N Sardinia, Italy, Fig. 1) in order to assess the

[1] Centro Marino Internazionale, Loc. Sa Mardini, 09072 Torregrande-Oristano, Italy
[2] Dipartimento di Scienze della Terra, Università di Cagliari, Via Trentino 51, Cagliari, Italy
[3] Dipartimento di Botanica ed Ecologia Vegetale, Università di Sassari, Via Muroni 25, Sassari, Italy
[4] Battelle-Medsar, Tramariglio-Alghero, Italy
[5] Eq.E.L., Université de Corse, Faculté des Sciences, BP 52, 20250 Corte, France

F.M. Faranda, L. Guglielmo, G. Spezie (eds)
Mediterranean Ecosystems: Structures and Processes

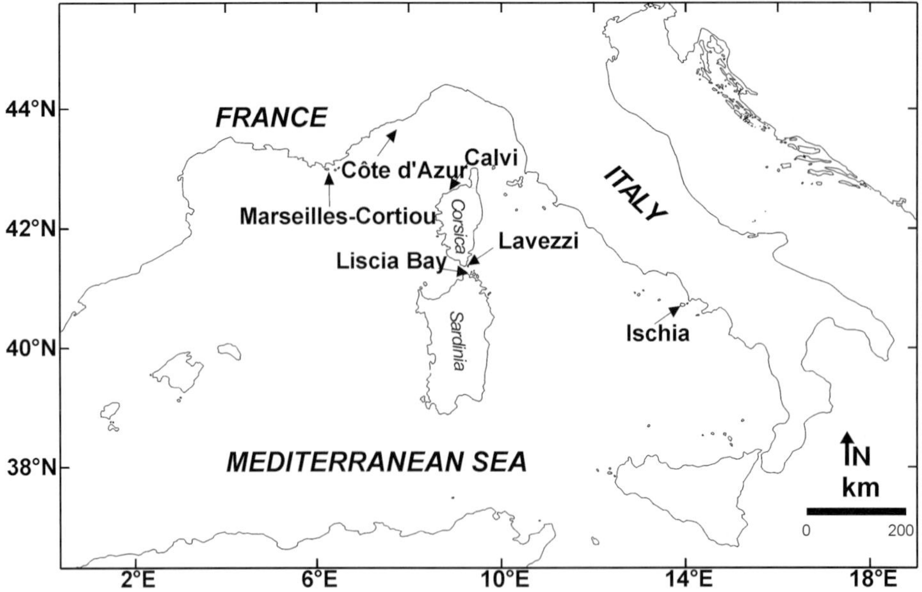

Fig. 1. Map of Western Mediterranean sea showing the study area (Liscia Bay) and the comparison sites

degree of metal pollution in the area. Concentrations of Cd, Cu, Fe, and Pb were determined in dated sheaths and segments of rhizomes collected from the study site and compared with other Western Mediterranean (WMed) reference sites.

Materials and Methods

The studied area is located on the northern coast of Sardinia (Italy) near the mouth of the Liscia river. The geological features of the catchment area of the river, of 510 km^2, involve granitic and metamorphic outcrops. Mineralisation, in the form of lenses and disseminated Cu, Pb and Zn sulphides, has been reported for the coastal sector. No anthropogenic activity is reported in the area, which may be considered a remote rural-maritime one.

Forty-five samples of *P. oceanica* were collected by scuba diving and immediately transported in a large volume of water to the laboratory. The samples were divided into sheaths and segments of rhizome, each one between two sheaths of minimum thickness, corresponding to a one-year period of growth by the lepidochronology technique. Epiphytes were carefully removed, with a PVC knife from every part of the plant. In order to remove sediments, specimens were then rinsed in double-distilled water.

Concentrations of Cd, Cu, Fe and Pb were determined in dated sheaths and rhizome segments of *P. oceanica*. Samples were dried (40°C, 12-24 hours) and solubilised in a microwave oven with HNO_3. The Fe concentration was determined by the inductively coupled plasma technique (ICP-OES), Cu, Pb, Cd concentrations were determined on a Perkin-Elmer 3030 Zeeman (GFAAS) Flameless Graphite furnace.

Results and Discussion

Metal distributions in the different tissues of the plant are due to many factors, such as the sorption mechanism (Ward 1989), the metabolic processes linked to plant physiology (Catsiki et al. 1987; Catsiki and Panayotidis 1993) and the physicochemical conditions of the environment (Malea and Haritonidis 1989). Figure 2 shows the temporal trends of metal concentrations in sheaths and rhizomes from the Liscia site: comparison according to lepidochronological years produces significant differences.

Cadmium concentrations do not show any time trend, either in sheaths or in rhizomes (Fig. 2), and the values are similar in the two tissues (Table 1). The enrichment of Cd in sheaths from different WMed sites is not clearly correlated with anthropogenic causes (Romeo et al. 1995): the highest values (1.2±0.5 μg • g^{-1}) were

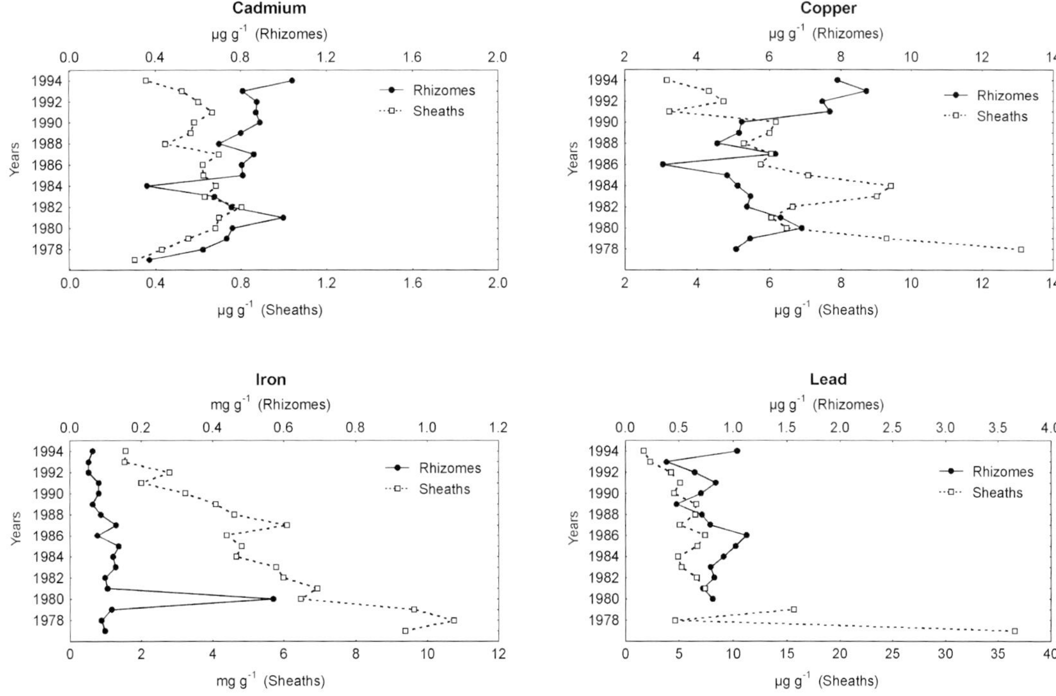

Fig. 2. Metal concentration records of sheaths and rhizomes of *Posidonia oceanica*. A double scale is used for Fe and Pb

found in clean areas of Corsica with respect to other areas of the Côte d'Azur (0.8±0.5 µg • g^{-1}). Pergent-Martini (1992) believes that the high Cd concentration detected in Corsica is linked to natural phenomena such as upwelling of deep waters rich in Cd. Liscia bay together with La Darse shows the lowest values in the WMed (Table 2).

Cu concentrations are similar in the two tissues, in agreement with data from the other WMed sites (Table 1). Time trends in sheaths and rhizomes are opposite, decreasing with time in the sheaths and increasing in rhizomes (Fig. 2). It is assumed that concentrations in sheaths are more representative of the level of Cu in the environment. As a matter of fact, Cu translocation from rhizome to leaves may occur, Cu being a micronutrient involved in the photosynthesis of the plant. Mean Cu concentrations in sheaths (6.0±2.5 µg • g^{-1}) are similar to those from clean areas in Corsica (8.1±2.1 µg • g^{-1}, Lavezzi, Romeo et al. 1995; 7.3±0.6 µg • g^{-1}, Calvi), and lower with

Table 1. Comparison of metal concentrations (median and standard deviation, below range) in dated rhizomes and sheaths of *Posidonia oceanica* in four western Mediterranean sites

	Cd (µg • g^{-1})		Cu (µg • g^{-1})		Fe (mg • g^{-1})		Pb (µg • g^{-1})	
	Rhizomes	Sheaths	Rhizomes	Sheaths	Rhizomes	Sheaths	Rhizomes	Sheaths
Liscia bay	0.8±0.2	0.6±0.1	5.4±1.4	6.0±2.5	0.09±0.12	4.75±2.69	0.8±0.2	5.2±8.0
(Sardinia)	(0.4-1)	(0.3-0.8)	(3.1-8.7)	(3.2-13.1)	(0.05-0.57)	(1.53-0.75)	(0.4-1.1)	(1.7-36.6)
Ischia	0.8±0.1	0.7±0.2	12.8±1.5	17.0±2.7	0.37±0.08	1.28±0.30	2.3±0.2	11.2±9.8
	(0.4-1.7)	(0.1-3.9)	(7.4-29.0)	(5.0-57.9)	(0.08-1.19)	(0.19-3.38)	(1.8-5.3)	(1.8-18.3)
Calvi	2.4±0.4	2.0±0.1	14.9±1.3	7.3±0.6	0.12±0.02	0.68±0.10	2.4±0.3	7.8±0.7
(Corsica)	(0.9-6.2)	(1.2-2.8)	(7.4-22.0)	(5.2-12)	(0.05-0.26)	(0.15-1.47)	(0.2-5.1)	(4.5-13.8)
Marseille	1.1±0.1	1.3±0.2	15.3±0.3	16.8±1.9	0.11±0.01	0.53±0.07	1.3±0.2	9.5±1.2
(Cortiou)	(0.1-2.3)	(0.1-3.7)	(4.0-45.6)	(4.4-35.5)	(0.03-0.14)	(0.06-1.37)	(0.04-3.5)	(1.2-28.3)

Table 2. Comparison of metal concentrations (median and standard deviation, below range) in dated sheaths of *Posidonia oceanica* in Western Mediterranean sites

	Cd ($\mu g \cdot g^{-1}$) Sheaths	Cu ($\mu g \cdot g^{-1}$) Sheaths	Fe ($mg \cdot g^{-1}$) Sheaths	Pb ($\mu g \cdot g^{-1}$) Sheaths
Liscia bay	0.6±0.1	6.0±2.5	4.75±2.69	5.2±8.0
(Sardinia)	(0.3-0.8)	(3.2-13.1)	(1.53-10.75)	(1.7-36.6)
Lavezzi[a]	1.2±0.2	8.1±2.1	0.35±0.21	5.0±2.9
(Corsica)	(0.7-1.5)	(6.2-14.6)	(0.11-0.95)	(1.7-12.0)
La Darse[a]	0.5±0.2	33.6±6.5	0.77±0.50	10.4±7.1
(Côte d'Azur)	(0.3-0.9)	(27.1-46.3)	(0.12-1.96)	(1.6-26.1)
Rochambeau[a]	0.9±0.4	17.2±10.8	1.52±0.95	39.6±20.1
(Côte d'Azur)	(0.3-1.4)	(7.4-40.1)	(0.43-2.87)	(4.6-54.7)
Cap-Martin[a]	0.7±0.3	12.5±8.7	0.87±0.41	9.9±6.7
(Côte d'Azur)	(0.6-1.4)	(11.1-35.8)	(0.47-1.77)	(2.5-22.1)

[a] Romeo et al. 1995

respect to the polluted sites of the Côte d'Azur (Table 2; Romeo et al. 1995).

The Fe concentration is higher in sheaths (1.5-10.8 mg • g^{-1}), with differences up to 2 orders of magnitude with respect to rhizomes (0.05-0.57 mg • g^{-1}). This has also been recorded in other Western Mediterranean sites (Table 1), but the difference is much higher at our site. Sheath concentrations decrease over time, whereas rhizome concentrations show no trend related to lepidochronological years. Several authors (Pergent-Martini 1992; Romeo et al. 1995) explain the decrease in iron concentration in sheaths, observed in different sites, as due to biochemical modifications linked to the biology of the species (liberation of metal binding sites) or to Fe^{++} diffusion from the reducing zone in sediments, and oxidation and adsorption at the sheath surface. This may explain the high concentration in basal sheaths, which are the oldest and nearest to sediments. The iron concentrations in sheaths at Liscia are much higher than those of the other sites considered for comparison (Table 2), probably linked to the high background values in sediments (unpublished data).

Mean Pb concentrations in sheaths at Liscia are 5.2±8.0 µg • g^{-1} and reach higher values (up to 36.7 µg • g^{-1}) in the first lepidochronological year (1977). Rhizomes show lower values (0.8±0.2 µg • g^{-1}) with respect to sheaths, as is also evident in the other comparison sites. Lead concentrations in the sheaths refer to the adsorption of lead released from sediments onto the cellulose cell wall (Flower 1982): this phenomena is probably less important in rhizomes than in sheaths, the latter being more exposed to sediments. Comparison of Mediterranean sites shows lower values in Liscia bay and Lavezzi.

Conclusions

Trace metal values detected in sheaths and rhizomes of *P. oceanica* in the Liscia bay are generally low, with the exception of Fe, compared to other WMed sites, and are similar to the values detected in the cleanest areas of Corsica. The trends of metal concentrations related to lepidochronological years also highlight the absence of high signals over the last twenty years.

Relative concentrations in the two tissues reflect the same trends as the comparison sites: Fe and Pb have lower values in rhizomes than in sheaths, whereas Cd and Cu are similar in both tissues. The factors which influence the distribution of the various elements in the tissues are still not well known and further studies are necessary on this issue.

The data confirm that our site is a clean area without anthropogenic enrichment, and metal concentration values detected in *P. oceanica* represent background values.

Acknowledgements. This study has been funded by the European programmes INTERREG and STRIDE and by the Regione Autonoma della Sardegna. The authors are grateful to Luigi Muscas for the chemical analysis. Sincere thanks to Dr. Stefano Guerzoni for his invaluable advice and for the critical review of the manuscript.

References

Augier H (1985) L'herbier à *Posidonia oceanica*, son importance pour le littoral méditerranéen, sa valeur comme indicateur biologique de l'état de santé de la mer, son utilisation dans la surveillance du milieu, les bilans écologiques et les études d'impact. Vie Mar 7: 85-113

Augier H, Gilles G, Ramonda G (1977) Utilisation de la phanérogame marine *Posidonia oceanica* Delile pour mesurer le degré de Contamination mercurielle des eaux littorales méditerranéennes. CR Acad Sci Ser III-Vie 285: 1557-1560

Boudouresque CF, Crouzet A, Pergent G (1983) Un nouvel outil au service de l'étude des herbiers à *Posidonia oceanica*: la lépidochronologie. Rapp P-V Réun Comm Int Explor Sci Médit 28: 111-112

Calmet D, Boudouresque CF, Meinesz A (1988) Memorisation of nuclear atmospheric tests by rhizomes and scales of Mediterranean seagrass *Posidonia oceanica* (Linnaeus) Delile. Aquat Bot 30: 279-294

Catsiki VA, Panayotidis P (1993) Copper, chromium and nickel in tissues of the Mediterranean seagrasses *Posidonia oceanica* and *Cymodocea nodosa* (Potamogetonacea) from Greek coastal areas. Chemosphere 26 (5): 963-978

Catsiki VA, Panayotidis P, Papathanassiou E (1987) Bioaccumulation of heavy metals by seagrasses in Greek coastal waters. Posidonia Newsl 1 (2): 21-30

Crouzet A (1981) Mise en évidence de variations cycliques dans le écailles de rhizomes de *Posidonia oceanica* (Potamogetonaceae). Trav Sci Parc Nat Port-Cros 7: 129-135

Fowler SW (1982) Biological transfer and trasport processes. In: Kullenberg G (ed) Pollutant transfer and transport in the sea. Vol 2 CRC Press, Boca Raton, pp 2-65

Giraud G (1977) Contribution à la description et à la phénologie quantitative des herbiers à *Posidonia oceanica* (L.) Delile. Thèse Doctorat 3e cycle, Univ Aix-Marseille II

Malea P, Haritonidis S (1989) Uptake of Cu, Cd, Zn and Pb in *Posidonia oceanica* (Linnaeus) from Antikyra Gulf, Greece: preliminary note. Mar Environ Res 28: 495-498

Maserti B E, Ferrara R, and Paterno P (1988) Posidonia as an indicator of mercury contamination. Mar Pollut Bull 19(8): 381-382

Pergent G (1987) Recherches Lépidochronologiques chez *Posidonia oceanica* (Potamogetonaceae). Fluctuations des paramètres anatomiques et morphologiques des écailles des rhizomes. Thèse Doctorat, Univ Aix-Marseille II

Pergent G (1990) Lepidochronological analysis in seagrass *Posidonia oceanica*: a standardized approach. Aquat Bot 37: 39-54

Pergent-Martini C (1992) Contribution à l'étude des stocks et des flux d'éléments dans l'écosystème à *Posidonia oceanica*. D.E.S.S. Ecosystèmes méditerranéens. Université de Corse, Corte

Pergent-Martini C (1998) *Posidonia oceanica*: a biological indicator of past and present mercury contamination. Mar Environ Res 45(2): 101-110

Romeo M, Gnassia-Barelli M, Juhel T, Meisnesz A (1995) Memorisation of heavy metals by scales of the seagrass *Posidonia oceanica*, collected in the NW Mediterranean. Mar Ecol Progr Ser 120: 211-218

Ward TJ (1989) The accumulation and effects of metals in seagrass habitats. In: Larkum AWD, McComb AJ, Shepherd SA (eds) Biology of Seagrass (Aquatic Plant Studies 2) Elsevier, Amsterdam, pp 787-820

CHAPTER 14

Mercury Concentrations in Sea Water Sampled with the New "Marikiki" Sampler

R. Capelli, V. Minganti, and R. De Pellegrini

ABSTRACT

During a series of oceanographic cruises (December 1993, June 1994, December 1994, and July 1997) in the area of the buoy "ODAS Italia 1" in the Ligurian Sea (43° 48.90'N 09° 06.80'E), a new sampling bottle for deep sampling of sea water for trace metals analysis at ng/l level was tested. The reactive mercury concentration was determined directly on board the oceanographic vessel, immediately after the sampling by reduction with tin(II) chloride, pre-concentration on gold, and atomic fluorescence detection. Mercury concentrations ranged between 0.06 and 0.90 ng/L with an average of 0.22 ng/l for samples collected in December 1993, June and December 1994. Higher values were found (mean 0.41 ng/l, range 0.27-0.61 ng/l) in samples collected in July 1997.

Introduction

In order to investigate the biogeochemical cycle of mercury, the concentration of both organic and inorganic forms of the metal in all components of the ecosystem (air, water, organisms and sediments) must be determined. In the environment mercury concentrations range from ng/m^3 (air) to ng/l (water) to µg/g (organisms and sediments), and different analytical methods with the required sensitivity and accuracy have been developed.

As far as sea water is concerned, mercury is present at concentrations of a few ng/l or even less. Regarding the Mediterranean, there are data for the presence of mercury at the surface (Ferrara and Maserti 1986; Copin-Montegut et al. 1986; Minganti et al. 1993), but very few data on mercury concentrations at different depths (Cossa et al. 1997; Ferrara and Maserti 1988).

At present, determination of pg or ng of mercury has been made possible by highly sensitive analytical methods such as atomic fluorescence and atomic absorption spectrometry and pre-concentration on metals such as silver and, especially, gold. However, there are still some difficulties connected with the purity of reagents and of all materials used during the sampling, storage and analytical procedures. It is possible to choose materials and reagents that allow to reduce sample contamination to a minimum, while sampling remains the most difficult procedure, particularly if it has to be carried out at depths in excess of 10-15 metres.

After examining the various available equipment to sample sea water at different depths, a new sampling device was designed and tested which combines a simple sampling procedure with the possibility to obtain contamination-free samples (Capelli et al. 1998). With this sampling equipment a series of sea water samples have been collected at different depths, near the buoy ODAS Italia -1, in the Ligurian Sea, in order to verify if reactive mercury concentrations can be used as a tracer of the different water masses.

Materials and Methods

Sampling

Sea water samples were collected during offshore investigations carried out in December 1993, June 1994, December 1994 and July 1997 in co-operation with C.N.R. (Italian National Research Council) on board M/V Urania. The

Dipartimento di Chimica e Tecnologie Farmaceutiche ed Alimentari, Università di Genova, Via Brigata Salerno, 16147 Genova, Italy

F.M. Faranda, L. Guglielmo, G. Spezie (eds)
Mediterranean Ecosystems: Structures and Processes

study of the water column was carried out in the proximity of the buoy ODAS Italia-1 in the Ligurian sea (43°48.90' N 09°06.80' E), anchored at a 1270 metre depth. During these off-shore investigations, density and salinity along the water column were measured by means of a CTD Profiler. Samples were collected at different depths, starting from the surface and down to 600 metres. Depths were chosen by taking into consideration the thermocline and temperature and salinity values. Sea water sampling down to a 600 meter depth was carried out by means of the new MARIKIKI sampling equipment designed and tested by the Department of Pharmaceutical and Food Chemistry and Technologies of the University of Genoa. This previously described sampling equipment (Capelli et al. 1998) basically consists of a nylon body holding a container, in the specific case a quartz tube (diameter 5 cm, thickness 1 mm, volume 300-500 ml). A vacuum can be created simultaneously inside the tube and the container through a glass capillary that connects them. The sampler is fastened to the cable attached to a winch on the research vessel and lowered to the desired depth. At this point a messenger slides down along the cable and breaks the glass capillary. Water is sucked inside and fills both the quartz tube and the external container. The sampler is retrieved on board and the quartz tube full of sea water is sealed with a cap and either stored or immediately analysed on board the ship. The "MARIKIKI" sampler has two nylon bodies and two quartz tubes in order to collect two samples in the same place and at the same time. The most significant characteristics of this kind of sampler can be summarised as follows:

1. Each sample is obtained with a previously washed quartz tube whose blank value is known.
2. The tube in which the vacuum is created does not have any contact with other water until the moment in which the capillary breaks.
3. It is possible to obtain two samples in the same place and, especially, at the same time.
4. The quartz tube can be heated up to 800°C to ensure its cleanliness and also sterilised in case of microbiological research.
5. Both the quartz tube and the container can be substituted with other more suitable materials to obtain water samples for the determination of other substances.

Analytical Methods

All analyses were carried out on board the ship immediately after sampling, to avoid the need to add acids to prevent loss of ionic mercury due to absorption onto the container surface. Mercury was chemically reduced to Hg° vapour by means of 10% (m/v) tin (II) chloride dihydrate in 20% (v/v) sulphuric acid. The Hg° vapours were stripped from the solution by a stream of argon and collected on a gold trap, with 200-300 mg of 0.1 mm diameter gold wire placed in a 4 mm (inner diameter) quartz tube. After pre-concentration of all mercury from the sample, the trap was electrically heated and the released Hg° vapours were detected with a mercury fluorescence detector (Merlin, PSA Analytical LTD) connected with a Hewlett-Packard HP3396 Series II Recorder/Integrator. The obtained concentrations of mercury correspond to "reactive mercury", i.e. the dissolved inorganic mercury species, labile organo-mercury associations and mercury that is easily leachable (Gill and Fitzgerald 1985).

Calibration was carried out by means of the standard additions method on each sample, adding 0.1 and 0.2 ng of mercury. Slope of calibrations was checked by comparison with a direct injection of known amount of mercury vapour obtained by gas sampling in a container with metallic Hg in equilibrium with Hg vapour, kept at a constant temperature.

Each quartz container was acid washed, heated at 600°C for 12 hours, and, after cooling, filled with ultra-pure (>18MOhm·cm) water. The water was then analysed with the procedure described, and the cleaning was repeated until a mercury level lower than 0.08 ng/L was achieved. The mean (n=37) blank value obtained is 0.06 ng/L with values ranging 0.01-0.08 ng/L and a standard deviation of 0.02 ng/L. The detection limit of the method (3 σ) is 0.06 ng/L. The blank value for each quartz container was recorded and subtracted from the mercury concentration measured in the sea water.

Results and Discussion

Reactive mercury concentrations obtained are reported in Fig. 1, together with salinity and temperature data. No correlation between mercury concentration and salinity, temperature or depth

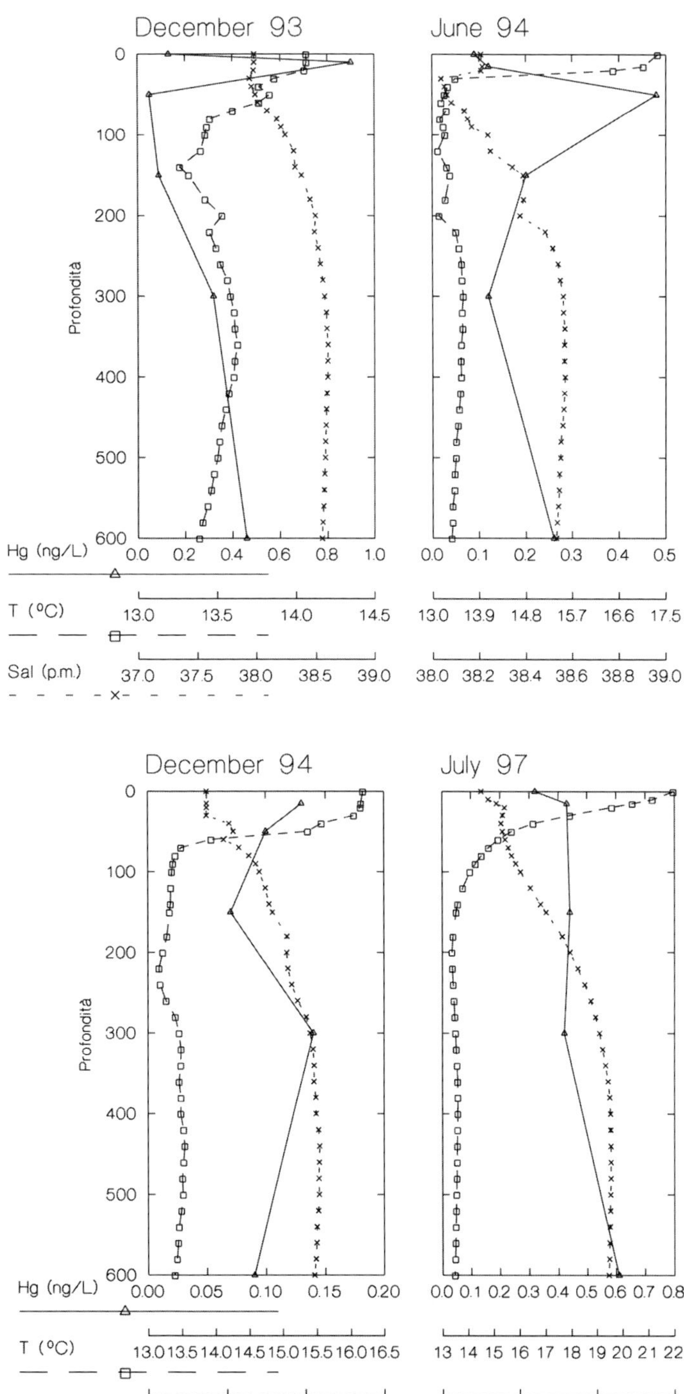

Fig. 1. Mercury concentration profiles obtained during four samplings (December 1993, June 1993, December 1994, July 1997) at the buoy "ODAS Italia 1" (43°48.90' N 09°06.80' E)

can be detected. Mercury concentration ranges between 0.06 and 0.90 ng/l, with a mean value of 0.22 ng/l, for samples collected in December 1993, June 1993 and December 1994. Higher values have been found for the July 1997 samples with values ranging from 0.27 to 0.61 ng/l with a mean value of 0.41 ng/l.

Oceanographic data (temperature and salinity) recorded during the various investigations at the same location and time at which sea water samples were collected show the presence of two different water masses:

1. Surface water of Atlantic origin (MAW, Modified Atlantic Water) usually extending from the surface to a depth of 200 metres.
2. Intermediate water of levantine origin (LIW, Levantine Intermediate Water) warmer and saltier than the Atlantic water, usually found down to a 700 metres depth.

A comparison between MAW (mean 0.21 ng/l) and LIW (mean 0.23 ng/l) was made for the first three samplings (Dec. 93-Dec. 94) but no significant differences have been detected. Data collected in July 97 show significantly higher reactive mercury concentrations (mean 0.41 ng/l, range 0.27-0.61 ng/l). No explication was found for such difference, and July 97 data have been excluded from the comparison. No relationship was found between the concentration of reactive mercury, salinity, temperature and depth.

It is difficult to compare these data with those found in the literature, because there are few data available for the Mediterranean Sea and especially for reactive mercury at great depth. Ferrara and Maserti (1986) reported reactive mercury concentrations of 3.5 ng/l in the Tyhrrenian Sea (range 2.0-5.9 ng/l) in off-shore waters and concentrations of 2.0 ng/l (range 0.5-2.5 ng/l) in samples of coastal waters. Significantly lower values recently found by Cossa et al. (1997), who recorded reactive mercury values below 0.14 ng/l, are in agreement with data found in this work.

Data referred to surface water samples in the Ligurian Sea show reactive values of 0.3±0.2 ng/l (n=33, range 0.2-0.8 ng/l, Minganti et al. 1993).

The use of reactive mercury concentrations, from data reported in this work, seems to be unable to differentiate water masses of different origin.

Acknowledgements. The authors thank Dr. Marco Orsi for his help in acquiring the oceanographic data and the crew of the M/V Urania for their assistance during sampling. This research was carried out in the framework of the project "ODAS Italia1" (Istituto per l'Automazione Navale, Genova, Consiglio Nazionale delle Ricerche).

References

Capelli R, Chiarini C, De Pellegrini R, Minganti V (1998) A new sampler (Marikiki) for deep water for the determination of mercury at ng L^{-1} level. Ann Chim 88: 413-417

Copin-Montegut G, Coureau P, Laumont F (1986) Occurrence of mercury in the atmosphere and waters of the Mediterranean. in: Pap FAO/UNEP/WHO/IOC/IAEA Meet Biogeochem Cycle Mercury Mediterr, Siena, Italy, 27-31 August 1984, FAO Fish Rep (325) Suppl, pp 51-75

Cossa D, Martin JM, Takayanagi K, Sanjuan J (1997) The distribution and cycling of mercury species in the western Mediterranean. Deep-Sea Res II 44: 721-740

Ferrara R, Maserti B (1986) Mercury in the Mediterranean Basin. Mar Pollut Bull 17: 533-534

Ferrara R, Maserti BE (1988) Mercury exchange between outflowing and inflowing water in the Strait of Gibraltar. Mar Pollut Bull 19: 387-388

Gill GA, Fitzgerald WF (1985) Mercury sampling of open ocean waters at the picomolar level. Deep-Sea Res 32: 287-297

Minganti V, Fiorentino F, De Pellegrini R, Capelli R (1993) Mercury concentrations in coastal and offshore waters of the Ligurian Sea (Mediterranean). Fresenenius Environ Bull 2: 65-69

The Use of Carbon Stable Isotopes to Investigate the Origin and Distribution of Suspended and Sedimentary Organic Matter in a Semi-enclosed Mediterranean Marine System

A. Mazzola[1], G. Sarà[1], and R.H. Michener[2]

ABSTRACT

The natural stable isotope values of different primary sources have been used to trace the fate of organic carbon that enters in the marine food webs of Stagnone di Marsala (Italy). Water and sediment samples were collected monthly (March 1996 – February 1997) at 3 stations, characterized by different amounts of vegetal cover and analysed to determine total organic matter (OM), phytopigments, biopolymeric organic carbon (BPC) and stable carbon isotopic composition ($\delta^{13}C$), the latter measured also in main primary producers. Sedimentary OM accumulated in summer, while total suspended organic matter reached highest concentrations in May and December. The concentration of chlorophyll-a carbon in the sediments and suspended material (on average 227 ± 165 µg C g^{-1} and 14 ± 8.7 µg C l^{-1} respectively) were highest in spring and autumn. Sedimentary BPC concentrations showed several peaks throughout the year, suspended BPC concentrations peaked in May and in summer. The $\delta^{13}C$ from suspended particulate matter and sediments (on average −19.5 ± 2.7‰ and −13.0 ± 3.1‰ respectively) followed a gradient of enrichment moving towards the inner part of the Stagnone. The $\delta^{13}C$ of primary sources ranged from −21.4‰ (phytoplankton) to −4.5‰ (seagrass detritus). The Dauby's model (1989) was used in order to estimate the contribution of each primary carbon source to sedimentary and particulate reservoirs. In the northern basin, OM in sediments was mainly influenced by macroalgae carbon (46%), while that in the water column was influenced by phytoplankton (60%) and microphytobenthic (30%) carbon. In the southern basin, main sedimentary carbon sources were seagrasses and their detritus (61%, mainly *Posidonia*), while in the water column were phytoplankton and microphytobenthos (50% and 24% respectively). The role of some physical constraints (wind and tidal energy) in influencing isotopic composition of organic matter is discussed.

Introduction

In recent years, several authors investigated organic matter flows and food web structures using stable isotope ratios as natural tracers (Lajtha and Michener 1994; Griffiths 1998). The bulk of organic matter (i.e. particulate and sedimentary) in shallow environments originates from different primary (such as phanerogames, macroalgae, phytoplankton and microphytobenthos) and secondary sources (such as plankton, benthos and fish). These sources provide organic matter with varying degrees of availability to consumers and with different extent as a function of environmental gradients (Valiela 1984).

In shallow coastal systems, resuspension, sedimentation and lateral drifting are the primary processes controlling the quality and quantity of organic matter (MacIntyre et al. 1996; Sarà et al. 1999). Sediment resuspension events, which are mainly driven by wind energy (Smaal and Haas 1997; Sarà et al. 1999), may contribute to sediment-water exchange. Hopkinson (1985) suggested that resuspension may control the relative amounts of organic carbon as well as the sites and rates of organic matter degradation in the water column and benthos. As a conse-

[1] Laboratorio di Biologia Marina e Ricerche, Dipartimento di Biologia Animale, Università di Palermo, Via Archirafi 18, 90123 Palermo, Italy
[2] Department of Biology, Boston University, Boston, Massachusetts, USA

F.M. Faranda, L. Guglielmo, G. Spezie (eds)
Mediterranean Ecosystems: Structures and Processes

quence, the distribution of organic matter is highly variable in these systems.

The Stagnone di Marsala is a Mediterranean semi-enclosed marine system in which the food chain is largely constrained by detritus originating from seagrasses (mainly *Posidonia oceanica* and *Cymodocea nodosa*) (Pusceddu et al. 1999; Sarà et al. 1999). Despite the oligotrophy of its waters, the Stagnone functions as a detrital trap (Pusceddu et al. 1999), although different zones can be individuated as a function of the different degrees of resuspension-sedimentation-lateral drifting processes. In these zones particulate and sedimentary organic matter seem to be available for consumers in different ways and at different times.

It is thus important to provide information on the role and contribution of each organic matter source to the trophic pathway and to investigate in what way this organic matter is distributed among the different loops of the trophic web. Since little information exists on Mediterranean areas (Dauby 1989; Jennings et al. 1997), in this study we used stable carbon isotopic analysis to determine the monthly contribution of carbon from each primary producer, both in the sediment and in the water column. By comparing the data from this analysis with those regarding sedimentary and particulate biopolymeric organic carbon, we aim at investigating the role of each carbon source in determining the temporal and spatial variability of organic matter available to consumers.

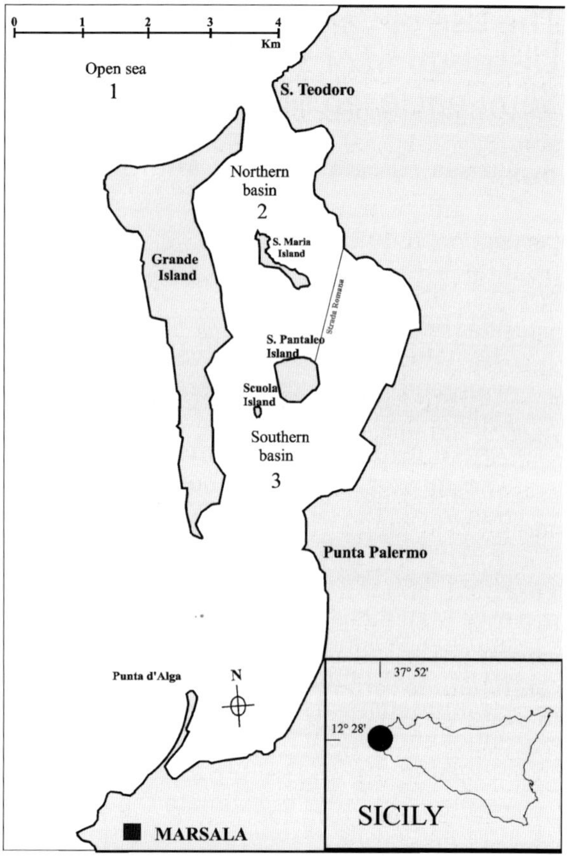

Fig. 1. The study site and the three investigated sites: station *1* – open-sea; 3 m depth and unvegetated; station *2* – Northern basin; 0.6-0.8 m depth and dominance of macroalgae (mainly *Caulerpa prolifera*) plus sparse *Cymodocea nodosa*; station *3* – Southern basin; 1.2-1.8 m depth and dominance of phanerogames (*Cymodocea nodosa* and *Posidonia oceanica*) plus sparse macroalgae (mainly *Caulerpa prolifera*)

Description of the Study Site

The study was carried out in a semi-enclosed marine system (15 km^2, 37° 52' N, 12° 28' E), characterised by two main communication channels with the open sea. A platform separates the rest of the basin from the open sea (Fig. 1). The basin is very shallow, with depths ranging from 2 m along the eastern shore of the platform to 0.50 m in the western area. Basin depth increases gradually to about 2.5 m in the southernmost area, close to the open sea. The northern channel is 450 m wide and is characterised by occasional turbulent inputs of seawater. In this area sediments are characterised by scant infrequent algal coverage. The southern mouth, 1450 m wide, is open to sea water inflow and is characterised by internal tides. South of this area, a luxuriant *Posidonia oceanica* meadow is present, influencing water circulation and silting. No continental inputs are present.

Materials and Methods

Water and sediment samples were collected monthly from March 1996 to February 1997 at 3 stations located along a north-south transect (Fig. 1), characterized by different amounts of vegetal cover (Sarà et al. 1999). Water samples were collected using a Niskin bottle and immediately processed in the laboratory to determine total organic suspended matter (OSM, mg·l^{-1}; Strickland and Parsons 1972), chlorophyll-a and phaeopigments (CHLa and Phaeo, µg·l^{-1}; Lorenzen and Jeffrey 1980), carbon isotopic composition (Susp-δ^{13}C, ‰; Lajtha and Michener 1994) and particulate biopolymeric organic carbon (Susp-BPC, µg C l^{-1}; Fabiano and Pusceddu 1998).

For sediment analysis, three replicate cores per station and period were collected randomly from three quadrats (20 cm length, 400 cm^2 surface area) belonging to a larger quadrat (1 m^2). The top 0-1 cm layer of each core was used for the analysis of total organic matter (SOM, mg·g^{-1}) according to Parker (1983), chlorophyll a (CHLa, µg·g^{-1}) and phaeopigments (Phaeo, µg·g^{-1}) were analysed according to Lorenzen and Jeffrey (1980) and sedimentary biopolymeric organic carbon (Sed-BPC, µgCl^{-1}) and isotopic composition (Sed-δ^{13}C, ‰) were analysed according to Pusceddu et al. 1999. Phytoplankton (Susp-CCHLa) and microphytobenthic (Sed-CCHLa) carbon was calculated by converting CHLa concentrations to carbon content (C-Chla) according to Nival et al. (1972). Zooplankton (>125 µm), macroalgae and seagrass isotopic analysis was performed according to Jennings et al. (1997).

The isotopic contribution of each carbon source to the organic matter bulk (sedimentary and suspended) was calculated according to Dauby (1989). Isotopic data were compared with frequency of resuspension calculated according CERC (1977) model reported in Smaal and Haas (1997). Temporal and spatial fluctuations were assessed by analysis of variance (ANOVA; Underwood 1997), while correlation between parameters was ascertained by means of Spearman-Rank correlation (Sokal and Rohlf 1981).

Results

Table 1 reports particulate and sedimentary organic matter descriptors. Total sedimentary organic matter (Fig. 2) mainly accumulated in summer (106 ± 0.9 mg·g^{-1}). Moving from open sea (5.7 mg·g^{-1}) to the southern station (156 mg·g^{-1}), a marked gradient in SOM was observed (ANOVA, $P<0.05$; Table 1). Total suspended organic matter showed two peaks in May and December (3.5 mg·l^{-1} and 3.1 mg·l^{-1}) (Fig. 2). Among stations no significant differences in OSM were observed (ANOVA, $P>0.05$; Table 1).

The concentration of chlorophyll-a carbon in the sediments (Fig. 3) ranged from 0 to 85 µg C g^{-1}, with the highest values in autumn and spring (814.0 ± 536 µgCg^{-1} and 112.0 µgCg^{-1}, respectively). No significant differences were observed between stations (ANOVA, $P>0.05$; Table 1). Sedimentary phaeopigments peaked in late summer (August-October) and in December (2.95 ± 0.6 and 5.2 µg·g^{-1}, respectively). The concentrations of sedimentary phaeopigments were significantly higher in the southern than in the open-sea station (ANOVA, $P<0.05$; Table 1).

Table 1. Statistics and ANOVA results of main trophic descriptors of sediments and suspended organic matter. See text for acronyms. *ns*, no significant difference ($P>0.05$)

Variable/Station	Open-sea				North				South				ANOVA
	Mean	± SD	Min	Max	Mean	± SD	Min	Max	Mean	± SD	Min	Max	$F_{(2,33)}/P$
Resuspension	1.7	1.3	0.4	4.4	4.4	2.6	1.9	11.5	2.9	1.7	1.1	6.4	6.0/***
Water column													
OSM (mg·l^{-1})	1.8	1.3	0.2	4.6	1.8	1.1	0.4	4.0	1.2	0.5	0.1	2.2	1.5/ns
CHLa (µg·l^{-1})	0.4	0.2	0.1	0.8	0.4	0.3	0.0	1.1	0.3	0.1	0.1	0.5	0.5/ns
CCHLa (µgCl^{-1})	15.6	8.1	3.2	31.0	15.1	12.5	0.4	45.9	12.1	4.0	3.7	18.6	0.6/ns
Phaeo (µg·l^{-1})	0.2	0.3	0.0	0.9	0.3	0.5	0.0	1.7	0.1	0.2	0.0	0.5	0.7/ns
BPC (µg·l^{-1})	223.3	168.4	75.0	596.4	246.3	157.3	86.7	542.9	179.7	103.4	77.7	357.0	0.6/ns
δ^{13}C (‰)	−21.4	2.2	−25.1	−16.4	−18.3	1.8	−20.5	−15.8	−18.6	2.8	−24.7	−15.4	6.4/***
Sediment													
SOM (mg·g^{-1})	5.7	3.9	1.4	15.0	93.5	35.8	35.0	144.8	156.3	43.7	71.8	201.1	64.1/***
CHLa (µg·g^{-1})	1.8	1.2	0.0	4.0	3.5	3.6	0.2	11.8	11.7	23.7	0.2	84.4	1.8/ns
CCHLa (µgCg^{-1})	72.4	46.5	0.0	159.6	139.4	142.2	9.6	473.9	467.9	947.2	8.7	3390.2	1.8/ns
Phaeo (µg·g^{-1})	0.0	0.2	0.0	0.5	2.3	1.0	0.0	3.8	4.6	3.8	0.0	13.0	12.0/***
BPC (µg·g^{-1})	1989.8	1926.9	104.5	7478.4	19011.1	11947.3	3907.9	42281.4	26619.2	18468.8	2749.9	59114.5	11.8/***
δ^{13}C (‰)	−15.8	3.1	−22.3	−10.6	−13.1	1.4	−15.3	−10.6	−10.0	1.2	−11.5	−8.0	22.9/***

* $P\leq0.05$; ** $P\leq0.01$; *** $P\leq0.001$

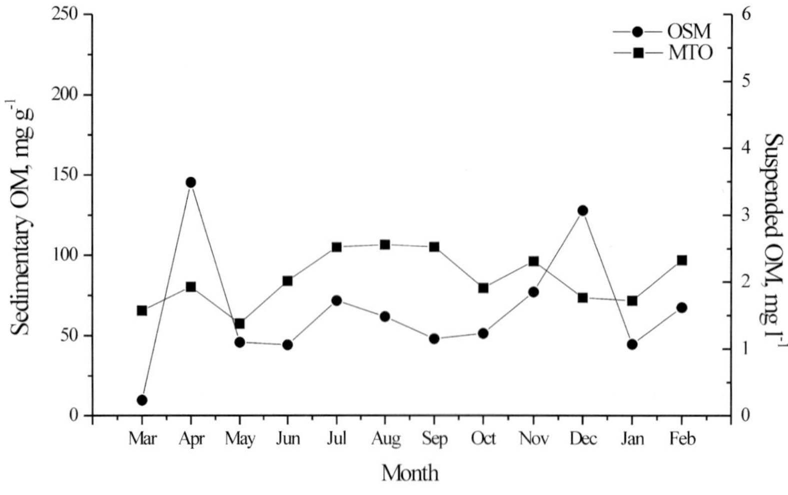

Fig. 2. Temporal changes in sedimentary (SOM, mg·g^{-1}; *left* axis, *solid square symbols*) and suspended (OSM, mg·l^{-1}; *right* axis, *solid circle symbols*) total organic matter in the three sites investigated

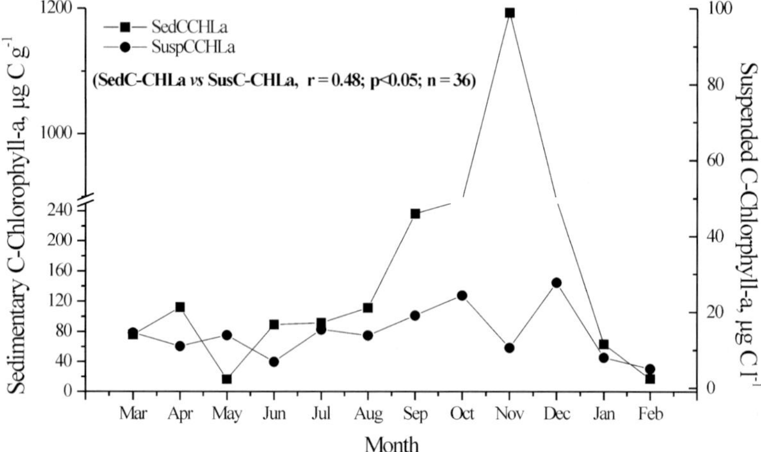

Fig. 3. Temporal changes in microphytobenthic (Sed-CCHLa, µg·g^{-1}; *left* axis, *solid square symbols*) and phytoplankton (Susp-CCHLa, µg·l^{-1}; *right* axis, *solid circle symbols*) carbon in the three sites investigated

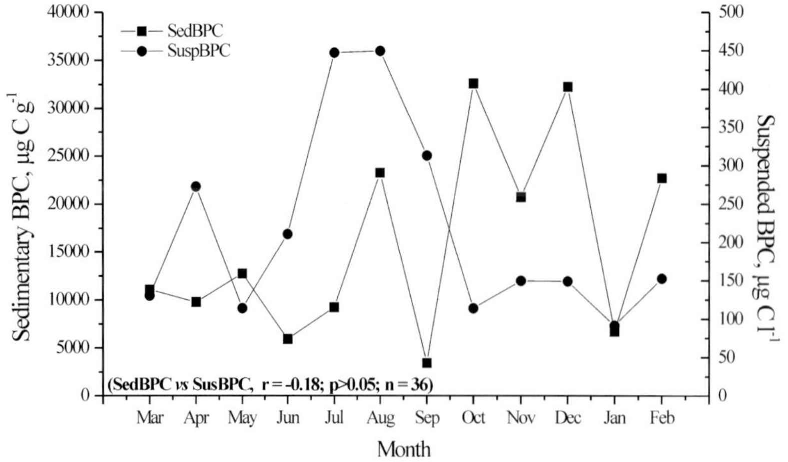

Fig. 4. Temporal changes in sedimentary (Sed-BPC, µg·g^{-1}; *left* axis, *solid square symbols*) and suspended (Susp-BPC, µg·l^{-1}; *right* axis, *solid circle symbols*) biopolymeric organic carbon in the three sites investigated

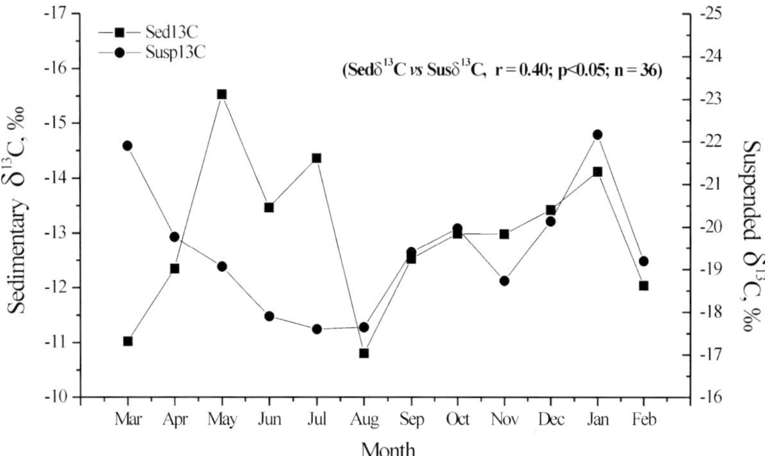

Fig. 5. Temporal changes in sedimentary (Sed-δ^{13}C, ‰; *left* axis, *solid square symbols*) and suspended (Susp-δ^{13}C, ‰; *right* axis, *solid circle symbols*) biopolymeric organic carbon in the three sites investigated

Suspended chlorophyll-a carbon values (Fig. 3) ranged from 0.4 to 46 μgCl^{-1} with main peaks in October and December (26.0 ± 2.4 μgCl^{-1}). Suspended phaeopigments showed the main peaks in May (0.6 $\mu g \cdot l^{-1}$) and January (0.8 $\mu g \cdot l^{-1}$). For both parameters no significant differences were observed between stations (ANOVA, $P>0.05$; Table 1).

Sedimentary biopolymeric organic carbon concentrations (Fig. 4) showed several peaks throughout the year in August, October, December and February. Comparison between stations was significant, with Sed-BPC higher (ANOVA, $P<0.05$; Table 1) in the southern station (26,620 ± 18,468 μgCg^{-1}) than at the other two stations (1,990 ± 1,926 μgCg^{-1} at the open-sea station and 19,011 ± 11,946 μgCg^{-1} at the north station).

Suspended biopolymeric organic carbon (Fig. 4) concentrations peaked in May (272 ± 51 μgCl^{-1}) and in summer (403 ± 78 μgCl^{-1}). No differences were observed between stations (ANOVA, $P>0.05$; Table 1).

The mean value of the isotopic composition of sedimentary carbon (Fig. 5) was −12.9 ± 3.1‰, ranging from −22.3 to −8.0‰. Depletion in ^{13}C was observed in the May-July period and in December-January (−14.3 ± 0.8‰), while carbon was enriched with ^{13}C in spring and in August and in autumn (−12.1 ± 0.9‰). A clear gradient of enrichment (ANOVA, $P<0.05$; Table 1) was observed moving from the open sea (−15.8 ± 3.1‰), through the northern station (−13.1 ± 1.4‰) to the southern station (−10.0 ± 1.8‰).

The isotopic composition of the particulate organic carbon (Fig. 5) ranged from −25.1 to −15.4‰ (mean value = −19.5 ± 2.7‰). Except in summer (−17.7 ± 0.2‰) the isotopic composition was depleted in ^{13}C (−20.3 ± 1.2‰). The δ^{13}C of particulate organic carbon followed a gradient of enrichment with ^{13}C moving towards the inner part of the Stagnone (−21.4‰ and about −18.4‰ at the open-sea station and at the southern station, respectively).

Carbon isotopic ratios determined from the main organic matter sources are summarised in Table 2.

Table 2. Carbon isotopic ratios in the main organic matter sources measured in the sites studied (*OC*, organic carbon). For comparison, mean values (*bold*) of suspended and sedimentary organic carbon measured at each of the three sites are reported

Source	$^{13}C/^{12}C$
Zooplankton	-23.4
Phytoplankton	-21.4
Microphytobenthos (mainly diatoms)	-16.5
Caulerpa prolifera	-13.5
Posidonia oceanica	-7.80
Cymodocea nodosa	-7.20
Seagrass detritus	-4.50
Suspended OC open-sea	**-21.4**
Suspended OC North	**-18.3**
Suspended OC South	**-18.6**
Sedimentary OC open-sea	**-15.8**
Sedimentary OC North	**-13.1**
Sedimentary OC South	**-10.0**

Discussion and Conclusion

Several studies have focused on identifying the origin of the organic matter entering food webs and quantifying the fraction actually available to consumers (Fry and Sherr 1984; Mann 1988; Michener and Shell 1994). Seagrass ecosystems (*Posidonia oceanica*) are highly productive, although *Posidonia* produced organic carbon is highly refractory (Pollard and Kogure 1993). It has been recently demonstrated that the main transfer route for seagrass carbon into the benthic food web is through bacterial decomposition (Pollard and Kogure 1993). Seagrass meadows function as "detrital traps", gathering detrital carbon from all sources (Dauby 1989).

In the study area, five main primary producers were identified. *Posidonia oceanica* and *Cymodocea nodosa* appear to be slightly enriched compared to the literature data (McMillan et al. 1980; Dauby 1989; Jennings et al. 1997). Such an enrichment could be the result of higher temperatures and PAR (Wiencke and Fisher 1990) due to the shallowness of the studied area compared to other investigated systems (Dauby 1989; Tufano 1991; Jennings et al. 1997). In shallow environments, sediments accumulate organic carbon from different sources, including the overlying function of water column. Exchange between sediments and the overlying water column are a function of water movement intensity, which depends on the pulsing of wind and tidal energy (Sarà et al. 1999). Thus, variability in the sedimentary and particulate organic matter content and composition can be more strongly affected by these physical phenomena than by seasonality of the biological life cycle of each primary producer. In this way, most of the sedimentary carbon can be resuspended, to settle again rapidly within a short time. Such processes can lead to continuous changes in organic carbon composition, affecting the overall availability of organic matter. In the Stagnone di Marsala a coexistence of two different zones has been demonstrated. These two systems (the northern and the southern basin) are constrained by different hydrodynamic regimes depending upon wind exposure (Mazzola et al. 1999; Pusceddu et al. 1999; Sarà et al. 1999). It has been estimated that the occurrence of resuspension phenomena as a function of depth, effective fetch and wind speed (Smaal and Haas 1997), is about double in the northern basin of the Stagnone (Sarà et al. 1999) in which sedimentation is prevalent.

Based on Dauby model calculations (Dauby 1989), our isotopic data indicate that in the northern basin, sediments are mainly affected by macroalgae carbon (46%), while carbon from *Cymodocea*, microphytobenthos and phytoplankton represented about 14.5% each one, whilst *Cymodocea* detritus contributed for about 10%. In the water column, phytoplankton (60%) and microphytobenthos (30%) represent the main sources of suspended organic carbon, while the contribution of carbon from *Cymodocea*, macroalgae and detritus is negligible. If we correlate the monthly occurrence of resuspension at this site (Sarà et al. 1999), it can be inferred that organic carbon derived from *Cymodocea* and macroalgae accumulated in sediments (resuspension factor *vs* *Cymodocea* r_s = 0.65; $P<0.05$; $n = 12$; resuspension factor *vs* macroalgae $r_s = 0.75$; $P<0.05$; $n = 12$). Carbon from these two sources is not transferred to the water column as suggested by the absence of a correlation with the occurrence of resuspension events. In contrast, when resuspension intensity increases carbon from microphytobenthos and *Cymodocea* detritus is largely transferred to the water column. However, the absence of a correlation between the isotopic signature of these two producers and resuspension could suggest that much of this carbon is probably transported outside.

In Fig. 6a and b, the monthly carbon contributions are reported in comparison with the monthly trend of sedimentary and particulate biopolymeric organic carbon. A significant correlation between sedimentary BPC and microphytobenthic carbon was observed (S-BPC *vs* microphytobenthos $r_s = 0.8$; $P<0.05$; $n = 12$), suggesting that bio-available organic carbon largely originates from this source, when present.

Between March and July, when the microphytobenthic signature in sedimentary carbon was negligible (Fig. 6a), the biopolymeric organic carbon was quite low, while from August to February, when the microphytobenthic signature was present, S-BPC increased. Although no significant correlation was shown with S-BPC, enriched *Cymodocea* detritus also seems to play a definite role.

In the southern basin, the frequency of resuspension is quite low and tidal movements (Sarà et al. 1999) continuously mix the water col-

Fig. 6a,b. Carbon contribution (%) by each primary producer: **a** in sedimentary; **b** in particulate organic matter at station 2

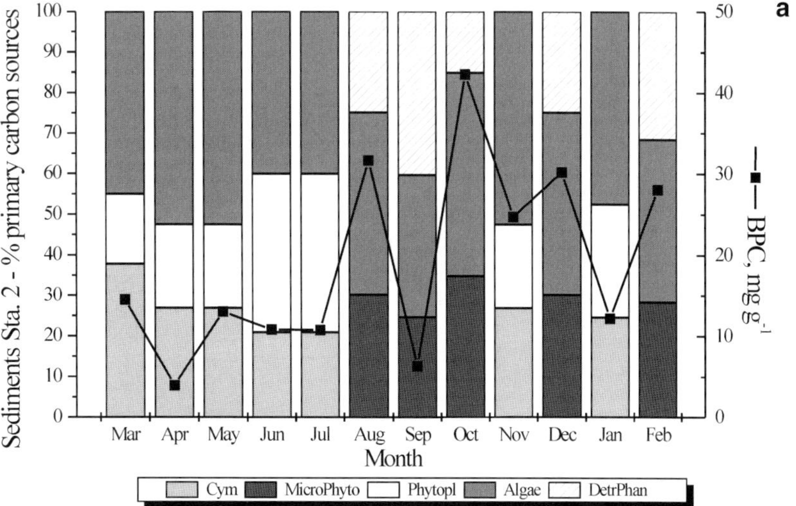

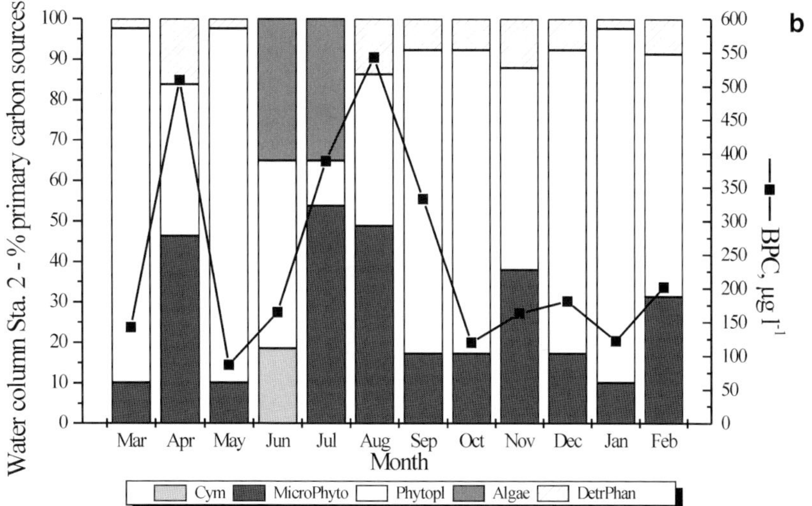

umn. At this site no significant correlation was shown between resuspension and the contribution of carbon from primary producers, but in December a main storm suggested a coupling between these variables.

Sediments in the southern basin were mainly affected (Fig. 7a) by seagrass organic carbon (35%; *Posidonia* and *Cymodocea*), seagrass detritus (26%) and microphytobenthic carbon (28%), while macroalgae carbon accounted only for 11%, finally phytoplankton contribution was negligible. No correlation was shown between the monthly occurrence of resuspension at this site and the isotopic signature of sedimentary and suspended carbon.

In the water column (Fig. 7b), microphytobenthic carbon accounted for about 24%, representing the second main carbon source after phytoplankton (50%), while macroalgae and detritus accounted for 15 and 10% respectively. The isotopic signature of living seagrasses was negligible.

Open-sea sediments were mostly affected by microphytobenthic and phytoplankton carbon (32 and 30%, respectively), by macroalgae (18%), detritus (14%), *Cymodocea* (6%). This station is a typical shallow unvegetated marine site with scant submerged vegetation, and in which seagrass detritus from other sites accumulates. The biopolymeric organic carbon concentrations,

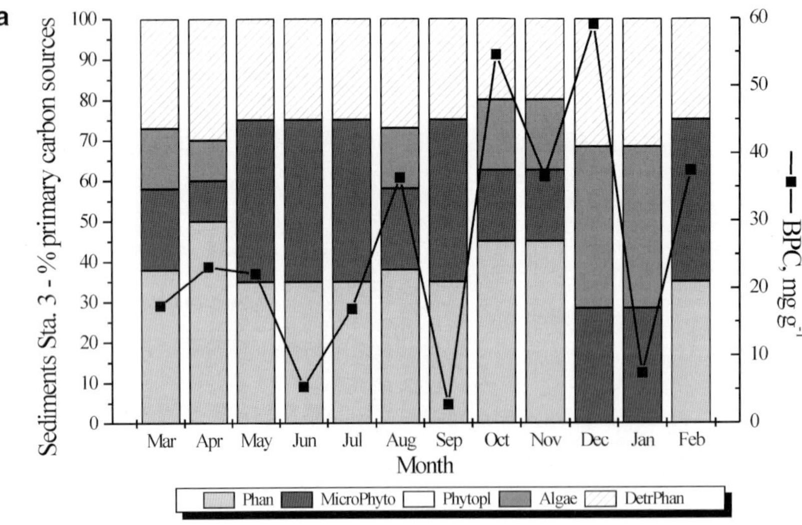

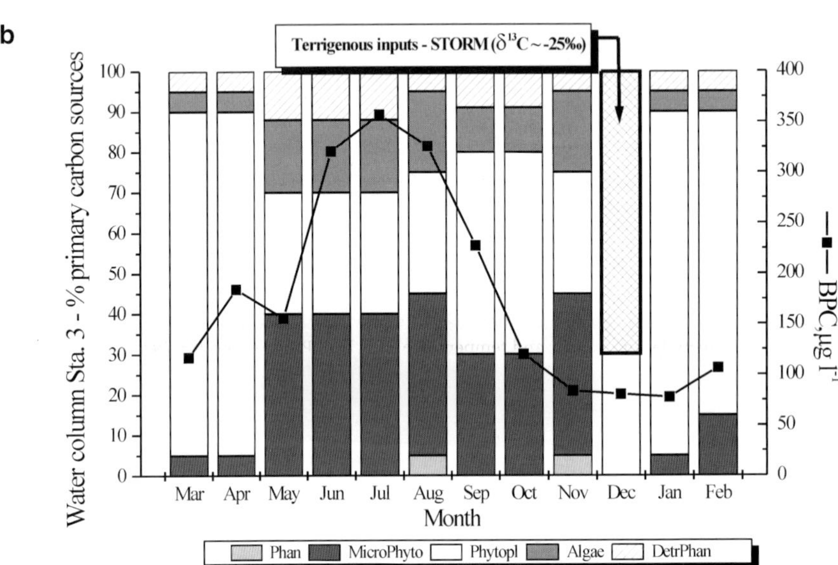

Fig. 7a,b. Carbon contribution (%) by each primary producer: **a** in sedimentary; **b** in particulate organic matter at station 3

both in sediments and suspended particulate matter, were 10 times lower than those of the sites within the Stagnone and were similar to those reported for other Mediterranean areas (see Pusceddu et al. 1999).

Acknowledgements. Authors would like to thank Dr. M. Caruso and Dr. David Catalano (University of Palermo) for help in chemical analyses and Dr. S. Vizzini for suggestions on improving the early version of the manuscript. The Ministero Politiche Agricole and Ministero Università Ricerca Scientifica Tecnologica (Italy) financially supported this work.

References

CERC (1977) Shore protection manual, 3d edn, vol 1. US Army Corps of Engineers, Coastal Engineering Centre, Washington, DC

Dauby P (1989) The stable isotope ratios in benthic food webs of the Gulf of Calvi, Corsica. Cont Shelf Res 9: 181-195

Fabiano M, Pusceddu A (1998) Total and hydrolizable particulate organic matter (carbohydrates, proteins and lipids) at a coastal station in Terra Nova Bay (Ross Sea, Antarctica). Polar Biol 19: 125-132

Fry B, Sherr EB (1984) δ^{13}C measurements as indicators of carbon flow in marine and freshwater ecosystems. Contrib Mar Sci Univ Texas 27: 13-47

Griffiths H (1998) Stable isotopes: integration of biological, ecological and geochemical processes. Bio Sci Publ, Washington, pp 270

Hopkinson CS (1985) Shallow water benthic and pelagic metabolism: evidence of heterotrophy in the nearshore Georgia Bight. Mar Biol 87: 19-32

Jennings S, Renones O, Morales-Nin B, Polunin NVC, Moranta J, Coll J (1997) Spatial variation in the ^{15}N and ^{13}C stable isotope composition of plants, invertebrates and fishes on Mediterranean reefs: implications for the study of trophic pathways. Mar Ecol Prog Ser 146: 109-116

Lajtha K, Michener RH (1994) Stable isotopes in ecology and environmental sciences. Blackwell Science, Oxford

Lorenzen CJ, Jeffrey SW (1980) Determination of chlorophyll and phaeopigments spectrophotometric equations. Limnol Oceanogr 12: 343-346

MacIntyre HL, Geider RJ, Miller DC (1996) Microphytobenthos: the ecological role of the "secret" garden of unvegetated, shallow water habitats. I. Distribution, abundance and primary production. Estuaries 19: 186-201

Mann KH (1988) Production and use of detritus in various freshwater, estuarine and coastal marine ecosystem. Limnol Oceanogr 33: 910-930

Mazzola A, Sarà G, Venezia F, Caruso M, Catalano D, Hauser S (1999) Origin and distribution of suspended organic matter as inferred from carbon isotope composition in a Mediterranean semi-enclosed marine system. Chem Ecol 16: 215-238

McMillan CP, Parker PL, Fry B (1980) $^{13}C/^{12}C$ ratios in seagrasses. Aquat Bot 9: 237-249

Michener R, Shell DM (1994) Stable isotope ratios as tracers in marine aquatic food webs. In: Lajtha K, Michener RH (eds) Stable isotopes in Ecology and environmental sciences. Blackwell Science, Oxford pp 138-157

Nival P, Charra R, Malara G, Boucher D (1972) La matière organique particulaire de la Méditerranée occidentale en mars 1970. Mission Mediprod du "Jean Charcot". Ann Ist Oceanogr Paris 48: 141-156

Parker JG (1983) A comparison of methods used for the measurement of organic matter in sediments. Chem Ecol 1: 201-210

Pollard PC, Kogure K (1993) The role of epiphitic and epibenthic algal productivity in a tropical seagrass, *Syringodium isoetifolium* (Aschers.) Dandy, community. Aust J Mar Freshwater Res 44: 141-154

Pusceddu A, Sarà G, Armeni M, Fabiano M, Mazzola A (1999) Seasonal and spatial changes in sediment organic matter composition of a semi-enclosed marine system (W-Mediterranean Sea). Hydrobiologia 397: 59-70

Sarà G, Leonardi M, Mazzola A (1999) Spatial and temporal changes of suspended matter in relation to wind and vegetation in a Mediterranean shallow coastal environment. Chem Ecol 16: 151-173

Smaal AC, Haas AA (1997) Seston dynamics and food availability on mussel and cockle beds. Estuarine Coast Shelf Sci 45: 247-259

Sokal RR, Rohlf FJ (1981) Biometry. WH, Freeman and Co, New York

Strickland JDH, Parsons TR (1972) A practical handbook of sea water analysis. Bull Fish Res Board Can 167: 310

Tufano M (1991) Il rapporto isotopico $^{13}C/^{12}C$ ($\delta^{13}C$) nello studio dell'ecosistema bentonico costiero (Mar Ligure). PhD Diss, Univ Genoa

Underwood AJ (1997) Experiments in ecology: Their logic and interpretation using analysis of variance. Cambridge University Press, Cambridge

Valiela I (1984) Marine ecological processes. Springer Verlag, Berlin Heildelberg New York Tokyo

Wiencke C, Fisher G (1990) Growth and stable carbon isotope composition of cold-water macroalgae in relation to light and temperature. Mar Ecol Prog Ser 65: 283-292

Synchronous Fluorescence Spectra as Chemical Tracers to Monitor the Organic Matter Dissolved in North Adriatic Waters

M. Mingazzini

ABSTRACT

The use of synchronous fluorescence spectra as chemical tracers of the DOM in Adriatic Sea waters is here proposed. Laboratory experiments using both marine diatom cultures and natural phytoplankton provided evidence for the ability of the technique to characterize and quantify the extracellular products. Fluorescence parameters defined on laboratory cultures were applied to the study of the natural DOM present in the mixing area of the Po River with marine waters. The results support the effectiveness of the parameters to detect the contribution of the terrestrial as well as of the newly produced components mixed in seawater DOM. Experimental evidence was provided for the potential of the technique to trace the accumulation of the EOM in marine waters.

Introduction

While the cycling of dissolved organic matter (DOM) in marine systems is the focus of intense studies, both the low concentration and the chemical complexity of the DOM components which are mixed in natural waters make difficult to balance the marine carbon budget. Due to the fluorescence properties of the DOM, many of the earliest studies concentrated on using fluorescence as a chemical tracer of the source, circulation and mixing of DOM in waters (Mopper and Schultz 1993; De Souza Sierra et al. 1994; Coble 1996).

In 1992 a global fluorescence budget was estimated (Chen and Bada 1992) to assess the major sources of DOM in seawater. The Authors concluded that, while rain, rivers and diffusion from sediments are local sources probably insignificant on a global scale, the *in situ* formation is the most likely source for the fluorescence of seawater. Nevertheless, the same Authors underlined a lack of information on the last source. The most recent and in-depth investigations on the nature of the seawater DOM still pointed out the research need on *in situ* biological production (Coble 1996). The information on phytoplankton sources are still related only to the fluorescence spectral characterization performed by Traganza 30 years ago (Traganza 1969). No more information has been provided by experimental evidence.

In the case of the Adriatic Sea, where great amounts of phytoplankton extracellular organic matter (EOM) are involved in the accumulation process of massive quantities of mucilage, with serious economic implications for the Italian country, the need of a better understanding of the marine DOM cycling is particularly urgent (Rinaldi et al. 1995; Mingazzini and Thake 1995). The use of synchronous fluorescence technique has recently been proposed to characterize the EOM produced by phytoplankton in Adriatic Sea waters (Mingazzini et al. 1995). The same technique was also applied to the study of the mixing DOM in a lagoonal system of the Po River delta (Ferrari and Mingazzini 1995).

In the present study the synchronous fluorescence technique was used to compare results obtained both from marine phytoplankton laboratory experiments and from natural Adriatic samples collected in the transition zone of the Po River delta. The aim was to determine fluorescence parameters able to characterize the naturally occurring components of DOM as well as to detect and trace the accumulation of the EOM in North Adriatic seawaters.

Istituto di Ricerca sulla Acque, CNR, Via della Mornera 25, 20047 Brugherio (Mi), Italy

F.M. Faranda, L. Guglielmo, G. Spezie (eds)
Mediterranean Ecosystems: Structures and Processes

Materials and Methods

Algal Cultures

Different algal species were used to characterize the organic matter produced and released in the extracellular medium. Four diatom species: *Skeletonema costatum*, *Nitzschia* (renamed *Cylindrotheca*) *closterium*, *Chaetoceros* sp. and *Navicula* sp., isolated from North Adriatic Sea, were cultured in the laboratory. While *S. costatum* and *N. closterium* were collected in summer 1991 from mucilage samples (Mingazzini et al. 1995), *Navicula* and *Chaetoceros* were isolated from water samples respectively collected in September 1994 and May 1995 at nearshore Cesenatico. The monospecific cultures were grown in MAAP medium (EPA 1974) prepared using oligotrophic seawater enriched with inorganic nutrients from the medium stock solutions (Mingazzini et al. 1995). Cultures were kept in controlled conditions at 20±1°C temperature, at 3000-lux light intensity, with a light:dark cycle of 14:10 hours. Natural phytoplankton from the marine sample stations (see below) was also used to monitor the extracellular production. The unfiltered seawater samples (100 ml) containing the natural community were incubated without any enrichment and kept in the same conditions used for the monospecific cultures.

Differently aged cultures of *S. costatum* (0 to 40 days from inoculum) were used to monitor the trend of production and accumulation of the fluorescing compounds released in the different growth phases. The same experimental framework was applied to the natural phytoplankton cultures, starting from the incubation time 0 of the samples.

For the fluorimetric characterization of the extracellular medium, 5 to 10 ml-subsamples were taken from the diatom as well as from the natural phytoplankton cultures. All samples (for EOM as well as for natural DOM determinations) were filtered on Millipore HA 0.45 μm filters under reduced pressure (<50 mm Hg) in order to separate the EOM from the biomass, avoiding breakage of cells. Each filter was rinsed in two ways before use. First, prefiltered Ultrapure water, and secondly, a part of the same sample were passed through each filter before collecting the filtrate.

Sampling Stations

The Po River delta, as the main contributor of terrestrial DOM, providing over 50% of the freshwater input to the North Adriatic basin, was investigated. The map in Fig. 1 shows the study area and the location of sampling stations. One riverine water sample (Po) was collected close to the river mouth, in Po della Pila, which is the main channel of the Po River delta. Seven marine water samples (1 to 7) were taken from an onshore-offshore transect, starting just outside from the river mouth, at salinities ranging from 8 to 32.2 PSU.

The samples were collected from surface waters (0.5 m) by the DAPHNE II oceanographic vessel of the Emilia-Romagna Region. The sampling was performed within one day (21 June 1995) and the water samples (for DOM determinations) were filtered (see above) and transferred to the laboratory in dark-refrigerated conditions. All the DOM sample analyses were performed within 36 h after sampling.

This sampling represents a part of the surveys carried out within the PRISMA 1 Project (Research Programme for the Adriatic Sea). A number of chemical-physical determinations was made within the project, and DOM was also investigated by DOC analyses performed on the same samples (Pettine et al. 1998).

Spectrofluorimetric Analyses

The synchronous fluorescence technique (Vo-Dinh 1978), which involves the simultaneous scanning of the excitation and emission monochromators keeping a constant wavelength offset ($\Delta\lambda$), was adopted in order to resolve the mixture of chromatophoric components of the DOM. This technique has been explored in the last years in different fields of application (Cabaniss and Shuman 1987; Vodacek 1989; Senesi 1990; De Souza Sierra et al. 1994; Ahmad and Reynolds 1995; Galapate et al. 1998). The choice of both the range of excitation wavelengths and the interval $\Delta\lambda$ should be determined for each application. In multicomponent mixtures of natural DOM samples the capacity to discriminate between fluorophores having different origin but similar fluorescence properties makes particularly difficult the choice of $\Delta\lambda$ (De Souza Sierra et al. 1994). In the present study the instrumental parameters

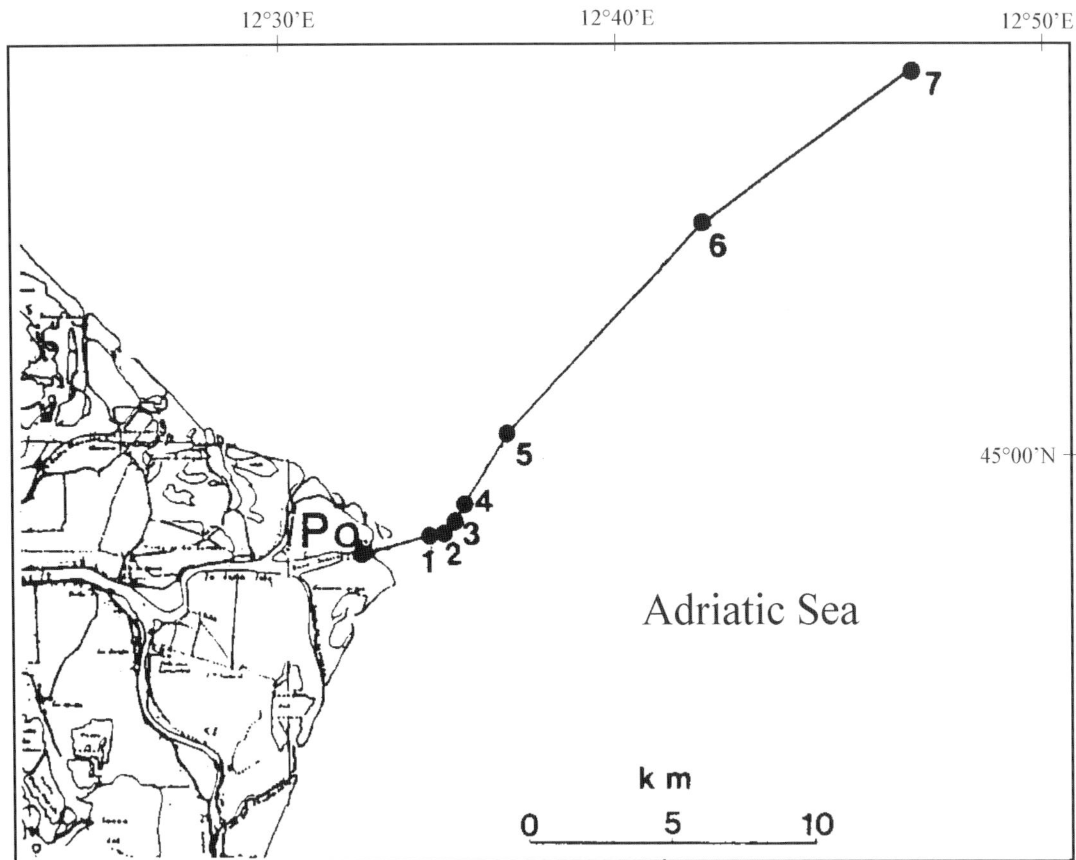

Fig. 1. Map of sampling stations

which provided spectral selectivity and sensitivity for the characterization of algal EOM, as well as of natural DOM samples (Mingazzini et al. 1995; Ferrari and Mingazzini 1995; Mingazzini and Onorato 1998), were employed. A SPEX FluoroMax spectrofluorimeter, equipped with a 150-W ozone-free Xenon lamp, was used. Spectra were recorded in an excitation wavelength range of 250-500 nm with 25 nm $\Delta\lambda$, 4.25 nm bandpass, 2 nm increment and 1s integration time. No correction factors were applied. All spectra were blank-subtracted using the Ultrapure water spectrum, recorded immediately before samples with the same instrumental parameters. Analyses were performed at room temperature using 1 cm path length cell (precision quartz cuvette) as sample holder. Synchronous spectra are presented in the figures on the excitation wavelengths scale, while the fluorescence intensities are expressed as fluorescence units (F.U. = cps x 10^3).

Results and Discussion

Laboratory Cultures

Typical fluorescence spectra of the EOM produced by different species of marine diatoms are illustrated in Fig. 2. Two main fluorescence signals are present in each of the four spectra. The two maxima, which are respectively labelled A and B, are referred to the two types of fluorescent DOM observed in seawater. The fluorescence peak A (276 nm Ex max) representing the "aminoacidic-like" component is referred to "relatively young" production by recent biological activity and was observed to be produced both in marine waters and in laboratory cultures (Traganza 1969). The fluorescence peak B (340 nm Ex max) representing the "humic-like" component is generally referred to "old" material, such as humic compounds of terrestrial origin (Mopper and Schultz 1993; De Souza Sierra et al.

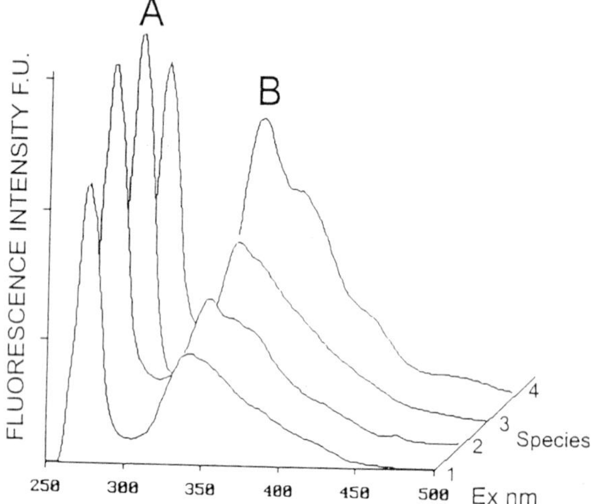

Fig. 2. Synchronous fluorescence spectra of the EOM produced by the marine diatoms *Navicula* sp. (*1*), *S. costatum* (*2*), *N. closterium* (*3*) and *Chaetoceros* sp. (*4*). *A*, aminoacidic-like component; *B*, humic-like component Ex max

1994). The production of the last by phytoplankton was just recently observed for the first time (Mingazzini et al. 1995). Some interspecific differences in the spectral features can be noticed in Fig. 2 as shoulders in additional bands (360 and 410 nm), but the B peak at 340 nm is always the Ex max in the visible region within the diatom EOM (Mingazzini and Onorato 1998).

While the production of the aminoacidic-like component gives an intense increase of peak A only during the exponential growth phase, the humic-like component is gradually produced also in stationary growth phase and progressively accumulates in the extracellular medium. According to what has been previously observed on a number of phytoplankton species (Mingazzini et al. 1995), Fig. 3 (*S. costatum*) gives a good representation of the EOM production trend over time. The ratio UV/visible Ex max A/B, which is high in the active growth phase, exponentially decreases with time. This fluorescence parameter is thus a good descriptor of the observed production trend, since high values of the A/B ratio can signify that the A component, as "recent production" prevails, while progressively lower A/B values can be referred to the progressive increase of the humic-like component, reflecting an increased contribution of the accumulated "old production" in the medium.

Natural DOM Samples

Field samples collected in the riverine to marine transition zone (Fig. 1) were analysed using the same technique in order to verify the effectiveness of the fluorescence parameters when applied to the study of the nature and distribution of the naturally occurring components of DOM in Adriatic seawater. In Fig. 4 all the synchronous spectra of natural DOM samples are shown. Compared to the EOM, the more noticeable differences in the spectral features were observed in the visible region. It is to be noted that the humic-like component exhibits the same Ex max (340 nm) as observed on diatom EOM spectra only in the most pelagic DOM (Station 7 spectrum). The riverine DOM (Station Po) spectrum is characterized by a broad band located between 300 and 400 nm, exhibiting a shift of the Ex max towards longer wavelengths, the humic-

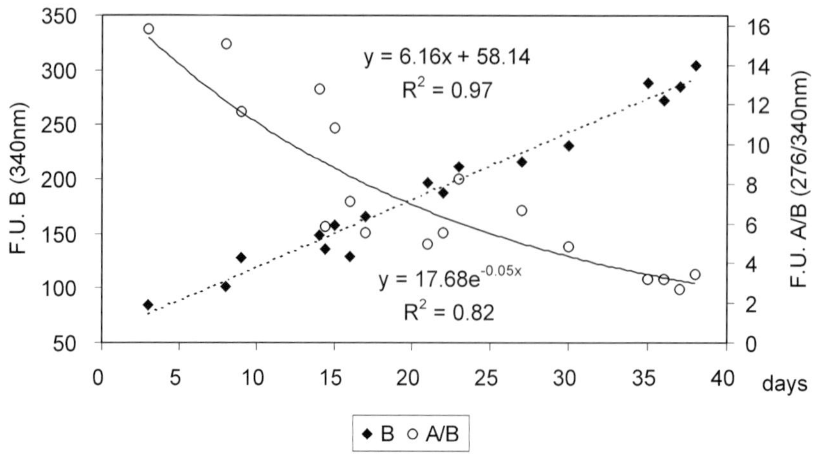

Fig. 3. Trend over time of the EOM produced by *Skeletonema costatum*: linear increase of B fluorescence intensity (*top* equation, n=17, P<0.001) and exponential decrease of the A/B intensity ratio (*bottom* equation, n=17, P<0.001) during a 40-days algal growth experiment

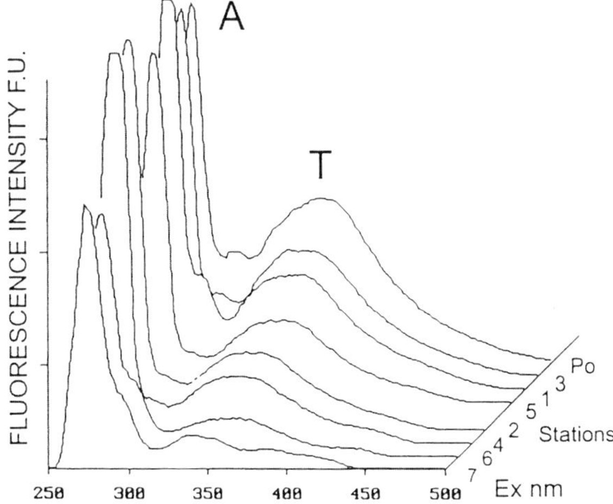

Fig. 4. Synchronous spectra of natural DOM from riverine (Po) and seawater (1 to 7) samples. The last are ranked according to the salinity order. *A*, aminoacidic-like component; *T*, Ex max of the riverine humic-like component

like peak being located at 350 nm. This red-shifted peak is labelled peak T (Fig. 4) to signify the difference in position of maximum fluorescence between the EOM peak B (Fig. 2) and the riverine DOM peak. The fluorescence intensity measured in T position was observed to decrease progressively as river water mixes with seawater. Thus, the use of band T as a fluorescent tracer of the humic terrestrial contribution makes possible to define accurately the dilution trend of the riverine DOM in Adriatic seawater. This is strongly supported by the good correlation found between fluorescence and salinity values (r^2 = 0.98) plotted in Fig. 5.

A linear decrease with salinity was also observed for peak A (not shown in figure) in marine samples (Stations 1 to 7) spectra, but its fluorescence intensity showed higher variability and decreased less steeply with salinity (A = 423.34 – 5.72 salinity, r^2 = 0.62). This behaviour, which was already found in a similar study undertaken on the Columbia River Estuary (Prahl and Coble 1994), seems to support the hypothesis that the UV-band can mostly reveal the presence of organic compounds which have not only an inland, but also an *in situ* origin. Results of a number of data sets from different environments suggested that UV and visible fluorescence peaks may vary independently of each other (Coble 1996).

In Fig. 5 the A/T ratios measured on Adriatic DOM samples are also plotted as a function of salinity. The well defined trend of the UV/visible ratio strongly supports the effectiveness of this parameter to describe changes in the relative contribution of the mixed DOM components having different nature and origin. While low values indicate that the contribution of remote DOM (terrestrial humic) prevails, increasing values signify a relative increase of newly produced DOM (marine phytoplankton EOM).

These results (Fig. 5) can be appreciated even more when compared with DOC results reported for the same samples (Pettine et al. 1998), the last being surprisingly unable to describe the dilution trend of the riverine DOM. The humic terrestrial DOM seems to be underevaluated by DOC measurements since they found higher DOC concentrations in marine than in riverine water samples. The hypothesized bacterial consumption of the terrestrial DOC in the prodelta

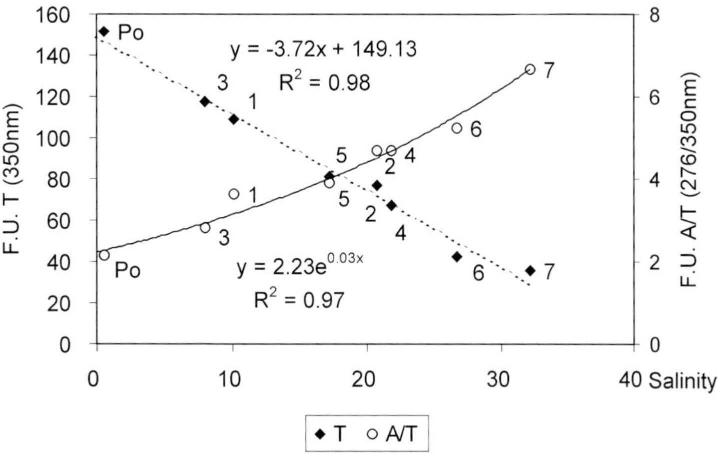

Fig. 5. Mixing gradient of natural DOM: linear decrease of the terrigenous (*T*) component (*top* equation, n=8, P<0.001) and exponential increase of the A/T intensity ratio (*bottom equation*, n=8, P<0.001) as a function of salinity

region seems to be unlikely since we clearly found the riverine DOM spectral features (Fig. 4, Station Po) unchanged in the marine Stations 1 to 6. On the other hand, the EOM production, which could explain the missing of a dilution trend, can not be detected using DOC measurements, which are unable to distinguish or trace it.

Natural Phytoplankton EOM

While fluorescence measurements performed on different estuaries over the world confirm the effectiveness of the fluorescence parameter – Ex max in the visible region – to describe the terrestrial DOM contribution, the behaviour of the UV/visible ratio is still poorly understood (De Souza Sierra et al. 1994; Coble 1996). The hypothesis (first suggested by Mingazzini et al. 1995) that recent biological activity (EOM production) can affect the UV/visible ratio, as well as the blue-shifted position of the humic Ex max, has never been supported by experimental evidence. The EOM production was therefore directly measured on the unfiltered seawater samples (Stations 1 to 7) where the natural phytoplankton was present and mostly represented by the diatom group.

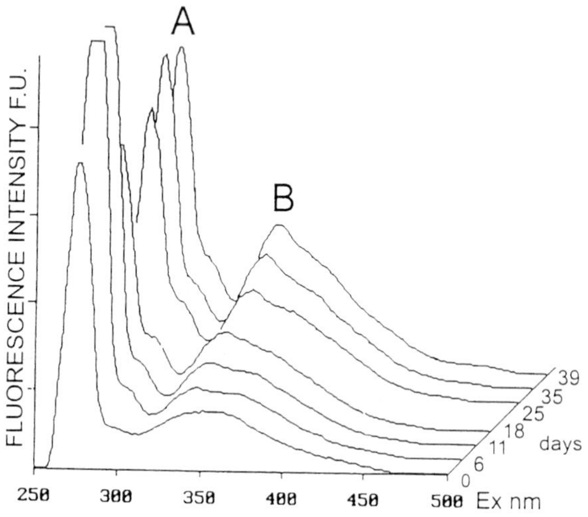

Fig. 6. Evolution trend of the DOM synchronous spectral features of the Station 2 sample incubated with the natural phytoplankton

Figure 6 gives a good representation of the results obtained on the 40 days incubation experiment performed on all seawater samples. The spectra (Station 2) are presented in chronological sequence, clearly showing how the time 0 spectral shape changes over time. Starting from day 6, it can be seen that two peaks are contemporaneously present in the humic band: the first of them (B, 340 nm) increases over time, the second (T, 350 nm) being progressively identified as just a shoulder. This change, giving a blue-shift of the Ex max in the visible region, is due to the accumulation of the EOM humic-like component B, which was observed (Fig. 7) to increase linearly in fluorescence intensity over time in all the incubated samples (Stations 1 to 7). The accumulation trend was ranging from 1.25 to 4.16 F.U./day (r^2 = 0.76 to 0.996, n = 7, P = 0.011 to 0.000), being of the same order of magnitude as the one measured on *S. costatum* cultures (6.16 F.U./day, Fig. 3). These can be considered interesting results since they also comprise the heterotrophic consumption by the bacterial compo-

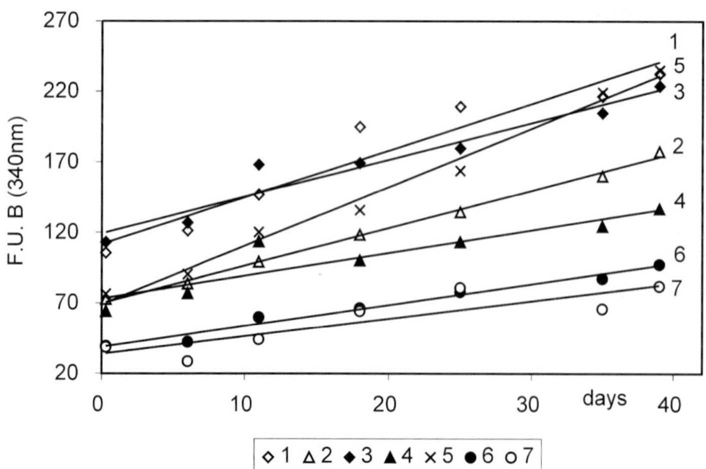

Fig. 7. Accumulation trend of the EOM humic-like component (*B*) produced by the phytoplankton in the marine samples 1 to 7 during a 40-days incubation

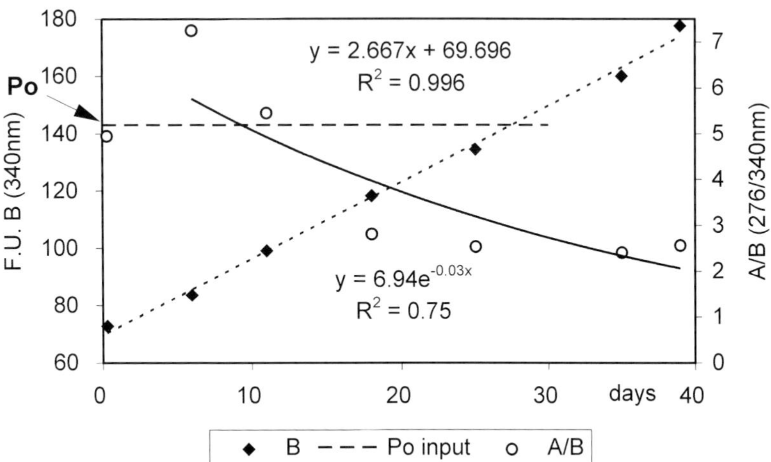

Fig. 8. Trend over time of the EOM produced by the natural phytoplankton (Station 2 sample): linear increase of B (top equation, n=7, P<0.001) and exponential decrease of the A/B ratio (bottom equation, n=6, P=0.026). The fluorescence intensity of the Po sample (*dotted line*) is plotted in figure for purpose of comparison

nent, which was naturally present in the unfiltered samples.

Figure 8 (Station 2, as representative sample) clearly shows how the trend over time of both the fluorescence parameters (B and A/B) closely follows the model (Fig. 3) defined on monospecific diatom cultures. At the beginning of the incubation time, the UV/visible ratio A/B increases, giving experimental evidence of what can happen in surface marine waters, due to a recent phytoplankton production. After a few days, the A/B ratio decreases as the B component accumulates. It is also to be noted that the humic-like B component, never considered before as a possible phytoplankton contribution to the marine DOM, can reach fluorescence intensities as high as the humic riverine contribution. While such massive quantities are unlikely to accumulate in surface waters, this EOM component can still represent an important contribution in shallow waters, pycnocline layers or reduced water exchange conditions.

Conclusions

These results provide experimental evidence of the ability of the two parameters: Ex max in the visible region and UV/visible Ex max ratio to monitor the nature and distribution of different components mixed in Adriatic seawater DOM. In particular:

1. The band T is a fluorescent tracer of the humic terrestrial contribution, effective to define the dilution trend of the riverine DOM;
2. The A/T ratio describes changes in the relative contribution of the terrestrial and newly produced components, respectively mixed in seawater DOM;
3. The trend model of the B and A/B parameters, as determined on algal cultures, is closely followed by the natural phytoplankton EOM, demonstrating for the first time the accumulation potential of the humic-like B component, as a phytoplankton contribution to the marine DOM.

The results from natural samples support the effectiveness of the spectrofluorimetric approach to trace and monitor the North Adriatic DOM.

Acknowledgements. The author is grateful to Dr. C.R. Ferrari and the crew of Daphne II (Emilia-Romagna Region) for sample collection and to Dr. L. Onorato for assistance in laboratory work. This research was in part supported by the Italian Ministry for University and Research within PRISMA 1 Project.

References

Ahmad SR, Reynolds DM (1995) Synchronous fluorescence spectroscopy of wastewater and some potential constituents. Water Res 29: 1599-1602

Cabaniss SE, Shuman MS (1987) Synchronous fluorescence spectra of natural waters: tracing sources of dissolved organic matter. Mar Chem 21: 37-50

Chen FR, Bada JL (1992) The fluorescence of dissolved organic matter in seawater. Mar Chem 37: 191-221

Coble PG (1996) Characterization of marine and terrestrial DOM in seawater using excitation-emission matrix spectroscopy. Mar Chem 51: 325-346

De Souza Sierra MM, Donard OXF, Lamotte M, Belin C, Ewald M (1994) Fluorescence spectroscopy of coastal and marine waters. Mar Chem 47: 127-144

EPA (1974) Marine algal assay procedure: bottle test (eutrophication and lake restoration branch). Pacific Northwest Environ Protect Lab, Corvallis, OR

Ferrari GM, Mingazzini M (1995) Synchronous fluorescence spectra of dissolved organic matter (DOM) of algal origin in marine coastal waters. Mar Ecol Prog Ser 125: 305-315

Galapate RP, Baes AU, Ito K, Mukai T, Shoto E, Okada M (1998) Detection of domestic wastes in Kurose River using synchronous fluorescence spectroscopy. Water Res 32: 2232-2239

Mingazzini M, Colombo S, Ferrari GM (1995) Application of spectrofluorimetric techniques to the study of mucilages in the Adriatic Sea: preliminary results. Sci Total Environ 165: 133-144

Mingazzini M, Onorato L (1998) Relation between phytoplankton exudates and producer species: differentiation indices (in Italian). Biol Mar Mediterr 5(1): 755-758

Mingazzini M, Thake B (1995) Summary and conclusions of the workshop on marine mucilages in the Adriatic Sea and elsewhere. Sci Total Environ 165: 9-14

Mopper K, Schultz CA (1993) Fluorescence as a possible tool for studying the nature and water column distribution of DOC components. Mar Chem 41: 229-238

Pettine M, Patrolecco L, Camusso M, Crescenzio S (1998) Transport of Carbon and Nitrogen to the Northern Adriatic Sea by the Po River. Estuarine Coast Shelf Sci 46: 127-142

Prahl FG, Coble PG (1994) Input and behaviour of dissolved organic carbon in the Columbia River Estuary. In: Dyer KR, Orth RJ (eds) Changes in fluxes in estuaries: implications from science and management. Olsen and Olsen, Fredensborg, pp 451-457

Rinaldi A, Volenweider RA, Montanari G, Ferrari CR, Ghetti A (1995) Mucilages in Italian seas: the Adriatic and Tyrrhenian Seas, 1988- 1991. Sci Total Environ 165: 165-183

Senesi N (1990) Molecular and quantitative aspects of the chemistry of fulvic acid and its interactions with metal ions and organic chemicals. II. The fluorescence spectroscopy approach. Anal Chim Acta 232: 77-106

Traganza ED (1969) Fluorescence excitation and emission spectra of dissolved organic matter in sea water. Bull Mar Sci 19:897-904

Vodacek A (1989) Synchronous fluorescence spectroscopy of dissolved organic matter in surface waters: application to airborne remote sensing. Remote Sensing Environ 30: 239-247

Vo-Dinh T (1978) Multicomponent analysis by synchronous luminescence spectrometry. Anal Chem 50: 396-401

CHAPTER 17

Atmospheric Deposition of Trace Metals in North Adriatic Sea

P. Rossini[1], S. Guerzoni[1], G. Rampazzo[2], G. Quarantatto[1], E. Garibbo[2], and E. Molinaroli[2]

ABSTRACT

Atmospheric bulk deposition of trace metals (Cd, Cu, Ni, Pb) was measured at two coastal urban sites (Cesenatico and Venice) in the north Adriatic Sea. Collection was carried out using polyethylene bulk passive samplers, samples being collected bi-weekly. Annual deposition fluxes of metals were calculated and compared with the Po river supply rates to the north Adriatic sea (surface area = 25000 km^2) and those of several rivers to the lagoon of Venice (surface area = 546 km^2). Atmospheric annual deposition at Venice was obtained by averaging data collected in the period 1993-97; Cesenatico values refer to 1995-96 depositions, measured in one coastal station and in one station 10 nautical miles offshore. Average north Adriatic fluxes (Venice + Cesenatico), in mg m^{-2} year^{-1}, range from 0.15 to 0.22 for Cd, 4.5 to 11.3 for Cu, 1.4 to 3.5 for Ni, and 3 to 21 for Pb, and are similar to French coastal sites data. The atmospheric deposition observed at Venice was 2- to 7-fold higher than that of Cesenatico for Cu, Ni and Pb, and comparable for Cd. Compared with riverine inputs of respectively 0.42, 4.0 and 5.5 tonnes year^{-1}, at Venice one-third of Cd (0.12 tonnes year^{-1}), nearly half of Ni (1.9 tonnes year^{-1}) and an almost equal amount of Cu (6.2 tonnes year^{-1}) are atmospherically derived, whereas the deposition flux of Pb is estimated to be 11.5 tonnes year^{-1}, i.e. 60% higher than the riverine input of 7 tonnes year^{-1}. Comparison with Po river data shows that the atmosphere delivers 70% of Cd, 10% of Ni, 70% of Cu and an amount of Pb 2-fold higher than riverine inputs. The magnitude of the atmospheric input indicates that eolian deposition is an important contribution which cannot be neglected in the study of biogeochemical cycles of anthropic elements introduced in the north Adriatic Sea.

Introduction

In the last decade, it has been demonstrated that the atmosphere is a significant pathway for the transport of many natural and polluting materials from the continents to the ocean (Duce et al. 1991). The atmospheric input of many of these elements may have an impact (either positive or negative) on marine geochemical cycling and effects on both terrestrial and aquatic ecosystems (Nriagu 1992). Atmospheric deposition of aerosols and gases occurs in two modes: wet and dry. The processes by which chemical elements are transferred in gas and solid phases are collectively referred to as "dry" deposition.

Dry deposition of particles occurs by direct impaction and gravitational settling on land or water surfaces. Liquid deposition, often referred to as "wet" deposition, comprises water and its dissolved gases and solutes, together with any insoluble particulate material contained therein (Duce et al. 1991; Golomb et al. 1997). To obtain total atmospheric loading, it is necessary to measure both wet and dry deposition.

As reported by Peters and Reese (1995), bulk deposition may not provide precise estimates of wet plus dry atmospheric deposition, and is further affected by local contamination. In said paper, deposition was lower and less variable in the wet precipitation than in the bulk. The higher variability of the latter was attributed to local contamination, particularly by dust and insects. In order to compare the validity of different sampling methods, in a previous work (Rossini et al.,

[1] Consiglio Nazionale delle Ricerche, Istituto di Geologia Marina, Via Gobetti 101, 40127 Bologna, Italy
[2] Dipartimento di Scienze Ambientali, Università di Venezia, Dorsoduro 2137, 30123 Venezia, Italy

F.M. Faranda, L. Guglielmo, G. Spezie (eds)
Mediterranean Ecosystems: Structures and Processes

personal communication) preliminary comparisons between located wet and dry and bulk samplers have been reported. It emerged that for Cd, Ni, Pb and Cu, wet+dry fluxes were ≅1.5 times higher than bulk. Therefore, as reported, in general there is a good correlation between all median fluxes. Moreover, bulk samplers are easier to operate than wet and dry, and may also be located at sites with no electrical power supply, where it is difficult to manage sampling, e.g. on a gas extraction base in the open sea.

Thus, in order to assess the importance of atmospheric loads and compare them with riverine inputs, in this paper results obtained from trace metal analysis on bulk samples collected during the period 1993-97 at two coastal urban sites (Cesenatico and Venice) in the north Adriatic Sea, and on an offshore rig (AGIP), are shown.

Materials and Methods

In order to obtain data on atmospheric transport and deposition of metals, useful for management policy, three atmospheric deposition sampling stations were chosen, located at two coastal urban sites (Cesenatico, 44°12' N, 12°24' E; Venice, 45°26' N, 12°21' E), and in one site 10 nautical miles offshore, on a gas extraction rig in the open Adriatic sea (AGIP; 44°31' N, 12°30' E), (Fig. 1). Cesenatico is a tourist beach resort, the gas rig is off the industrial area of Ravenna, and an important chemical plant is located close to Venice.

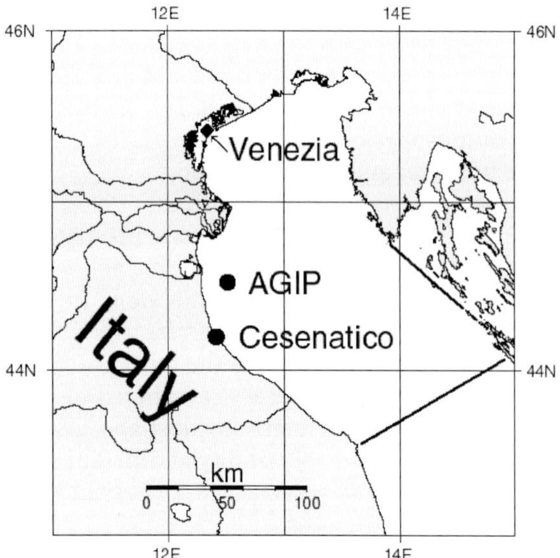

Fig. 1. Location of bulk sampling stations. Deposition area considered for atmospheric load calculations is indicated

Sampling was carried out following protocols adopted by the WMO (World Meteorological Organization), whose Marine Geology Institute is a scientific consultant (WMO/UNEP-MAP 1998). Atmospheric total deposition was measured by bulk samplers (Mosello et al. 1988). Bulk deposition collection was carried out using a polyethylene bucket of known diameter, with a polyethylene funnel, mounted on a stainless steel support painted with an inert varnish in order to avoid contamination. Bulk samples were collected bi-weekly, for a total of 106 samples (AGIP n=22, Cesenatico n=25, Venice n=59).

Bulk samples were analysed for Cd, Cu, Ni and Pb, in order to compare atmospheric and river contributions for these anthropogenic-derived elements. From 250 to 1000 ml of sample (depending on volume collected) were concentrated in Teflon beakers to a volume of approximately 15 ml by evaporation at 60 degrees centigrade in a laminar flow bench. Then, after addition of 3.5 ml of 65% HNO_3, 1.5 ml of 30% H_2O_2 and 1 ml of 40% HF, the solutions were dissolved in Teflon bottles in a microwave digestion unit (MILESTONE MLS 1200). Digested solutions were made up to 50 ml in tpx matrasses. Sampling blanks were collected at Venice by washing the polyethylene funnel of the sampler with 500 ml of Milli-Q water, and the obtained solutions were then treated and analysed following the same procedures used for other samples. Sampling blank values were always lower than 10% of sample concentrations. Laboratory blanks were also produced with Milli-Q water, both normal and pre-concentrated as for other samples by evaporation at 60 degrees centigrades in a laminar flow bench. Quality control of analyses was carried out by an external laboratory, qualified for UNI CEI EN 45001 and UNI EN ISO 9000 certifications. Differences between our analyses and certified values were under 5%.

Chemical analyses were conducted by AAS + ICP mass spectrometry, following the procedures of Guerzoni et al. (1987). Analytical error was between 5 and 10% for Cu, Ni and Pb, and 12% for Cd. All labware and glassware used were prewashed with diluted acid and rinsed with Milli-Q water. All reagents were of *suprapur* type.

Results and Discussion

Atmospheric Fluxes

As bulk collectors are open all the time, the amount of collected water is variable, being affected by evaporation due to weather conditions (e.g. dryness or high humidity). As the element concentrations of samples could not be compared, deposition fluxes in μg•m^{-2} had to be computed. Fluxes (F_x, μg•m^{-2}) were calculated according to the following equation:

$$F_x = \frac{C \times V_c \times V_r}{V_f} \Big/ S_a$$

where C is the element value in μg• ml^{-1}, V_c is the volume of digested solution expressed as ml, V_r is the amount of collected rain (ml), V_f is the concentrated volume of the sample (ml), and S_a is the sampler surface area (0.140 m^2).

Total daily fluxes of metals were very variable at all sites; in Figs. 2 (Pb) and 3 (Cd), deposition fluxes measured at the Venice and AGIP stations

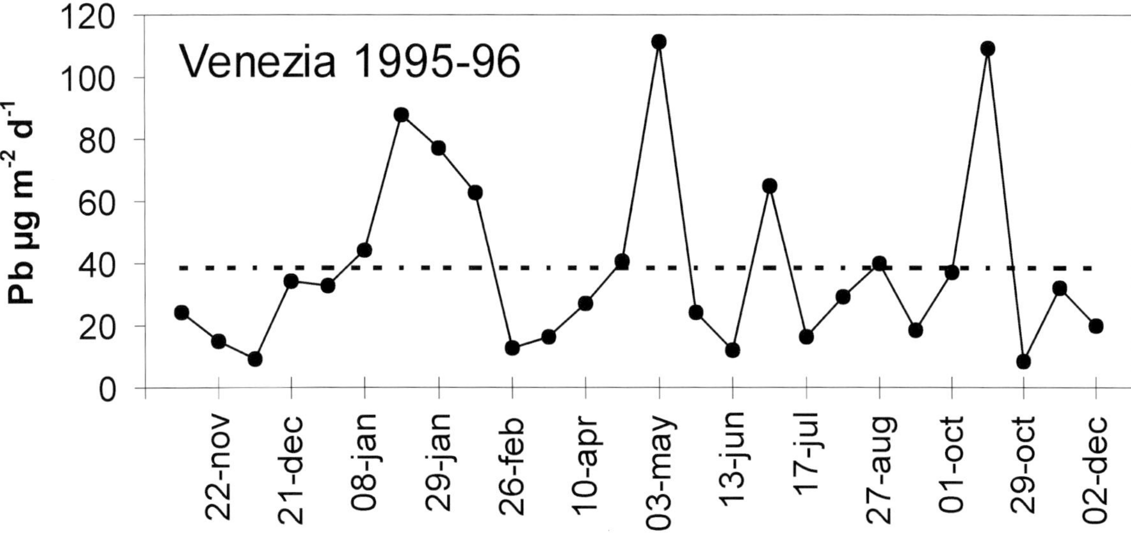

Fig. 2. Bulk daily fluxes of Pb collected at Venice during the period November 1995-December 1996. *Dashed line*, mean of 39 μg•m^{-2}•d^{-1}

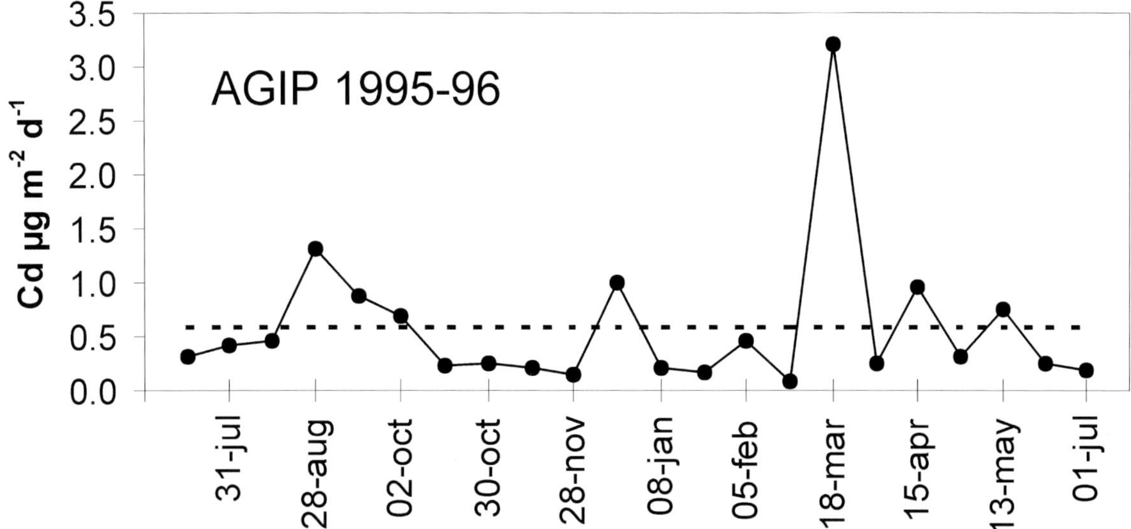

Fig. 3. Bulk daily fluxes of Cd collected at AGIP during the period July 1995-July 1996. *Dashed line*, mean of 0.6 μg•m^{-2}•d^{-1}

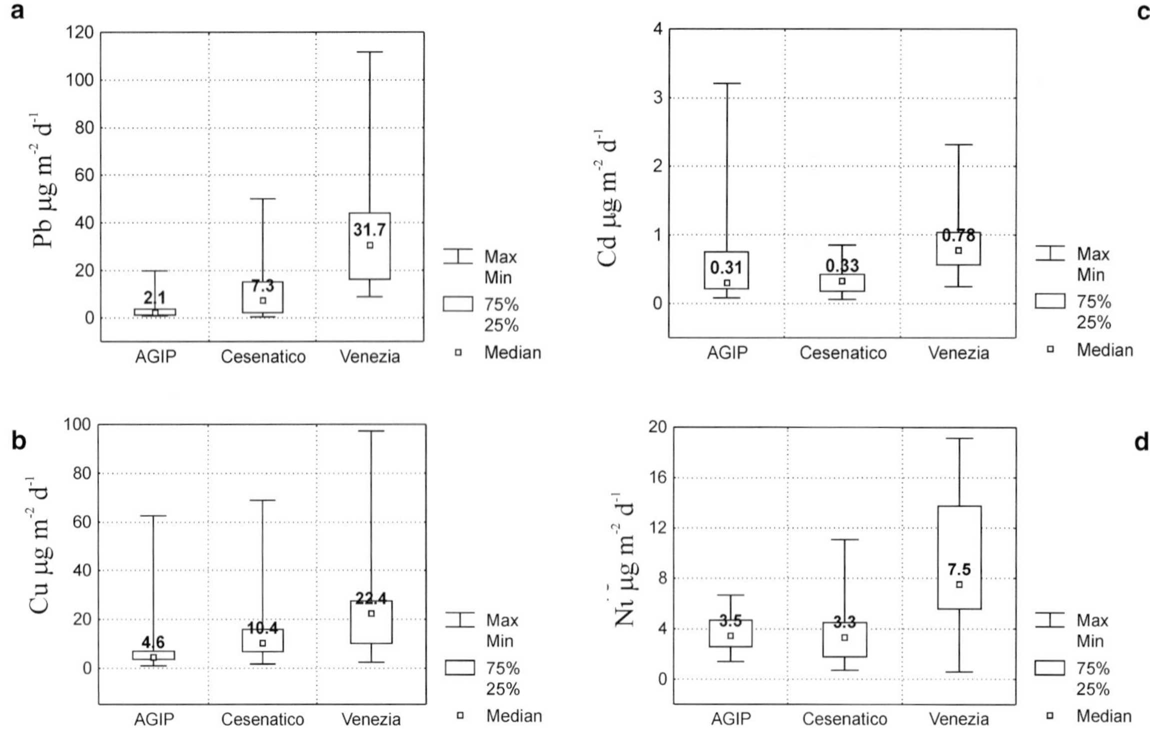

Fig. 4a-d. Descriptive statistics of metal fluxes measured at different sampling stations (period 1995-96). *Bold numbers*, median values

are shown. Relevant variations in Pb levels during the sampling period, ranging from 9 to 112 $\mu g{\bullet}m^{-2}{\bullet}d^{-1}$, with an average of 39 $\mu g{\bullet}m^{-2}{\bullet}d^{-1}$ were observed. With regard to AGIP Cd fluxes, levels varied from 0.1 to 3.2 $\mu g{\bullet}m^{-2}{\bullet}d^{-1}$, with an average of 0.6 $\mu g{\bullet}m^{-2}{\bullet}d^{-1}$. For all metals, seasonal variability could not be observed.

High variability may also be seen in the other stations. The box plots (Fig. 4) show that the distribution of data, and median values observed at AGIP, Cesenatico and Venice were, respectively, 2.1, 7.3 and 31.7 $\mu g{\bullet}m^{-2}{\bullet}d^{-1}$ for Pb (Fig. 4a), 4.6, 10.4 and 22.4 for Cu (Fig. 4b), 0.31, 0.33, and 0.78 for Cd (Fig. 4c), and 3.5, 3.3 and 7.5 for Ni (Fig. 4D). As expected, based on median values, the highest fluxes were observed at the urban site of Venice, whereas the lowest were observed at sea. In general, Cd fluxes at the AGIP site were comparable with those of Cesenatico (Fig. 4c), but fluxes at sea were more variable and reached high values, despite the origin of this element, wich is mainly anthropogenic.

A good correlation (r=0.6, P<0.001) was found between Pb and Cd in all sites (Fig. 5), with different Pb/Cd ratios. The observed Pb/Cd ratios at AGIP, Cesenatico and Venice were, respectively, 7, 26 and 39, the latter being comparable with values of emissions from anthropogenic sources in Europe (Pb/Cd = 46), as reported by Pacyna (1986). Differences in the Pb/Cd ratio between the sites were probably due to Cd enrichment caused by sea-spray at AGIP.

Table 1 lists yearly total deposition fluxes of metals measured at Venice in the 1993-97 period. Bulk sampling provides an estimate of both wet and dry depositions, as the bulk collectors are open all the time. Rainfall is reported in tables for comparison of different sampling periods. The Cd atmospheric flux increased 3-4 times from 1993 to 1995-97. As Cd is characterised by high solubility and is mainly deposited by wet flux (Chester et al. 1993; Guieu et al. 1997; Migon et al. 1997), differences between 1995-96 and 1996-97 were probably caused by rainfall variations. The observed fluctuations in Cu and Pb fluxes may also be due to changing meteorological conditions, Cu and Pb being mainly deposited by the dry mode, as also reported by Migon et al. (1997). On the contrary, Ni flux seemed to be stationary.

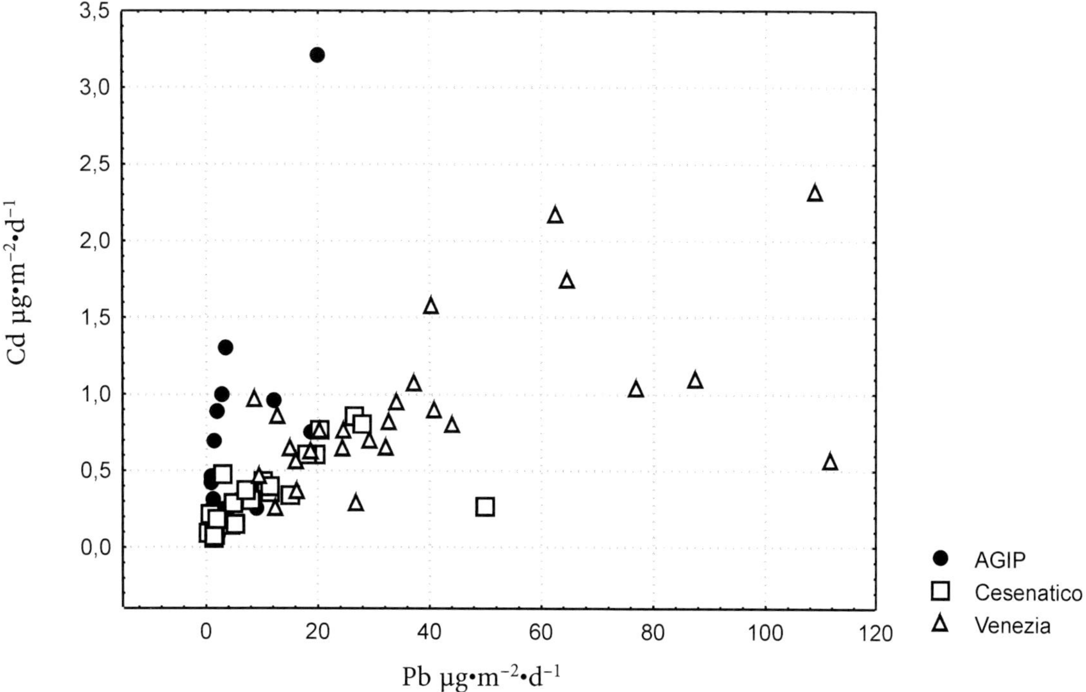

Fig. 5. Cd versus Pb bulk fluxes at AGIP (•; r=0.66, P<0.001), Cesenatico (□; r=0.59, P<0.001) and Venice (Δ; r=0.59, P<0.001) sampling stations

Atmospheric loads to the north Adriatic sea may be calculated from annual deposition fluxes, considering a surface area of 25000 km^2 (Table 1, Fig. 1), north of a W⇒E transect between Ancona (Italy) and Zadar (Croatia). As may be seen, atmospheric loads of Cd and Pb were comparable with estimates made by Migon (1993) for the Ligurian Sea. Lead emissions are diffuse because of the extensive use of motor-fuels in Europe, and atmospheric Cd mainly originates from industry (Nriagu and Pacyna 1988).

Table 2 lists metal fluxes measured at different sites. Fluxes of Cu, Ni and Pb were higher at Venice than at Cesenatico, whereas Cd deposition was comparable. In general, Pb and Cd levels fell in the flux range obtained by other authors in a coastal site in France (see references in table).

Table 1. Annual bulk fluxes at Venice. Bulk sampling provides an estimate of both wet and dry depositions. Rainfall is reported in tables for comparison of different sampling periods. Units: mg•m^{-2} year^{-1}

	Rain mm	Cd	Cu	Ni	Pb
1993-94	700	0.08 ± 0.02	12 ± 3	3.2 ± 0.8	30 ± 7.5
1995-96	950	0.33 ± 0.80	9 ± 2	3.3 ± 0.8	20± 5
1996-97	722	0.24 ± 0.06	13 ± 3	4.0 ± 1	20± 5
Load[a]		5.42	283	88	525
Ligurian Sea[b]		4.34	46	–	448

[a] Deposition area considered for northern Adriatic = 25000 km^2. Units: tonnes year^{-1}

[b] Migon (1993). Values recalculated for an area of 25000 km^2. Units: tonnes year^{-1}

Table 2. Metal fluxes expressed as mg•m^{-2}•year^{-1}. Bulk sampling provides an estimate of both wet + dry depositions. Rainfall is reported in tables for comparison of different sampling periods

	Rain mm	Cd	Cu	Ni	Pb
Venezia[a]	800	0.22 ± 0.13	11.3 ± 2.1	3.5 ±0.4	21.0 ± 8.5
Cesenatico[b]	650	0.15 ± 0.04	4.5 ± 1.1	1.4 ±0.4	3.0 ± 0.8
Cap Ferrat[c,d,e]	760	0.07 – 0.31	1.8 – 3.5	0.6 –1.4	1.8 –18.0

[a] Atmospheric deposition values for Venice obtained by averaging data collected in 1993-97

[b] For Cesenatico and AGIP in 1995-96

[c] Migon et al. (1997)

[d] Migon et al. (1991)

[e] Guieu et al. (1997)

Atmospheric versus Riverine Fluxes

In order to evaluate the role of atmospheric deposition to aquatic surfaces, we calculated the supply rates of metals to the lagoon of Venice (surface area = 546 km^2) and to the northern Adriatic Sea (Tables 3 and 4), and compared them with riverine loads. For Venice we examined data for several river tributaries of the lagoon (Silone, Dese, Osellino, Naviglio, Brenta, Loca, Taglio Novissimo, Montalbano and Trezze), and for the north Adriatic sea comparisons were made with the Po. As Table 3 shows, compared with riverine inputs of respectively 0.42, 4.0 and 5.5 tonnes $year^{-1}$, at Venice one-third of Cd (0.12 tonnes $year^{-1}$), nearly half of Ni (1.9 tonnes $year^{-1}$) and an almost equal amount of Cu (6.2 tonnes $year^{-1}$) is atmospherically derived, whereas the deposition flux of Pb is estimated to be 11.5 tonnes $year^{-1}$, i.e. 60% higher than the riverine input of 7 tonnes $year^{-1}$. For comparison with the Po, the atmospheric deposition was obtained by averaging data recorded in 1995-96 at the Cesenatico, AGIP and Venice stations. The deposition area of 25000 km^2 was chosen for budget purposes within the PRISMA project (a research programme on the Adriatic Sea from the Italian government). In this area the total amount of water load was estimated to be 20×10^9 $m^3 \cdot year^{-1}$, equivalent to 40% of the mean annual discharge of the Po River (Tartari et al. 1997). As Table 4 shows, compared with riverine inputs of respectively 7, 281 and 514 tonnes $year^{-1}$, in the northern Adriatic an almost equal amount of Cd (5 tonnes $year^{-1}$) and Cu (201 tonnes $year^{-1}$) and nearly 1/10 of Ni (63 tonnes $year^{-1}$) is atmospherically derived, whereas the deposition flux of Pb (300 tonnes $year^{-1}$) is twofold higher than riverine input (151 tonnes $year^{-1}$).

The extrapolation of these deposition data to calculate the atmospheric flux of metals must be considered with care. The high spatial and temporal variability of atmospheric transport and deposition, coupled with the short duration of oceanographic research cruises, made it difficult to obtain good estimates of atmospheric deposition to those areas of the open sea devoid of islands at which continuous sampling can be undertaken. Encouragingly, however, recent results from ship-board collections of aerosol metals over the North Sea, where a steep spatial gradient in aerosol concentration is found, are comparable with samples collected at the coast (Guerzoni et al. 1999). This finding suggests that atmospheric fluxes estimated from coastal sites may be extrapolated to adjacent marine areas. Thus, even though it would be important to have more sampling stations, the data derived from our study can usefully increase knowledge of eolian deposition in the North Adriatic.

Table 3. Supply rates of metals to the lagoon of Venice (tonnes $year^{-1}$). Atmospheric annual deposition obtained by averaging three years of data recorded in 1993-97. Deposition area = 546 km^2

	Cd	Cu	Ni	Pb
Rivers[a,b]	0.42	5.5	4.0	7.0
Atmosphere[c]	0.12	6.2	1.9	11.5
Atm/rivers	0.3	1.1	0.5	1.6

[a] Arcari et al. (1985)
[b] Bernardi et al. (1986)
[c] Atmospheric annual deposition obtained by averaging three years of data recorded in 1993-97. Deposition area = 546 km^2

Table 4. Supply rates of metals to the northern Adriatic (tonnes $year^{-1}$)

	Cd	Cu	Ni	Pb
Rivers Po[a]	7	281	514	151
Atmosphere[b]	5	201	63	300
Atm/Po	0.7	0.7	0.1	2.0

[a] Camusso et al. (1993)
[b] Atmospheric deposition is obtained by averaging data recorded in 1995-96 at Cesenatico, AGIP and Venice. Deposition area =25000 km^2

Conclusions

1. Bulk samplers may be useful to assess the importance of atmospheric loads at sea in sites where sampling is difficult to manage.
2. Total daily fluxes of metals were very variable at all stations; the highest fluxes were observed at urban sites, and the lowest at sea.
3. Atmospheric loads of Cd and Pb were comparable with estimates made by other Authors in different Mediterranean coastal sites.
4. The atmospheric deposition observed at Venice was 2- to 7-fold higher than that of Cesenatico for Ni, Cu, and Pb, and comparable for Cd.
5. The magnitude of atmospheric input indi-

cates that eolian deposition is an important contribution which cannot be neglected in the study of biogeochemical cycles of anthropic elements introduced in the northern Adriatic Sea.

Acknowledgements. This work was partially supported by the PRISMA project of MURST, Italy, and the MATER project of the European Community (contract n. MAS3-CT96-0051). This is IGM-CNR scientific contribution no. 1150. The authors thank Dr. P. Fonti of the Environmental Studies Centre, Rimini, Italy, for chemical work, Drs. C.R. Ferrari and C. Mazziotti of ARPAER – Daphne Oceanographic Structure, Cesenatico, Italy, for sampling at the Cesenatico and AGIP stations, and Ms. G. Walton for revision of the English text.

References

Arcari G, Bernardi S, Costa F et al (1985) Rapp Tec CNR 133: 1-30

Bernardi S, Cecchi R, Costa F, Ghermandi G, Vazzoler S (1986) Trasferimento di acqua dolce e di inquinanti nella laguna di Venezia. Inquinamento 1/2: 46-64

Camusso M, Martinotti W, Pettine M (1993) Trace metal distribution in the lower river Po and fluxes to the Adriatic Sea. In: Allan RJ, Nriagu JO (eds) Heavy metals in the environment, vol I. Int Conf, Toronto, Canada, Gordon & Brench Science Publishers, pp 308-311

Chester R, Murphy KJT, Lin FJ, Berry AS, Bradshaw GA, Corcoran PA (1993) Factors controlling solubilities of trace metals from non-remote aerosols deposited to the sea surface by the "dry" deposition mode. Mar Chem 42: 107-126

Duce RA, Liss PS, Merrill JT, Atlas EL, et al (1991) The atmospheric input of trace species to the world ocean. Global Biogeochem Cycles 3: 193-259

Golomb D, Ryan D, Eby N, Underhill J, Zemba S (1997) Atmospheric deposition of toxics onto Massachusetts Bay. I. Metals. Atmos Environ 31: 1349-1359

Guerzoni S, Chester R, Dulac F, Herut B, Loÿe-Pilot MD, et al (1999) The role of atmospheric deposition in the biogeochemistry of the Mediterranean Sea. Prog Oceanogr 44: 147-190

Guerzoni S, Rovatti G, Molinaroli E, Rampazzo G (1987) Total and "selective" extraction methods for trace metals in marine sediment reference samples (Mess-1, NBS 1646). Chem Ecol 3: 39-48

Guieu C, Chester R, Nimmo M, Martin JM, Guerzoni S, Nicolas E, Mateu J, Keise S (1997) Atmospheric input of dissolved and particulate metals to the north-western Mediterranean. Deep-Sea Res II 44: 655-674

Migon C (1993) Riverine and atmospheric input of heavy metals to the Ligurian Sea. Sci Total Environ 138: 289-299

Migon C, Journel B, Nicolas E (1997) Measurement of trace metal wet, dry and total atmospheric fluxes over the Ligurian Sea. Atmos Environ 31: 889-896

Migon C, Morelli J, Nicolas E, Copin-Montegut G (1991) Evaluation of total atmospheric deposition of Pb, Cd, Cu and Zn to the Ligurian Sea. Sci Total Environ 105: 135-148

Mosello R, Marchetto A, Tartari GA (1988) Bulk and wet atmospheric deposition chemistry at Pallanza (N Italy). Water Air Soil Pollut 42: 137-151

Nriagu JO (1992) Worldwide contamination of the atmosphere with toxic metals. In: Verry ES, Vermette SJ (eds) The deposition and fate of trace metals in our environment. General Tech Rep NC-150, US Dep of Agri, For Serv, North Central For Exp Stn, St Paul, MN

Nriagu JO, Pacyna JM (1988) Quantitative assessment of worldwide contamination of air, water and soils by trace metals. Nature 333: 134-139

Pacyna JM (1986) Atmospheric trace elements from natural and anthropogenic sources. In: Nriagu JO, Davidson CI (eds) Toxic metals in the atmosphere. Wiley and Sons, Pittsburgh, pp 145-171

Peters NE, Reese RS (1995) Variations of weekly atmospheric deposition for multiple collectors at a site on the shore of Lake Okeechobee, Florida. Atmos Environ 29: 179-187

Tartari G, Facchini U, Giuliacci M, Riva A (1997) Evaluation of northern Adriatic open sea rainfall. In: Lipiatou E, Mosetti R, Heussner S et al (eds) Research in Enclosed Seas Series. 3. Sci Rep First Conf Prog Oceanogr Mediterr Sea, 17-19 November, 1997, Rome, Italy, p 227. EUR 181312 EN

WMO/UNEP-MAP (1998) MED-POL Manual on sampling and analysis of aerosols and precipitation for major ions and trace elements. MAP Tech Rep Ser, Linep, Athens, No. 123

CHAPTER 18

Diversity and Vertical Migration of Euphausiids Across the Straits of Messina Area

G. Brancato, R. Minutoli, A. Granata, O. Sidoti, and L. Guglielmo

ABSTRACT

Samples used for this research were collected during the oceanographic cruises POP - EOCUMM '95 (15 to 30 July 1995, N/O Italica) and carried out by a BIONESS multinet. Samples were taken on 8 stations, in Ionian and in South Tyrrhenian Seas, across the Straits of Messina. All stations have been sampled at regular intervals of six hours (6.00 h; 12.00 h; 18.00 h; 24.00 h). The maximum sampled depth was 2030 m. A total of 5801 specimens of juvenile and adult euphausiids, belonging to 11 species (*Thysanopoda aequalis, Meganyctiphanes norvegica, Euphausia krohni, E. brevis, E. hemigibba, Nematoscelis megalops, N. atlantica, Stylocheiron suhmi, S. longicorne, S. abbreviatum* and *S. maximum*) were found and species composition of the two basins was related. *N. megalops* and *E. krohni* were the most common in the Ionian Sea-Straits of Messina area, while in South Tyrrhenian Sea, *T. aequalis, E. hemigibba* and *S. abbreviatum* were the dominant species. Different abundance values and occurrence mean depths for juveniles and adults were underlined in the two areas just for the most important species. Day/night vertical distributions of the most representative species, in both areas, were studied.

Introduction

The importance of euphausiids as an active vehicle of energy flux along the water column (Casanova 1970), as well as an important food source for many fishes of high commercial value (Froglia 1973, 1976), have led many researchers to be interested in this group of crustaceans for a very long time. A first check-list, in the Straits of Messina, was compiled by Claus (1863) and was subsequently revised by Thiele (1905). A more accurate study of the taxonomy and distribution was made later by Colosi (1916, 1922). From 1950 to nowadays the knowledge of euphausiids' ecology in these areas has been scarce and fragmentary. References on their presence and abundance, in Ionian and Tyrrhenian waters, were reported by meso- and bathypelagic stranded organisms (Guglielmo 1969; Guglielmo et al. 1973), by pelagic trawling net (Guglielmo et al. 1994), by visual observations with the mesoscaphe "F.A. Forel" (Genovese et al. 1985; Guglielmo et al. 1994), and also by studies on feeding behaviour of the most common and abundant mesopelagic fishes of the Straits of Messina, *Myctophum punctatum* and *Hygophum benoiti* (Scotto Di Carlo et al. 1982).

Today 12 out of 13 mediterranean species of euphausiids live in the Straits of Messina (*Thysanopoda aequalis, Meganyctiphanes norvegica, Euphausia krohni, E. brevis, E. hemigibba, Thysanoëssa gregaria, Nematoscelis megalops, N. atlantica, Stylocheiron suhmi, S. longicorne, S. abbreviatum* and *S. maximum*). In fact, only *Nyctiphanes couchi* has never been collected from the Straits of Messina even though this species has been reported very close by in the Ionian Sea (Ruud 1936; Casanova 1968; Wiebe and D'Abramo 1972; Guglielmo et al. 1995a). A more recent study about zooplankton and micronekton vertical distribution in South Tyrrhenian Sea (Aeolian Islands) showed that euphausiids represent about 1.7% of the whole community (Guglielmo et al. 1995c, 1996).

The main aim of this investigation is a study to improve our understanding of diversity and day/night vertical migration of euphausiids in Tyrrhenian and Ionian basins of the Straits of Messina hydrographic area.

Dipartimento di Biologia Animale ed Ecologia Marina,Università di Messina, Italy

F.M. Faranda, L. Guglielmo, G. Spezie (eds)
Mediterranean Ecosystems: Structures and Processes

Materials and methods

Study Area

The Straits of Messina (Fig. 1a) must be considered the site of intense and complex hydrodynamic phenomena caused both by the particular funnel-shaped geographical conformation and by the bottom profile which follows this conformation since there is a sharp decrease in depth northward, from about 1500 m off Capo dell'Armi to a minimum of 72 m along the Ganzirri-Punta Pezzo "sill" (Fig. 1b). After this sill the depth begins to increase again northward, but more gradually reaching 1000 m off Milazzo. This region represents the junction point between two basins, the Ionian and Tyrrhenian Seas, which have different physicochemical characteristics also due to the influences of Levantine intermediate waters and Atlantic ones, respectively. The Tyrrhenian Sea, in fact is on average colder and less salty than the Ionian Sea. However, where the two seas meet at the Straits, Ionian water is somewhat colder. This condition is due to the upwelling waters which are colder and saltier, thus the Ionian waters in the Straits are remarkably colder than the average for the entire Ionian basin. This phenomenon causes the onset of stationary currents which flow southward in the layer between the

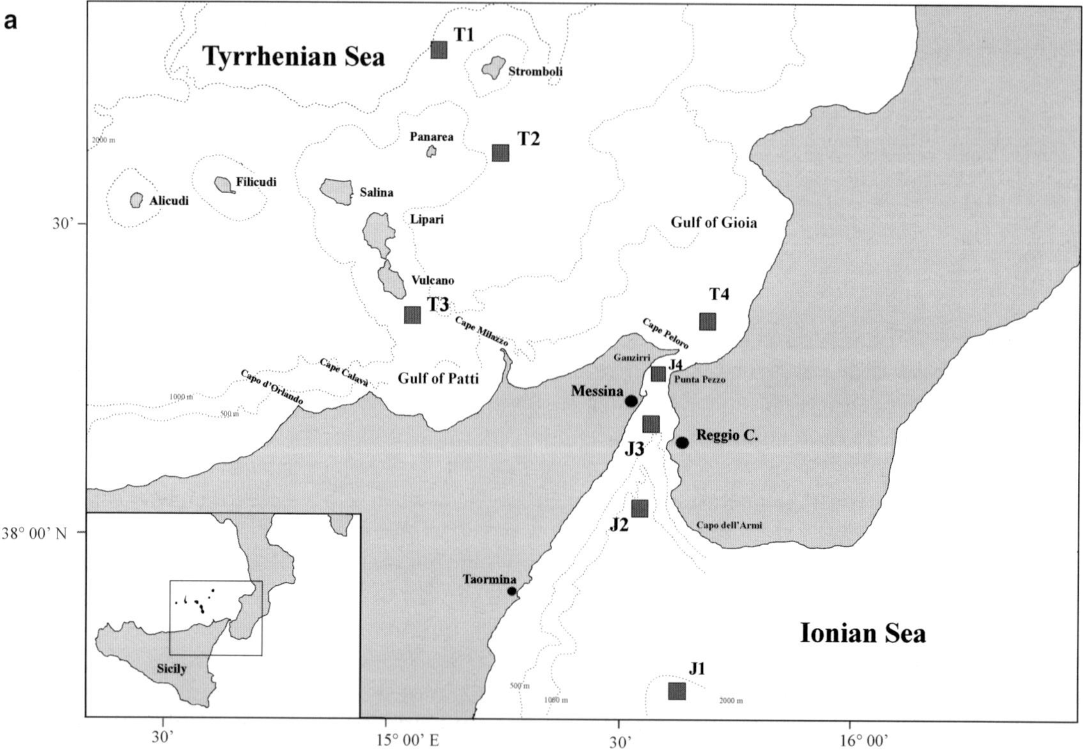

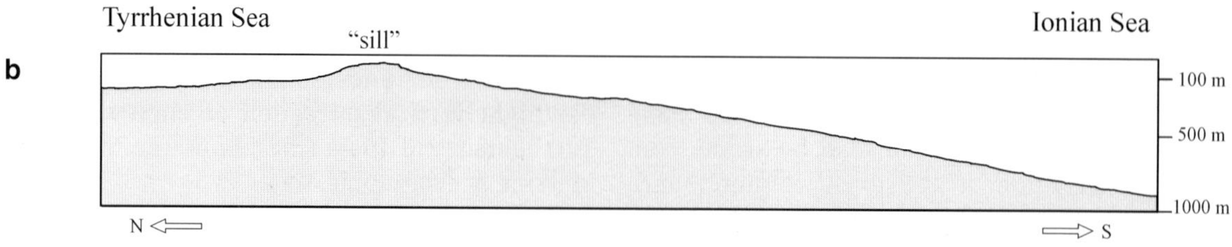

Fig. 1a,b. **a** Sampling area and station locations in Ionian and Tyrrhenian seas across the Straits of Messina; **b** Bottom profile section along the Straits of Messina

surface and the first 30 m of water, while the same underflow (30 m to the bottom) northward. However, the Tyrrhenian and Ionian Seas also exhibit oscillatory differences caused by the tidal currents, in fact, when there is low tide in the Tyrrhenian Sea at its southern border with the Straits, the Ionian Sea is in the phase of high tide at its northern border. This means that Ionian waters flow north, through the Straits, into Tyrrhenian waters forming the so-called "montante" current. While, when there is high tide in the Tyrrhenian and low tide in Ionian Sea, at their respective bounds with the Straits, the "scendente" current flows South. During the "scendente" current phase, the lighter Tyrrhenian waters flow on the top of the heavier Ionian ones until they fill completely the entire central part of the Straits, while in the "montante" current phase the Ionian waters sink into lighter Tyrrhenian ones, stratifing from 200 m down to the bottom (Magazzù and Andreoli 1971). These movements of waters occur with considerable velocity and turbulence leading to various phenomena that involve the water masses both horizontally ("cuts" and "sea steps") and vertically ("eddies" and "oil spots").

Furthermore, the movements of the stationary and tidal currents may cause, sometimes, the meeting of water masses which, on account of their respective physico-chemical characteristics, can mix with each other leading to the formation of "mixed" waters (Guglielmo et al 1995b). The study of these complex hydrodynamic processes are undoubtedly problematical, but the resolution of these could sometimes be ascribed to the use of biological indicators such as many euphausiid species (i.e. *Stylocheiron suhmi* as an indicator of Oriental waters or *Thysanoëssa gregaria* as an indicator of Atlantic ones, Casanova 1968, 1974) to know the features of a certain type of water.

Sampling Procedure

Samples used for this research were carried out during the oceanographic cruises POP – EOCUMM '95, from 15 to 30 July 1995 by N/O Italica. Samples were collected at 8 stations in Ionian and in Tyrrhenian Seas, across the Straits of Messina (Fig. 1a), by the BIONESS (Sameoto et al. 1980) electronic multinet (1 m^2 of mouth width and 12 nets of 500 μm mesh size). Each station has been sampled at regular intervals of six hours (6.00 h, 12.00 h, 18.00 h, 24.00 h). Simultaneously temperature, salinity, fluorescence and depth were measured by CTD and fluorometer sensors. BIONESS was hauled down, at reduced speed, to programmed maximum depth and then trawled up obliquely at a speed of 1.5-2 ms^{-1}. Nets were opened and closed by operator command at 50-100 m in profiles of 1000 m and 200-500 m between 1000 m down to maximum depth (2030 m) recorded profiles. Each net has filtered from 38 to 4225 m^3 of sea water.

Sample Analysis

On board samples were preserved in 5% buffered formaldehyde and sea water solution. In the laboratory euphausiids were sorted and identified to species level and divided into adults and juveniles according to Casanova (1968, 1974) and Mauchline and Fisher (1969). Abundances were combined values with respect to the total of sampled strata calculated from total specimens counted and total filtered volume at each station (or at each layer for vertical distribution patterns) as individuals per 10^{-3} m^{-3}. Similarity (*S*) and Shannon and Weaver diversity (*H'*) indices were calculated for specimens of both areas, moreover the cluster analysis by Euclidean distances method (with single linkage) for species and principal component analysis for stations were made. Finally, the weighted mean depth (WMD) of each species was determined according to Andersen and Sardou (1992).

Results

BIONESS Temperature-Salinity-Fluorescence Profiles

As shown in Fig. 2a, in Ionian Sea, temperature and salinity values of surface water masses were about 23-24°C and 37.7 PSU respectively, while, under the pycnocline (about 15-20 m) down to the bottom, they were colder and saltier (average 13.5°C and 38.5 PSU). At stations J3 and J4 there was not a clear pycnocline due to the presence of mixed waters in surface layers.

In South Tyrrhenian Sea, different, temperature and salinity values of about 27.5 °C and 37.8 PSU were found close to the surface layers

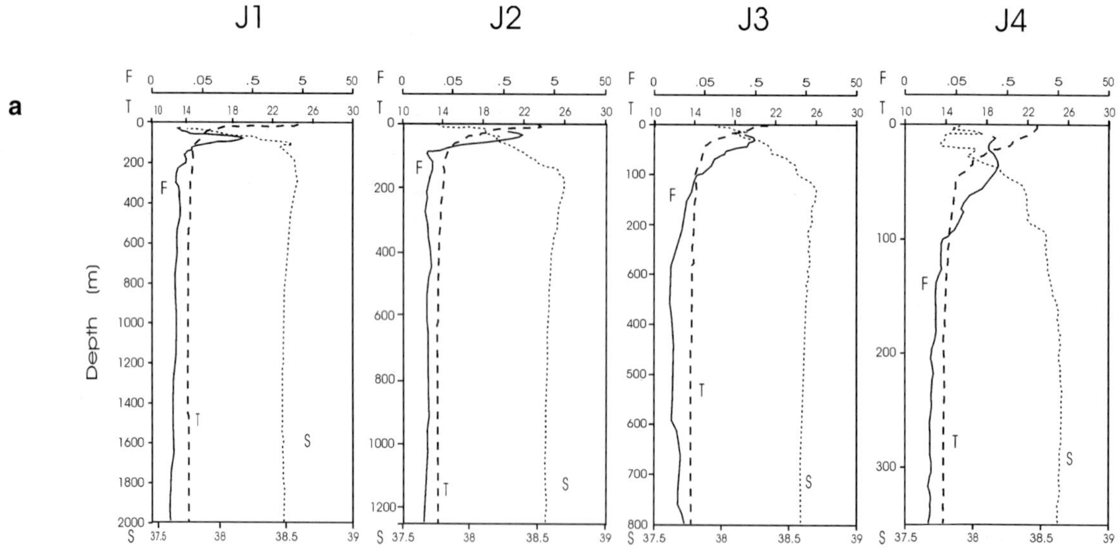

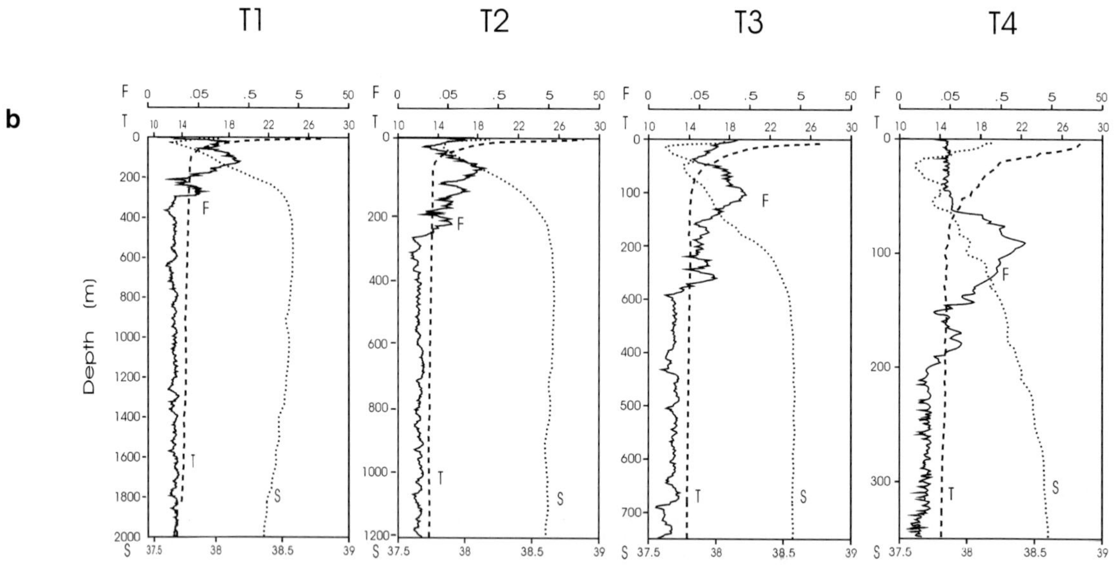

Fig. 2a,b. Temperature, salinity and fluorescence vertical profiles in Ionian Sea (**a**) and Tyrrhenian Sea (**b**)

until the pycnocline (about 10-15 m), but down at the bottom they reached values of about 13.7 °C and 38.5 PSU (Fig. 2b), very similar to Ionian water parameters, according to the described sinking waters phenomenon. There were not so many differences among physico-chemical and biological features at each station as in the Ionian Sea.

Fluorescence values showed a well defined maximum (about 1.5 mg/m^3 chl*a*) between 50 m and 80 m in Ionian Sea while in Tyrrhenian Sea it reached 0.25 mg/m^3 chl*a* between 80 and 120 m.

Species Composition and Abundance

In Table 1 the average abundance (ind./10^3 m^3), the relative percentage (F%) and the percentage of juveniles out of respective species total population (F% Juv/Tot) of all euphausiids were reported. The same number of euphausiid species (11) was found in both investigated areas and a total of 5801 specimens was sorted and divided into adults and juveniles, of which 2580 (2243 ad.+ 337 juv.) were in Ionian Sea and 3221 (1883 ad.+ 1338 juv.) were in South Tyrrhenian Sea. Higher abundances were found in southern Tyrrhenian Sea (84.69 ind./10^3 m^3) than in Ionian Sea (63.54 ind./10^3 m^3). A decreasing trend, in both areas, outward from the Straits (about 150 ind./10^3 m^3 at stations T4 and J3) to the outer stations J1 and T1 (27 ind./10^3 m^3) was found. In Fig. 3a, dendrograms for species showed three clusters of typical species of a certain basin (J = Ionian Sea and T = Tyrrhenian Sea) or both areas (B). In Fig. 3b, the plot of the first two principal factors showed that species composition recorded at Ionian stations were clearly different from those found at Tyrrhenian ones as confirmed by the percentage ratio of species composition at both basins. Particularly, some stations had an almost similar species composition such as T3 and T4 or J2 and J3 while outer ones such as T1, T2, J1 and J4 showed differente species composition.

In Ionian Sea-Straits of Messina (POP '95 Project) *Nematoscelis megalops* and *Euphausia krohni* were the most abundant species (25.1 and 23.5 ind./10^3 m^3) with a frequence percentage of 39% and 37%, respectively, while *N. atlantica* (4.26 ind./10^3 m^3), *Stylocheiron longicorne* (3.49 ind./10^3 m^3), *Thysanopoda aequalis* (2.42 ind./10^3 m^3), *S. abbreviatum* (2.21 ind./10^3 m^3) followed in decreasing order (Table). Rare individuals of *Meganycthiphanes norvegica* and *S. suhmi* were collected (0.12 ind./ 10^3 m^3 and 0.02 ind./ 10^3 m^3). Diversity index for each station was calculated: J1=1.93, J2=1.52, J3=1.13, J4=1.46 (mean = 1.51).

In the southern Tyrrhenian Sea (EOCUMM '95 Project), instead, the most important species, in order of abundance, were *T. aequalis* (22.1 ind./ 10^3 m^3; 26%), *E. hemigibba* (19.9 ind./ 10^3 m^3; 23%), *S. abbreviatum* (18.2 ind./ 10^3 m^3; 21%) and *N. atlantica* (6.6 ind./ 10^3 m^3; 8%) (Table). Just 3 specimens of *S. suhmi* were found during the whole trawling period (0.15 ind./ 10^3 m^3; 0.2%). The H' values were higher in the mean (1.84) and single values for each station were: T1=1.87, T2=1.95, T3=1.7, T4=1.82.

Vertical Distribution

Depth ranges and WMD of all euphausiid species were displayed in the Table. As regards vertical distribution, many species were caught in the

Table. Individuals per 1000 m^{-3} of filtered sea water (Ind./1000 m^{-3}), relative frequences (F%) for each species and Juveniles out of respective species total percentage (F% Juv./Tot), WMD and depth range of euphausiids in both areas

Species	Ionian Sea-Straits of Messina					South Tyrrhenian Sea					Depth range
				WMD (m)					WMD (m)		
	Ind./1000 m^3	F (%)	F% Juv/Tot	Adults	Juveniles	Ind./1000 m^3	F (%)	F% Juv/Tot	Adults	Juveniles	(m)
Thysanopoda aequalis	2.42	3.81	6	192	149	22.04	25.97	80	337	123	0-2000
Meganyctiphanes norvegica	0.12		0.19	0	50	1.37	1.61	0	176		0-700
Euphausia krohni	23.51	36.99	15	99	136	2.05	2.42	33	192	151	0-1250
E. brevis	0.86	1.35	0	147	5.33		6.28	40	94	78	0-800
E. hemigibba	1.38	2.17	27	115	423	19.98	23.54	33	187	298	0-2000
Nematoscelis megalops	25.10	39.50	2	155	1125	4.13	4.87	13	300	207	0-2000
N. atlantica	4.26	6.70	5	180	125	6.61	7.79	30	423	51	0-1500
Stylocheiron suhmi	0.02	0.03	0		0.15	0.18		0	75		0-150
S. longicorne	3.49	5.49	17	120	99	4.80	5.66	15	143	103	0-2000
S. abbreviatum	2.21	3.48	53	89	85	18.25	21.50	78	137	63	0-800
S. maximum	0.18	0.28	0	187	0.17	0.20		19	323	166	0-800

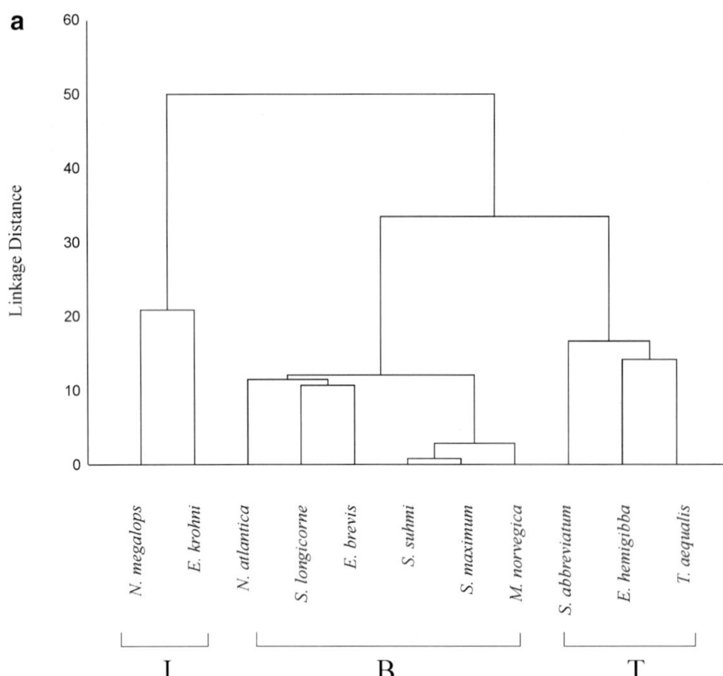

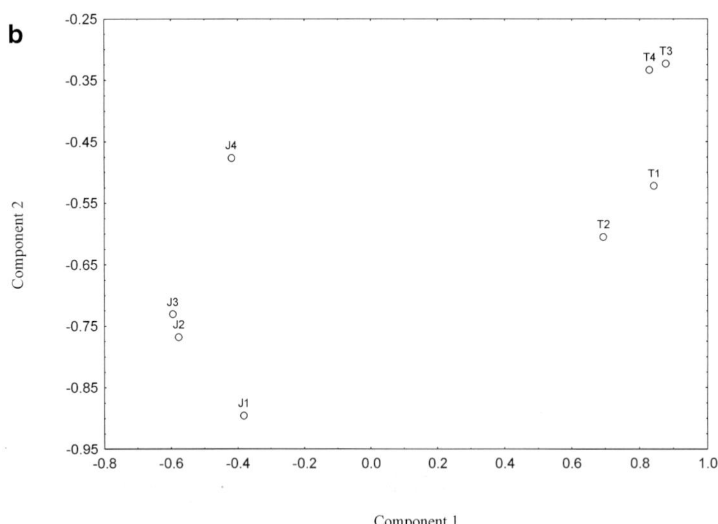

Fig. 3a,b. **a** Tree diagrams with Euclidean distances method for euphausiid species (*J*, Ionian Sea; *T*, Tyrrhenian Sea; *B*, Both areas); **b** Plot of the principal components for sampled stations

deepest strata (600-2000 m) such as *Thysanopoda aequalis, Euphausia krohni, Nematoscelis megalops, N. atlantica, Stylocheiron abbreviatum* and *S. longicorne* while greater densities were found in the upper layers often at 24.00 h, depending on their vertical day/night migratory habits, the presence of juveniles in their respective populations and the bottom depths of each station. For example, in Tyrrhenian Sea, adult euphausiids occurred in a deeper layer (205 m) than in Ionian Sea (136 m), but at station T4 they were concentrated in the upper layer (88 m), according to the hydrodynamic characteristics. The most important species of euphausiids, the adults of *Thysanopoda aequalis* concentrated at a WMD of 192 m in Ionian and 337 m in Tyrrhenian Sea, *Euphausia krohni* at 99 and 192 m, *E. brevis* at

146 and 94 m, *E. hemigibba* at 115 and 187 m, *Nematoscelis megalops* at 155 and 300 m, *N. atlantica* at 180 and 423 m, and *S. abbreviatum* at 89 and 137 m, while only juveniles of *E. hemigibba* (1750 m, at station J1) and *N. megalops* (1125 and 800 m at stations J2 and T1) were found deeper.

Day/Night Vertical Migration

Day/night vertical distribution of *E. hemigibba, N. megalops* and *E. krohni* in Ionian basin and of *T. aequalis, E. hemigibba, N. megalops* and *S. abbreviatum* in southern Tyrrhenian Sea are reported in Figs. 4 and 5. Some of these species

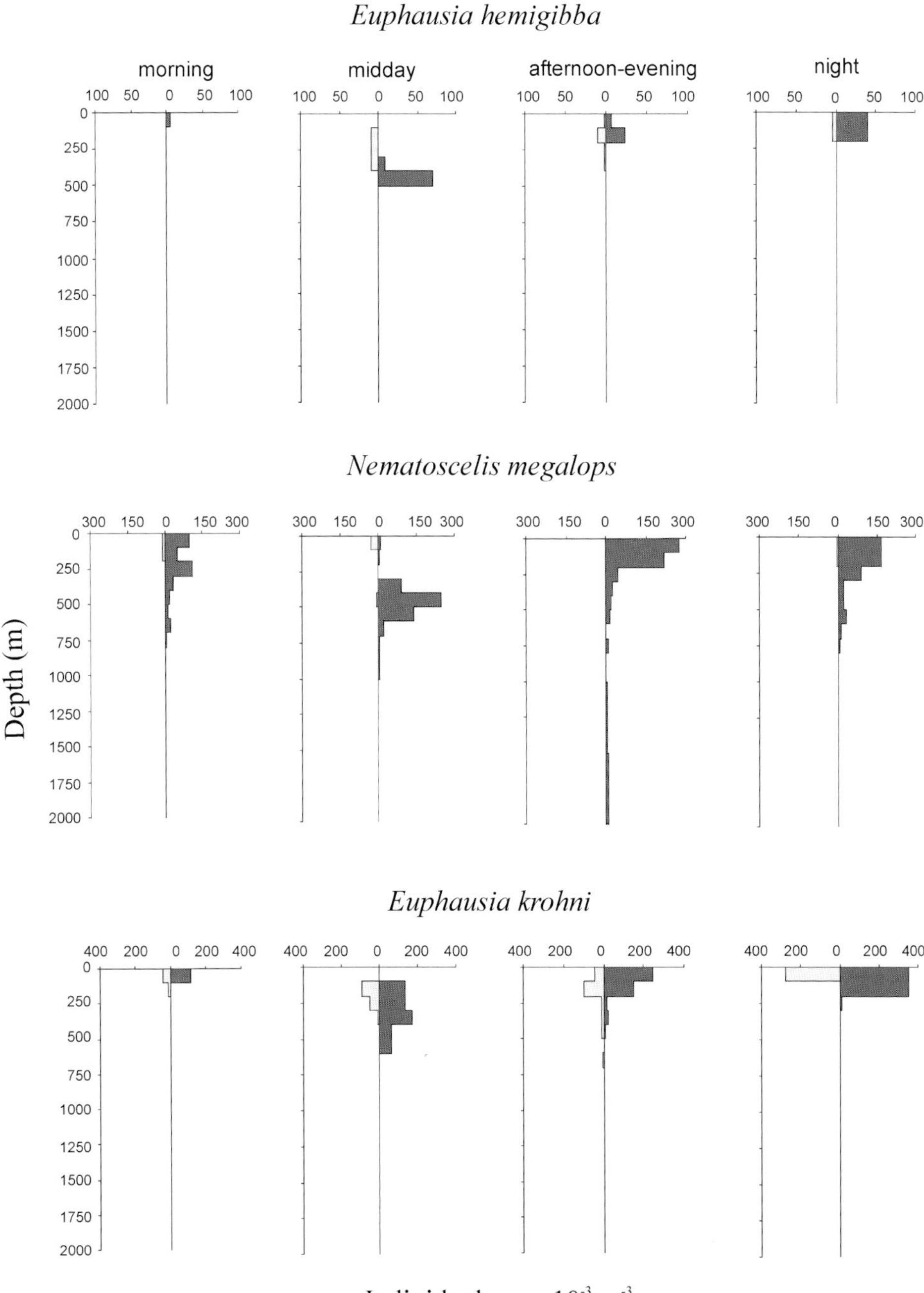

Fig. 4. Day/night vertical distribution of juveniles (*left*) and adults (*right*) of *E. hemigibba, N. megalops* and *E. krohni* in Ionian Sea

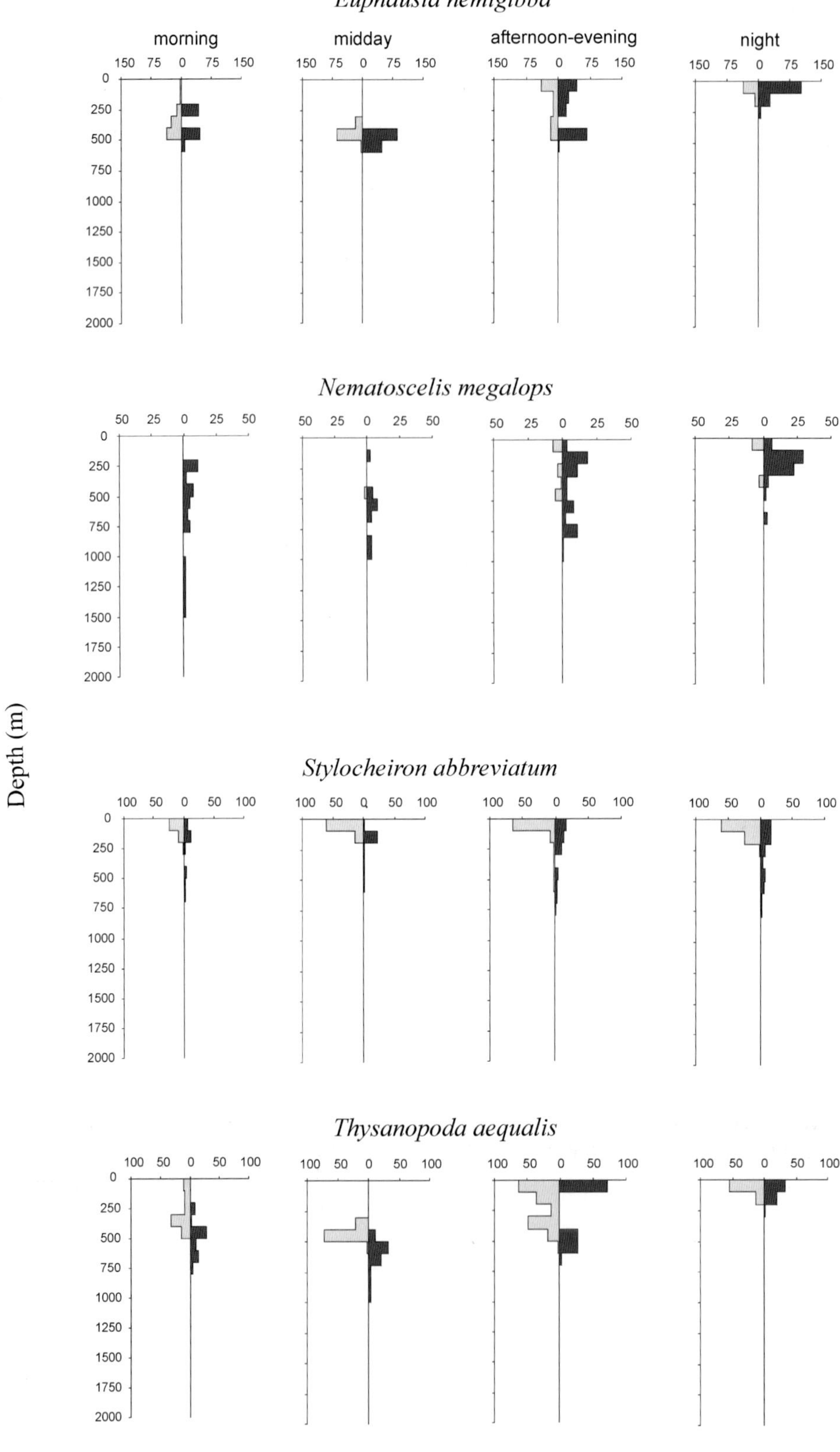

Fig. 5. Day/night vertical distribution of juveniles (*left*) and adults (*right*) of *E. hemigibba, N. megalops, S. abbreviatum* and *T. aequalis* in South Tyrrhenian Sea

can be utilized in a comparison of distribution in both areas.

During the daily cycle *Euphausia hemigibba,* in the Ionian basin, showed clear migratory behaviour. In the morning, this species was near the surface. The few caught individuals showed that the greatest part of the population started migrating down at daybreak, reaching at midday the maximum density for adults between 400 and 500 m, while the juveniles occurred up to 100 m. At 18.00 h they shifted upward reaching a peak of abundance between 100 and 200 m, while in the night time they concentrated in the upper 200 m.

Also *Nematoscelis megalops* had clear migratory behaviour. In the morning, its vertical distribution pattern showed that the maximum density was between 0 and 250 m, with the remaining population down to 1000 m. At midday, the greatest concentration of individuals was at 500 m depth and a lesser percentage of juvenile population remained near the surface down to 100 m. At 18.00 h species began assembling in surface layers between 0 and 200 m though some individuals were caught down to 2000 m. This condition remained the same also at night time. Patterns of vertical distribution of *Euphausia krohni* showed this species had higher abundance near the surface (0-200 m) during the evening and at 6.00 h, whereas at 12.00 h this species occupied a wide depth range down to 600 m.

The *E. hemigibba*, in South Tyrrhenian Sea, showed a more different day/night vertical migration than in Ionian Sea. In the morning it was found between 200 and 600 m, while its vertical distribution became sharper at midday when it assembled between 400 and 600 m. In the afternoon and in the evening, the population migrated upward, occurring in the 0-500 m strata, while at 24.00 h it showed a preference for the surface layers between 0 and 100 m.

Nematoscelis megalops didn't migrate, but had a wide homogeneous distribution, between 200 and 1000 m during daytime, while in the evening there was a slight migration upward, which was clearer at night with a maximum of density between 200 and 300 m. *Stylocheiron abbreviatum* didn't make day/night migrations but always occurred in the same bathymetric range down to 750 m, with its maximum concentration between 0 and 100 m, especially in the evening and at 24.00 h. *Thysanopoda aequalis* had a vertical profile that showed a separation of adults and juveniles. During daylight, adults were in the strata between 400 and 700 m, while juveniles occurred approximately in the upper 700 m layer to the surface; then they started to migrate, concentrating in late morning down to 500 m, adults instead occupied lower strata down to 1000 m. Population began again to reverse migration during afternoon to evening, inhabiting the surface layer down to 200 m.

Discussion

The number and species composition confirmed what was reported by Guglielmo et al. (1995a) in the Straits of Messina except that *Thysanoëssa gregaria* was considered an indicator species of Atlantic waters (Casanova 1974). Its absence in our samples could demonstrate that, during the sampling, Atlantic water influence in the study area was slight. Another reason could be ascribed to the fact that BIONESS didn't sample near the surface which is a typical habitat of this species. The number of euphausiids were lower during the day, indicating that avoidance of the sampler may be a problem (Sameoto 1980), while at J4 the lowest abundances could be due to the water turbulences of that zone which may negatively influence the sampling.

Lower diversity index value in the Ionian (1.51) than in Tyrrhenian Seas (1.84) and a similarity index of 30.5%, were probably due to different hydrological conditions which seemed to influence the faunistic composition and the dominance of certain species; this is similar to observations made by Casanova (1968) and Wiebe and D'Abramo (1972) at stations located close to the Straits of Messina area. Firstly, evident differences, showed that abundance values in an area didn't correspond with a similar area in the other basin. *Thysanopoda aequalis, Euphausia hemigibba* and *Stylocheiron abbreviatum* were more abundant in the South Tyrrhenian Sea (about 24%) than in Ionian Sea where they reached a percentage of 3%, while *Nematoscelis megalops* and *Euphausia krohni* were more abundant in the Ionian Sea than in Tyrrhenian Sea with percentages of about 38% and 3.6%, respectively. By analysing these values and the cluster graphs we can deduce that sets of species could be considered as characteristic of a single basin: *N. megalops* and *E. krohni,* in Ionian-Straits of Messina (cluster J), and *Thysanopoda aequalis, E. hemigibba, Stylocheiron abbreviatum, E. brevis,* in South Tyrrhenian Sea (cluster T). Lastly, only *N. atlantica, S. longicorne* and *S. maximum* showed a similar distribution (cluster B). A difference was noticed with regard to the observations of Casanova (1968) and Wiebe and D'Abramo (1972), about *T aequalis* and *E. hemigibba* inhabiting the Tyrrhenian Sea, which in our investigation, had an inverted abundance order, the former more abundant than the latter, as shown in the Table.

The abundance of juvenile stages in comparison with adults showed a marked difference between two investigated areas. Indeed, for almost all species, the respective percentages of juvenile individuals, out of total population, were higher in the South Tyrrhenian than in Ionian Sea such as *S. abbreviatum* (from 77 to 53%), *E. brevis* (from 40 to 0%) and *T. aequalis* (from 80 to 6%). This difference could be correlated with the different hydrological features of the two areas, the former being warmer than the latter, which allows us to suppose that the southern Tyrrhenian Sea is a nursery area.

Our results on the vertical distribution of the euphausiid species, confirmed those stated by Ruud (1936), Casanova (1968) and Andersen and Sardou (1992). At Tyrrhenian stations, except station T4 where they concentrated in the upper layers, all adult euphausiids occurred in deeper layers than in Ionian ones, according to either the different temperature and salinity values or the chlorophyll maximum, while juveniles were usually located at the surface layers. Moreover, during daylight, the highest concentration of euphausiids was found at the same depth as the main concentration of zooplankton (Guglielmo et al. 1996), suggesting that the maximum prey occurrence may be more important in determining their depth rather than some physical parameter as for *Thysanopoda aequalis* and *Euphausia* species. Indeed, during the day these animals usually fed on zooplankton at deeper layers, but during the night they fed on phytoplankton near the surface as described by Sameoto et al. (1987). *Nematoscelis megalops*, instead, concentrated in deeper layers of Tyrrhenian Sea than in Ionian Sea, maybe in relation to the presence of sinking dinoflagellates coming from Ionian Sea as suggested by Magazzù and Andreoli (1971); while *Stylocheiron abbreviatum*, a detritus feeder (Mauchline and Fisher 1969), did not migrate, occurring in surface layers down to 200 m.

Summing up, the presence of 11 out of 13 mediterranean species confirmed that the Straits of Messina hydrographic area has a very considerable richness of euphausiids. In fact the peculiarity of the principal tidal and upwelling currents determine, and maintain through time, the structure of the zooplankton and micronekton communities. Many species of euphausiids reach significant concentrations in the Straits of Messina which identify this area as a zone of accumulation that produces a subsequent "insemination" of the neighbouring Tyrrhenian and Ionian basins (Guglielmo et al. 1995a).

Euphausiids are important components of the pelagic food web, occurring in the diet of many fish species of high commercial value (Froglia 1973, 1976) and becoming a part of an energy flux which also involves mesopelagic fishes (Scotto Di Carlo et al. 1982) and cephalopod mollusks (Guglielmo et al. 1995a; Marabello et al. 1996). Moreover since euphausiids are a distinctive group of organisms that live mainly in the open deep sea, it thus confirms the pelagic nature of the Straits of Messina waters, the importance and the peculiarity of this ecosystem in the frame of Mediterranean Sea.

References

Andersen V, Sardou J (1992) The diel migrations and vertical distributions of zooplankton and micronekton in the Northwestern Mediterranean Sea. 1. Euphausiids, mysids, decapods and fishes. J Plankton Res 14(8): 1129-1154

Casanova B (1968) Les Euphausiaces de la Méditerranée. In: Commun Int Explor Sci Mer Méditerr Com Plancton, pp 23-49

Casanova B (1970) Répartition bathymétrique des euphausiacés dans le Bassin ocidentale de la Méditerranée. Rev Trav Inst Peches Marit 34: 205-21

Casanova B (1974) Les Euphausiacés de Méditerranée: systématique et développement larvaire: biogéographie et biologie. Thèse Doct Etat, Univ Provence, Marseille

Claus C (1863) Über einige Schizopoden und niedere Malakostraken Messinas. Z Wiss Zool 13: 422-454

Colosi G (1916) Contributo alla Conoscenza degli Euphausiacei dello Stretto di Messina. Monitore Zool Ital 27: 61-74

Colosi G (1922) Eufausiacei e Misidacei dello Stretto di Messina. Mem R Com Talassogr Ital, 98: 1-22

Froglia C (1973) Osservazioni sull'alimentazione del merluzzo (*Merluccius merluccius* L.) nel medio Adriatico. In: Atti V Congr Nazi Soc Ital Biol Mar, Nardò, pp 327-341

Froglia C (1976) Observations on the feeding of *Helicolenus dactylopterus* (Delaroche) (Pisces, Scorpaenidae) in the Mediterranean Sea. Rapp Comm int Mer Méditerr 23: 47-48

Genovese S, Guglielmo L, Ianora A, Scotto Di Carlo B (1985) Osservazioni biologiche con il mesoscafo "Forel" nello Stretto di Messina. Arch Oceanogr Limnol, 20: 1-30

Guglielmo L (1969) Spiaggiamenti di Eufausiacei lungo la costa messinese dello Stretto dal dicembre 1968 al dicembre 1969. Boll Pesca Piscic Idrobiol 24: 71-77

Guglielmo L, Arena G, Granata A, Sidoti O, Bonanzinga V, Soraci F (1996) Distribuzione verticale e migrazione giornaliera dello zooplancton e del micronecton nel Tirreno meridionale (Isole Eolie). In: Faranda FM, Povero P (eds) Caratterizzazione ambientale marina del sistema Eolie e dei bacini limitrofi di Cefalù e Gioiosa. EOCUMM 95, Data Rep, pp 217-246

Guglielmo L, Costanzo G, Berdar A (1973) Ulteriore contributo alla conoscenza dei crostacei spiaggiati lungo il litorale messinese dello Stretto. Atti Soc Peloritana 19:129-156

Guglielmo L, Crescenti N, Vanucci S (1994) Lo Stretto di Messina visto dal mesoscafo "F. A. Forel". Boll Soc Adriat Sci 75(1): 209-227

Guglielmo L, Crescenti N, Costanzo G, Zagami G (1995a) Zooplankton and micronekton communities in the Straits of Messina. In: Guglielmo L, Manganaro A, De Domenico E (eds) The Straits of Messina ecosystem. Proc Symp Messina 4-6 April 1991, Messina pp 247-270

Guglielmo L, Marabello F, Vanucci S (1995b) The role of the mesopelagic fishes in the pelagic food web of the Straits of Messina. In: Guglielmo L, Manganaro A, De Domenico E (eds) The Straits of Messina Ecosystem. Proceeding of the symposium held in Messina 4-6 April 1991, Messina pp 223-246

Guglielmo L, Zagami G, Sidoti O, Granata A (1995c) Distribuzione e migrazione giornaliera dello zooplancton nel Tirreno meridionale (Isole Eolie). In: Faranda FM (ed) Caratterizzazione ambientale marina del sistema Eolie e dei bacini limitrofi di Cefalù e Gioiosa. EOCUMM 94, Data Rep, Genova pp 167-190

Magazzù G, Andreoli C (1971) Trasferimenti fitoplantonici attraverso lo Stretto di Messina in relazione alle condizioni idrologiche. Boll Pesca Piscic Idrobiol 26: 125-193

Marabello F, Guglielmo L, Granata A, Sidoti O (1996) Studi preliminari sulle abitudini alimentari di *Todarodes sagittatus* (Cephalopoda) nel Tirreno meridionale. In: Albertelli G, Demaio A, Piccazzo M (eds) Atti XI Congr AIOL (Sorrento, 26-28 Ottobre 1994) Genova-AIOL, pp 271-278

Mauchline J, Fisher LR (eds)(1969) The biology of euphausiids. Academic Press, New York

Ruud JT (1936) Euphausiacea. Rep Dan Oceanogr Exped 1908-10, 11(2): 1-86

Sameoto DD (1980) Relationships between stomach contents and vertical migration in *Meganyctiphanes norvegica*, *Thysanoëssa raschii* and *T. inermis*, Crustaceans (Euphausiacea). J Plankton Res 2: 129-143

Sameoto DD, Guglielmo L, Lewis MK (1987) Day/night vertical distribution of euphausiids in the eastern tropical Pacific. Mar Biol 96: 235-245

Sameoto DD, Saroszynsky LO, Fraser WB (1980) BIONESS a new design in multiple net zooplankton sampler. J Fish Res Board Can 37: 722-724

Scotto Di Carlo B, Costanzo G, Fresi E, Guglielmo L, Ianora A (1982) Feeding ecology and stranding mechanisms in two lanternfishes, *Hygophum benoiti* and *Myctophum punctatum*. Mar Ecol Prog Ser, 9: 13-24

Thiele J (1905) Über einige stielaugige Krebse von Messina. Zool Jahrb Neapel, 8: 443-474

Wiebe HP, D'Abramo L (1972) Distribution of euphausiid assemblages in the Mediterranean Sea. Mar Biol 15: 139-149

Relationship between Fish Larval Biomass and Plankton Production in the South Tyrrhenian Sea

R. Bruno, A. Granata, A. Cefali, L. Guglielmo, G. Brancato, and P. Barbera

ABSTRACT

Fish larval biomass was compared with plankton abundance and production in the coastal south Tyrrhenian Sea from November 1994 to October 1995. *Sardina pilchardus* was the most representative species in biomass, followed by *Gymnammodytes cicerellus*. *S. pilchardus* showed a larger spawning period (January to May) together with *Engraulis encrasicolus* (July to November). Ichthyoplankton standing stocks peaked in February (6.28 gm^{-2}), preceeding both the first primary production peak and the spring maximum zooplankton biomass in May. The results demonstrate that no temporal overlapping occurred among fish larval species in the coastal waters of the southern Tyrrhenian Sea.

Introduction

Our understanding of the structure and dynamics of coastal marine food webs has progressed considerably in recent years. Cushing (1967, 1969) assumed that the spawning period of the pelagic fishes was fixed and did not change with environmental conditions. On the contrary, we now know that the timing of spawning and successfull larval fish recruitment are strongly related to plankton production and zooplankton species composition (Sameoto 1982; Fortier 1995; Mousseau et al. 1998). This has also been demonstrated for both the sardine and anchovy from the coastal Adriatic Sea (Vucetic 1975; Casavola et al. 1998). We therefore undertook a similar study to better understand the relationship between fish larval biomass, phytoplankton abundance and production, and zooplankton cycles and species composition in the coastal waters of the south Tyrrhenian Sea.

Material and Methods

The investigated area extends from S. Agata di Militello to Capo D'Orlando (Tyrrhenian Sea), about two miles from the coast along the 5-10 m isobaths (Fig. 1). This zone was identified as a probable nursery area for many pelagic and coastal fishes (Cefali et al. 1997). Ichthyoplankton was captured by a trawl net with a cod end of 1 mm mesh size from November 1994 to October 1995. Fish larvae, from a significative fraction of each sample (about 10%) were sorted, weighed, identified to species level and measured to nearest mm (TL). A total of 27 species were classified (Table 1). Detailed information on sampling strategy and methods are reported in Cefali et al. (1997). The present results on fish larval biomass were compared with data on production and phytoplankton dominance (Magazzù and Andreoli 1972, 1973; Magazzù et al. 1975), microzooplankton abundance (Sparla and Guglielmo 1994), zooplankton biomass and species composition in the coastal south Tyrrhenian Sea (Scotto di Carlo 1985; Zagami et al. 1996; L. Guglielmo, unpublished data).

Results

The results show that during the investigated period *Sardina pilchardus* was the most representative species (60%) followed by *Gymnammodytes cicirellus* (25%). Other speciessuch as *Aphia minuta* (6%), *Engraulis encrasicholus* (4%), *Sardinella aurita* (2%), *Pomatoschistus marmora-*

Dipartimento di Biologia Animale ed Ecologia Marina, Università di Messina, Italy

F.M. Faranda, L. Guglielmo, G. Spezie (eds)
Mediterranean Ecosystems: Structures and Processes

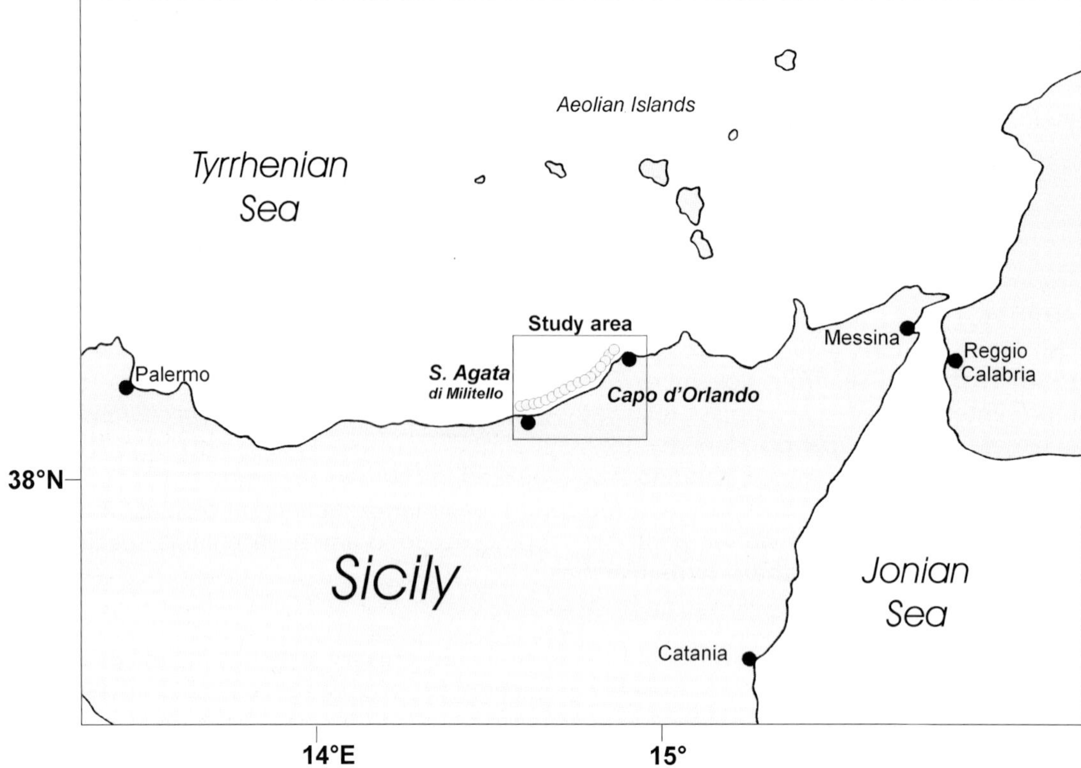

Fig. 1. Sampling area of fish larvae

Table 1. List of identified fish larvae species

Aphia minuta (Risso 1810)
w imberbis (Linnaeus 1758)
Boops boops (Linnaeus 1758)
Boops salpa (Risso 1826)
Bothus podas (Delaroche 1809)
Callionymus phaeton (Gunther 1861)
Caranx fusus (Geoffroy Saint Hilaire 1809)
Caranx rhonchus (Geoffroy Saint-Hilaire 1817)
Deltentosteus quadrimaculatus (Valenciennes 1837)
Diplodus annularis (Linnaeus 1758)
Engraulis encrasicolus (Cuvier 1817)
Gobius sp.
Gymnammodytes cicerellus (Rafinesque 1810)
Lithognathus mormyrus (Linnaeus 1758
Maena maena (Linnaeus 1758)
Mullus barbatus (Linnaeus 1758)
Mullus surmuletus (Linnaeus 1758)
Pagellus acarne (Risso 1826)
Pagellus centrodontus (Delaroche 1809)
Pagellus erythrinus (Linnaeus 1758)
Pomatoschistus marmoratus (Risso 1810)
Sardina pilchardus (Walbaum 1792)
Sardinella aurita (Valenciennes 1847)
Sphyraena sphyraena (Linnaeus 1758)
Spicara smaris (Linnaeus 1758)
Spondyliosoma cantharus (Linnaeus 1758)
Trachurus trachurus (Linnaeus 1758)
Trigla lucerna (Linnaeus 1758)

tus (1%) and *Lithognathus mormyrus* (1%) constituted the remaining ichthyoplankton biomass (Fig. 2).

Table 2 and Figure 3 report monthly values for species composition, and temporal changes in biomass distribution of captured fish larval species in the ichthyoplankton community. In January, *S. aurita* was the only abundant species in terms of biomass, while in February *S. pilchardus* was more abundant (87%); this latter species maintained higher percentage values also in April, May and June. In March, the dominant species was *G. cicerellus* (93%). The month of July was characterized by the co-occurrence of many fish larval species of which *S. pilchardus* (34%), *E. encrasicholus* (29%) and *P. marmoratus* (28%) were the most important species.

Sardina pilchardus showed a longer spawning period (January to May) together with *E. encrasicholus* (July to November). In August, *L. mormyrus* was the only representative species in terms of total numbers (77%) and biomass (65%) while *E. encrasicholus* dominated from October to December. It is evident that no overlapping occurred among the different species

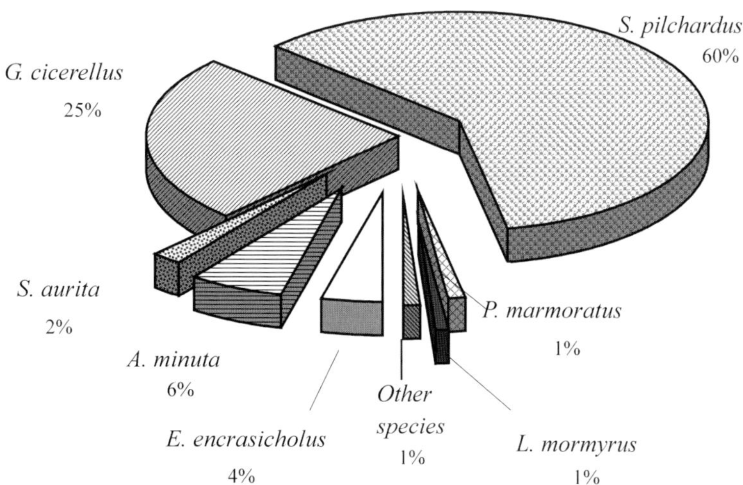

Fig. 2. Relative frequency percentage of fish larvae community during the investigated period

Table 2. Monthly species composition percentage of the captured fish larvae

Species	January	February	March	April	May	June	July	August	October	November	December
E. encrasicholus	0.47	-	-	-	-	-	28.63	7.67	69.25	98.20	77.07
A. minuta	6.88	11.57	0.03	-	1.22	-	3.31	-	-	0.30	6.66
S. aurita	81.49	-	-	-	-	-	0.38	-	-	0.05	13.31
G. cicerellus	0.13	1.14	92.85	2.50	-	-	-	-	-	-	-
S. pilchardus	11.00	87.27	7.09	95.60	65.26	97.00	34.27	-	-	-	-
P. marmoratus	-	-	-	-	-	-	27.65	15.05	10.02	-	-
L. mormyrus	-	-	-	-	-	-	0.04	77.28	19.99	-	-
Other species	0.03	0.02	0.03	1.90	33.52	3.00	5.72	-	0.74	1.45	2.96

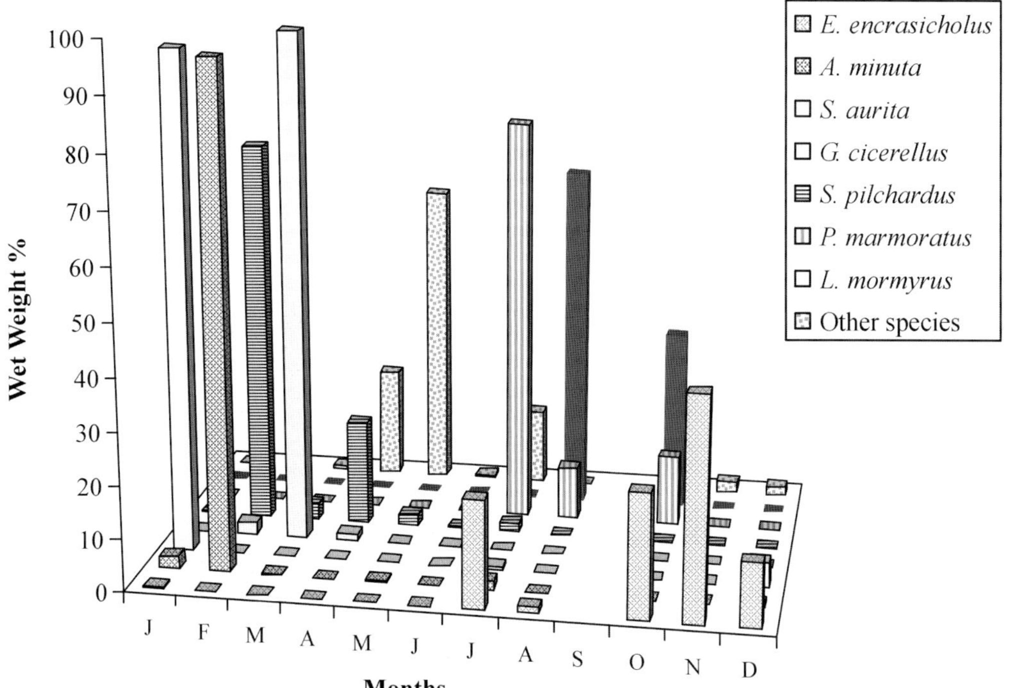

Fig. 3. Temporal biomass distribution of the dominant fish larvae species in the investigated area

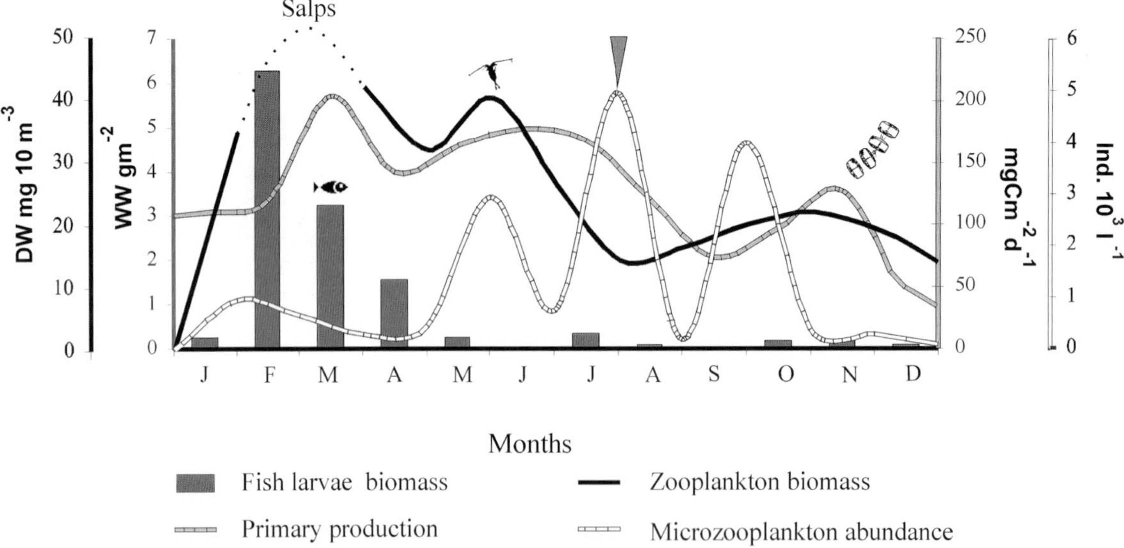

Fig. 4. Temporal fluctuation of primary production, micro-zooplankton abundance and zooplankton biomass in the upper 50 m layer of the coastal Tyrrhenian Sea compared with fish larvae biomass in the investigated area

which seemed to show a regular alternation of spawning periods.

Fish larval biomass was compared with temporal fluctuations in primary production, microzooplankton abundance and zooplankton biomass in the coastal southern Tyrrhenian Sea (Fig. 4). Ichthyoplankton standing stocks peaked in February (6.28 gm^{-2}), preceeding both the first primary production peak and the spring maximum zooplankton biomass in May. Corresponding values for microzooplankton densities were very low. From March to June fish larval biomass values decreased compared to primary production and zooplankton biomass which had much higher values.

A second peak of fish larval biomass was evident in July when phytoplankton production was low and zooplakton biomass started to decrease. On the other hand, microzooplankton increased in the same period. The low peak of fish larval biomass in November coincides with a second primary production and zooplankton biomass peaks. This was followed by higher values of microzooplankton densities in September.

Discussion

Our preliminary data confirm the results of previous authors (Casavola et al. 1998) that there is a significant relationship between the temporal distribution of ichthyoplankton biomass and primary production, phytoplankton and zooplankton seasonal cycles and species composition. The spawning period of some fish larval species appears to precede plankton blooms. Gamulin (1954) found that spawning of the sardine in the coastal Adritic Sea was related to both zooplankton biomass and composition. In fact, these larvae attained maximal biomass values in the coastal zooplankton community when primary production and phytoplankton standing crop reached maximum abundances and preceded the zooplankton maximum in May. The timing and biomass of the spring phytoplankton bloom are of fundamental importance for the survival of copepods and fish larvae. Sameoto (1982) stated that the total number of fish larvae (anchovy and sardine) m^{-2} showed a linear increase with the biomass of zooplankton. A significant correlation was found between the numbers of Peruvian anchovy *E. ringens* larvae m^{-2} and maximum of chlorophyll *a* values.

Our results agree with Vucetic (1971) that sardine and anchovy have a regular alternation of spawning seasons. The spawning period of the sardine occurred from November to April and the maximum of eggs occurred in the Adriatic Sea in January and March at different locations (Vucetic 1975). This coincides with the large numbers of sardine larvae caught in February, April, May and June in coastal waters of the southern Tyrrhenian Sea. Compared with plankton fluctuations, the peaks from February to May parallel phyto- and zooplankton standing stocks. Many phytoplankton species participate in the spring bloom and *Sardina pilchardus* and *Gymnammodytes cicerellus* in March may feed on a wide size range of particles, comprising eggs, nauplii and larval stages.

Dinoflagellates are the most important group in the size range 20-50 mm (*Ceratium furca, Prorocentrum micans, Goniaulax polygramma*) while Diatoms and Cyanophycae were mainly represented by *Rhizosolenia alata* and *Nostoc planctonicum*, respectively. In June, when phytoplankton standing stock decrease, young sardine was the only species occurring in the ichthyoplankton and this species seems to feed on the micro- and zooplankton communites. Anchovies will feed successfully on dinoflagellates (Blaxter and Hunter 1982), and indeed these motile phytoplankters seem to be the best food for early larvae of *Engraulis ringens* (Walsh et al. 1980).

Tintinnids and copepods are very important components of spring-summer zooplankton. The copepods *Acartia margalefi* and *A. clausi, Clausocalanus* and *Paracalanus* copepodites (Badalamenti et al. 1990; Zagami et al. 1996) and the tintinnids *Eutintinnus tubulosus, Metacylis annulifera, Helicostemella subulata* (Fonda Umani and Monti 1993, Sparla and Guglielmo 1994) are typical species of the coastal south Tyrrhenian community.

When the spawning of sardine is completed, anchovy eggs occur in large numbers (Vucetic, 1975). *Engraulis encrasicholus* larvae appeared in July and reached higher densities from October to December. In July, many fish larval species contribute to the ichthyoplankton community but in August *Lithognathus mormyrus* is the most abundant species. In summer and autumn, the diatoms *Nitzchia seriata* and *Rhizosolenia alata,* the dinoflagellates *Prorocentrum micans, Exuviaella compressa* and *E. apora,* and the ciliate *Strombidium* may constitute the primary food for anchovy larvae together with nauplii, copepodites and adults of the autumn zooplankton community (the copepods *Centropages kroyeri* and *Temora stylifera*). The abundance of *S. aurita* in January may be correlated with the diatoms *Thalassionema nitzschoides, Rhizosolenia alata* and small copepods such as *Clausocalanus furcatus*. The newly hatched larvae are sustained by remaining yolk, but while this is still being used they begin feeding on diatoms. At the end of yolk sac phase, there is a fairly abrupt transition in the zooplankton, with larvaceans such as *Oikopleura* being particularly important (Shelbourne 1953, 1957). Two short trophic pathways from primary producers to fish larvae were indentified in the Scotian Shelf, NW Atlantic (Mousseau et al. 1998): the herbivorous food chain (large phytoplankton ⇒ calanoid copepods ⇒ fish larvae) and the large-microphage shunt of the microbial food web (small phytoplankton ⇒ appendicularians/pteropods ⇒ fish larvae).

These preliminary results demonstrate that no temporal overlapping occurred among fish larval species in the coastal waters of the southern Tyrrhenian Sea. Differences in the timing of spawning and different larval development imply that each species occupies each a characteristic trophic niche. The strong relationship between fish larval abundance and phyto-micro- and zooplankton cycles and species composition depends on many factors including water temperature, egg size and amount of larval yolk. Its is evident that r-strategist filter feeders species, as sardine and anchovy, have a longer spawning season, different from K-strategist species such as *S. aurita* and *L. mormyrus,* that occur over a much shorter time interval.

References

Badalamenti F, Zagami G, Manganaro A, Guglielmo L (1990) Variazioni giornaliere della comunità zooplanctonica litorale nei pressi di Capo Peloro (Me): relazioni con la fauna ittica ed il regime idrodinamico. In: 53° Congr UZI Palermo 1-5 Ottobre 1990, pp 79-80

Blaxter JHS, Hunter JR (1982) The biology of the clupeoid fishes. Adv Mar Biol 20: 1-223

Casavola N, Hajderi E, Marano G (1998) Relationship between the density of *Engraulis encrasicholus* eggs and zooplanktonic biomass in the southern Adriatic sea. Biol Mar Mediterr 5 (1): 56-62

Cefali A, Potoschi A, Bruno R, Cavallaro G, Manganaro A, Costa F (1997) The qualitative-quantitative analysis of the first juvenile stages of fish in Sicilian Tirrenic coast, and observations about the reproductive periods of some species. Biol Mar Mediterr 4 (1): 211-216

Cushing D (1967) The grouping of herring population. J Mar Biol Ass UK 47 (1): 193-208

Cushing D (1969) The regulatory of the spawning season of some fishes. J Cons Perm Int Expl Mer 33 (1): 81-82

Fonda Umani S, Monti M (1993) Distribuzione dei popolamenti micronectonici nell'arcipelago toscano. In: Università di

Firenze (ed) Progetto mare: ricerca sullo stato biologico, chimico e fisico dell'Alto Tirreno Toscano. Univ Firenze, Regione Toscana, pp 157-258

Fortier L (1995) Export production and the production of fish larvae and their prey at hydrodynamic singularities. In: Guglielmo L Manganaro A, De Domenico E (eds) The Straits of Messina ecosystem. Proc Symp Messina, 4-6 April 1991, pp 213-222

Gamulin T (1954) Mrijescenje j mrijestiliste srdele (*Sardinia pilchardus Walb*) u Jadranu u 1947-1950, Izvjesca, Reports Rib, Biol Eksp "HVAR" 1948-1949, 4: 1-65

Magazzù G, Andreoli C (1972) Contributo alla Conoscenza del fitoplancton e della produzione primaria delle acque costiere siciliane (Canale di Sicilia e Tirreno Occidentale). Mem Biol Mar Ocean 2 (1): 1-30

Magazzù G, Andreoli C (1973) Ciclo annuale della produzione primaria e del fitoplancton in una zona di avamporto (Milazzo). In: Atti 5° Coll Int Oceanogr Mediterr Messina, pp 379-398

Magazzù G, Andreoli C, Munaò F (1975) Ciclo annuale del fitoplancton e della produzione primaria del Basso Tirreno (1969-1970). Mem Biol Mar Ocean V (2): 25-48

Mousseau L, Fortier L, Legendre L (1998) Annual production of fish larvae and their prey in relation to size-fractionated primary production (Scotian Shelf, NW Atlantic). ICES J Mar Sci 55: 44-57

Sameoto D (1982) Vertical distribution and abundance of the Peruvian anchovy, *Engraulis ringens*, and sardine, *Sardinops sagax*, larvae during November 1977. J Fish Biol 21: 171-185

Scotto Di Carlo B (1985) Appunti sullo zooplancton del Mediterraneo. Nova Thalassia 7 Suppl 3: 83-97

Shelbourne JE (1953) The feeding habits of plaice post-larvae in the southern bight. J Mar Biol Assoc UK 32: 149-159

Shelbourne JE (1957) The feeding and condition of plaice larvae in good and bad plankton patches. J Mar Biol Assoc UK 36: 539-552

Sparla MP, Guglielmo L (1994) Distribuzione del microzooplancton nello Stretto di Messina (estate 1990). In: Atti X Congr AIOL, Alassio 4-6 Novembre 1992, pp 307-325

Vucetic T (1971) Long term zooplankton standing crop fluctuation in the Central Adriatic coastal region. VI Simp EMBS 1970, Thalassia Jugosl Thjuap 7 (1): 419-428

Vucetic T (1975) Synchronism of the spawning season of some pelagic fishes (sardine, anchovy) and the timing of the maximal food (zooplankton) production in the central Adriatic. Pubbl Staz Zool Napoli 39 Suppl: 347-365

Walsh JJ, Whitledge TE, Esaias WE, Smith RL, Huntsman SA, Santander H, DeMendolia BR (1980) The spawning habitat of the Peruvian anchovy, *Engraulis ringens*. Deep-Sea Res 27: 1-27

Zagami G, Badalamenti F, Guglielmo L, Managanaro A (1996) Short-term variations of the zooplankton community near the Straits of Messina (North-eastern Sicily): relationships with the hydrodynamic regime. Estuarine Coast Shelf Sci 42: 667-681

CHAPTER 20

First Data on the Mysid Community in the "Stagnone di Marsala" (Western Sicily)

M. Campolmi[1], A. Vaccaro[2], and A. Mazzola[2]

ABSTRACT

First data on the spatial and temporal distribution of the mysid community in the "Stagnone di Marsala" (western Sicily) from January to December 1996 are reported. The assemblage consists mainly of five species which are common in the Mediterranean Sea: *Diamysis bahirensis* (55.2%), *Siriella armata* (18.6%), *S. clausii* (15.8%), *Mysidopsis gibbosa* (9.8%) and *Mesopodopsis slabberi* (0.6%). Population structure is described for each species and the reproductive period inferred. A sharp increase in abundance occurred in the spring, while the minima occurred during autumn-winter. The *D. bahirensis* dominated from March to October and *S. clausii* in the winter. A large difference, which was mainly quantitative, was recorded between the population from the inner area and that from the area more influenced by the sea. Clear seasonal patterns and spatial segregation in the inner part of the sampling area were observed.

Introduction

Mysids are usually demersal organisms, living near the bottom during the day and rising up into the water column at night, following a typical daily migratory pattern. Their spatial and temporal distribution in a given habitat appears to be affected by a number of abiotic (mainly temperature, light and salinity) and biotic (e.g. predation, food availability) factors (Clutter 1967; Mauchline 1980; Williams and Collins 1984; Wittmann 1984; Wooldridge 1989; Webb and Wooldridge 1990; Mees et al. 1993). Mysid distribution is also influenced by their typical aggregative behaviour (Clutter 1969; Wittmann 1977; Mauchline 1980; Omori and Hamner 1982; Carleton and Hamner 1989; Mees et al. 1993).

As reported by Mauchline (1980), mysids can contribute significantly to animal biomass in lagoons, where they use organic detritus as a food source. The mysid community can play an important role in remineralizing detritus in shallow environments (Carleton and Hamner 1989). These animals may act as a major link between pelagic and benthic food webs since they undergo vertical migrations and are preyed on by fish (Ariani and Spagnuolo 1975; Mauchline 1980; Turpen et al. 1994).

Many studies have been carried out on mysid communities along the Italian coasts, mainly on their taxonomy (Colosi 1929; Genovese 1956a, b, 1963; Ariani 1966, 1967; Ariani and Spagnuolo 1975; Wittmann 1977).

In the present paper we provide first data on species composition, population structure and distribution of the mysid assemblage in a shallow coastal sound (Stagnone di Marsala, western Sicily).

This study was carried out in the framework of a research programme which focused on zooplankton. Although the sampling method was not specifically designed to collect mysids (only a plankton net was used), the data presented can provide first quali-quantitative information on the mysid assemblage occurring in this environment, where they constitute a food source for many fish species such as *Mullus surmuletus* (La Rosa et al. 1997), *Atherina boyeri* (Scilipoti 1998), *Syngnathus abaster* and especially *S. typhle* (Campolmi et al. 1996; Scilipoti 1998), and where they may play an important role as detritivores (Vaccaro 1998).

[1] Dipartimento di Biologia Animale ed Ecologia Marina, Università di Messina, Salita Sperone 31, 98166 Messina, Italy
[2] Dipartimento di Biologia Animale, Università di Palermo, Via Archirafi 18, 90123 Palermo, Italy

F.M. Faranda, L. Guglielmo, G. Spezie (eds)
Mediterranean Ecosystems: Structures and Processes

Materials and Methods

The "Stagnone di Marsala" (37°52'N 12°28'E) is a shallow coastal sound on the western coast of Sicily (Fig. 1). It is characterized by low depth (0.5-2.5 m) and large fluctuations in physical-chemical parameters, especially temperature and salinity. The bottom is mainly sandy-muddy and is covered by *Cymodocea nodosa* (Ucria) Ascherson, *Caulerpa prolifera* (ForsskÂl) Lamouroux *and Posidonia oceanica* (Linneus) Delile (restricted to the central and southern basins).

Samples were collected monthly from January to December 1996 at nine sampling stations (Fig. 1): stations 1-6 (0.5-1 m depth) are located in the inner shallow area; stations 7, 8 and 9 (1.5-2 m depth) are located near the southern sea mouth in an area which is under greater influence from the sea.

Samples were collected during the day, using a plankton net (mouth diameter 40 cm; mesh size 125 (m) equipped with a flowmeter (Hydrobios) and towed horizontally at the surface in the layer between the surface and 1 m depth.

The specimens were fixed in 4% neutralized formalin/sea-water immediately after capture. Mysids were identified to species level and divided into six categories sensu Mauchline (1980): mature and immature males; ovigerous, empty and immature females; juveniles.

Physical-chemical parameters (temperature and salinity) were recorded by means of a multiparameter probe.

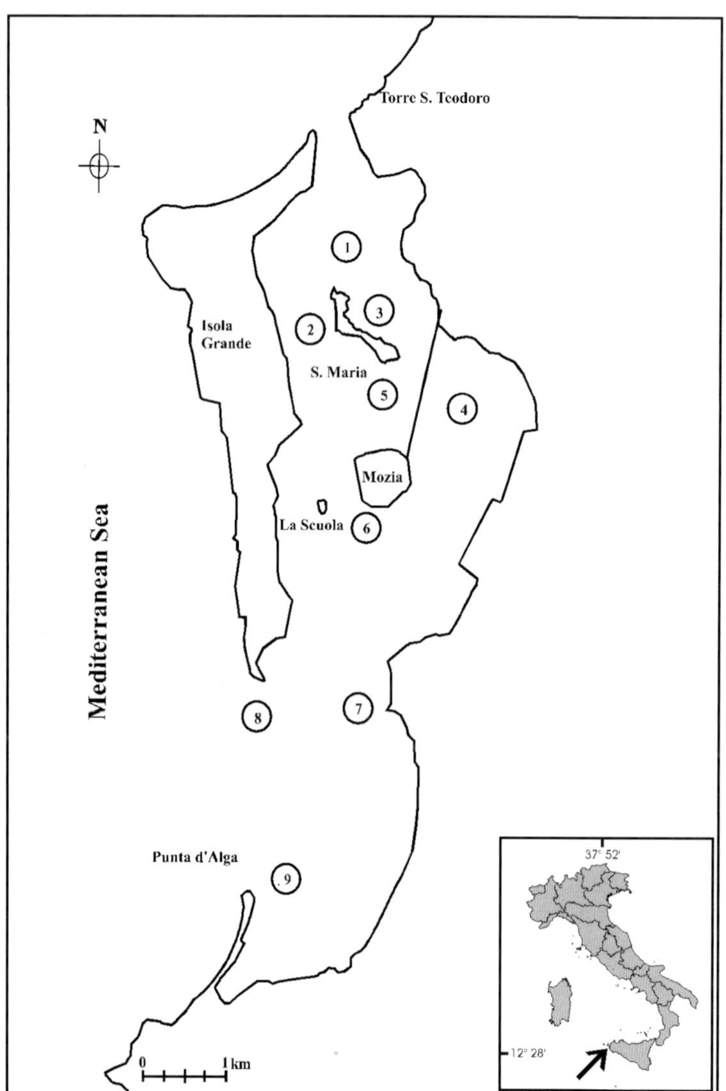

Fig. 1. Map of the sampling area (Stagnone di Marsala, Western Sicily) with the sampling stations

Results

Mysid abundance in the whole area averaged 192.2 ± 171.6 ind./100 m^3, with the highest peak in May (672.5 ind./100 m^3) (Fig. 2). Low densities occurred in July and in winter, when temperature and salinity were at their highest and lowest values respectively. The mysid assemblage consisted mainly of five species which are common in the Mediterranean Sea: *Diamysis bahirensis* (55.2%), *Siriella armata* (18.6%), *S. clausii* (15.8%), *Mysidopsis gibbosa* (9.8%) and *Mesopodopsis slabberi* (0.6%). Throughout the year, the highest abundances were recorded at the inner, shallow stations (Fig. 3). The *D. bahirensis* showed an average yearly abundance of 106.1 (137.2 ind./100 m^3 and two abundance peaks in late spring and late summer which were mainly due to the presence of juveniles (Fig. 4). This species reproduces almost all year except during the winter months, as confirmed by the absence of ovigerous and empty females during this period. The two species belonging to Siriella (*S. armata* and *S. clausii*) showed similar patterns of abundance throughout the year, on average 35.7 (39.7 ind./100 m^3) and 30.3 (22.5 ind./100 m^3) respectively, with a spring peak which was mainly due to juveniles (Fig. 4). The *S. clausii* also occurred in the area in winter, in contrast with the cogeneric species (Fig. 4). The *S. armata* appeared to reproduce from March to August, while *S. clausii* did so from March to October. The average yearly abundance of *M. gibbosa* was rather low, 18.8 (15 ind./100 m^3), with a peak in May (111.6 ind./100 m^3) (Fig. 4). Juveniles were consistently the most represented group. This species appeared to reproduce mainly during the spring. The *M. slabberi* showed an extremely low average yearly abundance, 1.2 (2.3 ind./100 m^3), with slight

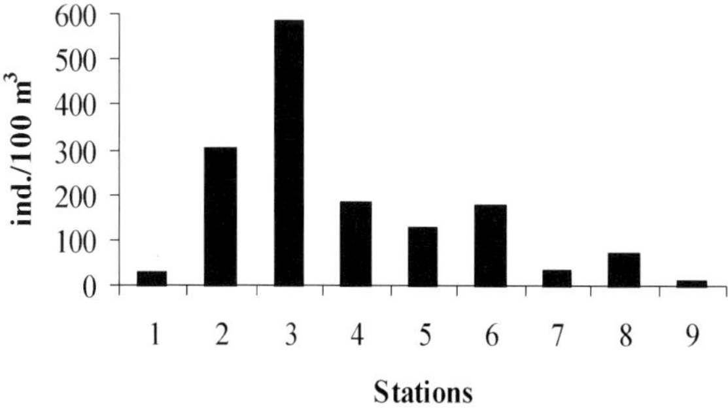

Fig. 3. Average annual abundance (ind/100 m^3) of mysid community at the sampling stations

increments during the winter (Fig. 4). Specimens of this species were never found during spring or summer. The sex ratio (male : female) was 0.9 : 1 for *D. bahirensis*, 0.8 : 1 for *S. clausii* and *M. gibbosa*, and 1.1 : 1 for *S. armata*. Mature females of *M. slabberi* were never collected.

Discussion

The mysid community in the "Stagnone di Marsala" showed a clear seasonal pattern in abundance. A sharp increase in density was observed in the spring and minima during the autumn-winter period. The maximum observed was due to the real increase in the size of the populations (except for *M. slabberi*), owing to active breeding and the production of young (Mauchline 1971). The low numbers in winter were probably caused by natural mortality and/or active migration from coastal areas towards deeper areas off-shore (Mees et al. 1993).

A large quantitative difference between the population of the inner stations and that of the area near the open sea was observed. During

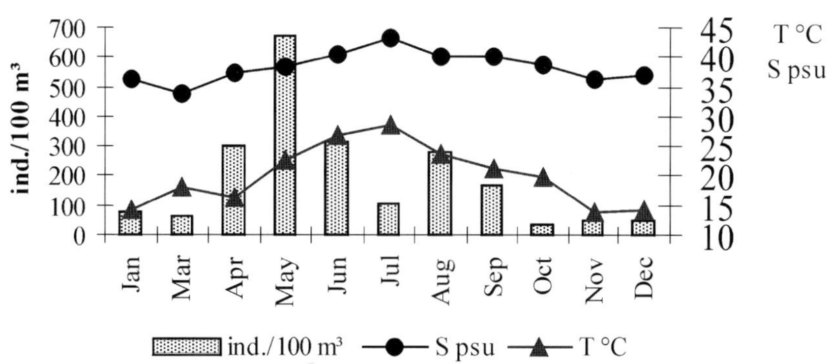

Fig. 2. Monthly distribution of the spatially averaged abundance (ind/100 m^3) for the mysid community, temperature (°C) and salinity (psu) in the "Stagnone di Marsala"

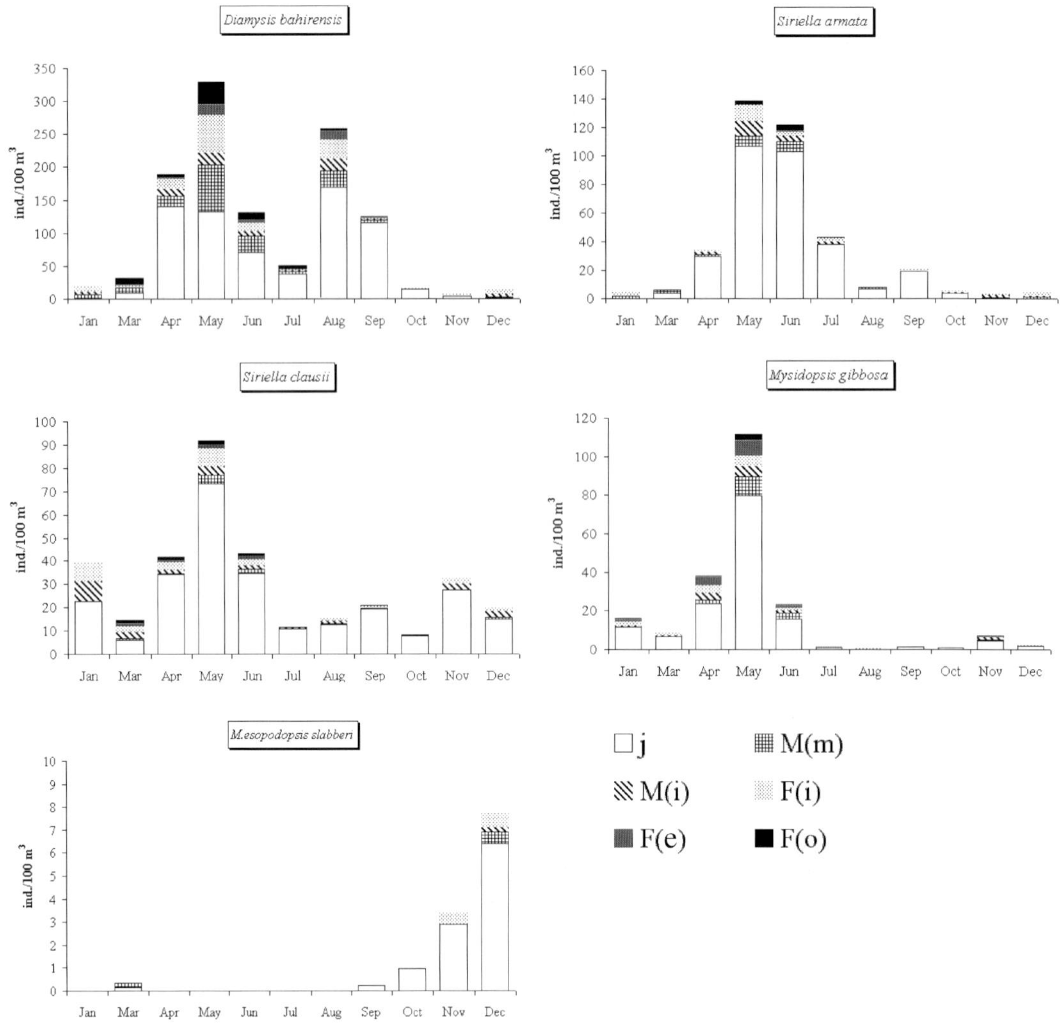

Fig. 4. Monthly distribution of the spatially averaged abundance for the main mysid species (ind./100 m^3) in the "Stagnone di Marsala". [*j*, juveniles; *M(m)*, mature males; *M(i)*, immature males; *F(i)*, immature females; *F(e)*, empty females; *F(o)*, ovigerous females]. Note different scale size

the year higher average densities were recorded in the shallower and more sheltered stations, while low abundances were observed in the deeper stations which are more influenced by the sea. The real mysid abundance at the sea mouth stations was probably underestimated due to the sampling method (only a plankton net was employed). The higher density in the inner area could be caused by the vicinity of the net to the bottom and by the resuspension produced by wind action which in these shallow environments often gives rise to a notable water mixing. It is also necessary to consider the possible underestimation caused by the "escape response" of these organisms. Different authors have noted that mysid schools are able to avoid the zooplankton net in an organised and co-ordinated way (Emery 1968; Mauchline 1980).

Diamysis bahirensis dominated the mysid community in the Stagnone, except in winter when the population structure is less diversified and *S. clausii* becomes more abundant. The *D. bahirensis*, a typically euryhaline species (Genovese 1956a, 1963), occurs mainly in the inner stations and reproduces from March to September, a slightly longer period in comparison to that reported by Genovese (1956a, 1963) for the lake of Ganzirri. Specimens belonging to Diamysis found in this lake have in fact been erroneously ascribed to the species *D. bahirensis* (A.P. Ariani, personal communication).

The *S. armata*, a large shallow water species

(Tattersall and Tattersall 1951), and *S. clausii*, a littoral species (Genovese 1956a), seemed to be associated mainly with the inner stations. *S. armata* has its most abundant period during late spring and early summer, differently from that observed by Mees et al. (1993) for coastal and estuarine waters of the south-western Netherlands, where a bimodal abundance pattern with maxima in spring and autumn was recorded. This species is not very abundant in the Stagnone during summer, as it probably departs for colder and deeper areas, as reported by Cuzin-Roudy and Tchernigovtzeff (1985) for the Ligurian Sea. It reproduces from March to August, confirming observations made by Mauchline (1980) for other littoral habitats; whereas Ariani (1967) reported a longer reproductive period (March-October) for the Adriatic coasts of Apulia. This species is the only one in the Stagnone to show a sex ratio which is slightly biased towards males. This could be caused by an underestimation of the actual number of females, which show secondary sexual features which can be recognised when a bigger body size is reached in comparison with that of males (Mauchline 1980), with the result that females may be erroneously classified as juveniles. The *S. clausii* also shows its maximum density between late spring and early summer, as observed by Genovese (1956a) for the Lake of Faro. Differently *from S. armata*, this species also shows a moderate abundance in the winter months. In the "Stagnone di Marsala" *S. clausii* prefers the inner area and, in particular, the stations characterized by *P. oceanica*, as pointed out by Furnestin (1959) for the Moroccan waters of the Atlantic Sea. It shows a single reproductive period (from March to October), as observed by Genovese (1956a, 1963) for the Lake of Faro (where it reproduces from March to July), and in contrast with the marine forms which have two separate reproductive periods in spring and autumn (Colosi 1929; Tattersall and Tattersall 1951).

The *M. gibbosa*, a neritic species most frequently found in shallow waters (Ariani 1967), shows a spring peak in abundance. It is present at a fairly low density in winter and disappears almost completely in late summer to early autumn. As reported by Mauchline (1970, 1980), this species seems to prefer areas with a lot of organic terrigenous fall-out material. In the Stagnone *M. gibbosa* reproduces from April to June, the same period reported by Mauchline (1980) for British coasts of the Northeast Atlantic.

The *M. slabberi* is a typically euryhaline species (Tattersall and Tattersall 1951; Mees et al. 1993) and is little represented in the "Stagnone di Marsala". It is present at a very low density and mainly during the autumn-winter period. This is in contrast with what has been reported by Vaan der Baan and Holthuis (1971), according to whom this species moves towards off-shore waters during the winter period. In the Stagnone *M. slabberi* is more abundant near areas of urban discharge (Stn 9), confirming what has been noted by Ariani (1967) for the Adriatic coast of Apulia. Reproductive females were never collected in the Stagnone but, as reported by Mauchline (1980) for British coasts, *M. slabberi* seems to reproduce throughout the year, particularly in spring and summer.

Analysis of the results obtained shows clear distribution patterns which vary from species to species. These distribution patterns, such as the relative abundance of different species, may, in such a complex environment, be ascribed to the synergy of a number of factors, both biotic and abiotic, as has already been observed by Williams and Collins (1984) for the Bristol Channel. Among these, temperature, salinity, vegetal covering, quantity of organic detritus and water quality seem to be the most important. Some merely biotic factors such as predation, space-time variations in food sources and interspecific competition for food resources may also be determinant.

In the "Stagnone di Marsala", predation by the "resident" (Quignard 1984) fish species, especially pipefishes and sandsmelt, is very important. In fact *Syngnathus typhle* feeds almost exclusively on mysids, which constitute more than 95% of the diet of the adult specimens and more than 40% of that of the juveniles (percentages refer to the total number of prey) (Scilipoti 1998). Mysids also form a considerable part of the diet of *S. abaster* (up to 12%) and *Atherina boyeri* (up to 10%) adults (Scilipoti 1998), and are preyed on to a lesser extent (up to about 3% of diet) by juveniles of *M. surmuletus*, a transient species in the Stagnone (La Rosa et al. 1997).

The array of known predators of mysids indicates their importance in the food webs of this environment.

Their contribution to the trophodynamics is probably also through the remineralization of detritus. Mysids are generally mixed feeders and in coastal areas utilize organic detritus to a considerable extent as a food source (Mauchline 1980). It is possible, therefore, that in the Stagnone, where a large amount of unconsumed organic matter is deposited on the bottom (Pusceddu 1999), mysids mobilize the organic particles, being involved in a short food chain: organic detritus – mysid-fish (Vaccaro 1998).

Acknowledgements. We would like to thank Prof. Giuseppe Costanzo and Dr. Nunzio Crescenti of the Department of Animal Biology and Marine Ecology, University of Messina, and Prof. Antonio Ariani (Department of Zoology, University of Naples) for their collaboration in taxonomic identification.

References

Ariani AP (1966) Su una forma di Diamysis bahirensis (G. O. Sars) rinvenuta in territorio pugliese. Boll di Zool 33: 227-229

Ariani AP (1967) Osservazione su Misidacei della costa adriatica pugliese. Ann Ist Mus Zool Univ Napoli 18(5): 1-38

Ariani AP, Spagnuolo G (1975) Ricerche sulla misidiofauna del Parco di Santa Maria di Castellabate (Salerno) con descrizione di una nuova specie di Siriella. Bull Soc Nat Napoli 84: 441-481

Campolmi M, Franzoi P, Mazzola A (1996) Observations on pipefish (Syngnathidae) biology in the Stagnone Lagoon (North-West Sicily). Publ Espec Inst Esp Oceanogr 21: 205-209

Carleton JH, Hammer WM (1989) Resident mysids: community structure, abundance and small-scale distributions in a coral reef lagoon. Mar Biol 102: 461-472

Clutter RI (1967) Zonation of nearshore mysids. Ecology 48: 200-208

Clutter RI (1969) The microdistribution and social behaviour of some pelagic mysids shrimps. J Exp Mar Biol Ecol 3: 125-155

Colosi G (1929) I Misidacei del Golfo di Napoli. Pubbl Staz Zool Napoli 9(3): 405-441

Cuzin-Roudy J, Tchernigovtzeff C (1985) Chronology of the female moult cycle in Siriella armata M. Edw. (Crustacea: Mysidacea) based on marsupial development. J Crust Biol 5(1): 1-14

Emery AR (1968) Preliminary observations on coral reef plankton. Limnol Ocean 13: 293-303

Furnestin ML (1959) Mysidacés du plancton marocain. Rev Trav Inst Peches Marit 23: 297-316

Genovese S (1956a) Su due Misidacei dei laghi di Ganzirri e di Faro (Messina). Boll Zool 23(2): 117-197

Genovese S (1956b) Analisi biometrica di una popolazione di Siriella clausii (Mysidacea, Siriellinae) vivente nel lago di Faro. Boll Pesca Piscic Idrobiol 11: 246-263

Genovese S (1963) Osservazioni preliminari sullo zooplancton degli stagni salmastri di Ganzirri e di Faro. Arch Bot Biogeogr Ital 39: 111-114

La Rosa T, Lopiano L, Sarà G, Mazzola A (1997) Osservazioni sulla dieta di forme giovanili di Mullus surmuletus (Linneo, 1758) nello Stagnone di Marsala (Sicilia occidentale). Biol Mar Mediterr 4(1): 530-532

Mauchline J (1970) The biology of Mysidopsis gibbosa, M. didelphis and M. angusta (Crustacea: Mysidacea). J Mar Biol Assoc UK 50: 381-396

Mauchline J (1971) Seasonal occurrence of mysids (Crustacea) and evidence of social behaviour. J Mar Biol Assoc UK 51: 809-825

Mauchline J (1980) The biology of Mysids. Adv Mar Biol 18: 1-369

Mees J, Cattrijsse A, Hamerlynck O (1993) Distribution and abundance of shallow-water hyperbenthic mysids (Crustacea, Mysidacea) and euphasiids (Crustacea, Euphasiacea) in the Voordelta and the Westerschelde, southwest Netherlands. Cah Biol Mar 34: 165-186

Omori M, Hamner WM (1982) Patchy distribution of zooplankton: behaviour, population assessment and sampling problems. Mar Biol 72: 193-200

Pusceddu A, Sarà G, Armeni M, Fabiano M, Mazzola A (1999) Seasonal and spatial changes in the sediment organic matter of a semi-enclosed marine system (W-Mediterranean Sea). Hydrobiologia 397: 59-70

Quignard JP (1984) Les caractéristiques biologiques et environementales des lagunes en tant que base biologique de l'aménagement des pêcheries. FAO-GFCM Etud Rev 61(1): 3-38

Scilipoti D (1998) Studio della comunità ittica residente allíinterno dello Stagnone di Marsala (Sicilia occidentale): distribuzione delle specie e ripartizione delle risorse in dipendenza di habitat a diversa complessità strutturale. PhD thesis, Sci Amb Univ Messina

Tattersall WM, Tattersall OS (1951) The British Mysidacea. The Ray Soc, London, p 460

Turpen S, Hunt J, Anderson BS, Pearse JS (1994) Population structure, growth and fecundity of the kelp forest mysid Holmesimysis costata in Monterey Bay, California. J Crust Biol 14(4): 657-664

Vaccaro AM (1998) Distribuzione di Misidacei (Crustacea, Malacostraca) nello Stagnone di Marsala. Graduation thesis, Biol Sc, University of Palermo

Van der Baan SN, Holthuis LB (1971) Seasonal occurrence of Mysidacea in the plankton of the southern North Sea near the "Texel" lightship. Neth J Sea Res 5(2): 227-239

Webb P, Wooldridge TH (1990) Diel horizontal migration of Mesopodopsis slabberi (Crustacea: Mysidacea) in Algo Bay, southern Africa. Mar Ecol Prog Ser 62: 73-77

Williams R, Collins NR (1984) Distribution and variability in abundance of Shistomysis spiritus (Crustacea: Mysidacea) in the Bristol Channel in relation to environmental variables, with comments on other mysids. Mar Biol 80: 197-206

Wittmann KJ (1977) Modification of association and swarming in north Adriatic Mysidacea in relation to habitat and interacting species. In: Keegan BF, Ceidigh PO, Boaden PJS (eds), Biology of benthic organism. Pergamon Press, Oxford, pp 605-612

Wittman KJ (1984) Effects of size and ambient conditions on development and reproduction. Oceanogr Mar Biol Annu Rev 22: 394-428

Wooldridge TH (1989) The spatial and temporal distribution of mysid shrimps and phytoplankton accumulations in a high energy surfzone. Vie Milieu 39(3/4): 127-133

Short-term Variability of Mesozooplankton in a Mediterranean Coastal Sound (Stagnone di Marsala, Western Sicily)

M. Campolmi[1], G. Zagami[1], L. Guglielmo[1], and A. Mazzola[2]

ABSTRACT

Short-term variations in mesozooplankton were studied during four 24-h periods at one station in a Mediterranean coastal sound (Stagnone di Marsala, western Sicily) in May, July and October 1996 and March 1997. Zooplankton samples were collected every three or six hours with a plankton net (mesh size 125 μm) towed horizontally at the surface. Hydrometric data (height of tide, speed and direction of current) were also recorded to assess the influence of the hydrodynamic conditions on fluctuations in zooplankton abundance and changes in population structure. The zooplankton assemblages were markedly dominated by copepods and their naupliar stages, which consistently accounted for more than 95% of the total numbers. Peaks in abundance were always recorded at night, due to the increase in coastal forms (mainly of the genus Acartia) and to a lesser extent in autochthonous forms (harpacticoids). The short-term variability of zooplankton seemed to be related both to hydrodynamic factors (tidal and wind-driven currents) and to the nocturnal upward migration of organisms.

Introduction

The zooplankton of shallow-water systems, coastal brackish lagoons and estuarine areas (mainly represented by calanoid copepods) is characterised by higher abundances and lower values of species diversity than in offshore communities. This is mainly due to the great temporal variability of the environmental parameters (Alcaraz 1983), which determines short- and medium-term variability of zooplankton abundance.

Knowledge of diel variation patterns is very important in understanding the complex dynamics of zooplankton assemblages and their role in the trophic webs of coastal and shallow-water systems. Zooplankton structure and space-time distribution can significantly be influenced by hydrodynamic factors (Haury et al. 1978, 1990; Alldredge and Hamner 1980; Lagadeuc et al. 1997), which contribute to a patchy distribution, but also by the daily vertical migration (Frost 1988; Lampert 1989; Ohman 1990; Checkley et al. 1992).

There have been many studies of short-term zooplankton variability in near-shore waters or estuaries (Sameoto 1975, 1978; Youngbluth 1980; Gagnon and Lacroix 1981; Maranda and Lacroix 1983) and lagoons (Ferrari et al. 1985a; Madhupratap et al. 1991; Cervetto et al. 1993). However, there is very little or no information about coastal environments of the central (Zagami et al. 1996) and southern Mediterranean Sea.

Brackish coastal habitats, generally characterised by high continental inflow and a wide tidal range, occur in the north and centre of the Mediterranean basin. Lagoons in the south (especially along the North African coasts) show small tidal effects and typical marine features, with salinity values close to that of sea water or even higher due to negligible runoff and a high level of evaporation (Riggio and Chemello 1992).

The "Stagnone di Marsala" is a shallow coastal sound on the western coast of Sicily (central Mediterranean). This hypersaline environment has many climatic and faunal features common to the lagoons along the North African coast (Riggio and Chemello 1992) and represents an environment of transition between open sea and typical lagoon systems. Until now, no information has

[1] Dipartimento di Biologia Animale ed Ecologia Marina, Università di Messina, Salita Sperone 31, 98166 Messina, Italy
[2] Dipartimento di Biologia Animale, Università di Palermo, Via Archirafi 18, 90123 Palermo, Italy

F.M. Faranda, L. Guglielmo, G. Spezie (eds)
Mediterranean Ecosystems: Structures and Processes

been available on the structure and dynamics of the zooplankton community in this ecosystem. Indeed research on the zooplankton was carried out only recently (Campolmi 1998); that study revealed that rare hyperbenthic calanoid copepods belonging to plesiomorphic families (including Pseudocyclopidae, Arietellidae and Ridgewayiidae) or poorly known families (such as the Stephidae) occur in the "Stagnone". These findings are important from a biogeographical point of view because some of the species have never been recorded for the Mediterranean (Campolmi et al. 1999), while others (currently under study) are completely new to science.

The aims of the present research were: (1) to analyse the diel and seasonal changes in zooplankton abundance and composition in this shallow-water environment over 24-hour cycles in four seasons; (2) to evaluate the effects of variability in physical-chemical and hydrodynamic conditions on the temporal distribution and structure of the zooplankton.

Materials and Methods

Study Area

The "Stagnone di Marsala" (37°52'N 12°28'E) is a large (2000 hectares), shallow (average depth about 1.5 m) semi-enclosed basin located on the western coast of Sicily (Fig. 1). It is connected to the open sea by two mouths. The southern mouth (1450 m wide and about 2.5 m deep) is open to a continuous inflow of sea water. The northern mouth is narrower (450 m wide) and

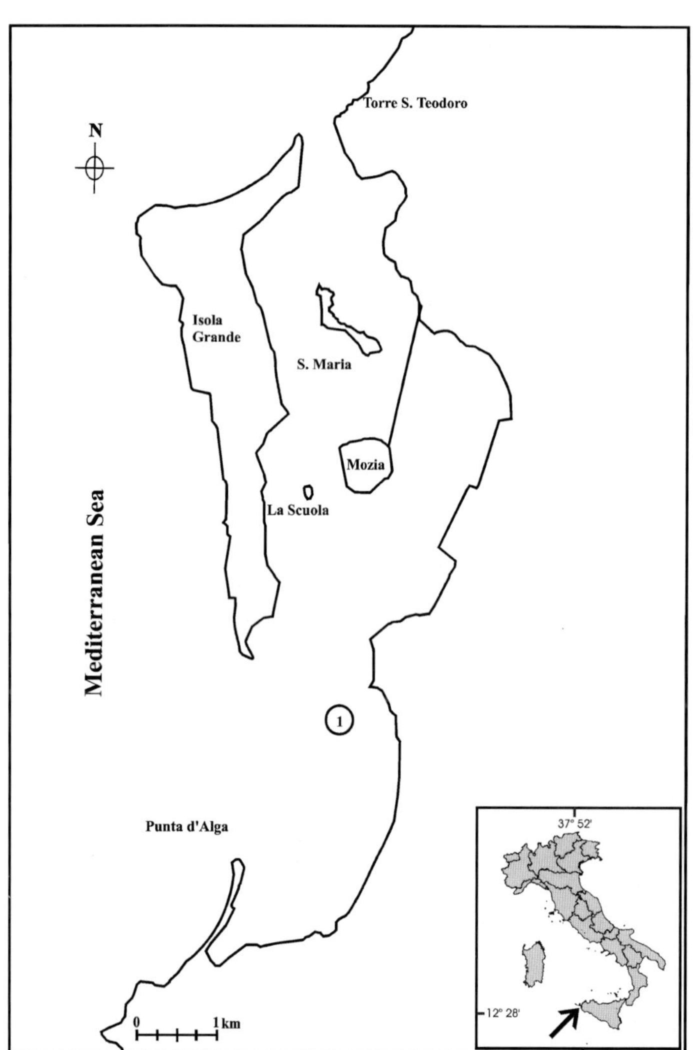

Fig. 1. Stagnone di Marsala (Western Sicily): sampling station

shallower (less than 0.5 m deep), being partially obstructed by sedimentary material, and allows occasional turbulent inputs of marine water (Pusceddu et al. 1999).

The "Stagnone" is characterised by large fluctuations in the values of physical-chemical parameters, especially temperature and salinity; in the inner part of the sound, they range respectively from 11.2 to 29.1°C and from 32.8 to 47.1 psu (Campolmi 1998). The bottom is sandy-muddy and covered mainly by *Cymodocea nodosa* (Ucria) Ascherson, and *Caulerpa prolifera* (Forsskål) Lamouroux. Meadows of *Posidonia oceanica* (Linneus) Delile are restricted to the central and southern basins where they are particularly luxuriant and greatly affect currents and silting (Pusceddu et al. 1999).

Sampling

Zooplankton samples were collected at a station (1.5-2 m deep) located near the southern sea mouth during four 24-hour periods on 9-10 May, 29-30 July and 22-23 October 1996, and on 14-15 March 1997. Samples were taken about every six (May and July) or three (October and March) hours with a plankton net (mouth opening diameter 40 cm; mesh size 125 µm) equipped with a fluximeter (Hydrobios). The net was towed horizontally in the layer between the surface and a depth of 1 m. About 50 m^3 of water were filtered each time and three replicates of each sample were taken. Samples were fixed immediately in 4% neutralised formalin/sea water solution. Zooplankton organisms were counted in subsamples ranging from 1/40 to 1/10 of the entire sample, depending on abundance. Whole samples were analysed for rare species. Adult copepods were identified at the species level (except Harpacticoida, Monstrilloida and Cyclopidae), while copepodid stages were identified at the genus level. Specimens belonging to other zooplankton groups were classified at higher taxonomic levels. Abundance values were expressed as individuals per cubic metre (ind. m^{-3}).

Contemporaneously with the zooplankton sampling, physical-chemical parameters were recorded from the surface layer by means of a multiparameter probe. Data of hydrometric height and profiles of current speed and direction were recorded at 5 min. intervals with a SENSOR DATA 6000 current meter.

Data Processing

The correlation between tide and chemical-physical parameters was calculated with the Spearman ρ coefficient. Statistical differences in zooplankton abundance between day and night were assessed with the Mann-Whitney U-test (Siegel 1956). The samples were classified by cluster analysis of copepod abundance using the UPGMA method on a Euclidean distance matrix of quantitative data (Pielou 1984). Factor correspondence analysis (FCA) (Benzecri 1982) was used as an ordination technique and performed exclusively on the correlation matrix obtained from the abundances of copepods, the most important taxon (>95%). The significance of the axes was evaluated using Lebart tables (1975). "Statistica" software (Statsoft Inc.; rel. 5.1) was used for the statistical analysis.

Results

Environmental Parameters

The average value of the diel tidal range was 25 cm. The values of physical-chemical parameters recorded during the four 24-h survey cycles are shown in Table 1. Temperature values ranged from 14.1 (March) to 29.2°C (July). The maximum diel range (2.4°C) was observed in July (26.8-29.2°C) and March (14.4-16.8°C). The highest values were always recorded during the after-

Table 1. Physical-chemical parameters during the four 24-h survey cycles

	9-10 May 1996		29-30 July 1996	
	Average	± SD	Average	± SD
Temperature (°C)	22.9	0.7	27.5	1.0
Salinity (psu)	38.6	0.2	41.8	1.1
pH	8.7	0.0	8.9	0.1
Dissolved oxygen (mg•l^{-1})	7.6	0.9	6.9	1.9
Dissolved oxygen (%)	112.4	15.2	115.4	24.8
	22-23 October 1996		**14-15 March 1997**	
	Average	± SD	Average	± SD
Temperature (°C)	19.0	0.3	15.6	0.8
Salinity (psu)	38.5	0.6	38.4	0.2
pH	9.0	0.1	9.6	0.1
Dissolved oxygen (mg•l^{-1})	7.1	0.3	8.2	1.3
Dissolved oxygen (%)	96.8	4.4	104.6	18.3

noon. Temperature did not show any correlation with tide.

Salinity values ranged from 37.8 (October) to 43.8 psu (July). Significant ($P<0.05$) diel variations (2.6 psu) were recorded in July, while the fluctuations were very low in May (0.5 psu). In March, the salinity values averaged 38.4 psu and the day-night variations were significant ($P<0.05$). The diel variations in salinity values appeared to be correlated negatively with tide in March ($\rho = -0.84$; $P<0.05$) and October ($\rho = -0.67$).

In almost all the surveys, dissolved oxygen showed a clear diel cycle with maximum saturation in the afternoon and a minimum in the early morning. The variations in oxygen concentration and saturation levels over the 24-h period were significant ($P<0.05$) only in July (3.9-9.1 mg/l; 81.9-146.1%). In general, these parameters appear to be affected by two different phenomena: the day-night cycle of photosynthesis and respiration processes and the movement of water masses with different oxygen contents through the sea mouth (Ferrari et al. 1985a).

The pH values were fairly high in all the 24-h sampling surveys. Significant ($P<0.05$) diel variations were recorded in March (9.4-9.7).

The speed and direction of the current recorded during the four surveys are shown in Fig. 2. A clear periodicity of current direction was

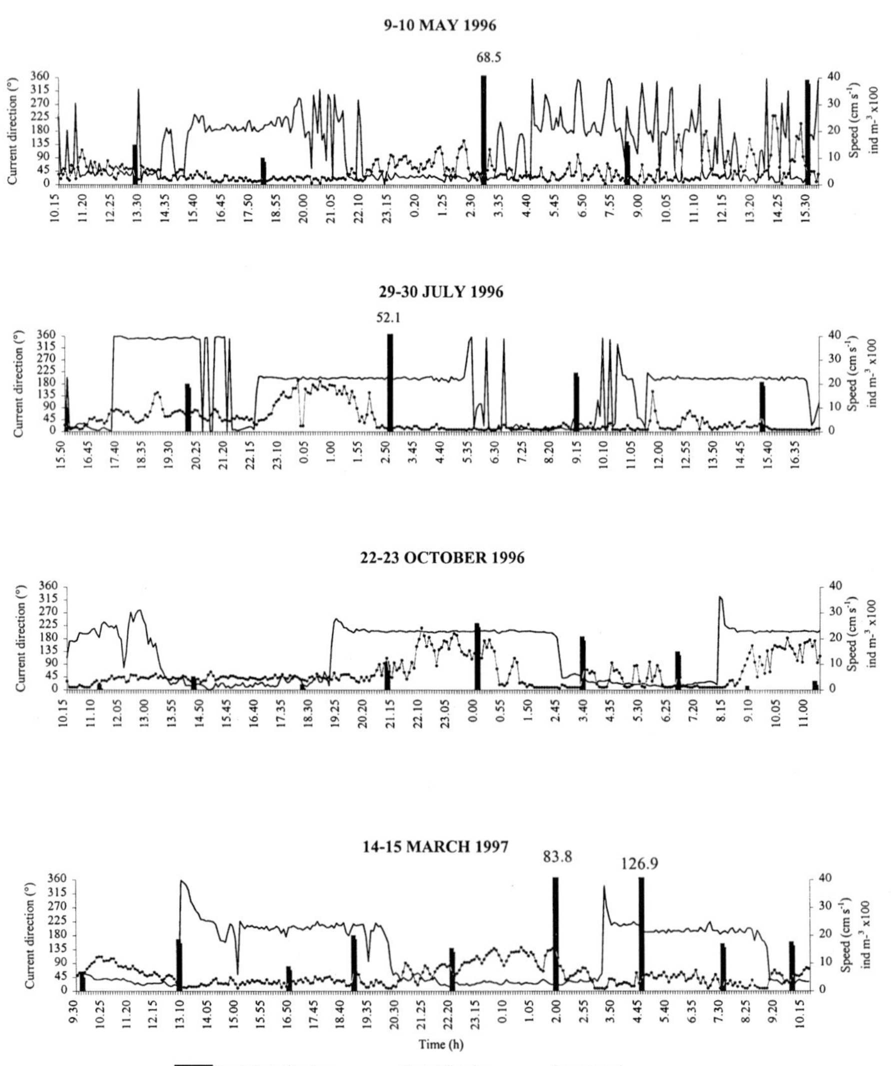

Fig. 2. Current speed (cm•s^{-1}) and direction (degrees), and zooplankton abundance (ind. m^{-3} x 100) during the four 24-h cycle surveys

evident in all the surveys and appeared to be influenced by the tidal cycle. The current direction rotated approximately 180° at 6-h intervals, so that four current phases took place over the 24-h period, two at about 45° (N/E) and the other two at about 225° (S/W). The incoming current (N/E) was generally recorded during high tide while the outgoing current (S/W) was recorded during low tide, except in May (15:00 h) when an outgoing current was recorded during high tide. This was probably due to the effect of the S/E wind, pushing the waters out of the basin. The opposition of the two driving forces (tide and wind) gives rise to marked turbulence and water mixing.

The current speed was very variable and did not exhibit a clear cyclic component. The highest current speed (26 cm/s) was recorded in a N/E direction, but high speed (24.2 cm/s) was also recorded in the opposite direction (S/W).

Zooplankton

Were identified 22 zooplankton taxa (including numerous demersal organisms such as amphipods, cumaceans, isopods, mysids, nematodes, ostracods and tanaids) and 30 copepod species (Table 2).

Copepods (mainly calanoids) and their naupliar stages were the dominant component of the zooplankton, forming up to 99% of the total number of individuals. Holoplankton predominated over meroplankton (consisting mainly of decapod, polychaete and gastropod larvae), which accounted for a very low percentage of the total population (about 1-4%). The zooplankton assemblages showed a clear seasonal pattern of abundance with a maximum in May (on average 3979 ± 4957 ind. m^{-3}) and a minimum in October (on average 1041 ± 918).

May 1996

Large variations in zooplankton abundance (938-6846 ind. m^{-3}) were recorded throughout the 24-h cycle (Fig. 2; Table 3a). A night-time (03:00 h) peak in abundance was evident and was due to the increase in copepods, mainly *Acartia* copepodids, and to a lesser extent harpacticoids and *Isias* copepodids (Table 3b). Mysids and decapod larvae were not abundant during the day, showing peaks at night (Table 3a). Sixteen copepod species were found: coastal-neritic species were abundant, while the pelagic ones (*Haloptilus longicornis* and *Paracandacia bispinosa*) were sporadic. Among the adults, *Acartia margalefi and Isias clavipes* were the most abundant species; the former showed peaks at 03:00 h (during high tide) and 15:00 h (during high tide and outgoing current), while the latter

Table 2. Mesozooplankton categories in samples from "The Stagnone di Marsala"

Acari	Calanoida	Cyclopoida
Amphipoda	*Acartia clausi*	Cyclopidae
Appendicularia	*Acartia discaudata*	*Oithona atlantica*
Ascidiacea (larvae)	*Acartia josephinae*	*Oithona brevicornis*
Chaetognatha	*Acartia latisetosa*	*Oithona nana*
Cirripedia (nauplia, larvae)	*Acartia margalefi*	*Oithona similis*
Copepoda	*Centropages typicus*	
Cumacea	*Clausocalanus furcatus*	Harpacticoida
Decapoda (larvae)	*Haloptilus longicornis*	*Porcellidium Sp.*
Echinodermata (larvae)	*Isias clavipes*	
Gastropoda (larvae)	*Labidocera brunescens*	Monstrilloida
Hydromedusae	*Metacalanus acutioperculum*	
Isopoda	*Paracalanus nanus*	Poecilostomatoida
Mysidacea	*Paracalanus parvus*	*Corycaeus flaccus*
Nematoda	*Paracandacia bispinosa*	*Corycaeus giesbrechti*
Ophiuroidea (larvae)	*Parapontella brevicornis*	*Corycaeus latus*
Ostracoda	*Pseudocyclops spp.*	*Corycella rostrata*
Osteichthyes (larvae and eggs)	*Ridgewayia marki minorcaensis*	*Oncaea media*
Pantopoda	*Stephos spp.*	*Oncaea sp.*
Polychaeta (larvae)	*Temora stylifera*	
Siphonophora		
Tanaidacea		

occurred exclusively at night (Table 3b). Some rare hyperbenthic calanoid species were collected: *Pseudocyclops spp.* and *Stephos spp.* The former were not abundant during the day, showing a clear peak at night, while the latter were collected only at night (Table 3b).

July 1996

The zooplankton community showed fairly large day-night variations in abundance (110 - 5211 ind. m^{-3}) (Fig. 2; Table 4a). There were relatively low values during the day (average 185 ind. m^{-3}) and a sharp peak at night (03:00 h; during low tide), mainly due to the increase in Acartia copepodids, *A. margalefi* and harpacticoids (Table 4b). Hydromedusae and nematodes showed significant variations in abundance ($P<0.05$), increasing respectively at night and during the day, while both mysids and decapod larvae exhibited a clear peak at night (Table 4a). Eighteen copepod species were identified, six of which were found only occasionally. Among these, *Ridgewayia marki minorcaensis*, a hyperbenthic calanoid, was found only at night. The *I. clavipes* adults and copepodids were more abundant at night ($P<0.05$) (Table 4b). The same trend was shown by *Pseudocyclops* species (Table 4b).

October 1996

The diel variation in abundance was lower in October (139-2531 ind. m^{-3}) than in the other months. However the pattern of a sharp increase during the dark period (during both low and high tide) was the same (Fig. 2; Table 5a). Mysids showed two significant ($P<0.05$) nightly peaks (Tab. 5a). Decapod larvae increased slightly at 21:00 h (Table 5a). The structure of the copepod assemblage was characterised by coastal species, mainly *Acartia* copepodids and *A. margalefi*, which were significantly ($P<0.05$) more abundant from 21:00 to 7:00 h (Table 5b). Harpacticoids and *I. clavipes* followed in rank order, both showing a significant ($P<0.05$) increase during the dark period (Table 5b). Among the hyperbenthic

Table 3a,b. a May 1996: diel variation of average abundance (ind. m^{-3}) for the main systematics groups. **b** May 1996: diel variation of average abundance (ind. m^{-3}) for the main copepod species (*SD*, standard deviation)

Time (h)	13:00		18:00		3:00		8:00		15:00	
	Average	± SD	Average	± SD	Average	± SD	Average	± SD	Average	± SD
a										
Copepoda	1067.29	574.9	806.86	319.6	6368.33	1375.8	1399.58	391.3	3506.30	1359.2
Copepoda nauplia	157.00	133.5	54.79	32.6	145.33	98.4	121.53	57.6	332.28	235.7
Ostracoda	117.95	143.9	21.94	16.4	96.18	161.6	9.60	5.3	846	5.5
Polychaeta larvae	33.98	37.5	8.60	5.3	23.60	22.8	4.80	2.6	0.73	0.6
Mysidacea	0.79	1.0	2.14	3.7	55.70	29.9	0.37	0.4	0.46	0.7
Amphipoda	5.40	3.2	13.36	19.1	20.66	12.9	2.50	2.7	2.76	2.6
Decapoda larvae	3.53	3.3	1.10	0.5	24.35	15.0	8.00	5.4	5.38	1.1
Total abundance	1426.2	783.7	938.3	340.3	6845.8	1509.9	1561.0	391.8	3864.1	1593.2
b										
Acartia copepodites	615.49	442.8	301.11	136.7	5009.97	1183.4	1115.40	442.0	3084.50	1238.8
Harpacticoida	353.53	163.4	460.69	429.5	897.01	163.3	234.45	67.8	239.00	45.2
Isias copepodites	58.94	46.3	21.74	10.1	198.12	106.4	20.75	15.9	86.80	0.8
Acartia margalefi	14.58	10.1	8.08	8.4	63.80	55.4	10.75	9.1	76.10	75.2
Isias clavipes					120.96	80.6				
Pseudocyclops spp.	2.45	3.3	0.57	0.9	29.30	11.8	0.07	0.1		
Pseudocyclops copepodites					2.18	3.8				
Stephos spp.					0.18	0.3				
Total abundance	1067.3	574.9	806.9	319.6	6368.3	1375.8	1399.6	391.3	3506.30	1359.24

Table 4a,b. **a** July 1996: diel variation of average abundance (ind. m^{-3}) for the main systematics groups. **b** July 1996: diel variation of average abundance (ind. m^{-3}) for the main copepod species

Time (h)	15:00		20:00		3:00		9:00		15:00	
	Average	± SD	Average	± SD	Average	± SD	Average	± SD	Average	± SD
a										
Copepoda	79.34	31.9	152.54	198.9	5103.16	1862.1	157.41	65.6	133.37	56.6
Copepoda nauplia	10.47	10.0	35.67	6.7	48.62	50.4	37.88	21.0	4.15	1.8
Polychaeta larvae	4.88	8.2	0.05	0.04	0.59	0.1	6.88	9.4	17.18	20.2
Mysidacea	0.09	0.1			14.63	7.6	1.41	0.8	2.11	1.1
Nematoda	2.18	3.7					6.56	10.6	8.88	5.8
Decapoda larvae	0.17	0.1	0.03	0.05	11.90	9.3	2.64	1.8	0.08	0.1
Hydromedusae	0.15	0.2	0.36	0.4	13.13	10.2	0.52	0.9	0.35	0.3
Gastropode larvae	1.19	1.7	0.86	0.9	4.23	5.8	1.50	1.4	2.88	3.0
Total abundance	109.9	52.3	192.6	195.5	5211.0	1858.2	239.2	89.3	201.8	92.0
b										
Acartia copepodites	55.23	25.8	121.25	191.6	4111.35	1175.0	53.60	7.4	63.81	46.0
Acartia margalefi	0.75	0.1	0.61	0.9	774.85	823.9	13.31	14.9	17.02	10.8
Harpacticoida	11.31	10.3	3.90	3.7	121.75	61.2	59.03	87.8	38.86	17.6
Clausocalanus copepodites	2.78	2.4	9.13	2.2	13.30	14.4	5.09	4.2	0.31	0.5
Cyclopidae	2.53	3.4	0.34	0.4	10.35	10.2	3.18	3.1	8.43	1.9
Pseudocyclops spp.	0.07	0.1			17.35	8.4	0.71	0.4	2.49	3.2
Jsias copepodites	1.69	1.0	2.22	1.7	13.87	8.6	0.84	0.9	0.31	0.5
Pseudocyclops copepodites	0.27	0.5			2.39	4.1	1.67	1.0	0.19	0.3
Isias clavipes					4.46	2.4	0.05	0.1		
Ridgewayia marki minorcaensis					0.08	0.1				
Stephos spp.					0.07	0.1				
Total abundance	79.3	31.9	152.5	198.9	5103.2	1862.1	157.4	65.6	133.4	56.6

calanoids, *Pseudocyclops spp.*, *Metacalanus acutioperculum* and *Stephos spp.* specimens were collected mainly at night (Table 5b).

March 1997

During the spring 24-h cycle, the greatest diel variation (570-12691 ind. m^{-3}) and the highest nightly peak ($P<0.05$; during high tide) were observed (Fig. 2; Table 6a). Copepod nauplii were particularly abundant, with a sharp increase at night (Table 6a). Decapod larvae and mysids showed significant diel variations ($P<0.05$) with maxima at night (Table 6a).

The structure of the copepod assemblage was characterised by greater homogeneity among the species. The sharp peak in abundance at night was due to an increase in coastal forms, mainly *Acartia* copepodids, *I. clavipes* (adults and copepodids) and *A. margalefi*. The last two were both significantly ($P<0.05$) more abundant at 05:00 h (Table 6b). Harpacticoids and Cyclopidae increased significantly ($P<0.05$) from 19:00 h to 05:00 h (Table 6b). Among the hyperbenthic calanoids, specimens (adults and copepodids) of *Pseudocyclops spp.* and *Stephos spp.* were found almost exclusively at night ($P<0.05$) (Table 6b).

Cluster and Factor Analysis

In each of the four 24-h cycles, the classification and ordination techniques show two distinct clusters of samples, which are homogeneous with regard to the relative abundance of the different copepod species (Fig. 3). The first cluster includes samples collected at night, while the second includes those collected during the day,

Table 5a,b. a October 1996: diel variation of average abundance (ind. m^{-3}) for the main systematics groups. **b** October 1996: diel variation of average abundance (ind. m^{-3}) for the main copepod species

Time (h)	12:00		15:00		18:00		21:00		0:00		4:00		7:00		9:00		11:00	
	Average	±SD	Average	±SD	Average	±SD	Average	±SD	Average	± SD	Average	±SD	Average	± SD	Average	± SD	Average	± SD
a																		
Copepoda	131.27	3.2	432.42	39.5	114.56	53.3	992.78	87.0	2476.53	405.8	1897.30	149.6	1392.17	228.1	58.86	24.9	244.95.	6.1
Copepoda nauplia	5.82	4.7	13.27	4.8	22.47	6.8	38.45	1.5	4.48	6.3	19.20	10.9	38.58	0.3	25.99	7.3	21.03	5.1
Mysidacea	0.03	0.0					21.19	22.4	2.49	3.1	24.94	2.8	0.07	0.1	1.02	0.8		
Ostracoda	0.27	0.2	0.05	0.1	0.03	0.0	0.26	0.2	21.13	29.7	0.52	0.2	0.15	0.0	1.82	1.7	0.04	0.1
Polychaeta larvae	0.21	0.3	0.49	0.5	0.76	0.0	2.87	2.9	6.92	3.4	6.64	6.8	0.29	0.2	0.40	0.6	0.08	0.1
Decapoda larvae	0.23	0.2	0.49	0.5	0.02	0.0	2.97	0.7	0.19	0.1	0.79	0.3	0.07	0.1	0.08	0.0	0.77	0.1
Total abundance	140.7	9.6	450.6	35.3	139.1	61.8	1061.0	114.7	2530.8	453.2	2011.5	161.0	1434.3	225.5	92.9	20.2	270.9	10.9
b																		
Acartia copepodites	109.85	0.5	390.26	44.2	81.64	49.0	578.00	213	1867.34	151.2	1144.96	20.1	1074.69	283.5	21.60	5.7	188.92	2.6
Acartia margalefi	5.67	1.0	27.47	4.6	1634	3.4	104.45	18.5	235.71	149.6	360.40	90.9	159.95	75.3	5.22	0.6	28.1	0.6
Harpacticoida	5.24	2.7	3.54	1.2	4.54	0.2	171.09	30.4	208.93	194.1	154.85	72.9	85.03	23.7	18.91	13.2	1.59	2.2
Isias clavipes	0.66	0.8					84.96	82.4	10636	77.6	122.92	92.7	2.29	2.8	0.11	0.2		
Clausocalanus copepod	2.86	1.2	5.74	3.0	6.04	0.8	15.62	8.1	6.72	9.5	31.41	44.4	28.51	28.3	4.24	2.6	18.72	1.8
Isias copepodites	0.48	0.5	0.05	0.1	0.09	0.1	12.55	10.2	22.70	12.2	45.39	0.7	25.88	29.7	0.29	0.2	0.54	0.3
Pseudocyclops spp							5.47	6.2	2.45	0.5	17.95	25.4	0.51	0.5	0.24	0.3		
Metalanus copepodites							6.58	5.8	0.15	0.2	4.54	63	0.07	0.1	0.62	0.3	0.04	0.1
Pseudocyclops copepodites							2.74	3.1	2.02	2.4	4.49	6.3	0.07	0.1			0.04	0.1
Stephos spp.							1.43	0.6	0.22	0.3	4.72	6.0	0.28	0.4				
Metacalanus copepodites					0.03	0.0	0.27	0.4			0.06	0.1			0.22	0.3		
Stephos copepodites							0.07	0.1										
Total abundance	131.3	3.2	432.4	39.5	114.6	53.3	992.8	87.0	2476.5	405.8	1897.3	149.6	1392.2	228.1	58.9	24.9	244.9	6.1

Table 6a,b. a March 1997: diel variation of average abundance (ind. m^{-3}) for the main systematics groups. **b** March 1997: diel variation of average abundance (ind. m^{-3}) for the main copepod species

Time (h)	9:00		13:00		17:00		19:00		22:00		2:00		5:00		8:00		10:00	
	Average	± SD	Average	± SD	Average	± SD	Average	± SD	Average	± SD	Average	± SD	Average	± SD	Average	± SD	Average	± SD
a																		
Copepoda	336.21	174.7	1300.12	623.9	500.10	125.3	1822.28	616.3	1307.42	498.8	7016.91	956.0	11348.24	4081.1	1066.31	159.2	1304.25	473.2
Copepoda nauplia	225.29	72.9	485.05	146.4	309.22	15.5	91.20	31.3	94.65	62.8	1183.54	332.5	991.05	921.4	508.54	253.3	385.63	122.6
Decapoda larvae	6.93	3.6	15.33	5.2	5.68	2.1	12.86	4.8	30.61	26.2	47.51	13.0	278.55	75.4	54.81	24.6	10.10	3.7
Mysidacea							3.02	2.4	5.41	4.8	56.91	20.5	47.25	53.8			0.04	0.1
Polychaeta larvae	0.55	0.2	1.21	1.5	0.28	0.2	3.80	1.2	10.41	11.6	13.86	20.6	10.48	17.1	6.14	5.9	0.30	0.3
Total abundance	570.1	244.9	1806.2	746.3	829.0	139.5	1945.2	596.1	1487.5	546.1	8386.7	1228.3	12691.6	5044.1	1643.0	378.0	1708.3	579.8
b																		
Acartia copepodites	234.06	140.1	972.39	489.9	335.34	73.3	464.22	162.6	243.96	177.0	3565.48	464.6	3937.39	1813.3	773.37	79.4	1065.08	444.1
Isias copepodites	2.24	1.0	12.44	6.7	0.76	0.6	295.50	94.4	137.36	98.5	1342.95	261.4	2398.71	1164.6	110.13	30.6	84.73	31.4
Isias clavipes					0.03	0.0	309.18	153.6	141.91	121.7	633.77	224.0	2897.76	680.0				
Harpacticoida	14.81	4.0	834	5.1	41.76	22.0	418.28	111.1	576.82	43.2	741.57	160.8	896.89	318.7	71.77	41.0	56.03	5.5
Acartia margalefi	47.62	21.6	135.18	89.8	60.73	29.8	216.78	73.1	110.37	116.2	543.77	96.9	962.88	226.5	82.49	37.3	90.27	32.1
Clausocalanus copedod	26.33	13.4	155.78	45.5	49.43	33.0	44.79	10.0	17.32	12.3	63.72	37.6	27.68	26.5	4.04	3.5	1.95	3.4
Cyclopidae											74.77	17.2	121.75	112.9	6.31	6.0		
Pseudocyclops spp.							0.24	0.1	0.32	0.3	4.47	1.0	21.01	18.3				
Pseudocyclops copepodites									0.13	0.2	1.08	0.7	12.55	20.3	0.05	0.1		
Stephos spp.							0.40	0.4	4.26	6.3	0.33	0.3						
Stephos copepodites							1.82	2.8	0.57	0.6	0.43	0.2	0.38	0.3				
Total abundance	336.2	174.7	1300.1	623.9	500.1	125.3	1822.3	616.3	1307.4	498.8	7016.9	956.0	11348.2	4081.1	1066.3	159.2	1304.3	473.2

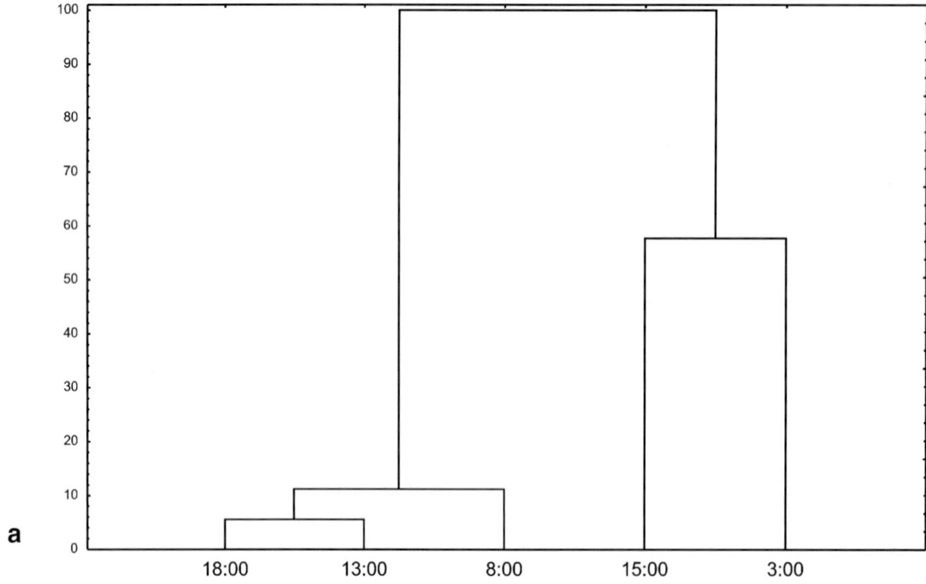

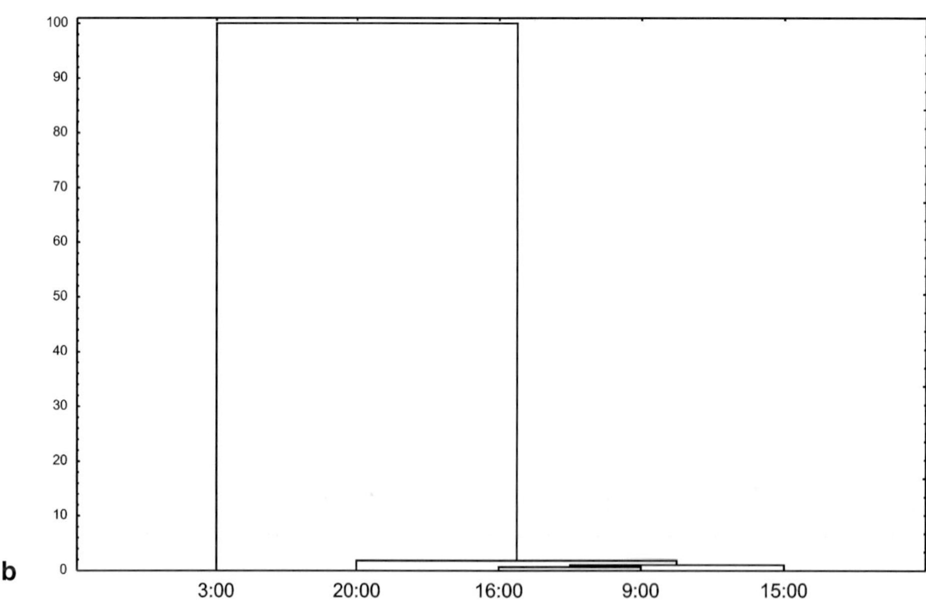

Fig. 3a-d. Cluster analysis of the abundance of copepod species. Sampling time is reported on horizontal axis and the relative dissimilarity index (based on Euclidean distance) is reported on the vertical axis. (A May; B July; C October; D March)

100
90
80
70
60
50
40
30
20
10
0
21:00 15:00 11:00 18:00 12:00 9:00 7:00 4:00 0:00
c
100
90
80
70
60
50
40
30
20
10
0
22:00 19:00 17:00 9:00 13:00 10:00 8:00 5:00 2:00
d

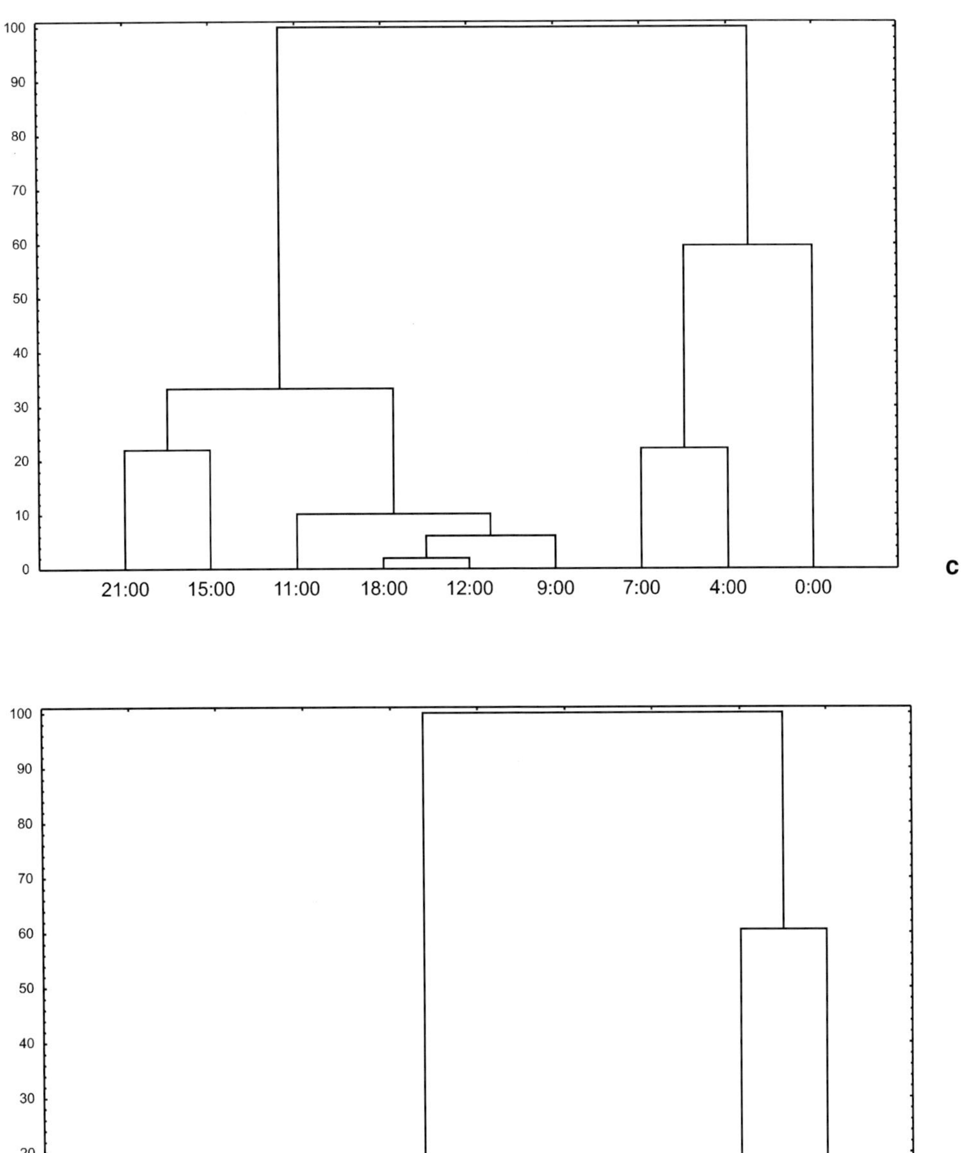

Fig. 3a-d. (Cont.)

a

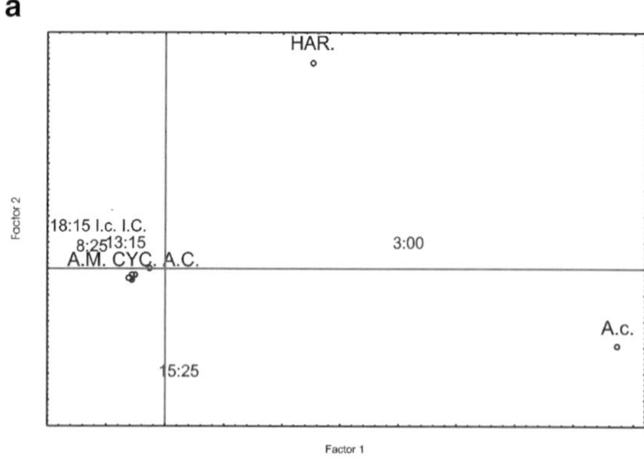

b

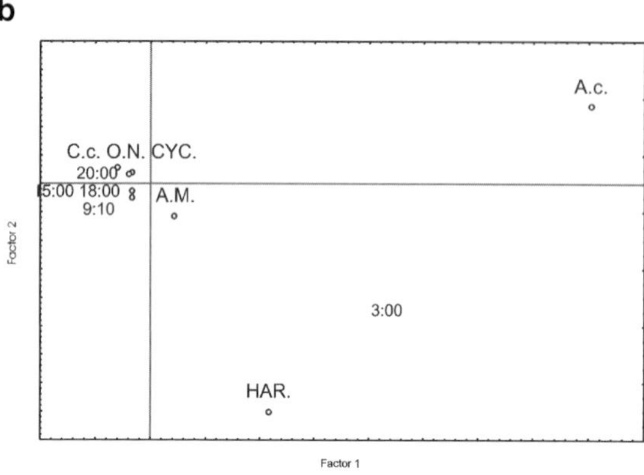

c

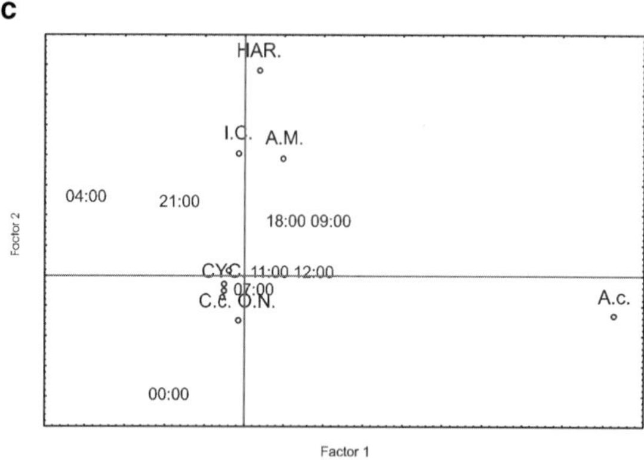

d

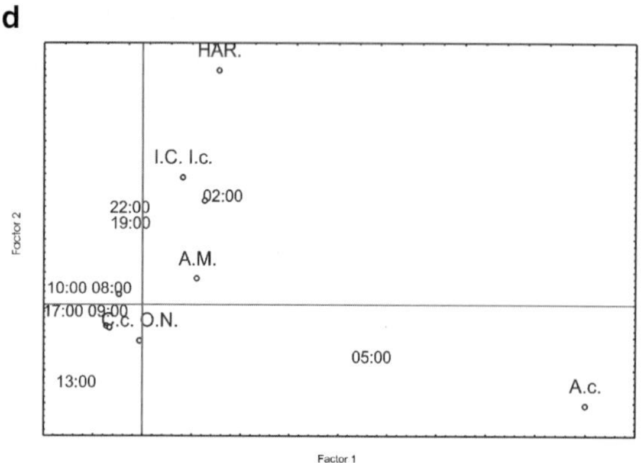

except in May when the 15:00 h sample is associated with that of 03:00 h (probably due to the peak in abundance recorded in the early afternoon).

The FCA (Fig. 4) performed on the correlation matrix has three significant axes ($P<0.05$), with the first two explaining most of the total variance (May 97.5%; July 97.1%; October 99.3%; March 95.2%). In May and July, *Acartia* copepodids and harpacticoids are separated at the extremes of the two axes, associated with the night sample (3:00 h), while all the other species are associated with the day samples. In October and March, *A. margalefi* and *I. clavipes* (adults and copepodids) also characterise the night samples, in addition to *Acartia* copepodids and harpacticoids.

Discussion

The zooplankton community at the station close to the southern mouth of the "Stagnone di Marsala" sound consists primarily of coastal-neritic calanoids (mainly of the genus *Acartia*) and to a lesser extent autochthonous forms, since this area is greatly influenced by the sea. The native component is represented principally by benthic harpacticoids (mainly Thalestridae, Tisbidae and Harpacticidae), cyclopoids (Cyclopidae) and some calanoid species of the hyperbenthic genus *Pseudocyclops*, often found in shallow coastal environments (Bowman and Gonzàlez 1961; Ohtsuka et al. 1999). The structure of the copepod assemblages differs from that observed in the inner part of the basin, where the zooplankton community is quantitatively poor and composed mainly of harpacticoids (Campolmi 1998).

The presence of numerous demersal groups (amphipods, cumaceans, isopods, mysids, nematodes, ostracods and tanaids) in the water column could be due primarily to hydrodynamic factors. These are related mainly to wind action, which in these shallow environments often gives rise to strong water mixing, so that sediments

Fig. 4a-d. Factorial correspondence analysis of the abundance of copepod species. (***a***, May; ***b***, July; ***c***, October; ***d***, March) (*AM, Acartia margalefi*; *AC, Acartia clausi*; *Ac, Acartia copepodites*; *Cc, Clausocalanus copepodites*; *CYC*, Cyclopidae; *HAR*, Harpacticoids; *IC, Isias clavipes*; *Ic, Isias copepodites*; *ON, Oithona nana*) and bottom organisms are re-suspended. Secondly, the presence of these demersal groups in our samples could be due to specific daily migratory behaviour (Alldredge and King 1980). In all the 24-h surveys, mysids (mainly *Diamysis bahirensis*, *Siriella armata*, *S. clausii* and *Mysidopsis gibbosa*) showed one or two clear abundance peaks during the dark period, as found in other shallow-water environments (Genovese 1963). This phenomenon could be related to the vertical migration performed by these organisms (Mauchline 1980).

Isolated abundance peaks of decapod larvae were observed during the dark hours in all the 24-h cycles, confirming reports by Ferrari et al. (1985a) for a lagoon of the Po River Delta. The diel abundance pattern of these organisms appears to be due to endogenous circatidal rhythms in their behaviour (Cronin and Forward 1979).

The copepod community was characterised by low diversity and the marked dominance of one or two species (*Acartia margalefi* and *Isias clavipes*), as observed by numerous authors for brackish-water environments of temperate climates (Alcaraz 1983; Colombo et al. 1983-84; Ferrari et al. 1985b; Belmonte and Bianchi 1992; Zagami and Guglielmo 1995). Other coastal-neritic species (mainly *Oithona nana*, *A. clausi*, *A. josephinae*, and *Labidocera brunescens*) entered the basin but showed very low abundance and frequency of occurrence, suggesting that they were unable to stay and reproduce in the area. This confirms that a certain degree of adaptability to the frequent temporal variations in the hydrological parameters is required to establish a population in this environment; this is the case for *A. margalefi*, an "r" strategist species (Alcaraz 1983; Ferrari and Carrieri 1985; Zagami and Guglielmo 1995) which is able to enter the inner basin and spend its entire life cycle there (Campolmi 1998). The temporal variations in physical-chemical parameters have a greater importance than their range of values in determining the spatial distribution of species; the species that are more "r" strategist occupy the less stable areas (Alcaraz 1983).

The seasonal succession of species is different from that reported for coastal waters of the Mediterranean Sea (Scotto di Carlo 1985). In the Stagnone mouth area, two species that usually characterise the Mediterranean neritic community, *A. clausi* and *Paracalanus parvus*, show very low abundances throughout the year. This is probably due to the location of the area between the coastal waters and the inner basin. The structure of the mesozooplankton assemblage of this area is intermediate between the two adjacent environments (Campolmi 1998), following a gradient of abundance and species diversity which overlaps with that of the hydrological features, as also reported in the Sacca di Scardovari (Ferrari and Colombo 1988). This transition environment, although influenced by the nearby sea water, shows high instability and significant diel and seasonal fluctuations in physical-chemical parameters.

The finding of rare hyperbenthic calanoid species (*Pseudocyclops spp.*, *Stephos spp.*, *Metacalanus acutioperculum* and *Ridgewayia marki minorcaensis*) is noteworthy, and their greatest occurrence at night confirms the hypothesis of demersal behaviour. As reported for other shallow-water environments, these species probably migrate at various times between the bottom and the water column (Damkaer 1970; Alldredge and King 1985; Ohtsuka 1985; Jacoby and Greenwood 1988).

The variability of the zooplankton structure throughout the 24-h period seems to be affected by hydrodynamic factors (tidal and wind-driven currents). The diel tidal range, although low in comparison with other Mediterranean shallow-water environments (Ferrari et al. 1985a, b), seems to affect the currents; they show a clear periodicity of about six hours, especially when the winds are weak or absent. Incoming (N/E) and outgoing (S/W) currents alternate through the sea mouth during the 24-h period, which contribute to the water-mass turnover and strengthen the effects of wind action. These phenomena contribute to the transportation of coastal-neritic species into the basin, but are not sufficient to explain the large diel fluctuations in zooplankton abundance. Independently of the tidal phase and the presence of outgoing currents (which transport the zooplankton organisms out of the basin), high peaks in abundance were observed at night in all the time series. These were due mainly to an increase in coastal-neritic species and to a lesser extent autochthonous forms. Thus it can be hypothesised that these increases in abundance are due to a specific daily migratory behaviour, which seems to be related principally to the light and dark periods. These species may stay near the bottom

during the day to avoid the strong solar radiation in the surface layers (Zaret and Suffern 1976) and then move towards the surface layers during the night probably to avoid predation (Fulton 1984). The only peak in abundance recorded during the day (May, 15:00 h) may have been due to water mixing and re-suspension (caused by opposing effects of tide and wind action) of the near-bottom water layer, in which the highest zooplankton abundance probably occurs during the day.
A sampling strategy specifically designed to study daily vertical migration (using a benthic sledge and plankton net contemporaneously) would be necessary to confirm this hypothesis.

References

Alcaraz M (1983) Coexistence and segregation of congeneric pelagic copepods: spatial distribution of the Acartia complex in the rÌa of Vigo (NW of Spain). J Plankton Res 5: 891-900

Alldredge AL, Hamner WM (1980) Recurring aggregation of zooplankton by a tidal current. Estuarine Coast Mar Sci 10: 31-37

Alldredge AL, King JM (1980) Effects of moonlight on the vertical migration patterns of demersal zooplankton. J Exp Mar Biol Ecol 44: 133-156

Alldredge AL, King JM (1985) The distance demersal zooplankton migrate above the benthos: implications for predation. Mar Biol Berlin 84: 253-260

Belmonte G, Bianchi CN (1992) Zooplankton structure and distribution in a brackish-water basin. Oebalia 18: 1-15

Benzecri JP (1982) L'analyse des donnés. II. L'analyse des correspondances. 3rd edn. Dunod, Paris

Bowman TE, Gonzàlez JG (1961) Four new species of Pseudocyclops (Copepoda: Calanoida) from Puerto Rico. Proc US Nat Mus 113: 37-59

Campolmi M (1998) Studio della comunità zooplanctonica di un bassofondo costiero mediterraneo (Stagnone di Marsala, Sicilia occidentale). PhD thesis, Sci Amb, Univ Messina

Campolmi M, Costanzo G, Crescenti N, Zagami G (1999) First record of the hyperbenthic Calanoid Copepod Metacalanus acutioperculum in the Mediterranean Sea. In: Schram FR and von Vaupel Klein JC (eds) Crustaceans and the Biodiversity Crisis. Proceedings of the 4th Int Crust Cong, Amsterdam, The Netherlands, July 20-24 (1998) 1: 559-567

Cervetto G, Gaudy R, Pagano M, Saint-Jean L, Verriopoulos G, Arfi R, Leveau M (1993) Diel variations in Acartia tonsa feeding, respiration and egg production in a Mediterranean coastal lagoon. J Plankton Res 15: 1207-1228

Checkley DM, Uye S, Dagg MJ, Mullin MM, Omori M, OnbË T, Zhu M-Y (1992) Diel variation of the zooplankton and its environment at neritic stations in the inland sea of Japan and the north-west Gulf of Mexico. J Plankton Res 14: 1-40

Colombo G, Ceccherelli VU, Ferrari I (1983-84) Lo zooplancton delle lagune. Nova Thalassia 6: 185-200

Cronin TW, Forward RB (1979) Tidal vertical migration: an endogenous rhythm in estuarine crab larvae. Science 205: 1020-1022

Damkaer DM (1970) Parastephos occatum, a new species of hyperbenthic copepod (Calanoida: Stephidae) from the inland marine waters of Washington state. Proc biol Soc Washington 83: 505-514

Ferrari I, Carrieri A (1985) Distribuzione e dinamica stagionale dei Cladoceri e dei Copepodi planctonici della Sacca di Scardovari. Nova Thalassia 7(3): 157-162

Ferrari I, Cantarelli MT, Mazzocchi MG, Tosi L (1985a) Analysis of a 24-hour cycle of zooplankton sampling in a lagoon of the Po River Delta. J Plankton Res 7: 849-865

Ferrari I, Carrieri A, Coen R (1985b) Distribuzione delle taxocenosi planctoniche a Copepodi e Cladoceri nella Sacca di Scardovari. Oebalia 11: 187-201

Ferrari I, Colombo G (1988) Struttura e dinamica dello zooplancton nelle lagune salmastre italiane. In: Carrada GC, Cicogna F, Fresi E (eds) Le lagune costiere. Ricerca e gestione CLEM, Massa Lubrese (Napoli), pp 107-117

Frost BW (1988) Variability and possible adaptive significance of diel vertical migration in Calanus pacificus; a planktonic marine copepod. Bull Mar Sci 43: 675-694

Fulton RS (1984) Distribution and community structure of estuarine copepods. Estuaries 7: 38-50

Gagnon M, Lacroix G (1981) Zooplankton sample variability in a tidal estuary: an interpretative model. Limnol Oceanogr 26: 401-413

Genovese S (1963) Osservazioni preliminari sullo zooplancton degli stagni salmastri di Ganzirri e di Faro. Arch Bot Biogeogr Ital 39: 111-114

Jacoby CA, Greenwood JG (1988) Spatial, temporal, and behavioural patterns in emergence of zooplankton in the lagoon of Heron Reef, Great Barrier Reef, Australia. Mar Biol 97: 309-328

Haury LR, McGowan JA, Wiebe PH (1978) Patterns and processes in the time-space scales of plankton distribution. In: Steele J (eds) Spatial patterns in plankton communities. Plenum Press, New York, pp 277-327

Haury L, Yamazaki H, Itsweire EC (1990) Effects of turbulent shear flow on zooplankton distribution. Deep-Sea Res 37: 447-461

Lagadeuc Y, Boulé M, Dodson JJ (1997) Effect of vertical mixing on the vertical distribution of copepods in coastal waters. J Plankton Res 19(7): 1183-1204

Lampert W (1989) The adaptive significance of diel vertical migration of zooplankton. Funct Ecol 3: 21-27

Lebart (1975) Validité des resultats en analyse des données. Centre de Recherches et de Documentation sur la Consommation, Paris. L. L/cd No 4465, pp 1-157

Madhupratap M, Achuthankutty CT, Sreekumaran SR (1991) Zooplankton of the lagoons of the Laccadives: diel patterns and emergence. J Plankton Res 13(5): 947-958

Maranda Y, Lacroix G (1983) Temporal variability of zooplankton biomass (ATP content and dry weight) in the St. Lawrence Estuary: advective phenomena during neap tide. Mar Biol 73: 247-255

Mauchline J (1980) The biology of Mysids. Adv Mar Biol 18: 1-369

Ohman MD (1990) The demographic benefits of diel vertical migration by zooplankton. Ecol Monogr 60: 257-281

Ohtsuka S (1985) Calanoid copepods collected from the near-bottom in Tanabe Bay on the Pacific coast of the middle Honshu, Japan. II. Arietellidae (cont). Publ Seto Mar Biol Lab 30 (4/6): 287-306

Ohtsuka S, Fosshagen A, Putchakarn S (1999) Three new species of the demersal calanoid copepod Pseudocyclops from Phuket, Thailand. Plankton Biol Ecol 46(2): 132-147.

Pielou EC (1984) The interpretation of ecological data. A primer on classification and ordination. John Wiley and Sons, New York

Pusceddu A, Sarà G, Armeni M, Fabiano M, Mazzola A (1999) Seasonal and spatial changes in the sediment organic matter of a semi-enclosed marine system (W-Mediterranean Sea). Hydrobiologia 397: 59-70

Riggio S, Chemello R (1992) The role of coastal lagoons in the emerging and segregation of new marine taxa: evidence from the Stagnone di Marsala Sound (Sicily). Bull Inst Océanogr, Monaco 9: 1-19

Sameoto DD (1975) Tidal and diurnal effects of zooplankton sample variability in a nearshore marine environment. J Fish Res Board Can 32: 347-366

Sameoto DD (1978) Zooplankton sample variation on the Scotian Shelf. J Fish Res Board Can 35: 1207-1222

Scotto di Carlo B (1985) Appunti sullo zooplancton del Mediterraneo. Nova Thalassia 7(3): 83-97

Siegel S (1956) Nonparametric statistics for the behavioural sciences. McGraw-Hill, New York

Youngbluth MJ (1980) Daily, seasonal and annual fluctuations among zooplankton populations in an unpolluted tropical embayment. Estuarine Coast Mar Sci 10: 265-287

Zagami G, Badalamenti F, Guglielmo L, Manganaro A (1996) Short-term variations of the zooplankton community near the Straits of Messina (North-eastern Sicily): relationships with the hydrodynamic regime. Estuarine Coast Shelf Sci 42: 667-681

Zagami G, Guglielmo L (1995) Distribuzione e dinamica stagionale dello zooplancton nei laghi di Faro e Ganzirri. Biol Mar Mediterr 2(2): 83-88

Zaret TM, Suffern JS (1976) Vertical migration in zooplankton as a predator avoidance mechanism. Limnol Oceanogr 21: 804-813

CHAPTER 22

Ultraplankton of the Gulf of Naples by Flow Cytometry

R. Casotti[1], C. Brunet[2], and A. Graziano[1]

ABSTRACT

Ultraplankton composition and distribution were investigated as a function of physical structures and water masses of the Gulf of Naples in November 1995. Cyanobacteria were distributed all over the Gulf, but mostly in the surface mixed layer. Prochlorophytes were more abundant offshore and at depth. Small eukaryotes were typical of the surface of coastal stations. Prochlorophytes vertical distribution showed a peculiar profile, with two populations at depth, where the Modified Atlantic Water (MAW) was present. A periodic sampling of a fixed station in the Gulf of Naples started in June 1998 in order further to investigate environmental parameters ruling this vertical distribution. Preliminary data interpretation is discussed. During the November cruise, bacterial abundances were correlated to chlorophyll *a* concentrations, and, despite the strong oligotrophy of the area, bacterial carbon never exceeded phytoplankton carbon.

Introduction

Phytoplankton represents the first step in the trophic marine structure and it is the main producer of organic matter in the oceans, even in coastal areas, where micro and macro benthonic algae dominate (Charpy-Roubaud and Sournia 1990). Phytoplankton is generally studied using chlorophyll *a* as biomass indicator. Values are obtained by filtering variable amounts of seawater, in order to obtain significative measurements even in oligotrophic regions. Although, such values represent a global phytoplankton community, without considering the algal composition, the trophic and physiological state of the algae, which are the parameters driving primary production. Therefore, valuable information is lost. Recently, discriminant analyses have received increasing attention, and among these, flow cytometry. Flow cytometry allows a fast and precise analysis of particles based on their scattering and fluorescence properties at a rate of thousands of cells per second (Shapiro 1988). Commercially available instruments, though, are limited by the maximum sample volume (fraction of a ml), and by the maximum size of the particles to be analysed (50 mm). For these reasons, flow cytometry of natural samples has been applied mainly to the study of small-sized, very abundant plankton cells, the ultraplankton *sensu* Li (1997), which includes both heterotrophic bacteria and algae (ultraphytoplankton). Ultraphytoplankton (including picophytoplankton, < 2 μm) is composed of photoautotrophic organisms smaller than 5 μm ESD (Equivalent Spherical Diameter) (Shapiro and Guillard 1986), and includes prokaryotic and eukaryotic cells. The prokaryotic component is dominated by unicellular Cyanobacteria (around 1 μm ESD, Waterbury et al. 1986) and Prochlorophytes (around 0.7 μm ESD, Chisholm et al. 1988). The eukaryotic component is mainly represented by small flagellates (2-5 μm ESD, Olson et al. 1989; Longhurst and Pauly 1987). Algae of this size range are very abundant in the oceans and can generate up to 80% of primary production in oligotrophic areas (Li et al. 1983). They represent, together with bacteria, a very important component of microbial food webs, and their discovery unveiled a whole new word on the complex interactions between organisms of different size (Pomeroy 1974). Small autotrophs can use regenerated forms of nitrogen

[1] Stazione Zoologica "A. Dohrn", Villa Comunale, Napoli, Italy
[2] Université du Littoral, Laboratoire Interdisciplinaire des Sciences de l'Environnement (LISE), Wimereux, France

F.M. Faranda, L. Guglielmo, G. Spezie (eds)
Mediterranean Ecosystems: Structures and Processes

remineralized by bacteria or else excreted by heterotrophs, therefore contributing to regenerated production. Indeed, thanks to their high surface-to-volume ratio, they can tolerate strong nutrient limitation and grow in conditions otherwise prohibitive for other algae (Vaulot et al. 1995). As a matter of fact, ultra-autotrophs are dominant in oligotrophic areas of every ocean and also in areas where iron and not nitrogen is the limiting element (e.g. Binder et al. 1996). Thanks to their particular pigment, divinyl chlorophyll *a*, Prochlorophytes can live at the very low light irradiances of the ocean's depths, but thanks to their exceptional photoacclimation properties, they span from the surface to the bottom of the euphotic zone and have been found also in the surface at very high light intensities. This exceptional photoacclimation ability is made possible by the physiological characteristics of prochlorophytes (Partensky et al. 1993), but also by the presence of different populations with different growth optima replacing one the other along the water column (Moore et al. 1995; 1998). Much less information is available on small eukaryotes, due to the heterogeneity of taxonomic groups represented and to the technical difficulties of their study, since they are very delicate cells and very difficult to culture (Andersen et al. 1993; Simon et al. 1994). The very few species cultured are probably not very representative of *in situ* species (Campbell et al. 1994).

Heterotrophic bacteria are also a very important component of microbial food webs and contribute significantly to biogeochemical cycles in the ocean. Their study has also very much improved in precision, speed and ease of analysis since flow cytometry has been used (Li et al. 1995). Much of the information is available from coastal systems and from species easy to grow on plates, but little is known about the importance of free-living planktonic bacteria in pelagic environments. Some authors observed that in oligotrophic areas bacterial biomass often exceeds phytoplankton biomass (Cho and Azam 1990), and termed the phenomenon "the inverted pyramid". The constancy of such a feature is under debate (Li et al. 1992; Robarts et al. 1996). Also in the Mediterranean Sea cyanobacteria and prochlorophytes dominate oligotrophic areas, averaging 71% of total primary production (e.g. Decembrini and Magazzù 1990; Maugeri et al. 1992; Magazzù and Decembrini 1995), but they are very abundant and growing actively also in neritic areas (Vaulot and Partensky 1992).

This study aims at describing and quantifying ultraplankton by flow cytometry in order to describe its general features and assess its contribution to total biomass in a coastal area of the Mediterranean Sea, the Gulf of Naples. The Gulf of Naples is a coastal eutrophic area open to the general circulation of the oligotrophic Tyrrhenian Sea (e.g. Scotto di Carlo et al. 1985 and references therein). From a biological point of view, phytoplankton production shows a surface peak in the spring, and a second development in the autumn, distributed in a thicker layer of the water column (Scotto di Carlo et al. 1985). When meteorological conditions are favourable, a minor peak occurs around mid-November (Zingone et al. 1995). The ultraplankton of the Gulf has been poorly studied. Modigh et al. (1996) investigated cyanobacteria distribution for one year at a coastal station, concluding that their concentrations are maintained stable by grazers control. No data are available on prochlorophytes, neither on planktonic heterotrophic bacteria.

Materials and Methods

The inner part of the Gulf has been sampled on 9th and 10th November 1995 along three coast-offshore transects (Fig. 1), in order to assess general distribution of ultraplankton as a function of the water masses and physical structures present. Successively, several aspects of ultraplankton features have been investigated through the weekly sampling of a fixed coastal station (named "MareChiara", Fig. 1), from which preliminary results will be presented. During the November 1995 cruise, temperature and salinity have been measured by using a CTD probe (Seabird SBE 16). Discrete samples have been taken at 2, 10, 20, 40 and 70 m by Niskin bottles for analysis of different parameters. Chlorophyll *a* concentrations have been estimated by HPLC (Beckman Gold System), using the protocol described in Brunet et al. (1993).

Flow cytometric parameters (FALS, RALS, red from chlorophyll and orange from phycoerythrin fluorescence) have been measured by a FACScalibur instrument (Becton-Dickinson) equipped with an Argon laser emitting at 488 nm and 15 mW. Fluorescence beads of 1 µm (Polyscience, USA) were used as an internal stan-

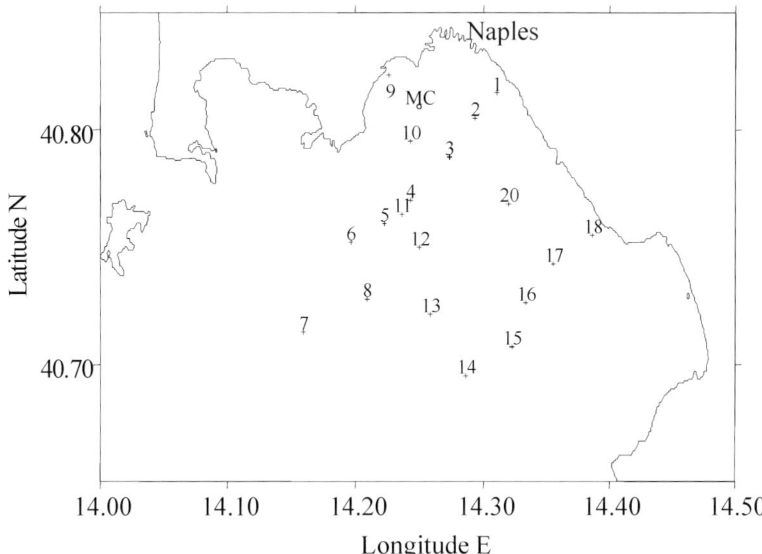

Fig. 1. Gulf of Naples with sampling sites (*MC*, MareChiara station)

dard. Bacteria were counted by flow cytometry after staining with the fluorescent DNA-stain TOTO-1 (Molecular probes, USA), according to the protocol of Li et al. (1995). The sampling of the fixed station "MareChiara" started on June 16th, 1998, and is still in course with a weekly period. Analytical methods are as described above, and discrete samples are taken at 0, 2, 5, 10, 20, 30, 40, 50, 60 and 70 m.

Results and Discussion

Hydrography and Chlorophylla

In November 1995 several water masses occupied the inner part of the Gulf of Naples. Both salinity and temperature increased from the coast to offshore (Fig. 2a,b). These data together with AVHRR satellite image (not shown) helped in identifying a warmer and saltier water mass in the middle of the sampled area, around st. 13, which was interpreted as an eddy of 8 km in diameter, which is in the size range of such mesoscale structures in the Mediterranean Sea (Robinson et al. 1987).

A salinity profile along the transect from station 9 to 14 crossed several physical structures, and is representative of the water column structure at the time (Fig. 3). Apart from the eddy, several fronts were found at the interface with surrounding waters, both vertical and horizontal. A transition layer separates the surface layer (0-50 m), from a low salinity water mass, present below

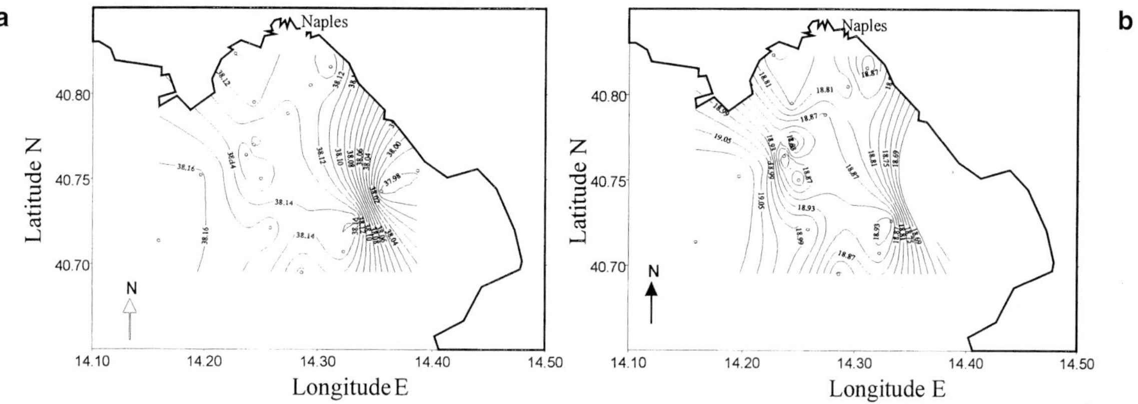

Fig. 2a,b. Surface salinity (psu) (**a**) and temperature (°C) (**b**) distribution in the Gulf of Naples, 9th and 10th November 1995

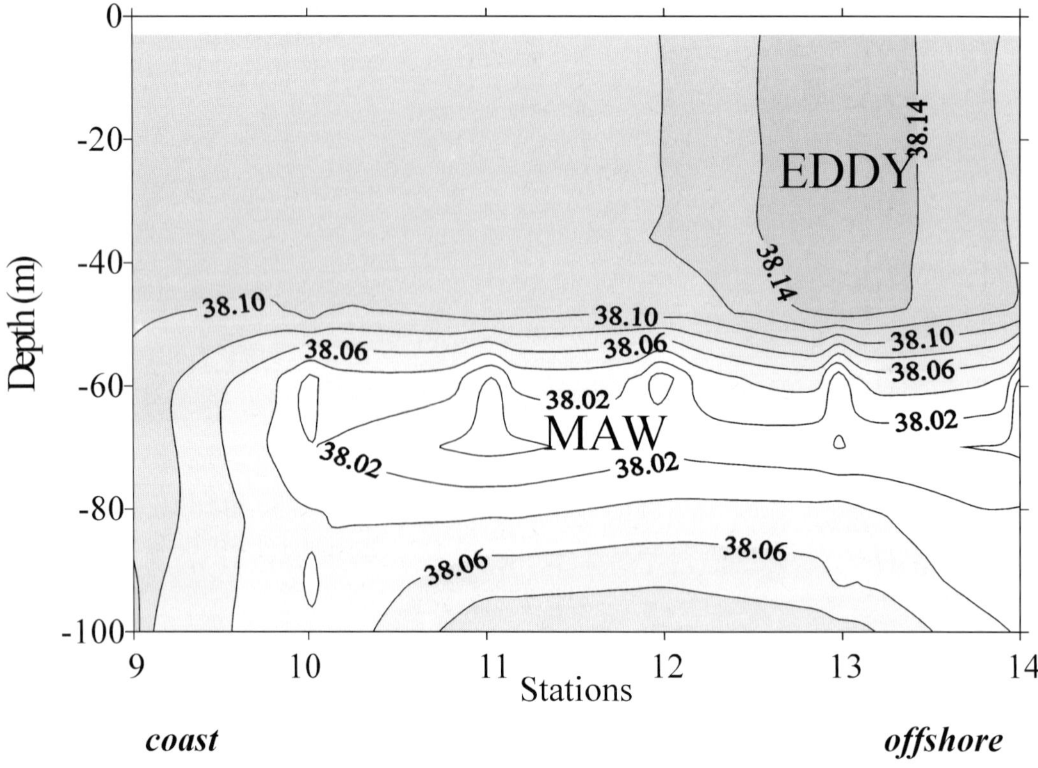

Fig. 3. Salinity distribution (psu) along a coast-to-offshore transect, from station 9 to 14, November 1995

60 m. It occupied the depth range of 70 m (±10 m) and had a water type of ~15.60°C and ~38.00 psu. Such a water mass is referred to as the Modified Atlantic Water (MAW), which is known to intrude the Tyrrhenian Sea at times (Fedorov 1972; Povero et al. 1990). Chlorophyll concentrations ranged from 0.26 to 1.16 µg/l, and were lowest inside the eddy (st. 13). Sixty percent of values lie below 0.5 µg/l, conferring to the area a character of oligotrophy (data not shown).

Ultraphytoplankton

Cyanobacteria were distributed all over the Gulf and were more abundant in the upper 40 m (Fig. 4). Concentrations averaged 10^4 cell/ml. Prochlorophytes were more abundant offshore and in the eddy, attaining concentrations of 10^3 cell/ml (Fig. 5).

No correlation was found between Prochlorophytes and Cyanobacteria abundances or cell parameters (P>0.05), due to a different vertical distribution of these two cell types. Cyanobacteria were limited above the thermocline while Prochlorophytes extended all over the water column, slightly increasing at depth. The lack of a DCM and the complementary distribution of Cyanobacteria and Prochlorophytes more resembles distribution in true oligotrophic regions, such as the Sargasso Sea (Chisholm et al. 1988; Li and Wood 1988) or the Pacific Ocean (Campbell et al. 1997), than what is found in more coastal areas, such as the northwestern Mediterranean Sea (Vaulot et al. 1990) or the Catalan Sea (Bautista and Gomez 1996), where both groups decrease with depth or subsurface peaks are noted. Also, no correlation was evident between cyanobacteria and prochlorophytes cell parameters (light scatter or fluorescence), and from this we must conclude that the two cell types did not respond to the same environmental constraints at the time of sampling.

The vertical distribution of Prochlorophytes shows a typical profile. In fact, multiple populations were present in single water samples, as evidenced by the bimodal distribution of the chlorophyll fluorescence (Fig. 6). Recent studies have shown that these populations are genetically distinct, adapted for optimal growth under different conditions but co-occurring in environ-

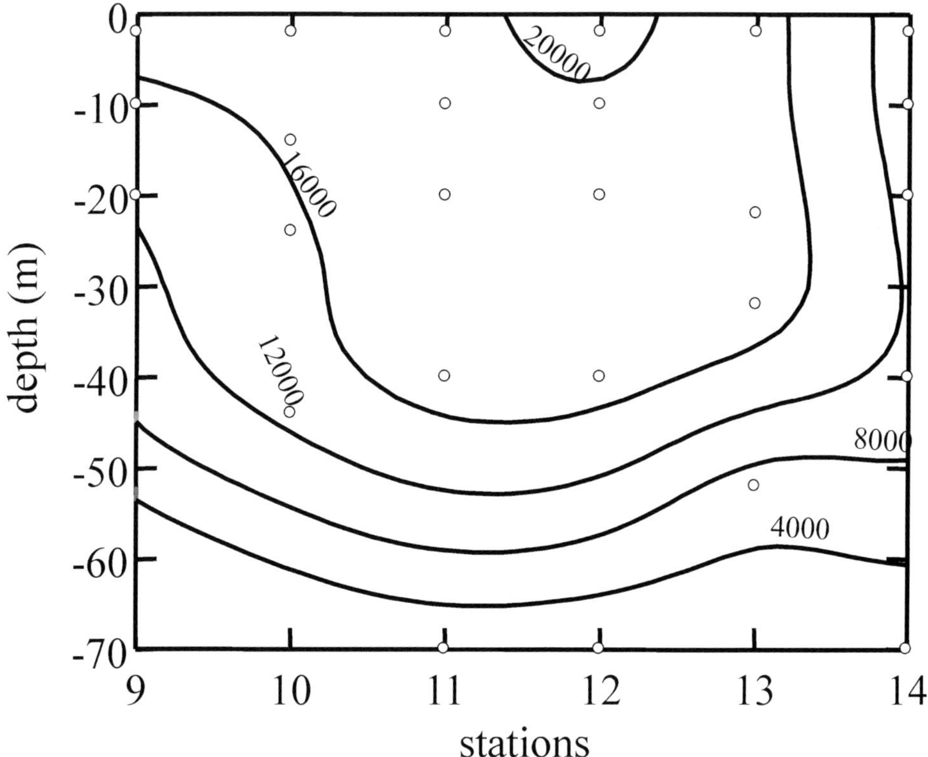

Fig. 4. Cyanobacteria distribution (cell/ml) along a coast-to-offshore transect, from station 9 to 14, November 1995

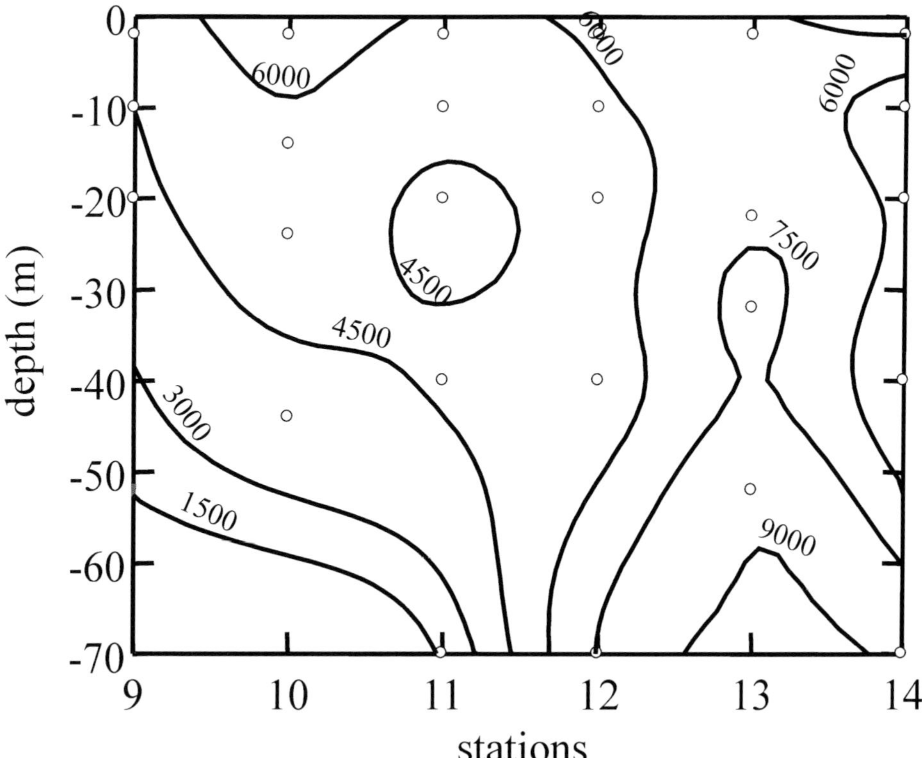

Fig. 5. Prochlorophytes distribution (cell/ml) along a coast-to-offshore transect, from station 9 to 14, November 1995

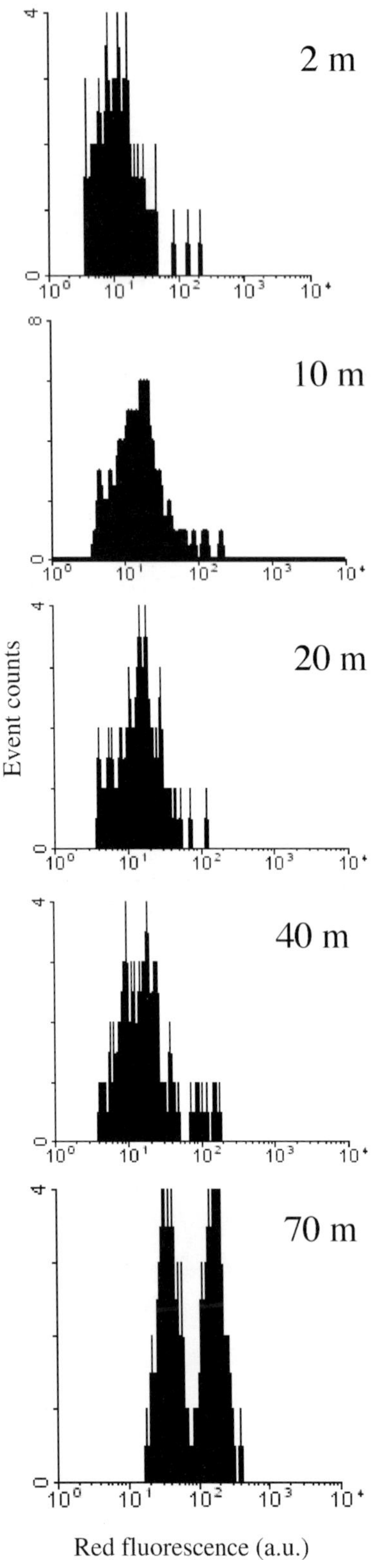

Fig. 6. Vertical profile of prochlorophyte red fluorescence (arbitrary units, relative to fluorescent beads) from station 13. Note the bimodal distribution at 70 m, where the MAW is present

ments where conditions are favourable for both (Moore et al. 1995, 1998). In our case, the contemporary presence of these populations occurred exclusively at 70 m, inside the MAW, and, together with other parameters, contributed to its characterization. Such coexistence of populations has been observed by other authors in the Pacific Ocean (Campbell and Vaulot 1993), the Red Sea (Veldhuis and Kraay 1993) or the Atlantic Ocean (Partensky et al. 1996), while this is the first report for the Mediterranean Sea.

In order to investigate the factors causing this bimodal distribution, a more systematic study of a coastal stations started in June 1998. Preliminary data from June 16th to September 15th will be presented here. So far, the two ecotypes have always been observed at depth. The upper limit of the layer of coexisting ecotypes varies with time (Fig. 7). Only on one date, September 15th (arrow in Fig. 7), could we observe both limits of the "coexistence layer". This sampling followed the first strong wind event after the summer stability, and it induced the breaking of the surface stratification and a slight cooling of surface waters (Fig. 8a,b). The mixing brought deep and saltier water in the upper levels of the water column, pushing the "coexistence layer" up in the water column. A significative correlation has been found between salinity and the presence of the two populations of Prochlorophytes (Spearman, r=0.65, P<0.001; n=78). Average value of salinity of the "coexistence layer" is 37.868 psu (SD=0.056, n=23), while values corresponding to only one population (the dimmer) average 37.657 psu (SD=0.261, n=55). A comparison of salinity means (non-parametric test of Mann-Whitney 1947) shows that the two values are highly significantly different (U=120, P<0.001).

The co-occurrence of the two populations of Prochlorophytes appears to be a characteristic of deep waters, with salinity acting as a threshold. Although we do not know whether the co-presence is due to passive mixing of waters with different salinities or to active growth of one or the other type of cells. Further analyses will try to address the question whether physics or also biology causes the observed distribution. As a matter of fact, if culture observations can be applied to these Mediterranean populations, then light must play a main role in limiting the deeper population (Moore et al. 1995). Small eukaryotes were never very abundant, rarely

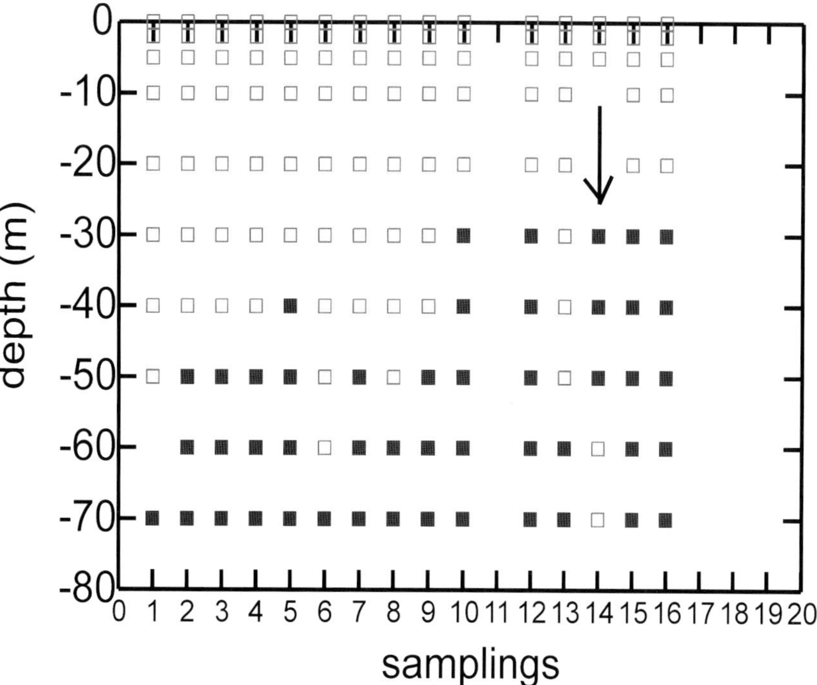

Fig. 7. Presence of two populations of Prochlorophytes (*closed symbols*) at station MareChiara in the Gulf of Naples. Sampling started on June 16th, 1998, and proceeded with a weekly period. Last sampling was on September 29th. *Open squares* indicate the presence of only one population of prochlorophytes (*the dimmer*). The *arrow* indicates the sampling on September 15th, where the two populations of prochlorophytes were present in a discrete layer between 30 and 50 m depth

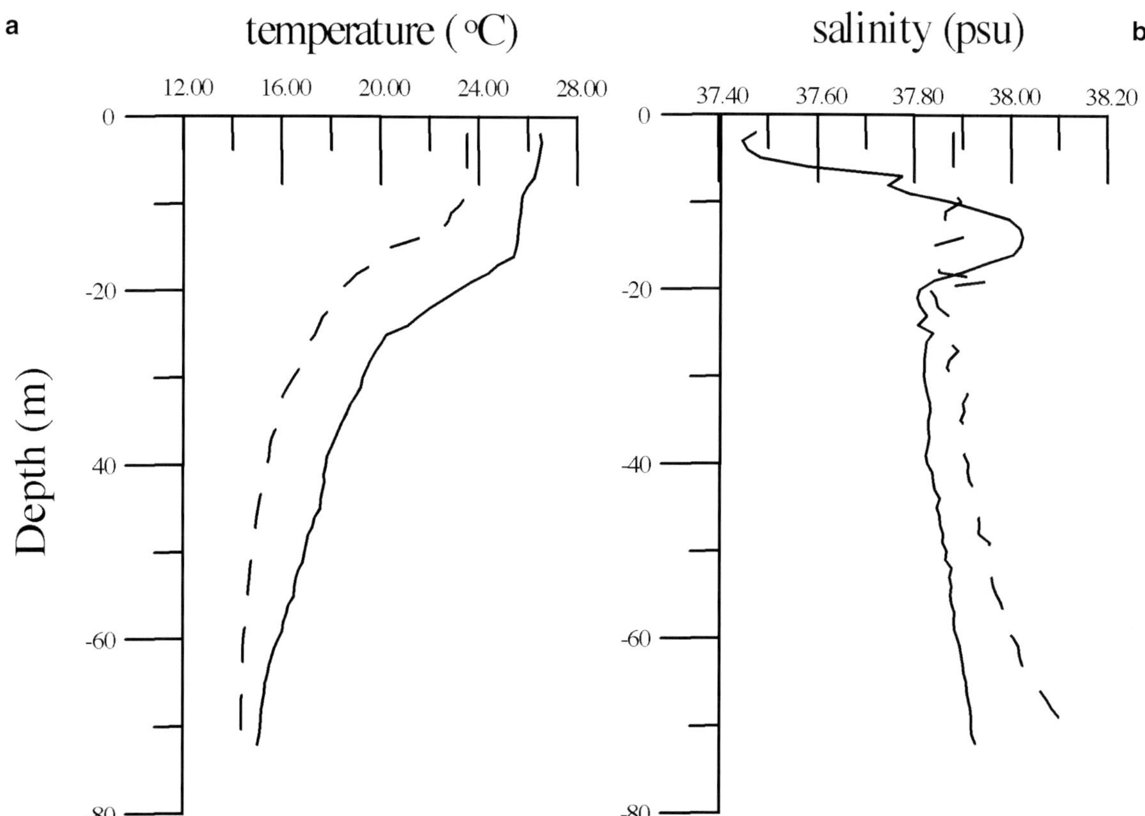

Fig. 8a,b. Comparison of temperature (**a**) and salinity (**b**) profiles at MareChiara station in the Gulf of Naples on September 7th (*solid line*) and September 15th (*broken line*)

attaining 10^3 cell/ml, and were present mostly at the surface of coastal stations (data not shown). No parameter allowed us to identify them at the taxonomic level. Cyanobacteria represented 65% (SD ±24%) of ultraphytoplankton Carbon, Prochlorophytes 15% (SD ±22%), while picoeukaryotes accounted for 27% (SD ±16%).

Bacteria

Heterotrophic bacteria were more abundant at surface and at coastal stations. Concentrations range from 2 x 10^5 cell/ml inside the eddy to 1.9 x 10^6 cell/ml at stations 17 and 18, located in the eastern part of the Gulf, an area of intense city and river runoffs (Fig. 9). A significative correlation with chlorophyll *a* concentrations has been found (Fig. 10), suggesting that bacterial biomass is controlled by phytoplankton production and is ruled by the same environmental variables as phytoplankton, in agreement with what was observed by Ducklow (1992) and Li et al. (1995). When C concentrations are estimated from cell number and chlorophyll concentrations (Buck et al. 1997), we find that bacterial C never exceeded phytoplankton C. Integrated values over the first 20 m averaged 617 µgC/m^2 for phytoplankton (SD 287, 74% of total), while bacterial carbon averaged 204 µgC/ m^2 (SD 71, 26% of total). Therefore, in spite of the general oligotrophy of the area, no inverted pyramid is observed, as opposite to what is found by other authors (e.g. Cho and Azam 1990).

From our data we may agree with Li et al. (1992) that in oligotrophic conditions bacterial biomass does not always dominate phytoplankton biomass. On the other hand, the oligotrophy of the Gulf is not a steady state situation, and is ruled by the external water mass penetrating the inner part of the Gulf (eddy). Although we do not know how often these phenomena affect the

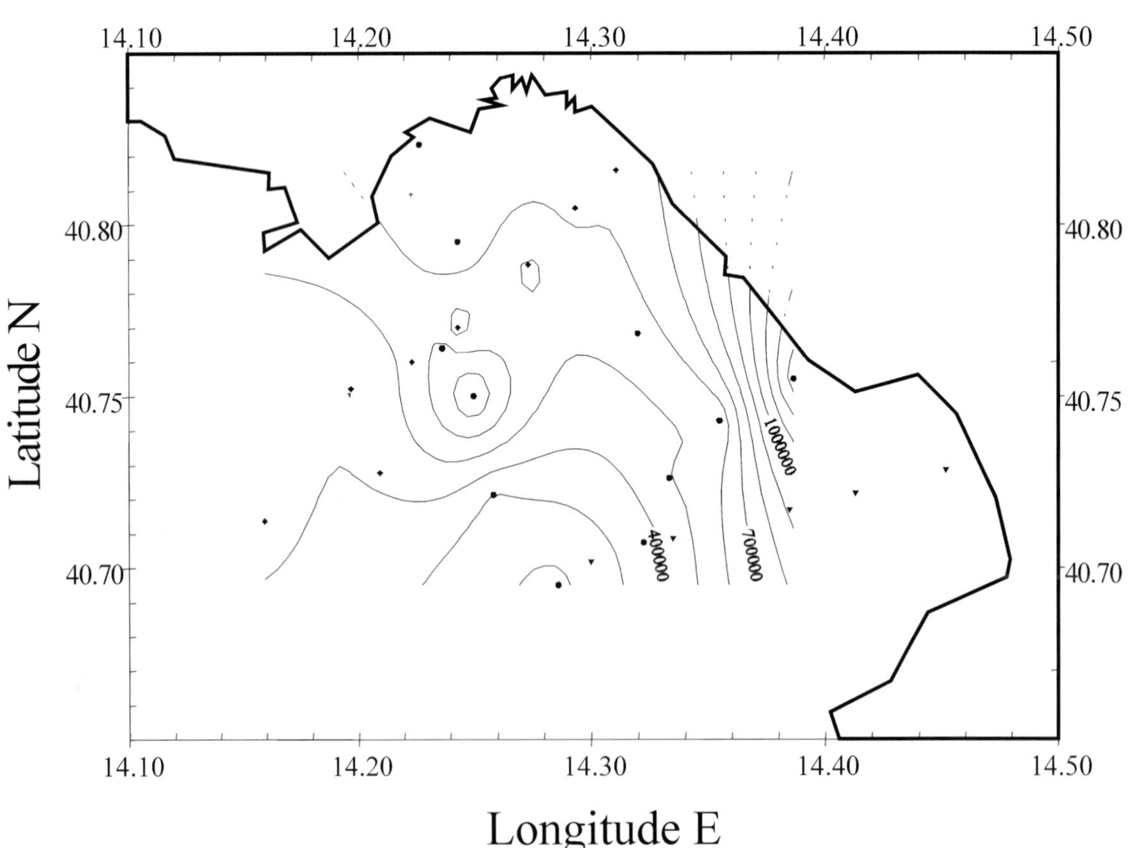

Fig. 7. Distribution of bacteria (cell/ml) at surface in the Gulf of Naples in November 1995

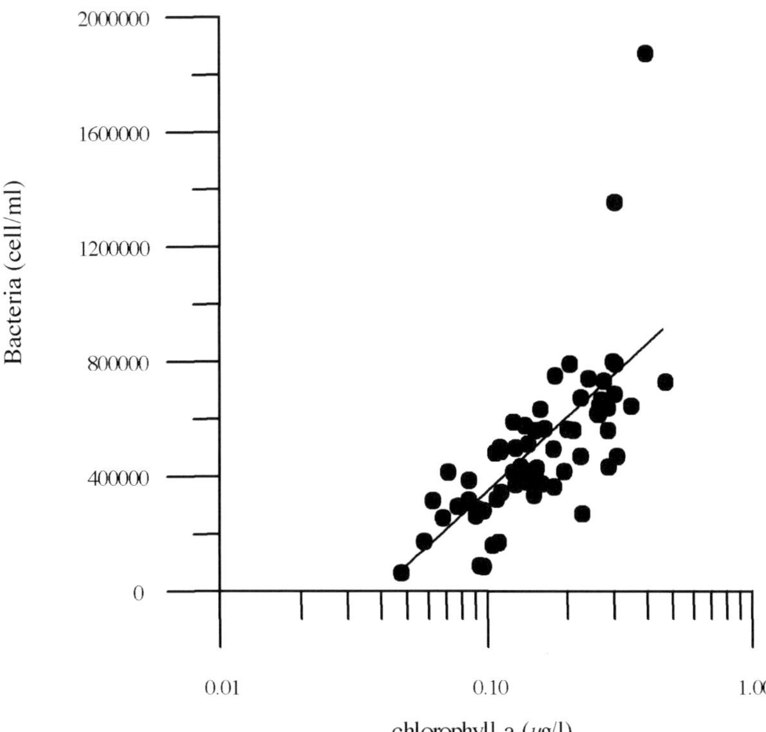

Fig. 10. Correlation between bacteria (cell/ml) and chlorophyll *a* concentrations (μg/l). The *line* is based on a log-linear regression. Bacteria = 371128x log(chlorophyll)+1.205x10^6. r=0.69, n=65 p<0.001

Gulf, we can reasonably expect that the oligotrophy of the area is a transient and unstable condition.

Conclusions

This study shows that the ultraplankton features are ruled by the physics of the area at mesoscale. The physical variability determines the biological variability and, in turn, biological features can be used as markers of water masses or physical structures. An example of this is ultraplankton, due to its small size and lack of independent motion. The Gulf of Naples appears to be at an alternating balance between the oligotrophic offshore system and the coastal meso-eutrophic waters, similar to that observed by Scotto di Carlo et al. (1985) and Zingone et al. (1995). In general, we conclude that, although chlorophyll concentrations are typically oligotrophic, and so are nutrient concentrations (data not shown), the planktonic compartment of the Gulf of Naples maintains its mesotrophic characteristics, even at seasons not favourable to algal development.

Acknowledgements. The authors wish to thank Federico Corato and Gianluca Zazzo for their technical help with sampling and data handling and the captain and the crew of R/V Vettoria.

References

Andersen RA, Saunders GW, Paskind MP, Sexton J (1993) Ultrastructure and 18S rRNA gene sequence for *Pelagomonas calceolata* gen. Et sp. nov. and the description of a new algal class, the Pelagophyceae classis nov. J Phycol 29: 701-715

Bautista B, Jimenez-Gomez F (1996) Ultraphytoplankton photoacclimation through flow cytometry and pigment analysis of mediterranean coastal waters. Sci Marina 60 (Suppl I): 233-241

Binder BJ, Chisholm SW, Olson RJ, Frankel SL, Worden AZ (1996) Dynamics of picophytoplankton, ultraphytoplankton and bacteria in the central equatorial Pacific. Deep-Sea Res II 43(4-6): 907-931

Brunet C, Brylinski JM, Lemoine Y (1993) *In situ* variations of the xanthophylls diatoxanthin and diadinoxanthin: photoadaptation and relationships with an hydrodynamical system in the eastern English Channel. Mar Ecol Prog Ser 120(1): 69-77

Buck KR, Chavez FP, Campbell L (1997) Basin-wide distributions of living carbon components and the inverted trophic pyramid of the central gyre of the North Atlantic Ocean, summer 1993. Aquat Micro Ecol 10: 283-298

Campbell L, Liu H, Nolla HA, Vaulot D (1997) Annual variability of phytoplankton and bacteria in the subtropical North Pacific Ocean at Station ALOHA during the 1991-1994 ENSO event. Deep-Sea Res I 44(2): 167-192

Campbell L, Shapiro LP, Haugen E (1994) Immunochemical characterization of eukaryotic ultraplankton from the Atlantic and Pacific Oceans. J Plankton Res 16: 35-51

Campbell L, Vaulot D (1993) Photosynthetic picoplankton community structure in the subtropical North Pacific Ocean near Hawaii (station ALOHA). Deep-Sea Res I 40(10): 2043-2060

Charpy-Roubaud C, Sournia A (1990) The comparative estimation of phytoplanktonic, microphytobenthic and macrophytobenthic primary production in the oceans. Mar Microb Food Webs 4(1): 31-57

Chisholm SW, Olson RJ, Zettler ER, Goericke R, Waterbury JB, Welschmeyer NA (1988). A novel free-living prochlorophyte abundant in the oceanic euphotic zone. Nature 334: 340-343

Cho BC, Azam F (1990) Biogeochemical significance of bacteria biomass in the ocean's euphotic zone. Mar Ecol Prog Ser 63: 253-259

Decembrini F, Magazzù G (1990) Clorofilla, numero di assimilazione e ATP del picoplancton fotosintetico nei mari italiani. Oebalia 16(1): 443-456

Ducklow HW (1992) Factors regulating bottom-up control of bacteria biomass in open ocean plankton communities. Ergeb Limnol 37: 207-217

Fedorov KN (1972) Temperature inversions in the Red and Mediterranean Sea. Oceanology 12 (6): 795-803

Li WKW (1997) Cytometric diversity in marine ultraphytoplankton. Limnol Oceanogr 42 (5): 874-880

Li WKW, Dickie PM, Irwin BD, Wood AM (1992) Biomass of bacteria, cyanobacteria, prochlorophytes and photosynthetic eukaryotes in the Sargasso Sea. Deep-Sea Res 39(3/4): 501-519

Li WKW, Jellet JF, Dickie PM (1995) DNA distributions in planktonic bacteria stained with TOTO or TO-PRO. Limnol Oceanogr 40(8): 1485-1495

Li WKW, Subba Rao DV, Harrison WG, Smith JC, Cullen JJ, Irwin B, Platt T (1983) Autotrophic picoplankton in the tropical ocean. Science 219: 292-295

Li WKW, Wood AM (1988) Vertical distribution of North Atlantic ultraphytoplankton: analysis by flow cytometry and epifluorescence microscopy. Deep-Sea Res 35 (9): 1615-1638

Longhurst AR, Pauly D (1987) Ecology of Tropical Oceans. Academic Press, London

Magazzù G, Decembrini F (1995) Primary production, biomass and abundance of phototrophic picoplankton in the Mediterranean Sea: a review. Aquat Microb Ecol 9: 97-104

Mann HB, Whitney DR (1947) Ann Math Stat 18: 50-54

Maugeri TL, Acosta Pomar MLC, Bruni V, Salomone L (1992) Picoplankton e picophytoplankton in the Ligurian Sea and in the Straits of Messina (Mediterranean Sea). Bot Mar 35: 493-502

Modigh M, Saggiono V, Ribera d'Alcalà M (1996) Conservative features of picoplankton in a Mediterranean eutrophic area, the Bay of Naples. J Plankton Res 18(1): 87-95

Moore LR, Goericke R, Chisholm SW (1995) Comparative physiology of *Synechococcus* and *Prochlorococcus*: influence of light and temperature on growth, pigments, fluorescence and absorptive properties. Mar Ecol Prog Ser 116: 259-275

Moore LR, Rocap G, Chisholm SW (1998). Physiology and molecular phylogeny of coexisting *Prochlorococcus* ecotypes. Nature 393: 464-467

Olson RJ, Zettler ER, Andersen OK (1989) Discrimination of eukaryotic phytoplankton cell types from light scatter and autofluorescent properties measured by flow cytometry. Cytometry 10: 636-643

Partensky F, Blanchot J, Lantoine F, Neveux J. Marie D (1996) Vertical structure of picophytoplankton at different trophic sites of the tropical northeastern Atlantic Ocean. Deep-Sea Res I 43(8): 1191-1213

Partensky F, Hoepffner N, Li WKW, Ulloa O, Vaulot D (1993) Photoacclimation of *Prochlorococcus sp.* (Prochlorophyta) strains isolated from the North Atlantic and the Mediterranean Sea. Plant Physiol 101: 285-296

Pomeroy LR (1974) The ocean's food web, a changing paradigm. Bioscience 24: 499-504

Povero P, Hopkins TS, Fabiano M (1990) Oxygen and nutrient observations in the Southern Tyrrhenian Sea. Oceanol Acta 13(3): 229-305

Robarts RD, Zohary T, Waiser MJ, Yacobi YZ (1996) Bacterial abundance, biomass and production in relation to phytoplankton biomass in the Levantine Basin of the southeastern Mediterranean Sea. Mar Ecol Prog Ser 137:273-281

Robinson AR, Hecht A, Pinardi N, Bishop J, Leslie WG, Rosentroub Z, Mariano AJ, Brenner S (1987) Small synoptic/mesoscale eddies and energetic variability of the eastern levantine Basin. Nature 327: 131-134

Scotto di Carlo B, Tomas CR, Ianora A, Marino D, Mazzocchi MG, Modigh M, Montresor M, Petrillo L, Ribera d'Alcalà M, Saggiomo V, Zingone A (1985) Uno studio integrato dell'ecosistema pelagico costiero del Golfo di Napoli. Nova Thalassia 7: Suppl.3: 99-128

Shapiro HM (1988) Practical flow cytometry, 2nd edn Alan R. Liss, New York

Shapiro LP, Guillard RRL (1986) Physiology and ecology of the marine eukaryotic ultraplankton. Can J Fish Aquat Sci 214: 371-389

Simon N, Barlow RG, Marie D, Partensky F, Vaulot D (1994) Characterization of oceanic photosynthetic picoeukaryotes by flow cytometry. J Phycol 30: 922-935

Vaulot D, Partensky F (1992) Cell cycle distributions of prochlorophytes in the northwestern Mediterranean Sea. Deep-Sea Res 39, 727-742

Vaulot D, Marie D, Olson RJ, Chisholm SW (1995) Growth of *Prochlorococcus*, a photosynthetic prokaryote, in the Equatorial Pacific Ocean. Science 268: 1480-1482

Vaulot D, Partensky F, Neveux J, Mantoura RFC, Llewellyn C (1990) Winter presence of prochlorophytes in surface waters of the northwestern Mediterranean Sea. Limnol Oceanogr 35: 1156-1164

Veldhuis MJW, Kraay GW (1993) Cell abundance and fluorescence of picoplankton in relation to growth irradiance and nitrogen availability in the Red Sea. Neth J Sea Res 31: 135-145

Waterbury JB, Watson SA, Valois FW, Franks DG (1986) Biological and ecological characterization of the marine unicellular cyanobacterium Synechococcus. In: Platt T, Li WKW (eds) Photosynthetic picoplankton. Can Bull Fish Aquat Sci 214: 71-120

Zingone A, Casotti R, Ribera d'Alcalà M, Scardi M, Marino D (1995) "St. Martin's Summer": the case of an autumn phytoplankton bloom in the Gulf of Naples (Mediterranean Sea). J Plankton Res 17(3): 575-593

CHAPTER 23

Distribution and Ecology of Mesozooplankton in the Northern and Central Adriatic Sea

O. Sidoti, G. Zagami, A. Granata, G. Brancato, L. Guglielmo, and M. Campolmi

ABSTRACT
Within the framework of the Prisma 2 Project, four oceanographic cruises were carried out in the central and northern Adriatic Sea from June 1996 to March 1997. Samples were collected both by BIONESS electronic multinet (204 samples from 54 sites) and by WP2 (101 samples from 19 sites) along inshore-offshore sections. The spatial-temporal distribution of the zooplankton community was analysed in relation to the variability of physico-chemical and biological parameters. The seasonal succession of the zooplankton community was characterized by an inversion of the dominance ratio between copepods and cladocerans. In early June, copepods and cladocerans represent on average 52 and 20% of the zooplankton community, respectively, while in late summer they represent 17 and 66%. During late spring-summer, the cladoceran population was clearly dominated by *Penilia avirostris*, which in some coastal sites constituted more than 90% of the zooplankton. In the coastal zone, the copepod population was characterized by low species diversity and greater dominance of *A. clausi*, *P. parvus* and *T. stylifera* (73% of the population). In the offshore zone of the neritic system, there was instead a more homogeneous copepod species composition. Because of their spatial distribution patterns, *P. elongatus* and *T. longicornis*, typical of estuarine environments, can be considered as hydrological indicator species of different water masses of the Adriatic neritic system. In winter, the zooplankton community was characterized by strong dominance of copepods (31 species identified), on average constituting 75% of the zooplankton, followed by invertebrate larvae, appendicularians, cladocerans, siphonophores and chaetognaths. From our analysis of the spatial distribution of the zooplankton community, we have formulated a preliminary trophic model concerning the association of ecologically similar species in the water masses of the Adriatic neritic system.

Introduction

The zooplankton of the Adriatic Sea has been studied since the first half of the last century (Claus 1881; Car 1890; Graeffe 1900; Steuer 1902a, b, 1910a, b). In particular, many studies have been conducted on the distribution of copepods (Hure et al. 1980; Specchi and Fonda Umani 1987, 1996; Ghirardelli et al. 1989; Fonda Umani et al. 1992) and on their medium and long term variations (Regner 1985).

Particular morphological, hydrological and hydrodynamic features of the northern Adriatic Sea influence the abundance and distribution of mesozooplankton. In fact, this basin is characterized by abundant biomass (Benovic et al. 1984) which decreases southward. However, the structure of the mesozooplankton community exhibits low diversity in the northern zone, which increases toward the southern zone and offshore (Fonda Umani et al. 1992). Hure et al. (1980) have identified three copepod communities: estuarine, coastal and oceanic.

The main aim of this study was to analyse the spatial distribution of the mesozooplankton in relation to the variability of physico-chemical and biological factors. In addition, we studied the structure of the community in order to formulate a preliminary trophic model of the northern and central Adriatic neritic system.

Dipartimento di Biologia Animale ed Ecologia Marina, Università di Messina, Salita Sperone 31, 98166 Messina, Italy

F.M. Faranda, L. Guglielmo, G. Spezie (eds)
Mediterranean Ecosystems: Structures and Processes

Materials and Methods

Study Area

The northern Adriatic basin is delimited in the south off Ancona, with varying depths from around 30 m to about 70 m (Franco 1983). The circulation and distribution of the water masses are strongly influenced by the bottom morphology (Franco et al. 1982), by meteorological conditions (Buljan and Zore-Armanda 1976; Franco 1973, 1983; Franco et al. 1982) and by large fresh water contributions from Italian rivers (Fonda Umani et al. 1994). In the northern basin, a frontal system is established throughout most of the year (Franco 1983; Fonda Umani et al. 1992), which clearly separates the neritic eutrophic waters from the oligo-mesotrophic offshore ones, producing two independent and ecologically different systems (Fonda Umani et al. 1994). These characteristics determine cyclic variations of the vertical stability of the water column; in winter, there is complete instability, while a clear stratification is evident during the rest of the year (Fonda Umani et al. 1992).

The central basin, from Ancona south to a transverse line off Pescara (Artegiani et al. 1993), exhibits a north to south transition in its features. It is characterized by increasing depths of the bottom until the meso-Adriatic depths. In this area, the characteristics of the neighbouring basins alternately prevail during the year.

Sampling Procedures

Within the framework of the Prisma 2 Project, four oceanographic cruises were carried out – two in the northern Adriatic Sea (June 1996 and February 1997) and two in the central Adriatic (August-September 1996 and February-March 1997). In the northern Adriatic, two sampling zones were chosen: the first north off Ravenna (northern area), the second north off Ancona (central-northern area). Two hundred and four samples were collected by the electronic multinet BIONESS from 54 sites located along inshore-offshore sections (Fig. 1a,b). The BIONESS (0.25 m^2 mouth area and 5 nets of 230 µm mesh size) was towed at a speed of 1-1.5 m/s. Depending on the bottom depth, samples were collected at 5-10 m intervals of the water column. More details of BIONESS features can be found in Guglielmo et al. (1998). In the central Adriatic, 101 samples were collected with a WP2 net from 19 fixed sites situated in three sections perpendicular to the

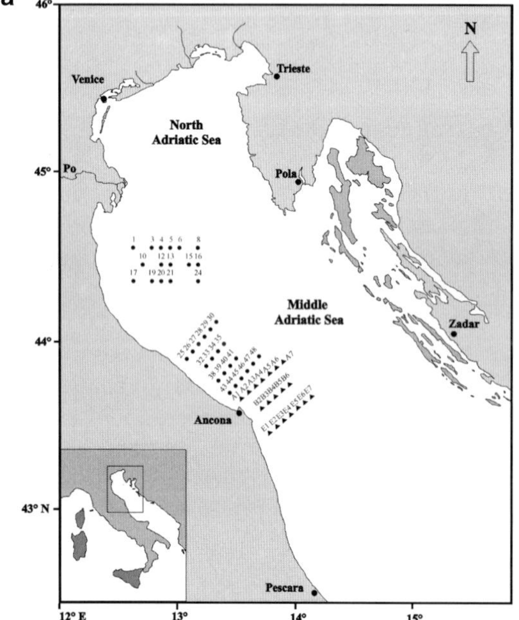

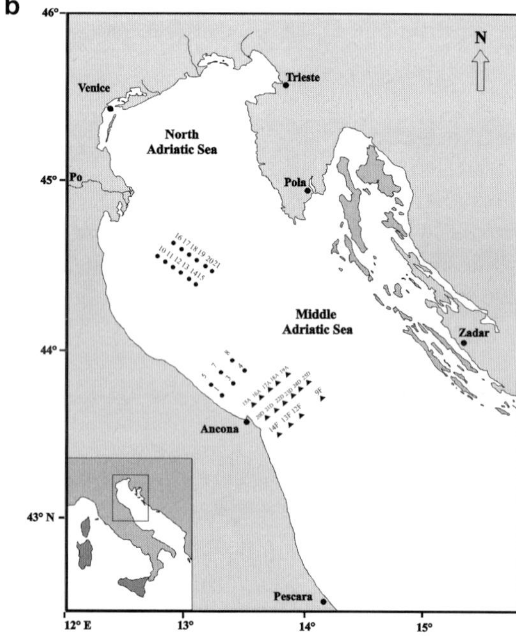

Fig. 1a,b. Sampling area and location of sites during the northern and central Adriatic cruises. **a** Summer (• June '96, ▲ August-September '96). **b** Winter (• February '97, ▲ March '97)

coast (Fig. 1a,b). At the sites with a clear thermocline, two vertical samplings were carried out, the first from the bottom up to the surface, the other one from the thermocline up to the surface. On board, the samples were preserved in a 4% buffered formaldehyde and sea water solution.

Analysis of the Samples

In the laboratory, a qualitative-quantitative analysis of the mesozooplankton was performed on subsamples from 1/10 to 1/20, depending on the total sample richness, while the whole samples were observed in their entirety for the identification of rare species. The specimens were sorted and divided per taxon. The copepods (adults) and cladocerans were identified at the species level, while the remaining groups were classified at higher taxonomic levels. Biomass values (as dry weight) were estimated from 500 ml of the whole sample, according to Lovegrove's method (1966). Abundance was expressed as ind/m^3 and biomass as mg/m^3. These parameters were calculated according to the bottom depth along onshore-offshore sections; the groups of sites were re-named as no. 1 to no. 6.

Data Processing

The data were log-transformed to avoid the excessive dominance of more abundant species in the analysis. The Bray Curtis dissimilarity coefficient was used to construct the between-sites distance matrix (Bakus 1990); using the furthest neighbour amalgamation rule, we obtained a classification of sites shown by dendrograms.

Results

BIONESS: Temperature-Salinity-Fluorescence Isolines

In June, temperature and salinity values (recorded by the BIONESS) revealed a strong vertical thermohaline water stratification, which determined a frontal system separating neritic ecosystem water masses (close to 40 m depth) with a pycnocline from 10-15 m depth (Fig. 2a,b). During this period, two sub-systems were identified in the neritic system and defined as coastal and offshore zones. The former included the warmer and less salty water masses from the coast to 20 m depth, while, in the layer over the thermocline, it spread outward until 40 m depth. The latter included the water masses under the pycnocline spreading from 20 m depth to the whole neritic system. Mean values of temperature and salinity of the water masses over and under the thermocline were 20.-13°C and 35-37.4 PSU, respectively.

The spatial distribution of fluorescence exhibited a decreasing inshore-offshore trend, ranging from 0.45-1.2 mg/m^3 chl*a* over the pycnocline. Another maximum (0.75-1.3 mg/m^3 chl*a*) was found close to the bottom at the offshore sites (Fig. 2c). During winter, a persistent unstable frontal system was caused by large fluctuations of meteo-marine conditions and river flows. The neritic system was characterized by a minor extension of the less salty water of the coastal zone and by a shift of the halocline up to the surface (Fig. 2d, e). The spatial distribution of mean temperature and salinity ranged from 9.38-11.64°C and from 35.46-37.85 PSU, respectively. During this period, the fluorescence values were lower (0.1-0.45 mg/m^3 chl*a*), with decreasing inshore-offshore and surface-bottom trends (Fig. 2f). Similar patterns, but with slightly different ranges, were recorded during August-September and March in the central Adriatic Sea.

Zooplankton Abundance and Spatial Distribution

Late Spring-Summer '96

The zooplankton community showed wide quantitative oscillations, both of abundance (501-25292 ind/m^3) and biomass (3.80 - 173.25 mg/m^3); the means ± SD were 3759 ± 3767 ind/m^3 and 27.06 ± 20 mg/m^3, respectively. The spatial distribution of abundance exhibited strongly decreasing inshore-offshore (Fig. 3) and surface-bottom (Fig. 4a,c) trends, mainly determined by cladocerans and especially by *Penilia avirostris*. The copepods, instead, showed higher values in the layer under the pycnocline.

The seasonal succession of zooplankton was characterized by a temporal inversion of the

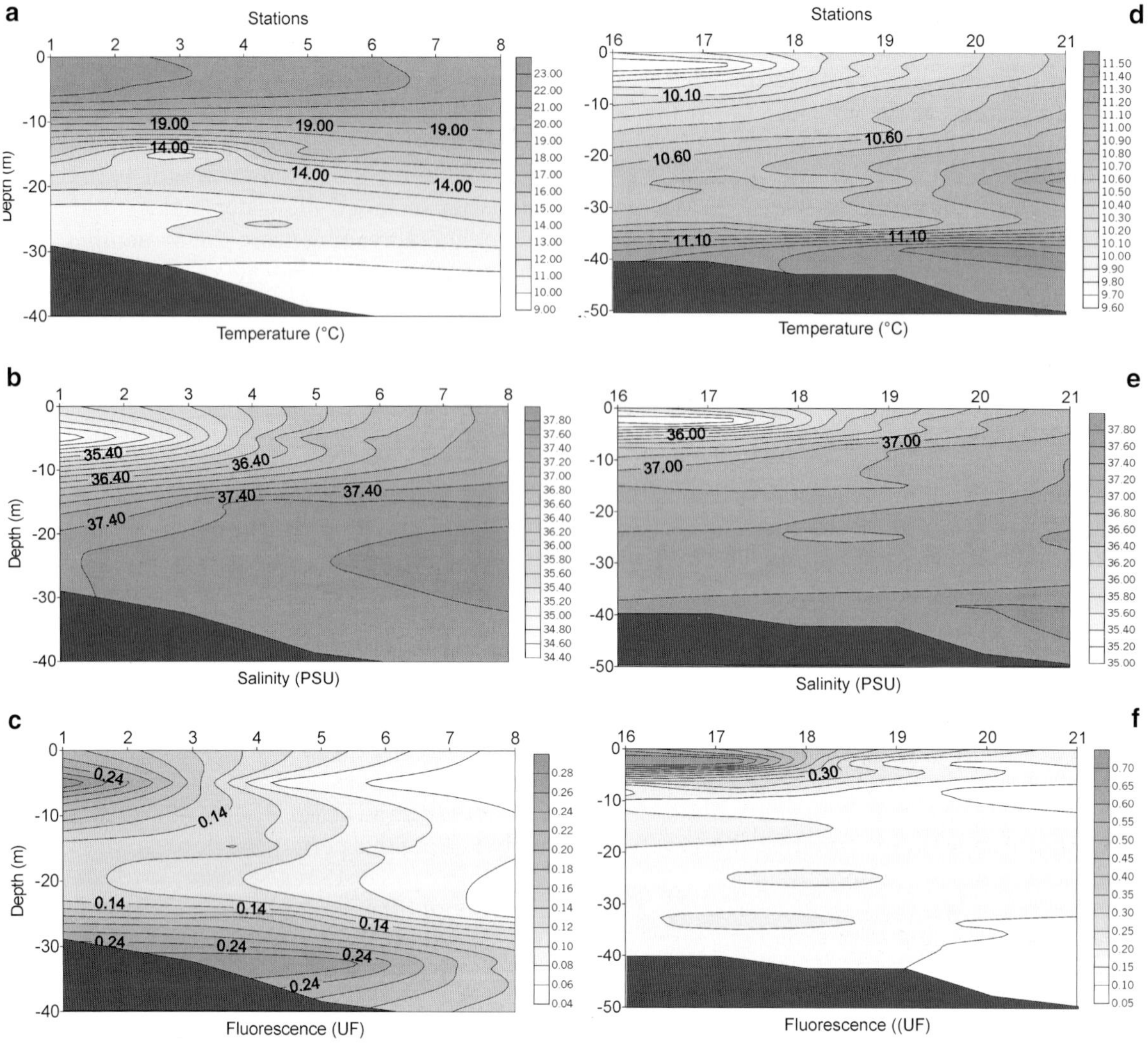

Fig. 2a-f. Spatial distribution of salinity, temperature and fluorescence. **a-c** Summer (sections 1-8); **d-f** Winter (sections 16-21)

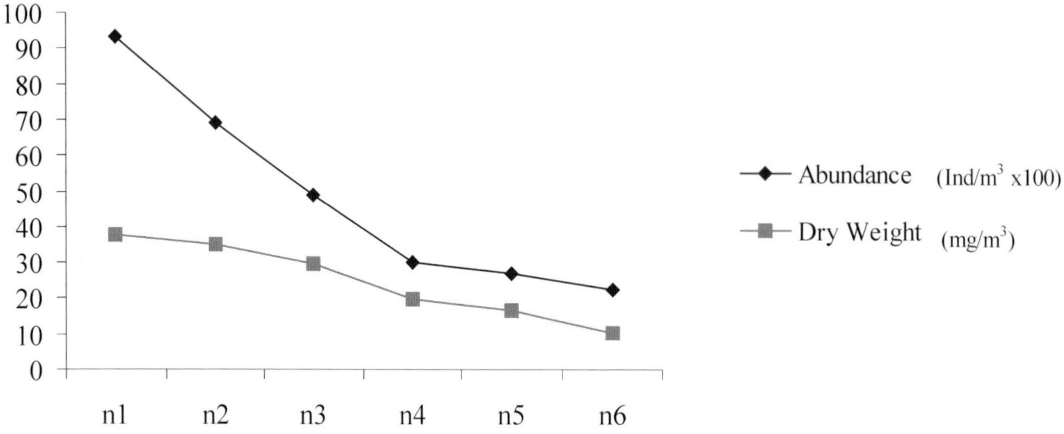

Fig. 3. Distribution of mean abundance (ind/m^3) and biomass (mg/m^3) of zooplankton in inshore-offshore sections (June and August-September '96)

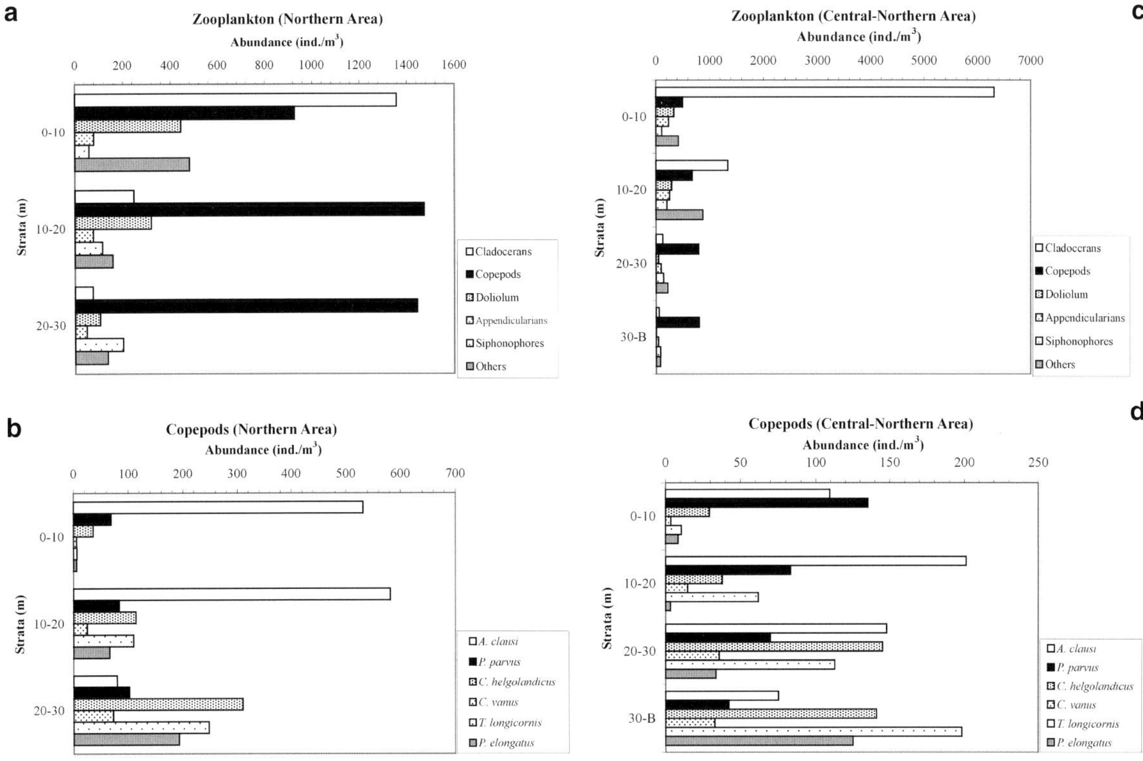

Fig. 4. Vertical distribution of mean abundance (ind/m^3) and main copepod species (ind/m^3) of zooplankton in the northern area (**a,b**) and in central-northern area (**c,d**) (northern Adriatic Sea, June '96)

dominance ratio between copepods and cladocerans. At the sites sampled in early June and located in the northern part of the study area, copepods (1354±808 ind/m^3) and cladocerans (517±922 ind/m^3) represented an average of 52 and 20% of the zooplankton community, respectively. During the remaining summer period, at central-northern and central Adriatic sites, copepods (745±329 ind/m^3) and cladocerans (2091±4071 ind/m^3) represented 17 and 66% of the zooplankton, respectively.

The cladoceran population was dominated by *Penilia avirostris*, which constituted more than 90% of the zooplankton at some coastal sites. *Evadne tergestina* and *E. spinifera* followed in decreasing order of abundance. There were only low abundances and rare occurrences of *Podon intermedius*, *Evadne nordmanni* and *Podon polyphemoides*. With regard to the spatial distribution of zooplankton, cladocerans were the dominant group in the coastal waters (particularly *P. avirostris*), whereas copepods were the dominant group in the offshore zone of the neritic system.

We identified 36 copepod species. A few species were dominant (*Acartia clausi, Paracalanus parvus, Temora stylifera, Centropages typicus, Calanus helgolandicus, Ctenocalanus vanus, Oithona plumifera, Temora longicornis, Pseudocalanus elongatus*), constituting on average (adults and copepodites) more than 93% of the entire population. In the coastal area, the copepod population was characterized by low species diversity and greater dominance of *A. clausi*, *P. parvus* and *T. stylifera* (73% of the whole population). In the offshore zone, instead, the percentage compositions of the various dominant species were more homogeneous (Fig. 5).

In late spring-summer, the vertical distribution of zooplankton (BIONESS samples) exhibited different associations of species in relation to different water masses along the water column (Fig. 6). Tunicates (*Oikopleura, Salpa* and *Doliolum*), cladocerans (*P. avirostris, E. spinifera* and *E. tergestina*) and copepods (*A. clausi* and *P. parvus*) occurred in the layer above the pycnocline, while *E. nordmanni*, *P. polyphemoides*, *P. intermedius*, *Oithona similis*, *O. plumifera*,

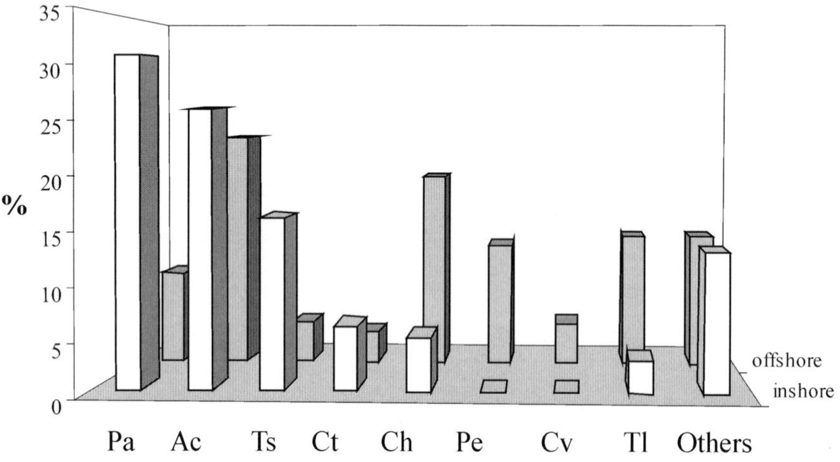

Fig. 5. Mean percentage composition of copepod species in late spring-summer in coastal and offshore zones. (*Pp*, *Paracalanus parvus*, *Ac*, *Acartia clausi*; *Ts*, *Temora stylifera*; *Ct*, *Centropages typicus*; *Ch*, *Calanus helgolandicus*; *Pe*, *Pseudocalanus elongatus*; *Cv*, *Ctenocalanus vanus*; *Tl*, *Temora longicornis*)

Centropages typicus, *Clausocalanus furcatus*, *Ctenocalanus vanus*, *Corycaeus* spp., *P. elongatus*, *T. longicornis* and *C. helgolandicus* were mainly found in the water mass under the pycnocline.

In June, there was a very high abundance of the copepod estuarine species *P. elongatus* and *T. longicornis* (235 and 268 ind/m^3, respectively) in the deepest samples at the offshore sites of the central and north-central areas (Fig. 4b,d), where high fluorescence values were recorded. In August-September, in the central Adriatic, *P. elongatus* and *T. longicornis* were found almost exclusively in the offshore zone of the neritic system, with a mean abundance of 30 and 45 ind/m^3, respectively, in the layer under the pycnocline where the North Adriatic Deep Water (NAdDW) was recorded. In June, *Acartia clausi* was the most abundant species; during August, in the central area, the abundance of *P. parvus* increased; in September, with the beginning of its annual biological cycle, *Temora stylifera* was the dominant species (Fig. 7).

The classification of the samples collected during the late spring-summer cruises revealed two main clusters, called A and B. The former was subdivided into a further three clusters, named A_1, A_2 and A_3 (Fig. 8). Cluster B was composed of the samples from the offshore zone of

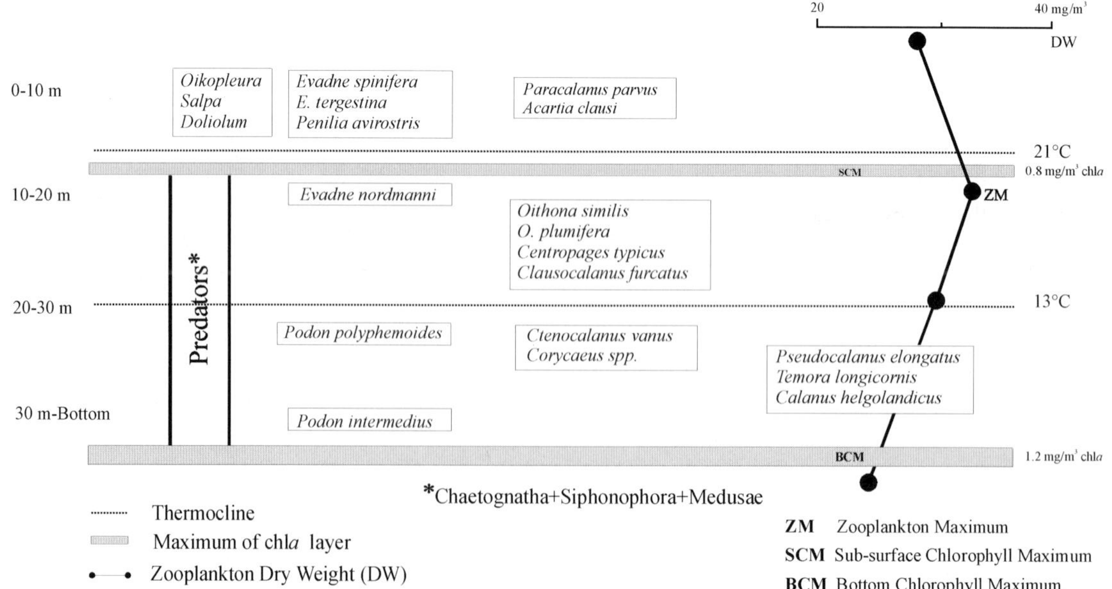

Fig. 6. Diagram of the depth distribution of mesozooplankton species with taxonomic and trophic similarities (June 1996)

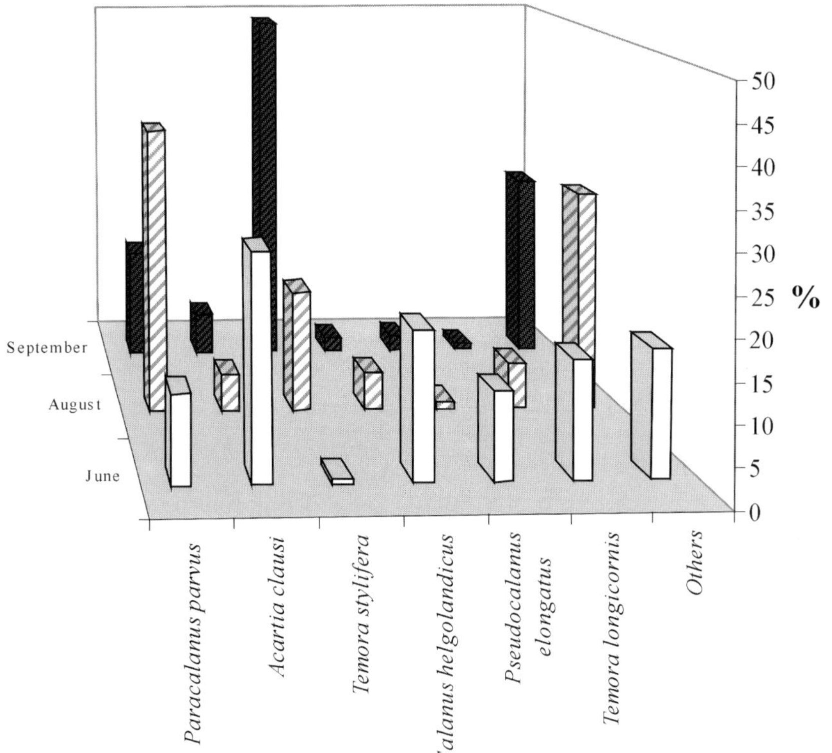

Fig. 7. Mean percentage composition of dominant copepod species in June-August-September '96

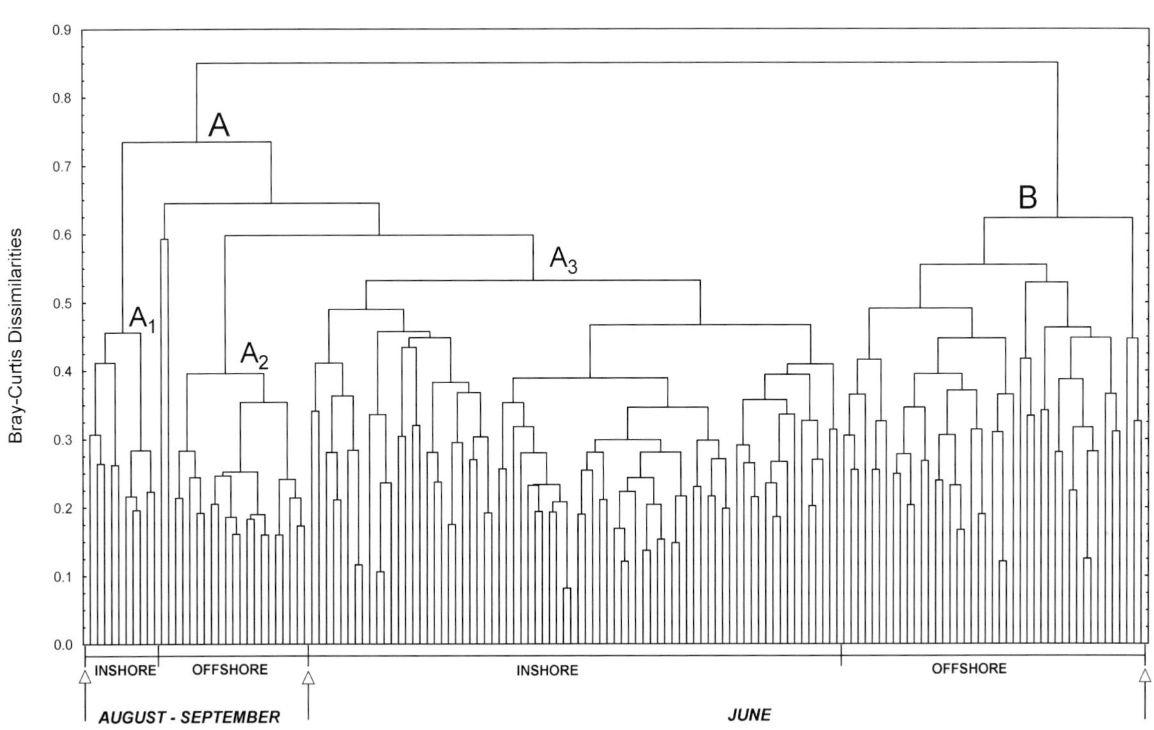

Fig. 8. Cluster analysis of the samples collected in late spring-summer '96

the neritic system, collected in June. Cluster A_3 included all coastal samples collected in June. Clusters A_2 and A_1 included, respectively, offshore and coastal samples collected in August-September.

Winter '97

The abundance and biomass of zooplankton ranged from 263-8825 ind/m^3 and 2.19-51.58 mg/m^3, respectively; the means ± SD were 1760±1472 ind/m^3 and 14.87±9.04 mg/m^3. There was greater homogeneity of the vertical (Fig. 9a,c) and horizontal abundance of zooplankton in this season than in summer. The zooplankton community was strongly dominated by copepods (31 identified species), representing on average 75% of all zooplankton, followed by invertebrate larvae, appendicularians, cladocerans, siphonophores and chaetognaths.

Copepods were mainly represented by juveniles and adults of *P. parvus*, *O. similis*, *C. helgolandicus*, *A. clausi*, *C. vanus*, *Clausocalanus* spp. copepodites and *O. media*, which represented 86% of the population. The mean percentage compositions of these species were more equally distributed in winter than in summer. The vertical (Fig. 9b,d) and horizontal distribution of copepod species mainly exhibited homogeneous patterns in the neritic system. Moreover, an inversion of the vertical distribution model of *C. helgolandicus*, with respect to that in summer, was caused by the decreasing surface-bottom gradient of abundance in winter (Fig. 9b,d). In this season, the most abundant cladoceran species were *P. intermedius* and *E. nordmanni*, accounting for about 96% of the population. Some juvenile copepods of pelagic species (*Lucicutia flavicornis*, *Pleuromamma gracilis*, *Clausocalanus pergens* and *Neocalanus gracilis*), typical of sub-surface waters and with a wide vertical distribution, were recorded at offshore sites, though with low values of abundance and percentage frequency.

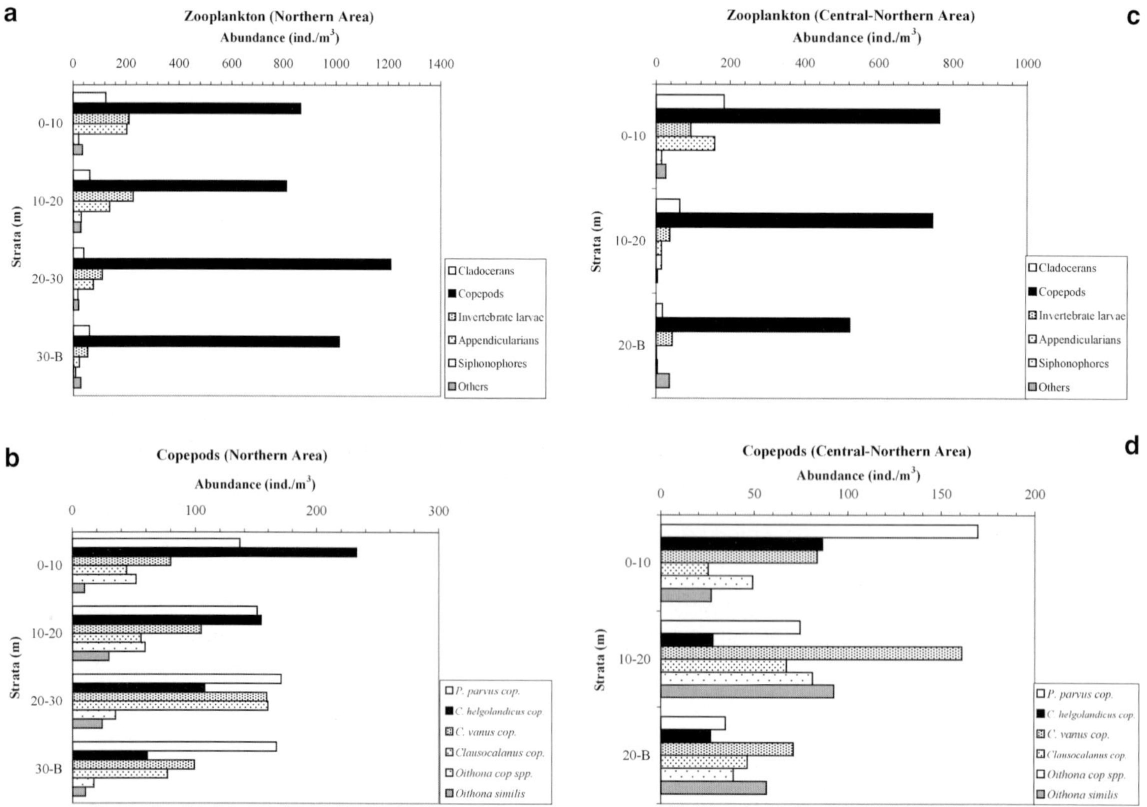

Fig. 9. Vertical distribution of mean abundance (ind/m^3) and main copepod species (ind/m^3) of zooplankton in the northern area (**a**,**b**) and in central-northern area (**c**,**d**) (northern Adriatic Sea, February '97)

Conclusions

Our analysis of the space-time distribution of zooplankton in the north and central Adriatic neritic system revealed two sub-systems in the late spring-summer period. Defined as coastal and offshore zones, they exhibit different physico-chemical and biological features. The two zones are separated by a frontal system, caused by thermohaline stratification of the water masses, with a pycnocline at about 10-15 m depth.

The zooplankton community of the neritic system shows a space-time variability in relation to the different reproductive strategies and ecological niches of the species. From late spring to summer, the zooplankton community of the coastal zone exhibits high abundance and low species diversity, as a result of the dominance of a few species. The north-central Adriatic Sea is characterized by a high inter-annual variability of the zooplankton community (Benovic et al. 1984; Fonda Umani et al. 1992, 1994; Fonda Umani 1996). The coastal zooplankton is characterized by the ecological association of opportunistic species of cladocerans (*P. avirostris*) and copepods (*A. clausi* and *P. parvus*). Their reproductive rate increases contemporaneously with the higher primary production in the spring-summer period (Fonda Umani et al. 1992; Zoppini et al. 1995; Scotto di Carlo et al. 1985); thus they come to dominate the structure of the coastal community.

The offshore zone is different from the coastal one, both in the lesser abundance of zooplankton and the more equal distribution of species composition. This is caused by the reduced abundance of *P. avirostris*, *A. clausi* and *P. parvus* and by the increasing numbers of *C. helgolandicus*, *C. vanus*, *O. plumifera* and *C. jobei*. The latter species are widespread mainly in waters with smaller oscillations of the physico-chemical parameters.

During late spring-summer, the copepod population structure changes significantly in relation to seasonal changes in the dominance ratio among *A. clausi*, *P. parvus* and *T. stylifera*. Their distribution patterns show a certain degree of temporal segregation of abundance peaks, which would reduce the negative aspects of interspecific competition. In summer, the spatial distribution of *P. elongatus* and *T. longicornis*, typical euryoecious species of estuarine environments and well adapted to higher temperature and lower salinity (Hure et al. 1980), assume particular importance. Their wide distribution in the neritic system depends on both the interaction of several biological and hydrological factors and their vertical migration pattern. In this season, the abundance of *P. elongatus* and *T. longicornis* is higher and their distribution range extends from coastal to offshore zones and from north to south (Hure et al. 1980). Their marked spreading into the offshore waters is caused by the very strong outflow of the Po River in this period (Marchetti et al. 1985). In the neritic system, the change from coastal hypertrophic surface waters to offshore oligo-mesotrophic ones could cause the vertical migration to the bottom of the above-mentioned species, either in the same water layer with high fluorescence values or in the NAdDW current which flows southward.

These observations on the spatial distribution pattern of *P. elongatus* and *T. longicornis* suggest some considerations on the circulation of water masses of the western Adriatic neritic system. These species could be considered as hydrological indicators both of the coastal waters flowing south along the western coastal zone (Hure et al. 1980) and of the NAdDW, which originates in the area between the Po delta and Istria and flows deeply from north to south in the offshore zone of the neritic system (Russo et al. in press).

During winter, the zooplankton community has a clearly different pattern of species composition than in summer. The low abundance and biomass of zooplankton are mainly due to the lack of *Penilia avirostris*, *Acartia clausi*, *Paracalanus parvus* and *Temora stylifera*. Copepods are the dominant zooplankton group, even though they strongly decrease in number. Moreover, there is an increased abundance of pelagic species, which have a wide vertical distribution and are typical of sub-surface waters of the southern Adriatic Sea. Their spreading toward north-central areas is mainly related to the seasonal ontogenetic migration to the surface (Hure et al. 1980). In this period, lower sunlight and temperature values, together with greater water turbulence, determine a strong reduction of the primary production. The low primary production resources (Fonda Umani et al. 1992) are used by a high number of species, so that the copepod population structure is characterized by a more equal distribution among the species of the neritic system. No species becomes so dominant that it characterizes the zooplankton community. In winter, a persistent unstable frontal system, caused by wide fluctuations of meteo-marine conditions, prevents a clear differentiation of the mean percentage species composition between the two sub-systems.

From our analysis of the spatial distribution of the most representative zooplankton groups, we formulated a preliminary trophic model concerning the association of ecologically similar species occupying the same habitat. In the coastal zone of the neritic system, the zooplankton biocoenosis is characterized by the dominance of the fine filter feeders, e.g. tunicates (*Oikopleura, Salpa* and *Doliolum*) and cladocerans (*P. avirostris, E. spinifera* and *E. tergestina*) which feed on nanoplankton (Fonda Umani et al. 1992), and mixtivorous copepods (*A. clausi* and *P. parvus*) able to feed on nano-, microzooplankton and particulate organic matter. In the offshore zone, in addition to the above-mentioned species but with less abundance, we found both herbivorous copepods, e.g. *Ctenocalanus vanus* and *Clausocalanus furcatus* which mainly feed on diatoms, and carnivorous species, e.g. *Oithona similis, O. plumifera* and *Centropages typicus* (Fonda Umani et al. 1992). Close to the bottom, the zooplankton population is characterized by herbivorous species and/or omnivorous ones, e.g. *Pseudocalanus elongatus, Temora longicornis* and *Calanus helgolandicus* which generally feed on diatoms, together with typical carnivorous species of the genera *Corycaeus, Oncaea* and *Podon*. The predator group, formed by chaetognaths, siphonophores and medusas, occurs all along the water column, but is more abundant in the layer under the pycnocline.

References

Artegiani A, Gacic M, Michelato A, Kovacevic V, Russo A, Paschini E, Scarazzato P, Smircic A (1993) The Adriatic Sea hydrography and circulation in spring and autumn (1985-1987). Deep-Sea Res II 40(6): 1143-1180

Bakus GJ (1990) Quantitative ecology and marine biology. A.A. Balkema, Rotterdam

Benovic A, Fonda Umani S, Malej A, Specchi M (1984) Net zooplankton biomass of the Adriatic Sea. Mar Biol 49: 265-275

Buljan M, Zore-Armanda M (1976) Oceanographic properties of the Adriatic Sea. Oceanogr Mar Biol Am Rev 14: 11-98

Car L (1890) Ein Beitrag zur Kenntnis der Copepoden von Triest. Glas Drus 5: 313-332

Claus C (1881) Neue Beitrage zur Kenntnis der Copepoden unter besonderer Berucksichtigung der Triester Fauna. Arb Zool Inst Wien 3: 313-332

Fonda Umani S (1996) Pelagic production and biomass in the Adriatic Sea. Sci Mar 60 Suppl 2: 65-77

Fonda Umani S, Franco P, Ghirardelli E, Malej A (1992) Outline of oceanography and the plankton of the Adriatic Sea. In: Olsen and Olsen (eds) Marine eutrophication and population dynamics, Proc 25th EMBS, pp 347-365

Fonda Umani S, Specchi M, Cataletto B, De Olazabal A (1994) Distribuzione stagionale del mesozooplancton nell'Adriatico settentrionale e centrale. Boll Soc Adriat Sci LXXV, 1: 145-176

Franco P (1973) L'influenza del Po sui caratteri oceanografici e sulla distribuzione della biomassa planctonica nell'Adriatico settemtrionale. Ann Univ Ferrara 1: 95-117

Franco P (1983) L'Adriatico Settentrionale: caratteri oceanografici e problemi. In: Atti V Congr AIOL, Stresa, pp 1-27

Franco P, Jeftic L, Malanotte-Rizzoli P, Michelato A, Orlic M (1982) Descriptive model of the Northern Adriatic. Oceanol Acta 5: 379-389

Ghirardelli E, Fonda Umani S, Specchi M (1989) Lo zooplancton dell'Adriatico. In: Il Mare Adriatico: problemi e prospettive. SOGESTA Urbino 24 maggio 1989, pp 47-74

Graeffe E (1900) Ubersicht der Fauna des Golfes von Triest. Arb Zool Inst Wien 13: 33-80

Guglielmo L, Granata A, Greco S (1998) Distribution and abundance of postlarval and juvenile *Pleuragramma antarcticum* (Pisces, Nototheniidae) off Terra Nova Bay (Ross Sea, Antarctica). Polar Biol 19: 37-51.

Hure J, Ianora A, Scotto di Carlo B (1980) Spatial and temporal distribution of Copepod communities in the Adriatic Sea. J Plankton Res 2 (4): 295-316

Lovegrove T (1966) The determination of the dry weight of plankton and the effect of various factors on the value obtained. In: Barnes H (ed) Some contemporary studies in marine science. Allen and Unwin, London, pp 429-467

Marchetti R, Pacchetti G, Provini A (1985) Tendenze evolutive della qualità delle acque del Po. Nova Thalassia 7 Suppl 2: 311-340

Regner D (1985) Seasonal and multiannual dynamics of copepods in the middle Adriatic. Acta Adriat 26: 11-99

Russo A, Totti C, Zagami G, Barletta D, Brancato G, Campolmi M, Galassi M, Pariante R, Solazzi A (2000) Caratterizzazione fisica e biologica delle masse d'acqua Nord-adriatiche nell'area Anconetana. In: XIII Congr AIOL, Ancona, 28/30 Settembre 1998 (in press)

Scotto di Carlo B, Tomas CR, Ianora A, Marino D, et al (1985) Uno studio integrato dell'ecosistema pelagico costiero del Golfo di Napoli. Nova Thalassia 7: 99-128

Specchi M, Fonda Umani S (1987) Influenza del Po sul sistema pelagico dell'Adriatico. Bull Ecol 18 (2): 135-144

Steuer A (1902a) Beobachtunger uber das Plankton des Trieser Golfes in Jabre 1901. Zool Anz 25: 369-371

Steuer A (1902b) Quantitative Planktonstudien im Golfes von Triest. Zool Anz 25: 372-375

Steuer A (1910a) Plankton Copepoden aus dem Hafen von Brindisi. Sitzb K Akad Wiss Wien Math Naturwiss KL Abt 119: 591-598

Steuer A (1910b) Adriatische Plankstoncopepoden. Ibid, 1005-1039

Zoppini A, Pettine M, Totti C, Puddu A, Artegiani A, Pagnotta R (1995) Nutrients, standing crop and primary production in western costal waters of the Adriatic Sea. Estuarine Coast Shelf Sci 41: 493-513

Distribution of Representative Oligotrophic Bacteria in a Pelagic Marine Environment (Ligurian Sea) by in situ Hybridization with rRNA-targeted, Fluorescently Labelled Oligonucleotides

L. Giuliano[1], M. De Domenico[1], and M.M. Yakimov[2]

ABSTRACT

The distribution of specific autochthonous oligotrophic bacteria in a surface layer (30 m depth) of a pelagic seawater station (DYFAMED, 28 miles off-shore, in the northwestern Mediterranean Sea) was studied by in situ hybridization with rRNA-targeted fluorescently labelled oligonucleotide probes (FISH). The data reported here extend a previous analysis of 16S rDNA clone libraries obtained by PCR from enriched oligotrophic bacterial populations in order to relate the previous enrichment results to population abundance and diversity in the original sample. For such a purpose, the frequency of matching of six previously designed group- and clone-specific probes with the naturally occurring 16S rRNA sequences was analysed. The results of in situ hybridization with the rRNA-targeted fluorescently labelled oligonucleotide probes support reports of *Vibrio* and *Rhodobacter* groups in the surface layer. The high percentage of coverage of *Vibrio*- and *Rhodobacter*- specific probes (24 and 19%, respectively) resulting from in situ experiments indicates the suitability of the previously described method to obtain enrichment of numerically representative autochthonous marine bacteria.

Introduction

Knowledge of microbial diversity has increased dramatically in recent years, in part as a result of sequencing of rRNA genes from DNA obtained directly from uncultured microbiota, often by the use of PCR and rRNA-specific primers. Fluorescence in situ hybridization (FISH) with rRNA-targeted oligonucleotide probes selectively visualizes bacterial cells with defined phylogenetic affiliations (Amann et al. 1995; 1997). Based on a rapidly growing set of 16S rRNA sequence data, it is probably the phylogenetically most sophisticated approach of a whole-cell in situ identification (Ludwig and Schleifer 1994). In contrast to other identification approaches, FISH largely maintains the original features of the targeted microorganisms, i.e. their morphologies, cell sizes (Pernthaler et al. 1996; Ramsing et al. 1996), and cellular rRNA content (Boyle et al. 1995; Poulsen et al. 1993). Although these approaches have produced a diverse collection of sequences and expanded our view of microbial diversity, analysis of microbial 16S rDNA sequences has limitations in relating specific rDNA sequences to other aspects of organisms in the environment under study. We have already described a method to obtain seawater enrichments of numerically representative indigenous oligotrophic marine bacteria within the naturally occurring microbial populations (Giuliano et al. 1998). Such a method provided the possibility to test physiologically the taxonomically characterized targeted populations representative of the original communities. In the present study, the density of specific populations from seawater enrichments in the natural pelagic marine environment was determined by in situ hybridizations with the previously obtained rRNA-targeted oligonucleotide probes. Based on the described hybridization protocol, the total density of the checked populations accounted for a large fraction of the bacterioplankton, thus suggesting the suitability of the method for the proposed aim. By relating the previous enrichment results to the population abundance and diversi-

[1] Dipartimento di Biologia Animale ed Ecologia Marina, Università di Messina, Salita Sperone 31, 98166 Messina, Italy
[2] Centro Siciliano per la Ricerca Atmosferica e la Fisica Applicata (CSRAFA), C. da Papardo 31, 98166 Messina, Italy

F.M. Faranda, L. Guglielmo, G. Spezie (eds)
Mediterranean Ecosystems: Structures and Processes

ty in the original sample, the data presented give a further contribution for checking the previously proposed method.

Materials and Methods

Sampling Site and Procedure

The DYFAMED station (43○ 25' N and 07○ 52' E), is an oligotrophic pelagic station located in the Ligurian Sea (northwestern Mediterranean Sea), 28 miles from the coast of Nice. Despite the general oligotrophic conditions, some eutrophication events usually occur in the euphotic layers of the sampled area in spring time (Minas and Minas 1990; Morel and André 1991; Prieur 1981). During the DYFABAC 07 cruise (15 to 21 May 1996), aboard the Tethys II research vessel, water samples were collected with 6l Niskin bottles attached to a CTD (conductivity-temperature-depth) rosette from 30 m depth layer. Processing of samples on board the research vessel was previously described (Bianchi and Giuliano 1996).

Direct Bacterial Counts

Formalin-fixed subsamples of 5 to 20 ml were filtered with black membrane filters (pore-size, 0.22 μm; diameter, 25 mm; Nuclepore) and stained with 4',6'-diamidino-2-phenylindole (DAPI), and bacterial abundances were determined by epifluorescence microscopy (Porter and Feig 1980).

Design and Characterization of Oligonucleotide Probes

Original samples were diluted in unamended, previously filtered, sterilized seawater to the extent of extinction in order to obtain enrichments of numerically representative oligotrophic marine bacteria. Detailed information about the utilized dilution culture technique, the DNA extraction, amplification, cloning and the analyses of the amplified products were previously reported (Giuliano et al. 1998). Briefly, water samples for DNA extraction were obtained from the highest positive seawater dilution of the natural samples. After filtration of the diluted sample, the total genomic DNA was extracted by following the protocol of Fuhrman et al. (1988). Almost full-length bacterial 16S rDNA fragments were amplified by PCR from the extracted DNA by using two general bacterial 16S rDNA primers, 16F27 and 16R1492 (Lane 1991). After a previous analysis of restriction fragments (RFLP), eight clones were selected and their 16S rDNA inserts were fully sequenced. The clone identifications, EMBL accession numbers, and nearest relatives in the database are presented in the Table. Six either clone- or group-specific oligonucleotide probes were constructed and evaluated with the RDP programme CHECK_PROBE (Larsen et al 1994). The sequences and target positions of the designed probes are given in the Table. After optimization and specificity study, the digoxigenin-11-ddUTP (DIG)-labelled probes, purchased from Gibco BRL (Germany) were utilized for the screening of the clone libraries. For in situ hybridization, the same probes were labelled with the indocarbocyanine dye Cy3.

FISH

The FISH of filter sections from the 30 m depth sample with oligonucleotide probes, counterstaining with DAPI, mounting for microscopic evaluation and epifluorescence microscopy were performed following the protocol of Glöckner et al. (1996). Cells were observed using an Axioplan epifluorescence microscope (Zeiss, Jena, Germany) equipped with a 50W mercury high pressure bulb and specific filter sets (DAPI: Zeiss 01, CY3: Zeiss 15). Colour photomicrographs were done on Kodak Ektachrome 1600 (Rochester, NY, USA). Exposure times were 15 s for DAPI and 20 s for CY3.

Results

Cell density determined by DAPI staining was 9.6×10^5 cells ml^{-1}. An increase of mean cell number ml^{-1} at a second survey, to 1.98×10^6 cells ml^{-1}, in agreement with an increase of the chlorophyll a values (data not shown) indicated a bloom event evolution in the sampled area, at the sampling time.

Between 15 and 23% of all DAPI-stained objects (mean, 19%) were detected by the group-specific G Rb probe, and between 20 and 28%

Table 1. Probe designed from bacterial 16S rDNA sequences that were retrieved from the highest positive dilutions of seawater sample from 30 m of depth (station DYFAMED)

Clone	EMBL accession no.	Nearest relative in EMBL database (% similarity)	Probe sequence (5'-3')	rRNA position[a]
SRF1	AJ002563	*Roseobacter algocolus* (95.0)	GTCAGTATCGAGCCAGTGAG	645-626
SRF2	AJ002564	*Roseobacter algocolus* (95.0)	GTCAGTATCGAGCCAGTGAG	645-626
SRF3	AJ002565	*Silicibacter lacustaerulensis* (94.7)	GTCAGTATCGAGCCAGTGAG	645-626
			CTCAAGACTACCAGTATTAG	806-787
DPT1.1	AJ002566	*Vibrio sp.2P44* (98.0)	AAATCCTCCGAAGATTCAAT	47-28
			AGGCCACAACCTCCAAGTAG	841-822
DPT1.2	AJ002567	*Vibrio sp.2P44* (98.0)	AAATCCTCCGAAGATTCAAT	47-28
			AGGCCACAACCTCCAAGTAG	841-822
DPT1.3	AJ002569	*Flavobacter salegens* (87.5)	GGTTCTTCCTCTGTAAAAGC	502-483
			CTGGCAACTAACCACAGGGG	1128-1109
DPT1.4	AJ002570	*Flavobacter salegens* (87.3)	GGTTCTTCCTCTGTAAAAGC	502-483
			CTGGCAACTAACCACAGGGG	1128-1109
DPT2.1	AJ002568	*Vibrio sp.2P44* (93.0)	AAATCCTCCGAAGATTCAAT	47-28
			AGGCCACAACCTCCAAGTAG	841-822

[a] *Escherichia coli* numbering

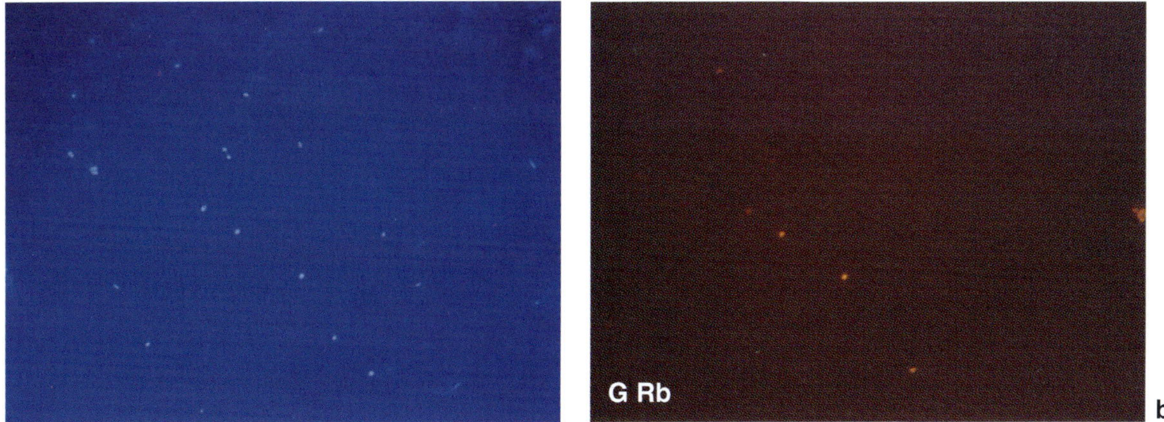

Fig. 1a,b. a DAPI stained cells and **b** cells hybridized with CY3-G Rb probe from natural microbial community of 30 m depth sample under epifluorescence microscopy examination

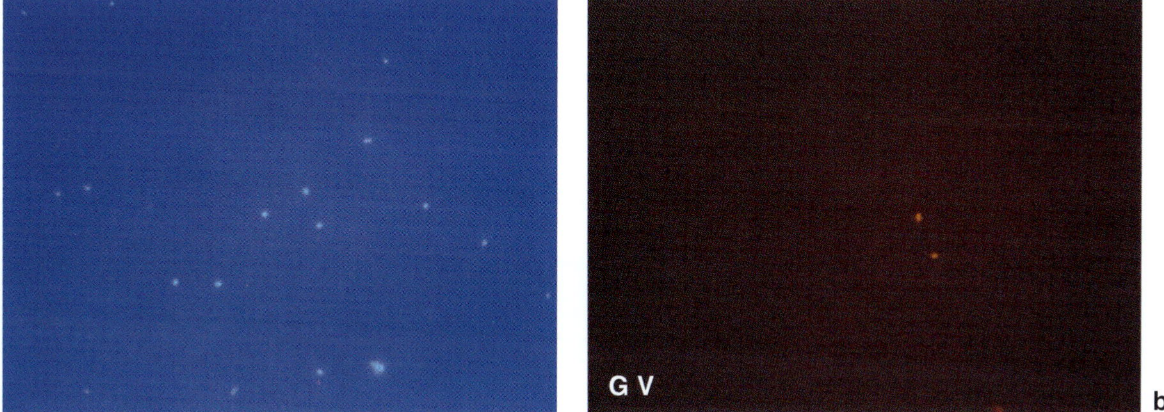

Fig. 2a,b. a DAPI stained cells and **b** cells hybridized with CY3-G V probe from natural microbial community of 30 m depth sample under epifluorescence microscopy examination

(mean, 24%) of them were detected by the group-specific G V probe (Figs. 1a,b; 2a,b).

Discussion

The high percentage of coverage of the group-specific probes indicated the suitability of the previously described method to obtain enrichments of numerically representative autochthonous marine bacteria. Moreover, because the two different group-specific probes were targeting enriched bacterial populations from the same dilution of the original sample, the equivalent size of detected populations agreed with the predicted occurrence of populations within the natural community. The presence of bacterial populations representative of *Vibrio* group in pelagic waters is well documented (Pinhassi et al. 1997). According to previous works in different oceans (Gonzalez and Moran 1997; Mullins et al. 1995), the *Rhodobacter* group, affiliated to the α-*Proteobacteria*, is best adapted to the conditions of the euphotic zone when the water column is highly stratified (e.g. sampling conditions), and some representatives of this group have been isolated from the surfaces of unicellular algae (Lafay et al. 1995). Because of the bloom event detected at the sampling time in the DYFAMED station, we could speculate about the possible existence of some spatial relations between SRF- targeted populations and some dominant algae. Monitoring by the clone-specific probes did not show any detectable signal under epifluorescence microscopy analysis. One of possible reasons for the absence in the natural sample of populations hybridizing with the obtained clone-specific probes could be explained by the high frequency of mutational events usually involving the most variable regions of 16S rRNA operons (i.e. minor nucleotide substitutions). Such events could cause the loss of the original sequences of some hypervariable regions during succession of bacterial generations and, therefore, the design of probes complementary to mutated regions.

Acknowledgements. We would like to thank Dr. A. Bianchi (CNRS, Marseilles, France) for generous hospitality, providing us the possibility to collect the probes during the DYFABAC cruise of r/v 'Tethys II' (Programme DYFAMED, CNRS-INSU) and Prof. K.N. Timmis (GBF, Braunschweig, Germany) for helpful discussions. The assistance of Prof. E. De Domenico is also gratefully acknowledged. This research work was supported in part by grants from University of Messina.

References

Amann R, Glöckner OF, Neef A (1997) Modern methods in subsurface microbiology: in situ identification of microorganisms with nucleic acid probes. FEMS Microbiol Rev 20: 191-200

Amann RI, Ludwig W, Schleifer KH (1995) Phylogenetic indentification and in situ detection of individual microbial cells without cultivation. Microbiol Rev 59: 143-169

Bianchi A, Giuliano L (1996). Enumeration of viable bacteria in the marine pelagic environment. Appl Environ Microbiol 62: 174-177

Boyle M, Ahla T, Molin S (1995) Application of a strain-specific rRNA oligonucleotide probe targeting *Pseudomonas fluorescens* Agl in a mesocosm study of bacterial release into the environment. Appl Environ Microbiol 61: 1384-1390

Fuhrman JA, Comeau DE, Hagström Å, Chan AM (1988) Extraction from natural planktonic microorganisms of DNA suitable for molecular biological studies. Appl Environ Microbiol 54: 1426-1429

Giuliano L, De Domenico M, De Domenico E, Höfle MG, Yakimov MM (1998) Identification of culturable oligotrophic bacteria within naturally occurring bacterioplankton communities of the Ligurian Sea by 16S rRNA Sequencing and Probing. Microb Ecol 37: 77-85

Glöckner FO, Amann R, Alfreider A, Pernthaler J, Psenner R, Trebesius K, Schleifer KH (1996) An in situ hybridization protocol for detection and identification of planktonic bacteria. Appl Microbiol 19: 403-406

Gonzalez JM, Moran MA (1997) Numerical dominance of a group of marine bacteria in the "-subclass of the class Proteobacteria in coastal seawater. Appl Environ Microbiol 63: 4237-4242

Lafay B, Ruimy R, Rausch de Traubenberg C, Breittmayer V, Gauthier MJ, Christen R (1995) *Roseobacter algicola sp.* nov., a new marine bacterium isolated from the phycosphere of the toxin-producing dinoflagellate *Prorocentrum lima*. Int J Syst Bacteriol 45: 290-296

Lane DJ (1991) 16/23 rRNA sequencing. In: Stackebrandt E, Goodfellow M (eds) Nucleic acids techniques in bacterial systematics. John Wiley and Sons, Chichester, UK, pp 115-175

Larsen N, Overbeek R, McCaughey MJ, Woese CR (1994) The ribosomal RNA database project (RDP). Nucleic Acid Res 24: 82-85

Ludwig W, Schleifer KH (1994) Bacterial phylogeny based on 16S and 23S rRNA sequence analysis. FEMS Microbiol Rev 15: 155-173

Minas HJ, Minas M (1990) New production considerations in the Mediterranean Sea. EOS 71: 1395-1396

Morel A, André JM (1991) Pigment distribution and primary production in the western Mediterranean as derived and modelled from Coastal Zone Colour scanner observations. J Geophys Res 96: 12685-12698

Mullins TD, Britschgi TB, Krest RL, Giovannoni SJ (1995) Genetic comparisons reveal the same unknown bacterial lineage in Atlantic and Pacific bacterioplankton communities. Limnol Oceanogr 40: 148-158

Pernthaler J, Sattler B, Simek K, Schwarzenbacher A, Psenner R (1996) Top-down effects on the size-biomass distribution of a freshwater bacterioplankton community. Aquat Microb Ecol 10: 255-263

Pinhassi J, Zweifel UL, Hagstrom Å (1997) Dominant marine bacterioplankton species found among colony-forming bacteria. Appl Environ Microbiol 63: 3359-3366

Porter KG, Feig YS (1980) The use of DAPI for identifying and counting aquatic microflora. Limnol Oceanogr 25: 943-948

Poulen LK, Ballard G, Stahl DA (1993) Use of rRNA fluorescence in situ hybridization for measuring the activity of single cells in young and established biofilms. Appl Environ Microbiol 59: 1354-1360

Prieur L (1981) Heterogéneité spatio-temporelle dans le bassin liguro-provençal en 1981-1982. Rapp Comm Int Mer Méditerr 27: 177-179

Ramsing NB, Fossing H, Ferdelman TG, Andersen F, Thamdrup B (1996) Distribution of bacterial population in a stratified fjord (Marianger Fjord, Denmark) quantified by in situ hybridization and related to chemical gradients in the water column. Appl Environ Microbiol 62: 1391-1404

Energy Flux in the South Tyrrhenian Deep-sea Ecosystem: Role of Mesopelagic Fishes and Squids

A. Granata, G. Brancato, O. Sidoti, and L. Guglielmo

ABSTRACT

We here review various studies concerning the ecology of mesopelagic organisms in the South Tyrrhenian Sea and other oceans to direct attention to the importance of mesopelagic animals within the context of meso- and epi-pelagic food webs. The results of recent studies on the micronekton in the South Tyrrhenian Sea were examined (Marabello 1994; Guglielmo et al. 1995), together with the relationship between immense populations of vertically migratory fishes and squids and the fixed resident population of more surface and deeper waters. The review summarizes the present state of knowledge on migratory activity related to trophic purposes of mesopelagic fishes and on the diet of squids, tunas and other large predators. The conclusions seem to confirm the hypothesis of a trophic connection between cephalopods and mesopelagic fishes. They also elucidate the fundamental role of cephalopods as a top link in the trophic chain of mesopelagic fishes. The hypothesis of a trophic chain characterised by cephalopods, mesopelagic fishes and crustaceans, with cephalopods acting as a connection between surface and mesopelagic trophic webs, is discussed.

Introduction

The so-called Deep Scattering Layer (DSL) has been known for some time in all of the major oceans. During the day, it is located at a depth of 450 down to 750 m, but at night time it splits into a component that rises towards the surface and another that remains stationary. The organisms responsible for this phenomenon are mainly midwater fishes and crustaceans, in addition to siphonophores (Farquhar 1977; Guglielmo and Zagami 1985). The DSL is discontinuous since an incalculable number of organisms moves towards the surface for trophic purposes (Mann 1984).

Marine organisms are linked to each other by complex trophic food webs, knowledge of which is important for understanding energy flows that are established within pelagic communities. The trophic dynamics of certain "key groups" delineate several basic features of the functioning of pelagic ecosystems (Fig. 1). Within the micronekton, mesopelagic fishes feeding on zooplankton are important components in several oceanic ecosystem models (Hopkins and Baird 1985). This group is the main component, in terms of biomass, of the mesopelagic zone (200-1000 m of depth). Due to their high density and wide diffusion in all oceans, mesopelagic fishes can be considered the most abundant vertebrates on earth (Mann 1984). Their importance in this context stems from the fact that each day they perform large vertical migrations, for trophic purposes, which make possible the continuous flow of energy between surface and underlying waters.

At present, the most accepted hypothesis regarding trophic migrations is that these organisms move at night to feed in upper layers, where presumably food can be found in higher concentrations, and then descend to greater depths during the day to avoid predation (Marshall 1960) (Fig. 2). The daily vertical migrations associated with feeding have been well documented in different species of mesopelagic fishes (Scotto di Carlo et al. 1982) and squids (Marabello et al. 1996).

The aim of this study was to address the problem of what relationship exists between these immense populations of vertically migra-

Dipartimento di Biologia Animale ed Ecologia Marina, Università di Messina, Salita Sperone 31, 98166 Messina, Italy

F.M. Faranda, L. Guglielmo, G. Spezie (eds)
Mediterranean Ecosystems: Structures and Processes

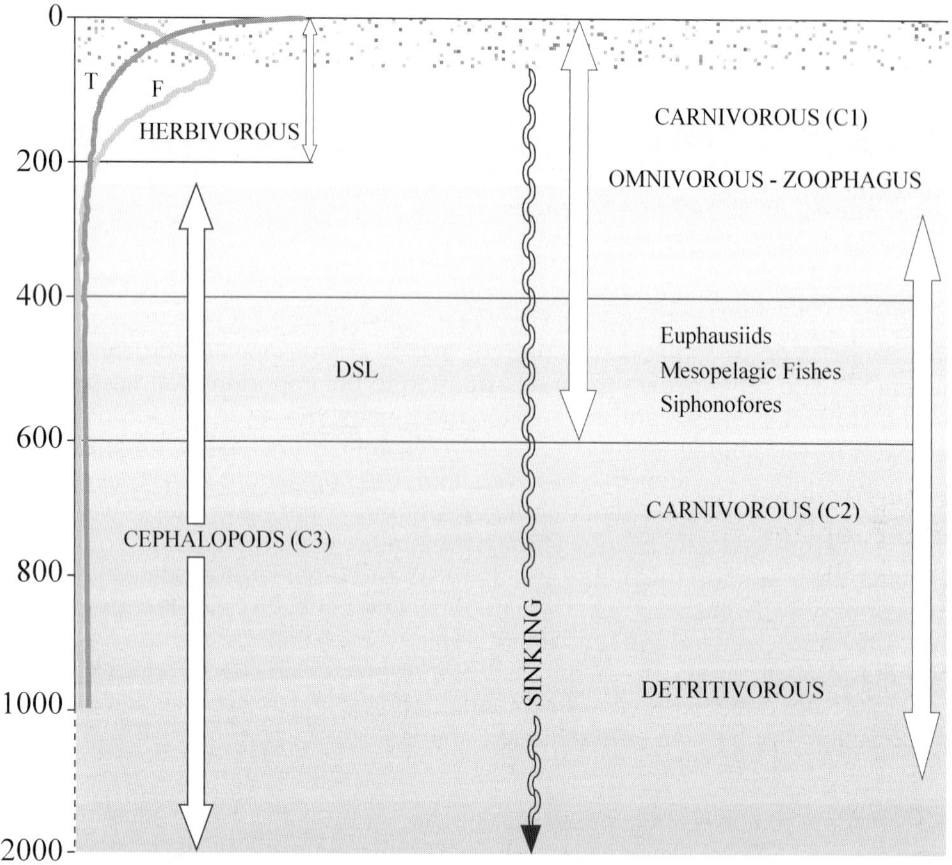

Fig. 1. Schematic representation of the biological pump of carbon flux in the pelagic food web (*sensu* from Longhurst 1989)

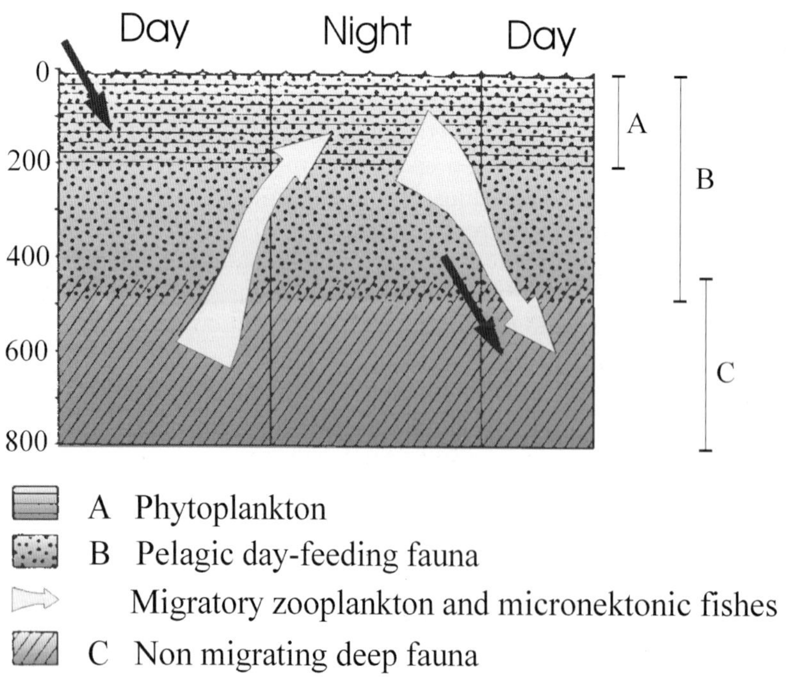

Fig. 2. Relationship between migratory and non-migratory fauna

tory animals, and the standing stocks of surface and deeper waters. The data examined in this paper derive from different literature sources and collections of samples from the Straits of Messina ecosystem in the Mediterranean Sea. These include: (1) trawling samples collected with classical fine-mesh nets such as the Juday-Bogorov, WP2, and Ori-Net 1.6 at fixed stations and along different sampling grids (Guglielmo et al. 1995, 1996); (2) collections of stranded organism (Berdar et al. 1979; Genovese et al. 1971); (3) observations with the deep-diving vehicle "FA Forel" (Genovese et al. 1985).

Relationship Between Mesopelagic Fishes and Zooplankton: Vertical Distribution and Feeding Ecology

Mesopelagic fishes dominate the mesopelagic zone of all oceanic regions (Dalpadado and Gjøsæter 1988). In recent years, several studies have been conducted better to understand various aspects of their biology and ecology. As shown in Fig. 3, the main families belonging to this group are: the Myctophidae (lanternfishes), Gonostomatidae, Sternoptychidae (hatchetfishes), Chauliodontidae and Stomiatidae.

Organisms of this group do not have a very hydrodynamic shape, thereby rendering them poorly adapted for long-range displacements as vertical migrators. Nonetheless, these organisms do carry out ample vertical migrations within the water column, between 1000 m up to the surface (Table 1). Underwater observations have demonstrated that they often spend long periods in an immobile position, head downward, often occurring in dense aggregations, but if disturbed, they are capable of rapid movements and quick scatterings (Mann 1984; Genovese et al. 1985).

A rich literature concerning the migratory and trophic activities of mesopelagic fishes has stated that they can be split into two categories, based on vertical distribution, with a deep non-migrating and mesopelagic migrating component (Legand et al. 1972). The former category contains only a few species while, species of the latter category are more numerous. These occur at the same depth as non-migrating species during the day, but clearly climb to surface layers at night. Depending on the species, the depth range of migration ranges from 200 and 700 m.

The diet of several myctophids exhibits various feeding rhythms and the composition of the stomach contents varies from species to species. The relationship between predator and prey size shows that smallest fishes are essentially cope-

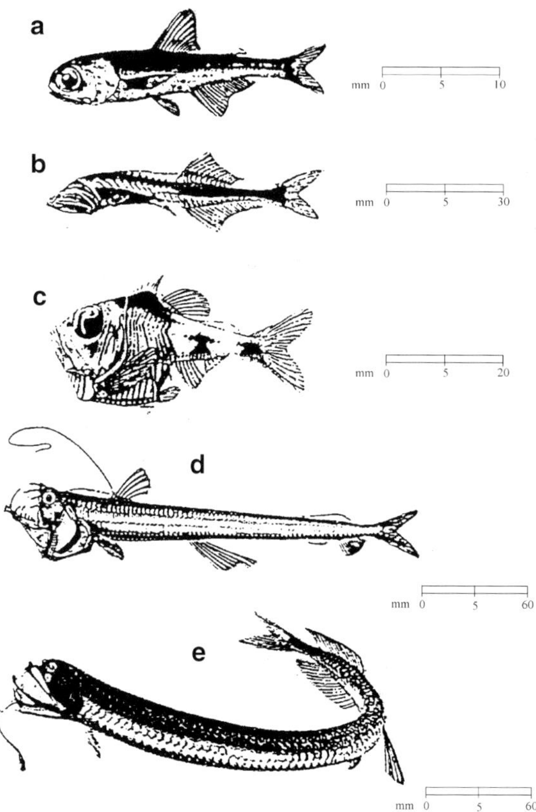

Fig. 3a-e. Main families of mesopelagic fishes: **a**, *Myctophidae*; **b**, *Gonostomatidae*; **c**, *Sternoptychidae*; **d**, *Chauliodontidae*; **e**, *Stomiatidae*

Table 1. Percentage occurrence (*PO*) and range of vertical distribution, by day and night, of the most important prey of *Todarodes sagittatus* (Modified from Marabello 1994)

Species	PO (%)	Depth range Day	Night
Onychoteuthis banksi	15.1	50-800	0-100
Histioteuthis sp.	2.5	500-700	50-400
Todarodes sagittatus	2.2	70-80	0-150
Chauliodus sloani	3.6	500-700	100-700
Maurolicus muelleri	3.2	400-450	
Vinciguerria sp.	2.5	300-500	50-400
Argyropelecus hemigymnus	1.8	300-500	200-500
Stomias boa	1.4	500-700	100-700
Lampanyctus crocodilus	0.7	700-1000	45-1000
Phrosina semilunata	8.2	40-80	50-220

pophagous, medium-sized fishes prey on euphausiids and large ones on other fishes. Guglielmo and Zagami (1985) estimated the dry weight and energetic value (as kJ) of both zooplankters and euphausiids, and the composition of the feeding categories in the South Tyrrhenian Sea, better to understand the flux of energy along the Tyrrhenian food chain, as shown in Table 2. Legand et al. (1972) argued that the agility of the predator and the size of its buccal parts and dentition also influenced the size and type of prey. The availability of food at the surface and the quantity of O_2 in the lower water layers can also condition the feeding behaviour of these animals.

The feeding behaviour of *Lampanyctus mexicanus* (Myctophidae) was examined by Holton (1969), during the day. This species is located at a depth of 600-700 m, but performs wide vertical migrations that bring it up to the surface at night. During its rise, the fish actively feeds on high quantities of ostracods, copepods and decapods, whereas upon its descent to deep waters, it regurgitates the same food items, probably to reduce oxygen consumption while passing through water layers with minimal O_2 levels.

Some species also show selective feeding. Merrett and Roe (1974) studied the diet of various mesopelagic species from the surface down to 2000 m, relating these to observations of day-night vertical distribution of the zooplankton and micronekton collected in the same area. The feeding selectivity of 7 species was compared to the composition of the marine zooplankton population at the time, and the composition of zooplankton prey items found in their stomachs was analysed. They found feeding selectivity in 3 species: *Valenciennellus tripunctulatus* feeds on calanoid copepods, *Argyropelecus aculeatus* on ostracods and *Lampanyctus cuprarius* on amphipods and euphausiids. *Argyropelecus hemigymnus* does not seem to exhibit a particular feeding strategy. Gorelova (1974) observed a difference between the relative proportions of the main taxonomic groups found in the stomachs of lanternfishes and their relative proportions in the marine environment. In the stomach contents, the zooplankton was characterized by greater percentages in weight of amphipods, appendicularians, chaetognaths and mysids than, for example, of copepods, which constituted the largest component in the ocean. This can be explained both by the selective ability of the fishes and by the patchy distribution of plankton.

Kinzer and Schulz (1985) confirming the relationship between daily vertical migrations and trophic activity of mesopelagic fishes, found an opportunistic predatory activity in 7 species of myctophids (*Lampanyctus alatus, Lepidophanes guentheri, Ceratoscopelus warmingii, Diaphus brachycephalus, D. lutkeni, D. dumerilii, Notolychnus valdiviae*), which ate mainly calanoid copepods at night.

The hatchetfishes (*Sternoptyx diaphana, S. pseudobscura, Argypelecus sladeni, A. affinis, A. hemigymnus*) show modest migratory activity, moving only 100-200 m during the night time and feeding mainly on copepods and ostracods. The larger forms prefer euphausiids and amphipods (Kinzer and Schulz 1988).

The vertical distribution and trophism of the same family (Sternoptychidae), in the Gulf of Mexico (Hopkins and Baird 1985), showed evidence of spatio-temporal and trophic partitioning of the four main species. The *A. aculeatus* appears to feed in the epipelagic zone early in the night, preferring ostracods, copepods and pteropods. The *A. hemigymnus* feeds in the late afternoon at somewhat lower depths, in the zone between 300 and 500 m, preferring ostracods and copepods. These two species coexist in the same layer during the day, the former migrating into

Table 2. Mean biomass estimated for zooplankton and euphausiids and mean percentage composition of the major constituents of zooplankton and micronekton in DSL (0-100 m, South Tyrrhenian Sea at 18:00-22:00 h)

Zooplankton	**Feb-March**	**July**
Dry weight (g/1000m^3)[a]	2.55	1.59
KJ[b]	58.65	36.57
Euphausiids		
Dry weight (g/1000m^3)[c]	0.51	2.41
KJ[d]	10.45	49.40
RATIO E/Z	0.20	1.50
	Mean % (min - max)	
Fishes	27.2 (0.3-72.6)	
Siphonophores	20.8 (2.3-72.1)	
Euphausiids	19.9 (0.2-41.7)	
Copepods	14.9 (2.2-32.9)	

[a] Measured to be 15±2% of wet weight (Sameoto 1982)
[b] Measured to be 23±0.4 kJ/g of dry weight (Sameoto 1982)
[c] Measured to be 22±2% of wet weight
[d] 1 g of dry weight = 20.5 kJ (Sameoto 1982)

the epipelagic zone at night. The latter species remains below 300 m where it feeds in the late afternoon, maybe to reduce trophic competiton. The *S. diaphana*, concentrated between 500 and 800 m, feeds on copepods, ostracods and amphipods (small individuals). The young stages of *S. pseudobscura*, occurring below 800 m, prefer copepods, polychaetes and euphausiids, while adults preferentially feed on amphipods and fishes.

Sameoto (1988) demonstrated how *Benthosema glaciale*, which dominates the mesopelagic zone of Nova Scotian waters in the North Atlantic selects certain species of copepods, but not the most abundant *Calanus finmarchicus*, despite the fact that *B. glaciale* has a feeding preference for prey species larger than 1 mm. During the day, the population is concentrated at a depth from 300 and 550 m, with a maximum from 320-360 m. At night, the population moves below the main concentration of copepods that occupy the upper 30 m. This could either imply that the predator occurs exactly in correspondence with the maximum concentration of zooplankton, and feeds on it there, or, that the predators remain below their prey with only brief excursions into the upper stratum to eat.

Goodyear et al. (1972) studied the ecology and distribution of mesopelagic fishes in the Mediterranean Sea, sampling more than 20,000 specimens from 36 species of 10 different families. Mesopelagic fishes occurred in the upper 1000 m during the day, and were mainly sampled between 100 and 1000 m. On the other hand, at night they were captured between 0 and 1000 m, with a maximum concentration between 0 and 100 m and 400 and 800 m. Depending on the size, a vertical stratification depending on size was evident for the majority of species, both during the day and at night. The authors subdivided this group into a broad scale, involving distinct and widely separated depths, and a narrow scale, involving continuous distributions usually of limited depth.

Matallanas (1982) studied the diet of *Lampanyctus crocodilus* collected in the western Mediterranean between 300 and 700 m, concluding that the species exhibits a high coefficient of fullness which indicates its voracity. The mean number of prey per stomach was higher during spring at the same time when euphausiids were more abundant, while the size of the prey increased proportionally to that of the predator. The species feeds also on copepods, mysids, amphipods, pelagic decapods and fishes (*Cyclothone braueri, L. crocodilus*) without any relation between trophism and daily vertical distribution.

The diet of *Cyclothone braueri* collected between 300 and 900 m in the Ligurian Sea includes zooplankton organisms with wide vertical distribution ranges. The most important groups are copepods (mainly *Pleuromamma gracilis, Euchaeta marina, Chiridius poppei* and *Calanus helgolandicus*) and ostracods, followed by larvae of euphausiids, amphipods and decapods (Palma 1990).

The two most abundant myctophids in the Straits of Messina, the lanternfishes *Hygophum benoiti* and *Myctophum punctatum* (Scotto di Carlo et al. 1982), were distributed at a depth between 100 and 1000 m, with a maximum concentration between 700 and 1000 m. Dietary variations were related to age, both as regards size and taxonomic composition of the prey (Fig. 4), indicating differences in feeding habits between the two species. The *H. benoiti* has a trophic behaviour related to the different size classes of prey. The young ate mainly copepods whereas the prey diversity and biomass of the other zooplanktonic taxa such as amphipods, mysids and chaetognaths were greatest in the stomachs of medium-sized individuals. On the other hand, adults had a more euphausiid-based diet. This model implies not only a greater efficiency in the capture of larger prey, but also selective feeding during all stages of the life cycle. *Myctophum punctatum*, instead, exhibited a mixed diet during all the developmental stages, with the exception of small individuals that fed almost exclusively on copepods. Prey diversity and biomass in the stomach contents generally increased with size. *Myctophum punctatum* became a more efficient and opportunistic predator with age. This type of strategy, which gradually adds a new type of prey to the diet as the fish matures, is the most common for mesopelagic fishes (Tyler and Pearcy 1975). *Myctophum punctatum* was also the most active predator, since it fed on a wide range of prey, such as gasteropods, thaliaceans and appendicularians, which were absent in the diet of *Hygophum benoiti*. The composition of the diet of both species very much resembled a nocturnal zooplanktonic community composed of both

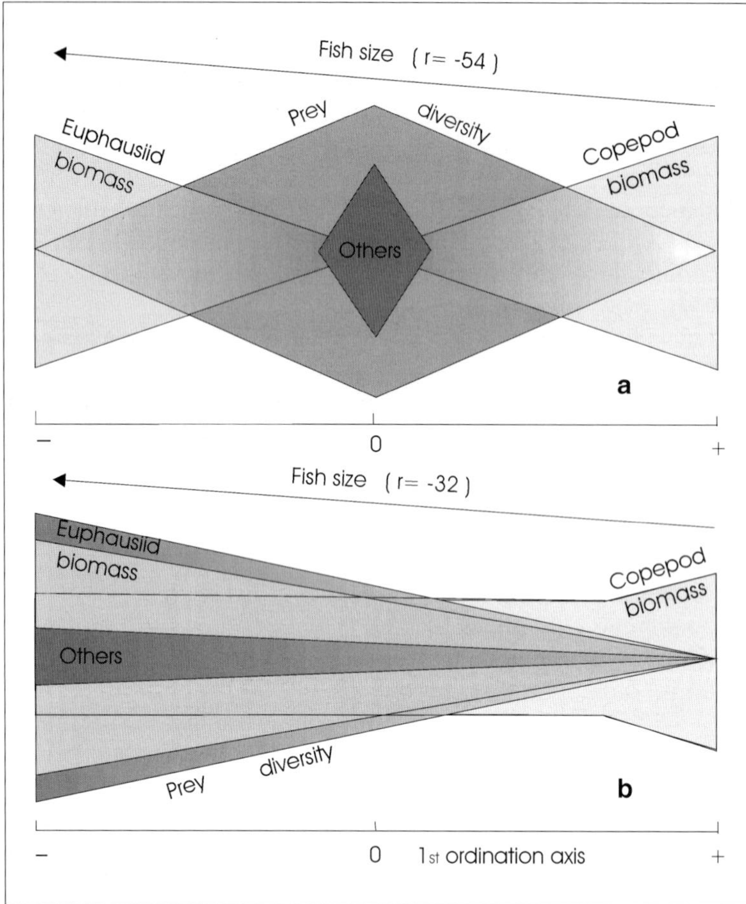

Fig. 4a,b. Conceptual model of the relationship between fish size and prey size, biomass and diversity in: **a** *Hygophum benoiti*; **b** *Myctophum punctatum* (modified from Scotto di Carlo et al. 1982)

surface and sub-surface migrating species. The *H. benoiti* ascended only in the open sea while *M. punctatum* also moved into more near-shore environments as demonstrated by the presence in the stomach contents of benthoplankters inhabiting shallow, soft-bottom waters

Genovese et al. (1985) carried out numerous dives in the Straits of Messina with the mesoscaph "Forel" to study the vertical distribution of some characteristic mesopelagic species. Among these, *Lestidiops sphyraenoides* (family Paralepididae) was distributed almost continuously between 350 and 500 m (mean depth of 450 m) and did not perform vertical migrations. *Maurolicus muelleri* was observed in the Tyrrhenian Sea, at a depth of 350 to 500 m, concentrated mainly in the demersal environment where it sometimes also assumed benthonic habits. Great versatility probably assures the species a more efficient utilization of the available food resources in levels characterized by high concentrations of zooplankton and micronekton.

Relationship Between Squids and Mesopelagic Fishes

Cephalopods seem to be a very important link in the trophic chain of the mesopelagic ecosystem. The lack of studies on their feeding regime is due to the difficulty in identifying the prey since these are reduced to small fragments in the stomach contents (Legand et al. 1972). The most complete studies on the diet of cephalopods have been carried out on the family Ommastrephidae in all oceans, also in the light of their commercial importance (Lipinski and Linkowski 1988).

Akimushkin (1963) and Fields (1965), in studying the feeding regime of pelagic cephalopods, were in agreement on the diversity of prey items during the ontogeny of squids. Thus, for the coastal species *Loligo opalescens,* the crustacean/fish ratio in the diet changed from 3/1 for young individuals to 1/1 for older ones, to 1/3 for reproductive animals. Morphological modifications of the radula were in direct relation to variation in feeding. During their growth, young squid gradually captured prey of diverse zoological groups and of increasing size, as the structure of the ever more-armed tentacle suggests. As their capturing apparatus became gradually perfected, the young squid began feeding on copepods, euphausiids, amphipods and eventually fishes. Medium-sized pelagic cephalopods (Onychoteuthidae, Ommastrephidae) fed mainly on prey items that were proportional to their dimensions, always concentrating on fishes. In the adult stage, reproductive animals exhibited, in some of the species studied, a high percentage of empty stomachs, which leads to the inference that feeding stops at the time of reproduction (Legand et al. 1972). Breiby et al. (1984) analysed the feeding habits of *Todarodes sagittatus* (Ommastrephidae) and found variations in the composition of the diet related to ontogeny. The stomach contents had a broth-like consistency, which precluded the division of the food remains for analysis of the volume and mass of the various prey categories. Prey items were identified from "key" fragments of skeletal tissue such as jaw elements, parapodia, otoliths and shell remains. The recovered otoliths were compared directly with a reference sample taken from fishes common to the study area. The authors noted a general trend toward a decrease of crustaceans in the diet and an increase of fishes as the squids increased in size. The size of the prey items ranged from a few millimetres for the calanoid copepods to more than 15 cm for some fishes, among which were mesopelagic species like *Maurolicus muelleri.*

With regard to the South Tyrrhenian Sea, a study was recently carried out, based on the annual sampling of *Todarodes sagittatus*, in the hydrographic area adjacent to the Straits of Messina (Marabello 1994; Marabello et al. 1996). The purpose of the study was to investigate the trophic relations between cephalopods and mesopelagic fishes. The results have provided further confirmation of the hypothesis of a close trophic coupling between cephalopods and mesopelagic fishes. Figure 5 illustrates that young squid, with DML (Dorsal Mantle Length) up to 23 cm, prefer above all cephalopods, followed by fishes and crustaceans with equal frequency. As squid increase in size, the predator's preference turns more and more to fishes rather than cephalopods and crustaceans. The composition of fishes in the diet is very distinct, with the families Gonostomatidae, Myctophidae, Chauliodontidae, Paralepididae, Sternoptychidae and Stomiatidae, all characteristic of the mesopelagic environment. The species, found in order of importance in the diet, are: *Chauliodus sloani, Maurolicus muelleri, Argyropelecus hemigymnus, Vinciguerria attenuata, Stomias boa, Vinciguerria poweriae, Lampanyctus crocodilus, Myctophum punctatum, Hygophum benoiti, Diaphus rafinesquei, Ichthyococcus ovatus, Sudis hyalina.* Table 1 reports the most important prey species in the diet of *T. sagittatus,* with their diurnal and nocturnal ranges of depth, while Fig.

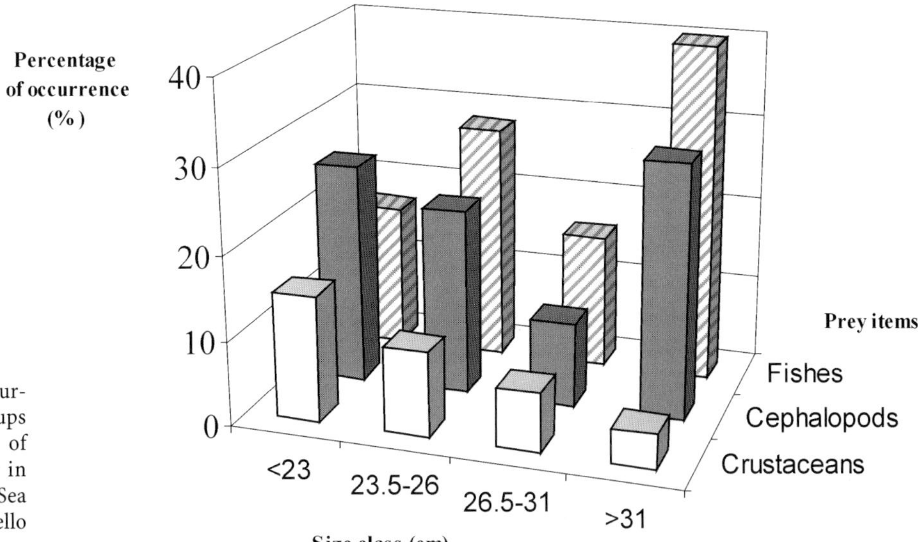

Fig. 5. Percentage occurrence of main prey groups with increasing size of *Todarodes sagittatus* in southern Tyrrhenian Sea (modified from Marabello 1994)

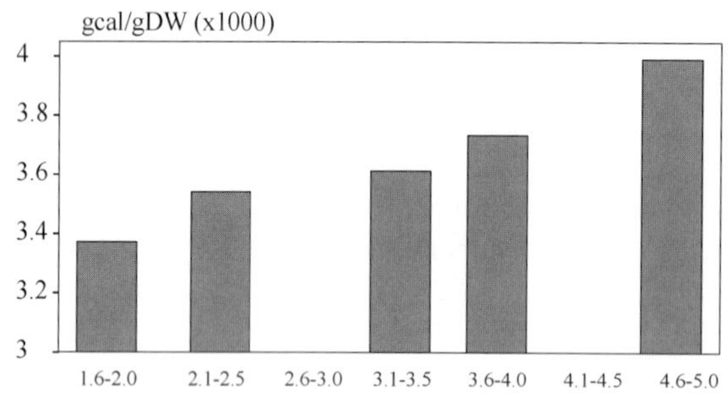

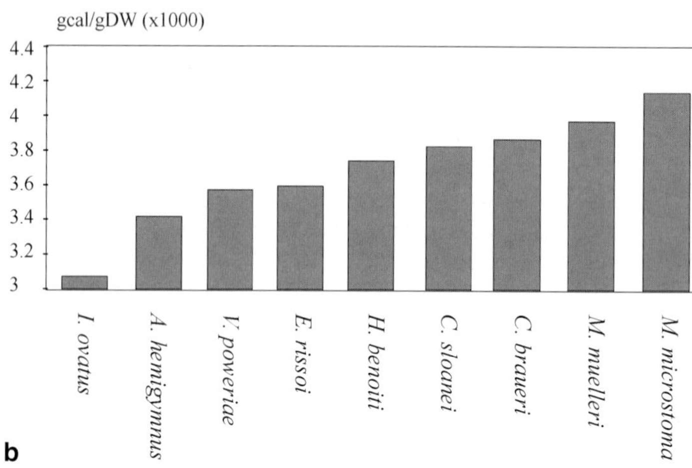

Fig. 6a,b. Energetic value: **a** per size class (mm) in *Argyropelecus hemigymnus*; **b** in the most representative midwater fishes (modified from Zagami et al. 1992)

6 reports the energetic value of the most important mesopelagic fish species (Zagami et al. 1992). Therefore, the data reported in the literature and our direct observations in the South Tyrrhenian Sea confirm the fundamental role of cephalopods as a top link in the trophic chain of mesopelagic fishes.

The diet of *Ommastrephes bartrami* was studied in the South-West Atlantic (Lipinski and Linkowski 1988), again with only qualitative results for the same reasons mentioned above. There was a high percentage of stomachs containing food in varying stages of digestion. Prey size varied markedly, although it was not measured directly on account of its fragmentation. The diet included three main groups: cephalopods (82%), fishes (34%) and crustaceans (18%). Squids comprised 77% of the diet (14% other *O. bartrami* individuals) while fishes were the second most important component, particularly species of myctophids *(Symbolophorus* sp., *Diaphus metopoclampus, Diaphus dumerilii, Diaphus effulgens, Protomyctophum gemmatum* and *Protomyctophum luciferum).* The *O. bartrami* was an opportunistic predator; the index of stomach fullness, the relatively small number of empty stomachs, the constituents of the diet and the majority of cephalopod species found suggest that *O. bartrami* probably feeds at night.

Sanchez (1982) investigated the feeding regime of *Illex coindetii,* a very common species in the Mediterranean with a vertical distribution between 45 and 575 m. The author found that this species was closely associated with the bottom during the day but exhibited a certain dispersion at night. With regard to the composition of the diet, three main groups of prey were observed: fishes, crustaceans and cephalopods. In autumn, fishes were most important (in order, *Sudis sp., Ceratoscopelus maderensis, Maurolicus muelleri, Myctophum punctatum, Engraulis encrasicholus, Gadiculus argenteus).* In winter, there was a presence of fishes in 57% of the stomachs among adults; *Sudis. sp.* disappeared while the occurrence of *Engraulis encrasicholus, Notoscopelus elongatus, Lampanyctus crocodilus* and *Ceratoscopelus maderensis* increased. In spring, fishes were again the preferred diet, although *Engraulis encrasicholus* did not appear and was substituted by *Micromisistius poutassou* and *Maurolicus muelleri.* In summer, fishes constituted a very high percentage of the diet (83%), predominantly the myctophids. Like most of the pelagic and bathypelagic cephalopods, *I. coindetii* is basically a fish predator, although its diet includes a sizeable quantity of crustaceans and sometimes also cephalopods. Young individuals exhibit less alimentary selectivity, but it appears that, with increasing size, the preference moves ever more toward fish.

Discussion and Conclusion

For an adequate understanding of pelagic trophic webs, various of problems must be investigated thoroughly. For example, it is necessary to understand the relationship between large surface predators and mesopelagic fishes that migrate at night into surface layers, the connection between these two important trophic food chains, and the distribution of available energy in the environment. The structure of the so-called epipelagic zone is well known, and is dominated during the day by tunas, swordfishes and sharks. These feed on a wide variety of smaller fishes, pelagic cephalopods and, to a lesser degree, on crustaceans (euphausiids). Tunas *(Thunnus thynnus, T. albacares, T. obesus, T. alalunga)* pursue their prey very rapidly, performing true migrations in search of adequate food sources only during the day and in the upper layers of water. Sharks generally occupy the same trophic level, except for the very large species that can occupy a higher trophic level, assuming the role of final predators. Legand et al. (1972) stated that the organisms found in the stomachs of tunas inhabiting tropical regions were very different from those caught with the IKMT (Isaacs-Kidd Midwater Trawl). Cephalopods were more numerous and larger in the stomachs than in the nets, in contrast to crustaceans which were numerous in the nets but virtually absent in the stomachs. Epipelagic fishes, which are rapid swimmers and thus not easily caught with pelagic nets, constitute the main resource of food for tunas, with a frequency percentage of 60%. A study by Roger (1973) on the relation between the euphausiids and organisms that occupy higher trophic levels *(Thunnus alalunga, T. albacares, T. obesus, Alepisaurus ferox)* confirmed that tunas do not feed directly on euphausiids. Instead, euphausiids constituted only 10% of the total food ingested by micronektonic epipelagic fishes, which in turn are preyed upon by tunas. These fishes normally measure 30 to 130 mm, are rapid swimmers and, consequently, are rarely caught in pelagic nets. Their preferred euphausiid prey belong to the genera *Stylocheiron* and *Nematoscelis*. This implies that epipelagic fishes preyed upon by tunas are able to utilize only organisms that inhabit sub-surface waters during the day and ascend toward the surface only at night *(Euphausia, Thysanopoda)*.

The same observations of daytime feeding were made for tunas. Roger and Grandperrin (1976) studied the diet of *Thunnus alalunga* and *T. albacares* and confirmed the previous data. They observed that euphausiids constituted 12% by volume of the food of micronektonic fishes; euphausiids were exclusively resident in the epipelagic zone and in turn formed 60% of the tuna diet. Cephalopods formed 40% of the tuna diet, whereas mesopelagic fishes that migrated to the sub-surface zone at night were virtually never found in tuna stomachs. Also in this case, euphausiids found in the stomachs of micronektonic fishes were non-migratory epipelagic species of the genus *Stylocheiron*. Thus, both tunas and their prey, micronektonic fishes, feed during the day. Therefore, large predators in surface waters have a diet composed of two thirds smaller epipelagic fishes and one third cephalopods, and do not have a direct trophic relationship with mesopelagic fishes. Hence, the trophic chain "Tunas ⇨ Prey fishes ⇨ Euphausiids" is based exclusively on epipelagic energy production (Mann, 1984). Roger and Grandperrin (1976) maintained that vertically migrating mesopelagic fishes subtract food, and thus energy, from the surface layers without returning the removed portion, since they do not form part of the epipelagic trophic chain. If this were so, there would be an important and active net transport of energy and materials toward lower depths, with the function of supporting meso- and bathypelagic production. However, there is another parallel chain that involves cephalopods, which form 40% of the tuna diet. It is important to know the role of cephalopods as a link that joins the epi- and mesopelagic trophic chains. This question is tied to the hypothesis that energy flow to lower layers is partly compensated by a return towards the surface, by means of a chain that has as its key organisms the cephalopods. In fact, recent studies (Marabello, 1994; Marabello et al., 1996) on the feeding ecology of *Todarodes sagittatus* in the South

Tyrrhenian Sea confirm the data found in the literature with regard to other species and geographical locations. There is substantial evidence of a return of energy to the upper layers of the ocean, with cephalopods forming a rather large component of the diet of tunas (40%) and preying on mesopelagic fishes to a conspicuous degree, thus restoring a portion of the energy that is taken from the surface levels.

The question of energy flow is not as simple as depicted in this schematic outline. There are a several unsolved problems: the low variability of prey species depending perhaps on the ability of the prey to avoid capture using the mechanism of bioluminescence of prey fishes (Baguet et al., 1983) or escaping from the predator by sudden darting movements (Mann 1984). Alternatively, the diet may be composed only of those species characterized by high biomass, and in which there is a lack of the most common species in the diet. For example, *Todarodes sagittatus*, *Myctophum punctatum* and *Hygophum benoiti* have a low percentage occurrence (0.4%) even though they are very abundant in this sea (Scotto di Carlo et al. 1982). These species may migrate to surface levels at night. These authors stated that *T. sagittatus* feeds at night and not in strictly surface layers supporting the greater importance in the diet of the squid *Maurolicus muelleri* and *Argyropelecus hemigymnus* (percentage occurrence of 3.2 and 1.8%, respectively, Table 1). Both are non-migratory species, the former remaining near the bottom (400-450 m in the southern Tyrrhenian Sea; Genovese et al. 1985) and the latter in the mesopelagic zone (300-600 m; Goodyear et al. 1972). Therefore, *T. sagittatus* should be considered a good vehicle of energy transfer from the zone ranging from sub-surface layers down to the bottom (100-800 m). Nevertheless, its diet lacks the mesopelagic species that populate the surface level (0-100 m) at night and which are responsible for the subtraction of energy from the euphotic zone.

This review raises questions concerning the complex trophic organization of the epi - and mesopelagic zones and the flow of energy between the different vertical trophic levels of the oceans. In other words, for a study like this, thorough investigations of the unresolved questions listed above are necessary. Moreover, the hypothesis of a trophic chain composed of cephalopods, mesopelagic fishes and crustaceans, with cephalopods acting as the connection between the surface and mesopelagic trophic chains, seems plausible.

References

Akimushkin II (1963) Cephalopods of the seas of the URSS. In: Izdat Akad Nauk SSSR Moskva (Translated 1965 by Israel Progr Sci Transl) IPST 1384

Baguet F, Piccard J, Christophe B, Marechal G (1983) Bioluminescence and luminescent fish in the Straits of Messina from the mesoscaph "Forel". Mar Biol 74: 221-229

Berdar A, Costanzo G, Guglielmo L, Ianora A, Scotto di Carlo B (1979) Some aspects of the feeding habits of two species of midwater fishes stranded along shores of the Straits of Messina. Rapp P-v Reun Comm Int Explor Sci Mer Mediterr 25/26: 209-210

Breiby A, Jobling M, Sundet JH (1984) Feeding habits ot the squid *Todarodes sagittatus* in north Norwegian waters. SCR Doc NW Atl Fish Org 84/IX/103: 11-99

Dalpadado P, Gjøsæter J (1988) Feeding ecology of the lanternfish *Benthosema pterotum* from the Indian Ocean. Mar Biol 99: 555-567

Farquhar GB (1977) Biological sound scattering in the oceans: a review. In: Andersen NR, Zahuranec BJ (eds) Oceanic sound scattering prediction. Plenum Press, New York, pp 493-527

Fields GW (1965) The structure, development, food relations, reproduction and life history of the squid *Loligo opalescens*, Berry. Fish Bull 131: 1-108

Genovese S, Berdar A, Guglielmo L (1971) Spiaggiamenti di fauna abissale nello Stretto di Messina. Atti Soc Peloritana 17: 331-370

Genovese S, Guglielmo L, Ianora A, Scotto Di Carlo B (1985) Osservazioni biologiche con il mesoscafo "Forel" nello Stretto di Messina. Arch Oceanogr Limnol 20: 1-30

Goodyear RH, Zahuranec BJ, Pugh WL, Gibbs RH (1972) Ecology and vertical distribution of mediterranean midwater fishes. In: Smits Inst Was Mediterr Biol Stud Final Rep 1: 91-229

Gorelova TA (1974) Zooplankton from the stomach of juvenile lantern fish of the family Myctophidae. Okeanologiya 14: 575-580

Guglielmo L, Arena G, Granata A, Sidoti O, Bonanzinga V, Soraci F (1996) Distribuzione verticale e migrazione giornaliera dello zooplancton e del micronecton nel Tirreno meridionale (Isole Eolie). Caratterizzazione ambientale marina del sistema Eolie e dei bacini limitrofi di Cefalù e Gioiosa (EOCUMM 95) In: Faranda FM, Povero P (eds) Data Rep, Genova, pp 217-246

Guglielmo L, Zagami G (1985) Role of Euphausiids in DSL of western Mediterranean Sea. Mem Biol Mar Ocean 15: 191-206

Guglielmo L, Zagami G, Sidoti O, Granata A (1995) Distribuzione e migrazione giornaliera dello zooplancton nel Tirreno meridionale (Isole Eolie): caratterizzazione ambientale marina del sistema Eolie e dei bacini limitrofi di Cefalù e Gioiosa (EOCUMM 94) In: Faranda FM (ed) Data Rep, Genova, pp 167-190

Holton AA (1969) Feeding behaviour of a vertically migrating lanternfish. Pac Sci 23: 325-331

Hopkins TL, Baird RC (1985) Feeding ecology of four hatchetfishes (Sternoptychidae) in the Eastern Gulf of Mexico. Bull Mar Sci 36: 260-277

Kinzer J, Schulz K (1985) Vertical distribution and feeding patterns of midwater fish in the central equatorial Atlantic. I. Myctophidae. Mar Biol 85: 313-322

Kinzer J, Schulz K (1988) Vertical distribution and feeding patterns of midwater fish in the central equatorial Atlantic. II. Sternoptychidae. Mar Biol 99: 261-269

Legand M, Bourret P, Fourmanoir P, Grandperrin R, Gueredrat JA, Michel A, Rancurel P, Repelin R, Roger C (1972) Relations trophiques et distributions verticales en milieu pelagique dans l'Ocean Pacifique intertropical. Cah ORSTOM Ser Oceanogr 10: 303-393

Lipinski MR, Linkowski TB (1988) Food of the squid *Ommastrephes bartrami* (Lesueur 1821) from the south-west Atlantic Ocean. S Afr J Mar Sci 6: 43-46

Longhurst AR (1989) The biological pump: profiles of plankton production and consuption in the upper ocean. Prog Oceanogr 22: 47-123

Mann KH (1984) Fish production in open ocean ecosystems. In: Fasham MJR (ed) Flows of energy and materials in marine ecosystems: theory and practice, Plenum Press, New York, pp 435-458

Marabello F (1994) Comportamento alimentare di *Todarodes sagittatus* Lamarck (1798) nell'area idrografica dello Stretto di Messina. Ph D Thesis, Univ Messina, Italy

Marabello F, Guglielmo L, Granata A, Sidoti O (1996) Studi preliminari sulle abitudini alimentari di *Todarodes sagittatus* Lamarck (1798) nel Tirreno Meridionale. In: Albertelli G, De Maio A, Picazzo M (eds) Atti XI Congr Naz Assoc Ital Oceanol Limnol, Sorrento (Na), 26-28 Ottobre 1994, pp 271-278

Marshall NB (1960) Swimbladder structure of deep-sea fishes in relation to their systematics and biology. Discovery Rep 31: 1-122

Matallanas J (1982) Estudio del regimen alimentario de *Lampanyctus crocodilus* (Risso, 1810) (Pisces: Myctophidae) en las costas catalanas (Mediterraneo Occidental). Tethys 10: 254-260

Merret NR, Roe HSJ (1974) Patterns and selectivity in the feeding of certain mesopelagic fishes. Mar Biol 25: 115- 126

Palma S (1990) Ecologie alimentaire de *Cyclothone braueri* Jespersen et Taning, 1926 (Gonostomatidae) en mer Ligure, Méditerranée occidentale. J Plankton Res, 12: 519-534

Roger C (1973) Recherches sur la situation trophique d'un groupe d'organismes pelagiques (Euphausiacea). V. Relations avec les thons. Mar Biol 19: 61-65

Roger C, Grandperrin R (1976) Pelagic food webs in the tropical Pacific. Limnol Oceanogr 21: 731-746

Sameoto DD (1982) Zooplankton and micronekton abundance in acoustic scattering layer and its significance to surface swarms. J Plankton Res 5: 129-143

Sameoto DD (1988) Feeding of lantern fish *Benthosema glaciale* off the Nova Scotia shelf. Mar Ecol Prog Ser 44: 113-129

Sanchez P (1982) Regimen alimentario de *Illex coindetii* (Verany, 1837) en el Mar Catalan. Environ Pesq 46: 443-449

Scotto Di Carlo B, Costanzo G, Fresi E, Guglielmo L, Ianora A (1982) Feeding ecology and stranding mechanisms in two lanternfishes, *Hygophum benoiti* and *Myctophum punctatum*. Mar Ecol Prog Ser 9: 13-24

Tyler HR, Pearcy WG (1975) The feeding habits of three species of lanternfishes (Family Myctophidae) off Oregon, USA. Mar Biol 32: 7-11

Zagami G, Badalamenti F, Guglielmo L (1992) Dati preliminari sul valore energetico dei più comuni pesci mesopelagici dello Stretto di Messina. OEBALIA 17: 165-168

CHAPTER 26

Diel Feeding Features of Juveniles of Two Sparids in the Stagnone di Marsala Coastal Sound (Western Sicily, Italy)

L. Lopiano, S. Mirto, D. Scilipoti, and A. Mazzola

ABSTRACT

Diet composition, feeding rhythms, gastric evacuation rates, daily rations of *Diplodus puntazzo* (Gmelin 1789) and *Sarpa salpa* (Linneo 1758) juveniles were studied in order to investigate their trophic ecology, niche breadth and overlap. These Sparids were collected in the Stagnone di Marsala (Western Sicily), in April 1997, in a 24 hour sampling period. Gastric content of 86 specimens of *D. puntazzo* (standard length = 31.9 ± 6.6 mm; body wet weight = 1.0 Ī 0.5 g) and of 100 specimens of *S. salpa* (standard length = 32.3 ± 2.9 mm; body wet weight = 0.6 Ī 0.2 g) were examined. Amphipoda and Tanaidacea were the most frequent prey items in stomach content of *D. puntazzo*, while Hydrozoa, Bryozoa and Rhodophycaea were the most frequent prey items in that of *S. salpa*. Both species displayed high frequency of occurrence (% Fr) of Copepoda and Ascidiacea larvae. A unimodal trend of daily rhythm of food consumption was deduced for both species. Gastric evacuation of juvenile *D. puntazzo* was best described by a square root model, with a gastric evacuation rate $R = 0.011 \pm 0.001$ ($r = 0.84$; $n = 28$; $P < 0.001$). Similarly, gastric evacuation of *S. salpa* followed a square root model, with a gastric evacuation rate $R = 0.014 \pm 0.001$ ($r = 0.90$; $n = 42$; $P < 0.001$). The amount of food consumed daily, calculated according to Pennington's model, was 0.0133 g dry weight g^{-1} body dry weight for *D. puntazzo* and 0.0159 g dry weight g^{-1} body dry weight for *S. salpa*, equal to 1.3% and 1.6% of the average body dry weight, respectively. Standardised Levins' index (B_A), calculated on percent volume (% V), suggested that juveniles of *D. puntazzo* have a niche breadth (mean standardised Levins' index = 0.2 ± 0.02) larger than the one of *S. salpa* (mean standardised Levins' index = 0.1 ± 0.01) at most sampling times. Juvenile specimens of these two Sparids presented a niche overlap at 4 pm, in correspondence with their feeding peak (Simplified Morisita's index $C_H = 0.9$). Feeding habits of *D. puntazzo* and *S. salpa* and their niche overlap are discussed in comparison with literature available on the same field.

Introduction

Diplodus puntazzo and *Sarpa salpa* are two species of Sparids very common along the Mediterranean coasts. They can also be found in the Black Sea and in the Atlantic Ocean, from the Bay of Biscay to South Africa, where they live on rocky bottoms covered by vegetation. Adults of *D. puntazzo* usually live alone or in small groups down to a depth of 50 metres, while adults of *S. salpa* live in the upper water layer and have strong gregarious habits (Tortonese 1975). Juveniles of these two species usually live near the coast (Macpherson 1998), in shallow and sheltered waters, that they selected as "nursery areas". Several studies are available on feeding habits and microhabitat requirements of these two Sparids, in relationship with other fish species (Porcile et al. 1987; Antolic et al. 1994; Mirto et al. 1994; Harmelin-Vivien et al. 1995; Sala and Ballesteros 1997; Planes et al. 1999).

The aims of this study are: (1) to point out diel feeding habits, niche breadth and overlap of juveniles of *D. puntazzo* and *S. salpa* in a typical Mediterranean shallow coastal environment; (2) to study feeding rhythm and daily food consumption of young *D. puntazzo* and *S. salpa*.

Dipartimento di Biologia Animale, Università di Palermo, Via Archirafi 18, 90123 Palermo, Italy

F.M. Faranda, L. Guglielmo, G. Spezie (eds)
Mediterranean Ecosystems: Structures and Processes

Materials and Methods

The study was carried out in the Stagnone di Marsala (western Sicily), a shallow coastal sound where *D. puntazzo* and *S. salpa* are transient species. Their larger recruitment, as juveniles, is included between last winter and early spring (Sarà et al. 1996). Samples were collected in April 1997, in the central part of the Stagnone di Marsala. This site ranged in depth between 0.5 m and 1.0 m and was characterised by submerged vegetation composed of *Cymodocea nodosa* and *Caulerpa prolifera*. Fish were caught using a beach seine (estimated opening wide = 6 m; length = 15 m; mesh size = 10 mm), every 4 hours over a 24 h cycle, then narcotised in chloroform and fixed in a (10%) formalin buffered water solution. In laboratory, the standard length (SL) and the body wet weight (BWW) of each fish were measured to the nearest 0.1 mm and 0.1 g, respectively. In total, 86 specimens of *D. puntazzo* (mean SL = 31.9 ± 6.6 mm; mean BWW = 1.0 ± 0.5 g) and 100 specimens of *S. salpa* (mean SL = 32.3 ± 2.9 mm; mean BWW = 0.6 ± 0.2 g) were examined. At each sampling time, about 10 specimens of both species were used in the qualitative analysis of the diet, the remaining part being used in the study of the gastric evacuation pattern and the calculation of daily food ration. For the qualitative analysis, stomachs were dissected under the stereomicroscope and the main items of the diet classified. The contribution to the total volume due to each item was calculated for each stomach using the subjective method proposed by Hellawell and Abel (1971). The principal simple indices such as the Coefficient of Vacuity (% CV), frequency occurrence (% Fr) (Kennedy and Fitzmaurice 1972) and the percentage in volume of prey (% V) were calculated. Among the composite indices only the Feeding Index (IA = % Fr * % V / 100), according to Lauazanne (1975) was calculated. For the quantitative dietary analysis, the relative weights of the dry eviscerated fish (BDW, g) and of the dry gastric content (DWsc, g) were measured after desiccation (60°C to constant weight). The latter was expressed as g dry weight g^{-1} BDW (Arrhenius and Hansson 1994). For both species, the feeding rhythm was derived plotting the mean corrected DWsc at different sampling hours. The values of gastric contents at different times, expressed as corrected DWsc were also compared using the Wilcoxon test (Sokal and Rohlf 1981). For both species evacuation rates (*R*) were calculated from the decreased stomach contents during non-feeding periods (i.e., assuming that no feeding occurred in darkness). Food consumption for each sampling interval (C_t) was calculated following Pennington (1985). Total consumption over an entire diel cycle was obtained by summing the positive values of C_t (Ruggerone 1989). Niche breadth of both species was calculated according to Levins (Krebs 1989). Niche overlap between juveniles of the two species, based on % V values, was calculated using simplified Morisita's index, C_H, (Krebs 1989).

Results

At four of the six sampling times stomachs appeared to be full (%CV = 0), in both species; at 12 pm % CV was equal to 22 % for *D. puntazzo*, and 10 % for *S. salpa*. All specimens caught in the early morning (4 am) presented empty stomachs (% CV = 100). The analysis of mean corrected DWsc, revealed, for both species, a unimodal diel feeding rhythm (Figure). *Diplodus puntazzo* feeding phase began between 8 and 12 am (8 am vs 12 am $P < 0.01$); it continued during the following hours (12 am vs 4 pm $P > 0.05$; 4 pm vs 8 pm $P > 0.05$), and ended from 8 to 12 pm (8 pm vs 12 pm $P < 0.05$), without any real feeding peak. A similar feeding rhythm was observed for *S. salpa* juveniles, whose feeding phase started between 12 am and 4 pm (12 am vs 4 pm $P < 0.01$) and ended from 8 to 12 pm (8 pm vs 12 pm $P < 0.01$).

The values of % Fr and % V in the different sampling times are reported in Table 1, for *D. puntazzo*, and in Table 2 for *S. salpa*.

The analysis of Feeding Index (Table 3) showed that, for *D. puntazzo* juveniles, Amphipoda were "dominant prey" (*sensu* Lauazanne 1975; 50 < IA < 100) at 8 pm and "fundamental prey" (25 < IA < 50) at 12 pm. Ascidiacea larvae were "fundamental prey" at the beginning of the feeding phase (IA = 45.3 at 12 am and 42.1 at 4 pm). Copepoda, despite their high % Fr values, were "not negligible prey"(10 < IA < 25) only at 8 pm, due to their low volume contribution to the stomach content. All other food items represented "secondary prey" (0 < IA < 10). For *S. salpa*, Phycophyta were "dominant prey" at most sampling hours, except at 4 pm, when they were decreased to the level of "not

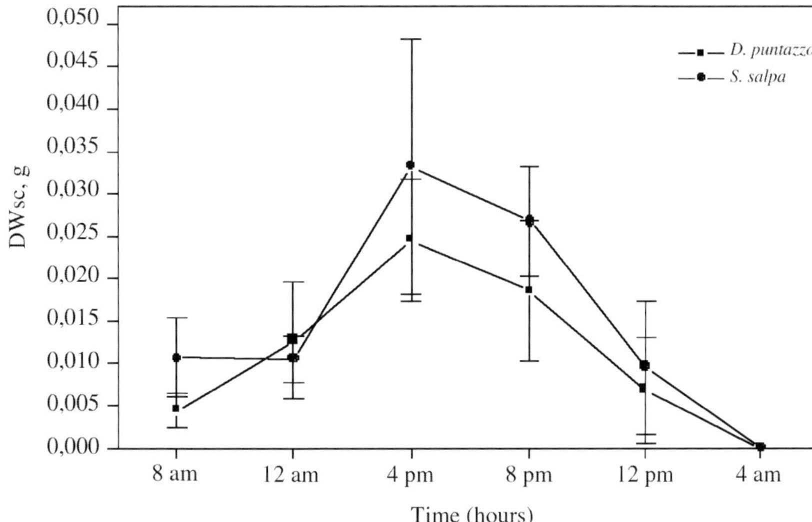

Figure. Diel feeding rhythm of *D. puntazzo* and *S. salpa* (*DWsc*, dry weight of stomach content)

Table 1. Frequency occurrence (*% Fr*) and percent volume (*% V*) of food items in the diet of *D. puntazzo*

Items	8 am		12 am		4 pm		8 pm		12 pm		4 am	
	% Fr	% V	% Fr	% V	% Fr	% V	% Fr	% V	% Fr	% V	% Fr	% V
Amphipoda	50.0	18.7	55.5	13.4	80.0	11.9	100.0	56.9	60.0	49.0	0.0	0.0
Ascidiacea larvae	0.0	0.0	77.8	58.2	80.0	52.6	44.0	7.8	40.0	27.1	0.0	0.0
Bryozoa	0.0	0.0	0.0	0.0	0.0	0.0	0.0	0.0	0.0	0.0	0.0	0.0
Copepoda	20.0	1.0	77.8	3.2	80.0	5.3	89.0	16.7	40.0	3.5	0.0	0.0
Fecal pellets	0.0	0.0	0.0	0.0	0.0	0.0	0.0	0.0	0.0	0.0	0.0	0.0
Hydrozoa	0.0	0.0	0.0	0.0	0.0	0.0	0.0	0.0	0.0	0.0	0.0	0.0
Isopoda	0.0	0.0	11.1	12.4	10.0	12.8	0.0	0.0	0.0	0.0	0.0	0.0
Mollusca larvae	20.0	23.1	0.0	0.0	50.0	15.4	11.0	2.6	0.0	0.0	0.0	0.0
Ostracoda	0.0	0.0	0.0	0.0	0.0	0.0	0.0	0.0	0.0	0.0	0.0	0.0
Phycophyta	0.0	0.0	11.1	2.1	10.0	1.8	11.0	5.2	10.0	2.9	0.0	0.0
Polychaeta	0.0	0.0	0.0	0.0	0.0	0.0	0.0	0.0	0.0	0.0	0.0	0.0
Tanaidacea	30.0	13.5	27.8	2.8	10.0	0.2	44.0	5.4	30.0	10.3	0.0	0.0
Unidentified eggs	20.0	43.7	38.9	7.9	0.0	0.0	22.0	5.4	20.0	7.2	0.0	0.0

Table 2. Frequency occurrence (*% Fr*) and percent volume (*% V*) of food items in the diet of *S. salpa*

Items	8 am		12 am		4 pm		8 pm		12 pm		4 am	
	% Fr	% V	% Fr	% V	% Fr	% V	% Fr	% V	% Fr	% V	% Fr	% V
Amphipoda	16.7	1.3	0.0	0.0	0.0	0.0	20.0	0.8	0.0	0.0	0.0	0.0
Ascidiacea larvae	16.7	0.7	30.0	2.4	80.0	67.4	20.0	7.4	40.0	11.0	0.0	0.0
Bryozoa	66.7	15.8	10.0	0.6	80.0	4.3	20.0	4.0	0.0	0.0	0.0	0.0
Copepoda	83.3	1.2	50.0	0.7	70.0.	1.9	90.0	1.0	50.0	0.7	0.0	0.0
Fecal pellets	16.7	0.3	30.0	2.2	0.0	0.0	40.0	0.3	10.0	0.2	0.0	0.0
Hydrozoa	66.7	10.7	100.0	31.2	80.0	5.3	60.0	8.0	40.0	3.4	0.0	0.0
Isopoda	0.0	0.0	0.0	0.0	0.0	0.0	0.0	0.0	0.0	0.0	0.0	0.0
Mollusca larvae	0.0	0.0	0.0	0.0	10.0	1.6	0.0	0.0	10.0	15.3	0.0	0.0
Ostracoda	16.7	0.7	0.0	0.0	30.0	0.1	20.0	0.1	0.0	0.0	0.0	0.0
Phycophyta	100.0	68.2	100.0	61.7	90.0	16.7	100.0	72.0	70.0	62.3	0.0	0.0
Polychaeta	16.7	0.4	10.0	0.2	10.0	0.1	40.0	0.6	10.0	0.1	0.0	0.0
Tanaidacea	0.0	0.0	0.0	0.0	0.0	0.0	0.0	0.0	0.0	0.0	0.0	0.0
Unidentified eggs	16.7	0.7	20.0	1.0	30.0	2.6	40.0	5.9	20.0	7.0	0.0	0.0

Table 3. Feeding Index (IA) of main food items in the diet of *D. puntazzo* (*D.p.*) and *S. salpa* juveniles (*S.s.*)

Items	8 am		12 am		4 pm		8 pm		12 pm		4 am	
	D. p.	*S. s.*	*D. p.*	*S. s.*	*D. p.*	*S. s.*	*D. p.*	*S. s.*	*D. p.*	*S. s.*	*D. p.*	*S. s.*
Amphipoda	9.4	0.2	7.4	0.0	9.5	0.0	56.9	0.2	29.4	0.0	0.0	0.0
Ascidiacea larvae	0.0	0.1	45.3	0.7	42.1	53.9	3.5	1.5	10.8	4.4	0.0	0.0
Bryozoa	0.0	10.5	0.0	0.1	0.0	3.4	0.0	0.8	0.0	0.0	0.0	0.0
Copepoda	0.2	1.0	2.5	0.3	4.2	1.3	14.8	0.9	1.4	0.4	0.0	0.0
Fecal pellets	0.0	0.1	0.0	0.7	0.0	0.0	0.0	0.1	0.0	0.0	0.0	0.0
Hydrozoa	0.0	7.1	0.0	31.2	0.0	4.2	0.0	4.8	0.0	1.4	0.0	0.0
Isopoda	0.0	0.0	1.4	0.0	1.3	0.0	0.0	0.0	0.0	0.0	0.0	0.0
Mollusca larvae	4.6	0.0	0.0	0.0	7.7	0.2	0.3	0.0	0.0	1.5	0.0	0.0
Ostracoda	0.0	0.1	0.0	0.0	0.0	0.0	0.0	0.1	0.0	0.0	0.0	0.0
Phycophyta	0.0	68.2	0.2	61.7	0.2	15.0	0.6	72.0	0.3	43.6	0.0	0.0
Polychaeta	0.0	0.1	0.0	0.0	0.0	0.1	0.0	0.2	0.0	0.0	0.0	0.0
Tanaidacea	4.0	0.0	0.8	0.0	0.0	0.0	2.4	0.0	3.1	0.0	0.0	0.0
Unidentified eggs	8.7	0.1	3.1	0.2	0.0	0.8	1.2	2.4	1.4	1.4	0.0	0.0

negligible prey", and at 12 pm, when they represented a "fundamental prey". Other "fundamental prey" were taxa Hydrozoa at 12 am and Ascidiacea larvae at 4 pm, while the other items may be considered "secondary prey".

Gastric evacuation of *D. puntazzo* juveniles fitted a square root model with a rate R of 0.011 ± 0.001 g DW h^{-1} ($r = 0.84$; $P < 0.001$; $n = 28$). The amount of food consumed daily, calculated according to Pennington's model, was 0.0113 g DW g^{-1} body DW, equal to 1.3% of the mean dry weight of the fish. Also gastric evacuation *of S. salpa* juveniles was best described by the square root model, with R values of 0.014 ± 0.001 g DW h^{-1} ($r = 0.89$; $P < 0.001$; $n = 42$). Daily food ration was 0.0159 g DW g^{-1} body DW, equal to 1.6% of the mean BDW.

Values of standardised Levins' index of niche breadth (B_A) and of simplified Morisita's index niche overlap (C_H) between *D. puntazzo* and *S. salpa* juveniles, in the different sampling times, are reported in Table 4. Juveniles of *D. puntazzo* had a niche breadth larger that *S. salpa* at most sampling times. The analysis of C_H values showed that juveniles of two species presented a niche overlap at 4 pm ($C_H = 0.9$).

Table 4. Standardised Levins' index (B_A) and simplified Morisita's index (C_H), in the different sampling times

Time	B_A		C_H
	D. p.	*S. s.*	
8 am	0.2	0.1	0.1
12 am	0.1	0.1	0.1
4 pm	0.2	0.1	0.9
8 pm	0.1	0.1	0.1
12 pm	0.2	0.1	0.1
4 am	0.0	0.0	0.0

Discussion

The qualitative analysis of the gastric contents of *D. puntazzo* in the Stagnone di Marsala allowed to identify these juveniles as micro-carnivorous mainly feeding on small Amphipoda and Tanaidacea, that in turn are associated to *Cymodocea* and *Caulerpa*. Another important contribution to their diet was provided by planktonic prey, such as Ascidiacea larvae, planktonic eggs and Copepoda Harparcticoida. The *D. puntazzo* diet appeared to be highly selective, due to the proportion of Ascidiacea in their stomachs compare to their density in the Stagnone di Marsala (Campolmi 1998). Copepoda were intensively preyed during the entire diel cycle. Similarly, several studies carried out in coastal and estuarine areas (Knox 1986; Coull 1990) have shown that juveniles of different fish species largely feed on Harpacticoida which, due to their high fatty acid content (Volk et al. 1984) are characterised by high calorific values (Watanabe et al 1983). Our results confirmed previous results on food preference of juvenile *D. puntazzo* from the same area (Mirto et al 1994). Algae and marine seagrasses did not represent important food items in the diet of these juveniles, and this is in contrast with results on food preferences for adults (Porcile et al. 1987; Sala and Ballesteros

1997). By contrast, macroalgae were the most important food item in the diet of *S. salpa* juveniles. In particular, they fed on Rhodophycaea (*Laurencia* sp. and *Polisiphonia* sp.), that have been reported in this environment (Scilipoti 1998). Our results are in agreement with Antolic et al. (1994) who found adults of *S. salpa*, in the southern Adriatic Sea, feeding mainly on Rhodophycaea and marine phanerogams. Among other food items, only Ascidiacea larvae and Hydrozoa provided an important contribution to the diet of this species.

Juveniles of *D. puntazzo* always presented a niche breadth larger than that of *S. salpa* juveniles. In fact, their diet was composed of largely diversified prey items, while macroalgae represented the dominant fraction of gastric contents of *S. salpa* juveniles. Niche overlap of the two species at 4 pm is clearly due to the intensive ingestion of Ascidiacea larvae by both *D. puntazzo* and *S. salpa*. As the availability of this item in the environment was scarce, it hypothesised a competition between the two species for this kind of resource.

The unimodal trend in the daily feeding rhythm of both species (without a real feeding peak), indicates that they were continuously feeding during the day, but not during the night. This habit allows them to avoid competition for food resources with other small-size residential species. In fact, additional studies carried out in the same area showed that residential species, such as *Pomatoschistus tortonesei*, *Aphanius fasciatus* and *Atherina boyeri*, fed largely on same resources (Copepoda, Amphipoda, Tanaidacea) (Scilipoti et al. 1997; Scilipoti 1998). Yet, a comparison between diel feeding cycles of these species suggests that there is a shift in feeding times in order to reduce trophic competition. Both *D. puntazzo* and *S. salpa* juveniles followed a square root model of gastric evacuation with a gastric evacuation rate of 0.011 and 0.014 g DW h^{-1}, respectively. These low rates of gastric evacuation may be due to a number of factors such as their very small size, the scant amount of food ingested, water temperature, and food quality. It is well known that all these factors may influence gastric evacuation rates (Jobling 1987; Dos Santos and Jobling 1988; Karjalainen et al. 1990; Parrish and Margraf 1990). The *D. puntazzo* diet was made up of small crustaceans with relatively hard chitin esoskeleton, and *S. salpa* gastric content was mainly composed of macroalgae, with strong cellulose content. Nonetheless evacuation rates reported here are very low when compared to those of the juveniles of *Mullus surmuletus* in the same area, who presented an evacuation rate of 0.66 g DW h^{-1} (Mazzola et al. 1999).

Daily food rations, calculated following Pennington (1985) were 1.3% BDW for *D. puntazzo* and to 1.6% for *S. salpa*. These results are lower than those reported by other authors, for juveniles of different species but based on Eggers' and Elliott and Persson's models (Lagardère 1987; Sagar and Glova 1988; Ruggerone 1989; Mazzola et al. 1999). So the differences observed might be due to the fact that the Pennington method, based on square root model, tends to underestimate food consumption, when compared with Eggers' (1979) and Elliott and Persson's exponential models, (Ruggerone 1989).

Acknowledgements. The authors are particularly indebted to Dr. G. Sarà (University of Palermo) for precious collaboration during sampling. Thanks are due to Prof. R. Danovaro (University of Ancona) and Dr. T. La Rosa (University of Palermo) for suggestions and support during ms revision. This work was supported by a grant of the Ministero dell'Università e Ricerca Scientifica e Tecnologica and the Ministero per le Politiche Agricole, Italy.

References

Antolic B, Skaramuca B, Musin D, Sanko-Njire J (1994) Food and feeding habits of a herbivore fish *Sarpa salpa* (L.) (Teleostei, Sparidae) in the southern Adriatic (Croatia). Acta Adriat 35: 45-52

Arrhenius F, Hansson S (1994) *In situ* food consumption by young-of-the-year Baltic Sea herring *Clupea harengus*: a test of predictions from a bioenergetics model. Mar Ecol Prog Ser 110: 145-149

Campolmi M (1998) Studio della comunità zooplanctonica di un bassofondo costiero mediterraneo (Stagnone di Marsala, Sicilia occidentale). Ph D thesis, Univ Messina.

Coull BC (1990) Are members of the meiofauna food for higher trophic levels? Trans Am Microsc Soc 109(3): 233-246

Dos Santos J, Jobling M (1988) Gastric emptying in cod, *Gadus morhua* L.: effects of food particle size and dietary energy content. J Fish Biol 33: 511-516

Eggers DM (1979) Comments on some recent methods for estimating food consumption by fish. J Fish Res Board Can 36: 1018-1019

Harmelin-Vivien ML, Harmelin JG, Leboulleux V (1995) Microhabitat requirements for settlement of juvenile sparid fishes on Mediterranean rocky shores. In: Balvay G (ed) Space partitionig within aquatic ecosystems, Hydrobiologia, vol 300-301, pp 309-320

Hellawell JM, Abel R (1971) A rapid volumetric method for the analysis of the food of fish. J Fish Biol 3: 29-37

Jobling M (1987) Influences of food particle size and dietary energy content on pattern of gastric evacuation in fish: test of a physiological model of gastric emptying. J Fish Biol 30: 299-314

Karjalainen J, Koho J, Viljanen M (1990) The gastric evacuation rate of vendace (*Coregonus albula* L.) larvae predating on zooplankters in the laboratory. Aquaculture 90: 343-351

Kennedy M, Fitzmaurice P (1972) Some aspects of the biology of gudgeon *Gobio gobio* (L.) in Irish waters. J Fish Biol 4: 425-440

Knox GA (1986) Estuarine ecosystem: a systems approach. CRC Press, Boca Raton, Florida

Krebs CJ (1989) Ecolological methodology. Harper and Row Publishers, New York

Lagardère JP (1987) Feeding ecology and daily food consumption of common sole, *Solea vulgaris* Quensel, juveniles on the French Atlantic coast. J Fish Biol 30: 91-104

Lauazanne L (1975) Régimes alimentaires d'*Hydrocyon forskalii* (Pisces, Characidae) dans le lac Tchad et ses tributaires. Hydrobiologia 9(2): 103-121

Macpherson E (1998) Ontogenic shifts in habitat use and aggregation in juvenile sparid fishes. J Exp Mar Biol Ecol 220 (1): 127-150

Mazzola A, Lopiano L, La Rosa T, Sarà G (1999) Diel feeding habits of juveniles of *Mullus surmuletus* (Linneo, 1758) in the lagoon of the Stagnone di Marsala (Western Sicily, Italy). J Appl Ichthyol 15: 143-148

Mirto S, Scilipoti D, Lopiano L, Badalamenti F, Mazzola A (1994) Primi dati sul ritmo alimentare giornaliero di tre specie ittiche nello Stagnone di Marsala (Sicilia occidentale). Biol Mar Mediterr 1(1): 335-336

Parrish DL, Margraf FJ (1990) Gastric evacuation rate of white perch, *Morone americana*, determined from laboratory and field data. Environ Biol Fish 29: 155-158

Pennington M (1985) Estimating the average food consumption by fish in the field from stomach contents data. Dana 5: 81–86

Planes S, Macpherson E, Biagi F, Garcia-Rubies A, Harmelin J, Harmelin-Vivien M, Jouvenel JY, Tunesi L, Vigliola L, Galzin R (1999) Spatio-temporal variability in growth of juvenile sparid fishes from Mediterranean littoral zone. J Mar Biol Assoc UK 79 (1): 137-143

Porcile P, Repetto N, Wurtz M (1987) Feeding behaviour of young Sparidae of the Ligurian Sea. Oebalia 1: 311-314

Ruggerone GT (1989) Gastric evacuation rates and daily ration of piscivorous coho salmon, *Oncorhynchus kisutch* Walbaum. J Fish Biol 34: 451-463

Sagar PM, Glova GJ (1988) Diel feeding periodicity, daily ration and prey selection of riverine population of juvenile chinook salmon, *Oncorhynchus tshawytscha* (Walbaum). J Fish Biol 33: 643-653

Sala E, Ballesteros E (1997) Partitioning of space and food resources by three fish of the genus *Diplodus* (Sparidae) in a Mediterranean rocky infralittoral ecosystem. Mar Ecol Prog Ser 152: 273-283

Sarà G, Mirto S, Mazzola A (1996) Analisi della diversità biologica del popolamento ittico di una laguna costiera italiana (Stagnone di Marsala, Sicilia occidentale). Biol Mar Mediterr (in press)

Scilipoti D (1998) Studio della comunità ittica residente all'interno dello Stagnone di Marsala (Sicilia occidentale): distribuzione delle specie e ripartizione delle risorse in dipendenda di habitat a diversa complessità strutturale. PhD thesis, Univ Messina.

Scilipoti D, Franzoi P, Nardo R, Mazzola A (1997) Primi dati sull'alimentazione di *Aphanius fasciatus* (Nardo, 1827) nello Stagnone di Marsala (Trapani). Biol Mar Mediterr 4 (1): 564-566

Sokal RR, Rohlf FJ (1981) Biometry. WH Freeman and Co, New York

Tortonese E (1975) Ostheicthyes, fauna d'Italia, vol 2. Calderini, Bologna, pp 104-118

Volk EC, Wissman RC, Simenstad CA, Eggers DM (1984) Relationship between otolith microstructure and growth of juvenile chum salmon *Oncorhynchus keta* under different rations. Can J Fish Aquat Sci 41: 126-133

Watanabe T, Kitajima C, Fujita S (1983) Nutritional values of live organisms used in Japan for mass propagation of fish: a review. Aquaculture 34: 115-143

CHAPTER 27

Fish Species Biodiversity on Trawlable Bottoms of South Adriatic Basin (Mediterranean Sea)

N. Ungaro and G. Marano

ABSTRACT

Groundfish biodiversity on trawlable bottoms of the South Adriatic basin was analysed by means of univariate ecological indices. Raw data came from three experimental trawl surveys carried out in the mentioned area during the summer 1996, 1997 and 1998. The surveys (112 samples per survey) provided a total list of 113 groundfish species (98 teleosts, 15 selachians). Margaleff (richness), Shannon (diversity) and Pielou (evenness) indices were estimated; the index values per sample were mapped by kriging method, and mean values for each survey were computed in order to detect yearly and/or spatial differences (shelf and slope zone, East and West coast).

The global analysis of the diversity of the groundfish species on the trawlable bottoms of the South Adriatic Sea did not show any particular trend over the triennium 1996-98, in spite of some significant differences among the years. The main changes concern the "shelf" ecological indices, while the "slope" situation appears to be reasonably stable; East (Albania) "shelf" values are generally higher and more homogeneously distributed than the West shelf (Italy) ones.

Introduction

The Adriatic Sea basin is characterised by typical geomorphologic and oceanographic features; the bottom structure (sediments) and slope, the water current flows and chemical-physical characteristics follow both a latitudinal (Northern, Middle and Southern Adriatic) and a longitudinal (East and West side) gradient (Buljan and Zore-Armanda 1979; Alfirevic 1981; Artegiani et al. 1981). These environmental features could affect the distribution of a large number of species, including groundfishes.

As regards the Southern Adriatic Sea, until 1996 scientific information about diversity of groundfish species was referred mainly to the western side (Italian waters) (Bello and Rizzi 1988; Ungaro et al. 1996; 1998a), while data were scarce from the eastern side (Albanian waters). During the last years additional information about the whole Southern Adriatic Sea came from the European Community research project "Mediterranean International Trawl Survey" (Bertrand 1996). The trawl surveys carried out within the above mentioned project provided new data on fish species presence and distribution in the basin; herein the collected data were analysed by means of univariate ecological indices in order to update groundfish biodiversity on South Adriatic trawlable bottoms during the years 1996, 1997 and 1998.

Materials and Methods

Data about groundfish species were obtained during three trawl-surveys carried out in South Adriatic Sea (Mediterranean Sea), from 10 to 800 m depth (trawlable bottoms), by using a 10 mm cod-end otter trawl net during the summer 1996, 1997 and 1998. One hundred and twelve stations (1 haul/60 square nautical miles) were sampled for each survey (1 hour of trawling on the slope, half an hour of trawling on the shelf) (Fig. 1). The sampling design referring to the first survey was random stratified (five bathymetric strata: 10-50 m, 51-100 m, 101-200 m, 201-500 m, 501-800 m) (Bertrand 1996) and the chosen points were re-sampled the following years.

Laboratorio di Biologia Marina, Molo Pizzoli (Porto), 70123 Bari, Italy

F.M. Faranda, L. Guglielmo, G. Spezie (eds)
Mediterranean Ecosystems: Structures and Processes

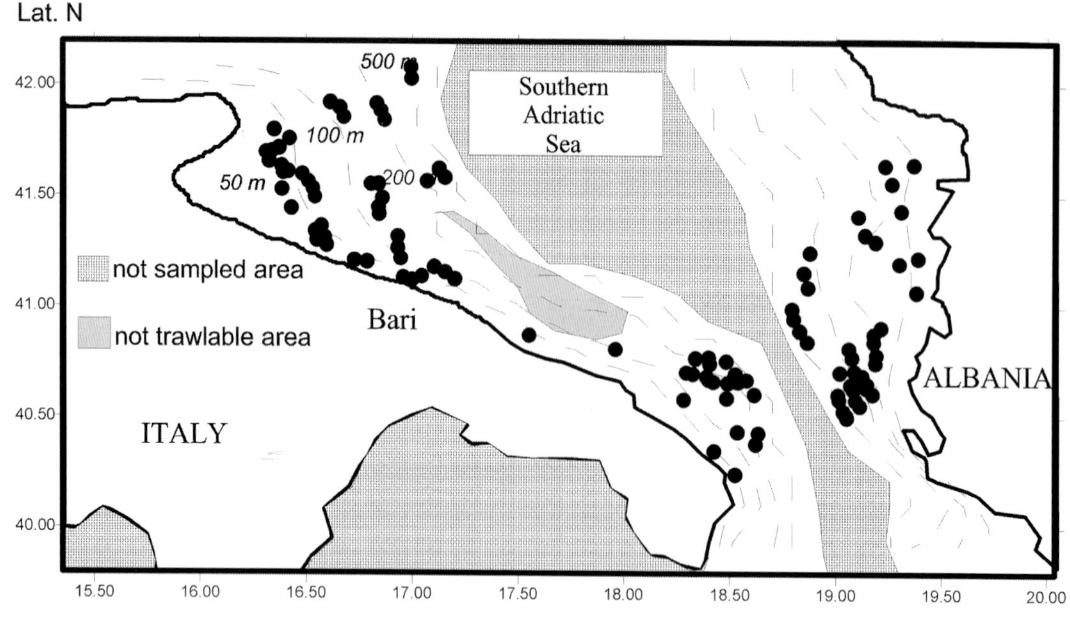

Fig. 1. Investigated area

All caught fish species were identified and a species list per haul was recorded; strictly pelagic and epi-mesopelagic species (mostly belonging to the Families Clupeidae, Engraulidae, Myctophidae, Carangidae, Scombridae), occasionally collected by the trawl net, were excluded from the list.

Biodiversity was evaluated by using univariate ecological indices (Margaleff richness, Shannon diversity, Pielou's evenness) (Magurran 1991). During processing the results were mapped per year by kriging method (Oliver 1990) and analysed per year, per bathymetric (shelf, slope) and geographic zone (East and West coast). The shelf and slope distinction was adopted because of the presence of "well defined" fish assemblages by depth (0-200 m; > 200 m) (Ungaro et al. 1998a). Index mean values were compared by Tukey and Tukey-Kramer statistical tests (Castino and Roletto 1991).

Results

The three surveys provided a list of 113 groundfish species (98 teleosts, 15 selachians) (Table 1); the number of caught and identified species were 93, 96 and 99, for the year 1996, 1997 and 1998 respectively. The Margaleff index per sample (D, by using Ln transformation) ranged, over the triennium, from 0.66 (year 1998) to 4.48 (year 1998), while Shannon index (*H*) ranged from 0.41 (year 1986) to 2.54 (year 1998) and Pielou's index (*J*) from 0.18 (1996) to 1 (1998) (Table 2). The results per sample were processed by ordinary kriging (linear variogram model) and mapped per year. Generally, the index highest values resulted more homogeneously distributed on the East side of the investigated area (Figs. 2-4).

Finally, the index mean values per year and geographic - bathymetric area (shelf, slope) were compared. The "slope" ecological indices per year (over the whole South Adriatic area) did not show significant differences, while some different values were found for the "shelf" indices (mostly richness index) among the years (Tukey test, a = 0.05). Furthermore, the Margaleff index mean values for the East coast shelf (Albania) were significantly higher than the West coast (Italy) ones in the surveys 1997 and 1998, as well as the Shannon index for the surveys 1996 and 1998 (Tukey-Kramer test, a = 0.05) (Table 3).

Discussion

The global analysis of groundfish species diversity on the trawlable bottoms of the South Adriatic

Table 1. Demersal fish species caught during MEDITS '96, '97 and '98 surveys in the Southern Adriatic Sea (depth range 10-800 m) (*A*, found in Albanian waters only; *I*, found in Italian waters only; *S*, found on shelf bottoms only)

	Teleosteans				
	Antonogadus megalokynodon (Kolombatovic 1894)		*Helicolenus dactylopterus* (Delaroche, 1809)	A	*Peristedion cataphractum* (Linnaeus 1758)
SI	*Aphia minuta* (De Buen 1931)		*Hoplostethus mediterraneus* (Cuvier 1829		*Phycis blennoides* (Brünnich 1768)
	Argentina sphyraena (Linnaeus 1758)		*Hymenocephalus italicus* (Giglioli 1884)	A	*Phycis phycis* (Linnaeus 1766)
	Ariosoma balearicum (Delaroche 1809)		*Lappanella fasciata* (Cocco 1833)		*Polyprion americanus* (Scheneider 1801)
	Arnoglossus laterna (Walbaum 1792)		*Lepidopus caudatus* (Euphrasen 1788)	SI	*Pomatoschistus marmoratus* (Risso 1810)
	Arnoglossus rueppelli (Cocco 1844)		*Lepidorhombus boscii* (Risso 1810)	SI	*Pomatoschistus minutus* (Pallas 1770)
S	*Arnoglossus thori* (Kyle 1913)		*Lepidorhombus whiffiagonis* (Walbaum 1792)	SA	*Psetta maxima* (Linnaeus 1758)
	Aspitrigla cuculus (Linnaeus 1758)		*Lepidotrigla cavillone* (Lacepede 1801)		*Scorpaena elongata* Cadenat 1943
S	*Blennius ocellaris* (Linnaeus 1758)	I	*Lestidiops sphyrenoides* (Risso 1820)	SI	*Scorpaena loppei* (Cadenat 1943)
S	*Boops boops* (Linnaeus 1758)		*Lesuerigobius friesi* (Malm 1874)	S	*Scorpaena notata* (Rafinesque 1810)
SI	*Callionymus fasciatus* (Valenciennes 1837)		*Lesuerigobius sueri* (Risso 1810)	S	*Scorpaena porcus* (Linnaeus 1758)
	Callionymus maculatus (Rafinesque Schmaltz 1810)		*Lophius budegassa* (Spinola 1807)	S	*Scorpaena scrofa* (Linnaeus 1758)
	Capros aper (Linnaeus 1758)		*Lophius piscatorius* (Linnaeus 1758)	S	*Serranus cabrilla* (Linnaeus 1758)
SA	*Carapus acus* (Brunnich 1768)		*Macrorhamphosus scolopax* (Linnaeus 1758)	S	*Serranus hepatus* (Linaeus 1766)
S	*Cepola macrophthalma* (Linnaeus 1758)	SI	*Merlangius merlangus euxinus* (Nordmann 1840)	S	*Solea vulgaris* (Quensel 1806)
	Chauliodus sloanei (Bloch and Schneider 1801)		*Merluccius merluccius* (Linnaeus 1758)	S	*Spicara flexuosa* (Rafinesque 1810)
A	*Chlopsis bicolor* (Rafinesque 1810)	SI	*Microchirus variegatus* (Donovan 1802)	S	*Spicara maena* (Linnaeus 1758)
	Chlorophthalmus agassizii (Bonaparte 1840)		*Micromesistius poutassou* (Risso 1810)	S	*Spicara smaris* (Linnaeus 1758)
SA	*Citharus linguatula* (Linnaeus 1758)		*Molva dipterygia macrophtalma* (Rafinesque 1810)	S	*Spondyliosoma cantharus* (Linnaeus 1758)
	Coelorhynchus coelorhynchus (Risso 1810)	SI	*Monochirus hispidus* (Rafinesque 1814)		*Stomias boa* (Risso 1810)
	Conger conger (Linnaeus 1758)		*Mora moro* (Risso 1810)	A	*Symphurus ligulatus* (Cocco 1844)
SI	*Dalophis imberbis* (Delaroche 1809)		*Mullus barbatus* (Linnaeus 1758)		*Symphurus nigrescens* (Rafinesque 1810)
S	*Deltentosteus quadrimaculatus* (Valenciennes 1837)		*Mullus surmuletus* (Linnaeus 1758)	A	*Synodus saurus* (Linnaeus 1758)
SA	*Dentex dentex* (Linnaeus 1758)	A	*Nemichthys scolopaceus* (Richardson 1848)	S	*Trachinus draco* (Linnaeus 1758)
S	*Diplodus annularis* (Linnaeus 1758)		*Nettastoma melanurum* (Rafinesque 1810)		*Trigla lucerna* (Linnaeus 1758)
I	*Echiodon dentatus* (Cuvier 1829)		*Nezumia sclerorhynchus* (Valenciennes 1838)		*Trigla lyra* (Linnaeus 1758)
	Epigonus denticulatus (Dieuzeide 1950)		*Notolepis rissoi* (Bonaparte 1840)	S	*Trigloporus lastoviza* (Bonnaterre 1788)
I	*Epigonus telescopus* (Risso 1810)		*Notacanthus bonapartei* (Risso 1840)		*Trisopterus minutus capelanus* (Lacepede 1800)
S	*Eutrigla gurnardus* (Linnaeus 1758)	SI	*Ophichthus rufus* (Rafinesque 1810)	S	*Uranoscopus scaber* (Linnaeus 1758)
	Gadella maraldi (Risso 1810)	S	*Pagellus acarne* (Risso 1826)		*Zeus faber* (Linnaeus 1758)
	Gadiculus argenteus (Guichenot 1850)		*Pagellus bogaraveo* (Brünnich 1768)		
	Glossanodon leioglossus (Valenciennes 1848)	S	*Pagellus erythrinus* (Linnaeus 1758)		
	Gnathophis mystax (Delaroche 1809)	SA	*Pagrus pagrus* (Linnaeus 1758)		
S	*Gobius niger jozo* (Linnaeus 1758)	I	*Paralepis coregonoides* (Risso 1820)		
	Selachians				
	Centrophorus granulosus (Bloch and Schneider 1801)		*Raja alba* (Lacepede 1803)	A	*Raja oxyrinchus* (Linnaeus 1758)
	Chimaera monstrosa (Linnaeus 1758)		*Raja asterias* (Delaroche 1809)		*Scyliorhinus canicula* (Linnaeus 1758)
	Etmopterus spinax (Linnaeus 1758)		*Raja circularis* (Couch 1838		*Squalus acanthias* (Linnaeus 1758)
	Galeus melastomus (Rafinesque 1809)		*Raja clavata* (Linnaeus 1758)	A	*Squalus blainvillei* (Risso 1826)
SA	*Mustelus mustelus* (Linnaeus 1758)		*Raja miraletus* (Linnaeus 1758)	S	*Torpedo marmorata* (Risso 1810)

Table 2. Margaleff richness index (*D*), Shannon-Wiener diversity index (*H*) and Pielou's evenness index (*J*): mean values (mean, SD, min., max.) for the Southern Adriatic Sea (shelf, slope, whole area). * , significantly different ($P<0.01$) from other years results, Tukey test

		Southern Adriatic Sea											
Depth strata	**Samples**	**Dmean**	**SD**	**Dmin**	**Dmax**	**Hmean**	**SD**	**Hmin**	**Hmax**	**Jmean**	**SD**	**Jmin**	**Jmax**
Year 1996													
10-200 m	*75*	2.14*	0.61	0.82	3.21	1.49	0.44	0.41	2.34	0.58*	0.15	0.18	0.84
200-800 m	*37*	2.59	0.67	0.96	3.64	1.77	0.39	0.90	2.46	0.72	0.14	0.30	0.95
10-800 m	*112*	2.29*	0.66	0.82	3.64	1.58	0.44	0.41	2.46	0.63*	0.16	0.18	0.95
Year 1997													
10-200 m	*75*	1.88*	0.52	0.70	3.78	1.53	0.39	0.44	2.42	0.68	0.17	0.19	0.93
200-800 m	*37*	2.48	0.45	1.35	3.28	1.84	0.29	0.83	2.32	0.72	0.11	0.43	0.88
10-800 m	*112*	2.08*	0.57	0.70	3.78	1.63	0.39	0.44	2.42	0.69	0.15	0.19	0.93
Year 1998													
10-200 m	*75*	2.48*	0.53	1.36	4.48	1.84*	0.34	0.67	2.54	0.569	0.12	0.26	0.89
200-800 m	*37*	2.54	0.55	0.66	3.86	1.86	0.37	0.64	2.40	0.75	0.09	0.51	1.00
10-800 m	*112*	2.50*	0.54	0.66	4.48	1.85*	0.35	0.64	2.54	0.71	0.11	0.26	1.00

Sea did not show a clear trend over the triennium 1996-98, in spite of some differences among the years. The main changes concern the "shelf" ecological indices, while the "slope" situation appears to be reasonably stable. East (Albania) "shelf" values are generally higher and more homogeneously distributed than West shelf (Italy) ones.

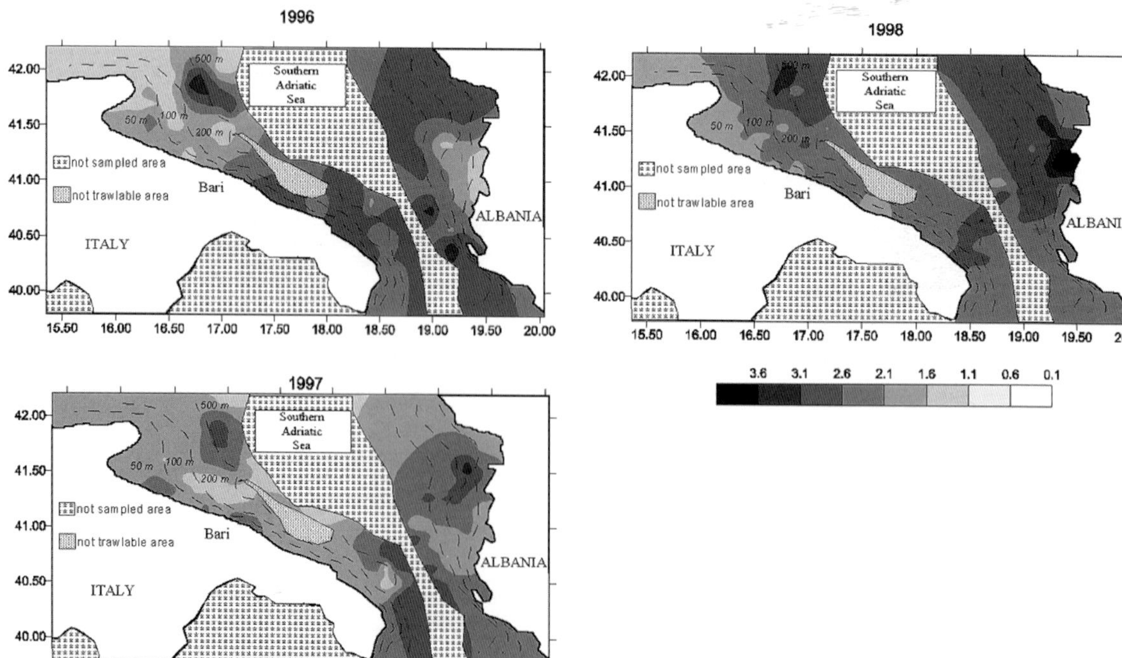

Fig. 2. Distribution per year of richness values (Margaleff index)

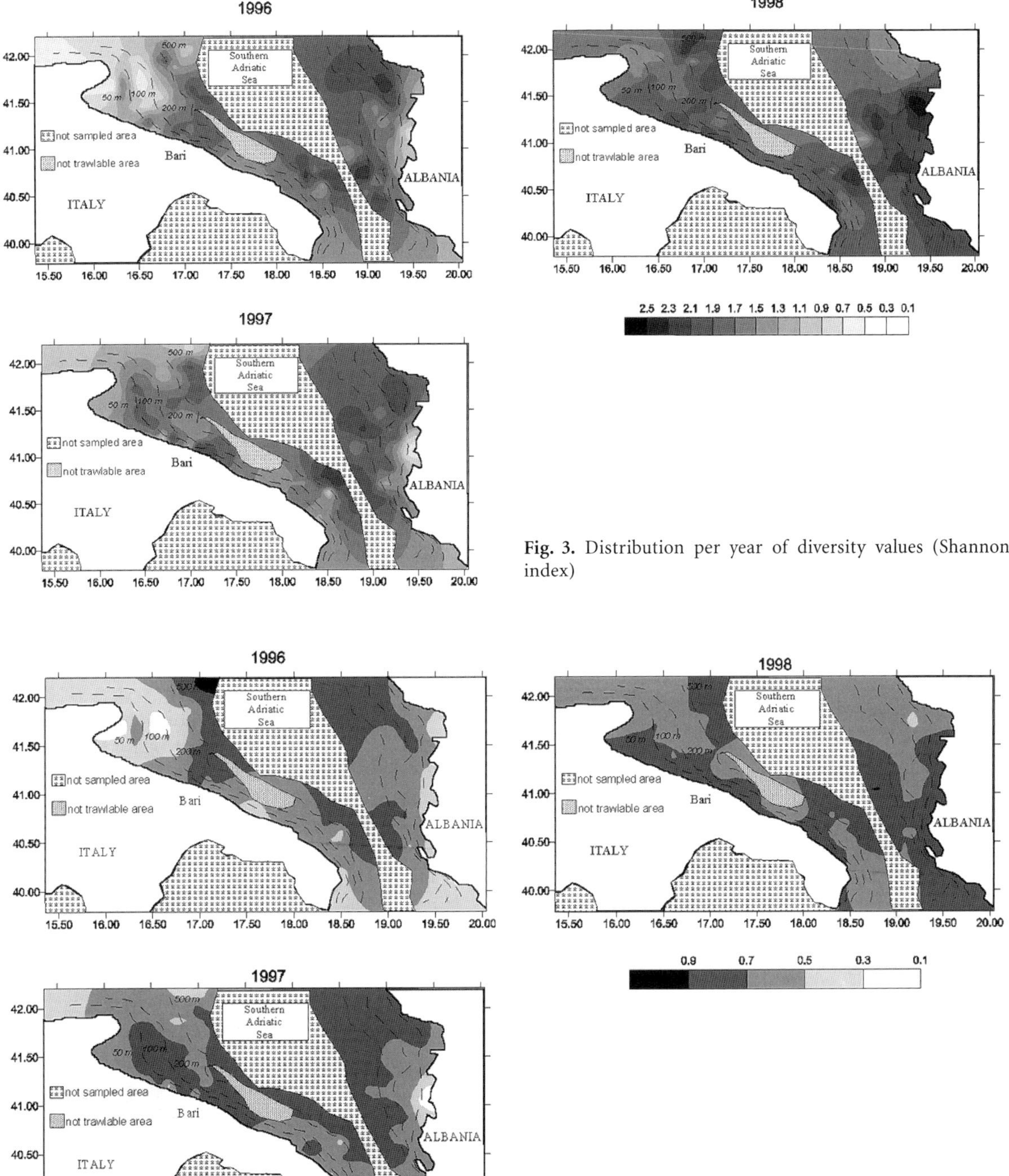

Fig. 3. Distribution per year of diversity values (Shannon index)

Fig. 4. Distribution per year of evenness values (Pielou index)

Species occurrence, as well as relative abundance, was also different between the West and East side of the basin; elasmobranch specimens were collected more frequently on "Albanian" bottoms and relative abundance of most of teleost commercial species (i.e. hake and red mullet) were higher in the same area (data not reported here). These last results could be due to a high fishing pressure on Italian shelf, as Ungaro et al. (1998b) suggested; as regard to the elamobranchs, the high "sensitivity" with respect to fishery (low fecundity, large length at maturity,

Table 3. Margaleff richness index (D), Shannon-Wiener diversity index (*H*) and Pielou's evenness index (J): mean values (mean, SD, min., max.) for Albanian and Italian shelves. *, significantly different ($P<0.01$) between the shelves, Tukey-Kramer test

Year 1996	Samples	Dmean	SD	Dmin	Dmax	Hmean	SD	Hmin	Hmax	Jmean	SD	Jmin	Jmax
Albanian shelf	*23*	2.28	0.56	1.00	3.14	1.69*	0.36	0.78	2.34	0.63	0.10	0.44	0.79
Italian shelf	*52*	2.08	0.63	0.82	3.21	1.40	0.45	0.41	2.25	0.56	0.16	0.18	0.84
Year 1997	**Samples**	**Dmean**	**SD**	**Dmin**	**Dmax**	**Hmean**	**SD**	**Hmin**	**Hmax**	**Jmean**	**SD**	**Jmin**	**Jmax**
Albanian shelf	*23*	2.10*	0.57	1.36	3.78	1.52	0.46	0.52	2.12	0.60*	0.16	0.19	0.89
Italian shelf	*52*	1.78	0.47	0.70	2.95	1.53	0.37	0.44	2.42	0.72	0.16	0.22	0.93
Year 1998	**Samples**	**Dmean**	**SD**	**Dmin**	**Dmax**	**Hmean**	**SD**	**Hmin**	**Hmax**	**Jmean**	**SD**	**Jmin**	**Jmax**
Albanian shelf	*23*	2.80*	0.60	1.86	4.48	2.00*	0.39	0.67	2.54	0.71	0.14	0.26	0.88
Italian shelf	*52*	2.34	0.44	1.36	3.34	1.77	0.30	0.78	2.44	0.69	0.11	0.29	0.89

large size at birth) (Holden 1974; Walker and Hislop 1998) could explain some differences between the two Adriatic sides, but local environmental features should be considered as well.

The sampling methodology could also strongly affect the results, and the use of univariate ecological indices is a matter of debate (Huston 1979) for marine resources (Caddy and Sharp 1986; Caddy and Griffiths 1995).

South Adriatic demersal fish diversity updating (1996-1998) could however be useful for the development of research (if the same methodology is used) in order to describe future marine environment changes due to ecological or anthropical factors.

References

Alfirevic S (1981) Contribution à la connaissance des caractéristiques bathymétriques et sédimentologiques de l'Adriatique. FAO Fish Rep 253: 43-52

Artegiani A, Azzolini R, Paschini E (1981) Seasonal thermocline evolution in the northern and middle Adriatic Sea. FAO Fish Rep 253: 55-64

Bello G, Rizzi E (1988) I teleostei raccolti nell'Adriatico meridionale nelle campagne sperimentali di pesca a strascico 1985-87. Quad Ist Ric Pesca Marit 5: 77-90

Bertrand J (1996) Campagne internationale de chalutage démersal an Méditerranée (MEDITS): campagne 1996. Rapp Final Rapp Contrat CEE-IFREMER-IEO-SIBM-NCMR (MED/95/19/54/65/27) 1-2: 1-71

Buljan M, Zore-Armanda M (1979) Hydrographic properties of the Adriatic Sea in the period from 1965 through 1970. Acta Adriat 20: 1-368

Caddy JF, Griffiths RC (1995) Living marine resources and their sustainable development. FAO Fish Tech Pap 353

Caddy JF, Sharp GD (1986) An ecological framework for marine fishery investigations. FAO Fish Tech Pap (283): 152 p.

Castino M, Roletto E (1991) Statistica applicata. Piccin, Padova

Holden MJ (1974) Problems in the rational exploitation of elasmobranch populations and some suggested solutions. In: Harden Jones FR (ed) Sea fisheries research. Halsted Press, J. Wiley and Sons, New York, pp 117-137

Huston M (1979) A general hypothesis of species diversity. Am Nat 113: 81-101

Magurran AE (1991) Ecological diversity and its measurement. Chapman and Hall, London

Oliver MA (1990) Kriging: a method of interpolation for geographical information systems. Int J Geog Inf Syst 4: 313-332

Ungaro N, Marano G, Rizzi E, Marzano MC (1996) Demersal Squaliformes and Rajiformes in the South-Western Adriatic Sea: trawl-surveys 1985-1994. FAO Fish Rep 533 Suppl: 87-96

Ungaro N, Marano G, Vlora A, Martino M (1998a) Space-time variations of demersal fish assemblages in South-Western Adriatic Sea. Vie Milieu 48: 191-201

Ungaro N, Marano G, Marsan R, Osmani K (1998b) Demersal fish assemblages biodiversity as an index of fishery resources exploitation. Ital J Zool 65 Suppl: 511-516

Walker PA, Hislop JRG (1998) Sensitive skates or resilient rays? Spatial and temporal shifts in ray species composition in the central and north-western North Sea between 1930 and present day. ICES J Mar Sci 55: 392-402

Notes on the Biology and Ecology of *Opeatogenys gracilis* (Canestrini 1864) (Pisces: Gobiesocidae) from Coastal Environments in Sicily (Mediterranean)

S. Vizzini, A. Mazzola, and D. Scilipoti

ABSTRACT

A particularly large number of specimens of the Mediterranean coastal fish *Opeatogenys gracilis* (Canestrini 1864) (Pisces: Gobiesocidae) were collected from *Posidonia oceanica* mats of the "Stagnone di Marsala" sound (western Sicily) and in the Gulf of Carini (northern Sicily). A study was made of biometric and meristic characters and, for the first time for this species, diet composition. The biometric and meristic data did not distinguish between specimens from the two study sites. In accordance with its benthic habit, *O. gracilis* feeds on small phytal prey living on *Posidonia oceanica* leaves. The preferential food items throughout the year were Copepoda Harpacticoida.

Introduction

Gobiesocidae is a cosmopolitan family of small marine fish, typical of shallow and warm temperate waters. Specimens are characterised by an adhesive structure located on the ventral body surface formed by two contiguous disks and flanked by pelvic fins. This adhesive structure determines the benthic habit of this species. The family includes 8 subfamilies, 43 genera and 150 species. Only 5 genera, all belonging to the subfamily Lepadogastrinae, are present in the Mediterranean Sea. *Opeatogenys gracilis* (Canestrini 1864) (Pisces: Gobiesocidae) is an endemic species of this area. It distinguishes itself from other Mediterranean Gobiesocidae by an acute and strong spine from the subopercolum. The literature data concerning *O. gracilis* are very poor, lacking information on both its distribution and ecology. This species is indicated as uncommon in that it has been recorded only at Nizza, Messina and in the Adriatic Sea (Bini 1968; Tortonese 1975). It has also been found recently along the southern coast of Spain (Reina-Hervás and Núñez-Vergara 1985; Cardona and Guerao 1992). Furthermore, the existing information regarding *O. gracilis* is derived from an exiguous number of specimens captured (two reported in Reina-Hervás and Núñez-Vergara 1985 and five in Cardona and Guerao 1992).

This paper reports the finding, distribution and feeding ecology of *O. gracilis* from the "Stagnone di Marsala". The first occurrence of this species in the Gulf of Carini (northern Sicily) is reported, verifying its distribution in *Posidonia oceanica* beds.

Materials and Methods

The "Stagnone di Marsala" is a semi-enclosed coastal sound on the western coast of Sicily. It is characterised by a wide range of salinity and temperature values. Its bottoms are covered by phanerogames (*Cymodocea nodosa* and *Posidonia oceanica*) associated with macroalgae (*Caulerpa prolifera* and *Cystoseira spp.*) (Calvo et al. 1980, 1982). The Gulf of Carini is located on the northern coast of Sicily. In this area the physical parameters are those typical of western Mediterranean shallow environments with little variation during the year (Giaccone and Sortino 1964).

Samples from the "Stagnone di Marsala" were collected monthly during one year by means of a trawling net (3.5 metres long and 3 millimetres mesh diameter). In the Gulf of Carini, samples were collected exclusively over

Dipartimento di Biologia Animale, Università di Palermo, Via Archirafi 18, 90123 Palermo, Italy

F.M. Faranda, L. Guglielmo, G. Spezie (eds)
Mediterranean Ecosystems: Structures and Processes

one month along a *Posidonia oceanica* transect (−1, −3, −5, −15 metres depth) by means of a hand-towed net (Russo et al. 1985).

Samples were fixed in formalin (10% buffered in sea-water) and the classic biometric (total length, standard length, head length, adhesive disk length, ocular diameter, maximum width) and meristic (dorsal fin rays, anal fin rays, caudal fin rays) measurements were carried out. Differences in the biometric and meristic relationships between *O. gracilis* from the two study sites were tested using t-test, after the verification of the normality.

Feeding and foraging habits were investigated in the specimens from the "Stagnone di Marsala". A subsample of 20 specimens for each season was dissected under a stereomicroscope. The prey were counted and identified. The percentage contribution of each prey taxon to the total prey volume was estimated visually. Quantitative and qualitative stomach content analysis was carried out using both simple (percentage of number, percentage of occurrence and percentage of volume of prey) (Hyslop 1980) and compound indexes (IRI: index of relative importance) (Pinkas et al. 1971). The IRI was standardised to 100% by calculating the percentage of the total IRI contributed by each prey type and ranged from 0 (absent from diet) to 100 (the only prey consumed).

Results

Biometric and Meristic Characters

Throughout the year the total catch of *O. gracilis* in the first study site was two hundred specimens, while in the second one the catch was 75. The mean standard length of specimens from "Stagnone di Marsala" was 14.51 ± 1.36mm, while in the Gulf of Carini it was 16.7 ± 2.41mm. The biometric and meristic characters are reported in Tables 1 and 2 respectively. The results of the t-test are shown in Table 3. Only some characters (adhesive disk length, maximum width, number of dorsal fin rays) were found to be significantly different in specimens from the two study sites (Table 3).

Table 1. Biometric characters of *Opeatogenys gracilis* in the "Stagnone di Marsala" sound and Gulf of Carini (*TL*, total length; *SL*, standard length; *HL*, head length; *AL*, adhesive disk length; *OD*, ocular diameter; *MW*, maximum width)

Biometric	Stagnone di Marsala		Gulf of Carini	
character	Mean ± SD (mm)	% SL	Mean ± SD (mm)	% SL
TL	17.51 ± 1.62	–	19.69 ± 2.78	–
SL	14.51 ± 1.36	100	16.70 ± 2.41	100
HL	5.28 ± 0.47	36.45 ± 1.86	6.11 ± 0.81	36.68 ± 1.51
AL	3.57 ± 0.41	24.63 ± 2.11	1.34 ± 0.16	8.13 ± 1.04
OD	1.18 ± 0.09	8.17 ± 0.64	3.70 ± 0.47	22.28 ± 1.71
MW	3.50 ± 0.42	24.12 ± 2.07	3.75 ± 0.58	22.53 ± 2.48

Table 2. Meristic characters of *Opeatogenys gracilis* reported in the literature and in this paper (*D*, dorsal fin rays; *A*, anal fin rays; *C*, caudal fin rays)

Authors	Meristic characters
Bini 1968	D 3; A 3/4; C 9/10
Briggs 1986	D 3/5; A 3/5; C 8/10
Cardona and Guerao 1992	D 3/4; A 3; C 9/10
Reina-Hervás and Núñz Vergara 1985	D 3; A 4; C 9
Soljan 1975	D 3; A 3
Tortonese 1975	D 3; A 4
This paper (Gulf of Carini)	D 2/4; A 2/4; C 9/13
This paper (Stagnone di Marsala)	D 2/3; A 2/3; C 10/14

Table 3. The significance levels of the t-test (*SL*, standard length; *HL*, head length; *AL*, adhesive disk length; *OD*, ocular diameter; *MW*, maximum width)

	P-level
HL/SL	0.29 (ns)
OD/SL	0.41 (ns)
AL/SL	0.00 (***)
MW/SL	0.00 (**)
Dorsal fin rays	0.02 (*)
Anal fin rays	0.32 (ns)
Caudal fin rays	0.06 (ns)

* $P < 0.05$; ** $P < 0.01$; *** $P < 0.001$ – *ns*, non-significant difference ($P>0.05$)

Table 4. Food preferences of *Opeatogenys gracilis* throughout the year. Percentage of number (*N%*) and percentage of occurrence (F%) are reported for each food item

	Spring		Summer		Autumn		Winter	
	N%	F%	N%	F%	N%	F%	N%	F%
Acarina	0.38	10	0.00	0	0.53	5	0.00	0
Amphipoda	0.96	20	2.67	10	0.00	0	0.00	0
Copepoda Colonoida	0.00	0	0.67	5	0.00	0	0.00	0
Chironomidae	0.38	10	0.00	0	4.28	55	2.44	25
Cumacea	0.38	10	0.67	5	5.88	35	0.00	0
Copepoda Harpacticoida	95.79	100	77.33	100	64.44	100	95.12	100
Isopoda	0.19	5	0.00	0	0.00	5	0.00	0
Ostracoda	0.19	5	0.00	0	0.80	15	0.00	0
Polychaeta	0.38	15	0.67	10	3.48	45	2.44	25
Tanaidacea	1.34	25	18.00	80	20.59	95	0.00	0.00
d.m.[a]	0.00	5	0.00	10	0.00	15	0.00	0.00

[a] (*d.m.*, digested material)

Stomach Content Analysis

The simple index values for each food item are reported in Table 4. In the Figure the relative importance index values are shown. Prey richness was highest in spring where 9 prey taxa were identified and declined to 3 taxa in winter (Table 4). Copepoda Harpacticoida represent the preferential and most important food item throughout the year (F%=100%; IRI% ranges from 53% to 97%). Tanaidacea are also important components of the diet in summer and autumn (F%=80%, 95%; IRI% = 28.27% and 34.33%). Polychaeta worms, Cumacea and Amphipoda were a lesser component of the diet.

Discussion and Conclusions

Biometric and Meristic Characters

When compared with the literature data, a large number of specimens of *Opeatogenys gracilis* were collected from both study sites. The small sizes of the specimens, benthic habit and the high habitat complexity of *Posidonia oceanica* beds make it difficult to assess the distribution and abundance of this species along western Mediterranean coasts. Although *O. gracilis* is considered to be an uncommon species (Bini 1968; Tortonese 1975), the fact that such a large number of individuals was collected leads us to

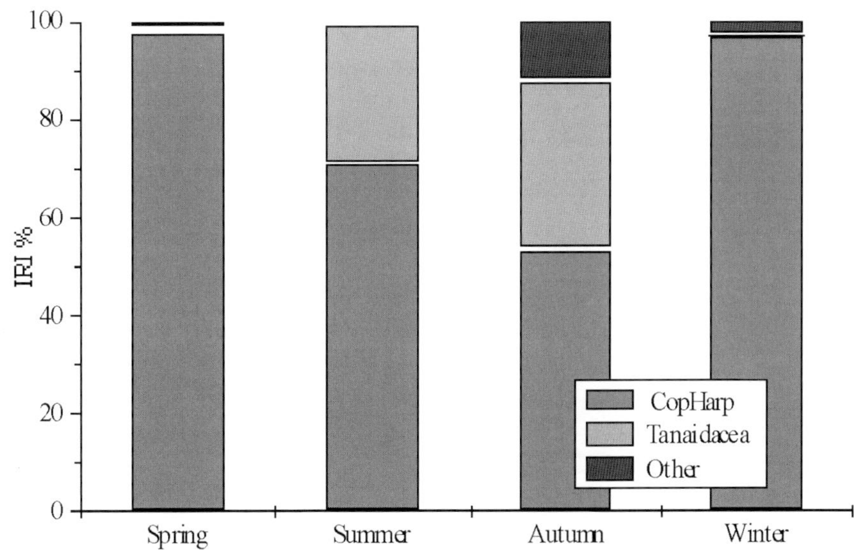

Figure. Index of relative importance (*IRI%*) of the main food items

suggest that its geographical distribution should be widened.

As regards the biometric and meristic characters, from the data reported it was not possible to distinguish between specimens from the two study sites: only two biometric and one meristic parameter showed statistically significant differences (Table 3). Generally speaking, biometric measurements are a tool for identifying different fish populations; differences in the average number of meristic parameters indicate that stocks are genetically different (Taning 1951). The meristic phenotype of fish depends on genetic and environmental variation. The main environmental factors are: temperature, salinity and light during egg incubation (Beacham and Murray 1986). In the case of *O. gracilis* further insights (karyotype characterisation, electrophoresis analysis of isoenzyme, nDNA and mDNA) is needed to make it possible to distinguish between two populations.

Stomach Content Analysis

Data on the feeding ecology of *O. gracilis* are reported for the first time in this paper. In accordance with its benthic habit *O. gracilis* feeds on small phytal prey living on *Posidonia* leaves. Throughout the year Copepoda Harpacticoida are the most common and abundant prey. Their importance as a food source is well known for other fish species living in the "Stagnone di Marsala" (Scilipoti 1998). The dominance of this systematic group is not surprising since Copepoda Harpactioida are normally the most numerous group in the phytal habitat of "Stagnone di Marsala". In summer and autumn Tanaidacea also become preferential prey. The feeding preferences of *O. gracilis* follow the annual trend of trophic resources in the study site. Copepoda Harpacticoida show a peak in abundance in the winter and spring months, while their number decrease during summer and autumn concomitantly with the increase in the Tanaidacea resource (although the latter are numerically fewer than the Copepoda Harpacticoida) (Campolmi 1998). In conclusion *O. gracilis* can be defined as a micro-carnivore specialised on epifaunal Crustacea especially Copepoda Harpacticoida and Tanaidacea.

References

Beacham TD, Murray CB (1986) The effect of spawning time and incubation temperature on meristic variation in chum salmon (*Oncorhynchus keta*). Can J Zool 64: 45-48

Bini G (1968) Atlante dei pesci delle coste italiane. In: Mondo sommerso, Roma 8: 129-130

Briggs JC (1986) Gobiesocidae. In: Whitehead PJP, Bauchot ML, Hureau JC, Nielsen J, Tortonese E (eds) Fishes of the north-eastern Atlantic and Mediterranean. UNESCO, Paris, pp 1351-1359

Calvo S, Drago D, Sortino M (1980) Winter and summer submersed vegetation maps of the Stagnone (western coast of Sicily). Rev Biol-Ecol Mediterr VII 2: 89-96

Calvo S, Giaccone G, Ragonese S (1982) Tipologia della vegetazione sommersa dello Stagnone di Marsala (TP). Nat Sicil S IV VI(2): 187-196

Campolmi M (1998) Studio della comunità zooplanctonica di un bassofondo costiero mediterraneo (Stagnone di Marsala, Sicilia occidentale). PhD thesis, Univ Messina

Cardona L, Guerao G (1992) Primera cita de Opeatogenys gracilis (Canestrini 1864) (Osteichthyes: Gobiesocidae) en el litoral catalán (Mediterráneo NO). Misc Zool 16: 243-245

Giaccone G, Sortino M (1964) Flora e vegetazione algale di Isola delle Femmine. Lav Ist Bot Giardino Colon Palermo 21: 3-27

Hyslop EJ (1980) Stomach contents analysis-a review of methods and their application. J Fish Biol 1: 491-429

Pinkas L, Oliphant MS, Iverson ILK (1971) Food habits of albacore, bluefin tuna and bonito in California waters. Calif Dep Fish Game Fish Bull 152

Reina-Hervás JA, Núñez-Vergara JC (1985) *Opeatogenys gracilis* (Canestrini 1864) (Gobiesocidae, Osteichthyes) en el Mediterráneo español. Misc Zool 9: 405-407

Russo GF, Fresi E, Vinci D (1985) The hand-towed net method for direct sampling in Posidonia oceanica beds. Rapp Comm Int Mer Mediterr 29(6): 175-177

Scilipoti D (1998) Studio della comunità ittica residente all'interno dello Stagnone di Marsala (Sicilia occidentale), con particolare riferimento alla distribuzione delle specie e alla ripartizione delle risorse in dipendenza di habitat a diversa complessità strutturale. PhD thesis, Univ Messina

Soljan T (1975) I pesci dell'Adriatico. In: Arnoldo Mondadori, Verona, pp 159-163

Taning AV (1951) Experimental study of meristic characters in fishes. Biol Rev Cambridge Philos Soc 27: 169-193

Tortonese E (1975) In: Fauna d'Italia: Osteichthyes. Pesci ossei. Calderini, Bologna, 11, pp 546-555

Benthic Fluxes of Oxygen, Ammonium and Nitrate and Coupled-uncoupled Denitrification Rates within Communities of Three Different Primary Producer Growth Forms

M. Bartoli[1], G. Castaldelli[2], D. Nizzoli[1], L.G. Gatti[2], and P. Viaroli[1]

ABSTRACT

Inorganic nitrogen and oxygen fluxes together with coupled-uncoupled denitrification were studied in sediments covered by different primary producers (benthic microalgae, the floating macroalga *Ulva rigida,* and the rooted phanerogam *Ruppia cirrhosa*). High DIN (Dissolved Inorganic Nitrogen) assimilation rates were measured for all the primary producers and resulted in low denitrification rates, in particular at the *Ulva* and *Ruppia* colonised sites. The competition for NH_4^+ and NO_3^- between phototrophic organisms and nitrifiers-denitrifiers was particularly strong when DIN concentrations in the water column were low. Despite algal uptake denitrification rates were appreciable (>200 μmol N $m^{-2}h^{-1}$) in the site covered by benthic diatoms due to high availability of NO_3^- in the water column and efficient coupling between nitrification and denitrification. In the sites with macrophytes losses of N due to coupled-uncoupled denitrification were negligible compared to assimilation rates. Most of the organic nitrogen pool in the *Ulva* biomass is probably recycled in the water column while a consistent part of the N stored in *Ruppia* may be buried in the sediment.

Introduction

In estuaries and shallow water impoundments primary producers can markedly affect N-cycling in many direct and indirect ways. Benthic microalgae can control at the sediment-water interface the diffusion of combined nitrogen to or from the sediment through their assimilation activity and thus compete with nitrifiers and denitrifiers for inorganic nitrogen. The influence of benthic microalgae on N-cycling at the sediment-water interface can be attributed to a combination of factors including: ammonium limitation, high pH and O_2 concentrations, CO_2 limitation and organic excretion products (Henriksen and Kemp 1988; Sundback and Graneli 1988; Nielsen et al. 1990; Nielsen and Sloth 1994). Photosynthesis at the sediment-water interface can expand the horizon of oxic sediment (Revsbech et al. 1981) with a positive effect on nitrifiers-denitrifiers but conversely remove NH_4^+ and CO_2 from the porewater and increase the pH to values up to 9-10 units, thus inhibiting the activity of nitrifiers and consequently denitrifiers (Rasmussen et al. 1983; Focht and Verstraete 1977; Henriksen et al. 1984).

Floating macroalgae have very high primary production rates which are sustained by significant amounts of DIN (Dissolved Inorganic Nitrogen) (Sand-Jensen and Borum 1991; Borum 1996). Whilst, part of the nitrogen requirement of macroalgal biomass production is probably due to internal recycling of N, most of it comes from the water column. Macroalgal mats can strongly influence the pool of DIN in the water column and may act as a significant sink of inorganic nitrogen during the growth season. Assimilation of NO_3^- is considered to be a much more important sink for N compared to denitrification, in particular where macroalgal mats attain biomass values over 200-300 $g_{dw}m^{-2}$ (Naldi 1994). Although, macroalgal uptake can be a major sink for DIN it is likely that the biomass becomes a N-source for the system when it is decomposed and during the frequent collapses occurring in summer, the so called "dystrophic crises" (Viaroli et al. 1996a).

Rooted macrophytes can control sediment-

[1] Dipartimento di Scienze Ambientali, Università di Parma, Parco Area delle Scienze 33A, 43100 Parma, Italy
[2] Dipartimento di Biologia Evolutiva, Università di Ferrara, Via Borsari, 44100 Ferrara, Italy

F.M. Faranda, L. Guglielmo, G. Spezie (eds)
Mediterranean Ecosystems: Structures and Processes

water fluxes of DIN through their assimilation activity (Rysgaard et al. 1996) and nitrification-denitrification processes within the sediment (Caffrey and Kemp 1992; Risgaard-Petersen and Jensen 1997). The roots of submerged phanerogams can excrete organic carbon (Sondergaard 1983) and transport O_2 to the sediment (Sand-Jensen and Prahl 1982; Christensen et al. 1994; Risgaard-Petersen and Jensen 1997). This can create an oxic layer around the roots resulting in optimal nitrification area (the availability of NH_4^+ is higher in the deep sediment); indirectly O_2 transport can stimulate denitrification when the produced NO_3^- diffuses to the anoxic sediment. This was clearly demonstrated in freshwater phanerogams (Risgaard-Petersen and Jensen 1997). However, aquatic plants also assimilate and incorporate DIN from the leaves and from the roots (Pedersen and Borum 1992) and thereby compete with nitrifiers and denitrifiers for inorganic nitrogen. The net effect of phanerogams on nitrification-denitrification processes depends on the balance between DIN uptake, DIN availability in the water column, NH_4^+ production in the sediment, O_2 release and the redox state of the sediment.

The role of primary producers and nitrifying-denitrifying bacteria on nitrogen transformations can be particularly strong in semi-enclosed bays and lagoons receiving high freshwater N-loads and where the water renewal is slow. In these areas the retention of DIN in the biomass and N-burial in the sediment as well as the conversion of NO_3^- to N_2 can quantitatively attenuate the flux of nitrogen through the coastal environment to the sea.

In this paper we consider oxygen and DIN fluxes, and coupled-uncoupled denitrification rates in three coastal areas of the North Adriatic Sea during three critical phases of the seasonal evolution of the benthic system. The investigated areas were colonised respectively by benthic diatoms (station Giralda), the macroalga *Ulva rigida* (station Gorino) and the phanerogam *Ruppia cirrhosa* (station Smarlacca).

Study Area

Station Giralda is located in the western part of the Sacca di Goro (Po River Delta, Northern Italy), close to the freshwater inlet of the Po di Volano Canal. The station is micro-tidal with water depths ranging between 30 and 100 cm with a mean depth of approximately 50 cm. and, although shallow, the water column at high tide is stratified with freshwater overlaying the salt water. The surficial sediments consist of a soft mud colonised by benthic diatoms, salinity of the bottom water varies between 0 and 25‰ while inorganic nitrogen concentrations range between 5 and 200 µM. This area of the lagoon receives an annual load of about 1,000 tons of N of which 50% is NO_3^- and NH_4^+ (Orlandi 1998). Previous studies, have demonstrated a relationship between sediment oxygen demand, macrofaunal densities and suspended matter concentrations in the Po di Volano canal (Bartoli 1996; Castaldelli 1998). *Corophium,* for example, may attain densities of 10,000 ind. m^{-2} in early spring (Ceccherelli V.U., personal communication), but completely disappear during the summer.

Station Gorino is located in the eastern area of the Sacca di Goro, where dense mats of the floating macroalga *Ulva rigida* develop. The sediment is muddy sand and is disturbed due to the intense harvesting of molluscs, in particular of clams. During summer this station suffers severe episodes of anoxia due to the decomposition of huge amounts of *Ulva*. Long term monitoring of this site shows consistent annual cycles, with nitrates disappearing from the water column during the entire growth season of *Ulva* and oxygen concentrations varying from supersaturated in spring to zero during the summer dystrophic crises (Viaroli et al. 1996a, b).

Station Smarlacca is a small basin (about 1.9 km^2) in the Valli di Comacchio (North-Adriatic Sea), a wide complex (about 100 km^2) of shallow water impoundments connected to the Adriatic Sea via a network of canals. Valle Smarlacca is surrounded by embankments but receives fresh water inputs from the adjacent Reno River through man-regulated sluices. Salinity is relatively stable (20 to 22 ‰) but can rise to 25-30 ‰ due to evaporation in summer. The sediment is colonised by a dense meadow of the aquatic phanerogam *Ruppia cirrhosa*. These seagrass meadows are often patchy and within the larger seagrass meadows, areas of different sizes may be devoid of plants due to disturbance events. Therefore, it is possible to recognise a mosaic of patches of bare sediments separated by seagrass beds.

Materials and Methods

Fluxes of DIN (Dissolved Inorganic Nitrogen) and O_2 across the sediment-water interface and coupled-uncoupled denitrification rates were measured during light and dark incubations of intact sediment cores. Distinct incubation systems were used for the different sites: relatively small cores were used for microalgae-covered sediments while larger cores and chambers were used respectively for phanerogam and macroalgae dominated systems. All incubation methods were tested and standardized before starting the investigation. All sediment cores, whether containing microphytes, rooted macrophytes or floating macroalgae were incubated the day after the sampling using *in situ* water at *in situ* oxygen concentration and temperature. Before starting the incubations the cores were maintained submersed in large tanks without the top lid and with the stirring system on to allow renewal of the water inside the cores. With the microalgae-covered sediment, Plexiglas cores (i.d. 8 cm, height 40 cm) were used for flux and denitrification measurements; a 4 cm long Teflon-coated magnetic stirring bar was suspended 6 cm above the sediment surface. The stirring was due to an external rotating magnet. In the macroalgae-covered sediment flux and denitrification measurements were performed in Plexiglas chambers (base 20x20 cm, working height 40 cm), stirring was created by drawing water out through a water dispersal unit placed at one side of the chamber and pumping it back at the other side. Undisturbed sediment with macroalgal mat (when present) was obtained with a hand held box corer pushed through the algal mat into the sediment. At the site covered by rooted macrophytes Plexiglas cores (i.d. 20 cm, height 40 cm) were used; stirring of the water column was obtained with a small aquarium pump placed inside the core. In all the incubations water-sediment fluxes of O_2, NH_4^+ and NO_3^- were measured as concentration changes in the water with time according to Eq. (1)

$$\text{Flux} = (a * V) / A \qquad (1)$$

where:
flux is expressed in µmol of the species of interest m^{-2} h^{-1};
a = slope of the linear regression of concentration versus time;
A = area of sediment surface in core/chamber (m^2);
V = volume of water in core/chamber (l).

Denitrification rates were measured by the isotope pairing technique (Nielsen 1992) adding $^{15}NO_3^-$ to the water column and then measuring the produced labelled N_2; this method allows to determine coupled (Dn) and uncoupled denitrification (Dw). Uncoupled denitrification (Dw) is denitrification of nitrate diffusing to the anoxic sediment from the water column while coupled denitrification (Dn) removes nitrate produced within the sediments via nitrification. Details of the experimental setup and analytical methods can be found in the protocol handbook of the NICE project (Dalsgaard 1999).

Results

Primary Producer Biomass and Nutrient Concentration in the Water Column

In 1997 sampling campaigns were carried out approximately every 40 days at each site; here we present results from three dates which represent critical phases in the seasonal evolution of the primary producer communities (Table). At station Giralda, chlorophyll *a* concentrations in the upper few millimeters of the sediment were relatively constant and ranged between 24 and 48 mg m^{-2}; the highest value was determined in summer. Ammonium and nitrate concentrations were minimum in July (Table) while higher values were measured in March (32 µM of NH_4^+ and 50 µM of NO_3^-) and November. At station Gorino *Ulva* biomasses of 191±5 and 311±15 g_{dw} m^{-2} were determined respectively in March and November; no macroalgae were present in August due to the biomass collapse associated with the dystrophic crisis. The collapse of *Ulva* production is a phenomenon well described which occurs regularly in the Gorino area (Viaroli et al. 1999) and is associated with water column hypoxia, NO_3^- deficiency and NH_4^+ release from the sediment. Oxygen and DIN concentrations in the water column are in a good agreement with this cycle, with an ammonium peak of 35 µM and a nitrate minimum of 11 µM in August. In this month the water column was hypoxic (110 µM O_2, about 40% saturation) during the day and anoxic at night.

Table. Primary producer biomass, oxygen and DIN concentrations in the water column at the three sites. At station Giralda microalgal biomass is expressed as mg Chl*a* m^{-2}; at station Smarlacca *Ruppia* biomass is expressed as g dry weight of roots and leaves. At station Gorino on 21.08.97 no *Ulva* thalli were found

Station	Date	Biomass	Oxygen (µM)	Ammonium (µM)	Nitrate (µM)
Giralda	03/26/97	39±12	330	32	50
(microalgae)	07/29/97	48±10	340	14	16
	11/11/97	24±3	290	36	45
Gorino	03/4/97	191±5	500	18	43
(Ulva)	08/21/97	0	110	35	11
	11/26/97	311±15	400	22	46
Smarlacca	03/19/97	*Roots* 20±2	315	15	7
(Ruppia)		*Leaves* 31±2			
	07/22/97	*Roots* 96±15	430	6	5
		Leaves 318±26			
	11/20/97	*Roots* 18±3	324	15	4
		Leaves 75±6			

At station Smarlacca, the *Ruppia* biomass reached a summer peak of 414±29 g_{dw} m^{-2} in July 77% of which was due to the leaves. At this station, the water column was about 100% O_2 saturated in March and in November whilst in July O_2 oversaturation reached 196%. The DIN concentrations were generally very low with a summer minimum coinciding with the biomass peak. The NO_3^- concentrations were almost negligible.

Oxygen and DIN Fluxes

Light and dark oxygen and ammonium fluxes determined at station Giralda are shown in Fig. 1. Dark respiration ranged between −3.42±0.18 and −2.38±0.11 mmol O_2 $m^{-2}h^{-1}$ and was not related to water temperature. A positive net efflux of O_2 was determined in July (740±250 µmol O_2 $m^{-2}h^{-1}$) when chlorophyll *a* attained the highest concentration in the surficial sediment (Table). Gross primary productivity values (1.35 to 4.04 mmol $m^{-2}h^{-1}$) demonstrate a high microalgal photosynthetic activity at the water–sediment interface (Fig. 1). During light incubations a net ammonium uptake by the benthic microalgal mat was observed. For example, in July an influx of 197±104 µmol NH_4^+ m^{-2} h^{-1} was measured in the light and an efflux of 515±189 µmol NH_4^+ m^{-2} h^{-1} was determined in the dark. Nitrate fluxes (values not shown) were quite variable even if mostly directed from the water column to the sediment. A net efflux of nitrates was determined in the light in July (783± 345 µmol NO_3^- m^{-2} h^{-1}) coinciding with the ammonium uptake. Nevertheless, nitrate fluxes were not correlated with NO_3^- concentrations in the water column or ammonium exchanges between sediment and water.

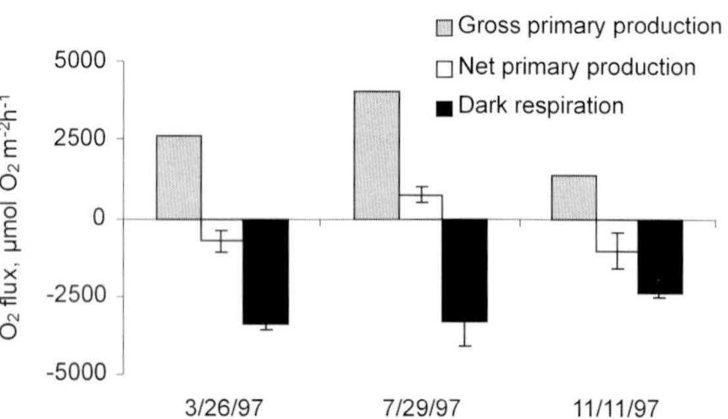

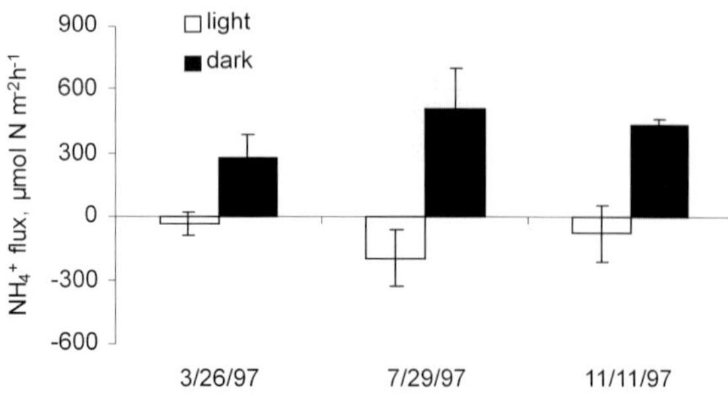

Fig. 1. Oxygen and ammonium fluxes determined at Station Giralda in cores incubated in light and dark conditions

At Station Gorino the presence of *Ulva* mats in March and November resulted in a high efflux of oxygen in the light and in a high oxygen uptake in the dark (Fig. 2). Dark oxygen respiration was also high in August, after the *Ulva* collapse, due to a combination of the high water temperature and the availability of fresh organic matter settled on the sediment surface. In August, ammonium fluxes were highly variable in light incubated cores of unvegetated sediments, whilst in the dark incubation 326 µmol NH_4^+ m^{-2} h^{-1} was released from the sediment to the water column. Differences between light and dark incubations were probably due to the assimilation by benthic microalgae, rapidly developing on the sediment surface. *Ulva* actively assimilated and retained nitrate and ammonium both in the dark and in the light conditions

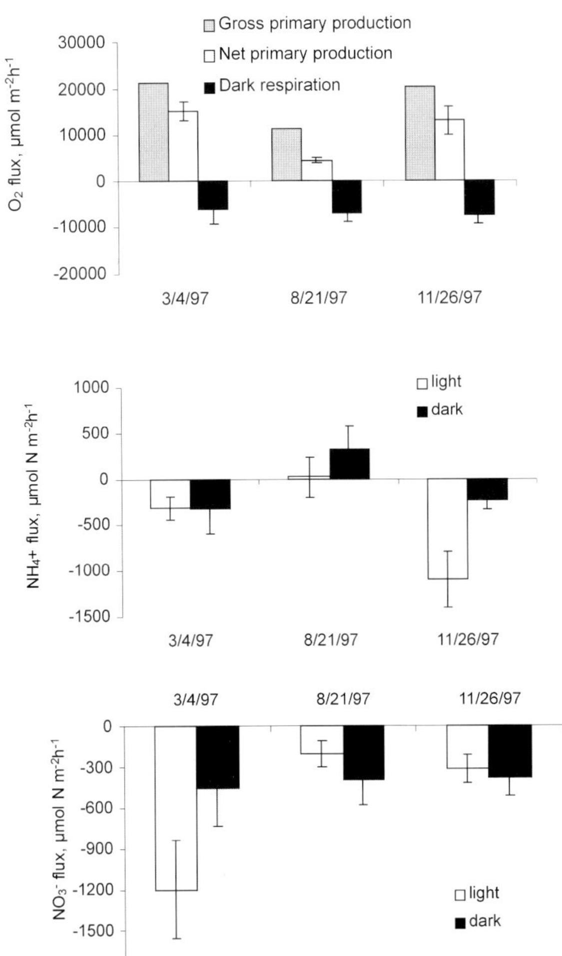

Fig. 2. Oxygen, ammonium and nitrate fluxes determined at Station Gorino in light and dark chamber incubation

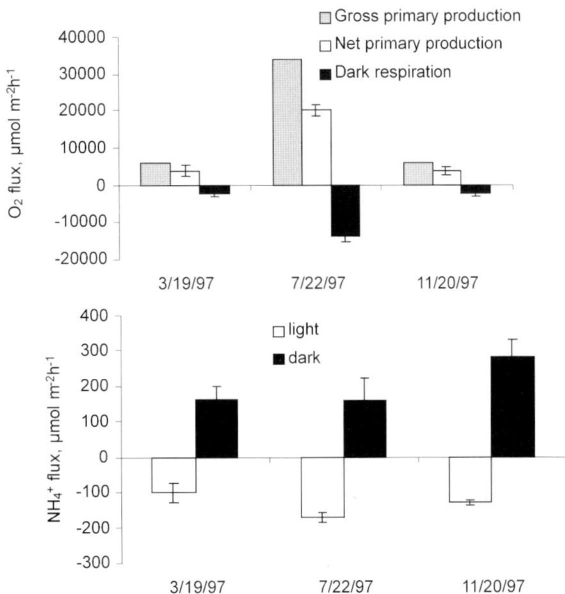

Fig. 3. Oxygen and ammonium fluxes determined at Station Smarlacca in light and dark core incubations

with a net DIN removal of 1.51 mmol m^{-2} h^{-1} in March and 1.54 mmol m^{-2} h^{-1} in November. In March the *Ulva* sediment system removed mostly nitrate, whilst in November ammonium uptake was dominant.

At Station Smarlacca, the phanerogam meadow supplied a net oxygen production on all the sampling dates and attained a maximum in July (Fig. 3) coinciding with the highest biomass of *Ruppia* (Table); dark respiration was also highest in July. There were significant (Fig. 3) differences between light and dark ammonium fluxes, with a net release in the dark and a net uptake in the light incubations. Dark ammonium regeneration from the *Ruppia* beds peaked in November (284 ± 48 µmol NH_4^+ m^{-2} h^{-1}), whilst the maximun light uptake (171 ± 15 µmol NH_4^+ m^{-2} h^{-1}) was determined in July. Nitrate fluxes (not reported) were always negative with a maximum in March (−223 ± 9 and −257 ± 19 µmol NO_3^- m^{-2} h^{-1} respectively in the light and in the dark incubations), whilst NO_3^- fluxes were negligible in July and November (< 40 µmol NO_3^- m^{-2} h^{-1}).

Coupled and Uncoupled Denitrification

Denitrification rates of nitrates diffusing from the water column (Dw) and of nitrates produced within the sediment (Dn) are shown in Fig. 4. At station Giralda the highest total denitrification rates (Dn+Dw) were determined in March (256 ± 15 and 267 ± 22 μmol N m^{-2} h^{-1} respectively in the light and in the dark). The Dn represented 67 and 78% of the total light and dark denitrification, respectively indicating that nitrification is an important N-process in the microphytobenthic system. Total denitrification was lower in July and in November (about 100 μmol N m^{-2} h^{-1}) due to a combination of lower nitrate concentrations in the water column and lower populations of benthic infauna.

At station Gorino total denitrification rates ranged between 63 and 119 μmol m^{-2} h^{-1}. The Dw was low in March in both light and dark incubations probably due to the high nitrate uptake by *Ulva* (–1197 ± 460 and –454 ± 280 μmol NO_3^- m^{-2} h^{-1}). The Dn was generally low with a peak of 49 ± 8 μmol N m^{-2} h^{-1} determined in the light incubation of August when *Ulva* was not present and it was not detectable in the dark

Station Giralda

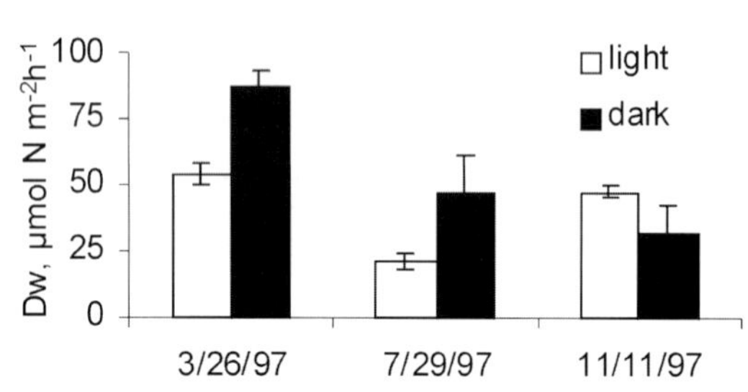

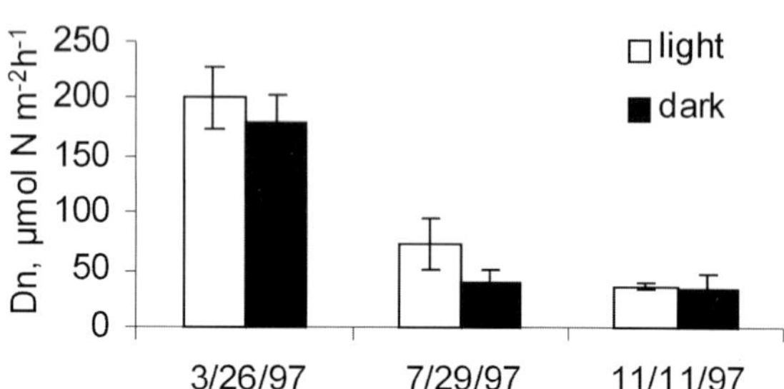

Station Gorino

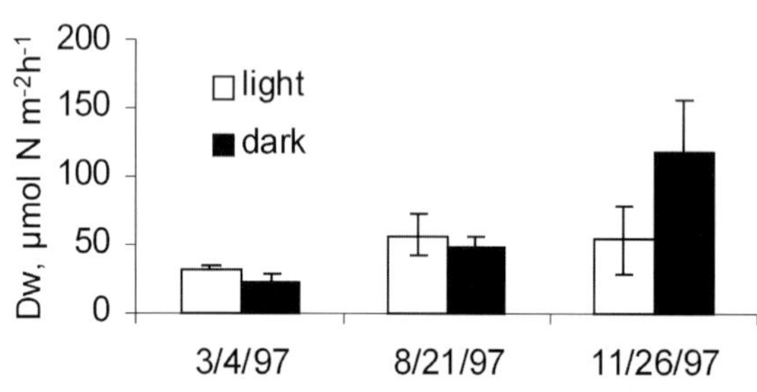

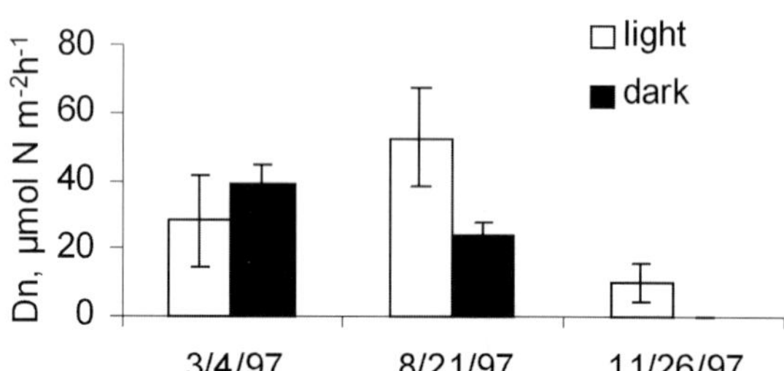

Station Smarlacca

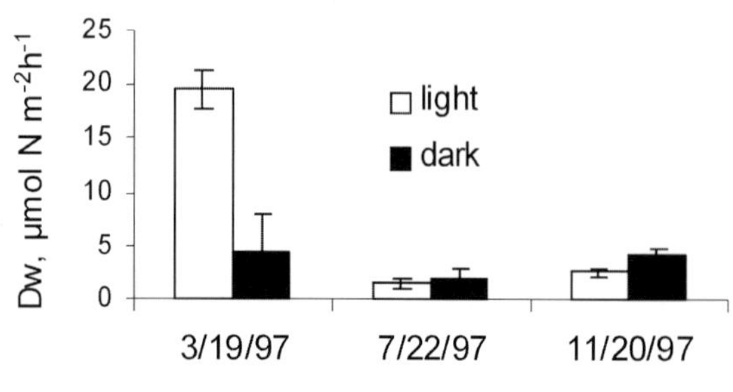

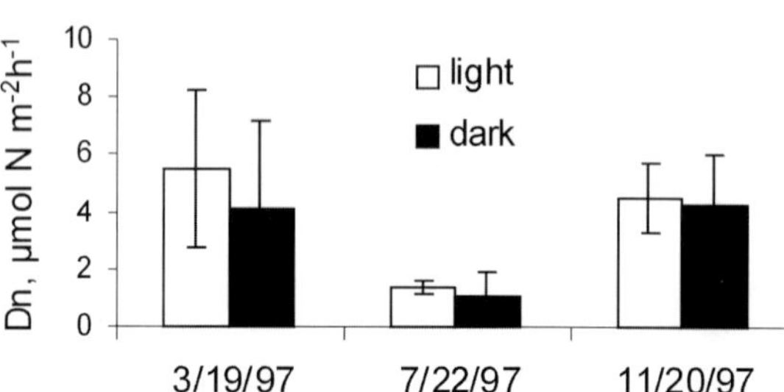

Fig. 4. Uncoupled (*Dw, left*) and coupled (*Dn, right*) denitrification rates determined at the three sites during light and dark incubations. Note different scales of *ordinate* axes

incubation of November, due probably to oxygen deficiency in the uppermost sediment layers.

At station Smarlacca denitrification rates were extremely low (between 3 and 25 µmol N m^{-2} h^{-1}) due to low nitrate concentrations in the water column (Table). The Dn rates were between 1 and 4 µmol N m^{-2} h^{-1} indicating that nitrification was not important. Low rates of Dn are probably due to the low oxygen penetration into the sediments. At Station Smarlacca the root system of *Ruppia* was not an efficient oxygen transport system to the sediment, or chemical oxidation due to reduced environment around the rhizosphere outcompeted nitrifiers for oxygen released (Barbanti et al. 1997). Potential nitrification rates have demonstrated that the population of nitrifiers around the rhizosphere of *Ruppia* is not significantly different from that of the sediment far from the roots and nitrification does not seem to be an important process at this site.

Discussion and Conclusion

At station Giralda nitrogen fluxes are influenced by external inputs from the Po di Volano as well as by water column transparency and organic matter inputs. These factors also regulate the structure and activities of benthic microalgal and macrofaunal communities influencing N-cycle processes. Additionally, tidal changes in water transparency, nutrient concentrations and salinity resulted in dramatic changes in both activities and fluxes over short time scales. Microalgae efficiently take up the regenerated NH_4^+ (Fig. 1), although high dark respiration rates in the oxic layer (>2300 µmol O_2 $m^{-2}h^{-1}$) and high sulphate reduction rates at 4-6 cm depth (>700 µmol SO_4^{2-} m^{-2} h^{-1}, Giordani 1997) maintained high NH_4^+ concentrations in the sediment. Pore water concentrations up to 700 µM (data not shown) are generally found at 5 cm depth. The availability of ammonium partially explains the high coupled nitrification-denitrification rates (Dn) determined in March (Fig. 4). No significant difference was found between Dn values determined in light and dark incubations meaning that the oxygen used by nitrifiers was not derived from microalgae photosynthesis but probably from burrow ventilation by benthic macrofauna (Kristensen 1984). Indeed, oxygen penetration in the sediment of station Giralda is generally limited to the upper 3 mm (Bartoli 1996) whereas potential nitrification rates up to 12 µM N g^{-1} d^{-1} have been estimated at this site at 2 cm depth (Castaldelli 1997) and active nitrifiers have been found down to 10 cm.

At station Gorino, DIN fluxes and denitrification rates were mostly controlled by macroalgal growth. In March, nitrate and ammonium availability results in a high assimilation and retention of DIN by the macroalgal biomass; the NO_3^- lost by the system via total denitrification in the light incubation is just 5% of the NO_3^- assimilated by Ulva (Figs. 2, 4). In the summer nitrate deficiency occurs in the water column and *Ulva* growth collapses: the decomposition of macroalgal biomass causes ammonium release and prolonged hypoxia (days to weeks) in the water column (Table, Fig. 2). In August, despite the absence of *Ulva*, denitrification of nitrate from the water column is limited by its low concentration whilst Dn is limited by the effect of oxygen deficiency for nitrification due to respiration processes. At this site the samplings of March and November were doubled and bare sediments (without *Ulva* in the chambers) were incubated in parallel (data not shown). Oxygen production and consumption as well as DIN fluxes were dramatically affected by the presence of the macroalgae. In the chambers with bare sediment, for example, net O_2 fluxes in the light were similar in March and November (–800 µmol O_2 $m^{-2}h^{-1}$) and about 15 times smaller than those measured in chambers with *Ulva*. Without macroalgae light and dark NH_4^+ fluxes were always from the sediment to the water column (with values between 90 and 260 µmol N $m^{-2}h^{-1}$), the opposite of those measured in the chambers with *Ulva* where ammonium was assimilated. In the bare sediment furthermore NO_3^- fluxes were negligible. The Dw in the chambers without *Ulva* was two times higher than for *Ulva* colonised sediments indicating the competition for inorganic nitrogen between denitrification and assimilation by primary producers.

At station Smarlacca nitrogen fluxes and water column DIN concentrations were regulated throughout the year by N-assimilation and storage by the seagrasses and their epiphytes, which are the dominant N-cycling processes. Thus, losses of N via denitrification are small in terms of the overall N-budget. The March and November samplings were doubled and bare sediment was incubated in parallel with *Ruppia* colonised sediment (data not shown). Oxygen and DIN fluxes were strongly affected by the presence of *Ruppia* and exchange rates were up to 10-15 fold higher. No significant differences were found comparing denitrification rates with or without macrophytes probably because of the extremely low NO_3^- concentrations in the water column and the absence of O_2 in the rhizosphere (Barbanti et al. 1997).

Overall, comparison of the three sites indicates the effects of the dominant primary producers on the balance between N-assimilation and denitrification. The seagrasses control the N-cycle throughout the year, whereas *Ulva* controls processes during the spring growth phase and in the early decomposition period. At the microalgal site rates have a high variability due to the low and rapidly changing biomass of the microalgae and the inherently high variability of physicochemical parameters and macrofaunal populations.

Acknowledgements. This research has been jointly supported by the EU projects NICE (contract MAS3-CT96-0048) and ROBUST (contract ENV4-CT96-0218).

References

Barbanti A, Bartoli M, Cavicchini F, Giordani G, Viaroli P, Varallo G, (1997) Nictemeral changes of redox equilibrium at the sediment-water interface of a *Ruppia cirrhosa* meadow determined by in situ microprofiling. In: VIII Congr Ital Soc Ecol Atti, pp 501-505

Bartoli M (1996) Nitrogen and phosphorus benthic fluxes in coastal environments characterised by different communities of primary producers (in italian). PhD thesis, Department of Environmental Sciences, Parma University, Italy

Borum J (1996) Shallow waters and land/sea boundaries. In: Jorgensen BB, Richardson K (eds) Eutrophication in coastal marine ecosystems. (Coastal and Estuarine Studies) Am Geophys Union 52, pp 179-203

Caffrey J, Kemp WM (1992) Influence of the submerged plant, *Potamogeton Perfoliatis*, on nitrogen cycling in estuarine sediments. Limnol Oceanogr 37: 1483-1495

Castaldelli G (1998) Nitrification and ammonium pools in the sediments of the Sacca di Goro, Po River Delta (in italian). PhD thesis, Department of Environmental Sciences, Parma University, Italy

Christensen PB, Revsbech NP, Sand-Jensen K (1994) Microsensor analysis of oxygen in the rhizosphere of the aquatic macrophyte *Littorella uniflora* (L.) ascherson. Plant Physiol 105: 847-852

Dalsgaard T (ed.) (1999). Protocol handbook for NICE (nitrogen cycling in estuaries). NICE contract MAS3-CT96-0048, Comm Eur Communities DC XII, Bruxelles, Appendix 25

Focht DD, Verstraete V (1977) Biochemical ecology of nitrification and denitrification. Adv Microbiol Ecol I: 135-214

Giordani G (1997) Sulphide speciation and inorganic phosphorus mobility in relation to iron availability in sediments of distrophic lagoon environments (in italian). PhD thesis, Department of Environmental Sciences, Parma University, Italy

Henriksen K, Kemp W (1988) Nitrification in estuarine and coastal marine sediment. In: Blackburn TH, Sorensen J (eds) Nitrogen cycling in coastal marine environments. John Wiley and Sons, New York, pp 210-249

Henriksen K, Jensen A, Rasmussen MB (1984) Aspects of nitrogen and phosphorus mineralization and recycling in the northern part of the Danish Wadden Sea. Neth Inst Sea Res Publ Ser 10: 51-69

Kristensen E (1984). Effect of natural concentrations on nutrient exchange between a polychaete burrow in estuarine sediment and the overlying water. J Exp Mar Biol Ecol 75: 171-90

Naldi M (1994) Benthic macroalgae and nutrient cycles in a lagoon of the Po River Delta (Sacca di Goro) (in italian). PhD thesis, Department of Environmental Sciences, Parma University, Italy

Nielsen LP (1992) Denitrification in sediment determined from nitrogen isotope pairing. FEMS Microbiol Ecol 86: 357-362

Nielsen LP, Sloth NP (1994) Denitrification, nitrification and nitrogen assimilation in photosynthetic microbial mats. In: Stahl LJ, Coumette P (eds) Microbial mats, structure, development and environmental significance. (NATO ASI Ser G Ecol Sci, vol 35) Springer Verlag, Berlin Heidelberg New York Tokyo, pp 319-324

Nielsen LP, Christensen PB, Revsbech NP, Sorensen J (1990) Denitrification and photosynthesis in stream sediment studied with microsensor and whole-core technique. Limnol Oceanogr. 35: 1135-1144.

Orlandi S (1998) Nutrient loads and restoration perspectives in a coastal lagoon of the Po River Delta (in italian). Bachelor thesis, Dep Environ Sci Univ Parma

Pedersen MF, Borum J (1992) Nitrogen dynamics of eelgrass during late summer period of high growth and low nutrient availability. Mar Ecol Prog Ser 80: 65-73

Rasmussen MB, Henriksen K, Jensen A (1983) Possible causes of temporal fluctuations in primary production of the microphytobenthos in the Danish Wadden Sea. Mar Biol 73: 109-114

Revsbech NP, Jorgensen BB, Brix O (1981) Primary production of microalgae in sediments measured by oxygen microprofile, $H^{14}CO_3$ fixation, and oxygen exchange methods. Limnol Oceanogr 26: 717-730

Risgaard-Petersen N, Jensen K (1997) Nitrification and denitrification in the rhizosphere of the aquatic macrophyte *Lobelia Dortmanna* L. Limnol Oceanogr 42: 529-537

Rysgaard S, Risgaard-Petersen N, Sloth NP (1996) Nitrification, denitrification and nitrate ammonification in two coastal lagoons in Southern France. In: Coumette P, Castel J, Herbert R (eds) Coastal lagoon eutrophication and anaerobic processes. Kluwer, Amsterdam, pp 133-144

Sand-Jensen K, Borum J (1991) Interactions among phytoplankton, periphyton, and macrophytes in temperate freshwaters and estuaries. Aquat Bot 41: 137-175

Sand-Jensen K, Prahl C (1982) Oxygen exchange with the lacunae and across leaves and roots of the submerged vascular macrophyte, *Lobelia Dortmanna* L. New Phytologist 91: 103-120

Sondergaard M (1983) Heterotrophic utilization and decomposition of extracellular carbon released by the aquatic angiosperm *Littorella uniflora* (L.) Aschers. Aquat Bot Mar 27: 547-555

Sunback K, Graneli W (1988) Influence of microphytobenthos on the nutrient flux between sediment and water: a laboratory study. Mar Ecol Prog Ser 43: 63-69

Viaroli P, Bartoli M, Bondavalli C, Christian RR, Giordani G, Naldi M (1996a) Macrophytes communities and their impact on benthic fluxes of oxygen, sulphide and nutrients in shallow eutrophic environments. Hydrobiologia 329: 105-119

Viaroli P, Bartoli M, Giordani G, Azzoni R, Tajè L, Agazzi G, Nizzoli D (1999) Growth and decomposition of *Ulva spp.* in relation to dystrophy, nutrient retention and the sedimentary buffering capacity in the Sacca di Goro lagoon. Ann Rep Robust Project (contract ENV4-CT96-0218)

Viaroli P, Naldi M, Bondavalli C, Bencivelli S (1996b) Growth of the seaweed *Ulva rigida* C. Agardh in relation to biomass densities, internal nutrient pools and external nutrient supply in the Sacca di Goro lagoon (Northem Italy). Hydrobiologia 329: 93-103

CHAPTER 30

The Role of Sponge Bioerosion in Mediterranean Coralligenous Accretion

C. Cerrano[1], G. Bavestrello[2], C.N. Bianchi[3], B. Calcinai[1], R. Cattaneo-Vietti[1], C. Morri[1], and M. Sarà[1]

ABSTRACT

Coralligenous biocoenoses developed almost 10,000 years ago in the Mediterranean Sea, forming environments characterized by high levels of biodiversity. They are complex biogenic structures growing through the continuous overlapping of organogenic layers, between 20 and 130 m depth. The coralligenous substrate, for the concurrent presence of bio-builders (algae, serpulid worms, bryozoans, and scleractinians) and destroying elements (clionid sponges, bivalves) is subject to dynamic evolution. Studies conducted on sections of coralligenous blocks show that algae are the main component responsible for the building of these concretions. Calcareous remains of benthic animals lack in mineral structures and are one of the most important detritus-forming elements at the bottom of vertical cliffs.

This seems to be due to the attack of clionids which selectively bore animal rather than plant carbonates. A study on the population dynamics of some anthozoans with a calcareous skeleton has shown that about 100% of the newly settled specimens are probably removed by clionids. In this way, the substratum is cyclically renewed by the continuous boring action of clionids. It is a sort of intermediate disturbance which controls the evolution and the structure of skiophilous zoocoenosis, through primary space regeneration. A high specific richness, typical of coralligenous biocoenosis, is developed by this action. From a geological point of view, it forms a kind of limestone, characteristic of biogenic constructions and detritic bottoms under the cliff.

Introduction

Coralligenous communities are, in the Mediterranean Sea, among the most complex and diversified assemblages living on hard-bottoms. In the last 10,000 years, they have contributed to creating significant organogenic reef-like bioconstructions, between 20 and 100 m depth (Laborel 1987; Sartoretto et al. 1996). These structures result from a multi-stratified accretion, made of a macroalgae and invertebrates complex, in dynamic equilibrium due to the simultaneous activities of builders (coralline rhodophythes, scleractinians, bryozoans, serpulids) and different disruptive agents, of which clionids and allied sponges are the most important ones (Sarà 1973). Whenever either of these actions prevails, it favours the accretion or the erosion of these bioconstructions.

Coralligenous biocoenoses have high biodiversity: for example, off Marseilles, they include more than 650 species of invertebrates (Hong 1982). Sarà (1973) listed one hundred sponges in the coralligenous biocoenosis platform off the Apulian coast. Their biomass, which can exceed 500 g dw m^{-2} (Bellan-Santini 1985), is also high. This figure is comparable to those measured in tropical coral reefs (Sorokin 1993).

The richness and diversity of species in coralligenous biocoenoses are partially due to their evident substrate heterogeneity, which has 3-dimensional features. Hong (1982) underlined the abundance of crevices and microcavities which, depending on their size and exposure, host large kinds of organisms, distributed according to light irradiance and water movement. Moreover, these cavities are subject to different rates of organic and inorganic sedimenta-

[1] Dipartimento per lo Studio del Territorio e delle Sue Risorse dell'Università di Genova, Via Balbi 5, 16126 Genova, Italy
[2] Istituto di Scienze del Mare dell'Università di Ancona, Via Brecce Bianche, 60131 Ancona, Italy
[3] Istituto di Ricerca sulle Acque, ENEA Santa Teresa, P.O Box 316, 19100 La Spezia, Italy

F.M. Faranda, L. Guglielmo, G. Spezie (eds)
Mediterranean Ecosystems: Structures and Processes

tion, inducing a strong microhabitat differentiation. Laubier (1966) defined these structures like eco-ethological cross-roads rather than a simple biocoenosis. According to Picard (1985), this is a common feature in what he called "complex climatic mesoecosystems", formed by a polybiocenotic species assemblage.

Radiocarbon dating has allowed to determine the age of these bioconstructions: the deeper ones date back to the early Holocene or to the late Pleistocene, during the last great transgression, caused by the general increase in temperature at the end of the Wurm period (Adey 1986). Sartoretto et al. (1996), dated the oldest Mediterranean reefs at around 8,500 B.C., suggesting they had formed at a depth not greater than 10-15 m. The coralligenous buildings growing on soft bottoms along the Apulian coasts are constituted by mounds of calcareous subfossil algae (*Neogoniolithon mamillosum*), developed probably 10,000 years ago in the littoral zone (Sarà 1973).

Laubier (1966) and Sarà (1969) observed the fast sessile fauna turnover in different coralligenous communities, and, in general, the balance existing between growth and erosion processes. Owing to this fast turnover, the calcareous skeleton contributes to the increase in detritic sediments lying below the cliff (Pérès and Picard 1964). However, no data are so far available to quantify the real impact of boring sponges on the dynamics of calcareous organisms involved in coralligenous accretion.

This paper is a first attempt to verify the impact of sponge boring on sessile fauna turnover, in order to provide an evaluation of its role in coralligenous building dynamics.

Materials and Methods

The study has been carried out on the Portofino Promontory cliff (Ligurian Sea) as follows:

1. Three blocks, from horizontal ledges between 35 and 45 m depth, have been cut by diamond saw in 1 cm thick sections, showing both lit and shady sides. Each section has been examined by stereo microscope to detect the organisms involved in its formation.
2. The boring sponge activity on the corallites of the scleractinian *Leptopsammia pruvoti* has been evaluated. Corallites have been collected in standard surfaces of 20 x 20 cm at 20, 25, 30, 35 m depth (three replicas per each depth). In bored corallites, the sponge tissue was collected directly from the boring chambers to prepare spicule slides in order to identify the different species.
3. The animal fraction present in the detritic bottoms below the cliff has been considered. The analysis has been conducted on the coarse fraction (> 2 mm) collected at 45 m depth. In this fraction, carbonatic remains of organisms have been sorted by taxa and weighed.

Results

On the Portofino Promontory cliff, the coralligenous ledges show an upper, lit surface occupied by coralline algae (*Mesophyllum lichenoides* and *Lithophyllum frondosum*) and a lower shady side characterised by the growth of an assemblage mainly composed of animal organisms with a calcified skeleton. The main organisms were anthozoans (*Corallium rubrum* and *Leptopsammia pruvoti*) and massive and erect bryozoans (*Smittina cervicornis*, *Myriapora truncata* and *Pentapora fascialis*). Among these elements, a cryptic fauna of serpulids, encrusting bryozoans, brachiopods, and bivalves, could be detected, all likely to add their calcareous remains to the bioconstruction.

The sections of coralligenous rocks (Fig. 1a) show that, although the animal component is very abundant on their lower surface, the inner rock structure is mainly constituted of coralline algae. The lower, shady side of the rocks shows a continuous layer of 1 cm thick perforations, caused by the action of boring sponges (Fig. 1b,e). This etching action is practically absent in the upper surface (Fig. 1c). Perforations by sponges, consisting of boring spherical chambers – 1-2 mm in diameter – penetrate from the substratum into the calcareous skeleton of benthic organisms (Fig. 1d,e). Insinuating sponges are often present in the large crevices produced by the coalescence of boring chambers inside the coral rock (Fig. 1f).

Quali-quantitative analysis of *L. pruvoti* corallites bored by sponges at different depths showed that the number of bored corallites ranges between 60 and 75% of the total (Fig. 2). The specific composition of the boring species inside the corallites varies according to depth.

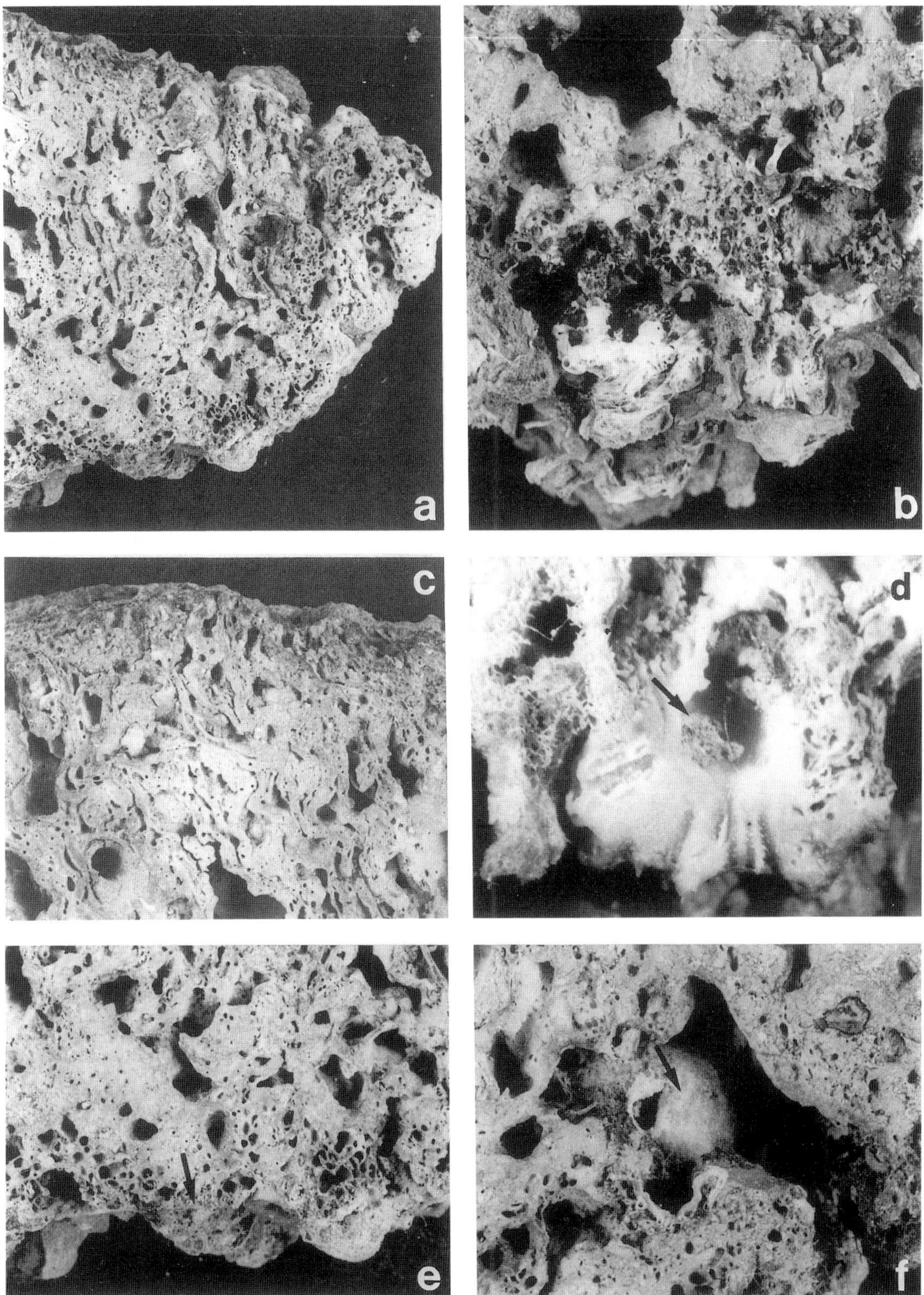

Fig. 1a-f. Section of a coralligenous ledge showing its structure mainly consisting of irregular layers of coralline algae. Under higher magnification, the absence of any sponge boring activity in the upper side of the ledge (**c**) and a layer of boring chambers immediately under the lower surface (**e**) can be observed. The boring chambers (*arrows*) penetrate from the substrate into the calcareous skeletons of *Leptopsammia pruvoti* (**d**) and *Madracis pharensis* (**e**). Large crevices of coralligenous structure are occupied by endobiotic sponge species like *Geodia cydonium* (*arrow*) (**f**)

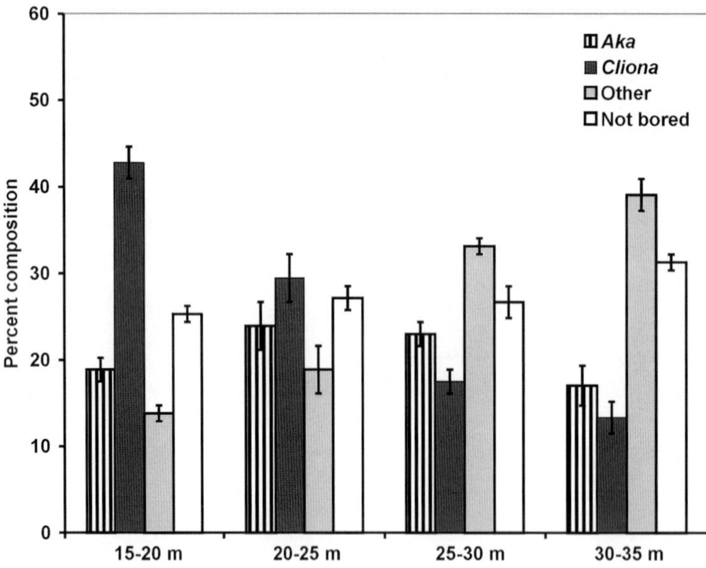

Fig. 2. Percentage of integer and bored corallites of *Leptopsammia pruvoti* (indicating only the genus composition of the boring sponge species) at four different depths. Data are the average (± SD) of five replicas for each depth

Cliona is the most represented genus in the 20 m depth specimens. This value decreases with depth. The decreasing percentage of *Leptopsammia* corallites bored by *Cliona* spp. is matched by a corresponding increase in corallites bored by other genera (mainly *Alectona, Delectona, Dercitus* and *Spyroxia*). The contribution of the genus *Aka* sponges is stable.

Studies of the detritus coarse fraction (> 2 mm) at the base of the cliff indicate that 85% of organogenous detritus is composed of animal debris subdivided as follows: about 6%, remains of organisms living inside the detritus itself (mainly mollusc shells and echinoid spines); about 40%, derives from the skeleton of *L. pruvoti* and *C. rubrum*; about 22%, consists of portions of erect bryozoan colonies; about 12% are tubes of serpulids; about 10%, indeterminable remains (Fig. 3).

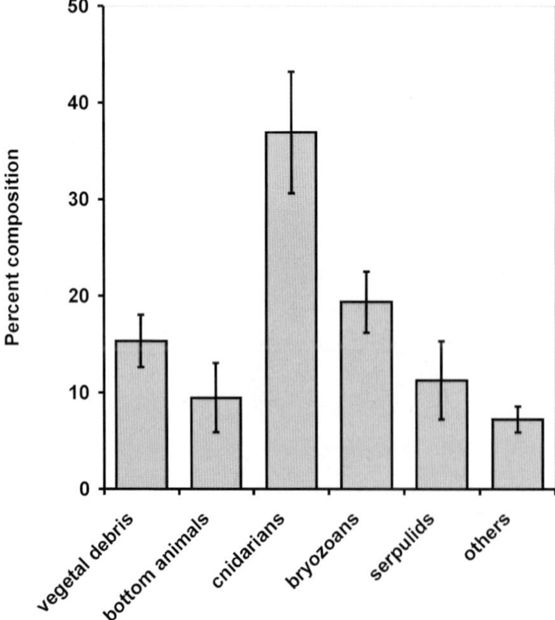

Fig. 3. Percentage composition of the coarse organic fraction of sediment at the base of the Portofino cliff. Data are the average (± SD) of five replicas, each formed of 1 litre of sediments

Discussion

By comparing the structure of coralligenous ledges of the Portofino Promontory cliff with the organogenous detritus composition accumulated at its base, the quick turnover of sessile zoobenthos living in the lower, shady side of coralligenous ledges was observed. The sub-fossil structure is, indeed, mainly consisting of coralline algae remains, while animal skeletons are very rare inside the sections and almost confined to cryptic elements. This shows that the animal fraction, prevailing in the shady side, does not take part in ledge building, since it is constantly being detached. The great abundance of coralligenous animal remains in underlying detritic bottoms is consistent with this observation.

This unexpected disappearance of the animal fraction from the diagenesised structure may be due to the erosion activity of clionid sponges and their allied species, which might be considered as the main controlling factor on animal populations settled on shady organogenous rocks. These sponges develop inside the first centimetre of substrate and attack the calcareous skeletons of sessile organisms, boring through their base. This action weakens the substrate as a whole, leading to the periodical detachment of a large part of the settled organisms. Since the animal fraction in coralligenous biocenosis generally lives by hanging from the ledges, its fall occurs by simple gravity.

This is in accordance with the population dynamics of *L. pruvoti* studied for a long period of time (Pronzato et al. 1994), which pointed to a massive mortality (about 80%) of the young polyps during the first year since their settlement. The severe impact of clionids on the newly settled scleractinians is due to the small diameter of the corallite bases (1-2 mm), compared to the diameter of the sponge boring chambers. Practically, one boring chamber inside the corallite is enough to cause its detachment. However, the larger the base becomes, the greater will be the polyp stability. Moreover, Harmelin (1990) suggested that a collar, consisting of an encrusting bryozoan at the base of the corallite, might preserve the polyp against clionid attack while the individuals are growing.

The mineral structure of the ledges may result out of clionids attacking carbonates deposed by animals and, in particular, by cnidarians, rather than those produced by algae. Among the 11 boring sponges recorded in the Portofino waters (Bavestrello et al. 1999), only *Cliona viridis* seems to be able to attack the crustose thalli of coralline algae. However, *C. viridis* is a photophilous species living at 10-15 m depth. Therefore, algae constituting the upper side of deeper coralligenous ledges cannot be attacked (Barbieri et al. 1995; Bavestrello et al. 1996). Consequently, living *Lithophyllum frondosum*, growing between 30 and 50 m depth, is the calcareous substrate least attacked by boring species along the cliff of the Portofino Promontory. Probably, when the coralligenous structure began to form in shallower waters, the boring activity of the sponges was offset by the quick growth of algae. After the rise in sea level, accretion of these bioconstructions, now in deeper waters, certainly decreased. However, at these depths, no clionid was able to continue its boring activity, hence these structures were preserved.

In conclusion, boring sponges may be considered as a key-stone group in coralligenous biocoenosis: eroding the substrate, influencing reef paleostructure and acting as intermediate disturbers (Connell 1978), they may represent one of the strongest forces modelling coralligenous communities, as they do in the coral reefs (Davies 1983; Sorokin 1993; Becker and Reaka-Kudla 1997).

Acknowledgements. This work was financially supported by Italian MURST funds.

References

Adey WH (1986) Coralline algae as indicators of sea-level. In: Van de Plaasche O (ed), Sea level research: a manual for the collection and evaluation of data. Geo Books 9: 229-280

Barbieri M, Bavestrello G, Sarà M (1995) Morphological and ecological differences in two electrophoretically detected species of *Cliona* (Porifera, Demospongiae). Biol J Linn Soc 54: 193-200

Bavestrello G, Calcinai B, Cerrano C, Pansini M, Sarà M (1996) The taxonomic status of some Mediterranean clionids (Porifera, Demospongiae) according to morphological and genetic characters. Bull Inst R Sci Nat Belg, Bruxelles 66: 185-195

Bavestrello G, Calcinai B, Cattaneo-Vietti R, Cerrano C, Pansini M (1999) Distribuzione e modalità d'erosione di alcune specie mediterranee di Clionidi (Porifera, Demospongiae) associati *Corallium rubrum*. In: Cicogna F, Cattaneo-Vietti R, Bavestrello G (eds) Mediterranean Red coral and Gorgonians biology. Ministero per le Politiche Agricole

Becker LC, Reaka-Kudla ML (1997) The use of tomography in assessing bioerosion in corals. In: Proc 8th Int Coral Reef Symp Panama 2, pp 1819-1824

Bellan-Santini D (1985) The Mediterranean benthos: reflections and problems raised by a classification of the benthic assemblages. In: Moraitou-Apostolopoulou M, Kiortsis V (eds) Mediterranean marine ecosystems. Plenum, New York, pp 19-48

Connell JH (1978) Diversity in tropical rain forests and coral reefs. Sci NY, 199: 1302-1310

Davies PJ (1983) Reef growth. In Barnes DJ (ed) Perspectives on coral reefs. Aust Inst Mar Sci, Townsville, pp 69-106

Harmelin JG (1990) Interactions between small sciaphilous Scleractinians and epizoans in the Northern Mediterranean, with particular reference to bryozoans. PSZNI Mar Ecol 11: 351-364

Hong JS (1982) Contribution à l'étude des peuplements d'un fond de concrétionnement coralligène dans la région marseillaise en Mediterranée Nord-occidentale. Bull KORDI 4: 27-51

Laborel J (1987) Marine biogenic constructions in the Mediterranean: a review. Sci Rep Port-Cros Natl Park Fr 13:97-126

Laubier L (1966) Le coralligène des Albères: monographie biocenotique. Thèse Doct, Fac Sci Univ Paris

Pérès JM, Picard J (1964) Nouveau manuel de bionomie benthique en Mediterranée. Rec Trav Stn Mar Endoume 31: 1-131

Picard J (1985) Réflexions sur l'écosystèmes marins benthiques: hiérarchisation, dynamique spatio-temporelle. Tethys 11: 230-242

Pronzato R, Manconi R, Bavestrello G, Parodi R (1994) Struttura e dinamica di popolazione di *Leptopsammia pruvoti* (Cnidaria, Madreporaria) del Promontorio di Portofino. Biol Mar Mediterr 1: 367-368

Sarà M (1969) Specie nuove di Demospongie provenienti dal coralligeno pugliese. Boll Mus Ist Biol Univ Genova 37: 89-96

Sarà M (1973) Sponge population of the Apulian coralligenous formations. Rapp Comm Int Mar Mediterr 21: 613-615

Sartoretto S, Verlaque M, Laborel J (1996) Age of settlement and accumulation rate of submarine "coralligène" (–10 to –60 m) of the northwestern Mediterranean Sea; relation to Holocene rise in sea level. Mar Geol 130: 317-331

Sorokin YI (1993) Coral reef ecology. Springer, Berlin Heidelberg New York Tokyo

Mortality of the Bryozoan *Pentapora fascialis* in the Ligurian Sea (NW Mediterranean) after Disturbance

S. Cocito and S. Sgorbini

ABSTRACT

Pentapora fascialis (Pallas) population at Tino Island (Ligurian Sea, NW Mediterranean) underwent total and partial mortality. A severe sea-storm occurred in December 1993, caused near total mortality of the colonies at the shallow station (11 m depth): seven out of eight colonies were swept away. Total mortality was not size-selective and was independent of density. The deeper station (22 m depth) resulted unaffected by the sea-storm. Partial mortality, i.e. necrosis and loss of portions of colony, affected exclusively the largest colonies subjected to long lasting epibiosis and high sedimentation. Changes in the morphological characters (large cross-sectional area and poor adhesion to the substratum), wich occurred in the largest colonies, enhanced their susceptibility to dislodgement.

Introduction

Ecologists have long been concerned with the processes that limit and regulate the distribution and abundance of natural benthic populations, and particular attention has been paid to the role of processes that determine and regulate the number of individuals in a population (Hughes 1990; Done and Potts 1992).

Sublittoral benthic populations are subjected to disturbance, which may fluctuate from disruptive and episodic events to relatively discrete and frequent events (Pickett and White 1985). The effects of disturbance may differ according to its intensity and periodicity resulting in total and partial mortality.

The effect of a disruptive event, such as a sea-storm, depends on the temporal and spatial scales under consideration, the life histories of the dominant species and its ability to recover afterward through recolonization and regrowth (Rogers 1993). Besides, the effect depends to a large extent on colony morphology, size and vulnerability to wave or current impact. Especially powerful storms are known to affect reefs in tropical regions down to 20-35 m (e.g. Woodley et al. 1981), but such storms are infrequent and have never been reported for the Mediterranean Sea.

The effect of relatively discrete and periodical events, such as sedimentation and epibiosis, affects the ability to capture food in suspension feeders thus reducing growth rate (Stebbing 1971; Cocito and Ferdeghini 1998).

Comprehensive studies on the effects of changes after disturbance are rare and need long time-series data. Among these very few concern bryozoans of temperate areas (Harvell et al. 1990; Hughes 1990; Cocito et al. 1998a), and usually the effects are mediated by competition for space and food (Buss 1979; Frechette et al. 1992).

The major traits of the population biology of *Pentapora fascialis* (Pallas)[Cheilostomatida: Ascophorina], a large erect bryozoan dominating a temperate rocky assemblage in the Ligurian Sea (north-western Mediterranean), have been under study since 1993 (Cocito et al. 1998b). We monitored over 4 years two stations at different depths through underwater photography. During monitoring, in December 1993, an exceptionally severe sea-storm struck the area, involving different effects on the *Pentapora fascialis* population.

This paper reports on how disturbance affected two stations situated at different depths. Partial and total colony mortality were analysed with respect to size, morphology, density of colonies and time.

Istituto di Ricerca sulle Acque, ENEA, 19100 La Spezia, Italy

F.M. Faranda, L. Guglielmo, G. Spezie (eds)
Mediterranean Ecosystems: Structures and Processes

Materials and Methods

The bryozoan *Pentapora fascialis* (Pallas) lives on hard and stable subtidal bottoms, down to 50 m or more (Gautier 1962); it is common on rocks exposed to strong currents (Hayward and Ryland 1979). It commonly develops massive erect colonies, with an encrusting base and convoluted bilamina sheets arising (Hayward and Ryland 1979; Jackson 1979), growing to a diameter of 20-30 cm (Sala et al. 1996). The *P. fascialis* is a perennial species with a growth-vertical rate up to 3.5 $cm \cdot y^{-1}$ when epibiosis is reduced (Cocito and Ferdeghini 1998). In the study sites, the population is formed by 7 ± 1.5 colonies$\cdot m^2$, growing closely adjacent and intertwined, and showing fusion among satellite colonies (Cocito et al. 1998b); at Medes Island the population has been reported to be made by 3.3 ± 2.5 colonies$\cdot m^2$ (Sala et al. 1996).

The study site, Tino Island (latitude 44°01' north, longitude 9°50' east), was located at the western border of the Gulf of La Spezia (eastern Ligurian Sea, Italy).

Current measurements were taken in the area for a two-year period (1993-1994) through Aanderaa and electromagnetic currentmeters. Currents had medium speed of 17 $cm \cdot s^{-1}$, ranging from 9 $cm \cdot s^{-1}$ during summer months (S. Sgorbini, unpublished data) to maximum speed of 80 $cm \cdot s^{-1}$ in autumn-winter (Cocito et al. 1997). In December 1993, during the sea-storm, current measurements were not registered. Wave measurements were registered for an eight-year period (1989-1996) and the maximum 'significant' wave (> 2 m) (Denny et al. 1985) was taken into account.

Two photographic stations were established on the rocky bottom with 1m^2 quadrats permanently marked with metal stakes at 11 and 22 m depth. The shallow station lay on a vertical cliff facing northwest; the deep station was located on a sloping substrate facing west. Photographs with Nikonos camera and 15 mm lens were taken periodically at 3-7 month intervals from November 1993 to November 1997. On images, vectorial digitisation and storage of computer-assisted drawings were produced with Autocad® software. Polygon areas represented the total projected area on the two-dimensional substrate. The complete procedure is described by Sgorbini et al. (1996).

Results and Discussion

The difference in relative frequency of factors that caused total mortality in the course of the whole investigation period at the 2 depth stations is a result of difference in exposure to a high wave-energy event, and thus to storm impact.

The sea-storm that occurred in December 1993 was anomalous and exceptionally severe. Wave measurements, recorded from 1989 to 1996, indicated numerous events in which waves reached 4 m up to a maximum of 6.5 m for a medium period of 3 days (Fig. 1). During December 1993, waves up to 5.5 m struck Tino Island for a long period (11 days with waves

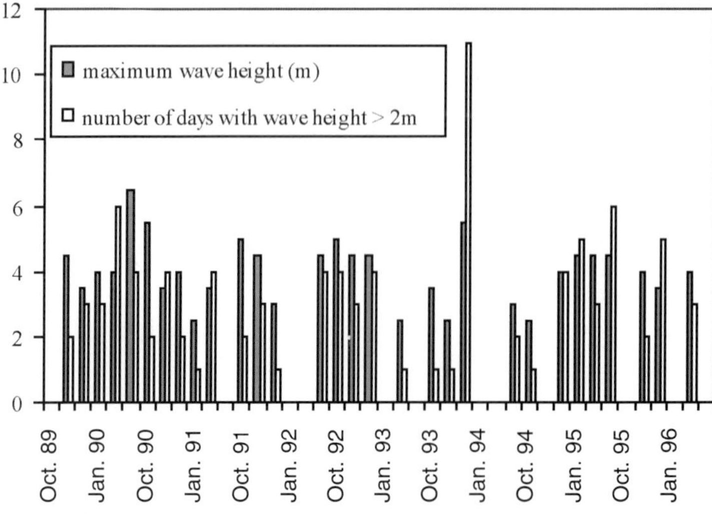

Fig. 1. Maximum wave height (m) and number of days with wave height > 2m measured in autumn-winter months (October to March) from 1989 to 1996

higher than 2 m, i.e. the maximum 'significant' wave according to Denny et al. 1985). Besides, whereas waves usually come from the southwest, on this occasion they changed from the west, directly striking the area under investigation.

Much of the rocky substrate down to about 10-12 m depth was scoured severely by the storm at Tino Island. Numerous *Pentapora fascialis* colonies were damaged or swept away; remnants were seen lying at the base of the cliff on horizontal substrates down to 25 m depth, adding carbonatic coarse debris to the muddy bottom. Dislodgement and dispersal of fragments and entire massive colonies under the influence of severe storm is commonplace for corals in tropical reefs where hurricanes are more frequent and severe (Woodley et al. 1981), but have never been reported for massive carbonatic organisms in the Mediterranean.

The sea-storm had different effects on the two monitored stations.

In the shallow station, at 11 m depth, one month before the storm eight differently sized colonies (from 150 to 2400 cm^2) were present. Seven out of eight colonies were dislodged away, the medium size (490 cm^2) survivor being sheltered in a rocky fissure. Colony removal, i.e. total mortality, was not size selective. Even if the most current- or wave-vulnerable colonies are those with a large cross-sectional area, hence large drag, and poor adhesion to the substratum (Cheetham 1986; McKinney and Jackson 1989), differently sized colonies of *Pentapora fascialis* were equally swept away by the storm.

Also mutual buttressing, that is the physical support among closely adjacent and intertwined colonies, was not effective in preventing mortality in the shallow station, where high density aggregation colonies cover up to 53% of the available substratum.

In the deeper station (22 m depth), one month before the storm, seven colonies, ranging from 80 to 2100 cm^2 were present, covering up to 52 % of the substratum. Only partial mortality was observed after the storm, exclusively affecting the largest colony. The outcome was an enlargement of the necrotic portion present in the central area of the colony and the loss of a portion of the colony, thus causing its splitting up into two. In this case, partial mortality was size-dependent as it was enhanced by the long-lasting pressure of epibiosis and fine sediment load on the central parts of the large sized colony.

The high sedimentation load, occurring in both stations as extensive thin sheet overlying a portion of the colonies and the substrate, contributed to the colony necrosis. Sediment, together with the overgrowth by encrusting algae and the overlying algal turf, as already observed in a shallow rocky coastal area in the Ligurian Sea (Airoldi et al. 1996), reduced the living bryozoan surface over several years prior to mortality.

The vulnerability of *P. fascialis* colonies to dislodgement by waves or current varied throughout their lifetime as the morphology changed with increasing size. In fact, colony's vulnerability has been demonstrated to be related to its erect growth form (Cheetham and Thomsen 1981; McKinney and Jackson 1989) and to the degree of exposure (Connell and Keough 1985). Similar observations were evidenced by Done (1992) in massive corals (genus *Porites*). In the years after the storm, in the deeper station where all colonies continued growing, we observed changes in the colonies morphology (Fig. 2). Smaller colonies approached a globular

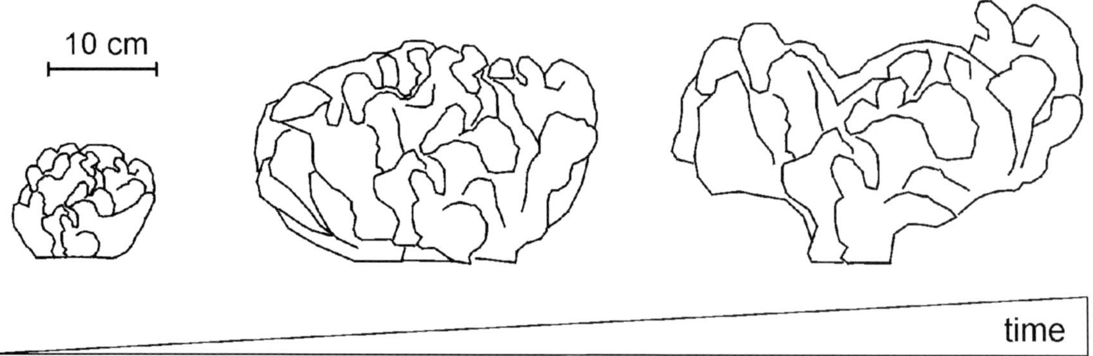

Fig. 2. Changes with time of morphological characters in *Pentapora fascialis* colonies

morphology with sub-circular perimeter, medium sized colonies developed a more enlarged shape with elliptic perimeter, and the largest colonies, after collapse of the central upper part and the periphery of the base of attachment, approximated a bowl perched on top of a narrow pedestal.

The vulnerability of a colony to water current impact is also a function of its age, strength of skeleton and tenacity of attachment (Denny et al. 1985). Consequently, changes in colony mass and morphology may be associated with changes in stability, hydrodynamic properties and susceptibility to dislodgement. In fact, in the deeper station, the three largest colonies, which exhibited the above mentioned morphology, and consequently a greater susceptibility to mechanical dislodgement, were no longer present two years after the storm.

The natural more routine disturbances, such as sedimentation and epibiosis, showed a longer-term effect to which *Pentapora fascialis*' colonies responded at the beginning with partial mortality and afterwards with total mortality.

Acknowledgements. We thank F. Ferdeghini (La Spezia) and G.M.R. Manzella (La Spezia) for suggestions. Research on *Pentapora fascialis* falls within the scope of the project 'Marine Biodiversity' (MURST) of the Marine Environment Research Centre of ENEA.

References

Airoldi L, Fabiano M, Cinelli F (1996) Sediment deposition and movement over a turf assemblage in a shallow rocky coastal area of the Ligurian Sea. Mar Ecol Prog Ser 133: 241-251

Buss LW (1979) Bryozoan overgrowth interactions - the interdependence of competition for space and food. Nature 281: 475-477

Cheetham AH (1986) Branching, biomechanics and bryozoan evolution. Proc R Soc London 228, pp 151-171

Cheetham AH, Thomsen E (1981) Functional morphology of arborescent animals: strength and design of cheilostome bryozoan skeletons. Paleobiology 7: 355-383

Cocito S, Ferdeghini F (1998) Marcatura con colorante ed etichettatura: due metodi per misurare la crescita in briozoi calcificati. In: Piccazzo M (ed) Atti XII Congr Assoc Ital Oceanol Limnol AIOL 2, pp 351-358

Cocito S, Sgorbini S, Bianchi CN (1997) Zonation of a suspension-feeder assemblage on a temperate rocky shoal: the influence of water current and bottom topography. In: Hawkins LE, Hutchinson S (eds) The response of marine organisms to their Environments. Univ Southampton, UK, pp 183-192

Cocito S, Ferdeghini F, Sgorbini S (1998a) *Pentapora fascialis* (Pallas) [Cheilostomata: Ascophora] colonisation of one sublittoral rocky site after sea-storm in the northwestern Mediterranean. In: Baden S, Pihl L, Rosemberg R, Strömberg, JO, Svane I, Tiselius P (eds) Recruitment, colonisation, and physical-chemical forcing in marine biological systems. Kluwer Acad Publ, Belgium, pp 59-66

Cocito S, Sgorbini S, Bianchi CN (1998b) Aspects of the biology of the bryozoan *Pentapora fascialis* in the north-western Mediterranean. Mar Biol 131: 73-82

Connell JH, Keough MJ (1985) Disturbance and patchy dynamics of subtidal marine animals on hard substrata. In: Pickett STA, White PS (eds) The ecology of natural disturbance and patch dynamics. Academic Press, London, pp 125-151

Denny MV, Daniel TR, Koehl MAR (1985) Mechanical limits to size in wave-swept organisms. Ecol Monogr 55: 69-102

Done TJ (1992) Effects of tropical cyclone waves on ecological and geomorphological structures on the Great Barrier Reef. Cont Shelf Res 12 (7/8): 859-872

Done TJ, Potts DC (1992) Influence of habitat and natural disturbance on contributions of massive *Porites* corals to reef communities. Mar Biol 114: 479-493

Frechette M, Aitken AE, Page L (1992) Interdependence of food and space limitation in benthic suspension feeder: consequence for self-thinning relationship. Mar Ecol Prog Ser 83: 55-62

Gautier YV (1962) Recherchés ecologiques sur les bryozoaires chilostomes en Méditerranée occidentale. Rec Trav Stn Mar Endoume 38: 163-168

Harvell CD, Caswell H, Simpson P (1990) Density effects in a colonial monoculture: experimental studies with a marine bryozoan (*Membranipora membranacea* L.). Oecologia 95: 595-598

Hayward PJ, Ryland JS (1979) British Ascophoran bryozoans. Kermack DM, Barnes RSK (eds) Synopses of the British Fauna. Academic Press, London, pp 1-312

Hughes TP (1990) Recruitment limitation, mortality, and population regulation in open system: a case study. Ecology 7 (1): 12-20

Jackson JBC (1979) Morphological strategies of sessile animals. In: Larwood G, Rosen BR (eds) Biology and systematics of colonial organisms. Academic Press, London, pp 499-555

McKinney FK, Jackson JBC (1989) Erect growth: problems of breakage and flow. In: McKinney FK, Jackson JBC (eds) Bryozoan evolution. Univ Chicago Press, Chicago, pp 167-189

Pickett STA, White PS (1985) Patch dynamics: a synthesis. In: Pickett STA, White PS (eds) The ecology of natural disturbance and patch dynamics. Academic Press, Orlando, pp 371-394

Rogers CS (1993) Hurricanes and coral reefs: the intermediate hypothesis revisited. Coral Reefs 12: 127-137

Sala E, Garrabou J, Zabala M (1996) Effects of diver frequentation on Mediterranean sublittoral populations of the bryozoan *Pentapora fascialis*. Mar Biol 126: 451-459

Sgorbini S, Cocito S, Bianchi CN (1996) Underwater photography as a tool to monitor the population dynamics of a clonal organism. In: Albertelli G, De Maio A, Piccazzo M (eds) Atti XI Congr Assoc Ital Oceanol Limnol AIOL, Genova, pp 819-826

Stebbing ARD (1971) Growth of *Flustra foliacea* (Bryozoa). Mar Biol 9: 267-273

Woodley JD, Chornesky EA, Clifford PA, Jackson JBC, Kaufman LS et al (1981) Hurricane Allen's impact on Jamaican corals. Science 214 (4522): 749-755

Spring Marine Vegetation on Rocky Substrata of the Tremiti Islands (Adriatic Sea, Italy)

M. Cormaci, G. Furnari, G. Alongi, M. Catra, F. Pizzuto, and D. Serio

ABSTRACT

Spring aspect of the benthic marine communities on rocky substrata from midlittoral to circalittoral zones of the Tremiti Islands (Adriatic Sea) is described. From a comparison with literature data the disappearance of species characterising well structured infralittoral photophilic communities on rocky substrata like *Cystoseira* spp. and *Sargassum* spp., resulted. This finding led us to suppose that a considerable structural deterioration of phytobenthic communities has occurred during the last three decades, probably because of an increase of water turbidity.

Introduction

The Tremiti Islands are a small archipelago situated in the Southern Adriatic Sea at a distance of about 22 km from the Italian coast (Fig. 1). There are three main islands: St Domino (2.08 km^2), St Nicola (0.41 km^2), Caprara (0.44 km^2) and an islet named Cretaccio (0.03 km^2). The islands are composed almost exclusively of Eocene dolomitic organic limestones and calcareous

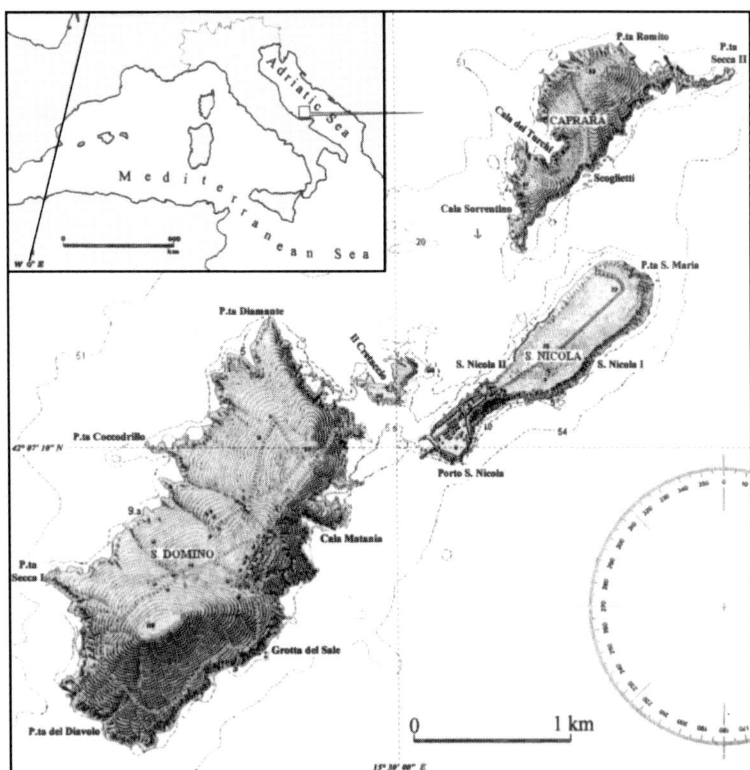

Fig. 1. Map of the Tremiti Islands with sampling sites

Dipartimento di Botanica, Università di Catania, Via A. Longo 19, 95125 Catania, Italy

F.M. Faranda, L. Guglielmo, G. Spezie (eds)
Mediterranean Ecosystems: Structures and Processes

arenites but the island of St Nicola is composed of more recent rocks like Pliocene dolomites (Rizzi Longo 1972).

The phytobenthos of the Tremiti Islands has, to date, been studied only from the floristic point of view and rather poorly too. The first publication on the area is that of Pignatti et al. (1967) who listed 70 species from the island of St Domino. Then, some more floristic data were reported by Giaccone (1969) who listed 75 species from the island of Caprara and 90 species from St Domino, Rizzi Longo (1972) who listed 260 species from the archipelago and Giaccone (1978) who, in his check-list of the Adriatic Sea, reported 3 new species. Finally, 18 additional new species from the archipelago were more recently reported by Solazzi et al. (1991-94). From the above literature data, appropriately updated from both taxonomic and nomenclatural points of view, the benthic marine flora of the Tremiti Islands consists of 281 accepted taxa at specific and infraspecific level, Cyanophyta excluded.

In order to evaluate changes that have possibly occurred in the phytobenthic communities in the last three decades, within the frame of the Project C.N.R. Prisma 2, sub-project 3, both benthic flora and vegetation of the above Islands were studied. In the paper by Cormaci and Furnari (1999), the first considerations on differences between the present flora and that from literature (both based on spring collections), consisting of 301 and 281 taxa respectively (Cyanophyta excluded), were reported. In the present paper the spring vegetational aspect of benthic communities of the archipelago is described.

Materials and Methods

Sampling was carried out on rocky substrata in May 1997 along 16 transects (Fig. 1) perpendicular to the shore, from the midlittoral zone to 40 m depth (circalittoral zone) for a total of 116 samples. Infra- and circalittoral samples were collected by SCUBA diving. Station data are reported in Table 1. Surfaces sampled were of 400 cm^2 in midlittoral communities and of 1600 cm^2 in both infra- and circalittoral communities. Such surfaces, wider than those generally estimated for analogous communities from different Mediterranean areas (Boudouresque 1974; Ballesteros 1992), were chosen in order to obtain data as significant as possible, since it wasn't possible to carry out any preliminary studies of qualitative minimal areas of different communities.

In Tables 2 to 6 every species was attributed a per cent cover value expressed according to the values of the scale of Braun-Blanquet (1928). In the above Tables species with per cent cover values < 1% and frequency < 50% are not reported

Table 1. Sampling sites

No. of transect	Name of transect	Latitude	Longitude	Exposure	Slope	Date
	St DOMINO					
1	Punta del Diavolo	42° 06' 40"	15° 28' 60"	270°	35°	18/05/97
2	Punta Secca I	42° 06' 75"	15° 28' 50"	270°	40°	21/05/97
3	Punta Coccodrillo	42° 07' 14"	15° 28' 95"	310°	35°	12/05/97
4	Punta Diamante	42° 07' 54"	15° 29' 45"	300°	30°	24/05/97
5	Cala Matania	42° 06' 88"	15° 29' 78"	110°	20°	13/05/97
6	Grotta del Sale	42° 06' 48"	15° 29' 45"	110°	25°	13/05/97
7	CRETACCIO	42° 07' 31"	15° 29' 95"	280°	35°	16/05/97
	CAPRARA					
8	Cala Sorrentino	42° 07' 83"	15° 30' 45"	250°	30°	20/05/97
9	Cala dei Turchi	42° 08' 18"	15° 30' 63"	300°	45°	17/05/97
10	Punta Romito	42° 08' 35"	15° 30' 96"	330°	50°	19/05/97
11	Punta Secca II	42° 08' 28"	15° 31' 47"	90°	25°	11/05/97
12	Scoglietti	42° 07' 95"	15° 30' 87"	120°	30°	15/05/97
	St NICOLA					
13	Punta S. Maria	42° 07' 66"	15° 31' 08"	40°	45°	14/05/97
14	San Nicola I	42° 07' 40"	15° 30' 88"	110°	20°	22/05/97
15	Porto San Nicola	42° 07' 05"	15° 30' 21"	130°	35°	22/05/97
16	San Nicola II	42° 07' 38"	15° 30' 46"	310°	15°	20/05/97

(lists of species not reported in Tables are available on request to the authors).

The material collected was preserved in sea water and Formalin (5%) for later study in the laboratory. Voucher specimens are kept in the Herbarium of the Department of Botany of the University of Catania.

Results

Midlittoral Zone

In the upper part of this zone the communities are characterised by a low mean number of species: 2.6 and 7 at +30 to +15 cm and at m.s.l. respectively; while the community occurring at 10 to 15 cm depth shows a higher mean number of species (20.2).

At +30 to +15 cm a community having *Polysiphonia sertularioides* as dominant species occurs (Table 2). To that species are associated, according to the stations, *Lithophyllum papillosum*[1] or *Nemalion helminthoides* and *Ralfsia verrucosa* or *Gelidiella lubrica* and, in a sheltered station of the island of Caprara (Scoglietti), *Cladophora dalmatica*. At the following two stations, Punta Coccodrillo (St Domino) and San Nicola II (St Nicola), *Gelidiella lubrica* is the dominant species instead of *P. sertularioides*. At m.s.l. (Table 2) a community with *Lithophyllum byssoides* and *L. papillosum* (both species with high per cent cover values) occurs. In some stations, however, only one of the two species is present. In the lower part of the midlittoral zone (–15 to –10 cm) a community with *Corallina elongata, Ceramium spp.* and *Callithamnion granulatum* as dominant species, accompanied by rhodomelacean species of the genera *Laurencia and Osmundea*, generally occurs (Table 2). At stations with a less steep slope of the substratum, the last species show high per cent cover values, e.g. *Osmundea truncata* and *O. verlaquei* at Cala Sorrentino and at Punta Secca (Caprara) respectively. At Grotta del Sale (St Domino), where *Corallina elongata* lacks, *Laurencia glandulifera* and *Osmundea verlaquei* are the dominant species.

Infralittoral Zone

The communities occurring in this zone are much richer of species than those of the midlittoral zone, having a mean number of species increasing from near the surface (22.3) to –10 m (58.4). Then, from 13 to 20 m depth, such a number reduces slightly (54.9) and more markedly from 22 to 30 m depth (50.8).

In the upper part of this zone (Table 3), a community with *Cystoseira amentacea* v. *stricta* (with high per cent cover values) and a mean number of species of 21.6, forms a rather continuous belt in all islands. At Cala Matania (St Domino) *C. amentacea* v. *stricta* is accompanied by *C. compressa* which, however, shows a low per cent cover value. Only at Punta Romito (Caprara) and at Porto San Nicola (St Nicola) the above vegetational belt is substituted by a community with *Corallina elongata* and by a community with *Ceramium* spp., respectively.

Below the belt with *Cystoseira amentacea* v. *stricta*, at about 1 m depth (Table 4), the vegetation is characterised by a community with *Corallina elongata* and *Laurencia glandulifera* accompanied by some other species (as *Osmundea truncata, O. verlaquei, Ceramium* spp.) with lower per cent cover value. Only at San Nicola I (St Nicola) a community with *Cystoseira crinitophylla* occurs, while some particular facies with *Ectocarpus siliculosus* and with *Sphacelaria cirrosa*, at Grotta del Sale (St Domino) and at Scoglietti (Caprara) occur respectively.

From 3 to 5 m depth (Table 4) a community with *Corallina elongata* and some Dictyotaceae (*Dictyota dichotoma* v. *dichotoma, D. fasciola, D. spiralis*), generally occurs. At Cala Matania (St Domino), where *Corallina elongata* lacks, a community with *D. fasciola* accompanied by *Amphiroa rigida* and *Haliptilon virgatum* occurs. From 6 to 10 m depth (Table 4) the vegetation is characterised by a community with *Dictyota dichotoma* v. *dichotoma* often accompanied by other Dictyotaceae (*D. dichotoma* v. *intricata, D. fasciola, Taonia atomaria* f. *ciliata, Padina pavonica*), some Sphacelariaceae (*Halopteris filicina* and *Stypocaulon scoparium*) and some large thalli of *Codium bursa. Peyssonnelia squamaria,*

[1] According to Woelkerling et al. (1998) the placement of this taxon in *Lythophyllum* requires confirmation

Table 2. Midlittoral communities

Number of transect	1	3	4	6	8	9	10	12	16	1	2	3	4	5	6	7	8	9	10	11	12	13	15	16	1	2	3	4	5	6	7	8	9	10	11	12	13	14	15	16
Depth in cm	+30	+30	+15	+15	+30	+30	+30	+15	+15	0	0	0	0	0	0	0	0	0	0	0	0	0	0	0	–15	–10	–15	–15	–15	–15	–15	–10	–10	–10	–10	–15	–15	–15	–15	–15
Per cent total cover (x 10)	5	1	1	2	2	1	2	2	2	7	7	7	7	3	8	7	6	8	8	4	7	4	7	6	10	10	10	10	10	10	10	10	10	10	10	10	10	10	10	10
Number of species per sample	3	1	2	2	3	2	3	4	3	5	6	3	9	8	8	7	10	8	10	5	5	9	6	6	13	25	18	24	32	22	22	15	14	19	19	26	19	19	15	21
Polysiphonia sertularioides (Grateloup) J. Ag.	3		5	3	5		3	5	+			+		+		+	+			+		+									+	+			+	+	+	+		
Gelidiella lubrica (Kütz.) J. Feldmann *et* Hamel		5		2					3		+							1																						
Nemalion helminthoides (Velley) Batters						3	2										+			3		+																		
Ralfsia verrucosa (Areschoug) Areschough			+		+	+	1										+			1											+	+								
Polysiphonia opaca (C. Ag.) Moris *et* De Not.														4			+			+									+						+			+		+
Cladophora dalmatica Kützing								5																					+		+									
Lithophyllum papillosum (Hauck) Foslie	3							+		2	3	3	3			3	4	2	2	4	5	3														+				
Lithophyllum byssoides (Lamarck) Foslie										3	3	3	3	5	5	3		3	4			3	5	5																
Corallina elongata Ellis *et* Solander															+			+			+		+	+	5	5	3	5	3		2		5	3	+	5	2	2	5	3
Ceramium secundatum Lyngbye										+	+		+		+	+	+	+	2		+	+	+		+	1	3	+	+		2	3	2	2	+	3	2		+	2
Callithamnion granulatum (Ducluzeau) C. Ag.					+			+						+			2		+			+			+	1	1	+	+	+	+	+		2	2	2	2	+	1	1
Osmundea verlaquei G. Furnari																		1								+		2	1	3	2		+	2	2	+		+	+	
Ceramium ciliatum (Ellis) Ducluzeau v. *robustum* (J. Agardh) G. Mazoyer																	+				+									1		2			2	+	+	3		2
Ceramium diaphanum (Lightfoot) Roth													+									+			+		+	5	+		+	+		+	+	+	+		+	2
Laurencia glandulifera (Kützing) Kützing														+															2	3						+			+	+
Cladophora laetevirens (Dillwyn) Kützing									+					+										+		+			+									3		+
Osmundea truncata (Kützing) K.W. Nam *et* Maggs																																3								+
Valonia utricularis (Roth) C. Agardh													+		+	+									+	+	+	+	+	+	+	+	+		2	+	+	+	+	+
Ceramium tenerrimum (Martens) Okamura																		+															+					+		2
Ceramium circinatum (Kützing) J. Agardh																											+	+						+						1
Crouania attenuata (C. Agardh) J. Agardh																										1	+	+	+		+	+	+		+	+	+		+	+
Ceramium flaccidum (Kützing) Ardissone																									+			+	+	+	+	+		+	+	+	+			
Gastroclonium clavatum (Roth) Ardissone																									+	+		+	+				+		+		+		+	+
Gelidium pusillum (Stackhouse) Le Jolis										+						+	+								+			+	+			+			+	+	+	+		

(cont.)

Table 2. (cont.)

Number of transect	1	3	4	6	8	9	10	12	16	1	2	3	4	5	6	7	8	9	10	11	12	13	15	16	1	2	3	4	5	6	7	8	9	10	11	12	13	14	15	16
Depth in cm	+30	+30	+15	+15	+30	+30	+30	+15	+15	0	0	0	0	0	0	0	0	0	0	0	0	0	0	0	-15	-10	-15	-15	-15	-15	-15	-10	-10	-10	-10	-15	-15	-15	-15	-15
Per cent total cover (x 10)	5	1	1	2	2	1	2	2	2	7	7	7	7	3	8	7	6	8	8	4	7	4	7	6	10	10	10	10	10	10	10	10	10	10	10	10	10	10	10	10
Number of species per sample	3	1	2	2	3	2	3	4	3	5	6	3	9	8	8	7	10	8	10	5	5	9	6	6	13	25	18	24	32	22	22	15	14	19	19	26	19	19	15	21
Chaetomorpha linum (O.F. Müller) Kützing																			+					+		+		+	+				+			+		+	+	+
Sphacelaria cirrosa (Roth) C. Agardh																									+				+	+	+		+		+		+			+
Bryopsis corymbosa J. Agardh																											+	+		+	+						+			
Lithophyllum pustulatum (Lamouroux) Foslie															+											+		+	+						+	+	+	+		
Pneophyllum fragile Kützing																							+					+						+				+	+	
Stylonema alsidii (Zanardini) Drew																			+							+		+					+	+						
Antithamnion cruciatum (C. Agardh) Nägeli																					+					+				+	+			+		+				
Lomentaria ercegovicii Verlaque *et al.*															+												+	+	+								+			+
Number of species per sample not reported in the table	1	0	0	0	0	0	0	0	0	1	2	0	4	2	2	1	1	1	4	0	0	2	2	2	4	13	9	7	14	13	8	4	4	9	4	10	4	6	4	4

Table 3. *Cystoseira amentacea* v. *stricta* community

Number of transect	1	2	3	4	5	6	7	8	9	10	11	12	13	14	15	16
Depth in cm (x10)	3	2	3	3	3	3	3	2	2	2	2	3	3	3	3	3
Per cent total cover (x10)	10	10	10	10	10	10	10	10	10	10	10	10	10	10	10	10
Number of species per sample	25	23	18	25	21	18	22	16	23	30	14	28	22	22	25	25
Cystoseira amentacea (C. Agardh) Bory v. *stricta* Montagne	2	5	4	5	5	5	3	5	3		5	3	5	5		4
Corallina elongata Ellis *et* Solander	+	+	+	+	+		1		2	5		+	+	+	+	1
Ceramium secundatum Lyngbye		+	+	+	+	+		+	3	1		+	+		3	1
Valonia utricularis (Roth) C. Agardh	2	+	+	2	+	+	+	+	+	+	+	2	+	+	+	+
Cystoseira compressa (Esper) Gerloff *et* Nizamuddin					1	1					+		2	+		
Ceramium diaphanum (Lightfoot) Roth	+		+	+		+	+	+		1		+		+	3	
Ceramium circinatum (Kützing) J. Agardh										1					3	+
Ceramium ciliatum (Ellis) Ducluzeau v. *robustum* (J. Agardh) G. Mazoyer	+											2	+	2		
Laurencia glandulifera (Kützing) Kützing	3	+			+							+		+		
Osmundea truncata (Kützing) K.W. Nam *et* Maggs				2						+		+			+	+
Callithamnion corymbosum (Smith) Lyngbye									+	1			+			
Ceramium flaccidum (Kützing) Ardissone	+	+	+	+	+	+	+	+	+	+	+	+	+	+	+	+
Lithophyllum pustulatum (Lamouroux) Foslie	+	+		+	+	+	+	+	+		+	+	+	+	+	+
Gastroclonium clavatum (Roth) Ardissone	+	1	+	+	+	+		+	+	+		+		+	+	1
Callithamnion granulatum (Ducluzeau) C. Agardh	+	+	+	+	+	+	+	+				+			1	1
Crouania attenuata (C. Agardh) J. Agardh	+	+		+	+	+	+		+	+		+		+	+	
Ceramium siliquosum (Kützing) Maggs *et* Hommersand	+	+		+	+		+	+	+	+	+	+		+		+
Ectocarpus siliculosus (Dillwyn) Lyngbye v. *siliculosus*	+	+		+				+	+	+			+	+		+
Gelidium pusillum (Stackhouse) Le Jolis				+	+	+					+	+		+	+	1
Osmundea verlaquei G. Furnari	+		+				+	+	+		+		1		+	
Hydrolithon boreale (Foslie) Chamberlain		+	+	+	+	+	+					+			+	
Sphacelaria cirrosa (Roth) C. Agardh	+	+	+				+		+	+	+				+	
Antithamnion cruciatum (C. Agardh) Nägeli	+			+			+			+		+			1	+
Bryopsis corymbosa J. Agardh	+			+			+		+	+			+			
Ectocarpus siliculosus (Dillwyn) Lyngbye v. *pygmaeus* (Areschough) Gallardo			+				+		+		+	+	+			
Stylonema alsidii (Zanardini) Drew			+	+		+				+		+				+
Ceramium tenerrimum (Martens) Okamura		+							+			+	+			1
Dasya baillouviana (Gmelin) Montagne	+	+		+						+					+	
Lomentaria ercegovicii Verlaque *et al.*		+	+				+	+								+
Pneophyllum fragile Kützing				+					+	+				+	+	
Colpomenia sinuosa (Mertens *ex* Roth) Derbès *et* Solier												1				
Number of species per sample not reported in the table	7	6	4	5	7	5	6	4	6	12	4	8	7	7	7	8

P. rubra and *P. bornetii* are dominant among the understory species.

From 13 to 20 m depth (Table 5) the vegetation is characterised by a community with *Dictyota dichotoma* v. *dichotoma*, *Codium bursa* and *Peyssonnelia squamaria*; at −20 m *Peyssonnelia bornetii* is more frequent, shows high per cent cover values and in some stations replaces *P. squamaria*. At the following two stations, Scoglietti (Caprara) at −14 m and Punta Diamante (St Domino) at −18 m, *Codium vermilara* is present too. *Lithophyllum stictaeforme* is very frequent and abundant among the understory species.

Table 4. Upper infralittoral communities

Number of transect	1	2	3	4	5	6	7	8	9	11	12	13	14	15	16	5	7	10	11	14	15	16	3	12	1	13	2	4	8	9	10	14	15
Depth in m	1	1	1	1	1	1	1	1	1	1	1	1	1	1	1	3	5	5	5	5	6	6	7	7	8	9	10	10	10	10	10	10	10
Per cent total cover (x10)	10	10	10	10	10	10	10	10	10	10	10	10	10	10	10	10	10	10	10	10	10	10	10	10	10	10	10	10	10	10	10	10	10
Number of species per sample	29	43	25	47	26	10	23	30	43	32	41	32	45	43	29	35	28	46	57	41	43	43	53	60	61	66	61	70	59	46	58	51	64
Corallina elongata Ellis *et* Solander		5	2	5			2	+	4		+	+	2	2	3		2	5	1			2	+		1	+	1	+	+	+			
Laurencia glandulifera (Kützing) Kützing	4	2	2		4		2	3	3	+	+	4		3	3																		
Ceramium ciliatum (Ellis) Ducl. v. *robustum* (J. Ag.) G. Mazoyer	2							+	+				+					+				+											
Valonia utricularis (Roth) C. Agardh	4	+	+	+	+		1	+	+	+	+	+	+	+	+	+	+	+	+	+	+	+	+	+	+	+		+	+		+	+	+
Sphacelaria cirrosa (Roth) C. Agardh	+	+	+	+	+		+	+	+	+	5	+	+	2	+			+	+	+		+	+		+	+	+	+		3	+	+	+
Ceramium secundatum Lyngbye	+	+		+			+	2	+	+				2	2			+				+									+	+	
Ceramium tenerrimum (Martens) Okamura	+		+	+			1	2	+	+		+	+		+							+			+				+			+	+
Osmundea truncata (Kützing) K.W. Nam *et* Maggs				1			+	1	+	3	+		+	1		+		+	+	+	+			+					+				
Osmundea verlaquei G. Furnari	+	+	1		+							+			+																		
Cystoseira crinitophylla Ercegovic													5																				
Ectocarpus siliculosus (Dillwyn) Lyngbye						5								+		+									+				+				
Ceramium siliquosum (Kützing) Maggs *et* Hommersand	+	+	+	+	+				+	+	+			2	+	+			+		+			+	+	+			+				
Dictyota dichotoma (Hudson) Lamouroux v. *dichotoma*		+						+		1			+		+			+	3	+	+	2		+	4	4	3	2	3	1	4	2	3
Dictyota fasciola (Roth) Lamouroux											+		+			4	+			5				5				+	+				
Codium bursa (L.) C. Agardh																	2	2	+		+	3		3	+	+	2	2	2			2	2
Stypocaulon scoparium (Linnaeus) Kützing							+						+			+	+		+	+	+					+	+			3	+	3	3
Peyssonnelia squamaria (Gmelin) Decaisne		+								+	+					+		+	2			+		+	+	+	2	2	+		2	+	2
Padina pavonica (Linnaeus) Lamouroux			+							1	+		+			+	+	+	+	+	+	+	+	+	+	2	+	2	+	2			+
Haliptilon virgatum (Zanardini) Garbary *et* Johansen		+		+				+	+	+	+			+		2	+	+	+	+	+	2	+	+	2	+	+		+	+	+		+
Halimeda tuna (Ellis *et* Solander) Lamouroux				+	+		+	+	+		+				+	+	2	+			+	1	+	+	+	2	+	+			+	+	+
Jania adhaerens Lamouroux				+							+		+			+				+			+		2	+		+	+				+
Polysiphonia elongata (Hudson) Sprengel		1							+					+					+			+	+	+			+	+	2				
Polysiphonia dichotoma Kützing				+		+										+					+	1		+			+	+	+				
Neogoniolithon brassica-florida (Harvey) Setchell *et* Mason				1																	+			2			+		3				
Amphiroa rigida Lamouroux												+	+			2	+		+	+		+	+	+		+		+	+			+	
Dictyota dichotoma (Hudson) Lamour. v. *intricata* (C. Ag.) Greville																						2								+		2	2
Dictyota spiralis Montagne																				1	3			+									
Taonia atomaria (Woodward) J. Agardh f. *atomaria*																				2			+					2					+
Peyssonnelia rubra (Greville) J. Agardh																			+						3	+				+			2
Peyssonnelia bornetii Boudouresque *et* Denizot																							2				1		+		2		
Taonia atomaria (Woodward) J. Ag. f. *ciliata* (C. Ag.) Nizamuddin																					2						2					+	
Rodriguezella pinnata (Kützing) Schmitz *ex* Falkenberg																							+		+	+	+	+	+	+	2		+
Halopteris filicina (Grateloup) Kützing																									+			+		2			+
Osmundea pelagosae (Schiffner) Nam																						1			+	+	+						+
Number of species per sample not reported in the Table	21	32	17	35	20	8	14	19	31	21	29	25	32	33	19	22	19	34	43	29	30	28	39	45	44	50	45	54	40	36	48	39	47

Table 5. Lower infralittoral communities

Number of transect	5	12	16	3	6	7	11	1	4	2	8	9	10	13	3	15	14	6	11	1	4	8	10	13
Depth in m	13	14	14	15	15	15	15	16	18	20	20	20	20	20	22	22	23	24	24	25	25	30	30	30
Per cent total cover (x10)	10	10	10	10	10	10	10	10	10	10	10	10	10	10	10	10	10	10	10	10	10	10	10	10
Number of species per sample	46	53	64	50	53	59	61	37	54	59	65	72	42	54	46	66	55	44	59	58	53	50	32	45
Dictyota dichotoma (Hudson) Lamouroux v. *dichotoma*	2	+	2	2	+	+	5	+	2	2	3	3	2	+		3			+		+	3	+	
Codium bursa (Linnaeus) C. Agardh	3	3	2	3	2	2	+	3			2	+		2	1	+	1			2	3			
Lithophyllum stictaeforme (Areschoug) Hauck		2	2	2		+						2			+		1	+			+	1		
Peyssonnelia squamaria (Gmelin) Decaisne	2	2	3		+	2	+	+	+	2	2			+	1								+	
Peyssonnelia bornetii Boudouresque *et* Denizot			+	+	4				+	3	+	2	4	4	2	3	4	4	3	+	+	3	+	4
Halimeda tuna (Ellis et Solander) Lamouroux	+	+	+	+	+	4	+	+	1	1	+	+	+	+	+	2	+	1	+	+	+	1	+	2
Womersleyella setacea (Hollenberg) R.E. Norris		+			+	+			+		+	1		1	2	+	2	3	+		+	+	3	+
Peyssonnelia rubra (Greville) J. Agardh							+	3			+	+			1				2	2		1		
Unidentified encrusting Corallinaceae						4						2	+							1				
Osmundaria volubilis (Linnaeus) R.E. Norris															1				3		2			
Codium vermilara (Olivi) Delle Chiaje		3							2															
Aglaothamnion sp.		+					+		+			+		+	+		+		+	2	2	+		+
Peyssonnelia polymorpha (Zanardini) Schmitz			1	+		+			3	+					+		+					+		
Neogoniolithon brassica-florida (Harvey) Setchell *et* Mason			3							+	1												+	
Botryocladia boergesenii J. Feldmann	+	+	3		+				+	+	+	+				+			+		+	+		+
Peyssonnelia rosa-marina Boudouresque *et* Denizot	1				+					+		2				+		+				+		
Sphaerococcus coronopifolius Stackhouse										+			+										3	
Sphacelaria cirrosa (Roth) C. Agardh			+		+	+	+	+		+	+	+	+	2	+	+			+	+	+			
Lithophyllum pustulatum (Lamouroux) Foslie			2	+	+	+			+	+	+	+	+			+	+		+		+			
Cladophora prolifera (Roth) Kützing	2										+	+			+				+		+	+		+
Botryocladia botryoides (Wulfen) J. Feldmann											+					2							+	
Polysiphonia furcellata (C. Agardh) Harvey		+	+	+	+	+	+	+	+	+	+	+	1	+	+	+	+	+	+	+	+	+		+
Herposiphonia secunda (C. Agardh) Ambronn			+	+	+	+	+	+		+		+	1	+	+	+	+	+	+	+	+	+		+
Polysiphonia denudata (Dillwyn) Greville												1	+			+							+	
Hypoglossum hypoglossoides (Stackhouse) Collins *et* Hervey	+	+	+	+	+	+	+	+	+	+	+	+	+	+	+	+	+	+	+	+	+		+	+
Rodriguezella pinnata (Kützing) Schmitz *ex* Falkenberg	+		+		+	+	+	+		+	+	+		+	+	+	+	+	+	+	+	+	+	+
Ceramium bertholdii Funk			+	+	+	+				+	+	+		+	+	+	+	+	+	+	+	+	+	+
Lobophora variegata (Lamouroux) Womersley *ex* Oliveira	+	+	+		+	+		+	+		+	+	+		+		+	+	+	+	+	+	+	
Valonia utricularis (Roth) C. Agardh	+		+		+		+	+	+	+	+		+	+	+	+	+	+	+	+			+	+
Antithamnion cruciatum (C. Agardh) Nägeli	+		+	+	+	+	+			+		+	+	+	+	+			+	+	+	+		+
Hydrolithon boreale (Foslie) Chamberlain	+	+	+	+	+	+	+	+		+	+				+	+	+	+	+	+	+			
Lomentaria chylocladiella Funk	+	+	+			+	+	+		+		+	+	+	+	+			+	+	+		+	+
Nereia filiformis (J. Agardh) Zanardini		+	+	+	+				+	+	+	+	+	+		+	+	+	+	+			+	+
Polysiphonia elongata (Hudson) Sprengel	+	+		+	+	+	+		+	+	+			+	+	+	+	+	+	+	+			
Number of species per sample not reported in the table	31	37	42	35	32	39	45	23	38	37	43	48	26	36	23	44	37	29	36	39	31	33	16	30

Below 20 m depth (Table 5), *Dictyota dichotoma* generally shows low per cent cover values except at Porto San Nicola (St Nicola) at -22 m and at Cala Sorrentino (Caprara) at -30 m. From 22 to 30 m depth (Table 5) the vegetation is more varied: the dominant species is *Peyssonnelia bornetii* accompanied, according to stations, by other species with high percent cover values: by *Peyssonnelia rubra* and *Osmundaria volubilis* at Punta Secca II (Caprara, -24 m); by *Halimeda tuna* at Porto S. Nicola (St Nicola, -22 m) and at Punta S. Maria (St Nicola, -30 m); by *Womersleyella setacea* at two stations of St Domino (Punta Coccodrillo at -22 m and Grotta del Sale at -24 m) as well as at San Nicola I (St Nicola) at -23 m. At Punta Romito (Caprara) at -30 m, a community with *Womersleyella setacea* and *Sphaerococcus coronopifolius* occurs.

Circalittoral Zone

Circalittoral communities (Table 6) were found only at the following three stations: Punta Secca I (St. Domino) at -33 m, Cala dei Turchi (Caprara) at -36 m and Punta Romito (Caprara) at -40 m. The number of species per sample is 56, 46 and 38 respectively with a mean value of 46.7. The communities are characterised by a dominance of sciophilous species: in the first station, *Peyssonnelia squamaria, P. bornetii* and unidentified encrusting Corallinaceae; in the second one, *Lithophyllum stictaeforme, Codium effusum, Neogoniolithon brassica-florida, Sebdenia rodrigueziana* and *Rodriguezella strafforelloi*; in the third station, *Peyssonnelia squamaria, P. crispata, Neogoniolithon brassica-florida, Codium effusum, Rodriguezella strafforelloi, R. pinnata, Codium bursa* and unidentified encrusting Corallinaceae.

Table 6. Circalittoral communities

Number of transect	2	9	10
Depth in m	33	36	40
Per cent total cover (x10)	10	10	10
Number of species per sample	56	46	38
Peyssonnelia squamaria (Gmelin) Decaisne	2	+	2
Peyssonnelia bornetii Boudouresque *et* Denizot	2		
Aglaothamnion sp.	1	+	+
Brongniartella byssoides (Goodenough *et* Woodward) Schmitz	1		
Chylocladia verticillata (Lightfoot) Bliding	1		
Polysiphonia elongata (Hudson) Sprengel	1		
Unidentified encrusting Corallinaceae	3		2
Lithophyllum stictaeforme (Areschoug) Hauck		3	
Codium effusum (Rafinesque) Delle Chiaje		2	2
Neogoniolithon brassica-florida (Harvey) Setchell *et* Mason		2	2
Rodriguezella strafforelloi Schmitz *ex* Rodriguez		2	2
Sebdenia rodrigueziana (J. Feldmann) Parkinson		2	
Rodriguezella pinnata (Kützing) Schmitz *ex* Falkenberg	+	+	2
Codium bursa (Linnaeus) C. Agardh			2
Peyssonnelia crispata Boudouresque *et* Denizot			2
Antithamnion cruciatum (C. Agardh) Nägeli	+	+	+
Ceramium bertholdii Funk	+	+	+
Ceramium codii (Richards) G. Mazoyer	+	+	+
Erythroglossum sandrianum (Kützing) Kylin	+	+	+
Herposiphonia secunda (C. Agardh) Ambronn	+	+	+
Lomentaria chylocladiella Funk	+	+	+
Pseudochlorodesmis furcellata (Zanardini) Börgesen	+	+	+
Sphacelaria plumula Zanardini	+	+	+
Hypoglossum hypoglossoides (Stackhouse) Collins *et* Hervey	+		+
Stictyosiphon adriaticus Kützing	+		+
Cladophora albida (Nees) Kützing	+	+	
Hydrolithon boreale (Foslie) Chamberlain	+	+	
Gulsonia nodulosa (Ercegovic) J. *et* G. Feldmann		+	+
Polysiphonia denudata (Dillwyn) Greville		+	+
Number of species per sample not reported in the table	36	26	17

Discussion and Conclusions

Apart from midlittoral communities and the shallow community with *Cystoseira amentacea* v. *stricta*, the benthic vegetation of the Tremiti Islands appears little structured. In particular, the most striking thing is the lack of other infralittoral photophilic communities with *Cystoseira* spp. and *Sargassum* spp. that, as known, characterise in the Mediterranean Sea the vegetation on rocky substrata in unpolluted environments (Giaccone and Bruni 1973; Ros *et al.* 1985). Even though it was impossible to carry out any comparisons between previous and present vegetational characteristics of that area, because of all the previous papers on those islands dealt only with flora, the occurrence in the sixties-seventies years of *Cystoseira crinita*, *C. schiffneri* f. *tenuiramosa*, *C. spinosa*, *Sargassum acinarium* and *S. hornschuchii* (Rizzi Longo 1972), leads us to suppose the occurrence, at that time, of the above mentioned photophilic communities in that area. Therefore, we can conclude that the benthic algal vegetation of the Tremiti Islands has undergone a considerable structural deterioration, during the last three decades, mainly in the infralittoral zone. Such a kind of deterioration might be due to different factors like the occurrence of more or less toxic pollutants (often linked to suspended inert terrigenous materials), eutrophication, overgrazing by sea-urchins, fishery activities and anchoring, etc. However, at the Tremiti Islands, neither the occurrence of dense populations of sea-urchins causing overgrazing, nor fishery activities and anchoring sufficiently important to cause heavy damage to the benthic vegetation, nor eutrophication (see below), were observed. Conversely, during our dives, a noticeable water turbidity due to the presence of suspended inert terrigenous materials was observed. A reduction in water transparency in the Southern Adriatic from sixties to eighties in both the coastal and open waters was recorded by Morovic and Domijan (1991). Such a reduction was related by Zore-Armanda (1991) to the eutrophication of the Adriatic waters. But, both the persistence of the above mentioned well structured midlittoral communities as well as of the *Cystoseira amentacea* v. *stricta* belt compared with the lack of communities with Ulvales near the surface, indicate the lack of eutrophication in the Tremiti Islands, at least in shallow waters. Therefore, agreeing with Cormaci and Furnari (1999), we think that the above mentioned structural deterioration of the benthic algal vegetation is mainly due to the increased turbidity of waters related to the presence of inert terrigenous suspended materials, even though the occurrence of toxic pollutants cannot be excluded.

Acknowledgements. This work was supported by a grant from the Italian C.N.R. within the Project PRISMA 2, sub-project 3. We are grateful to the Mayor of the Commune Isole Tremiti for facilities during sampling campaign.

References

Ballesteros E (1992) Els vegetals i la zonació litoral: espècies, comunitats i factors que influeixen en la seva distribució. Inst Estud Catalans Secció Cienc Biol, Barcelona

Boudouresque C-F (1974) Aire minima et peuplements algaux marins. Bull Soc Phycol Fr 19: 141-157

Braun-Blanquet J (1928) Pflanzensoziologie. Springer, Berlin Heidelberg New York Tokyo

Cormaci M, Furnari G (1999) Change of the benthic algal flora of the Tremiti Islands (Southern Adriatic) Italy. In: Kain (Jones) JM, Brown TM, Lehaye M (eds) Proc 16th Int Seaweed Symp. Hydrobiology 398/399, pp 75-79

Giaccone G (1969) Raccolte di fitobenthos sulla banchina continentale italiana. G Bot Ital 103: 485-514

Giaccone G (1978) Revisione della flora marina del mare Adriatico. Annu WWF Parco Mar Miramare Trieste 6: 5-118

Giaccone G, Bruni A (1973) Le Cistoseire e la vegetazione sommersa del Mediterraneo. Atti Ist Veneto Sci Lett Arti Venezia 131: 59-103

Morovic M, Domijan N (1991) Light attenuation changes in the middle and southern Adriatic. Acta Adriat 32: 621-635

Pignatti S, De Cristini P, Rizzi L (1967) Le associazioni algali della grotta delle Viole nell'isola di S. Domino (Is. Tremiti). G Bot Ital 101: 117-126

Rizzi Longo L (1972) La flora sottomarina delle Isole Tremiti. Atti Ist Veneto Sci Lett Arti Venezia 130: 329-376

Ros JD, Romero J, Ballesteros E, Gili MJ (1985) Diving in blue water: the benthos. In: Margalef R (ed) Key environments: Western Mediterranean. Pergamon Press, Oxford, pp 233-295

Solazzi A, Totti C, Marzocchi M (1991-1994) Epifitismo in *Halimeda tuna* (El. et Sol.) Lam. (Chlorophyceae, Siphonales) Isole Tremiti, Adriatico Meridionale. Nova Thalassia 12: 69-79

Woelkerling WJ, Lawson GW, Price JH, John DM, Prud'homme van Reine W (1998) Seaweeds of the western coast of tropical Africa and adjacent islands: a critical assessment. IV. Rhodophyta (Florideae) 6: genera [Q] R-Z, and an update of current names for non-geniculate Corallinales. Bull Br Mus Nat Hist Bot 28: 115-150

Zore-Armanda M (1991) Natural characteristics and climatic changes of the Adriatic sea. Acta Adriat 32: 567-586

Distribution of Meio- and Macrofaunal Assemblages in Anthropogenically Impacted Coastal Sediments (NW Mediterranean)

A. Covazzi Harriague, M. Chiantore, E. Piccione, L. Maliardi, and G. Albertelli

ABSTRACT

Meio- and macrofaunal assemblages in front of Genova and Savona were studied in order to evaluate their structure in anthropogenically impacted coastal sediments. In the study area, sediment particle size clearly decreases with depth, while their organic matter content shows a clearly increasing pattern. Hydrocarbons show top values at 20 m depth in front of Genova and at 50 m depth in front of Savona. In the whole considered area mollusc and echinoderm assemblages evidence higher densities and species number at 20 m depth. Herein a transition zone is observed where species characteristic of organic enrichment and sediment instability, such as the bivalves *Corbula gibba* and *Thyasira flexuosa*, reach top densities. Suspension-feeders dominate at 10 m, at 50 m deposit-feeders turn out dominant, while at 20 m these alternative feeding strategies coexist. Meiofauna abundance and number of taxa reached top values at 10 m depth in front of Genova as well as in front of Savona.

Introduction

For some 30 years much attention has been focused on man induced changes on soft-bottom macrobenthic communities in the Western Mediterranean Sea under chronic pollutant input, due to domestic and industrial wastes (Bellan 1967, 1984, 1985; Glemarec and Hily 1981; Bourcier 1982). Bottom sediments constitute a sort of record of environmental processes and, consequently, vehiculate man induced pollutants to benthic organisms (Reish 1972). This kind of pollution induces basic changes in structural parameters of benthic communities (species composition and number, number of individuals and biomass), more or less intensely according to the input characteristics, recovery capacity and pristine conditions (Zajac and Witlach 1982; Bellan 1985). While most sensible species disappear, they are replaced by few opportunistic species (Sandberg 1994; Heip 1995).

More recently, for a variety of reasons (short generation time, life cycles spent entirely within the sediments), meiofauna has been suggested as a pollution indicator with several implications in routine biological monitoring (Heip 1980; Platt and Warwick 1980; Vincx and Heip 1991), although literature dealing mainly with effects of hydrocarbons on marine meiofauna is generally contradictory and the pollution effects seem to be dependent on spill size and habitat characteristics (Renaud-Mornant et al. 1981; Fleeger and Chandler 1983; Danovaro et al. 1995) as well as taxonomic grouping (Heip et al. 1988). Generally, nematodes appear to be relatively insensitive to heavy oil conditions and, if affected, they seem to recover rapidly, while copepods seem to be more severely affected by hydrocarbons and to recover more slowly than nematodes (Wormald 1976; Giere 1979; Boucher 1980; Elmgren et al. 1980; Bonsdorf 1981; Fricke et al. 1981; Gee et al. 1992), although evidences from experimental oil spills reported an increase in copepod density (Naidu et al. 1978).

The aim of this work is to assess the structure of macro and meiobenthic littoral communities in front of Genova and Savona, trying to evaluate possible relationships among faunistical and biocoenotical features, natural and man

Dipartimento per lo Studio del Territorio e delle Sue Risorse, Università di Genova, Viale Benedetto XV 5, 16132 Genova, Italy

F.M. Faranda, L. Guglielmo, G. Spezie (eds)
Mediterranean Ecosystems: Structures and Processes

induced sediment characteristics (grain size, organic matter and hydrocarbon contents), by means of correlative methods.

Material and Methods

Study Site and Sampling

The sediments in front of Genova and Savona (NW Mediterranean, Fig. 1) were sampled, using a Van Veen grab (0.1 m^2), in June 1996 along 7 transects (4 in front of Genova and 3 in front of Savona) at 10, 20 and 50 m depth. Macrofauna was sieved through a 1 mm mesh size and fixed in 10% neutralised formalin. Due to sampling constraints only a single sub-core (11.3 cm^2 surface area) was taken for meiofaunal analysis. The samples were fixed with 4% neutralised formalin in 0.4 mm prefiltered seawater solution. An additional sub-core was taken for grain size, organic matter and hydrocarbons analyses and frozen at −20°C.

Sediment Texture, Organic Matter and Hydrocarbon Analyses

Sediment texture analysis was performed using the dry sieve technique after organic matter removal with hydrogen peroxide and organic matter content was determined with a gravimetric method (Buchanan and Kain 1971).

Hydrocarbons were hot-extracted: carbon tetrachloride (20 ml) was added as solvent, after organic matter extraction with hydrochloric acid. The solution was boiled for half an hour. The liquid was cooled and then leached on filter-paper and anhydrous sodium sulphate. The filtered solution was passed through a column filled with Florisil (silica gel) to retain fatty acids. Hydrocarbons were estimated spectrophotometrically (Perkin Elmer Mod. 983; CNR 1988). Concentrations were referred to the sediment dry weight (105°C).

Organic matter and hydrocarbon contents were analysed in three replicates.

Faunal Analyses

In order to retain the meiofaunal fraction, sediments were sieved through 1000 and 37 µm mesh. The fraction remaining on the 37 µm sieve was centrifuged three times with Ludox HS (density arranged to 1. 18 g/ml). All meiobenthic animals were counted and classified per taxon under a stereo microscope, after staining with Rose Bengal (0.5 g/1).

Macrofauna was sorted in the laboratory using a stereo microscope. Molluscs and echinoderms were identified to species level. Macrofauna trophic guilds and biocoenotical stocks were assessed according to literature (Jangoux and Lawrence 1982; Russell-Hunter 1983; Riedl 1991).

A Spearman-Rank correlation analysis was performed on the data set. Diversity and evenness indices for macrofauna parameters were calculated using the PRIMER programme

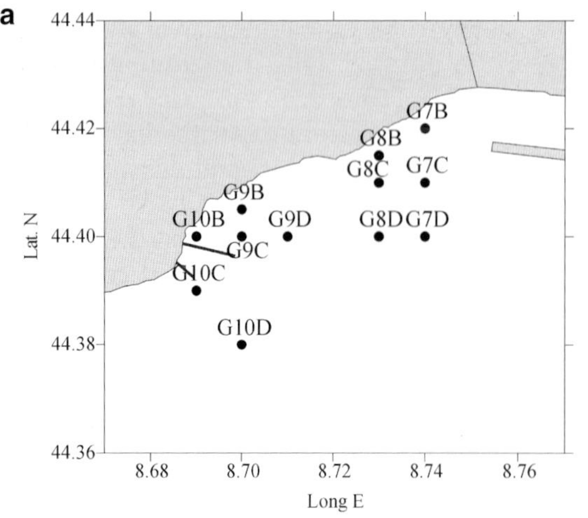

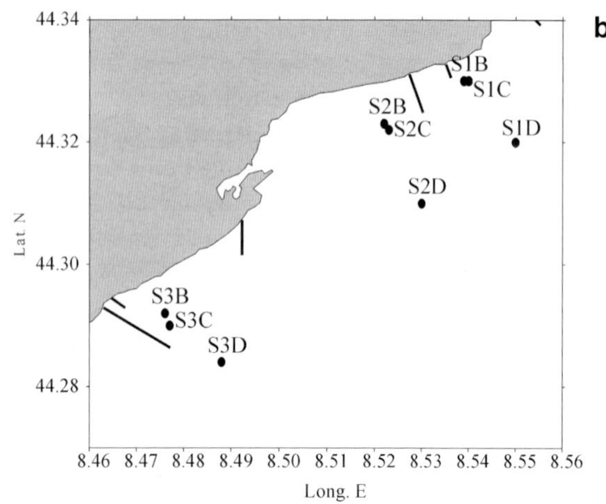

Fig. 1a,b. Sampling area: **a** Genova, **b** Savona

(Plymouth Marine Laboratory), as well as classification according to the Bray-Curtis similarity index (Bray and Curtis 1957).

Results

Sediment Features

Genova Area

Data relative to grain size, sediment total organic matter and hydrocarbon content in the Genova area are showed in Table 1a. Sediment grain size shows quite similar trends along the different transects, decreasing with depth. Sandy fraction dominated at the shallowest stations (B and C) representing, on average, 91.3 and 75.7% respectively, while silt/clay fraction reaches its maximum percentage at the deepest station (D), on average, 94.3%.

Total organic matter (TOM) content increases with depth, ranging on average between 3.3% (10 m) and 8.5% (50 m).

Hydrocarbon concentrations in the sediments in front of Genova are relatively high. Maximum values are observed values at 20 m depth (on average 0.3 mg/g) and minimum values at the shallowest stations (on average 0.1 mg/g).

Savona Area

Data relative to grain size, sediment total organic matter and hydrocarbon content in the Savona area are showed in Table 1b. Similar to the trend observed in front of Genova, sediment grain size decreases with depth: the sand fraction constitutes on average 90.5% at 10 m depth and 81.4% at 20 m depth.

Total organic matter content follows the same increasing pattern with depth observed in Genova.

Sediment hydrocarbon concentrations are sli-

Table 1a,b. Data relative to sampling depth, gravel, sand and silt/clay fractions, total organic matter (TOM: average ± standard deviation) and hydrocarbons (average ± standard deviation) concentrations: **a** Genova, **b** Savona

	Depth (m)	Gravel (%)	Sand (%)	Silt and Clay (%)	TOM (%)	Hydrocarbons mg/g
a						
G7B	10	0.1	94.2	5.7	2.4±0.3	0.27±0.18
G7C	20	0	63.1	36.9	6.4±0.4	0.08±0.01
G7D	50	0.1	3.2	96.7	9.6±0.1	0.13±0.04
G8B	10	8.7	90.2	1.1	2.3±0.2	0.13±0.10
G8C	20	0	81.4	18.6	6.1±0.2	0.37±0.20
G8D	50	0	6.4	93.6	9.0±0.1	0.27±0.18
G9B	10	1.4	94.3	4.3	2.3±0.2	0.03±0.02
G9C	20	0.3	76.9	22.8	4.7±0.6	ND
G9D	50	0.6	4.2	95.2	8.3±0.2	0.14±0.06
G10B	10	0.3	86.3	13.4	6.0±0.2	0.11±0.03
G10C	20	0.1	81.4	18.5	4.8±0.2	0.31±0.31
G10D	50	0.5	7.9	91.6	7.1±0.1	0.08±0.03
b						
S1B	10	0.6	91.7	6.7	2.9±0.3	0.05±0
S1C	20	2.3	88.7	9.0	8.3±0.7	0.06±0.03
S2B	10	0.7	96.8	2.5	0.7±0.1	0.22±0.01
S2C	20	7.8	87.4	4.8	3.4±0.4	0.08±0
S2D	50	0	21.4	78.6	6.4±0.2	0.29±0.14
S3B	10	12.8	83.0	4.2	3.6±0.4	0.19±0
S3C	20	5.9	67.9	26.2	6.0±0.5	0.18±0.10
S3D	50	2.8	38.0	59.2	ND	0.14±0.02

ghtly lower than those observed at Genova: maximum values are observed at the shallowest and deepest stations. A clear increasing pattern was observed from east to west (0.06 mg/g on average at transect S1, 0.2 mg/g on average at transect S3).

Mollusc and Echinoderm Density, Structure and Distribution

Genova Area

Data relative to mollusc and echinoderm abundance and number of species, diversity and evenness in the area of Genova are shown in Table 2a. Mollusc and echinoderm assemblages decrease with depth (top average values 388.3 and 52.5 ind/m^2 for molluscs and echinoderms respectively). On the contrary species number reaches the maximum average value for both assemblages at 20 m (18 and 4 species respectively).

Mollusc assemblage is characterised by the relatively high abundance of species indicator of organic enrichment and sediment instability (*Corbula gibba* and *Thyasira flexuosa*). These species are dominant at 20 m depth, reaching average values of 96 and 70 ind/m^2, respectively.

Suspension-feeders dominate at the shallowest stations (B), with an average percentage of 71%, while at 50 m deposit-feeders become dominant (61%). At 20 m suspension feeders (35%) and suspension/deposit-feeders (32%) coexist.

Diversity (*H'*) and evenness (*J*) values were relatively constant throughout the study area.

Savona Area

Data relative to mollusc and echinoderm abundance and species number, diversity and evenness in the area of Savona are showed in Table 2b. Abundance values of molluscs and echinoderms found in Savona are lower than those observed at Genova, while the species number was quite similar. An increasing abundance is observed from east to west. Similarly, species number reaches the maximum average value for both assemblages at transect S3. Echinoderms reach top

Table 2a,b. Mollusc and echinoderm abundance and species number, Shannon diversity (*H'*) and evenness (*J*): **a** Genova, **b** Savona

	Molluscs ind/m^2	Number of species	Echinoderms ind/m^2	Number of species	*H'*	*J*
a						
G7B	470	17	53	2	1.92	0.66
G7C	363	15	40	3	2.14	0.77
G7D	87	9	3	1	2.01	0.84
G8B	273	17	50	4	2.29	0.76
G8C	480	21	23	3	2.16	0.68
G8D	63	7	3	1	1.73	0.83
G9B	560	18	40	2	1.91	0.65
G9C	373	21	67	5	2.64	0.81
G9D	110	6	27	2	1.43	0.69
G10B	250	14	67	2	1.98	0.73
G10C	190	13	53	4	2.21	0.82
G10D	280	10	23	4	1.63	0.62
b						
S1B	160	11	23	3	2.01	0.76
S1C	110	9	20	1	2.20	1.0
S2B	40	2	0	0	0.69	1.0
S2C	230	26	63	3	2.93	0.87
S2D	155	13	45	3	2.46	0.89
S3B	390	24	77	5	2.38	0.71
S3C	170	17	63	6	2.64	0.84
S3D	250	4	100	2	1.75	0.98

Table 3a,b. Meiofauna composition, abundance (ind/10cm^2) and number of taxa: **a** Genova, **b** Savona

	Nematodes	Copepods	Polychaetes	Ostracods	Others	Total	N° of Taxa
a							
G7B	752.2	153.1	20.4	11.5	20.3	957.2	9
G7C	1056.6	403.5	130.1	15.0	62.8	1668.0	8
G7D	402.7	17.7	4.4	0	2.7	427.5	6
G8B	1229.2	1160.2	46.9	95.6	10.6	2542.5	12
G8C	629.2	116.8	1.8	3.5	31.8	783.1	8
G8D	170.8	21.2	1.8	5.3	5.3	204.4	8
G9B	2581.4	993.8	52.2	41.6	59.3	3728.3	11
G9C	2433.6	336.3	42.5	20.4	112.3	2946.1	13
G9D	405.3	45.1	0.9	15.9	13.3	480.5	8
G10B	882.3	139.8	13.3	5.3	18.6	1060.2	10
G10C	3623.9	481.4	65.5	24.8	56.5	4252.1	12
G10D	910.6	151.3	8.9	20.4	16.8	1108.0	9
b							
S1B	3060.2	574.3	58.4	28.3	52.2	3773.4	12
S1C	714.2	54.0	8.8	4.4	8.0	789.4	8
S2B	772.6	386.7	51.3	23.0	45.1	1278.7	13
S2C	1569.0	1570.8	161.9	94.7	149.6	3546.0	12
S2D	810.6	243.4	17.7	16.8	46.9	1135.4	15
S3B	892.0	131.9	15.0	4.4	25.7	1069.0	10
S3C	749.6	166.4	21.2	12.4	24.8	974.4	11
S3D	144.2	16.8	1.8	0	6.2	169.0	8

values at 50 m (on average 72.5 ind/m^2), while molluscs do not show a clear pattern with depth. For both assemblages species number reaches the top average value at 20 m (17 and 3 species for molluscs and echinoderms respectively).

In front of Savona, *Corbula gibba* and *Thyasira flexuosa* are found, but with much lower abundances than at Genova.

Suspension-feeders dominate both at stations B and C, with average percentages of 82 and 55% respectively, while deposit-feeders reach the top percentage at 50 m (45%).

Similar to Genova, diversity (H') and evenness (J) values were quite similar throughout the investigated area.

Meiofauna Density, Structure and Distribution

Genova Area

Data relative to meiofaunal composition, total abundance and number of taxa in front of Genova are showed in Table 3a. Meiofaunal abundance ranges between 2412.4 ind/10cm^2 (on average at 20 m) and 555.1 ind/10cm^2 (on average at 50 m). A generally increasing pattern is observed moving from east to west (transect G7: 1017.7 ind/10cm^2 on average; transect G9: 2385 ind/10cm^2 on average).

Number of taxa decreases with depth and from west to east, ranging between 11 on average at 10 m and 8 at 50 m depth and from 10.3 (transect G10) to 7.6 (transect G7).

Nematodes constitute the most represented taxon at all depths, followed by copepods. Nematodes display an increasing pattern with depth ranging on average from 65.4% (10 m) to 83.3% (50 m). By contrast, copepods percentage decreases with depth: highest values are found at 10 m (on average 29.4%) and lowest at 50 m (on average 10.4%).

Savona Area

Data relative to meiofaunal composition, total abundance, composition and number of taxa in front of Savona are showed in Table 3b. Meiofaunal abundance shows in Savona a clear decreasing pattern with depth and from east to west, average values ranging from 2040.4 ind/10cm^2 (10 m) to 652.2 ind/10cm^2 (50m) and between 2281.4 ind/10cm^2 (transect S1) and 737.5 ind/10m^2 (transect S3).

Number of taxa is quite constant with depth, ranging in average from 10 (20 m) to 12 (10 m). Along the transects the top value is found at transect S2 (on average 13) and the lowest at transect S3 (10).

Similar to Genova nematodes constitute the most represented taxon at all depths followed by copepods. Nematode abundance percentage of total meiofaunal community ranges on average from 56.1% (20 m) to 71% (50 m), while copepods display the maximum percentage at 20 m (on average 33.1%) and the minimum at 10 m (17.6%). Nematodes dominate also along the transects, the maximum percentage value is observed at transect S1 (on average 82%), while copepods reach their highest percentage at transect S2 (36.5%).

Table 4. Coefficients of the most significative correlations found in front of Genova

Parameters	Genova
TOM versus meiofauna taxa	−0.72 ($P < 0.01$)
TOM versus meiofauna abundance	−0.66 ($P < 0.05$)
TOM versus silt/clay fraction	0.89($P < 0.001$)
Sandy fraction versus meiofauna taxa	0.64 ($P < 0.05$)

Discussion and Conclusions

Sediments of both study areas are characterised by a clear increasing pattern of organic load with depth. This trend seems to be related with sediment texture: in fact, total organic matter, in the area of Genova, is significantly correlated with silt/clay fraction (Table 4). At Savona, although both parameters follow the same trend, insignificant correlations are found. In general, sediments in front of Genova are organically richer than those sampled at Savona, according to the increase of the silt/clay fraction. In particular, transects G7 and G10 display the highest organic contents due to their closeness to Voltri and Arenzano harbours.

Hydrocarbon concentrations are relatively high, the average value observed in front of Genova at 20 m is higher than those observed at Zoagli after "Agip Abruzzo" oil spill (Danovaro et al. 1995) and in the Bay of Morlaix and Lannion after "Amoco Cadiz" oil spill (Boucher 1980). The results found in this work seem to indicate a chronic hydrocarbon pollution in the examined areas, particularly evident in front of Genova. Similar concentrations were indicated by Bellan (1985) for the subnormal zones.

The area in front of Genova presents a clearly defined distribution of macro and meiobenthic assemblages following a depth gradient. Macrobenthic community distribution is strongly linked to sedimentary environment: a SFBC (Sable Fines Bien Calibrés) community is found at 10 m (Table 5), largely dominated by suspen-

Table 5. Biocoenosis succession and characteristic macrofaunal species for sediments in front of Genova and Savona along a depth gradient. In brackets average densities (ind/m^2)

	GENOVA		SAVONA	
	Biocoenosis	Discriminating species	Biocoenosis	Discriminating species
B	SFBC Suspension-feeders	*S. subtruncata* (164) *L. divaricata* (75) *C. gibba* (43) *T. papyracea* (30)	DC SFBC SD Suspension-feeders	*L. divaricata* (32) *P. papillosum* (21) *T. donacina* (9) *H. arctica* (58)
C	SFBC VTC Deposit/suspension-feeders	*C. gibba* (96) *T. flexuosa* (70) *S.subtruncata* (52)	DC Suspension-feeders	*P. papillosum* (18) *C. gibba* (11)
D	VTC Deposit-feeders	*T. flexuosa* (66) *T. alleni* (16)	VTC Deposit-feeders	*T. flexuosa* (33) *C. gibba* (27)

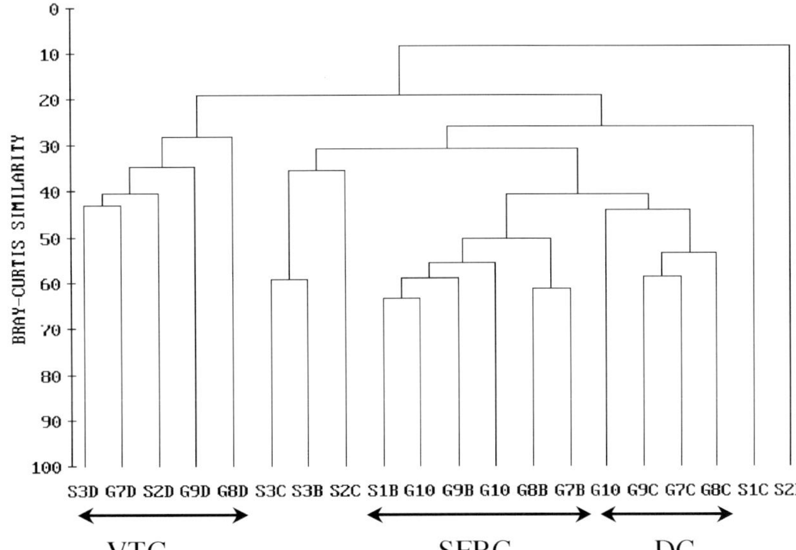

Fig. 2. Macrofauna cluster analysis based on Bray-Curtis similarities after double square root transformation of abundance data

sion-feeders, while at 50 m a VTC (Vases Terrigénes Côtieres) is widespread (Fig. 2). At the same depth deposit-feeders turn dominant in response to the increase in organic load and sediment stability (Weston 1990).

At 20 m depth a transition zone is found, characterised by a diversified community largely dominated by tolerant species such as *Corbula gibba* and *Thyasira flexuosa* (particularly hydrocarbon tolerant; López Jamar and González Mejuto 1987). Similar biocoenoses succession was observed in a boundary zone of Cogoleto by Drago and Albertelli (1978).

Similarly, meiofaunal abundance and number of taxa show maximum values at 20 m, inversely correlated with total organic matter and positively with the coarser grain size (Table 4), confirming the ecotone characteristics of this zone. A richer meiobenthic community is found on the western side particularly along transect 9, where lowest TOM and hydrocarbon concentrations are found. Anyway, total meiobenthic density, at 10 m depth, is lower than that recorded at the same depth in an unpolluted zone in the Ligurian Sea (Danovaro et al. 1994).

In terms of composition, meiobenthic assemblages are typical of coastal environments (Danovaro et al. 1994; Ólafsson and Elmgren 1997), except for an increased relevance of nematodes at 20 m depth, where hydrocarbons are higher. Conversely copepods are better represented in front of Savona, under less severe stress conditions.

In the area in front of Savona the transition between the different biocoenoses is less clear (Fig. 2), as previously observed (Albertelli and Rossi 1987), probably due to the coarser sediments and the presence of *Posidonia oceanica* mattes and *Cymodocea nodosa* meadows.

In conclusion, from the hydrocarbon concentrations measured in the whole study area we can identify general moderate and chronic perturbed conditions in relation to anthropic impact of the two towns. Such chronic but moderate pollution determines in front of Genova a decantation area characterised by high species and meiofaunal taxa richness, high nematodes abundance and strong presence of *C. gibba* and *T. flexuosa*, in accordance with the intermediate disturbance hypothesis (Connell 1975; Gray et al. 1990).

Acknowledgements. This work was carried out in the framework of the project SAReF. The authors are grateful to Mr. Nicolino Drago from Dip. Te. Ris for assistance on meiofaunal taxonomy.

References

Albertelli G, Rossi GL (1987) Studio del popolamento macrobentonico di un fondale posto in prossimità di un effluente urbano (1976-1980). Proc VII AIOL: 91-100

Bellan G (1967) Pollution et peuplements benthiques sur substrat meuble dans la région de Marseille. Rev Int Oceanogr Mediterr 8: 51-95

Bellan G (1984) Indicateurs et indices biologiques dans le domaine marin. Bull Ecol 15: 13-20

Bellan G (1985) Effects of pollution and man-made modifications on marine benthic communities in the Mediterranean: a review. In: Moraitou-Apostolopoulos M, Kiortsis V (eds) Mediterranean marine ecosystems. Plenum Publ Co, pp 163-188

Bonsdorf E (1981) The "Antonio Gramsci" oil spill: impact on the littoral and benthic ecosystem. Mar Pollut Bull 12: 301-305

Boucher G (1980) Impact of "Amoco Cadiz" oil spill on intertidal and sublittoral meiofauna. Mar Pollut Bull 11: 95-101

Bourcier M (1982) Évolution au course des quinze dernières années, des biocoenoses benthiques et de leurs faciès dans une Baie Méditerranèenne soumise à l'action lointaine de deux émissaires urbains. Tethys 10: 303-313

Bray JR, Curtis JT (1957) An ordination of the upland forest communities of Southern Wisconsin. Ecol Monogr 27: 325-349

Buchanan JB, Kain JM (1971) Measurement of the physical and chemical environment. In: Holme NA, McIntyre AD (eds) Methods for the study of marine benthos. (IBP Handbook, 16). Blackwell Scientific, Oxford, pp 30-58

CNR (1988) Metodi analitici per i fanghi. Quad Ist Ric Sulle Acque, Roma

Connell JH (1975) Some mechanisms producing structure in natural communities: a model and evidence from field experiments. In: Cody ML, Diamond JM (eds) Ecology and evolution of communities. Harvard Univ Press, Cambridge, pp 460-490

Danovaro R, Mees J, Vincx M (1994) Annual dynamics of meiobenthic communities in the Ligurian Sea (Northwestern Mediterranean): preliminary results. Proc AIOL X: 295-305

Danovaro R, Fabiano M, Vincx M (1995) Meiofauna response to the "Agip Abruzzo" oil spill in subtidal sediments of the Ligurian Sea. Mar Pollut Bull 30: 133-145

Drago N, Albertelli G (1978) Aspetti quantitativi del macrobenthos di Cogoleto (Golfo di Genova). Proc AIOL II: 187-192

Elmgren R, Hanson U, Sundelin B (1980) The Tsesis oil spill, report to the first year scientific study (26 October 1977 to December 1978). In: Kineman JJ, Elmgren R, Hanson S (eds). US Dep Commer Off Mar Pollut Assessment, Colorado, pp 97-126

Fleeger JW, Chandler GT (1983) Meiofauna response to an experimental oil spill in a Louisiana salt marsh. Mar Ecol Prog Ser 11: 257-264

Fricke AH, Henning H F-H O, Orren MJ (1981) Relationship between pollution and psammolittoral meiofauna density of two South-African beaches. Mar Environ Res 5: 57-77

Gee JM, Austin M, De Smet G, Ferraro T, McEvoy A, Moore S, Van Gausbeki D, Vincx M, Warwick RM (1992) Soft sediment meiofauna community responses to environmental pollution gradients in the German Bight and at a drilling site off the Dutch Coast. Mar Ecol Prog Ser 91: 289-302

Giere O (1979) The impact of oil pollution on intertidal meiofauna: field studies after the "La Coruña" spill, May 1976. Cah Biol Mar 20: 231-251

Glémarec M, Hily C (1981) Perturbations apportées à la macrofaune benthique de la Baie de Concarneau par les effluents urbains et portuaires. Acta Oecol Oecol Appl 2 (2): 139-150

Gray JS, Clarke KR, Warwick RM, Hobbs G (1990) Detection of initial effects of pollution on marine benthos: an example from the Ekofisk and Eldfisk oilfields, North Sea. Mar Ecol Prog Ser 66: 285-299

Heip C (1980) Meiobenthos as a tool in the assessment of marine environmental quality. In: McIntyre AD, Pearce JB (eds) Rapp P V Reun Cons Int Explor Mer 179: 182-187

Heip C (1995) Eutrophication and zoobenthos dynamics. Ophelia 41: 113-136

Heip C, Warwick RM, Carr MR, Herman PMJ, Huys R, Smol N, Van Holsbeke K (1988) Analysis of community attributes of the benthic meiofauna of Frierfjord/Langesundfjord. Mar Ecol Prog Ser 46: 171-180

Jangoux M, Lawrence JM (eds.) (1982) Echinoderm nutrition. A. A. Balkema Publishers, Rotterdam, p 1654

López Jamar G, González Mejuto J (1987) Ecology, growth and production of *Thyasira flexuosa* (Bivalvia, Lucinea) from Ría de La Coruña, North-West Spain. Ophelia 27(2): 111-126

Naidu AF, Feder HM, Norrel SA (1978) The effect of Prudhoe Bay crude oil on a tidal-flat ecosystem in Port Valdez Alaska. In: Proceedings of Tenth Offshore Technology Conference, Dallas, Texas, pp 97-103

Ólafsson E, Elrngren R (1997) Seasonal dynamics of sublitoral meiobenthos in relation to phytoplankton sedimentation in the Baltic Sea. Estuarine, Coastal and Shelf Sci 45: 149-164

Platt HM, Warwick RM (1980) The significance of free-living nematodes to the littoral ecosystem. In: Price JH, Ivine DEG, Farnham WF (eds) The shore environment. 2 Ecosystem. Academic Press, London, pp 729-759

Riedl R (1991) Fauna e flora del Mediterraneo. In: F. Muzzio & C. Editore (eds), Legoprint srl, Trento, pp 1-777

Reish DJ (1972) The use of marine invertebrates as indicators of varying degrees of marine pollution. In: Ruivo M (ed) Marine pollution and sea life. FAO Fishing News Ltd, pp 203-207

Renaud-Mornant J, Gourbault N, de Panafieu JB, Heleouet MN (1981) Effects de la pollution par hydrocarbures sur la meiofauna de la Baie de Morlaix. In: "Amoco Cadiz", consequences d'une pollution accidentale par les hydrocarbures. Actes Coll Int COB, pp 551-561

Russell-Hunter WD (ed) (1983) The mollusca. 6. Ecology. Academic Press, London, pp 1-695

Sandberg E (1994) Does short-term oxygen depletion affect predator-prey relationships in zoobenthos? Experiments with the isopod *Saduria enomon*. Mar Ecol Prog Ser 103: 73-80

Vincx M, Heip C (1991) The use of meiobenthos in pollution monitoring studies: a review. J Mar Sci 16: 50-67

Weston DP (1990) Quantitative examination of macrobenthic community changes along an organic enrichment gradient. Mar Ecol Prog Ser 61: 233-244

Wormald AP (1976) Effect of an oil spill of marine diesel oil on meiofauna of a sandy beach at Picnic Bay Hong Kong. Environ Pollut 11: 117-130

Zajac RN, Witlach RB (1982) Responses of estuarine infauna to disturbance. 1. Spatial and temporal variation of initial recolonization. Mar Ecol Prog Ser 10: 1-14

Population Dynamics and Secondary Production of *Owenia fusiformis* Delle Chiaje (Polychaeta Oweniidae) along the Coasts of Emilia-Romagna (Northern Adriatic, Mediterranean Sea)

E. Daelli and A. Occhipinti Ambrogi

ABSTRACT

The population structure and dynamics of *O. fusiformis* have been studied for one annual cycle along the coasts of Emilia-Romagna (Italy). Five seasonal samplings were carried out along two transects differently influenced by the Po river outflow, at the depth of 3, 8 and 12 m, using a 0.06 m^2 Van Veen grab. Secondary production has been evaluated according to the increment weight method. Densities and biomass values were higher at 8 m depth in both the considered localities, probably due to lower mud content of the sediment. The pattern of size-frequency histograms suggests an estimated life-span of at least two years. The growth rate of the youngest individuals was higher at 3 m depth, due to the higher hydrodynamic levels of the shallower waters. Total annual production ranged from a minimum of 0.12 g m^{-2} y^{-1} at the depth of 12 m off Cesenatico, to a maximum of 14.78 g m^{-2} y^{-1} at the depth of 3 m off Porto Garibaldi.

Introduction

Owenia fusiformis Delle Chiaje (Polychaeta, Oweniidae) is one of the most common species of sedentary worms with a wide geographical and bathymetrical distribution (Nilsen and Holthe 1985; Dauvin and Thiébaut 1994). It is present with high densities in the fine sand communities, ranging between depths of 0 and 40 m. The feeding behaviour of *O. fusiformis* exhibits high plasticity in relation to hydrodynamism (Fauchald and Jumars 1979; Gambi 1989; Desroy et al. 1997) and favours its colonisation in habitats subject to highly variable trophic fluxes such as the Bay of Seine (Dauvin and Gillet 1991) and along the western coast of the northern Adriatic Sea (Ambrogi et al. 1990; 1995).

This area, largely influenced by the eutrophic inputs of the Po River, shows intense algal blooms that can affect water oxygenation, producing a patchy distribution of anoxic states (Vollenweider et al. 1992; Rinaldi et al. 1997; Moodley et al. 1998), possibly influencing the life cycles of benthic populations (Diaz and Rosenberg 1995; Ambrogi 1997). The ecological importance of *O. fusiformis* in shallow sands and in highly variable environmental conditions is known, in stabilizing the bottom sediments, in enhancing recruitment by a reduction of post-larvae resuspension (Olivier et al. 1996; Desroy et al. 1997; Thiébaut et al. 1997), and in the trophic webs as a coupling element between the plankton and the nekton compartments (Woodin and Merz 1987; Gentil and Dauvin 1989).

In this work the population structure and dynamics of *O. fusiformis* have been studied for one year, in two transects located along the coasts of Emilia-Romagna (Italy), in order to evaluate its life strategy and secondary production.

Materials and Methods

Five seasonal samplings were carried out in July and October 1996, and in February, May and July 1997 along two transects located off Porto Garibaldi (stations 4, 304, 1004) and Cesenatico (stations 14, 314, 1014) at 3, 8 and 12-14 m depth, at the distance of 0.5, 3 and 10 km from the coast line, respectively (Fig.1, Table 1). The samplings were carried out on-board the oceanographic

Dipartimento di Genetica e Microbiologia, Sezione Ecologia, Università di Pavia, Via Sant'Epifanio 14, 27100 Pavia, Italy

F.M. Faranda, L. Guglielmo, G. Spezie (eds)
Mediterranean Ecosystems: Structures and Processes

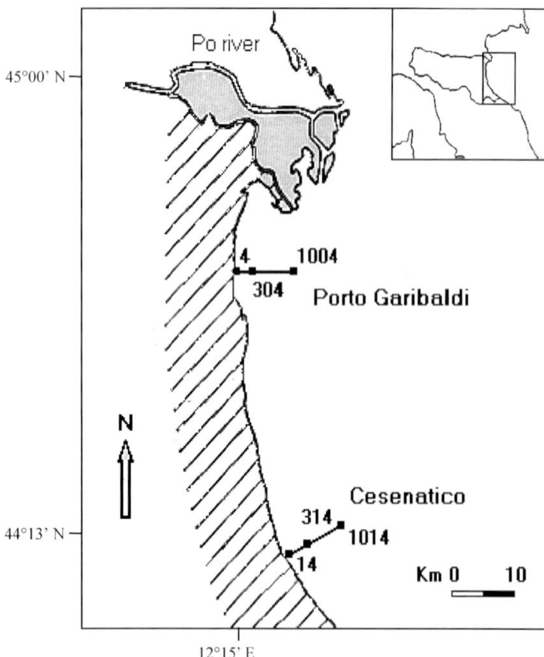

Fig. 1. Sampling stations along the two transects of Porto Garibaldi and Cesenatico

vessel Daphne II belonging to the Emilia-Romagna regional administration. In each station, four samples were collected together with a further sample of sediment for granulometric determinations (Buchanan 1984) using a 0.06 m^2 Van Veen grab. The collected material was sieved through 1mm mesh and preserved in a formol seawater solution (10%). Size-frequency histograms for each sample were obtained by measuring the width of the first abdominal segment (first uncinigerous) (Ambrogi et al. 1995) with an image analysis system (Optimas), after checking the allometric correlation between the total length and the width of the first uncinigerous segment of 50 individuals selected among all the samples ($r^2 = 0.98$).

Biomass was estimated by the regression equation between mean individual ash-free dry weight (AFDW) and width already calculated by Ambrogi et al. (1995) in a neighbouring area. Secondary production was calculated following the increment weight summation method (Crisp 1984). The physical-chemical parameters of the six sampling stations were provided by the oceanographic vessel Daphne II by means of the Idronaut multiparametric probe (mod. Ocean Seven 901).

Results

In the bottom layer, severe conditions of low oxygen concentration were recorded in November 1996 at stations 304, 1004 and 1014. Very low oxygen concentrations were also present in August of both years at stations 304 and 314 (Fig. 2). Stations at 3 m depth were characterised by a wide thermohaline range with the lowest temperature values around 5 -6 °C and the highest around 26 °C. Salinity ranged within values of 13.0‰-18.5‰ and 36.0‰-37.5‰ showing frequent fluctuations.

At the deepest stations (8 and 12-14 m), the range of variability of temperature and salinity of bottom water was much more limited, with the lowest values of temperature around 10 °C and the maximum around 23 °C and with an average value of salinity around 36‰ during the whole study period (Rinaldi et al. 1997; Montanari et al. 1998).

The sediment at 3 m depth was dominated by "very fine sand", according to the Wentworth scale and nomenclature. At 8 m depth, sand percentages changed according to season, ranging from "sand" to "silt", while at greater depth the percentage composition was rather constant with less than 6% sand (Table 2).

Table 1. Geographical coordinates, depth and distance from the coast-line of the sampling stations

Station	Latitude	Longitude	Depth (m)	Distance (km)
4	44° 39' 46"	12° 15' 06"	3	0.5
304	44° 39' 46"	12° 17' 14"	8	3
1004	44° 39' 46"	12° 22' 32"	14	10
14	44° 12' 35"	12° 24' 02"	3	0.5
314	44° 13' 14"	12° 25' 54"	8	3
1014	44° 14' 54"	12° 30' 40"	12	10

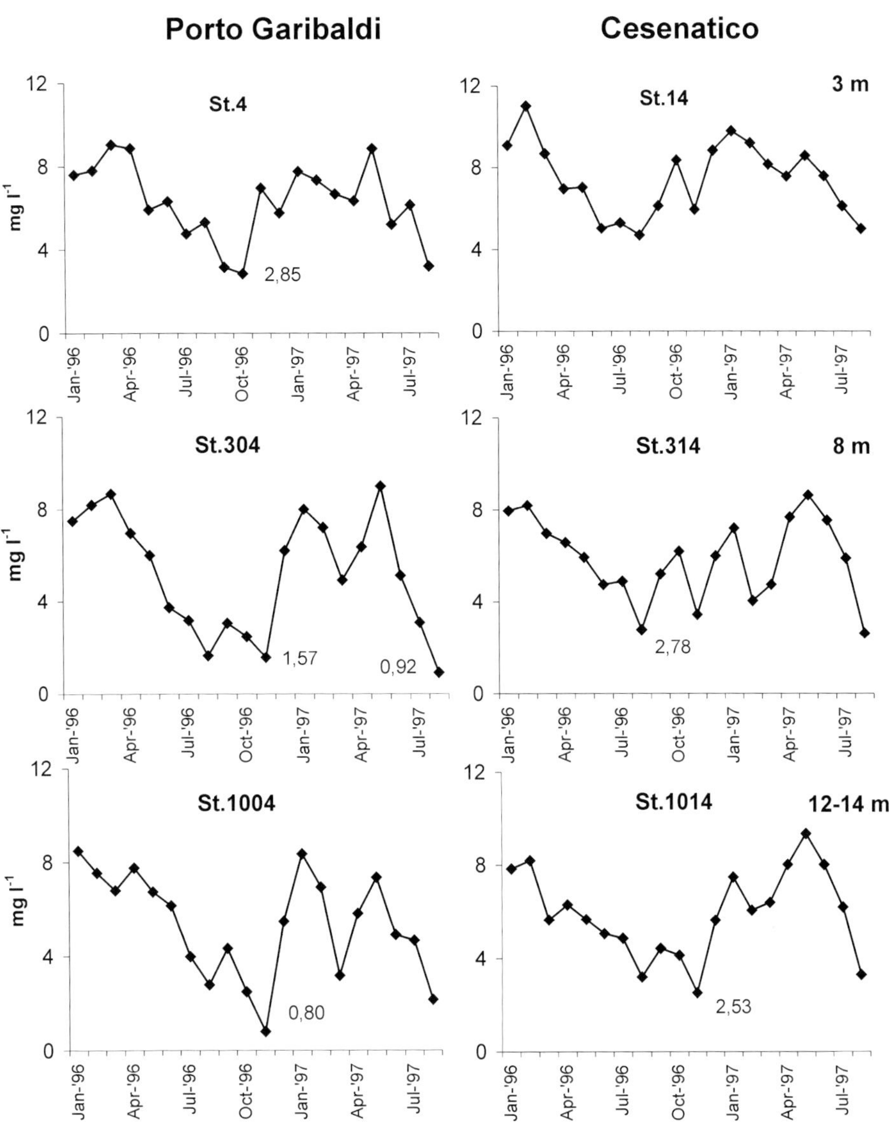

Fig. 2. Average values of oxygen concentration in the water layer close to the bottom in the six stations

Table 2. Sediment textures, according to the Wentworth nomenclature, in the six stations for each sampling date

	July 1996	October 1996	February 1997	May 1997	July 1997
Porto Garibaldi					
Station 4 (3 m)	Very fine sand	Very fine sand	Very fine sand	Very fine sand	Very fine sand
Station 304 (8 m)	Sandy silt	Silt	Silty sand	Sandy silt	Sandy silt
Station 1004 (14 m)	Silt	Silt	Silt	Silt	Silt
Cesenatico					
Station 14 (3 m)	Very fine sand	Very fine sand	Very fine sand	Very fine sand	Very fine sand
Station 314 (8 m)	Silt	Sandy silt	Sandy silt	Silt	Silty sand
Station 1014 (12 m)	Silt	Silt	Silt	Silt	Silt

The highest abundance of *O. fusiformis* was found at 8 m depth in the transect of Porto Garibaldi, with a top value of 27542 ± 1905 ind. m^{-2} at station 304 in July 1996 (Fig. 3).

In general, the Porto Garibaldi transect showed higher density values compared to Cesenatico. Stations at 3 and 8 m depth were more densely populated than those at 12-14 m. During the cold season, the whole community and all the most abundant species had very low densities. In July 1997 densities were much lower than in the corresponding sample of the previous year.

The population structure of *O. fusiformis* was analyzed by means of size-frequency distributions (Figs. 3 and 4). Three size classes were found at 3 m depth at both stations and also at 8 m depth at Porto Garibaldi. The remaining stations were characterised by a two size-class structure (only one in station 1014). The smallest size classes represented the new recruitment of 1996 (class 0); this class was probably underestimated due to the mesh size employed (1mm). The second (class I) and third (class II) size classes, when present, represented the overwintering individuals of the previous years.

At station 4 of Porto Garibaldi (Fig. 3), specimens of class "0" showed larger mean size than in other localities, probably due to an earlier recruitment compared to other stations. An earlier recruitment was confirmed in the second year (May 1997) at the same locality.

In all stations, the recruitment of July 1997 was not as successful as in 1996; it is represented by few individuals having a modal class smaller than that of the May 1997 sample.

During the year, a difference in the growth rate of the population was observed mainly in the class "0" of stations at 3 m depth, compared to those at 8, 12 and 14 m. At the shallowest depth, specimens showed a faster mean individual weight increment, highlighted by the growth

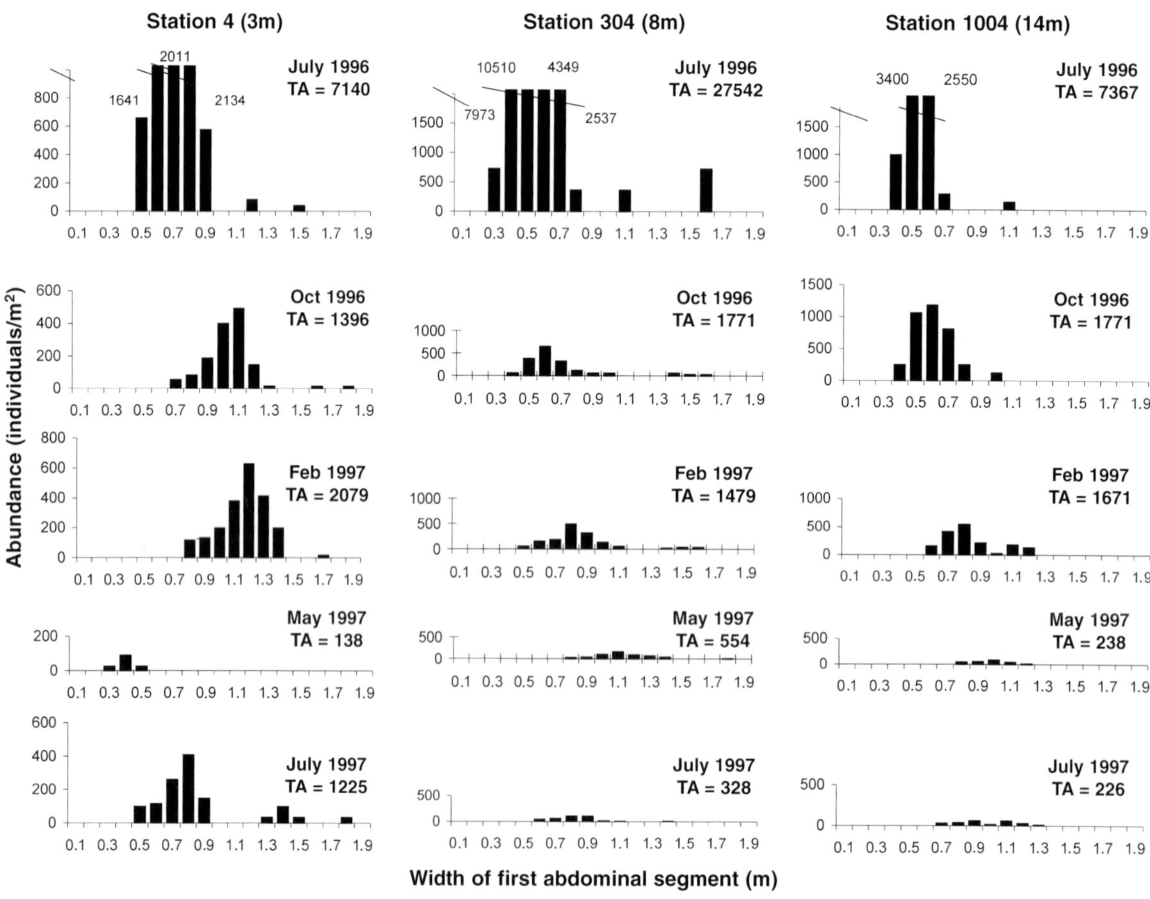

Fig. 3. Seasonal distribution of *O. fusiformis* size frequency at Porto Garibaldi stations. Total abundance (TA) is indicated for each sampling date

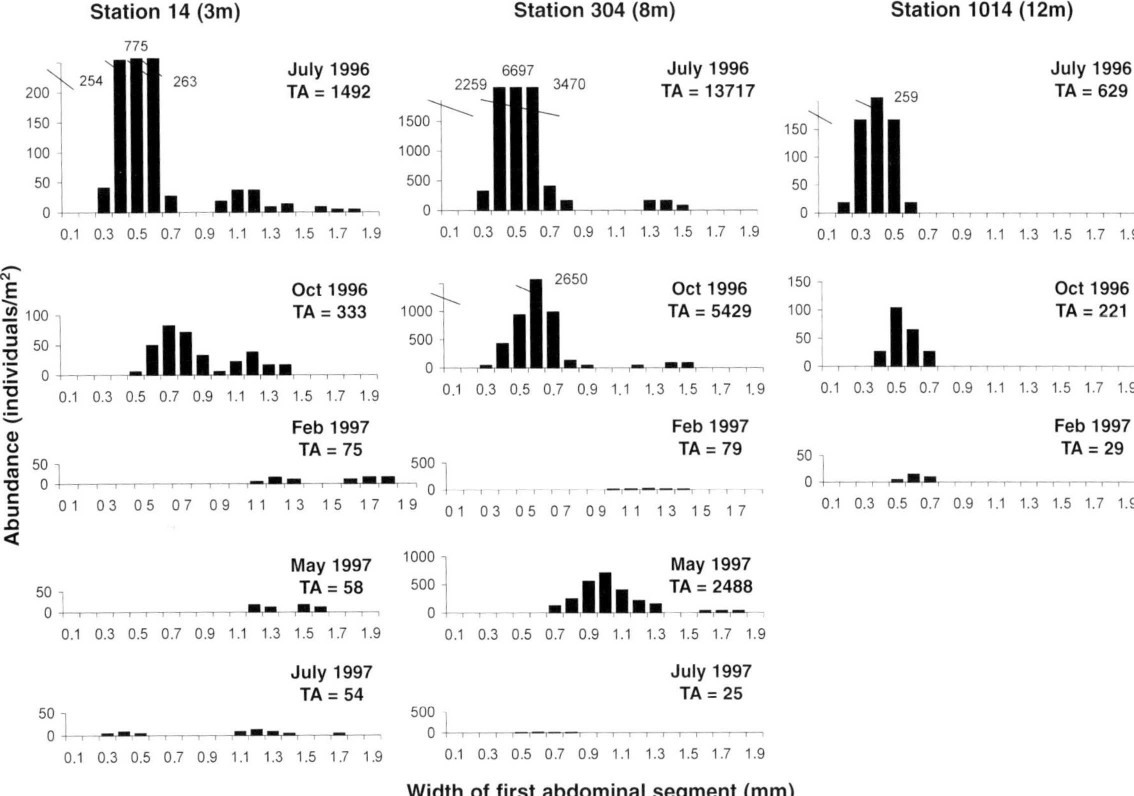

Fig. 4. Seasonal distribution of *O. fusiformis* size frequency at Cesenatico stations. Total abundance (TA) is indicated for each sampling date

rate values calculated for class "0" during the July-October and October-February periods (Table 3).

The estimation of the annual production for the six stations is shown in Table 4, together with the estimation of the mean biomass and production/biomass (P/B) ratio. Annual production values ranged from a minimum of 0.12 $g \cdot m^{-2} \cdot y^{-1}$ at the depth of 12 m off Cesenatico to a maximum of 14.78 $g \cdot m^{-2} \cdot y^{-1}$ at the depth of 3 m off Porto Garibaldi. The P/B ratio ranged from 2.18 to 3.62.

Table 3. Class "0" mean individual growth rate and total density variations of *O. fusiformis* during the summer and winter periods, evaluated at the different stations

Stations	Period	Mean individual growth rate (mg month^{-1})	Total density variation (ind m^{-2})
4 PG (3m)	Jul-Oct	0.48	7140-1396
	Oct-Feb	0.34	1396-2079
14 CE (3m)	Jul-Oct	0.17	1492-333
	Oct-Feb	0.91	333-75
304 PG (8m)	Jul-Oct	0.08	27542-1771
	Oct-Feb	0.18	1771-1479
314 CE (8m)	Jul-Oct	0.04	13717-5429
1004 PG (14m)	Jul-Oct	0.04	7367-3679
	Oct-Feb	0.15	3679-1671
1014 CE (12m)	Jul-Oct	0.06	629-221
	Oct-Feb	0.04	221-29

Table 4. Average biomass (*B*) determined as ash-free dry weight, annual values of secondary production (*P*) of *O. fusiformis* (sum of all age classes), and P/B ratio estimated in the six sampling stations

Stations	Average Biomass *B* ($g \cdot m^{-2}$)	Annual secondary production ($g \cdot m^{-2} \cdot y^{-1}$)	P/B
4 PG (3m)	4.53	14.78	3.26
14 CE (3m)	0.55	1.99	3.62
304 PG (8m)	3.34	10.60	3.17
314 CE (8m)	3.35	7.67	2.29
1004 PG (14m)	1.75	3.81	2.18
1014 CE (12m)	0.04	0.12	3.07

Discussion

The density and biomass values demonstrate that in the study area the polychaete *O. fusiformis* recruits preferentially at 8 m depths, particularly in Porto Garibaldi located closer to the Po River delta. The high abundance values recorded in July 1996 were caused by seasonal recruitment that did not take place with the same success in the following year.

The sediment texture may have played an important role in the settlement of larvae and their survival. In fact, station 304 in July 1996 showed a successful recruitment (Fig. 3) according to the composition of sediment (silt with 47% sand), while station 314, characterised in July by a silty sediment, showed a more limited recruitment (Fig. 4). In autumn, station 304 showed an abrupt increase of the silt fraction (silt = 95%) and a strong mortality of *O. fusiformis*. Differently, station 314 showed a mixed sediment composition in October and a lower mortality rate.

High density recorded in some stations in July 1996 (Table 3) and intraspecific competition may have reduced the individual growth rate at 8 and 12-14 m depths, compared to 3 m, as well as the sediment texture (Ambrogi et al. 1995). Nevertheless, in this case the lower growth rate observed in deeper stations can be better explained by the different hydrodynamic conditions which affect sedimentation rates. It should be noted that *O. fusiformis* can shift from the preferential suspension feeding behaviour to the less favourable one of a deposit feeder, according to hydrodynamic conditions (Gambi 1989). This is supported by the fact that also in winter the growth rate values were always higher at 3 m than at other depths, when the values of abundance were low and comparatively similar in all six stations.

The pattern of the size-frequency histograms suggests an estimated life-span of at least two years for *O. fusiformis* in this area, with a single reproduction and recruitment pulse, whose importance may vary in different years. Ambrogi et al. (1995) pointed out a similar pattern in a neighbouring population of the Po River delta.

O. fusiformis has been intensely studied by various authors also in other localities such as the Bay of Seine (France) (Ménard et al. 1989). As in the Po River delta, the population at the Bay of Seine shows a recruitment in summer and a large fluctuation in the recruits' abundance from year to year. Yet, this latter population is characterised by a more complex structure, with several year classes present at the same time and a four-year life-span.

The average biomass shows a wide range of variability among the six stations, mainly due to the success of recruitment in July 1996 (Table 4). In terms of annual production, stations located off Porto Garibaldi showed values higher than those in front of Cesenatico, probably because the former are most directly influenced by the Po River delta inputs and by the higher carbon content of the sediments (Albertelli et al. 1998). P/B values of stations located at 3 m depth, the highest in relation to the maximum growth rate observed at this depth, ranged from 3.6 to 2.2 in accordance with those already reported by Ambrogi et al. (1995) for populations of *O. fusiformis* living in this area.

In conclusion, total annual production showed great differences among the six stations, depending both on the recruitment success and on the individual growth rate in the summer period. The growth rate of class "0" was much higher at the shallower stations (3 m) in response to the high energy environment favouring the trophic input to the population and most favourable sediments texture. In the other stations, the growth rate may have been negatively affected by hypoxic conditions during the summer period, resulting also in a higher mortality.

Acknowledgements. This work was carried out as a part of a National Program (40% MURST). The authors thank the crew of Daphne II for the technical support during the sampling campaigns and the Emilia-Romagna Environmental Agency for the physical-chemical data.

References

Albertelli G, Bedulli D, Cattaneo-Vietti R et al (1998) Trophic features of benthic communities in the Northern Adriatic Sea. Biol Mar Medit 5(1): 136-143

Ambrogi R (1997) Tra Po e Adriatico: il sistema ecologico del Delta Padano. In: Prospettive di ricerca in ecologia delle acque. Quaderni IRSA 103, pp 188-208

Ambrogi R, Bedulli D, Zurlini G (1990) Spatial and temporal patterns in structure of macrobenthic assemblages: a three-year study in the Northern Adriatic Sea in front of the Po River Delta. PSZNI Mar Ecol 11:25-41

Ambrogi R, Fontana P, Gambi MC (1995) Population dynamics and estimate of secondary production of *Owenia fusiformis* Delle Chiaje (Polychaeta, Oweniidae) in the coastal area of the Po river Delta (Italy). In: Eleftheriou A et al (eds) Biology and ecology of shallow coastal waters. Olsen & Olsen, Fredensborg, pp 207-214

Buchanan JB (1984) Sediment analysis. In: Holme NA, McIntyre AD (eds) Methods for the study of marine benthos. 2nd ed. Blackwell, Oxford, pp 41-65

Crisp DJ (1984) Energy flow measurements. In: Holme NA and McIntyre AD (eds) Methods for the study of marine benthos, 2nd ed. Blackwell Sci Publ Oxford, pp 284-372

Dauvin JC, Gillet P (1991) Spatio-temporal variability in population structure of *Owenia fusiformis* Delle Chiaje (Anellida: Polychaeta) from the Bay of Seine (Eastern English Channel). J Exp Mar Biol Ecol 152: 105-122

Dauvin JC, Thiébaut E (1994) Is *Owenia fusiformis* Delle Chiaje a cosmopolitan species? In: Dauvin JC Laubier L Reish DJ (eds). Actes de la 4ème Conférence Internationale des Polychètes. Mém Mus Natn Hist Nat 162: 383-404

Desroy N, Olivier F, Retière C (1997) Effects of individual behaviours, inter-individual interactions with adult *Pectinaria koreni* and *Owenia fusiformis* (Annelida, Polychaeta), and hydrodynamism on *Pectinaria koreni* recruitment. Bull Mar Sci 60: 547-558

Diaz RJ, Rosenberg R (1995) Marine benthic hypoxia: a review of its ecological effects and the behavioural responses of benthic macrofauna. Oceanogr Mar Biol 33: 245-303

Fauchald K, Jumars PA (1979) The diet of the worms: a study of Polychaete feeding guilds. Oceanogr Mar Biol Annu Rev 17: 193-284

Gambi MC (1989) Osservazioni su morfologia funzionale e comportamento trofico di *Owenia fusiformis* delle Chiaje (Polychaeta, Oweniidae) in rapporto ai fattori ambientali. Oebalia 15: 145-155

Gentil F, Dauvin JC (1989) Etude allométrique de la Polychète *Owenia fusiformis* Delle Chiaje. J Rech Océanogr 1: 58-60

Ménard F, Gentil F, Dauvin JC (1989) Population dynamics and secondary production of *Owenia fusiformis* Delle Chiaje (Polychaeta) from the Bay of Seine (eastern English Channel). J Exp mar Biol Ecol 133: 151-167

Moodley L, Heip CHR, Middelburg JJ (1998) Benthic activit in sediments of the Northwest Adriatic Sea: sediments oxygen consumption, macro and meiofauna dynamics. J Sea Res 40: 263-280

Montanari G, Ghetti A, Ferrari CR (1998) Eutrofizzazione delle acque costiere dell'Emilia Romagna. Rapporto annuale 1997, Regione Emilia Romagna, Assessorato Territorio, Programmazione e Ambiente, ARPA, 220 pp

Nilsen R, Holthe T (1985) Arctic and Scandinavian Oweniidae (Polychaeta) with a description of *Myriochele fragilis* and comments on the phylogeny of the family. Sarsia 70: 17-32

Olivier F, Desroy N, Retière C (1996) Habitat selection and adult-recruit interactions in *Pectinaria koreni* (Malmgren) (Annelida, Polychaeta) post-larval populations – results of flume experiments. J Sea Res 36: 217-226

Rinaldi A, Montanari G, Ghetti A, Ferrari CR, Mazziotti C (1997) Eutrofizzazione delle acque costiere dell'Emilia Romagna. Rapporto annuale 1996, Regione Emilia Romagna, Assessorato Territorio, Programmazione e Ambiente, ARPA, 234 pp

Thiébaut E, Cabioch L, Dauvin JC, Retière C, Gentil F (1997) Spatiotemporal persistence of the *Abra alba* - *Pectinaria koreni* muddy fine sand community of the eastern Bay of Seine. J Mar Biol Ass UK 77: 1165-1185

Vollenweider RA, Rinaldi A, Montanari G (1992) Eutrophication structure and dynamics of a marine coastal system. Results of a ten-year monitoring along the Emilia-Romagna coast (Northwest Adriatic Sea). In: Vollenweider R A, Marchetti R, Viviani R (eds) Marine Coastal Eutrophication. Elsevier, Amsterdam, pp 63-106

Woodin SA, Merz RA (1987) Holding on by their hooks: anchors for worms. Evolution 41: 427-432

Comparison between Modern Clustering Techniques and Original Macrobenthic Community Classification on A. Vatova's Adriatic Data Set (1934-36)

P. Di Dato[1], E. Fresi[1], and M. Scardi[2]

ABSTRACT

We carried out a comparison between the results of several clustering procedures and the zoocoenoses defined by A. Vatova after his studies on the Adriatic macrozoobenthos (1934-36). The original data set was analysed by means of hierarchical and non-hierarchical clustering algorithms, with and without contiguity constraint, as well as with different combinations of distance/dissimilarity indexes and data normalization. None of the partitions obtained from these procedures matched satisfactorily the original classification, which had probably been based on information that is not available in the numerical abundance data reported by the Author. Therefore, we recommend that only binary information from this data set should be taken into account for further analyses involving recent data sets.

Introduction

The most comprehensive study on the benthic fauna of the Adriatic Sea is certainly the one carried out from 1934 to 1936 by Aristocle Vatova, an Italian marine biologist.

During four cruises he collected more than four hundred macrozoobenthos samples in 390 stations, both in the central and in the northern part of the Adriatic basin, using Petersen grabs (0.2 or 0.1 m^2). Almost three hundred species were identified and their abundance was reported both as number of individuals and as biomass. Qualitative information about the characteristics of the sea floor in each station was also reported (Vatova 1949).

On the basis of his data, A. Vatova defined fourteen different zoocoenoses. He classified the stations mainly according to a dominance criterion and named the zoocoenoses after their dominant species. This kind of classification, inherently subjective, was performed in order to outline large areas where the composition of the benthic fauna showed similar features.

The Author's concept of zoocoenosis was close to the one expressed by Clements and Shelford (1939), since he considered zoocoenoses as real and very sensitive entities, which were supposed to adapt to environmental changes by modifying their species composition.

The present paper provides a methodological framework for the exploitation of Vatova's data in view of future comparisons, with the objective to know:

1. If and how these data can be used as an historical reference for more recent data set.
2. Which technique provides results that more closely approximate Vatova's point of view in the description of the community.

In order to verify to what extent Vatova's classification was supported by the macrozoobenthos data, we compared it with the results of modern clustering techniques, which should provide, from a theoretical point of view, a more objective way to describe the spatial heterogeneity of the macrozoobenthic communities than as recorded more than 60 years ago. Of course, the purpose of our work is not to criticize Vatova's subjective approach to the classification of zoocoenoses. On the contrary, we think that Vatova's classification summarized valuable information and we want to check the possibility of comparing it to the classifications that can be obtained by means of the objective

[1] Dipartimento di Biologia, Università di Roma "Tor Vergata", Via della Ricerca Scientifica, 00133 Roma, Italy
[2] Dipartimento di Zoologia, Università di Bari, Via Orabona 4/A, 70125 Bari, Italy

F.M. Faranda, L. Guglielmo, G. Spezie (eds)
Mediterranean Ecosystems: Structures and Processes

algorithms that are used by present-day ecologists.

A new analysis of this data set is also useful because it provides a baseline for future studies on recent data aimed at effectively assessing the ecological changes that have occurred during this lapse of time.

Materials and Methods

Data analysis was carried out on a subset of the original data set. The subset included 256 stations and 274 taxa, as all the stations located above the 50 m isobath amidst the Dalmatian islands were excluded from the data analysis. These stations were not taken into account both because their benthic communities are much more heterogeneous in space than in other areas of the Adriatic basin and because the majority of the newer data sets focused on the central and western parts of the Adriatic Sea. The analysed subset included only nine of the zoocoenoses defined by Vatova (1949), which are listed in Table 1 and showed in Fig. 1.

Several clustering procedures were applied to these data, both hierarchical and non-hierarchical, and their results were compared with Vatova's classification. Only nine cluster partitions were considered, i.e. partitions in which the number of clusters matched the number of Vatova's zoocoenoses.

Hierarchical clustering was carried out with and without spatial contiguity constraint (Legendre and Legendre 1984; Legendre 1987), using the complete linkage algorithm on Euclidean, Manhattan (i.e. city block) and Canberra (Lance and Williams 1966) distance matrices and on a Jaccard (1900, 1901, 1908) dissimilarity matrix. All these distance and dissimilarity coefficients are presented in Table 2.

When applied, the contiguity constraint was defined on the basis of a Delaunay triangulation,

Table 1. Names of the zoocoenoses in the data subset. Each zoocoenosis is labelled with a different capital letter. Letters G and J were not used to avoid confusion in the following figures with letters C and I, respectively

A	*Amphioxus*
B	*Chione gallina (Chamelea gallina)* and local facies
C	*Lima hians*
D	*Nucula profunda*
E	*Schizaster chiajei* and local facies
F	*Syndosmya alba*
H	*Turritella*
I	*Tellina*
L	*Turritella profunda*

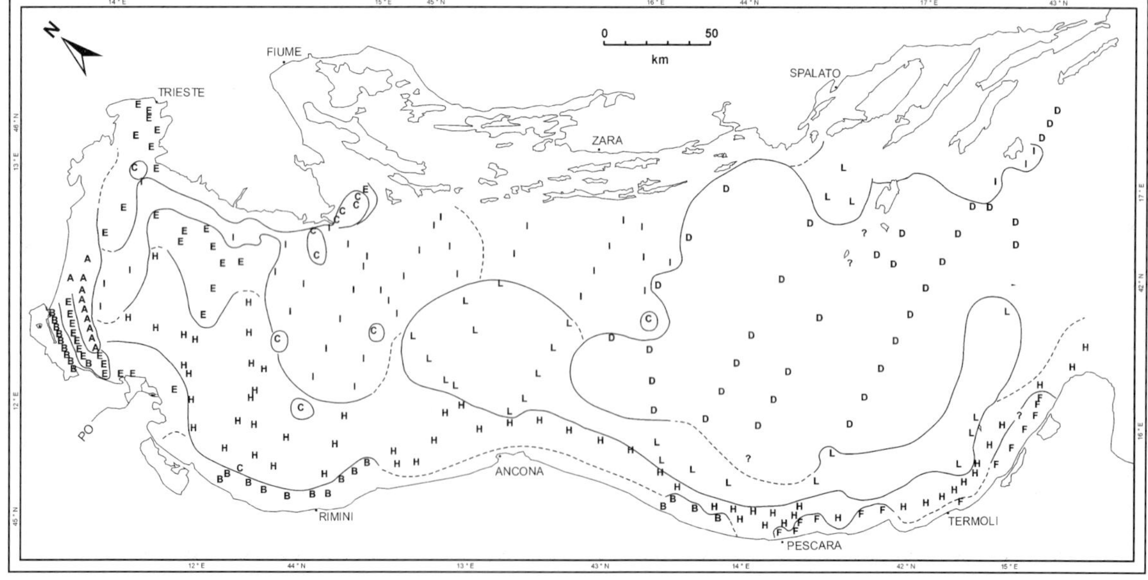

Fig. 1. The zoocoenoses defined by Vatova (1949) are labelled with a capital letter (see Table 1). Only four stations were not assigned to a zoocoenosis and are labelled with a question mark. *Continuous* and *dashed* boundaries between zoocoenoses were reproduced from Vatova's original drawings

Table 2. Distance and dissimilarity coefficients used for hierarchical clustering. D_{jk} is the distance or dissimilarity between stations j and k, considering p taxa, and x_{ij} and x_{ik} are the abundances of the i-th taxon in station j and k, respectively. In the Jaccard dissimilarity formula, a is the number of taxa that are present both in station j and k, whereas b and c are the number of taxa that are present in station j and k only, respectively

Euclidean distance	$D_{jk} = \sqrt{\sum_{i=1}^{p}\left(x_{ij} - x_{ik}\right)^2}$
Manhattan distance	$D_{jk} = \sum_{i=1}^{p}\left\lvert x_{ij} - x_{ik}\right\rvert$
Canberra distance	$D_{jk} = \sum_{i=1}^{p}\frac{\left\lvert x_{ij} - x_{ik}\right\rvert}{\left(x_{ij} = x_{ik}\right)}$
Jaccard dissimilarity	$D_{jk} = 1 - \frac{a}{a+b+c}$

i.e. all the stations that were Thiessen neighbours[1] were considered as connected.

All the distances and dissimilarities were computed using both absolute and relative abundance, i.e. using numerical abundance data as reported by Vatova (1949) and after normalization with respect to the station total, respectively.

The non-hierarchical clustering was performed on relative abundance data only and was based on a k-means algorithm (Sneath and Sokal 1973).

The comparison between Vatova's zoocoenoses and the partitions defined by the clustering procedures was performed by analysing 9 x 9 contingency tables in which row entries corresponded to Vatova's zoocoenoses and column entries corresponded to clusters of stations. Four stations, which had not been assigned to any zoocoenosis, were excluded from this analysis (they are labelled with a question mark in Fig. 1).

As a first step, all the expected values in the table were computed, i.e. the number of matches that should occur if the two classification criteria were independent of each other. Then, the largest positive deviation from expectation was selected and its row and column defined the most likely matching between the two classification criteria. All the values in the row and column pair were then excluded from the subsequent searches and other matches were defined in the same way on the remaining rows and columns of the table. An example of this matching procedure is shown in Table 3, where the largest positive deviations from expectation are evidenced by a gray background (it is important to notice that each of these cells is on a different row-column crossing).

It was not possible to perform statistical tests of independence on the 9 x 9 contingence tables because of the large number of either null or

Table 3. Vatova's zoocoenoses are compared to the partition obtained by clustering. The *gray cells* indicate the most likely matches between zoocoenoses and clusters (see text for explanation). This example refers to the hierarchical clustering with no spatial contiguity constraint performed on the relative abundance Manhattan distance matrix

Clusters		A	B	C	D	E	F	H	I	L
	A	−0.83	7.21	−1.19	−1.51	−0.56	−0.20	−2.26	0.48	−1.15
	B	9.62	−1.90	−2.86	−3.62	−1.33	−0.48	−3.43	−1.24	5.24
	C	−2.19	0.05	1.57	−0.81	−0.67	0.76	−2.71	5.38	−1.38
	D	−5.39	−2.78	0.83	−5.28	1.06	0.31	11.08	−1.81	1.97
zoocoenoses	E	−6.02	0.38	2.07	−0.98	3.17	0.35	−3.46	−0.70	5.20
	F	7.63	−1.03	−1.55	−1.96	−0.72	−0.26	−1.94	−0.67	0.50
	H	−5.77	−4.68	1.98	21.10	−2.28	−0.17	−1.35	−3.04	−5.79
	I	5.52	3.75	−0.88	−3.18	−2.28	0.19	0.73	−1.12	−2.72
	L	−2.56	−0.98	0.02	−3.77	3.61	−0.50	3.35	2.71	−1.88

[1] Given two points A and B, they are Thiessen neighbours if the circle whose radius is AB does not contain other points. A Delaunay triangulation can be obtained by connecting all the Thiessen neighbours

very low values. However, the overall degree of matching of each partition with Vatova's zoocoenoses was expressed according to a relative scale, where a unit value indicated a perfect matching. This "matching index" was obtained dividing the sum of the deviations from the expected frequencies of the matched zoocoenosis/cluster pairs by the value that would have been obtained in the case of perfect matching (i.e. comparing Vatova's zoocoenoses with themselves). It should be stressed that this index takes into account the effect of the number of items in each cluster when evaluating the degree of matching to the original classification, whereas the raw number of matches does not.

Results and Discussion

The comparison of the results of the different clustering procedures with the Vatova's zoocoenoses is summarized in Table 4. In spite of the variety of combinations of clustering techniques, distance indexes, data transformations and spatial constraints, none of the new partitions matched the original classification closely enough.

The "less independent" result was obtained by hierarchical clustering performed with no spatial constraints on a Manhattan distance matrix that was computed using relative abundance data (i.e. raw data divided by the station total abundance). The value of its "matching index" was 0.281, which is not only much lower than the theoretical maximum, but also quite close to the minimum values obtained by clustering approaches that are absolutely unrelated to Vatova's dominance criterion (e.g. those based on the binary Jaccard index). The best matching partition is shown in Fig. 2, where a gray circular background marks the stations (91 out of 252) that belong to clusters that correctly match Vatova's zoocoenoses. It is evident that most of the matching stations are close to the coastline. Deeper stations, which are usually more homogeneous in macrozoobenthos community structure, showed a less regular classification pattern.

If the results of the different clustering techniques are closely examined with respect to Vatova's dominance criterion in defining zoocoenoses, however, their ranking is not different from the expectation. In fact, partitions based on relative abundance performed better than those based on absolute abundance, which are more sensitive to sampling errors and to occasional maxima in species densities. Moreover, the binary Jaccard's index, which was used to represent the low end of the matching probability, provided the second and third lowest values of the "matching index".

Table 4. Comparison between station partitions obtained by different clustering procedures and by Vatova's zoocoenosis classification. The clustering procedures are ranked according to the "matching index", that summarizes the deviation from independence of the two classification criteria

Clustering criterion	Distance or dissimilarity	Contiguity constraint	Abundance coding	Number of matches	Matching index
Hierarchical	Manhattan	No	Relative	91	0.281
Hierarchical	Manhattan	Yes	Relative	83	0.257
Hierarchical	Canberra	Yes	Absolute	95	0.251
Non-hierarchical	Euclidean	No	Relative	74	0.245
Hierarchical	Canberra	No	Relative	83	0.221
Hierarchical	Euclidean	Yes	Relative	75	0.216
Hierarchical	Euclidean	No	Relative	56	0.163
Hierarchical	Canberra	No	Absolute	73	0.156
Hierarchical	Manhattan	No	Absolute	74	0.152
Hierarchical	Canberra	Yes	Relative	65	0.148
Hierarchical	Euclidean	No	Absolute	78	0.104
Hierarchical	Manhattan	Yes	Absolute	52	0.102
Hierarchical	Jaccard	No	Binary	66	0.077
Hierarchical	Jaccard	Yes	Binary	67	0.058
Hierarchical	Euclidean	Yes	Absolute	59	0.027

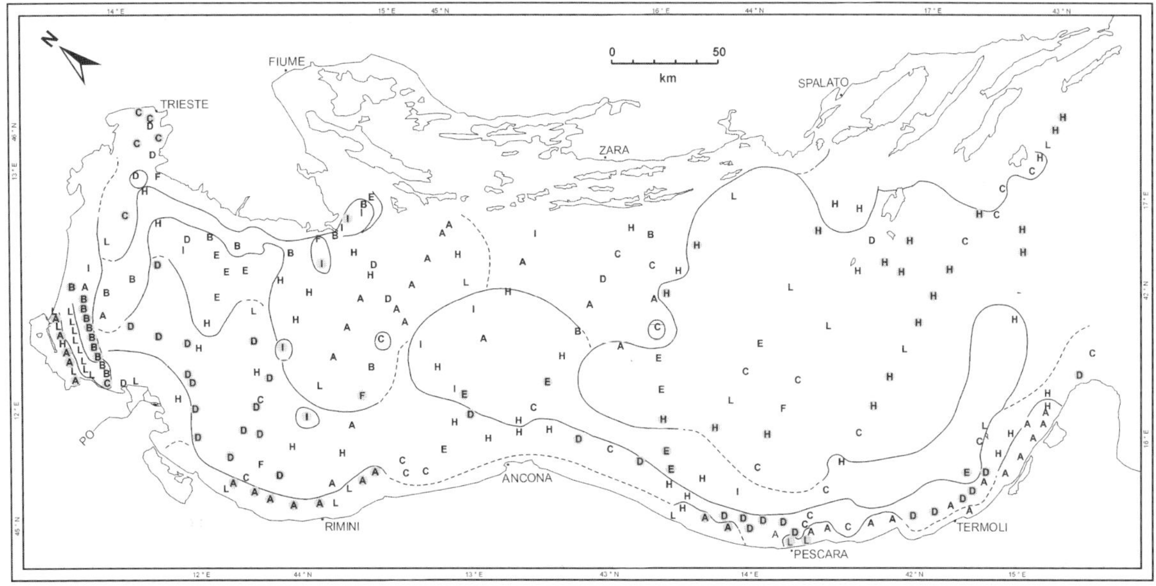

Fig. 2. The partition obtained by hierarchical clustering with no contiguity constraint from Manhattan distances computed on relative abundance data was the most similar to Vatova's classification (see Table 2, first line). A *gray circular background* marks the stations (91 out of 252) that belong to clusters that correctly match the Vatova's zoocoenoses

In conclusion, our results clearly indicate that Vatova's classification of Adriatic Sea zoocoenoses was defined using criteria that were not supported only by the information contained in the species abundance data set. If they had been, it should have been possible to approximate the original classification, but none of the clustering procedures we used succeeded in matching it.

This evidence implies that Vatova's data set does not include the whole set of descriptors and additional information that he used to classify zoocoenoses. Therefore, particular care is needed when Vatova's macrozoobenthos data are to be compared with more recent data. The only safe strategy in these cases is the one that takes into account the minimum level of information that is contained in Vatova's data set, i.e. presence data. The latter are more reliable as they are less influenced by the criteria that were used to collect samples and to record macrozoobenthos data.

References

Clements FE, Shelford VE (1939) Bio-ecology. Wiley, New York

Jaccard P (1900) Contribution au problème de l'immigration post-glaciaire de la flore alpine. Bull Soc Vaudoise Sci Nat 36: 87-130

Jaccard P (1901) Etude comparative de la distribution florale dans une portion des Alpes et du Jura. Bull Soc vaudoise Sci Nat 37: 547-579

Jaccard P (1908) Nouvelles recherches sur la distribution florale. Bull Soc Vaudoise Sci Nat 44: 223-270

Lance GN, Williams WT (1966) Computer programs for classification. In: Proc ANCCAC Conf, Canberra, May 1966, Pap 12/3

Legendre P (1987) Constrained clustering. In: Legendre P, Legendre L (eds) Developments in numerical ecology. (NATO ASI Series, vol. G14) Springer-Verlag, Berlin Heidelberg New York Tokyo, pp 289-307

Legendre P, Legendre V (1984) Postglacial dispersal of freshwater fishes in the Québec peninsula. Can J Fish Aquat Sci 41: 1781-1802

Sneath PHA, Sokal RR (1973) Numerical taxonomy – the principles and practice of numerical classification. In: Freeman WH (ed) San Francisco, XV+ pp 573

Vatova A (1949) La fauna bentonica dell'Alto e Medio Adriatico. Nova Thalassia 3: 3-110

CHAPTER 36

Deep-sea (250-1,550 m) Benthic Thanatocoenoses from the Southern Tyrrhenian Sea

I. Di Geronimo, A. Rosso, R. La Perna, and R. Sanfilippo

ABSTRACT

Bathyal benthic thanatocoenoses from 28 stations in the Southern Tyrrhenian Sea are studied, focusing on molluscs, bryozoans and serpulids. On the basis of faunal affinities, sampled stations were grouped into three bathymetrical belts: "A" (248-505 m), "B" (696-1,096 m) and "C" (1,139-1,536 m). Major faunal changes occurred between 500 and 600 m depths (i.e. between belts A and B), due to changes in the dominant molluscs (mainly *Benthonella tenella*, *Ennucula corbuloides*, *Ledella messanensis*, *Yoldiella micrometrica* and *Katadesmia cuneata*), bryozoans (e.g. *Tubulipora* sp. 1 and *Setosella folini*) and serpulids (e.g. *Filogranula stellata* and *Filograna* sp. 1), which were absent in belt A, and increased in abundance in the deepest belt (C). Faunas from stations below 500-600 m may define the typical Mediterranean deep bathyal assemblages. Differences between Pleistocene and Recent bathyal Mediterranean benthos are discussed. We hypothesise that the latter is "residual", i.e. composed mainly by species that endured changes that occurred from the psycrospheric Plio-Pleistocene to the warm-homothermic Recent conditions.

Introduction

The Mediterranean Sea is a threshold-basin whose peculiar hydrology is reflected in a poor deep-sea biota, in terms of both species richness and density (or biomass; Fredj and Laubier 1985). These features, coupled with a high spatial heterogeneity (Harmelin and d'Hondt 1993), represent problems in studying the deep Mediterranean communities. Dead assemblages (i.e. thanatocoenoses) from deep-sea bottoms are generally rather abundant and diversified. The origin of such thanatocoenoses cannot be referred to Atlantic "pseudopopulations" which, according to Bouchet and Taviani (1992), would inflow *via* Gibraltar as meroplankton, without reaching fertile conditions in adult populations settled into the Western-Mediterranean. Although larval inflow is likely, and "hard parts" of organism remains of Atlantic species (dating back to the Last Glacial) are present (Di Geronimo and Li Gioi 1981; Bouchet and Taviani 1989; Rosso 1990), most deep-sea thanatocoenoses should be referred to Recent and biogeographically autochthonous communities.

Studies on the deep Mediterranean macrobenthos are mainly taxonomic, often included in research focused on North Atlantic faunas. Few data are available on the deep-sea assemblages and the distribution pattern of species. Pérès and Picard (1964), Carpine (1970), Di Geronimo (1974) and Janssen (1989), mainly focusing on molluscs, have attempted to recognise depth-related assemblages.

In the present investigation, three benthic groups (molluscs, bryozoans and serpulids) were considered. Thanatocoenoses come from 25 cm deep cores from soft bottoms with high sedimentation rates. According to literature data from the westernmost Mediterranean (Loubere 1982) these thanatocoenoses can be assumed as being only two-three thousand years old and then wholly Holocene in age.

The present investigation focuses on composition and depth range of bathyal thanatocoenoses from the Southern Tyrrhenian Sea, discussing them within the evolution model of the deep Mediterranean benthos.

Dipartimento di Scienze Geologiche, Sezione di Oceanologia e Paleoecologia, Università di Catania, Corso Italia 55, 95129 Catania, Italy

F.M. Faranda, L. Guglielmo, G. Spezie (eds)
Mediterranean Ecosystems: Structures and Processes

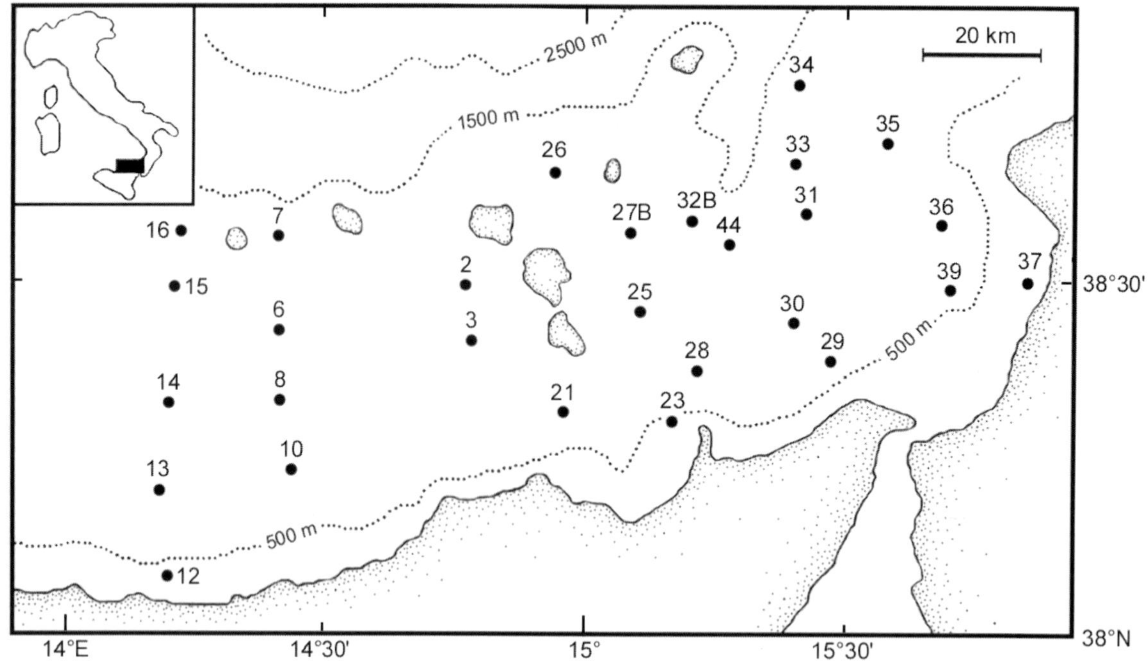

Fig. 1. Location of sampled stations (st)

Material and Methods

The investigated area (Fig. 1) is located in the Southern Tyrrhenian Sea, between the Aeolian Archipelago and the north-eastern coast of Sicily. The western part (Cefalù Basin) has a wide flat bottom, ca. 1,500 m deep and abrupt edges, while the eastern one (Gioia Basin) gradually declines to North and it is grooved by canyons.

Sediment samples (Table 1) were collected by means of a box-corer during the EOCUMM95 Cruise (summer 1995), onboard of "Ammiraglio Magnaghi" (Rosso and Corselli 1996). Sampled volume per station was ca. 6.5 dm^3 (core sampling surface 19x14 cm, depth 25 cm). The sediment was washed through 1 mm, 500 µm and 250 µm sieves.

Benthic assemblages from 28 stations were studied. Analysis was carried out on the fraction coarser than 500 µm, focusing on molluscs, bryozoans and serpulids.

Each core was preliminarily vertically subdivided into different layers, according to sediment colour and/or texture changes. No remarkable faunal difference was recorded through the layers, and each core was then treated as a single sample.

Specimens were scored as follows: whole

Table 1. Coordinates and depths of sampled stations

Station	Coordinates	(N, E)	Depth (m)
2B	38°30'27"	14°41'42"	1355
3	38°24'46"	14°44'57"	1410
7	38°34'09"	14°25'01"	1197
8	38°27'53"	14°23'10"	1510
9	38°20'08"	14°23'10"	1511
10	38°13'27"	14°24'50"	1357
12	38°04'40"	14°11'05"	411
13	38°12'54"	14°09'29"	1510
14	38°20'08"	14°10'47"	1139
15	38°29'06"	14°11'39"	1521
16	38°33'42"	14°13'59"	1340
21	38°18'00"	14°56'00"	1094
23	38°17'06"	15°10'06"	505
25	38°26'08"	15°04'55"	1018
26	38°38'30"	14°57'53"	945
27B	38°32'26"	15°05'02"	1096
28	38°20'40"	15°15'30"	820
29	38°23'47"	15°27'04"	897
30	38°26'49"	15°22'39"	1050
31	38°32'03"	15°28'18"	1061
32B	38°33'53"	15°05'33"	1219
33	38°35'23"	15°25'40"	875
34	38°42'36"	15°28'06"	1536
35	38°38'32"	15°38'21"	696
36	38°33'52"	15°41'12"	724
37	38°29'20"	15°50'19"	248
39	38°27'33"	15°34'26"	786
44	38°32'25"	15°18'53"	1210

shells, valves, apical and umbonal fragments for molluscs, whole colonies and colony fragments for bryozoans, whole tubes and tube fragments for serpulids. Based on faunal data, stations were clustered into three bathymetric groups (see Results). For each station group, three values are reported: *F* (species frequency) = number of stations in which a species occurs; *N* (number of specimens) = total amount of specimens of a species; *mA* (mean abundance) = number of specimens/number of stations within a bathymetric group.

Results

Except for samples 25 and 27B, that contained a remarkable sandy fraction (Amore et al. 1996), sediments were muddy (silty clay). Pumice (sands and small pebbles) occurred in the western part of the Gioia Basin. Siliciclastics (quartz, micas) and *Posidonia* fibres prevailed in the nearshore stations of both basins. Planktonic remains were encountered in the deepest and most distal stations.

Assemblages consisted only of dead specimens, although most of them were fresh-looking and some bivalves have joined valves and periostracum (particularly those from the core top).

The studied assemblages included 93 species: 76 molluscs (Table 2), 13 bryozoans (Table 3) and 4 serpulids (Table 4). Some additional species (20 bryozoans and 2 serpulids, proved to be downslope transported, being mainly associated with *Posidonia* grasses. Thus, they were excluded from the present study.

Molluscs were present in all samples and dominated assemblages (relative abundance 38.8-100%, mean 83.8%), whereas bryozoans and serpulids were scarce and present in 21 samples.

Molluscan assemblages fit the general deep-sea pattern (Rex 1981; Hickman 1984; Allen and Sanders 1996). Many typical deep-water genera were present, such as *Benthonella, Microgloma, Ledella, Yoldiella, Katadesmia, Bathyarca, Kelliella, Entalina*, etc. Main taxonomic groups were protobranchs (mean relative abundance 29.3%), gastropods (28.3%, mostly by the mesogastropod *Benthonella tenella*), heterodonts (16.5%, mostly by *Kelliella abyssicola* and some

Table 2. Mollusca: systematic list (see text for explanation)

	248-505 m 37, 12, 23		"A"	696-1096 m 35, 36, 39, 28, 33, 29, 26, 25, 30, 31, 21, 27B		"B"	1139-1536 m 14, 7, 44, 32B, 16, 2B, 10, 3, 8, 13, 9, 15, 34		"C"
	F	A	mA	F	A	mA	F	A	mA
Gastropoda									
Bathysciadium sp.							4	6	0.46
Copulabyssia corrugata (Jeffreys)		1	0.33	3	15	1.25	1	1	0.08
undet. cocculiniform					1	1	0.08		
Akrytogira conspicua (Monterosato)							3	3	0.23
Adeuomphalus ammoniformis (Seguenza				1	1	0.08			
Alvania cimicoides (Forbes)	2	4	1.33						
Alvania elegantissima (Monterosato)		1	0.33	5	24	2.00	2	1	0.08
Alvania subsoluta (Aradas)				5	7	0.58	1	8	0.62
Alvania testae (Aradas and Maggiore)	1	12	4.00						
Benthonella tenella (Jeffreys)				12	159	13.25	14	478	36.77
Talassia daugeneti (de Folin)	1	1	0.33						
Graphis gracilis (Monterosato)							1	1	0.08
Pagodula echinata (Kiener)	1	1	0.33	5	8	0.67	1	1	0.08
Trophonopsis muricata (Montagu)	1	5	1.67						
Nassarius lima (Dillwin)	3	16	5.33	3	7	0.58			
Granulina minusculina (Locard)	1	1	0.33						
Taranis moerchi (Malm)	1	1	0.33						
Drilliola loprestiana (Calcara)	1	1	0.33						
Drilliola emendata (Monterosato)				1	1	0.08			
Benthomangelia macra (Watson)				2	2	0.17	5	7	0.54
Lusitanops sp.							1	1	0.08
Melanella sp.				1	1	0.08			

(cont.)

Table 2. (cont.)

	248-505 m 37, 12, 23		"A"	696-1096 m 35, 36, 39, 28, 33, 29, 26, 25, 30, 31, 21, 27B		"B"	1139-1536 m 14, 7, 44, 32B, 16, 2B, 10, 3, 8, 13, 9, 15, 34		"C"
	F	A	mA	F	A	mA	F	A	mA
Haliella tyrrhena (Di Geronimo and La Perna)				1	1	0.08			
Pyramidella minuscula (Monterosato)	1	1	0.33						
Pyramidella octaviana (Di Geronimo)	1	1	0.33						
Chrysallida flexuosa (Monterosato)	2	5	1.67	1	1	0.08			
Turbonilla sp.				2	2	0.17			
Eulimella scillae (Scacchi)							2	2	0.15
Eulimella ventricosa (Forbes)	2	2	0.67	5	5	0.42	2	2	0.15
Odostomia sp.	2	2	0.67						
Tjaernoeia exquisita (Jeffreys)	1	1	0.33						
Acteon monterosatoi (Dautzenberg)	1	1	0.33						
Crenilabium exile (Jeffreys)	1	1	0.33	2	2	0.17			
Ringicula leptocheila (Brugnone)	1	5	1.67						
Roxania abyssicola (Dall)	1	2	0.67	1	1	0.08			
Diaphana cf. *marshalli* (Sykes)							2	2	0.15
Bivalvia									
Nucula sulcata (Bronn)	1	1	0.33						
Ennucula aegeensis (Forbes)	2	25	8.33	1	1	0.08			
Ennucula corbuloides (Seguenza)				12	119	9.92	13	192	14.77
Microgloma tumidula (Monterosato)	2	4	1.33	9	81	6.75	13	119	9.15
Saccella commutata (Philippi)	1	4	1.33						
Ledella messanensis (Jeffreys)				10	63	5.25	6	20	1.54
Yoldiella micrometrica (Seguenza)				4	28	2.33	9	106	8.15
Yoldiella nana (M. Sars)				1	1	0.08			
Yoldiella philippiana (Nyst)	1	2	0.67						
Yoldiella seguenzae (Bonfitto and Sabelli)							4	23	1.77
Katadesmia cuneata (Jeffreys)				2	3	0.25	8	52	4.00
Bathyarca pectunculoides (Scacchi)	2	34	11.33	3	7	0.58	12	51	3.92
Bathyarca philippiana (Nyst)	1	1	0.33	6	18	1.50			
Delectopecten vitreus (Gmelin)				2	3	0.25			
Cyclopecten hoskynsi (Forbes)				3	3	0.25			
undet. juv. Pectinidae							2	2	0.15
Pododesmus aculeatus (O.F. Mueller)				1	1	0.08			
Limatula gwyni (Sikes)	1	9	3.00	3	14	1.17			
Limatula sp.				1	1	0.08			
Notolimea crassa (Forbes)				8	138	11.50	5	46	3.54
Thyasira croulinensis (Jeffreys)	3	26	8.67	3	6	0.50	1	1	0.08
Thyasira ferruginea (Locard)	2	21	7.00	2	9	0.75	3	4	0.31
Thyasira flexuosa (Montagu)	1	3	1.00				1	2	0.15
Thyasira granulosa (Monterosato)				1	2	0.17			
Thyasira subovata (Jeffreys)	1	9	3.00	2	2	0.17	1	1	0.08
Abra longicallus (Scacchi)	2	63	21.00	6	28	2.33	3	11	0.85
Kelliella abyssicola (Forbes)	3	153	51.00	11	101	8.42	13	176	13.54
Xylophaga dorsalis (Turton)	1	1	0.33	6	24	2.00	8	25	1.92
Pholadomya loveni (Jeffreys)				2	2	0.17	1	2	0.15
Allogramma formosa (Jeffreys)				1	1	0.08			
Poromya granulata (Nyst and Westendorp)	1	1	0.33						
Verticordia trapezoidea (Seguenza)	2	2	0.67						
Polycordia gemma (Verrill)							2	2	0.15
Tropidomya abbreviata (Forbes)	1	1	0.33						
Cuspidaria rostrata (Spengler)	1	1	0.33						
Cardiomya costellata (Deshayes)	1	1	0.33				2	4	0.31
Scaphopoda									
Graptacme agilis (M. Sars in G.O. Sars)	1	8	2.67	9	27	2.25	5	16	1.23
Entalina tetragona (Brocchi)	2	37	12.33	12	94	7.83	8	37	2.85
Gadila subfusiforme (M. Sars)	1	1	0.33	4	8	0.67	6	22	1.69
Pulsellum lofotense (M. Sars),	1	7	2.33	3	3	0.25	1	4	0.31

Table 3. Bryozoa: systematic list (see text for explanation)

	248-505 m 37, 12, 23		"A"	696-1096 m 35, 36, 39, 33, 26, 25, 30, 31, 27B		"B"	1139-1536 m 14, 7, 32B, 16, 8, 13, 9, 15, 34		"C"
	F	A	mA	F	A	mA	F	A	mA
Bryozoa									
Crisia tenella Calvet *longinodata* (Rosso)	1	5	1.66	5	20	1.66	1	4	0.30
Anguisia verrucosa (Jullien)	3	8	2.66	9	82	6.83	7	50	3.84
Entalophoroecia gracilis (Harmelin)	1	36	12.00				2	8	0.61
Tubulipora sp. 1				4	9	0.75	7	35	2.69
Tervia irregularis (Meneghini)				1	1	0.08			
Hornera lichenoides (Linnaeus)				1	1	0.08			
Setosella folini (Jullien)				2	5	0.42	1	1	0.07
Palmicellaria cf. *elegans* (Alder)				2	2	0.16			
Gemellipora eburnea (Smitt)							1	1	0.07
Characodoma mamillatum (Seguenza)				3	5	0.42			
Tessaradoma boreale (Busk)	1	32	10.66				1	1	0.07
Jaculina tessellata (Hayward)	1	2	0.66		2	3	0.25		
Reteporella sparteli (Calvet)				5	8	0.66	5	16	1.23

Table 4. Serpulidae: systematic list (see text for explanation)

	248-505 m 37, 23		"A"	696-1096 m 35, 39, 33, 26, 25, 30, 27B		"B"	1139-1536 m 14, 7, 44, 32B, 16, 2B, 10, 3, 8, 9, 15, 34		"C"
	F	A	mA	F	A	mA	F	A	mA
Serpulidae									
Filogranula stellata (Southward)				3	8	0.66	10	70	5.38
Filograna sp. 1				3	5	0.42	6	6	0.46
Hyalopomatus variorugosus (Ben Eliahu and Fiege)	1	3	1.00	6	25	2.08	3	6	0.46
Hyalopomatus sp.	1	1	0.33	3	7	0.58	4	6	0.46

thyasirids), pteromorphs (10.5%) and scaphopods (9.1%). Many species were small-sized and thin-shelled and three dominant bivalves were micromorphic, i.e. *Kelliella abyssicola*, *Microgloma tumidula* (one of the smallest bivalves; Ockelmann and Warén 1998) and *Yoldiella micrometrica*. It is worth noting the presence of two species, *Xylophaga dorsalis* and *Copulabyssia corrugata*, known from organic substrates (sunken wood) (Janssen 1989).

Typical deep-sea genera are also the bryozoans *Anguisia*, *Gemellipora* and *Jaculina* (Lagaaij and Cook 1973; Hayward 1981). All the species have small-sized colonies, mostly with uni-to-pauciserial branches. Erect rigid morphoses (vinculariforms) largely prevail and only two species have flexible morphoses. Colonies often encrust small pteropod fragments or other bryozoans but they were rarely found on the pumice pebbles. Two species appeared specially adapted to fine-grained bottoms, i.e. *Characadoma mamillatum* with rod-like colonies bearing rhizoids (cellarinelliform, Rosso 1999), and *Setosella folini* exhibiting uniserial, scorpioid, probably free-living colonies.

Among serpulids, *Hyalopomatus variorugosus* and *Filogranula stellata* were the most frequent. *Hyalopomatus* is a typically bathyal genus (Zibrowius 1977). All tubes were small-sized, particularly of *H. variorugosus*, which did not exceed 200 µm in diameter. This species was generally present in samples with coarse sediments, where it was mainly found on pumice pebbles, sometimes cryptic within small cavities (Sanfilippo 1998).

The Cefalù Basin assemblages are slightly poorer than those from the Gioia Basin, where the serpulid *Hyalopomatus variorugosus* is markedly abundant, due to the presence of pumices.

On the basis of faunal affinities, stations were grouped into three bathymetric belts: "A" (248 to 505 m, 3 stns), "B" (696 to 1,096 m, 12 stns) and "C" (1,139 to 1,536 m, 13 stns) (Tables 2-4).

Several species were notably abundant and/or frequent all through the investigated depth range, such as the molluscs *Alvania elegantissima*, *Microgloma tumidula*, *Thyasira croulinensis*, *T. ferruginea*, *T. subovata*, *Abra longicallus*, *Graptacme agilis* and *Entalina tetragona*, the bryozoans *Crisia tenella longinodata*, *Anguisia verrucosa*, *Entalophoroecia gracilis* (deep-water morphotype) and the serpulid *Hyalopomatus variorugosus*. These species have a wide bathymetric distribution, starting from ca. 200 m. Other common species, *Bathyarca pectunculoides*, *Kelliella abyssicola* and *Xylophaga dorsalis*, displayed an even wider distribution, being also known from the mid-outer shelf. All these species can generally be regarded as eurybathic.

The molluscs *Alvania cimicoides*, *A. testae*, *Ringicula leptocheila*, *Ennucula aegeensis*, *Yoldiella philippiana* and *Verticordia trapezoidea* are only (or mostly, see *E. aegeensis*) present in the shallowest belt (A). They are actually known from the outer shelf to the upper bathyal, becoming less and less common with depth (e.g. Bouchet and Warén 1993; Salas 1996). Also the bryozoan *Tessaradoma boreale* (mostly found in the A belt) shared a similar distribution, although in the Atlantic it is much deeper (Rosso and Di Geronimo 1998). It is also worth noting the occurrence of some species, i.e. the molluscs *Nucula sulcata*, *Saccella commutata*, *Nassarius lima* and *Trophonopsis muricata*, typical of circalittoral communities and extending beyond the shelf break. Other molluscan species were restricted to the shallowest belt, or they partially extended to the deeper ones, but they were too scarce (or their distribution is poorly known) to be further commented upon.

Many species were abundant and/or frequent from 600-700 m to the deepest stations (belts B and C): the molluscs *Alvania subsoluta*, *Benthonella tenella*, *Benthomangelia macra*, *Ennucula corbuloides*, *Ledella messanensis*, *Yoldiella micrometrica*, *Katadesmia cuneata*, *Notolimea crassa*, the bryozoans *Tubulipora* sp. 1, *Setosella folini*, *Reteporella sparteli* and the serpulids *Filogranula stellata* and *Filograna* sp. 1. For *Reteporella sparteli*, a shallower range not exceeding 500-700 m in the Atlantic-Mediterranean area, was previously known (Hayward 1979; Harmelin and d'Hondt 1992). Differences between belts B and C are mainly due to a remarkable increase in abundance of *Benthonella tenella*, *Ennucula corbuloides*, *Yoldiella micrometrica*, *Katadesmia cuneata*, *Tubulipora* sp. 1 and *Filogranula stellata*. The nuculoid *Katademia cuneata* is one of the deepest species, its shallowest record being from station 29 (897 m) and becoming more and more common below 1,000 m.

A major faunal change occurred at ca. 500-600 m depth, where several species add to or replace the eurybathic ones. This change was more evident within molluscs. Some common species listed from belts B and C most probably define the typical deep bathyal assemblage through the Mediterranean, including the Eastern Basin (Janssen 1989). The gastropod *Benthonella tenella* is one of the most common species in 500-4,000 m in Mediterranean and Northeast Atlantic (Bouchet and Warén 1993). Di Geronimo (1974) recognised a "*Cithna* (=*Benthonella*) *tenella*" assemblage below 2,300 m in the Ionian Sea". *Ennucula corbuloides* is a common protobranch in Atlantic-Mediterranean bathyal bottoms below 500-700 m (Rhind and Allen 1992; Salas 1996, as *E. bushae*). *Ledella messanensis* is another widespread Atlantic-Mediterranean protobranch, common below 500 m (Warén 1989; Salas 1996). *Katadesmia cuneata* has an almost world-wide distribution, below 800-1,000 m (Knudsen 1970; Sanders and Allen 1985; Salas 1996). Among these common molluscan species, *Yoldiella micrometrica* deserves a special mention. It is probably present also in the Northeast Atlantic (Jeffreys 1879), where this species must be markedly rare as it was neither included by Allen et al. (1995) in their revision of Atlantic species of *Yoldiella*, nor reported by Salas (1996) from the Ibero-Moroccan Gulf. This species is present in other Tyrrhenian areas (Di Geronimo and Li Gioi 1981; Di Geronimo and Bellagamba 1985) and in the Eastern Basin (Janssen 1989), but it is unrecorded from the westernmost Mediterranean (Salas 1996). Other molluscan species from the "B" and/or "C" assemblages, such as *Copulabyssia corrugata*, *Adeuomphalus ammoniformis*, *Alvania subsoluta*, *Graphis gracilis*, *Benthomangelia macra*, *Yoldiella seguenzae*, *Delectopecten vitreus*, *Cyclopecten hoskynsi*, *Allogramma formosa*, *Polycordia gemma*, *Pholadomya loveni*, are also known from the Northeast Atlantic (e.g. Warén 1991, Bouchet and Warén 1993; Salas 1996).

The common serpulid *Filogranula stellata* shows a similar geographical and bathymetric distribution in the Northeast Atlantic and all through the Mediterranean (Zibrowius 1977; Ben Eliahu and Fiege 1996).

Among bryozoans, only *Setosella folini* is comparable in depth distribution with the species mentioned above, although it is much rarer. This species is known from a single Western Mediterranean site in 550 m (Calvet 1907). Conversely, it is frequent in the Northeastern Atlantic, including the Ibero-Moroccan Gulf, from 700 down to 3,700 m (Jullien 1882; Harmelin and d'Hondt 1992). The single shallower record (200 m) by Harmelin (1977) is from an upwelling area in the Canary Islands.

A small group of species deserves remarks about biogeography. The molluscs *Pyramidella minuscula* and *P. octaviana* (stn 37) in the Mediterranean, are only known from deep shelf and upper slope (Di Geronimo 1973). Also *Akrytogira conspicua* (stns 7, 10, 44) and *Alvania elegantissima* (stns 14, 23, 28, 29, 30, 35, 39), are only known from Mediterranean waters, the former in 100-2,400 (Warén 1992), the latter in 400-800 (Oliverio et al. 1992). *Diaphana* cf. *marshalli* (stns 8, 10) shows morphological differences from the Northeast Atlantic specimens and could be provisionally assumed as a distinct Mediterranean species. Another species (from stn 28), recently described as *Haliella tyrrhena* (Di Geronimo and La Perna in press), is very similar to an Atlantic species. Among bryozoans, *Characadoma mamillatum* (stns 35, 36, 39), an euribathyc species present with its deep morphotype, is known only from the Mediterranean (Rosso 1999). The serpulid *Hyalopomatus variorugosus*, known throughout the Mediterranean (Ben Eliahu and Fiege 1996; Sanfilippo 1998), is unknown from the Atlantic, except for a single unpublished record (see Sanfilippo 1998). It is also worth stressing the occurrence of *Gemellipora eburnea* (stn 14), a widely distributed species (Lagaaij and Cook 1973; Rosso and Di Geronimo 1998), so far unrecorded from the Recent Mediterranean.

Discussion

The deep Mediterranean faunas appear markedly impoverished compared to the Northeast Atlantic ones, especially when the numerous euribathyc species are excluded from the deep bathyal assemblages. Most euribathyc species actually show closer relations (taxonomic, ecological, etc.) to the shelf benthos, rather than to the deep-sea one. Marked differences are also evident from the Lower-Middle Pleistocene bathyal assemblages, recently studied by Barrier et al. (1996), Di Geronimo and La Perna (1996, 1997), Di Geronimo et al. (1996, 1997), Rosso and Di Geronimo (1998). The Pleistocene assemblages are more similar to the Recent Atlantic ones than to Recent Mediterranean ones. The Pleistocene assemblages comprise many species presently occurring only in the Atlantic, and extinct species with marked Atlantic, or Oceanic in general, taxonomic affinities. Atlantic species in the Mediterranean Pleistocene are the molluscs *Calliotropis ottoi* (Philippi), *Fissurisepta rostrata* Seguenza, *Puncturella noachina* (Linnaeus), *Torellia delicata* (Philippi), *Mitrolumna smithi* (Dauntzenberg and Fisher) and many others (see Di Geronimo and La Perna 1997; Di Geronimo et al. 1997), the bryozoans *Euginoma vermiformis* Jullien, *Caberea ligata* Jullien, *Scrupocellaria jullieni* Hayward, *Sertulipora guttata* Harmelin and d'Hondt (Rosso 1990; Di Geronimo et al. 1997; Rosso and Di Geronimo 1998), the serpulids *Neovernilia falcigera* (Roule) and *Vitreotubus digeronimoi* Zibrowius (Zibrowius 1979; Zibrowius and Ten Hove 1987; Di Geronimo et al. 1997). The extinct species are mostly represented by molluscs, such as *Fissurisepta papillosa* Seguenza, *Solariella marginulata* (Philippi), *Austrotindaria pusio* (Philippi), *Bathyspinula excisa* (Philippi), *Thestyleda cuspidata* (Philippi), *Katadesmia confusa* (Seguenza), *Cadulus ovulum* (Philippi), etc., all markedly abundant within the Pleistocene communities (Di Geronimo and La Perna 1997; Di Geronimo et al. 1997). The Pleistocene bryozoans *Tervia barrieri* Rosso, *Heliodoma angusta* Rosso, *Characadoma reclinatum* Rosso and *C. rostratum* Rosso have been recently described and added to the extinct stock (Rosso 1998, 1999). Most of these extinct species might have been palaeoendemics. However, most of the Recent species were also present in the Pleistocene. Preliminary data suggest a decrease in the species number of about one half.

In spite of such evident differences in composition, species dominance and diversity, the Recent molluscan assemblages are structurally comparable with the Pleistocene ones, in terms

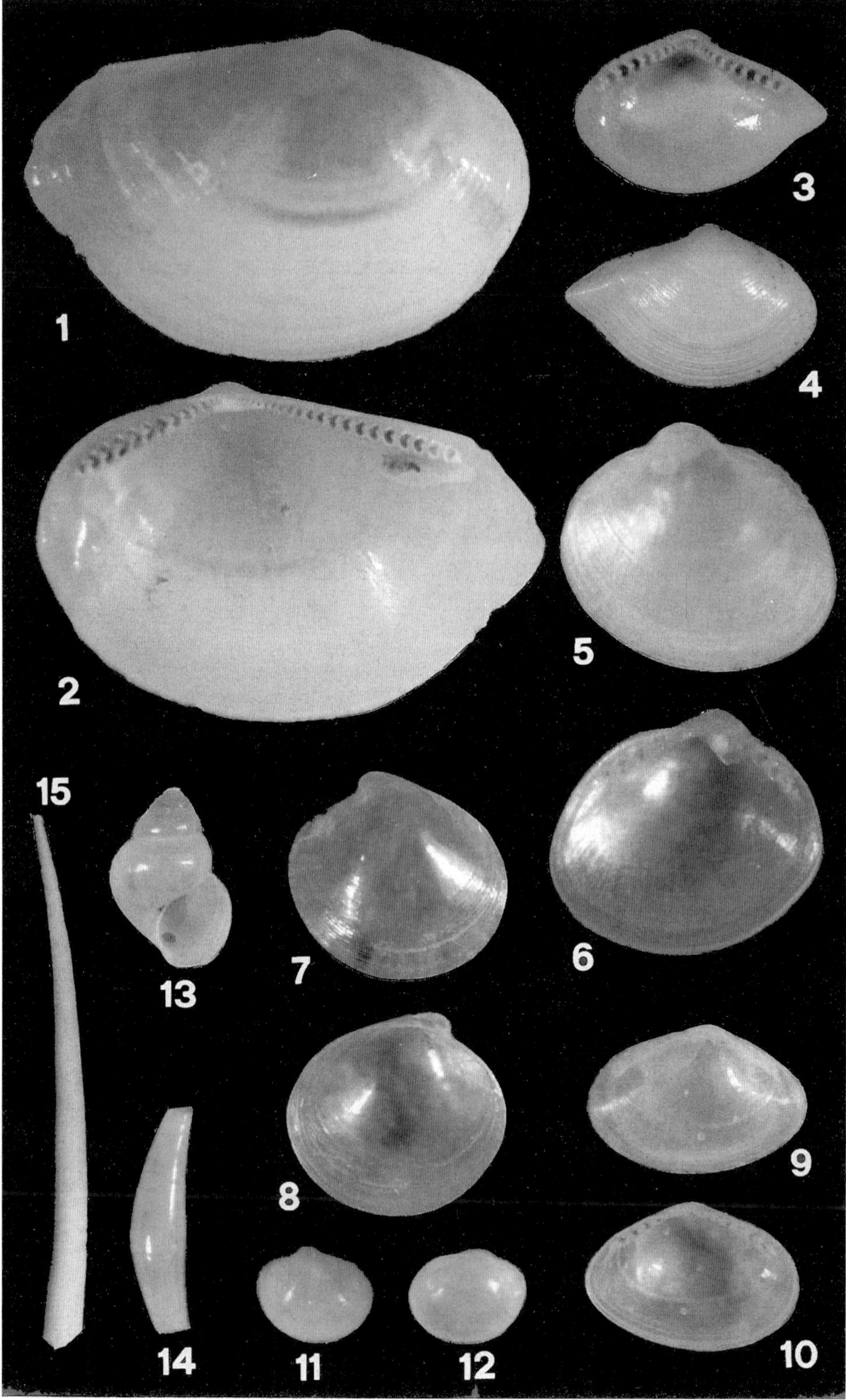

Plate 1. ***1, 2*** *Katadesmia cuneata* (Jeffreys), stn 10, shell length 7.7 mm. ***3, 4*** *Ledella messanensis* (Jeffreys), stn 35, shell length 3.8 mm. ***5, 6.*** *Ennucula corbuloides* (Seguenza), stn 10, shell length 2.0 mm. ***7, 8*** *Kelliella abyssicola* (Forbes), stn 37, shell length 1.5 mm. ***9, 10.*** *Yoldiella micrometrica* (Seguenza), stn 33, shell length 1.5 mm. ***11, 12*** *Microgloma tumidula* (Monterosato), stn 14, shell length 0.84 mm. ***13*** *Benthonella tenella* (Jeffreys), stn 16, shell height 2.6 mm. ***14*** *Gadila subfusiforme* (M. Sars), stn 14, shell height 1.6 mm. ***15*** *Graptacme agilis* (M. Sars), stn 16, shell height 23.6 mm

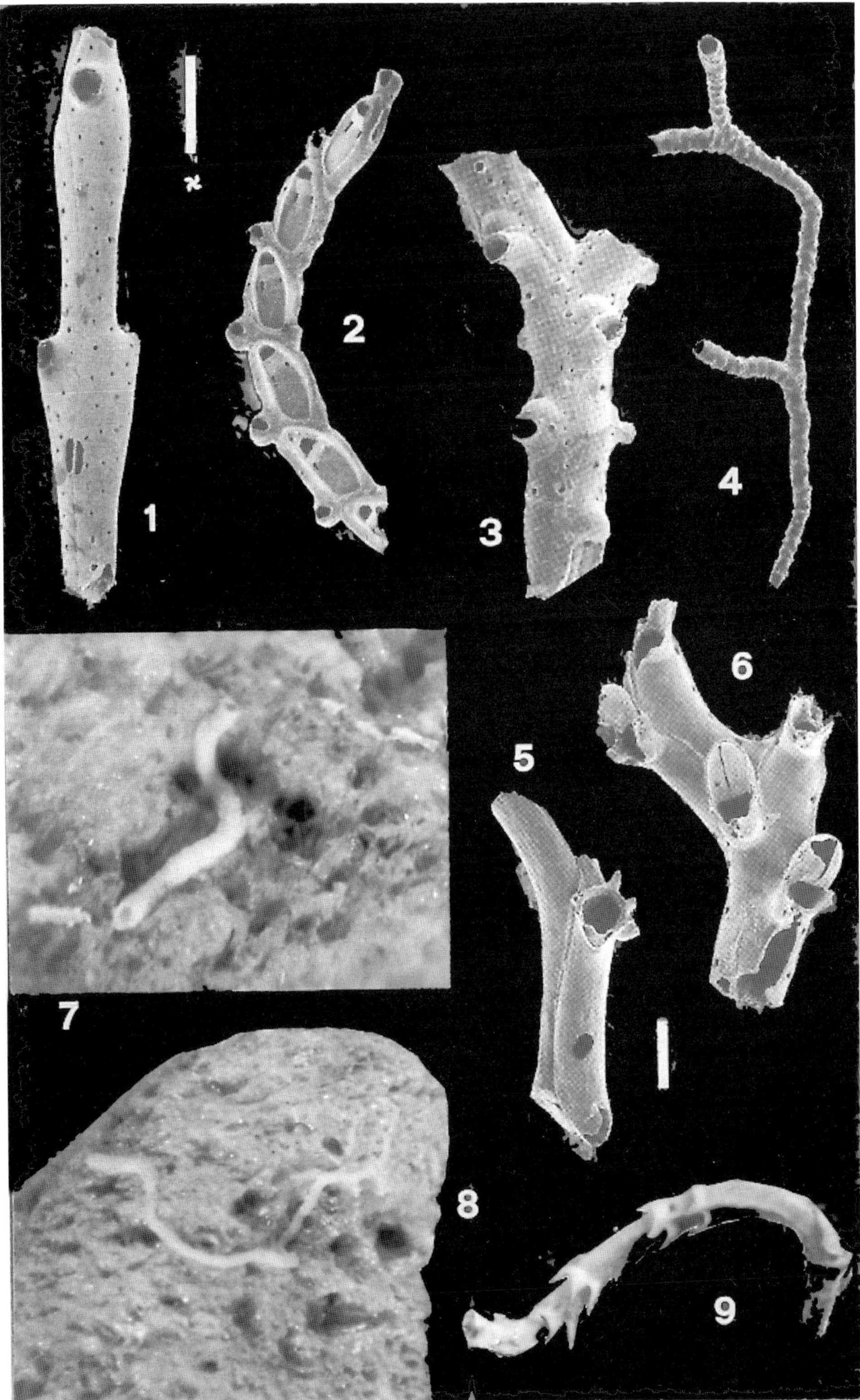

Plate 2. *1 Gemellipora eburnea* Smitt, stn 14, scale bar (*) 500 m. *2 Setosella folini* Jullien, fertile colony, stn 35, scale bar (*) 500 m. *3 Tessaradoma boreale* (Busk), fertile branch, stn 37, scale bar (*) 500 m. *4 Anguisia verrucosa* Jullien, stn 23, scale bar (*) 500 m. *5 Jaculina tessellata* Hayward, stn 27b, scale bar 250 m. *6 Reteporella sparteli* (Calvet), fertile branch, stn 7, scale bar 250 m. *7, 8 Hyalopomatus variorugosus* Ben Eliahu and Fiege on pumice, 7: stn 30, tube length 4 mm; 8: stn 27b, tube lenght 5.5 mm. *9 Filogranula stellata* (Southward), stn 9, tube length 3.5 mm

of dominant taxonomic groups (protobranchs, scaphopods, etc.) (Di Geronimo and La Perna 1997; Di Geronimo et al. 1997). Among bryozoans, a single anascan species is present in the examined material, while anascans are more diversified in the Pleistocene (Barrier et al. 1996; Di Geronimo et al. 1997). This rises the relative abundance of cyclostomatous, which parallel ascophorans as species number (46%) and exceed them (77%) at specimen level. The anascan/ascophoran ratio (0.16 and 0.09 respectively) is very low while values of about 0.6 can be valued from the Pleistocene assemblages (Rosso 1990; Di Geronimo et al. 1997).

The reported differences should be viewed within the model of changes in the deep Mediterranean benthos during the Quaternary (Di Geronimo et al. 1996; Di Geronimo and La Perna 1996, 1997; Rosso and Di Geronimo 1998). The deep-sea benthos underwent a progressive impoverishment due to the set of warm-homothermic conditions, which replaced the Plio-Pleistocene psycrospheric ones. The Gibraltar sill up-lift seems to have played a major role, turning the Mediterranean into a threshold-controlled basin, and cutting it of from the deep oceanic circulation. This led to the disappearance of most cold-stenothermic species (extinct species of Atlantic affinity, and extant Atlantic species), and to a drop in diversity and richness in general.

The "pseudopulation" model (Bouchet and Taviani 1992) does not fit the present data. The Recent deep Mediterranean benthos appear as "residual", i.e. composed of species which survived the hydrological changes. Many molluscs, usually encountered as empty shells, seem to have little chance of crossing the Gibraltar sill, as they lack long-living planktotrophic larvae and their depth range begins well beyond the sill depth (280 m). Further, several bryozoans are found as fertile colonies, testifying *in situ* reproduction.

Acknowledgements. Thanks are due to the master and the crew of "Ammiraglio Magnaghi" (Italy Navy), who greatly helped the cruise during the sampling; and to Dr M. Triscari (Messina University), for providing SEM facilities. This research was financially supported by M.U.R.S.T. (40% Di Geronimo and Rosso).

References

Allen JA, Sanders HL (1996) The zoogeography, diversity and origin of the deep-sea protobranch bivalves of the Atlantic: the epilogue. Prog Oceanogr 38: 95-153

Allen JA, Sanders HL, Hannah F (1995) Studies on the deep-sea Protobranchia (Bivalvia); the subfamily Yoldiellinae. Bull Nat Hist Mus London Zool 61(1): 11-90

Amore C, Curzi VP, Geremia F, Lanti E, Principato MS (1996) Risultati preliminari sulla dinamica sedimentaria attuale del sistema delle Isole Eolie e dei bacini limitrofi di Cefalù e di Gioia. In: Faranda FM, Povero P (eds) Caratterizzazione ambientale marina del sistema Eolie e dei bacini limitrofi di Cefalù e Gioia (EOCUMM95). Data Rep CoNISMa, Genova, pp 431-442

Barrier P, Di Geronimo I, La Perna R, Rosso A, Sanfilippo R, Zibrowius H (1996) Taphonomy of deep-sea hard and soft bottom communities: the Pleistocene of Lazzàro (Southern Italy). In: Melendez G, Blasco F, Pérez I (eds) II Meet Taphonomy Fossilization, Institución "Fernando el Católico", Zaragoza, pp 39-46

Ben Eliahu MN, Fiege D (1996) Serpulid tube-worms (Annelida Polychaeta) of the Central and the Eastern Mediterranean with particular attention to the levant basin. Senkenbergiana Marit 28(1-3): 1-51

Bouchet P, Taviani M (1989) Atlantic deep-sea gastropods in the Mediterranean: new findings. Boll Malac 25: 137-148

Bouchet P, Taviani M (1992) The Mediterranean deep-sea fauna: pseudopopulations of Atlantic species? Deep-Sea Res 39(2): 169-184

Bouchet P, Warén A (1993) Revision of the Northeast Atlantic bathyal and abyssal Mesogastropoda. Boll Malacol Suppl 3: 579-840

Calvet L (1907) Bryozoaires: expéditions scientifiques du "Travailleur" et du "Talisman" pendant les années 1880-1883. Masson C Paris 8: 355-495

Carpine C (1970) Ecologie de l'ètage bathyal dans la Méditerranée occidentale. Mem Inst Oceanogr 2: 1-146

Di Geronimo I (1973) *Tiberia octaviana* n. sp. di Pyramidellidae (Gastropoda, Opistobranchia) del Mediterraneo. Conchiglie 9(11-12): 217-222

Di Geronimo I (1974) Molluschi bentonici in sedimenti recenti batiali ed abissali dello Jonio. Conchiglie 10(5-6): 133-172

Di Geronimo I, Bellagamba M (1985) Malacofaune dei dragaggi BS 77-1 e BS 77-2 (Sardegna nord-orientale). Boll Soc Paleontol Ital 24(2-3): 111-129

Di Geronimo I, La Perna R (1996) *Bathyspinula excisa* (Philippi, 1844) (Bivalvia, Protobranchia): a witness of the Plio-Quaternary history of the deep Mediterranean benthos. Riv Ital Paleontol Stratigr, Milano, 102(1): 105-118

Di Geronimo I, La Perna R (1997) Pleistocene bathyal molluscan assemblages from Southern Italy. Riv Ital Paleontol Stratigr 103(3): 389-426

Di Geronimo I, La Perna R (2000) Some Quaternary bathyal eulimids from the Mediterranean, with description of two new species (Gastropoda, Eulimidae). J Conchol 36(6) (in press)

Di Geronimo I, Li Gioi R (1981) La malacofauna wurmiana della staz. BS 77/4 al largo di Capo Coda Cavallo (Sardegna nord-orientale). Ann Univ Ferrara Sez IX, (Suppl) 6: 123-151

Di Geronimo I, La Perna R, Rosso A (1996) The Plio-Quaternary evolution of the Mediterranean deep-sea benthos: an outline. In: La Méditerranée: variabilités climatiques, environnement et biodiversité. Colloq Sci Int 6-7 Avril 1995, Okeanos 95, Montpellier, pp 286-291

Di Geronimo I, D'Atri A, La Perna R, Rosso A, Sanfilippo R, Violanti D (1997) The Pleistocene bathyal section of Archi (Southern Italy). Boll Soc Paleontol Ital 36(1-2): 189-212

Fredj G, Laubier L (1985) The deep Mediterranean benthos. In: Moraitou-Apostoloupou M, Kiortis V (eds) Mediterranean marine ecosystems. Plenum Press, New York, pp 109-145

Harmelin J-G (1977) Bryozoaires du banc de la Conception (nord des Canaries) Campagne Cineca I du "Jean Charcot". Bull Mus Nat Hist Nat Ser 3 (492) Zool 341: 1057-1076

Harmelin J-G, d'Hondt J-L (1992) Bryozoaires des parages de Gibraltar (campagne océanographique BALGIM, 1984) 1 - Chéilostomes. Bull Mus at Hist Nat Ser 4, 14 (A) 1: 23-67

Harmelin J-G, d'Hondt J-L (1993) Transfers of bryozoan species between the Atlantic Ocean and the Mediterranea Sea *via* the Strait of Gibraltar. Oceanol Acta 16(1): 63-72

Hayward PJ (1979) Deep water Bryozoa from the coasts of Spain and Portugal. Cah Biol Mar 20: 59-75

Hayward PJ (1981) The Cheilostomata (Bryozoa) of the deep sea. Galathea Rep 15: 21-68

Hickman CS (1984) Composition, structure, ecology and evolution of six Cenozoic deep-water mollusc communities. J Paleontol 58: 1215-1234

Janssen R (1989) Benthos-Mollusken aus dem Tiefwasser des östlichen Mittelmeeres, gesammelt während der "METEOR"-Fahrt 5 (1987). Sencken Mar 20(5-6): 265-276

Jeffreys JG (1879) On the Mollusca procured during the "Lightning" and "Porcupine" expeditions, 1868-70. 2. Proc Zool Soc London (1878): 553-588

Jullien J (1882) Dragages du "Travailleur", Bryozoaires: espèces draguées dans l'Ocean Atlantique en 1881; espèces nouvelles ou incomplètement décrites. Bull Soc Zool Fr 7: 497-529

Knudsen J (1970) The systematics and biology of abyssal and hadal bivalvia. Galathea Rep 11: 1-241

Lagaaij R, Cook PL (1973) Some Tertiary to Recent bryozoa. In: Hallam A (ed) Atlas of palaeobiogeography, pp 498-498

Loubere P (1982) The Western Mediterranean during the last Glacial: attacking a non-analog problem. Mar Micropaleontol 7: 311-325

Ockelmann K, Warén A (1998) Taxonomy and biological notes on the Bivalve genus *Microgloma*, with comments on the Protobranch nomenclature. Ophelia 48(1): 1-24

Oliverio M, Nofroni I, Amati B (1992) Revision of the *Alvania testae* group of species (Gastropoda, Prosobranchia, Truncatelloidea, Rissoidea). Lav SIM 24: 249-259

Pérès JM, Picard J (1964) Nouveau manuel de bionomie benthique de la mer Méditerranée. Rec Trav Stat Mar Endoume 31(47): 1-137

Rex MA (1981) Community structure in the deep-sea benthos. Annu Rev Ecol Syst 12: 331-353

Rhind PM, Allen JA (1992) Studies on the deep-sea Protobranchia (Bivalvia): the family Nuculidae. Bull Br Mus Nat Hist Zool 58: 61-93

Rosso A (1990) Thanatocoenosis würmienne à Bryozoaires bathyaux en Mer Tyrrhénienne. Rapp Comm Int Explor Mer Mediterr 32(1): 23

Rosso A (1998) New deep-sea bryozoan species from the Pleistocene of Southern Italy. Riv Ital Paleontol Stratigr 104(3): 423-430

Rosso A (1999) Recent and fossil species of *Characadoma* Maplestone, 1900 (Bryozoa) from the Mediterranean with description of two new species. J Nat Hist 33: 415-437

Rosso A, Corselli C (1996) Evoluzione paleoecologica del sistema Eolie e bacini limitrofi: rapporto sulla campagna Giugno-Luglio 95. In: Faranda FM, Povero P (eds) Caratterizzazione ambientale marina del sistema Eolie e dei bacini limitrofi di Cefalù e Gioia (EOCUMM95). Data Rep CoNISMa, Genova, pp 385-412

Rosso A, Di Geronimo I (1998) Deep-sea Pleistocene Bryozoa of Southern Italy. Géobios 30(3): 303-317

Salas C (1996) Marine bivalves from off the Southern Iberian Peninsula collected by the Balgim and Fauna 1 expeditions. Haliotis 25: 33-100

Sanders HL, Allen JA (1985) Studies on deep-sea Protobranchia (Bivalvia); the family Malletiidae. Bull Br Mus Nat Hist Zool 49(2): 195-238

Sanfilippo R (1998) Tube morphology and structure of the bathyal mediterranean serpulid *Hyalopomatus variorugosus* Ben Eliahu and Fiege, 1996 (Annelida, Polychaeta). Riv Ital Paleontol Stratigr 104(1): 131-138

Warén A (1989) Taxonomic comments on some Protobranch bivalves from the Northeastern Atantic. Sarsia 74: 223-259

Warén A (1991) New and little known Mollusca from Iceland and Scandinavia. Sarsia 76: 53-124

Warén A (1992) New and little known "skeneimorph" gastropods from the Mediterranean Sea and the adjacent Atlantic Ocean Boll Malacol 27: 149-247

Zibrowius H (1977) Review of serpulidae (polychaeta) from depths exceeding 2000 metres. In: Reish DJ, Fauchald K (eds) Essays on polychaetus annelids in memory of Dr. Olga Hartmann. Allan Hancock Foundation, Los Angeles, pp 289-305

Zibrowius H (1979) *Vitreotubus digeronimoi* n.g., n.sp. (Polychaeta Serpulidae) du Pléistocène inférieure de la Sicile et de l'étage bathyal des Acores et de l'Océan Indien. Téthys 9(2): 183-190

Zibrowius H, Hove HA ten (1987) *Neovermilia falcigera* (Roule, 1898) a deep- and cold-water serpulid polychaete common in the Mediterranean Plio-Pleistocene. Bull Biol Soc Washington 7: 259-271

Seasonal Trends in the Adriatic Seagrass Communities of *Posidonia oceanica* (L.) Delile, *Cymodocea nodosa* (Ucria) Ascherson, *Zostera marina* L.: Plant Phenology, Biomass Partitioning, Elemental Composition and Faunal Features

P. Guidetti, M.C. Buia, M. Lorenti, M.B. Scipione, V. Zupo, and L. Mazzella

ABSTRACT

Three seagrass systems, represented by *Posidonia oceanica* (L.) Delile, *Cymodocea nodosa* (Ucria) Ascherson and *Zostera marina* L., were studied on a yearly cycle in their seasonal evolution in different sites of the Adriatic Sea. The aim of this study, conducted in the frame of the programme PRISMA 2 (*Programma di Ricerca e Sperimentazione per la Salvaguardia del mare Adriatico*) was to follow the temporal patterns of these seagrass ecosystems in the Adriatic Sea, in order to investigate their functional mechanisms. Structural attributes of the meadows (i.e. shoot density and phenological parameters) were measured and biomass contributions of the different plant compartments and associated communities (epiphytes and vagile fauna) were estimated. Analyses of macronutrients (carbon and nitrogen) of plant compartments and associated communities were aimed at identifying their relative proportion within the living part of the ecosystem. Shoot density showed a marked seasonality in *C. nodosa* and in *Z. marina*, while it was constant, as well known, in *P. oceanica*. Some phenological and structural parameters (i.e. shoot density, leaf length, LAI) varied seasonally, in the different seagrasses, following patterns in accordance with plant eco-physiology. Total biomass was highest in *P. oceanica*. Different plant compartments (above- vs below-ground) contributed differently to the total biomass in the investigated seagrass species. In *P. oceanica* and *C. nodosa*, below-ground parts accounted for the highest biomass through the year, while in *Z. marina* above- and below-ground contributions were more balanced. The study of carbon and nitrogen composition of different seagrass organs showed a comparative stability of carbon content over time, whereas nitrogen levels followed well-defined seasonal trends in each of the three species, with maxima generally occurring in winter, except for *C. nodosa*. The patterns observed point to the occurrence of different strategies among the three species at more than one level.

Introduction

The paramount importance of seagrasses for the dynamics and productivity of coastal waters has been largely described for several seas (Larkum et al. 1989). As regards the Adriatic Sea, early studies testify to their widespread and functional importance in coastal systems (Benacchio 1938). Marine phanerogams, in fact, are able to colonize different coastlines and types of substrata, according to their life requirements and adaptations to the marine and/or estuarine environments. It is interesting that all five species of marine phanerogams recorded in the Mediterranean Sea, *Posidonia oceanica* (L.) Delile, *Cymodocea nodosa* (Ucria) Ascherson, *Zostera noltii* Hornemann, *Zostera marina* L., *Halophila stipulacea* (Forsskal) Ascherson, are present in the Adriatic Sea, emphasizing the importance and peculiarity of the basin. The Adriatic Sea, in fact, is located at the centre of the Mediterranean basin and extends along a wide latitudinal gradient. The consequent variety of climatic conditions and marine environments highlights the importance of Adriatic seagrass systems in maintaining marine biodiversity.

On the basis of the spread or the peculiarity of their distributional patterns, the seagrasses *P. oceanica*, *C. nodosa* and *Z. marina* are of major interest. *P. oceanica* is distributed along the

Stazione Zoologica "A. Dohrn", Laboratorio di Ecologia del Benthos, Punta S. Pietro, 80077 Ischia (Napoli), Italy

F.M. Faranda, L. Guglielmo, G. Spezie (eds)
Mediterranean Ecosystems: Structures and Processes

coasts of the Southern and north-eastern Adriatic Sea (Gamulin-Brida 1967; Damiani et al. 1988). The Gulf of Trieste, where there are only few sparse patches of this plant (Caressa et al. 1995), represents the northernmost distributional area in the Mediterranean Sea. Furthermore, *C. nodosa* shows a widespread distribution in shallow Adriatic areas (Gamulin-Brida 1967), while *Z. marina* is confined to the northern portion of the basin (Buia and Marzocchi 1995).

In the frame of a general decay of the environmental quality in the Adriatic Sea, recent investigations revealed a massive regression of seagrasses in northern Adriatic (Ghirardelli et al. 1975; Zavodnik and Jaklin 1990; Caressa et al. 1995).

Therefore, the present investigation, carried out in the frame of the national research programme PRISMA 2 (Programma di Ricerca e Sperimentazione per la Salvaguardia del Mare Adriatico – Fase 2), was aimed at understanding some aspects of the dynamics and functional mechanisms of the ecosystems structured by different seagrass species.

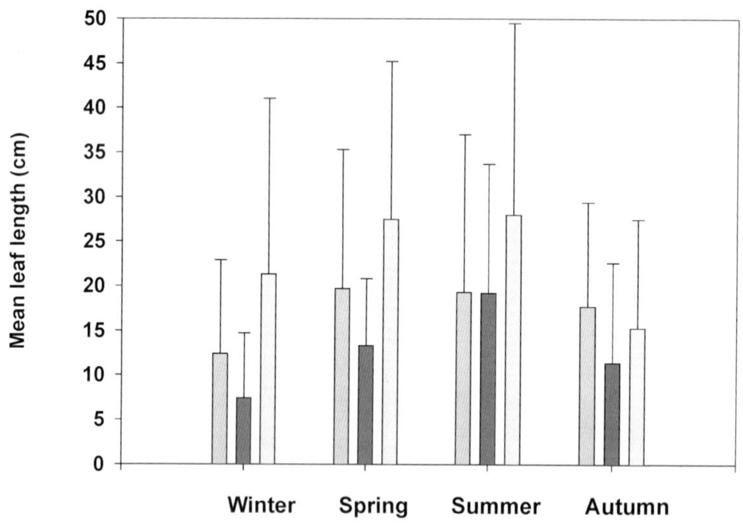

Fig. 1. Values of mean leaf length of the seagrass species studied, in the four seasons

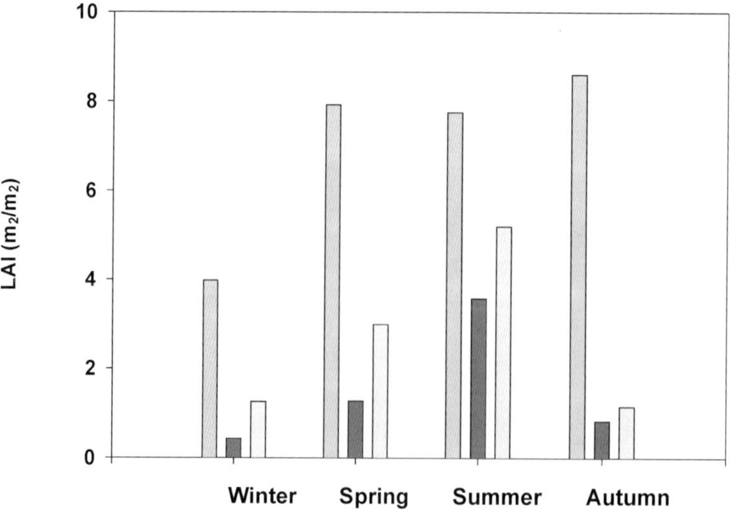

Fig. 2. Values of LAI of the seagrass species studied, in the four seasons

Materials and Methods

Marine phanerogams were sampled seasonally from February to November 1997 in two different areas located in the Southern and Northern Adriatic Sea, namely north of Otranto harbour (40°10'N, 18°29'E) for *P. oceanica* and outside Grado lagoon (45°41'N, 13°28'E) for *C. nodosa* and *Z. marina*. At Otranto, *P. oceanica* was sampled at 6.5 m depth, where the water temperature varied from 18°C (winter) to 25°C (summer), while salinity was about 38‰. The water temperature and salinity varied from 8.9°C (winter) to 24.4°C (summer), and from 28‰ to 34‰, respectively. At the meadow *C. nodosa* settled at 1.5 m depth farther away from the Grado lagoon than *Z. marina*. This latter was settled at 0.5 m depth near the entrance of the lagoon, where the water temperature and salinity ranged from 8.8°C (winter) to 24.4°C (summer), and from 25‰ to 33‰, respectively.

Samples for phenological analyses and biomass estimates were collected following the procedures reported by Mazzella et al. (1998). The above-ground (leaves) and below-ground (old sheaths, rhizomes and roots) compartments of plants were considered separately.

Total carbon and nitrogen contents of plant compartments, epiphyte and vagile fauna were obtained by means of a CHN analyser; results are reported as percentage of dry weight.

Results

Plant Phenology

In *P. oceanica*, a mean shoot density of 707.9 ± 65.2 shoots m^{-2} was measured. The shoot density is considered to be a constant parameter through the year in this plant (Mazzella and Buia 1989). In *C. nodosa*, the shoot density varied from 977.5 ± 158.3 (winter) to 1657.9 ± 85.3 (summer) shoots m^{-2} and in *Z. marina* from 277.8 ± 44.6 (winter) to 773.7 ± 90.5 (summer) shoots m^{-2}.

Higher seasonal variations of the leaf length were observed in *C. nodosa* and *Z. marina* than in *P. oceanica*, with maxima occurring in spring-summer (Fig. 1).

The Leaf Area Index (Fig. 2) was highest in *P. oceanica*, followed by *Z. marina* and, in turn, *C. nodosa*. This parameter varied according to season in *C. nodosa* and *Z. marina*, while in *P. oceanica* it was high and constant in all seasons but winter.

Biomass Partitioning

The *P. oceanica* accounted for the highest total biomass, with a striking difference in comparison to the other seagrass species (Fig. 3). The belowground compartment, mainly represented by old sheaths and living rhizomes, was responsible, in *P. oceanica*, for a higher fraction of plant biomass than the above-ground. In *C. nodosa*, a clear summer peak was observed both in the below-ground and above-ground, the former

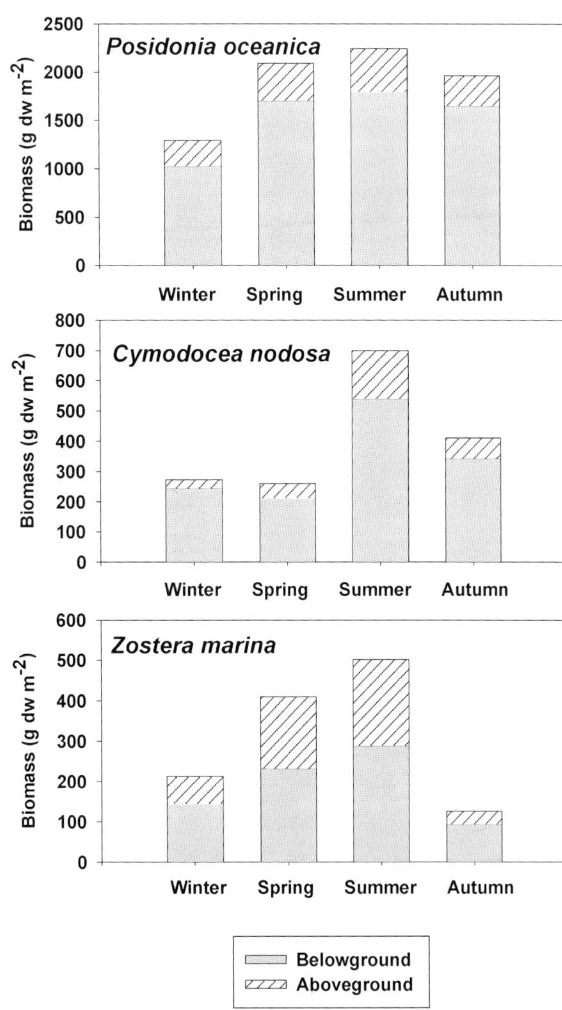

Fig. 3. Biomass partitioning in the plant compartments of the seagrass species studied, in the four seasons

accounting for the highest biomass in all seasons. Contributions of the two plant compartments to the total biomass in *Z. marina* were much more balanced, since they accounted for a comparable fraction in all seasons.

With regard to the biomass values of associated communities (Fig. 4), epiphytes were highly dominant in *P. oceanica* in comparison to the vagile fauna in all seasons, although a certain increase in the total biomass of vagile invertebrates was observed in winter. In *C. nodosa*, epiphytes and vagile fauna contributed equally to the biomass of associated communities, since the bulk of dry weight attributed to epiphytes in winter was represented by fine sediment trapped in filamentous microalgae. A slightly appreciable increase in biomass of the vagile fauna was observed in autumn. In *Z. marina*, the whole contribution of the vagile fauna to the total biomass of associated communities was much more evident than in other seagrasses. In particular, the biomass of vagile invertebrates was higher than epiphytes in winter and autumn, while epiphytes prevailed in spring and summer.

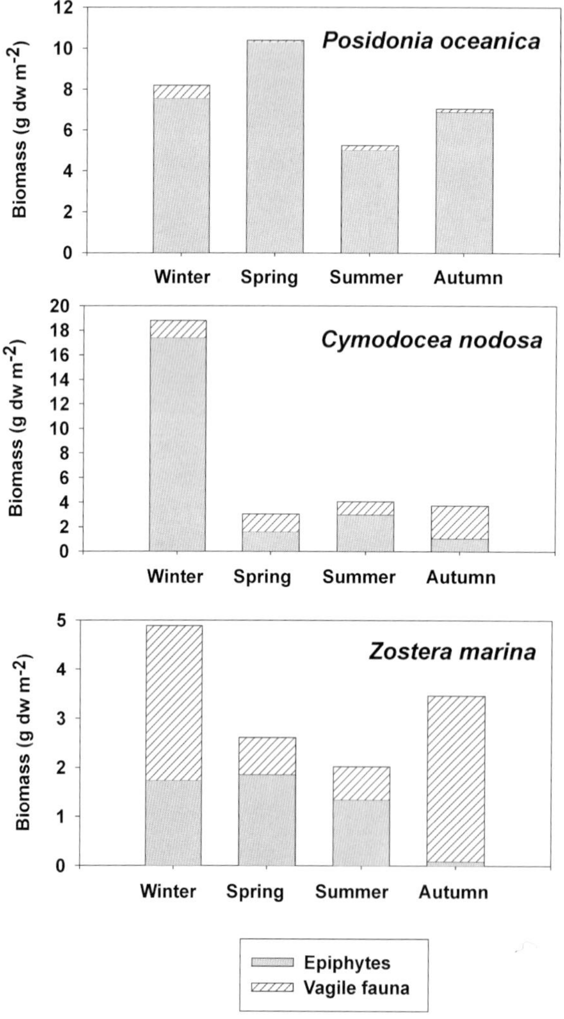

Fig. 4. Biomass partitioning of epiphytes and vagile fauna associated with the seagrass species studied, in the four seasons

Carbon and Nitrogen Content of Seagrass Tissues

From a general point of view, highest levels of carbon are reached in below-ground organs of *P. oceanica* and *C. nodosa* (up to 38.9% of dry weight in the former species), in particular in rhizomes, while minimum values, as low as 28.9% in summer, occur in *Z. marina* below-ground compartment. These figures, as well as those reported for nitrogen, do not take into account the fraction constituted by debris (*sensu* Velimirov et al. 1981); within this compartment, sheaths remaining still attached to the plant build up a large amount of biomass in *P. oceanica*. Considering the whole set of data, the average content of carbon is 35.3±3.3 (average percentage ± SD) in *P. oceanica*, 35.4±2.1 in *C. nodosa* and 32.6±2.9 in *Z. marina*. Trends of seasonal variations point to a decrease of carbon content in summer in all three species, notably as far as shoot tissues are concerned (Fig. 5). On the whole, differences in above- versus below-ground carbon content are more conspicuous in *P. oceanica*.

As for nitrogen, the highest relative content is reached in *P. oceanica* below-ground tissues in winter (4.12%), whereas at the shoot level, highest values occur in *C. nodosa* (up to 3.04% in spring). Contrasting patterns of above-ground versus below-ground N content are found in *P. oceanica* and *Z. marina*, with *P. oceanica* always accounting for higher levels in below-ground. It must however be pointed out that below-ground prevalence in *P. oceanica* is attributable to rhizomes, as roots have a comparatively low N content. In *C. nodosa*, although N content of shoot is generally higher than in below-ground, differences are not so sharp as in *Z. marina*, in which a relative decrease of 35-45%, according to the season, is found. Seasonal trends of N content show a summer decline in all three species. This is particularly evident in *P. oceanica*, in which shoot levels drop to about 40% of those measured in winter and also N content of below-ground in summer are the lowest recorded in the four seasons. By contrast, highest levels (both in below- and above-ground organs) are reached in winter in *P. oceanica*, in autumn-winter in *Z. marina* and in spring in *C. nodosa*.

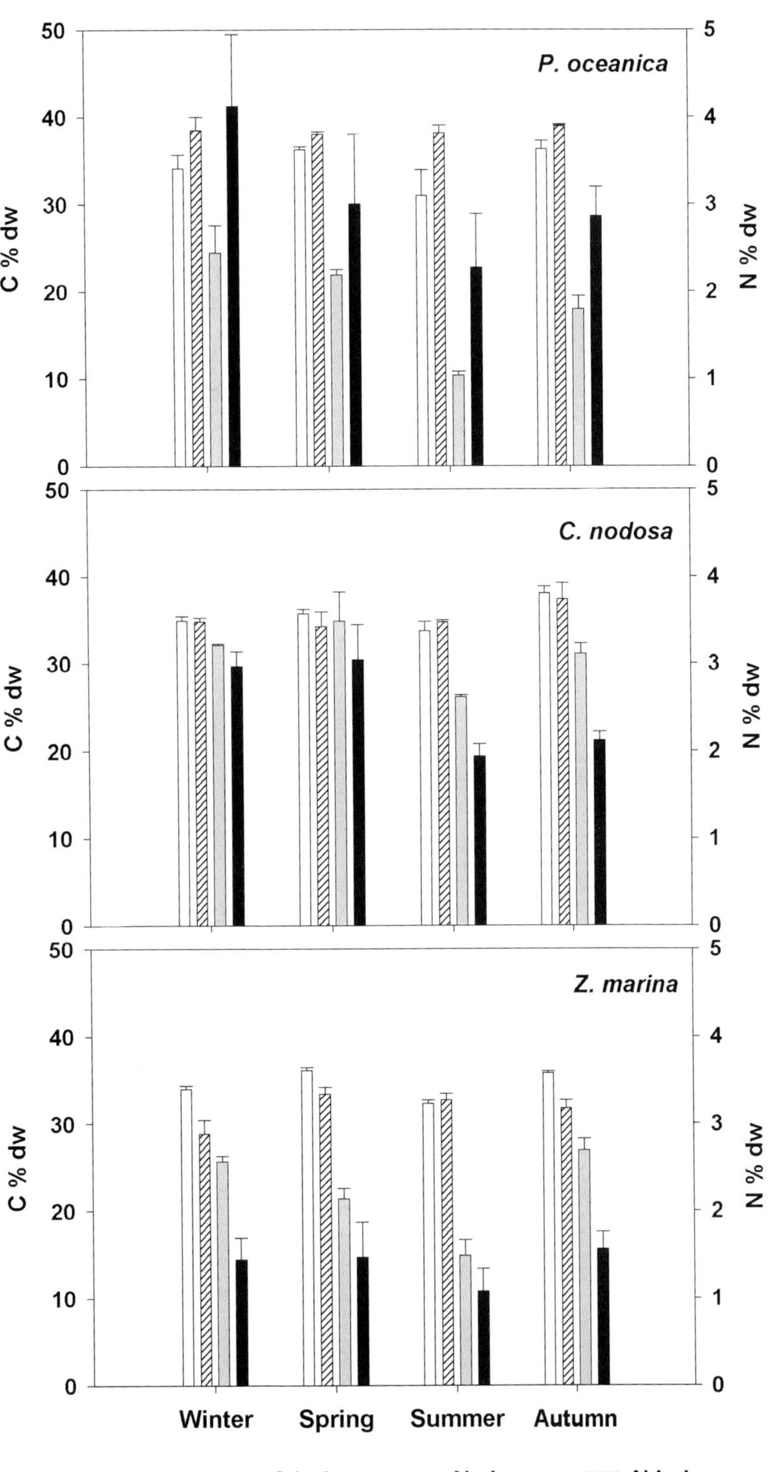

Fig. 5. Variations of carbon and nitrogen content in above- and below-ground parts of the seagrass species studied, in the four seasons

Carbon and Nitrogen Content of Epiphytes and Vagile Fauna

In percentage of dry weight, carbon average content of leaf epiphytes was 22.8±3.6 in *P. oceanica*, 21.9±7.0 in *C. nodosa* and 19.8±0.67 in *Z. marina*, while N content was 3.05±0.89, 2.32±1.42 and 1.85±0.38, respectively.

As for vagile fauna, yearly average content in carbon was 28.9±4.5 for *P. oceanica*, 32.4±3.9 for *C. nodosa* and 35.24±4.1 for *Z. marina*. Nitrogen content averaged 6.01±1.14 in *P. oceanica* fauna, 7.25±0.93 in *C. nodosa* and 7.99±0.93 in *Z. marina* one.

In both epiphytic and vagile fauna compartments, no obvious variations in C and N composition with season were detected.

Discussion

Differences among the three seagrass systems studied occur at more than one level. At plant level, a trend can be recognised in which *P. oceanica* shows the more complex structural architecture (as indicated e.g. by values of both leaf biomass and LAI), followed by *Z. marina* and, in turn, by *C. nodosa*. Partitioning of total biomass is also quite different among the three species, as well as patterns of nutrient allocation between the above- and the below-ground compartments. Seasonality in all considered structural parameters (shoot density, LAI and total biomass) is remarkable in *C. nodosa* and *Z. marina*, while in *P. oceanica* variations are not obvious. On the other hand, a clear seasonality is shown by N relative content in all three species; fluctuations in carbon content would be obscured, as is common in seagrasses, by the high contribution by structural compounds (Duarte 1990).

All the above observations point to marked differences among the three species in both strategies of resource allocation and structuring capacity with respect to the system. *Posidonia oceanica* appears fundamentally as a biomass-storer in which a key role is played by storage compartments mainly located in the below-ground parts. By contrast, *Z. marina* seems to maximise allocation of resources to the above-ground compartment. In *C. nodosa*, while below-ground biomass dominates in all seasons, a more even partitioning of nutrients occurs between the two compartments. Such strategy patterns compare well with those exhibited by other Mediterranean populations of the same species (Sfriso and Ghetti 1998; Pirc and Wollenweber 1988), while absolute values assumed by a number of parameters would indicate that the three Adriatic populations studied grow under non-limiting conditions (Sfriso and Ghetti 1998; Duarte 1990).

Both absolute values and seasonal variations of the epiphytic community seemed to be not strictly related to the available plant surface (LAI). The environmental stress (i.e. salinity and temperature variations) is thus invoked as a disturbance factor. In the same way, the biomass of vagile invertebrates, often related to the level of habitat complexity, was higher on the average in *Z. marina*. The above issues involve the existence of other superimposed factors which could explain our results, such as the influence of environmental disturbance (i.e. current tide, amplitude of salinity and temperature variations, nutrient availability) acting at different temporal scales as well as the amount and quality of detritus. As far as the nutrient composition of associated communities is concerned, one has to keep in mind that, dealing with mixed populations in which a seemingly varying taxonomic composition overlaps with seasonal effects, the values herein obtained are average figures which encompass differences within the two compartments across seasons.

Acknowledgements. This study, promoted by the regretted Dr. Lucia Mazzella, has been carried out in the framework of the national research programme PRISMA 2 (subproject 3: "Alterations of Communities"). Dr. P. Guidetti benefited from a grant founded by CNR. The authors are grateful to A. Rismondo and D. Curiel for their help during sampling at Grado, and to S. Caressa and P. Scaffidi for their field assistance.

References

Benacchio N (1938) Osservazioni sistematiche e biologiche sulle Zosteracee dell'Alto Adriatico. Thalassia 3 (3): 2-41

Buia MC, Marzocchi M (1995) Dinamica dei sistemi a *Cymodocea nodosa*, *Zostera marina* e *Zostera noltii* nel Mediterraneo. G Bot Ital 129 (1): 319-336

Caressa S, Ceschia C, Orel G, Treleani R (1995) Popolamenti attuali e pregressi nel Golfo di Trieste da Punta Salvore a Punta Tagliamento (Alto Adriatico). In: Cinelli, Fresi et al (eds). In: *Posidonia oceanica*. A Contribution to the preservation of a major mediterranean marine ecosystem. Rivista Marina 12: 160-187 (suppl)

Damiani V, Bianchi CN, Ferretti O, Bedulli D, Morri C, Viel M,

Zurlini G (1988) Risultati di una ricerca ecologica sul sistema marino costiero pugliese. Thalassia Salentina 18: 153-169

Duarte CM (1990) Seagrass nutrient content. Mar Ecol Prog Ser 67: 201-207

Gamulin-Brida H (1967) The benthic fauna of the Adriatic Sea. Oceanogr Mar Biol Annu Rev 5: 535-568

Ghirardelli E, Orel G, Giaccone G (1975) Esperienze sullo scarico a mare di Trieste - Metodologi: a e ricerche per la valutazione degli effetti sul benthos. Ing Ambientale 4: 413-418

Larkum AWD, McComb AJ, Shepherd SA (1989) The biology of seagrasses. Elsevier, Amsterdam, pp 841

Mazzella L, Buia MC (1989) Variazioni a lungo termine di alcuni parametri strutturali di una prateria a *Posidonia oceanica*. Nova Thalassia 10 (1): 533-542

Mazzella L, Guidetti P, Lorenti M, Buia MC, Zupo V, Scipione MB, Rismondo A, Curiel D (1998) Biomass partitioning in Adriatic seagrass ecosystems (*Posidonia oceanica*, *Cymodocea nodosa*, *Zostera marina*). Rapp Comm Int Mer Mediterr 35: 562-563

Pirc H, Wollenweber B (1988) Seasonal changes in nitrogen, free amino acids and C/N ratio in Mediterranean seagrasses. PSZN I: Mar Ecol 9 (2): 167-179

Sfriso A, Ghetti PF (1998) Seasonal variation in biomass, morphometric parameters and production of seagrasses in the lagoon of Venice. Aquat Bot 61: 207-223

Velimirov B, Ott JA, Novak R (1981) Microorganisms on macrophyte debris: biodegradation and its implications in the food web. Kiel Meeresforsch Sonderh 5: 333-344

Zavodnik N, Jaklin A (1990) Long-term changes in the Northern Adriatic marine phanerogam beds. Rapp Comm Int Mer Mediterr 32 (1): 15

Anatomical and Physiological Characteristics of Tentacular Nematocytes Isolated by Different Methods from *Aiptasia diaphana* (Cnidaria: Anthozoa) in the Brackish Pond Faro (Messina, Italy)

G. La Spada, A. Marino, and G. Sorrenti

ABSTRACT

We studied the effects of isolation on anatomical and functional characteristics of nematocytes isolated from tentacles of *Aiptasia diaphana*, an anthozoan living in the brackish pond Faro (Messina) where salinity can change for both climatic and anthropic reasons. The anatomical effects of isolation were investigated by scanning electron microscopy (SEM). To evaluate the functional effects, the capacity to regulate cell volume under a 35% hyposmotic shock (Regulatory Volume Decrease, RVD) was measured by image computer processing of the sagittal area. Furthermore the membrane integrity was revealed by applying Trypan blue. The isolation using a physical method was performed by heat dissociation (45°C for 20 min), while, as a chemical method, we used a treatment of tissue with SCN^-. The isolated nematocytes revealed different characteristics both anatomical and physiological, although the Trypan blue test showed that in both cases the cell membrane was not damaged. In particular nematocytes isolated by heat dissociation show the mechanosensorial apparatus while those isolated by SCN^- appeared to be deprived of this structure. As far as the volume regulation is concerned, the cells isolated by heat dissociation did not appear to regulate their volume until the hypotonic treatment was maintained. On the other hand the nematocytes isolated by SCN^- regulate volume in hypotonic conditions.

Introduction

The nematocytes are particular stinging cells of Cnidaria (Coelenterates). They have a mechanosensorial apparatus which makes them mechanoreceptive, while either support or sensorial cells, which have specific membrane receptors, act as chemoreceptors (Watson and Hessinger 1992; Mire-Thibodeaux and Watson 1994). A particular feature of the nematocytes is that of synthesizing the nematocyst, made up of a wall and an inverted tubule immersed in the capsule fluid (Mariscal 1974). The application of chemical and mechanical stimulation (Pantin 1942) results in a rapid extroflexion of the tubule which penetrates the tissues of the target, injecting toxins of the capsule fluid (Burnett 1991; Norton 1991; Long and Burnett 1994). This phenomenon is called discharge and it is the most rapid process of exocytosis known today (Holstein and Tardent 1984). It is regulated *in situ* not only by nematocytes, but also by support and sensorial cells and, as has recently been shown, a release of nitric oxide (Salleo et al. 1996); therefore it was deemed as useful to carry out research on isolated nematocytes in order to eliminate the possibility of interference by other cells.

So far, either chemical methods, such as either enzymatic (Anderson and McKay 1987; McKay and Anderson 1988) or non-enzymatic (Salleo et al. 1996; Barra et al. 1997), or physical, such as maceration and heat dissociation (Schmid et al. 1981; Weber 1995) have been attempted to isolate nematocytes from various species. Recently Weber (1995), on the basis of what Schmid reported (1981), proposed a new non fixative physical method using two labelling dyes, in order to investigate nematocyte turnover and incubating *Hydra* specimens at 40°C for 10 min in a hypertonic medium, obtain-

Istituto di Fisiologia Generale, Università di Messina, Salita Sperone 31, 98016 Messina, Italy

F.M. Faranda, L. Guglielmo, G. Spezie (eds)
Mediterranean Ecosystems: Structures and Processes

ing cellular elements of various types still organized in tissue fragments and, after shaking, isolated, capable of reaggregation. Alternatively, Salleo et al. (1996) and Barra et al. (1997) did isolate acontial nematocytes of *Aiptasia diaphana*, treating the tissue with an SCN^- isotonic solution. Thus isolated, the nematocytes appeared anatomically intact, but lacking in a mechanosensorial apparatus, and after being exposed to saline hypotonic solutions (35%), showed a capacity to regulate cell volume (La Spada et al. 1999), a feature which is vital in order to maintain cell volume constant in conditions of hyposmotic or hyperosmotic shock (respectively, Regulatory Volume Decrease RVD or Regulatory Volume Increase RVI). The mechanisms of regulation of volume are due to outflow of ions such as K^+ and Cl^-, to the H^+/K^+ exchange, to the cotransport of K^+ and Cl^-, or to the Na^+/Ca^{2+} exchange, and to the extrusion of aminoacids. The cause of this is, in many cases, an increase in intracellular concentration of Ca^{2+}, either flowing from outside through stretch-activated ion channels, or being released from intracellular stores. A dissertation on the mechanisms of cell volume regulation has been recently delivered by Lang et al. (1998). Volume regulation was investigated on cells of Mammals, Amphibians and Crustacea, but not on cells of Coelenterates (Morris 1990). In the present investigation, we observed the effects of isolation on the morphological and functional characteristics of nematocytes isolated from *Aiptasia diaphana*, an Anthozoan which lives in the euphotic zone of the brackish pond Faro (Messina, Italy), the salinity of which can vary for both anthropical and climatic factors suggesting the existence of osmoregulatory mechanism in species living there. We therefore decided to compare the morphological features of the isolated nematocytes both by heat dissociation as a physical method, and by treatment with SCN^- as a chemical method, evaluating morphological features by observation with scanning electron microscopy (SEM), and functional features by measuring the changes in volume of the nematocytes under hyposmotic stress of 35%.

Materials and Methods

Nematocytes were isolated from tentacles of *Aiptasia diaphana* collected in the brackish pond Faro (Messina), maintained in a closed-circuit aquarium at 19°-24°C and fed with prawn meat. The tentacles were excised using ophthalmic scissors from animals previously anaesthetized with a solution of $MgCl_2$ 0.6 M and artificial sea water (ASW) in a 1:1 ratio. The composition of ASW (in mM) used in all tests was NaCl 520, KCl 9.7, $MgCl_2$ 24, $MgSO_4$ 28, $CaCl_2$ 10, imidazol 5, pH 7.6, π=1100 mOsm/kg. Nematocytes were isolated either by tissue heat dissociation, or by treatment with an isosmotic solution (605 mM of SCN^- containing 0.01 mM of Ca^{2+}). In order to obtain isolation by heat dissociation two different procedures were adopted: the first consisted of the incubation of tentacles in ASW rendered hypertonic by adding mannitol (π=2200 mOsm/kg). The second consisted of incubation in an isotonic ASW. Tentacles were incubated in 0.3 ml of ASW at 40°C, or alternatively at 45°C in a thermostat-regulated double boiler for either 10 or 20 min. The test-tube was then returned to room temperature and delicately shaken manually, until the suspension appeared cloudy. A drop of the obtained suspension was then put on to a slide previously treated with polylysine (1% in distilled water). After a few minutes, in order to allow sedimentation of cellular elements, the medium was carefully absorbed using blotting paper, with accuracy being checked under the microscope. The hypertonic ASW used to incubate the suspension was replaced with isotonic ASW, in order to re-establish normal physiological conditions. The elements obtained, which included nematocytes, nematocysts and other cells, were counted using a Leitz-Dialux light microscope with an incorporated video-camera (JVC mod. TK-1180 E). About 200 elements were observed, almost exclusively nematocytes and nematocysts, immediately after isolation and at one hour intervals.

Isolation was obtained by the chemical method, as described by Salleo et al. (1996) and Barra et al. (1997). In short, the tentacles, after being washed several times with ASW Ca^{2+} 0.01 mM, were treated with a solution of SCN^- and Ca^{2+} 0.01 mM for a time not exceeding 5 min; this caused a contraction of the tentacle, followed by extrusion of nematocytes from the tissue. The solution was gradually replaced, first by ASW Ca^{2+} free and then ASW, to restore control conditions. The cells isolated by the two methods were then maintained in a damp room at

12°-16°C. In order to verify the anatomical integrity of nematocytes isolated with both methods, the Trypan blue test (0.4% in ASW) was applied, a dye that, owing to its molecular weight (1200 dal), does not penetrate through intact cell membrane.

The RVD test was carried out as follows: a suspension placed on a slide was left for about an hour in a damp room at 12°-16°C, in order to render adhesion of nematocytes on the surface possible; then, two strips of bi-adhesive tape (Scotch 3 M) (Hidaka 1993) were placed at the margins of the drop of suspension; these strips were used to fix a coverslip on the suspension to allow the total replacement of all experimental solutions. The test was carried out in three periods on a single nematocyte chosen for its strong adhesion to the slide during the flow of ASW. First period: isosmotic ASW for 5 min.; 2nd period: hyposmotic ASW (35%, 710 mOsm by a reduction of concentration of NaCl) for 10 min.; 3rd period: isosmotic ASW for 5 min. The osmotic pressure of all solutions was measured by an osmometer (Fiske OS). About 20 images of nematocytes, recorded at regular intervals, were then elaborated, using a Macintosh Performa 6200 computer, equipped with the suitable programme, to measure the variation of the sagittal area. The results were then expressed in area relative to the A/A_0 control. Statistical analysis was carried out using Student's *t*-Test ($P<0.01$). For SEM observation, the cells were then fixed with glutaraldehyde (1% in ASW) for 45 min, washed several times with ASW, dehydrated in a series of alcohols, coated with gold using an Edwards S150 metallizer, and observed with a DSM 950 Zeiss microscope.

Results

Isolation

The incubation of tentacles of *Aiptasia diaphana* in hypertonic ASW at 40°C for 10 min did not allow us to obtain isolated cellular elements. By extending treatment with hypertonic ASW to 20 min at 45°C, a few damaged cellular elements were seen. In particular, nematocytes showed disorganization of cytoplasm and the cell profile did not appear well defined, although control conditions had gradually been restored replacing the suspension with isotonic ASW even soon after isolation. After treating the tentacle with isotonic ASW at 40°C for 10 min with successive mechanical shaking the tissue still appeared intact and there were no isolated cellular elements whatsoever. At 45 °C for 10 min a few isolated cellular elements were observed, but many of these were still part of the tissue. After incubating the tentacles in isotonic ASW for 20 min at 45°C, many isolated cellular elements were obtained, and even if there were a few still attached to the tissue, they were detached by successive manual shaking. We observed nematocysts, nematocytes and other cells. Among the nematocytes it was possible to identify microbasic-mastigophore nematocytes, with a length of between 23-30 µm and basitric-isorhiza nematocytes, with a length of between 12-17 µm. For our investigation in RVD, we studied, taking into consideration their size, microbasic-mastigophore nematocytes, isolated by heat dissociation, at 45°C for 20 min.

As far as morphological features are concerned, microbasic-mastigophore nematocytes showed their membrane to be intact at SEM observation, and this was confirmed by cytological test with Trypan blue. Cell volume is almost completely occupied by the nematocysts, while the cytoplasm seemed to be relegated into an equatorial position. A peculiar feature of almost all nematocytes was that of showing a distinct cnidocil, (Fig. 1a) sometimes surrounded by stereocilia emerging from the same cell (Fig. 1b). In fact, immediately after isolation, nematocytes showing the cnidocil made up 81.8%±5 of the entire population of nematocytes. One hour after isolation, keeping the cells at 12-16°C, the value was 71%±15, which then further diminished, after two hours, to 60%±9.9.

Due to the treatment of tissue with SCN^- only microbasic-mastigophore nematocytes were isolated. These appeared intact on SEM observation, as also confirmed by the Trypan blue test, but they were lacking in the mechanosensorial apparatus (Fig. 2a), which was still present in those nematocytes which were partially extruded in the tentacle (Fig. 2b).

RVD

The exposure to a hypotonic stress (35%) of nematocytes isolated by heat dissociation induced the expected cell swelling, which after

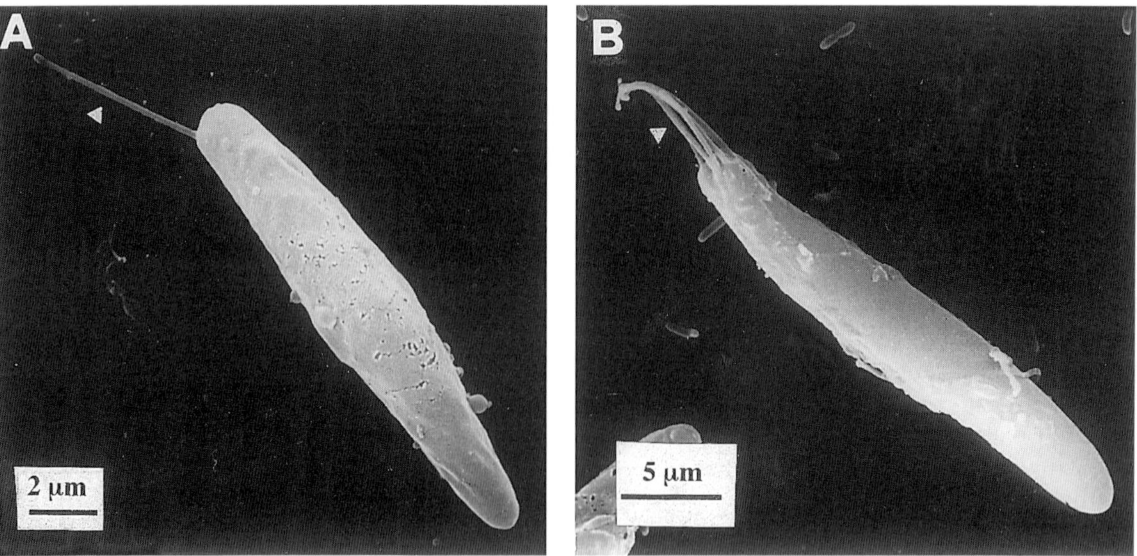

Fig. 1a,b. Nematocytes isolated by heat dissociation. Note in **a** the presence of a distinct cnidocil (*arrowhead*), that in **b** is surrounded by stereocilia (*arrowhead*)

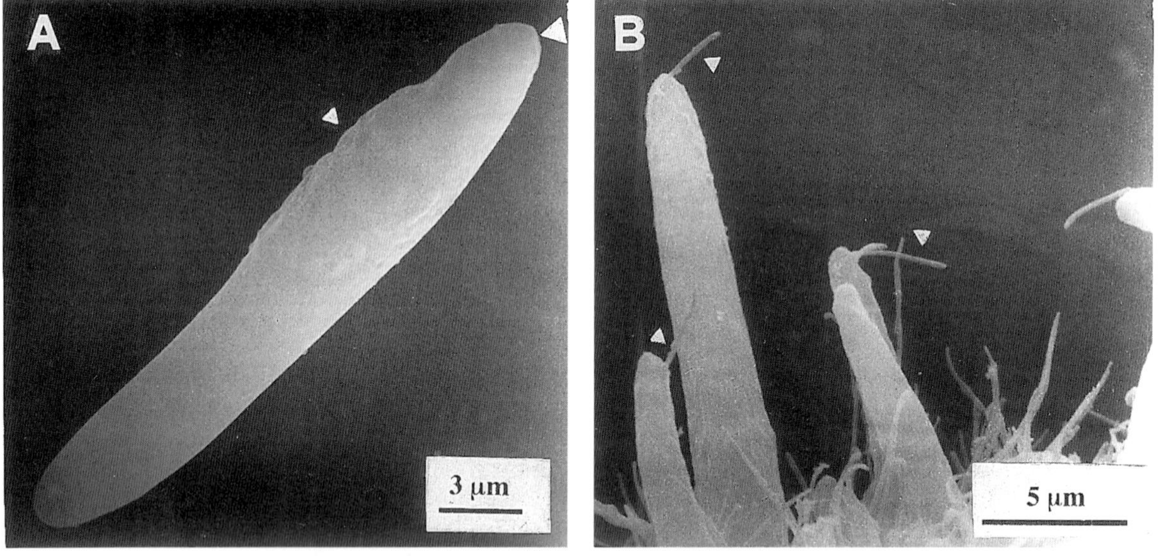

Fig. 2a,b. **a** Nematocyte isolated by SCN^- treatment. Note the absence of cnidocil at the apical end (*large arrowhead*) and the centrally placed girdle (small arrowhead). **b** Segment of tentacle in which the treatment of SCN^- was blocked once the protrusion of nematocytes was observed. Note that cnidocil was still present in partially extruded nematocytes (*arrowheads*)

3.5 min from the beginning of the hyposmotic shock corresponded to a maximum A/A_0 value of 1.041±0.008. Such increase, calculating the difference between the value immediately prior to the hyposmotic ASW and that corresponding to 3.5 min from the application of the solution, was highly significant (*P*<0.001). Such swelling, remained without any decrease for the entire duration of the hyposmotic shock, not showing RVD. Only after removing the hypotonic ASW and replacing it with the isotonic ASW was the value of the cell area restored to initial values (Fig. 3a).

The isolated nematocytes, treating the tentacles with SCN^- also showed the expected cell swelling, reaching their maximum value of 1.06±0.009 in 1.5 min. This result was highly significant (*P*<0.001). Shortly afterwards however a gradual decrease in cell volume was observed, and this continued for the whole duration of the hyposmotic shock reaching, in 5 min, values equal to 1.03±0.006, less than 50% compared to peak value, owing to RVD. Also in this case, after removing the hypotonic ASW, the restoration of the median sagittal area was observed, with values equal to those observed in the initial control (Fig. 3b).

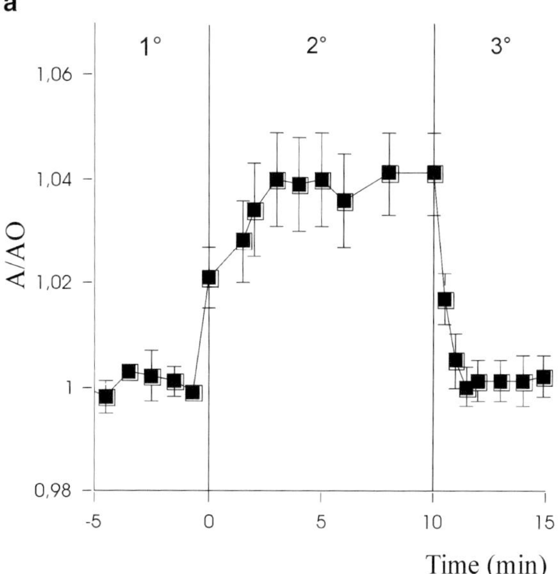

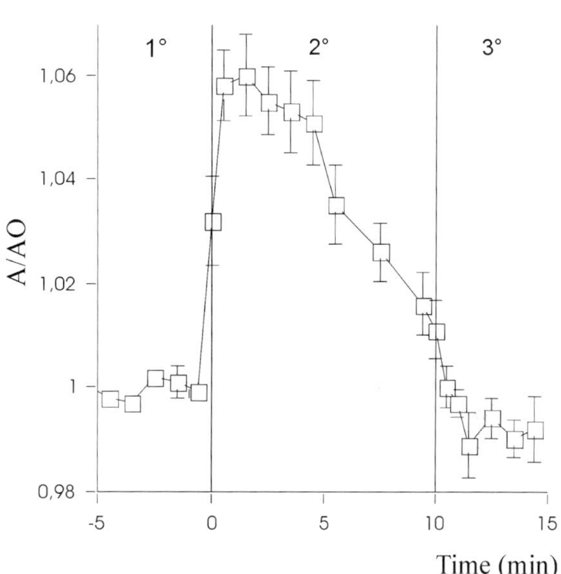

Fig. 3a,b. **a** RVD in nematocytes isolated by heat dissociation and **b** by SCN^- treatment. The test was carried out in three periods: isosmotic ASW for 5 min; hyposmotic ASW for 10 min; isosmotic ASW for 5 min. Note in **a** that the swelling remained unchanged for entire 2nd period, RVD was therefore absent, (means ± SE, *n*=10); while in **b** there was a gradual decrease in cell volume during the 2nd period, owing to RVD (means ± SE, *n*=10)

Discussion

The physical method of isolation by incubation of tentacles in isotonic ASW at 45°C for 20 min made it possible to obtain nematocytes isolated from *Aiptasia diaphana*. These nematocytes, with their membrane intact, showed their mechanosensorial apparatus, made of cnidocil and stereocilia, both had their origin in the same nematocyte. Watson and Hessinger (1992), Mire-Thibodeaux and Watson (1994) observed *in situ* two cellular complexes on the surface of the tentacle of the sea anemone *Haliplanella luciae*: CSCC (Cnidocyte-Supporting Cell Complex) and SNSC (Sensory Cell-Supporting Cell Complex). In the former the cnidocil belongs to the nematocyte, while the stereocilia belong to the supporting cells; in the latter the cnidocil belongs to the sensory cell and it is surrounded by the stereocilia. In the nematocytes isolated from *Aiptasia diaphana* both the cnidocil and the stereocilia, if present, would appear to belong to the nematocyte. It is in fact possible to identify the point of insertion both of the cnidocil and of the stereocilia. The difference between what we have observed and that described by Mire-Thibodeaux and Watson (1994) could be explained by the fact that these authors made their observations *in situ*, where a high number of ciliary protrusions is present, so that it is impossible to distinguish the origin of each of them. Alternatively, a difference between species could be hypothesized. Our results and those of Watson and Hessinger (1992) and Mire-Thibodeaux and Watson (1994) strongly contrast with those of Rifkin (1996) who indicated the absence of cnidocil in Anthozoa.

The method of isolation by treatment with SCN^- made it possible to obtain cellular elements which essentially consisted of microbasic-mastigophore nematocytes, anatomically intact but lacking in mechanosensitive apparatus and this differs notably from that observed in nematocytes isolated with the physical method. However, it must be considered that after treatment with SCN^-, when nematocytes are still *in situ* and only partially extruded, they do show their mechanosensitive apparatus, which afterwards disappears in the extruded nematocytes.

The exposure of the nematocytes, isolated with the two above-mentioned methods, to hyposmotic shock corresponding to 35% resulted in different responses. In fact, in nematocytes isolated by treatment of the tentacle with SCN^- the cell swelling gradually decreased, while still keeping hypotonic conditions unvaried, showing therefore mechanisms of RVD. Obviously the value of osmotic swelling is moderate, considering their particular anatomy. In fact, given that the cell volume is taken up by the nematocyst (~85%) (A. Salleo, unpublished), the volume of which does not vary under hyposmotic shock, the cell profile only varies to a moderate extent, as the cytoplasm is situated in a thin rim in an equatorial, and sometimes apical, position. It must also be considered that these cell have a higher capacity for volume regulation, compared to that seen in other cells (Park et al. 1994; De Smet et al. 1995). This result can be evaluated taking into account the great variability, observed in limited areas, of the seawater salinity of the brackish pond Faro, due to both the poor communication with the sea and anthropic intervention.

In nematocytes isolated by heat dissociation, cell swelling, which took place over a longer time than that in isolation with SCN^-, was constant for the entire duration of the hyposmotic shock. The absence of RVD in nematocytes isolated in this way could be attributed to some possible conformational modifications of the channel-proteins through which RVD mechanisms take place. Such modifications could be attributed to the degree of temperature and duration to which the tentacle was subjected.

From the present study we can conclude that: nematocytes isolated from *Aiptasia diaphana* by heat dissociation are anatomically intact, have mechanosensorial apparatus but lose a property, the RVD, that depends on integrity of ion transport mechanisms, while the general physiological integrity of nematocytes isolated with SCN^- is better maintained, with the exception of the mechanosensitive function.

Acknowledgements. We are grateful to Prof. Alberto Salleo of Institute of General Physiology (University of Messina) for helpful discussion and revision of the manuscript.

References

Anderson PAV, McKay MC (1987) The electrophysiology of cnidocytes. J Exp Biol 133: 215-230

Barra PFA, Musci G, Salleo A (1997) Control of discharge of acontial nematocysts in *Aiptasia diaphana. In:* Proc 6th Int Conf Coelenterate Biol (1995), pp 39-46

Burnett JW (1991) Clinical manifestations jellyfish envenomation. Hydrobiologia 216/217: 629-636

De Smet P, Simaels J, Van Driessche W (1995) Regulatory volume decrease in a renal distal tubular cell line (A6). I. Role of K^+. and Cl^- Pflügers Arch 430: G, Schaller H, Trenker E (1972) regeneration of *Hydra* from reaggregated cells. Nat New Biol 239 (91): 98-101

Hidaka M (1993) Mechanism of nematocyst discharge and its cellular control. Adv Comp Environ Physiol 15: 45-76

Holstein T, Tardent P (1984) An ultrahigh-speed analysis of exocytosis: nematocyst discharge. Sciences 223: 830-833

Lang F, Gillian L, Busch Ritter M, Völkl H, Waldegger S, Gulbins E, Haussinger D (1998) Functional significance of cell volume regulatory mechanisms. Physiol Rev 78: 247-306

La Spada G, Biundo T, Nardella R, Meli S (1999) Regulatory volume decrease in nematocytes isolated from acontia of *Aiptasia diaphana*. Cell Mol Biol 45: 249-258

Long KO, Burnett JV (1994) Characteristics of ialuronidase and hemolitic activity in fishing tentacle nematocyst venom of *Crhysaora quinquecirrha* . Toxicon 32: 165-174

Mariscal RN (1974) Nematocysts. In: Muscatine L, Lenhoff HM (eds) Coelenterate biology. Academic Press, New York, pp 129-178

McKay MC, Anderson PAV (1988) Preparation and properties of cnidocytes from the sea anemone *Anthopleura elegantissima* . Biol Bull Mar Biol Lab Woods Hole 174: 47-53

Mire-Thibodeaux P, Watson GM (1994) Morphodynamic hair bundles arising from sensory cell/supporting cell complexes frequency-tune nematocyst discharge in sea anemones. J Exp Zool 278: 282-292

Morris CE (1990) Mechanosensitive ion channels. J Membr Biol 113: 93-107

Norton RS (1991) Structure and structure function relationships of the sea anemone proteins that interact with the sodium channel. Toxicon 29: 1051-1084

Pantin CFA (1942) The excitation of nematocysts. J Exp Biol 19: 294-310

Park KP, Beck JS, Douglas IJ, Brown PD (1994) Ca^{2+} activated K^+ channels are involved in regulatory volume decrease in acinar cells isolated from the rat lacrimal gland. J Membr Biol 141: 193-201

Rifkin J (1996) Cnida structure and function. In: Williamson JA, Fenner PJ, Burnett JW, Rifkin JF (eds) Venomous and poisonous structures in marine animals. Univ New South Wales Press, Sidney 2502, Australia, pp 156-158

Salleo A, Musci G, Barra PFA, Calabrese L (1996) The discharge of acontial nematocytes involves the release of nitric oxide. J Exp Biol 199: 1261-1267

Schmid V, Stidwill R, Bally A, Marcum B, Tardent P (1981) Heat dissociation and maceration of marine cnidaria. Wilhem Roux Archiv 190: 143-149

Watson GM, Hessinger DA (1992) Receptors for N-acetylated sugars may stimulate adenylate cyclase to sensitize and tune mechanoreceptors involved in triggering nematocysts discharge. Exp Cell Res 198: 8-16

Weber J (1995) Novel tools for the study of development, migration and turnover of nematocytes (cnidarian stinging cells). J Cell Sci 108: 403-412

CHAPTER 39

The Oliveri-Tindari Lagoon (Messina, Italy): Evolution of the Trophic-sedimentary Environment and Mollusc Communities in the Last Twenty Years

M. Leonardi[1] and S. Giacobbe[2]

ABSTRACT

In the present study the evolution of the benthic ecosystem of Oliveri-Tindari (Messina, northern coast of Sicily), is described through the examination of data collected every ten years, from 1978 up to now. In particular three cycles of investigations (1978, 1987, 1997) provided comparable information on the edaphic, trophic and biological characteristics of the area. As shown, there has been a constant rise in the fine sedimentation occurring in the pools during these twenty years. With regard to the trophic characteristics, a general decrease of the organic carbon content in the sediments was recorded in the second cycle of study. The trophic-sedimentary evolution caused marked changes in the structure of the benthic communities, as pointed out by the qualitative and quantitative variations. In general a decrease in the density of the mollusc benthic communities was observed, because of the progressive isolation of the pools.

Introduction

The Oliveri-Tindari lagoon (Fig. 1), a small brackish system located along the Tyrrhenian coast of Sicily (Italy), is characterized by marked geomorphologic dynamism and high structural and hydrobiological complexity (Leonardi and Giacobbe in press). The particular anemological regime of this area determines the periodical formation of littoral bars, thus delimiting little coastal pools (Brambati 1988). At present, the system is constituted of six pools, formed in different periods of the last century (Amore and Randazzo in press).

The heterogeneity of allochthonous inputs (Azzaro 1995; Leonardi *et al.* in press), with the reciprocal isolation of the basins and the progressive confinement (sense of Guelorget and Perthuisot 1983; Giacobbe et al. 1990), have determined a very high hydrological differentiation among the pools, as shown by nutrients and salinity values. In particular the most confined pools (A, B, C), where the continental inputs are more important, have always shown lower salinity than the outermost ones (D, E, F) (CNR-I.T.M. 1990; Giacobbe and Leonardi 1986).

The depth is low in all basins, and the maximum value was found in the Marinello pool (4.20 m).

Because of its natural and ecological exclusiveness, this ecosystem has been the subject of numerous studies in the last fifty years (Abbruzzese and Aricò 1955; Crisafi 1961; Faranda et al. 1975; Genovese 1964; Giacobbe and Giordano 1975; Giacobbe et al. 1992).

In this paper we compare three cycles of investigations, which have provided useful information to describe the edaphic, trophic and faunal evolution of this area in the last twenty years. In particular we carried out the first investigation in 1978, the second in 1987, and the last one in 1997/'98, in the framework of a multidisciplinary project (Progetto Finalizzato Beni Culturali C.N.R.), aimed at protection, exploitation, and management of this complex ecosystem.

Materials and Methods

The sediment samplings were seasonal in 1978 and 1987, whereas they were monthly in 1997, but in this paper we refer only to the winter data of the three sampling series, since we have the most complete and comparable data in this peri-

[1] Istituto Sperimentale Talassografico, CNR, Spianata San Raineri 86, 98122 Messina, Italy
[2] Dipartimento di Biologia Animale ed Ecologia Marina, Università di Messina, Salita Papardo, 98100 Messina, Italy

F.M. Faranda, L. Guglielmo, G. Spezie (eds)
Mediterranean Ecosystems: Structures and Processes

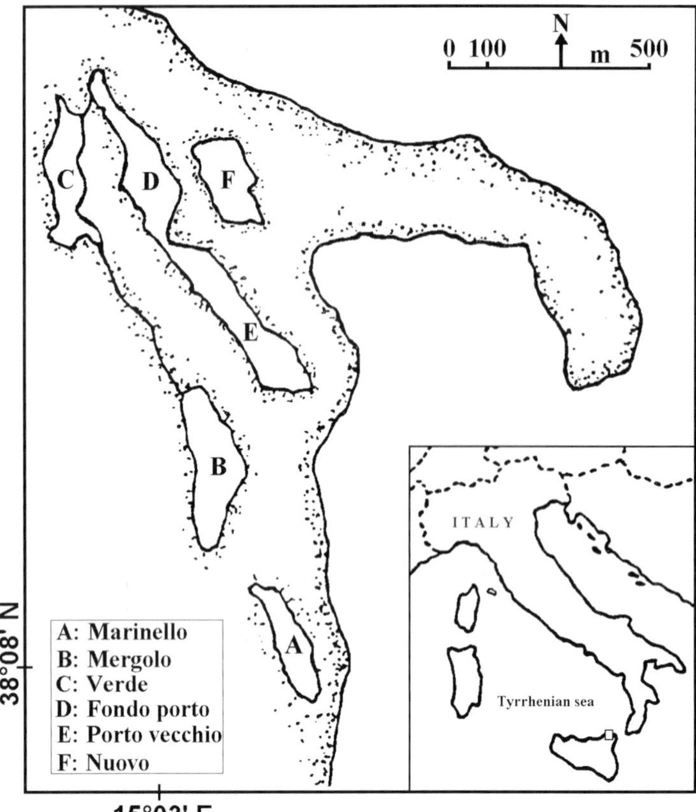

Fig. 1. The Oliveri-Tindari coastal system

od. Moreover, in this season benthic communities in the Oliveri-Tindari pools are less affected by anthropic disturbance.

The sediments by Van Veen grab from the middle of the pool and by box-corer along the edge were collected. In the first method samples were taken from the superficial layer (2 cm), for trophic-sedimentary characterization, while for the benthic fauna specimens the coastal sediments were collected at a depth of 0.5-1 m. This differentiation was necessary because in deeper parts of the pools, where the sedimentological evolution of the basin is better registered, the soft bottom macrozoobenthic communities do not occur.

For the benthic communities study, six samples were taken at every station, to obtain 50 dm^3 of sediment (Pérès and Picard 1964) and cover 0.5 m^2 of surface. Sampled sediment was 20 cm thick. The benthic macrofauna was picked up after washing through 1 mm sieves.

The textural analysis was carried out according to the Buchanan and Kain method (1971), using a sieve series with an interval of Φ ($\Phi = -\log_2 \varnothing$ mm).

The organic carbon content, according to progressive improvement of our analytical routine, was determined by different methods in the three investigation cycles (1978, 1987, 1998). In particular we used the method for the organic matter content suggested by Fanucci et al. (1973) in the first investigation, and the $K_2Cr_2O_7$ oxidation method by Walkley and Black (1934) in the second. Finally, in the last cycle we used a Perkin Elmer 2400 Autoanalizer, with Acetanilide as a standard at 950°C (Hiekel 1984). For a true comparison, we converted all values in TOC (Total Organic Carbon) by the conversion factor of Dunbar et al. (1989).

Results and Discussion

The textural features of the sediments sampled during the three investigation cycles, showed high variations, especially in the percentage content of the pelitic fraction. A constant and general enrichment of mud can be seen (Fig. 2), more or less marked, according to the source and rate of the sedimentary inputs, which are peculiar to each pool.

In particular, Marinello and Mergolo pools, which were geo-morphologically stable and well defined since ancient times (Abbruzzese and Aricò 1955), in 1978 already showed a pelitic substratum. The mud increase in these basins, which mainly receive continental inputs, was regular but low. Different sedimentary evolution occurred at Verde and Fondo Porto pools, the more north-westerly ones, where the effect of sea-storms has produced an important rise of mud, recorded in the second investigation cycle. In these pools, during the considered twenty years, the primary sandy substratum slightly covered by mud (Verde: 29.60%, Fondo Porto: 0.50%, Crisafi et al. 1981), became a sandy mud, with a pelitic fraction ranking 76.33% and 66.28%, respectively (Fig. 2). Finally, in Porto Vecchio and Nuovo pools, which have became established in the last decades, the substratum is still heterogranular and badly sorted, also in relation to the continuous modifications of the littoral bar separating it from the sea. For this reason the pelitic fraction content in these pools is also low in the third investigation cycle (64.60% and 40.97%), although a progressive increase has been observed.

It is important to point out that the noted fluctuations of the sedimentary budget, are linked to some human interventions executed in this area over the last thirty years, and have caused considerable modifications in the littoral dynamics of the system (Amore and Randazzo in press).

The total organic carbon content (TOC) is generally elevated in all pools (min: 0.99%; max: 3.18%), with remarkable variations during the period considered. In the Verde and Fondo Porto pools, the organic content greatly increased in 1987, in concomitance with a marked rise of the pelitic fraction. On the contrary, in the other pools, the higher TOC values were recorded in the first investigation cycle (1978), with a more or less noticeable decrease during the two successive periods. Therefore TOC space-temporal trend does not show a clear relationship with the pelitic fraction content, even if some significant coincidences have been observed. In fact the TOC content generally decreases from the first investigation cycle to present day, in opposition to the pelitic trend, save the case of marked increase of sedimentary rate (Fig. 2c,d).

As far as the soft bottom mollusc communities are concerned, the data collected describe a generalized oligotipic situation in all the periods under investigation. An extreme case is shown by the Marinello pool, where monospecific communities, observed in 1978, became extinct later on (Table). Also in the Nuovo pool, after a period of high oligospecificity (1978-97), the mollusc communities became extinct. The causes which determined such extinctions are different. In the Marinello pool this is probably linked to the marshy vegetation growth (Raimondo and Rossitto 1978), which has progressively invaded all the soft bottoms. However in the Nuovo pool the continuous morphological changing of the littoral bar, which determined the burying of the pre-existing communities (Table), now hinders a new colonization.

Setting aside these two extreme cases, the highest values of specific richness were recorded

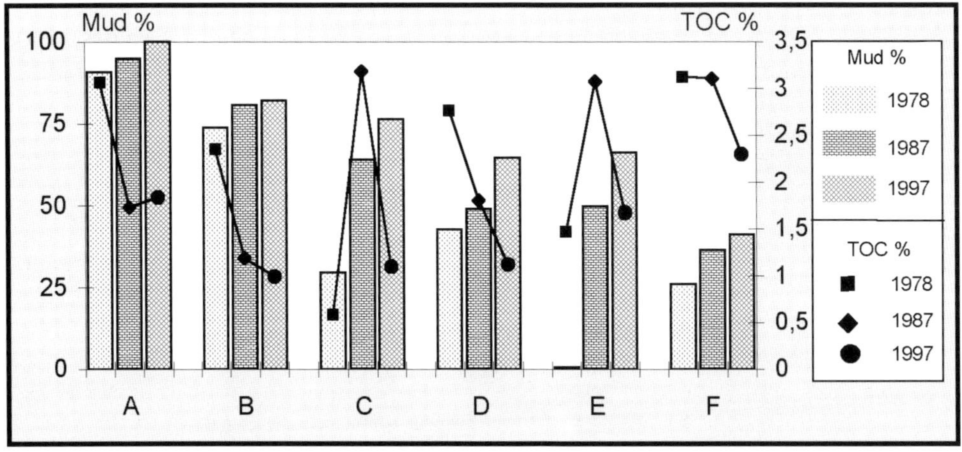

Fig. 2. Mud and total organic carbon (TOC) percentage evolution in the investigated pools

Table. Percentage of singular species at each pool

	Marinello			Mergalo			Verde			Porto V.			Fondo P.			Nuovo		
	1978 %	1987 %	1997 %	1978 %	1987 %	1997 %	1978 %	1987 %	1997 %	1978 %	1987 %	1997 %	1978 %	1987 %	1997 %	1978 %	1987 %	1997 %
Suspension feeders				72.5	100.0	60.3	39.0	65.0	23.6	82.5	73.0	33.2	68.0	63.0	66.3	100.0	100.0	
Mytilaster marioni (Locard 1889)						48.6						16.6						
Loripes lacteus (L. 1758)				8.3		2.7				10.2		16.6			26.9			
Cerastoderma glaucum (Poiret 1789)				44.2	78.2		10.7	65.0	3.3									
Tapes (Ruditapes) *decussatus* (L. 1758)							28.3		20.3				35.0	63.0	6.8	70.0	25.0	
Paphia aurea (Gmelin in L. 1791)			20.0	21.8	9.0			72.3	73.0		33.0		32.6		32.6	30.0	75.0	
Detritus feeders	100.0	100.0					52.5	10.0	52.5			16.6		18.0				
Abra segmentatum (Rechuz 1843)	100.0	100.0					52.5	10.0	52.5			16.6		18.0				
Carnivorous				27.5		39.7	8.5	25.0	23.8	17.5	27.0	50.0	32.0	19.0	33.7			
Centhium rupestre (Risso 1826)				27.5		38.7				17.5	27.0	50.0	32.0	7.0	3.8			
Cyclope neritea (L. 1758)						1.0	8.5	25.0	13.6				12.0	1.5				
Nassarius costulatus (Brocchi 1814)															0.8			
Hamminoea hydatis (L. 1758)									10.2						27.7			

in the third investigation cycle, after a decrease observed in the previous cycle. The highest number of species, seven, was observed in the Fondo Porto pool.

The population density in each pool varied over time, following two different trends. In particular, the first was observed in the Marinello, Mergolo, Nuovo and Porto Vecchio pools, where density was constantly decreasing (Fig. 3). Such reduction was at times strongly marked, as observed in the Porto Vecchio pool, in which mollusc communities shift from 576 specimens per square metre in 1978, to 12 specimens in 1998. In the Mergolo pool the reduction was less evident and the passage was from 460 to 222 specimens per square metre. Another typology was observed in the Verde and Fondo Porto pools, where the highest density of specimens was observed in the second investigation cycle, with 952 and 408 respectively.

As far as the specific composition of mollusc communities is concerned, significant temporal variations were registered in each pool (Table). The more evident variations concern the massive appearance of *Mytilaster marioni* in mobile bottoms of the Mergolo and Porto Vecchio pools, which is a result of an ongoing stabilization of the substratum. It is also interesting to point out the appearance of *Abra segmentum*, a highly typical euryhaline species, which, together with the already mentioned *Mytilaster marioni* shows the evolution of the pools towards a more brackish typology. Moreover, *Abra segmentum*, already present in the oldest pools (Marinello, Mergolo, Verde), settled in the still unoccupied trophic niche of deposit feeders in the mollusc community of the Porto Vecchio pool. Additional species which have more recently enriched the mollusc communities include the gastropods *Haminoea hyatis* and *Nassarius costulatus*, the former being an active carnivore and the latter mainly necrophagous.

Given the extreme poverty of species encountered, the appearance or disappearance of even a single one, may determine noticeable variations in the trophic structure of the mollusc communities. In our investigations the most evident variations concern the incidence of suspension feeders, which is generally decreasing (Table). The only exception is the Fondo Porto pool, which has more or less remained unaltered in time. However, while in the Porto Vecchio pool the incidence of the suspension feeders decreases constantly, in the Mergolo and Verde pools the present decrease follows a peak which was observed in 1987. In the Marinello and Nuovo pools, where molluscs have not been found at present, the original communities contained only deposit feeders in the former and suspension feeders in the latter.

Conclusions

Several studies have already shown the very high dynamism of the Oliveri-Tindari system. However, can this dynamism be interpreted as a real evolution of the ecosystem, or does it repre-

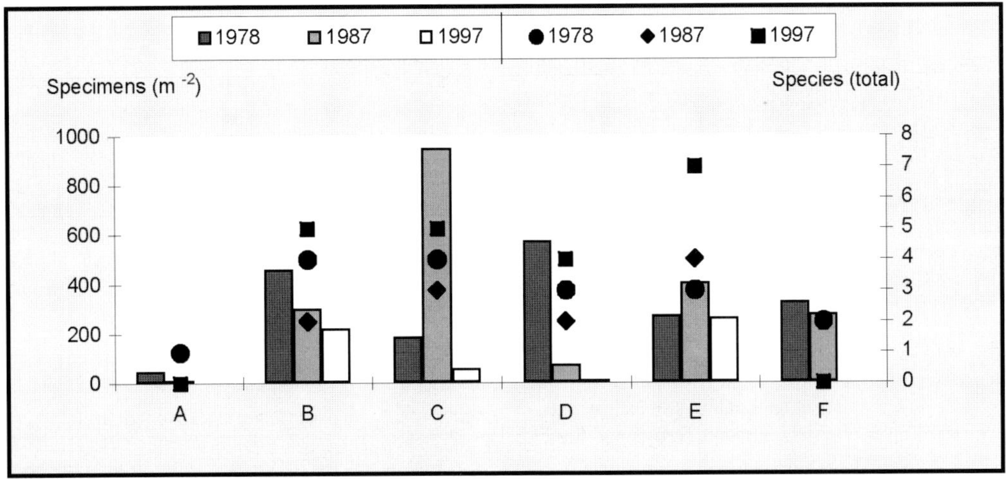

Fig. 3. Mollusc richness and total abundance at each pool

sent the effect of the human interventions on the updrift areas?

Some of the data we present here leads us to think that the six pools, during the time of their short existence, are actually evolving in the natural direction of these environments, that is, towards a progressive confinement, hypothesis actually confirmed by the progressive muddying of bottom sediments. However the organic load does not grow proportionally, but it seems to diminish, albeit with wide fluctuations.

The benthic communities, besides, are also undergoing these modifications, and they tend to organize themselves more and more in lagoon-type communities, as is evident, among other things by the growing diffusion of specific salt-marsh species (*Abra segmentatum* or *Cerastoderma glaucum)*, in comparison with those which are simply eurytopic (*Paphia aurea* or *Nassarius costulatus)*. Such a process of organization is also highlighted by the increase of specific richness as well as by the lower homogeneity of the trophic structure of the communities, which tend to become less numerous and more articulated.

At the current phase of the study we still have not clarified whether this quantitative decrease in the benthic communities is structural or occasional. It is however true that the density of the populations and the TOC content of the sediments have a parallel evolution.

Without drawing too premature a conclusion, it would seem reasonable to presume that both phenomena are manifestations of the same evolutionary process, which push the ecosystem towards a continuous reorganization through biological differentiation.

Also the interference of the anthropic compartment, although it has determined fluctuations in the evolutionary trend of the system, has not substantially modified the pathway, being that of a progressive confinement.

References

Abbruzzese D, Aricò F (1955) Osservazioni geomorfologiche e fisico-chimiche sui laghi di Oliveri-Tindari. Boll Pesca Piscic Idrobiol 10: 1 23

Amore C, Randazzo G (2000) Geographical and geomorphological Evolution of the Marinello-Tindari area (North Sicily). J Coast Conserv (in press)

Azzaro F (1995) Osservazioni biennali sull'ecosistema degli stagni costieri di Oliveri-Tindari (Messina): Nutrienti e clorofilla *a*. Atti Soc Ital Ecol 16: 495-497

Brambati A (1988) Lagune e stagni costieri: due ambienti a confronto. In: Carrada GC, Cicogna F, Fresi E (eds) Le lagune costiere: ricerca e gestione. CLEM, Massa Lubrense (Napoli), pp 9-33

Buchanan JB, Kain JM (1971) Measurement of the physical and chemical environment. In: Holme NA, McIntyre AD (eds) Methods for the study of marine benthos. IPB Handbook 16: 41-65

CNR-Istituto Sperimentale Talassografico, Messina (1990) Indagine interdisciplinare sul sistema degli stagni salmastri costieri di Oliveri-Tindari (Messina). Rapp 4: 1-34

Crisafi E, Giacobbe S, Leonardi M (1981) Nuove ricerche idrobiologiche nell'area lagunare di Oliveri-Tindari (Messina). 1. Morfologia dei bacini e caratteristiche fisico-chimiche delle acque e dei sedimenti. Mem Biol Mar Oceanogr 11: 139-186

Crisafi P (1961) Primo contributo alla conoscenza dei Copepodi degli stagni litorali di Oliveri (Messina). Rapp P V Reun CIESM 16: 841-843

Dunbar RB, Leventer AR, Stockton WL (1989) Biogenic sedimentation in McMurdo Sound, Antarctica. Mar Geol 85: 155-179

Fanucci F, Fierro G, Grosso F, Piacentino GB (1973) Contributo di una indagine sedimentologica e ricerche ecologiche nel Golfo di La Spezia. Ist Idrigr Mar 1050: 1-17

Faranda F, Gangemi G, Guglielmo L (1975) Nuove condizioni dell'arenile di Oliveri e dei laghetti salmastri Mergolo della Tonnara e Verde (Prime osservazioni). Atti Soc Peloritana 21: 15-31

Genovese S (1964) Sali nutritivi ed attività microbica in alcuni ambienti salmastri. Boll Zool 31(II): 409-421

Giacobbe MG, Maimone G, Azzaro F (1992) Distribuzione del fitoplancton in un'area salmastra della Sicilia nord-orientale. G Bot Ital 126: 531-547

Giacobbe S, Giordano R (1975) Prime osservazioni sugli insediamenti bentonici della zona lagunare di Oliveri-Tindari (Messina). Atti Soc Peloritana 21: 169-182

Giacobbe S, Leonardi M (1986) L'area lagunare di Oliveri-Tindari: Sue variazioni morfologiche recenti ed evoluzione dei popolamenti a molluschi. In: Atti VII Congr AIOL: 355-366

Giacobbe S, Leonardi M, Azzaro F, Rinelli P (1990) Descrizione di un esempio di "confinamento": l'area lagunare di Oliveri-Tindari (Messina). Oebalia XVI-2: 675-678

Guelorget O, Perthuisot JP (1983) Le domaine paralique: expressions géologiques, biologiques et économiques du confinement. Trav Lab Geol 16: 1-136

Hickel W (1984) Seston in the Wadden Sea of silt (German Bight, North Sea). Neth Inst Sea Res 10: 113-131

Leonardi M, Giacobbe MG (2000) Hydrobiological conditions of the Oliveri-Tindari lagoon: state of the art and study prospects. In: Adorno F, Leonardi M, Randazzo G (eds) Atti Conv Int "Salvaguardia e Sviluppo della Riserva Naturale Orientata Laguna di Oliveri-Tindari" 14-16/11/1996 Messina, Sicily, Fondazione Laboratorio Mediterraneo, Napoli (*in press*)

Leonardi M, Azzaro F, Azzaro M, Decembrini F, Monticelli LS (2000) Ciclo della sostanza organica nell'ecosistema lagunare di Tindari (ME). Biol Mar Mediterr (in press)

Pérès JM, Picard J (1964) Nouveau manuel de Bionomie benthique de la mer Méditerranée. Rec Trav St Mar Endoume 36: 1-160

Raimondo FM, Rossitto M (1978) La vegetazione della laguna e dell'arenile di Oliveri-Tindari (Messina) e problemi relativi alla sua tutela. G Bot Ital 112(4): 309-310

Walkley A, Black IA (1934) An examination of the Degtjarreff method for determining soil organic matter, and a proposed modification of the chromic acid titration method. Soil Sci 37: 29-38

Influence of Allochthonous Plant Detritus on *Gammarus insensibilis* Stock (Amphipoda) Occurrence in the Soft-bottom Epifauna of the Northern Adriatic Sea

G. Mancinelli and L. Rossi

ABSTRACT

Sediments and benthic epifauna were examined at a coastal station (ca. 26 m water depth) located in the northwestern Adriatic Sea. Sediment cores were taken every 90 days from April 1995 to January 1996. Epifauna was further investigated by means of trophic traps deployed on the sea bottom. Sediments were characterized by the occurrence of coarse plant detritus derived from aquatic angiosperms (*Zostera* sp., *Cymodocea nodosa*) commonly distributed in lagoons and salt marshes of the Po delta area. The epifauna was dominated by the brackish amphipod *Gammarus insensibilis* Stock, whereas other taxa occurred in negligible densities (< 2% quantitative dominance). A positive relationship was found between coarse organic matter in sediments and *G. insensibilis* spatial abundance. Analyses of population parameters from core samples and of size-frequency distributions from trophic traps permitted to exclude passive dispersal as the causative factor for the constant, even though seasonally variable, presence of the amphipod in the benthic system. Our results provide clear evidence of an "outwelling" flux of particulate plant detritus linking brackish coastal areas to adjacent marine environments in the northwestern Adriatic Sea. Moreover, they give support to the hypothesis that advection of allochthonous particulate detritus in marine environments may determine an "estuarine" characterization of epifaunal soft-bottom assemblages.

Introduction

Plant detritus export from salt marshes, estuaries and coastal lagoons plays a significant role in nutrients and energy dynamics of marine ecosystems (Nixon 1980 and literature cited therein; Dame et al. 1991; Taylor and Allanson 1995). On a global scale, macrophytal net primary production exceeds by ca. 40% total energy requirements of estuaries and salt marshes (Duarte and Cebrian 1996). In this perspective, angiosperms act as major sources of the organic carbon stored and subsequently buried in sediments of adjacent marine environments (ca. 30% of total ocean carbon storage). Furthermore, allochthonous macrophytal detritus may represent a relevant structuring factor at different levels of marine benthic systems, *e.g.* enhancing rates of microbial activity (Peduzzi and Herndl 1991), or affecting macrobenthic communities structure and secondary production (Vetter 1995).

In the northwestern Adriatic Sea, the distribution and productivity of macrophytes in coastal lagoons have been widely investigated (*e.g.* the Venice Lagoon: Scarton et al. 1995; den Hartog et al. 1996; Sfriso and Ghetti 1998). On the other hand, scarce information is available on the fate of macrodetritus, in terms of particulate matter fluxes from brackish to adjacent marine environments, and on effects of detritus advection on benthic assemblages.

Our study examines sediment characteristics and the epibenthic macrofauna at a coastal site located in northwestern Adriatic Sea, focusing on relationships between organic matter richness of sediment coarsest dimensional fractions and dominant epifaunal taxa.

Materials and Methods

A sampling station (45°08.96' N, 12°23.21' E; water depth: 26 m; Fig. 1) was located in the

Dipartimento di Genetica e Biologia Molecolare, Università di Roma I "La Sapienza", Sezione Ecologia, Via Lancisi 29, 00161 Roma, Italy

F.M. Faranda, L. Guglielmo, G. Spezie (eds)
Mediterranean Ecosystems: Structures and Processes

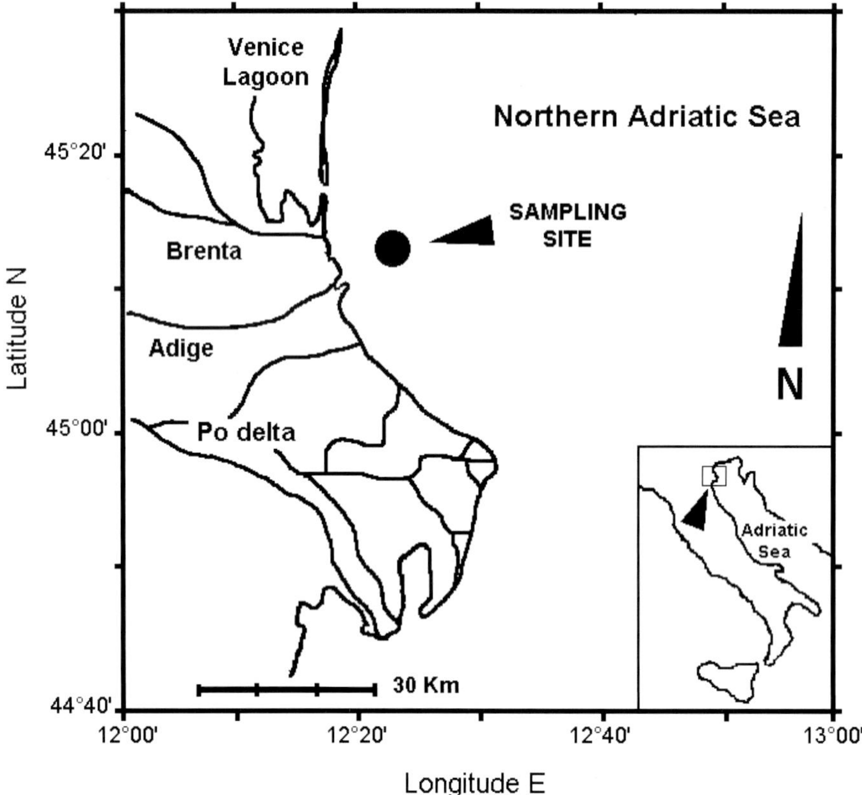

Fig. 1. The northwestern Adriatic Sea: location of the sampling site

northwestern Adriatic Sea 4.28 nautical miles offshore. River mouths (Adige River: Boldrin et al. 1992), lagoon outflows (e.g. the Venice Lagoon: Bergamasco et al. 1998), and tidal- and wind-induced currents (Franco and Michelato 1992) interact with the general counter-clockwise circulation of the basin, strongly influencing the dynamics of water masses in the area.

Sediment cores (superficial layer, upper 5 cm) were taken every 90 days using a Reineck box-corer (170 cm^2, 5 replicas) from April 1995 to January 1996. After retrieval, cores were placed in plastic containers with the original bottom water overlying the sediment, and transferred (at 4°C in the dark) within 12 h to the laboratory. The following substrate attributes were determined:

1. Frequency distribution of grain size diameters: sediment cores where wet-sieved through a graded series of standard ASTM sieves (ϕ units: –1.585; –0.585; 0.415; 1.415; 2.415; 3.415; 4.988). After removal of the macrofauna, every size class was oven-dried (60°C, 72 h) and weighed to the nearest 0.1 mg (dry weight; DW in the following text). Median grain size (Mdϕ in the text) was graphically determined on sediment size-frequency cumulative curves.
2. Sediment organic content: each grain size fraction was ashed (450°C, 6 h) and subsequently re-weighed. Organic matter content was determined as the loss of dry weight after ignition. Total organic content (TOC in the following text) was determined as the dry weight loss after ignition of sediment subsamples taken from three additional cores.
3. Epifauna: organisms retained in sieves > 0.5 mm ($\phi < 0.415$) were separated from inorganic particles and fixed in ca 5% buffered formalin. All epifaunal organisms were counted and identified to the lowest taxonomic level necessary for trophic strategy and motility determination.

Organisms were dried at 60°C for at least 72 h and weighed individually to the nearest 0.01 mg. Ash free dry weight (AFDW in the text) was determined after ignition. For every taxon, quantitative dominance (QD in the following text) was expressed as the percentage of the whole number of sampled epifaunal specimens; spatial density was expressed both in terms of *n* individual m^{-2} and mg AFDW m^{-2}.

Together with sediment cores were retrieved three trophic traps, deployed on the sea bottom seven days before sediment sampling. Each trap consisted of two plastic bags (40 × 80 cm, 10 × 3 mm mesh size) fixed to a concrete base (0.36 m^2) loosely tethered to a buoy. Bags were filled with 1000 ± 1 g DW of mixed *Cymodocea nodosa* Ascherson and *Zostera marina* Linnaeus leaf detritus previously collected in the Venice Lagoon. After retrieving, the detritus contained in each bag was washed on board on 0.5 mm-mesh sieves. Living specimens were separated from inorganic particles and preserved in ca. 5% buffered formalin. In the laboratory, specimens were identified and counted. Individual dry weight was determined according to the aforementioned procedures. Consequently, individuals were grouped on the basis of their DW in eleven size classes determined on a log-2 scale.

Results and Discussion

The sediment of the sampling station consisted of silt-clay, characterized by negligible seasonal variations in median grain size (Table; Mdϕ = 4.18 ± 0.02 annual mean ± ES). On the other hand, total organic content (TOC: Table) varied significantly among seasons (1-WAY ANOVA on arcsine-transformed data: $F_{3,8} = 8.77, P = 0.006$). The coarsest dimensional fraction ($\phi = -1.585$, > 2 mm) resulted significantly enriched in organic matter (Table; organic content in $\phi = -1.585$: 16.29% ± 3.5; TOC: 8.97% ± 2.21, annual means ± ES; Kolmogorov - Smirnov test, $P < 0.05$). Organic content of the size fraction showed significant among-season differences (1-WAY ANOVA on arcsine-transformed data: $F_{3,16} = 13.05, P < 0.001$), unrelated to TOC variations (r =0.09, 2 d. f., NS). Organic enrichment in the fraction $\phi = -1.585$ was due to the occurrence of particulate plant detritus, mainly constituted by *Zostera* sp. and *Cymodocea nodosa* (Ucria) Ascherson leaf fragments (Fig. 2). Both phanerophytes occur in brackish environments of the northwestern Adriatic coasts (e.g. the Venice Lagoon: den Hartog et al. 1996; Sfriso and Ghetti 1998), where they provide a significant fraction of the total macrophytal production (ca. 55-60% net production in the Venice Lagoon: Sfriso and Ghetti 1998). The low decay rates of marine angiosperms in lagoon and salt-marsh environments (Tenore et al. 1984), make angiosperm-derived particulate detritus susceptible to export and dispersion by tidal currents (or by other large-scale oceanographic factors) in adjacent sublittoral environments (Mann 1988). In this perspective, our study clearly indicates that in the northwestern Adriatic basin "outwelling" fluxes (*sensu* Odum 1980) of particulate detritus from brackish sites may represent a potentially significant source of organic matter for marine benthic systems, conventionally related to pelagic production or freshwater inputs (Faganeli et al. 1994; Pettine et al. 1998).

The epibenthic fauna observed in sediment cores and in trophic traps was dominated by the detritivorous amphipod *Gammarus insensibilis* Stock (QD = 98.78 and 99.73% in cores and in trophic traps, respectively. The QD in trophic traps refers only to spring and summer samples: both fall and winter traps were lost due to bad weather condition). Other taxa (e.g. crustaceans: *Corophium* sp., *Caprella* sp.; echinoderms: *Ophiura* sp.; see Mancinelli et al. 1998 for a complete list of species found in core samples) occurred in negligible densities and they were not included in further analyses.

In Europe, *Gammarus insensibilis* (Fig. 3, left) is widely distributed in brackish environments down to 15 m (Sheader and Sheader

Table. Seasonal variation of sediment parameters [*Mdϕ*, median grain size; *TOC*, total organic content (% dry weight); *OC in ϕ = -1.585*, organic richness (% dry weight) of the coarsest dimensional fraction (> 2 mm)]. ES *in brackets*

Parameter season	Md ϕ		TOC		OC in ϕ <-1.585	
Spring (April)	4.21	(0.01)	3.64	(0.54)	16.62	(5.07)
Summer (July)	4.17	(0.01)	11.74	(0.18)	25.77	(2.27)
Fall (October)	4.19	(0.02)	7.18	(2.35)	13.55	(1.47)
Winter (January)	4.14	(0.03)	13.33	(2.11)	9.25	(3.78)

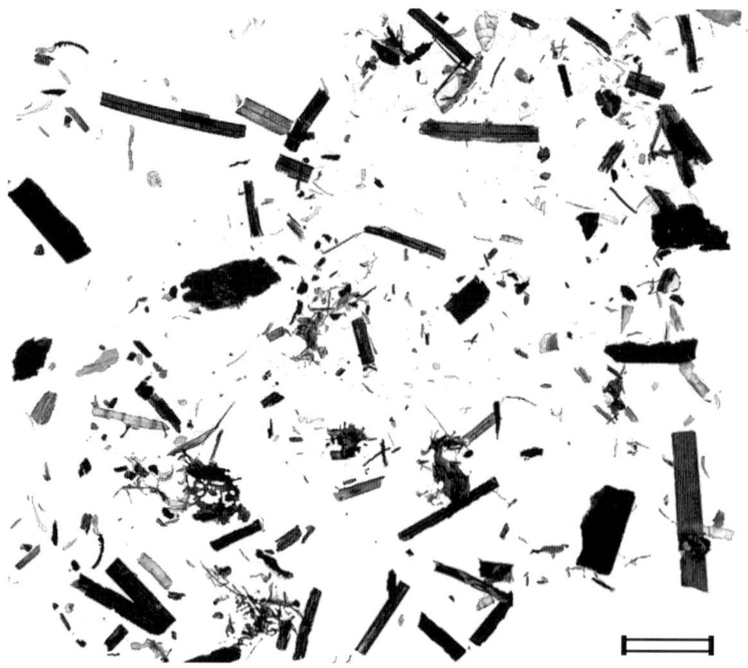

Fig. 2. Coarse detritus, mainly *Zostera sp.* e *Cymodocea nodosa* fragments, in sediment size fraction ϕ = -1.585. *Scale bar* = 1 cm

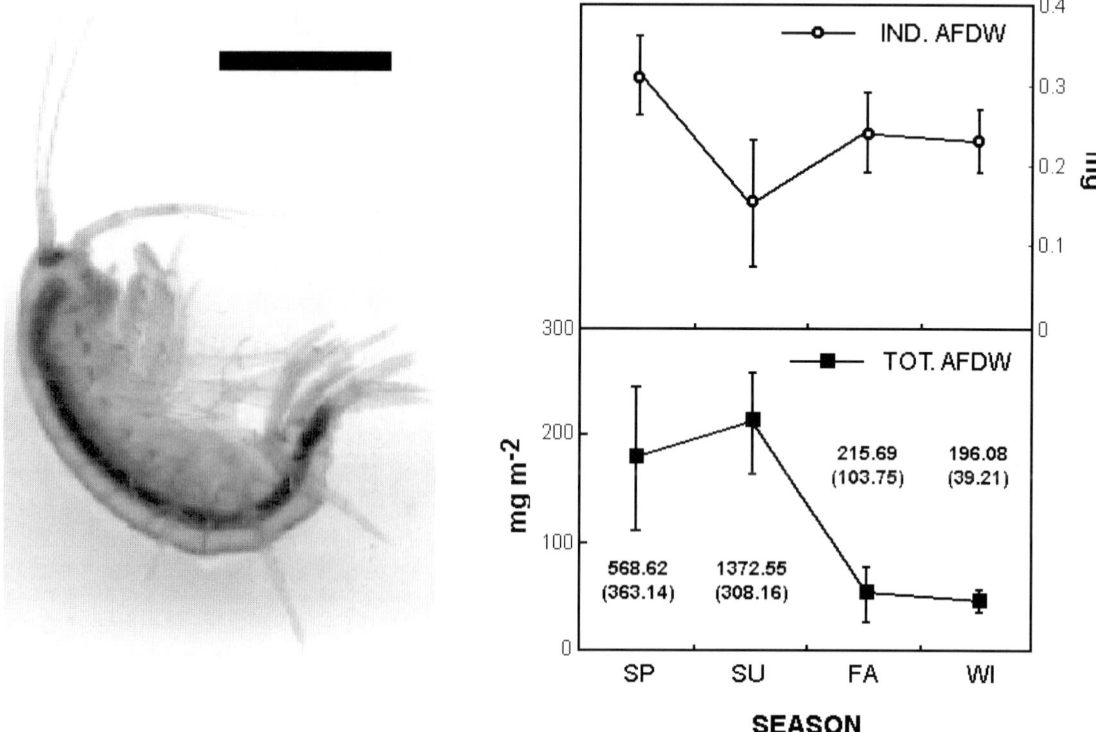

Fig. 3. *Left*: the amphipod *Gammarus insensibilis* (Stock). *Scale bar* = 1 mm. *Right*: seasonal patterns of the amphipod population parameters in core samples. Mean individual AFDW (top) and total AFDW density (*bottom*). *Numbers* in the *bottom graph* refer to *G. insensibilis* numerical density (*n* individual·m^{-2}, ES *in brackets*)

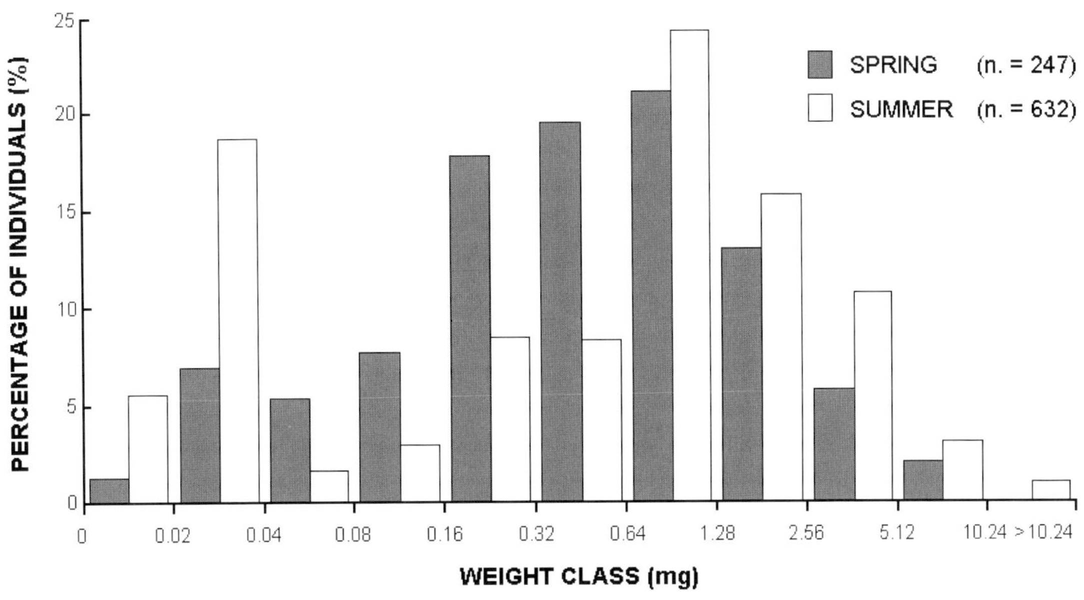

Fig. 4. Size-frequency histograms of *Gammarus insensibilis* specimens sampled in spring and summer trophic traps

1985), in association with macroalgae and macrodetritus (Stock 1967; Nicolaidou and Karakiri 1989; Procaccini and Scipione 1992). The species is particularly active (it can be enclosed in the "epifaunal free-living" ethological group, *sensu* Biernbaum 1979), euryhaline, and plurivoltine. Its reproductive phase is continuous during the whole year, even though the maximal release of juveniles usually occurs in summer (Brun 1975; Sheader 1996).

In our study, several evidences permit to exclude that the presence of the amphipod may be related to a phenomenon of passive dispersal of allochthonous individuals from brackish habitats. In fact: (1) population parameters from core samples (Fig. 3, right) are consistent with available information on populations of the amphipod in estuarine and salt-marsh environments (Brun 1975; Nicolaidou and Karakiri 1989); (2) both spring and summer size-frequency distributions of trophic traps-sampled specimens (Fig. 4) show features consistent with those recorded for autochthonous gammaridean populations (*Gammarus aquicauda*: Kevrekidis and Lazaridou-Dimitriadou 1988; *Gammarus duebeni*: Watt and Adams 1993); (3) in the Po delta area, *Gammarus insensibilis* is commonly associated to detritivorous isopods (e.g. *Idotea baltica* Pallas, *Sphaeroma* sp.) and/or other gammaridean amphipods (*Gammarus aequicauda* Martynov) (Fava et al. 1992; Patarnello et al. 1992). None of these brackish species were found either in cores nor in trophic traps.

The positive relationship found between *G. insensibilis* spatial density and organic content of the coarsest fraction ($y = 1.296x - 11.117$, where $y = n$ individuals • m^{-2}, x = % OM in the fraction $\phi = -1.585$: $r = 0.97$, 2 d. f., $P < 0.05$) may thus indicate the occurrence of a direct trophic linkage, where detritus, as the only trophic resource available in the soft-bottom environment, represents a limiting factor for the abundance of the amphipod.

In conclusion, our results support the hypothesis that particulate detritus advection from brackish coastal areas may represent in northwestern Adriatic Sea both a significant component of sediment organic matter and a structuring factor for benthic macrofaunal assemblages, determining an "estuarine" characterization of sublittoral benthic systems.

Further investigations are needed to assess the functional role played by the detritus – detritivores interaction in the Adriatic benthic system. In fact, macrodetritivores may affect particulate organic matter decay and nutrients biogeochemistry, a specific functional role already recognized in freshwater (Oertli 1993), estuarine, and salt-marsh environments (Robertson and Mann 1980; Peduzzi and Herndl 1991).

Acknowledgements. We wish to thank two anonymous reviewers for their valuable comments on the first draft of the manuscript. This research was supported by CNR-PRISMA1 and MURST funds.

References

Bergamasco A, Carniel S, Pastres R, Pecenik G (1998) A unified approach to the modelling of the Venice Lagoon-Adriatic Sea Ecosystem. Estuarine Coast Shelf Sci 46: 483-492

Biernbaum CK (1979) Influence of sedimentary factors on the distribution of benthic amphipods of Fishers Islands Sound, Connecticut. J Exp Mar Biol Ecol 38: 201-223

Boldrin A, Juracic M, Menegazzo M (1992) Sedimentation of riverborne material in a shallow shelf area: Adige River, Adriatic Sea. Mar Geol 103: 473-482

Brun B (1975) Quelques aspects des cycles biologiques de deux Gammares (Crustacés Amphipods) des eaux saumâtres du littoral méditerranée français. Arch Zool Exp Gen 116: 343-358

Dame RF, Spurrier J, Williams T, Kjerfve B, Zingmark RG, Wolaver TG, Chrzanowki TH, McKellar HN, Vernberg FJ (1991) Annual material processing by a salt marsh-estuarine basin in South Carolina, USA. Mar Ecol Prog Ser 72: 153-166

den Hartog C, Vergeer LHT, Rismondo AF (1996) Occurrence of *Labyrinthula zosterae* in *Zostera marina* from Venice Lagoon. Bot Mar 39: 23-26

Duarte CM, Cebrian J (1996) The fate of marine autotrophic production. Limnol Oceanogr 41: 1758-1766

Faganeli J, Pezdic J, Ogorelec B, Mišic M, Najdek M (1994) The origin of sedimentary organic matter in the Adriatic. Cont Shelf Res 14: 365-384

Fava G, Zangaglia A, Cervelli M (1992) Ecology of *Idotea baltica* (Pallas) populations in the lagoon of Venice. Oceanol Acta 15: 651-660

Franco P, Michelato A (1992) Northern Adriatic Sea: oceanography of the basin proper and of the western coastal zone. Sci Total Environ Suppl. 1992: 35-62

Kevrekidis T, Lazaridou-Dimitriadou M (1988) Relative growth and secondary production of the Amphipod *Gammarus aequicauda* (Martynov, 1931) in the Evros Delta (N. Aegerian sea). Cah Biol Mar 29: 483-495

Mancinelli G, Fazi S, Rossi L (1998) Sediment structural properties mediating dominant feeding type patterns in soft-bottom macrobenthos of the Northern Adriatic Sea. Hydrobiologia 367: 211-222

Mann KH (1988) Production and use of detritus in various freshwater, estuarine and coastal marine systems. Limnol Oceanogr 33: 910-930

Nicolaidou A, Karakiri M (1989) The distribution of Amphipoda in a brackish-water lagoon in Greece. PSZNI Mar Ecol 10: 131-139

Nixon SW (1980) Between coastal marshes and coastal waters - a review of twenty years of speculation and research on the role of salt marshes in estuarine productivity and water chemistry. In: Hamilton R, MacDonald KB (eds) Estuarine and wetland processes. Plenum Press, New York, pp 437-525

Odum EP (1980) The status of three ecosystem-level hypotheses regarding salt marsh estuaries: tidal subsidy, outwelling and detritus-based food chains. In: Kennedy V (ed) Estuarine Perspectives. Academic Press, New York, pp 437-525

Oertli B (1993) Leaf litter processing and energy flow through macroinvertebrates in a woodland pond (Switzerland). Oecologia 96: 466-477

Patarnello T, Battaglia B, Bisol PM (1992) Genetic differentiation among geographic populations of two species of the genus *Gammarus*: *G. insensibilis* and *G. aequicauda* (Crustacea, Amphipoda) Vie Milieu 42: 263-268

Peduzzi P, Herndl GJ (1991) Decomposition and significance of seagrass leaf litter (*Cymodocea nodosa*) for the microbial food web in coastal waters (Gulf of Trieste, Northern Adriatic Sea). Mar Ecol Prog Ser 71: 163-174

Pettine M, Petrolecco L, Camusso M, Crescenzio S (1998) Transport of carbon and nitrogen to the northern Adriatic Sea by the Po River. Estuarine Coast Shelf Sci 46: 127-142

Procaccini G, Scipione MB (1992) Observations on the spatio-temporal distribution of crustacean amphipods in the Fusaro coastal lagoon (Central Tyrrhenian Sea, Italy) and some notes on their presence in mediterranean lagoons. PSZNI Mar Ecol 13: 203-224

Robertson J, Mann KH (1980) The role of isopods and amphipods in the initial fragmentation of eelgrass detritus in Nova Scotia, Canada. Mar Biol 59: 63-69

Scarton F, Curiel D, Rismondo A (1995) Aspetti della dinamica temporale di praterie a fanerogame marine in Laguna di Venezia. Lav Soc Venice Sci Nat 20: 95-102

Sfriso A, Ghetti PF (1998). Seasonal variation in biomass, morphometric parameters and production of seagrasses in the lagoon of Venice. Aquat Bot 61 207-223.

Sheader M (1996) Factors influencing egg size in the gammarid amphipod *Gammarus insensibilis*. Mar Biol 124: 519-526

Sheader M, Sheader AL (1985) New distribution records for *Gammarus insensibilis* Stock, 1966, in Britain. Crustaceana 49: 101-105

Stock JH (1967) A revision of the European species of the *Gammarus locusta*-group. Zool Vehr Leiden 90: 1-56

Taylor DI, Allanson BR (1995) Organic carbon fluxes between a high marsh and estuary, and the inapplicability of the outwelling hypothesis. Mar Ecol Prog Ser 120: 263-270

Tenore R, Hanson RB, McClain J, MacCubbin AE, Hodson RE (1984) Changes in composition and nutritional value to a benthic deposit feeder of decomposing detritus pools. Bull Mar Sci 35: 299-311

Vetter EW (1995) Detritus based patches of high secondary production in the nearshore benthos. Mar Ecol Prog Ser 120: 251-262

Watt PJ, Adams A (1993) Adaptative sex determination and population dynamics in a brackish-water amphipod. Estuarine Coast Shelf Sci 37: 237-250

CHAPTER 41

Indirect Control of Vagile Epifauna on *Gracilaria verrucosa* (Hudson) Papenfuss* Growth: a Field Study in the Lesina Lagoon (FG), Italy

G. Mancinelli, P. Polimeno, and L. Rossi

ABSTRACT

The effect induced by vagile macrofauna on the growth of the Rhodophyta *Gracilaria verrucosa* (Huds.) was evaluated in the Lesina Lagoon (southern Italy) through an in situ manipulation experiment. Percent wet weight daily growth rates and percent ash content of pre-weighted *G. verrucosa* thalli were monitored for 28 days in enclosures made of nets with fine (220 μm, "F" treatment) and medium mesh (2000 μm, "M" treatment). Both partial and total faunal exclusion treatments showed a significant negative variation of daily growth rates together with evident fouling phenomena confirmed by the percent total ash content analysis on *Gracilaria* samples. A further analysis on ash content of thalli fragments free from fouling organisms excluded any effective variation of macroalgal tissue ash content. The study highlighted a significant effect induced by vagile epifauna on *G. verrucosa* growth, accomplished through an indirect control on fouling organisms. Our results support, besides regeneration phenomena of nutrients necessary for growth induced by particulate detritus processing (widely documented in the literature), the hypothesis of a further, positive action of herbivorous-detritivorous macrofauna on macroalgal populations in salt marsh and coastal environments. The results, moreover, suggest a potentially effective control method on the quality of *Gracilaria*, which is extensively used as an agar source.

Introduction

In estuarine and coastal environments, macroalgae are commonly associated with a diverse assemblage of small invertebrates, many of which are potential consumers of their host plants. Collectively referred to as "mesograzers" (Hay et al. 1987), these small herbivores include amphipods (Duffy 1990; Poore 1994) and isopods (Arrontes 1990). Despite their high abundance, often in the order of thousands per square metre of substratum (Healy and O'Neill 1984), the way in which vagile macrofauna affect algal growth is poorly understood (Brawley 1992). Mesograzers inhabiting macroalgae have the opportunity to feed not only on their host plant but also on a large spectrum of fouling organisms, usually exerting a detrimental effect on algal growth rates (Wahl 1989). Therefore, the trophic activity displayed by mesograzers has great significance for host plant fitness and development patterns: in fact, they may affect their host both indirectly through removal of the fouling colonisers, or directly through consumption of the host plant tissues.

The present study investigated the effects induced by vagile macrofauna on the growth of the Rhodophyta *Gracilaria verrucosa* (Huds.) in the Lesina Lagoon (FG), Italy. In an experimental site characterised by the co-occurrence of the red macroalga with high densities of benthic macrocrustaceans, the effective impact of epifauna on daily growth rates of pre-weighted *Gracilaria* living thalli was examined in an situ manipulation experiment involving the use of partial and total exclusion cages. Macrofaunal

* According to Steentoft et al. (1995) the correct name of this species is *Gracilariopsis longissima* (S.G. Gmelin) Steentoft, Irvine L.M. and Farnham W.F.

Dipartimento di Genetica e Biologia Molecolare, Sezione Ecologia, Università di Roma "La Sapienza", Via dei Sardi 70, 00185 Roma, Italy

F.M. Faranda, L. Guglielmo, G. Spezie (eds)
Mediterranean Ecosystems: Structures and Processes

effects were also evaluated by thalli visual analysis, percent total ash content and ash content of fragments free from fouling colonisers.

Materials and Methods

A survey of the epifaunal assemblage associated with *Gracilaria verrucosa* in the Lesina Lagoon at the time of the present study showed that three species of vagile mesograzers - the isopods *Idotea baltica* Pallas and *Sphaeroma serratum* Fabricius, and the amphipod *Gammarus insensibilis* Stock – together accounted for ca. 90% of the total number of epifaunal organisms found on the macroalga (Mancinelli, unpublished data). Our findings corresponded with previous investigations on the Lesina Lagoon epifauna (De Benedictis et al. 1996; Dinardo et al. 1997). The encrusting bryozoan *Conopeum seurati* Canu and the filamentous macroepiphyte *Enteromorpha intestinalis* (L.) Link. were observed as the prevailing fouling colonizers of the macroalga. At the site chosen for the manipulation experiment, *I. baltica*, *G. insensibilis* and *S. serratum* together acconted for 92.7 ± 2.8% of epifaunal organisms sampled on *Gracilaria*, corresponding to a total density of 5.3 ± 0.6 individuals g^{-1} algal wet weight. *C. seurati* represented the dominant fouling colonizer.

The manipulation experiment was done using cages. They consisted of cylindrical polyethylene frames (Ø 25 cm, 60 cm high), with the bottom covered with a 10-mm nylon mesh. One day prior to the experiment, fouling-free *G. verrucosa* turfs of similar branching and size were collected and stored overnight in oxygenated tanks containing defaunated lagoon water.

On 4 May 1998, randomly chosen turfs free from fouling bryozoans and grazing damage were blotted dry, singularly weighted (1.787 ± 0.1 g, mean wet weight ± 1 SE), and carefully tied with nylon threads to the bottom of the cages. Invertebrates were excluded or allowed access to the macroalga in three 'cage' treatments: (1) no exclusion, *Gracilaria* in open frames without mesh that allowed access to all invertebrates ("C" treatment); (2) total exclusion, *Gracilaria* in frames covered with small nylon mesh (0.22 mm mesh size) that prevented invertebrates access ("F" treatment) and (3) partial exclusion, *Gracilaria* in frames covered with large mesh (2 mm) allowing access to juveniles and immature stages of the three species dominating the epifauna ("M" treatment).

A total of 27 enclosures were randomly attached to wooden sticks embedded in the bottom mud at the same depth at which thalli fragments were sampled.

Wet weight and percent ash content of *Gracilaria* turfs were determined after 5, 15 and 28 day. At each sampling time, three replicates of each treatment were retrieved, and the algal samples were rinsed in deionized water to eliminate macrofauna and residual salts, tapped in blotting paper and wet-weighted. They were successively oven-dried for 96 h at 60 °C, weighted (dry weight, DW) and ignited (450 °C, 6 h, to avoid carbonate volatilization; Rosa et al. 1994). Fouling-free algal subsamples were taken from each turf and subjected to the same procedure.

Growth rates of *Gracilaria* were measured from increases in turf wet weight and presented as percent growth per day according to the formula $G = [(W_t - W_0)^{1/t} - 1] \times 100$ (Penniman et al. 1986), where G = percent increase in wet weight per day, W_0 = initial weight, and W_t = weight after t days. Percent ash content of samples was calculated as the percent ratio between DW after and before ignition. For parametric statistical analysis, all percentages were normalised as $x = arcsin \sqrt{P/100}$.

Results and Discussion

Gracilaria demonstrated in the control treatment no significant variations in G values during the study (1-way ANOVA, F = 0.67, NS; min, 10.49 ± 0.18%; max 12.23 ± 0.39 % d^{-1}). In contrast, in both partial and total faunal exclusion treatments, *Gracilaria* showed negative variations in "G" values (Fig. 1). In particular, whereas no difference was found among treatments after 5 d (1-way ANOVA, F = 0.92, NS), after 15 d and 28 d the observed negative variations were statistically significant (F=18.09, $p<0.01$ and F=21.43, $p<0.01$, respectively).

Additionally, in comparison to the control, no differences occurred between exclusion treatments at day 15 and 28 (1-way ANOVA, contrast analysis 15 d : "F" and "M" treatments vs. "C", F = 32.26, $p<0.01$ "F" vs. "M", F = 3.91, NS; 28 d: "F" and "M" vs. "C", F = 37.83, $p<0.001$, "F" vs. "M", F = 5.03, NS). After 15 and 28 days it was also possible in "M" and "F" treatments to detect a signif-

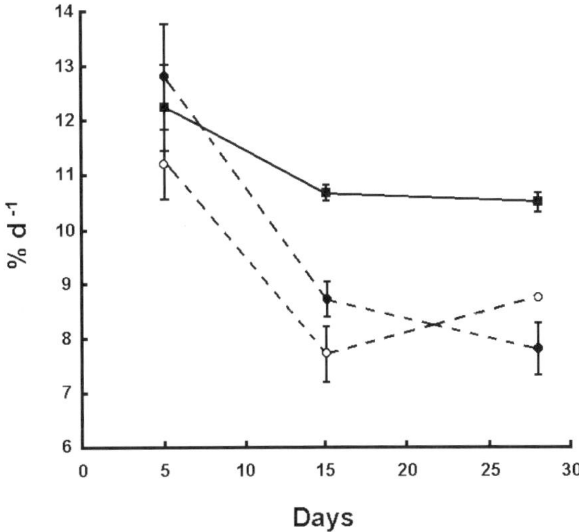

Fig. 1. Percent daily growth rates (% WW d^{-1}), G, under total ("F" treatment, ●), partial ("M" treatment, ❍), and no ("C" treatment, ■) macrofaunal esclusion conditions

icant fouling phenomenon (completely absent in the control treatment, Fig. 2), characterised by the strong prevalence of the encrusting bryozoan *Conopeum seurati* Canu (Fig. 3).

The three crustacean species dominating the epifaunal assemblage present in the experimental site may potentially exploit *G. verrucosa* as a trophic resource. In fact, *I. baltica* is an active mesograzer-detritivore (Robertson and Mann 1980; Costantini and Rossi 1995; Schaffelke et al. 1995), whereas representatives of the genus *Sphaeroma* are reported as scraping grazers-detritivores exploiting epibionts together with living and/or non-living plant substrates (Frier 1979; Kittlein 1991). Only scarce direct information is available on the diet of *G. insensibilis*. Sheader and Sheader (1985) and Sheader (1996) defined *G. insensibilis* as a mesograzer-detritivore on the basis of its association with macroal-

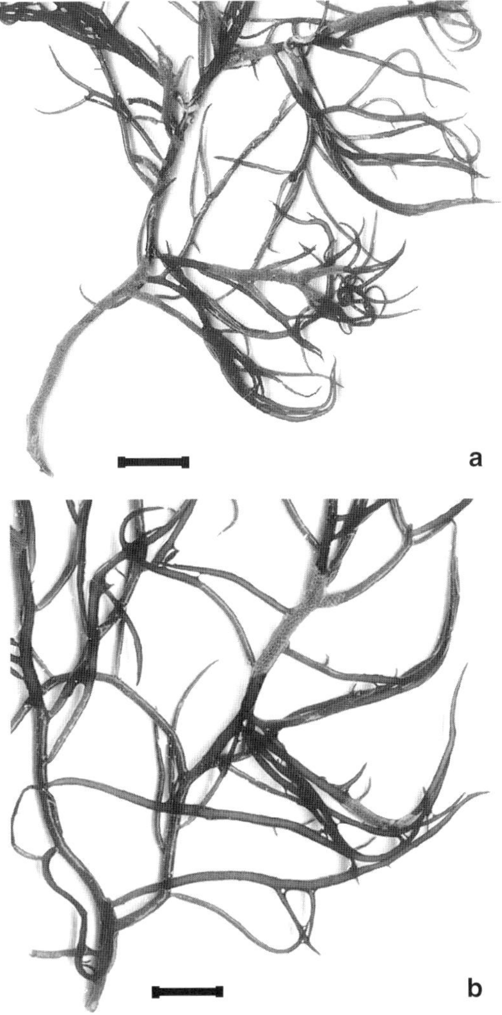

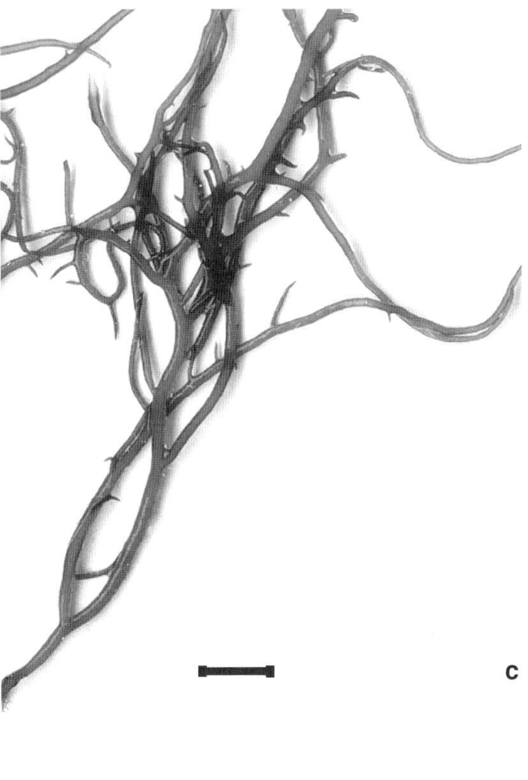

Fig. 2a-c. *Gracilaria verrucosa* thalli recovered at day 28. **a** "F" treatment; **b** "M" treatment; **c** "C" treatment. *Bars* = 1 cm

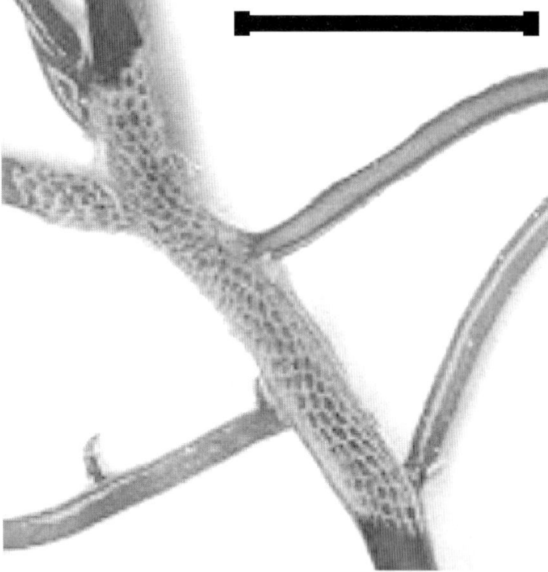

Fig. 3. Detail of the encrusting bryozoan *Conopeum seurati* Canu on *Gracilaria* thallus. *Bar* = 1 cm

gal and macrophytal beds. In Costantini and Rossi (1995), it is found strictly associated with phanerogam coarse detritus. Nevertheless, marine amphipods of the *Gammarus locusta* group are generally described as macrophagous mesograzers/detritivores (Harrison 1977; Lincoln 1979; Robertson and Mann 1980; Agnew and Moore 1986; Pavia et al. 1999). In our study, several lines of evidence exclude direct grazing as a significant controlling factor on algal growth: macrofaunal exclusion failed to show both significant short-term effects and body-size related differences; moreover, G values in control treatment fell within the reported ranges for other *Gracilaria* species cultivated under grazer-free conditions (Haglund and Pédersen 1993; Pickering et al. 1993). The significant, negative effect induced by fouling colonisers on *G. verrucosa* growth in "F" and "M" treatments seems to be strongly related to the exclusion of the vagile epifauna. These findings support the hypothesis that in the Lesina Lagoon the epifaunal assemblage associated with *G. verrucosa* is beneficial to the host plant exerting an indirect control on epibiont development, that reduce plant growth through shading, interference with nutrient uptake, and increased mechanical stress (Orth and Van Montfrans 1984; D'Antonio 1985; Howard and Short 1986; Wahl 1989).

In the present study, the negative effect of fouling development on daily growth rates probably represents a conservative estimation. In fact, the percent total ash content analysis on *Gracilaria* samples (Fig. 4) after 5 days showed no difference among treatments (1-way ANOVA, F = 1.90, NS), while after 15 and 28 days a significant increase was detected (F=15.01, $p<0.01$ and F=9.33, $p<0.05$, respectively). A further analysis on ash content of thalli fragments free from fouling organisms (Table 1) excluded any effective among-treatment variation of macroalgal tissues ash content (1-way ANOVA, always NS), and permitted to relate total ash content variation strictly to fouling organisms (i.e. bryozoan calcite skeletons, Fig. 3). Thus, G values determined for partial and exclusion treatments are probably biased by an overestimation due to the contemporary ash content increase.

Our results strongly support the hypothesis of a fouling-mediated, positive action of the invertebrate assemblage on macroalgal populations in salt marsh and coastal environments (Hay et al. 1987). Besides direct trophic exploita-

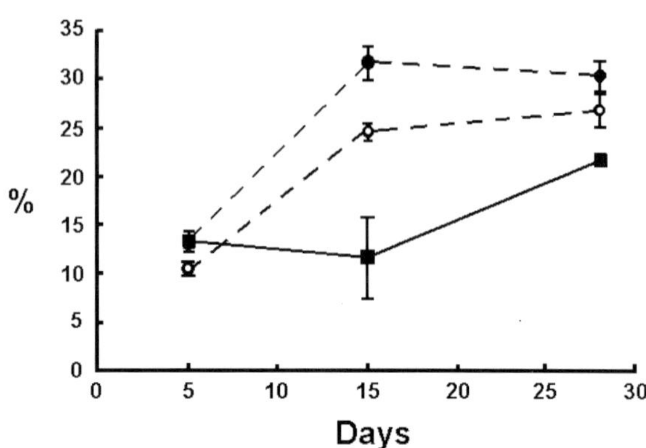

Fig. 4. Percent ash content of *Gracilaria* thalli under total ("F" treatment, ●), partial ("M" treatment, ❍), and no ("C" treatment, ■) macrofaunal esclusion conditions

Table 1. Percent ash content (mean, SE in parentheses) of *Gracilaria* thalli free from fouling colonisers under total ("F" treatment), partial ("M" treatment), and no ("C" treatment) macrofaunal esclusion conditions at different sampling times

Treatment	Day 5	Day 15	Day 28
F	**12.81** (0.49)	**11.35** (0.73)	**23.84** (2.58)
M	**10.46** (0.58)	**12.31** (2.20)	**23.67** (2.37)
C	**13.15** (1.00)	**11.68** (4.11)	**21.77** (0.52)

tion and nutrient regeneration phenomena induced through direct grazing of living macroalgal tissues and particulate detritus processing (widely documented in the literature), the present investigation highlights a further aspect of the functional relationships established among biotic components in brackish ecosystems. In fact, notwithstanding the large body of information available on the effect of water column chemical and physical characteristics on the development of *Gracilaria* spp., only scattered information is available on interactions with the macrofauna under field conditions (Brawley and Fei 1987).

In conclusion, our results suggest a potentially effective control method on the quality of *Gracilaria*, used worldwide as a commercial agar source. The control on fouling phenomena in seaweed mass cultivation through macro-organisms (i.e. fishes, Friedlander et al. 1996; amphipods, Shacklock and Doyle 1983; isopods, Shacklock 1992), in fact, is recognised as an effective, alternative method to chemical (Pickering et al. 1993; Friedlander 1992; Haglund and Pédersen 1993) and mechanical (Polne et al. 1980) treatments. Our results may be of interest for the development of highly productive, intensive and semi-intensive cultivations of *G. verrucosa* in the Lesina Lagoon.

Acknowledgements. This research was funded by the CNR ISEC - Insitute for the Research on Coastal Ecosystems - to G.M. We are grateful to Dr. Antonio Trombetta, Major of the City of Lesina, for his strong support; to Dr. Paolo Villani, director of the Institute for the Research on Coastal Ecosystems, and to all the personnel at CNR ISEC for the help provided at all stages of the investigation. This paper is dedicated to the memory of Prof. Giuseppe Magazzù.

References

Agnew DJ, Moore PG (1986) The feeding ecology of two littoral amphipods (Crustacea), *Echinogammarus pirloti* (Sexton and Spooner) and *E. obtusatus* (Dahl). J Exp Mar Biol Ecol 103: 203-215

Arrontes J (1990) Diet, food preference and digestive efficiency in intertidal isopods inhabiting macroalgae. J Exp Mar Biol Ecol 139: 231-249

Brawley SH (1992) Mesoherbivores. In: John DM, Hawkins SJ, Price JH (eds) Plant-animal interactions in the marine benthos. Clarendon Press, Oxford, pp 235-263

Brawley SH, Fei XG (1987) Studies of mesoherbivory in aquaria and unbarricaded maricolture farm on the chinese coast. J Phycol 23: 614-623

Costantini ML, Rossi L (1995) Role of fungal patchiness on vegetal detritus in the trophic interactions between two brackish detritivores, *Idotea baltica* and *Gammarus insensibilis*. Hydrobiologia 316: 117-126

D'Antonio C (1985) Epiphytes on the rocky intertidal red alga *Rhodomela larix* (Turner) C. Agardth: negative effects on the host and food for herbivores? J Exp Mar Biol Ecol 86: 197-218

De Benedictis A, Cardellicchio N, Guarino SM, De Nicola M (1996) Distribution of Fe, Zn, Cu and Cd in some organisms, water and sediment of the Lake of Lesina (Southern Italy). Ann Chim 86: 577-580

Dinardo C, Gambardella C, Guarino SM, De Nicola M (1997) Reproductive biology of a *Gammarus insensibilis* Stock population living in the Lesina Lagoon. Biol Mar Medit 4: 377-379

Duffy JE (1990) Amphipods on seaweeds: partners or pests? Oecologia 83: 267-276

Friedlander M (1992) *Gracilaria conferta* and its epiphytes: the effect of culture conditions on growth. Bot Mar 35: 423-428

Friedlander M, Weintraub N, Freedman A et al. (1996) Fish as potential biocontroller of *Gracilaria* (Rhodophyta) culture. Aquaculture 145: 113-118

Frier JO (1979). Character displacement in *Sphaeroma* spp. (Isopoda: Crustacea). I. Field Evidence. Mar Ecol Prog Ser 1: 159-163

Haglund K, Pedersén M (1993) Outdoor pond cultivation of the subtropical marine red alga *Gracilaria tenuistipitata* in brackish water in Sweden. Growth, nutrient uptake, co-cultivation with rainbow trout and epiphyte control. J Appl Phycol 5: 271-284

Harrison PG (1977) Decomposition of macrophyte detritus in seawater: effects of grazing by amphipods. Oikos 28: 165-169

Hay ME, Duffy JE, Pfister CA, Fenical W (1987). Chemical defense against different marine herbivores: are amphipods insect equivalents? Ecology 68: 1567-1580

Healy B, O'Neill M (1984) The life cycle and population dynamics of *Idotea pelagica* and *I. granulosa* (Isopoda: Valvifera) in south-east Ireland. J Mar Biol Assoc UK 64: 21-33

Howard RK, Short FT (1986) Seagrass growth and survivorship under the influence of epiphyte grazers. Aquat Bot 24: 287-302

Kittlein MJ (1991) Population biology of *Sphaeroma serratum* Fabricius (Isopoda, Flabellifera) at the Port of Mar del Plata, Argentina. J Nat Hist 25: 1449-1459

Lincoln RJ (1979). British Marine Amphipoda: Gammaridea. British Museum (Nat Hist), London

Orth RJ, Van Montfrans J (1984) Epiphyte-segrass relationships with an emphasis on the role of micrograzing: a review. Aquat Bot 18: 43-69

Pavia H, Carr H, Aberg P (1999) Habitat and feeding preferences of crustacean mesoherbivores inhabiting the brown seaweeed *Ascophyllum nodosum* (L.) Le Jol and its epiphytic macroalgae. J Exp Mar Biol Ecol 236: 15-32

Penniman CA, Mathieson AC, Penniman CE (1986) Reproductive phenology and growth of *Gracilaria tikvahiae* McLachlan (Gigartinales, Rodophyta) in the Great Bay Estuary, New Hampshire. Bot Mar 29: 147-154

Pickering TD, Gordon ME, Tong LJ (1993) Effect of nutrient pulse concentration and frequency on growth of *Gracilaria chilensis* plants and levels of epiphytic algae. J Appl Phycol 5: 525-533

Polne M, Gibor A, Neushul M (1980) The use of ultrasounds for the removal of macro-algal epiphytes. Bot Mar 23: 731-734

Poore AGB (1994) Selective herbivory by amphipods inhabiting the brown alga *Zonaria angustata*. Mar Ecol Prog Ser 107: 113-123

Robertson AI, Mann KH (1980) The role of isopods and amphipods in the initial fragmentation of eelgrass detritus in Nova Scotia, Canada. Mar Biol 59: 63-69

Rosa F, Bloesch J, Rathke DE (1994) Sampling the settling and suspended particulate matter. In: Mudroch A, McKnight SD (eds) Handbook of Techniques for Aquatic Sediment Sampling. Lewis Publishers, CRC Press, Boca Raton, Florida, pp 97-131

Shacklock PF (1992) Biology of *Idotea baltica* in *Chondrus* aquaculture. J World Aquacol Soc 23: 241-249

Shacklock PF, Doyle RW (1983) Control of epiphytes in seaweed coltures using grazers. Aquacolture 31: 141-151

Shaffelke B, Evers D, Walhorn A (1995) Selective grazing of the isopod *Idotea baltica* between *Fucus evanescens* and *F. vesiculosus* from Kiel Fjord (western Baltic). Mar Biol 124: 215-218

Sheader M (1996) Factors influencing egg size in the gammarid amphipod *Gammarus insensibilis*. Mar Biol 124: 519-526

Sheader M, Sheader AL (1985) New distribution records for *Gammarus insensibilis* Stock, 1966, in Britain. Crustaceana 49: 101-105

Steentoft M, Irvine LM, Farnham WF (1995). Two terete species of *Gracilaria* and *Gracilariopsis* (Gracilariales, Rhodophyta) in Britain. Phycologia 34: 113-127

Wahl M (1989) Marine epibiosis. I. Fouling and antifouling: some basic aspects. Mar Ecol Prog Ser 58:175-189

CHAPTER 42

Colonisation, Gene Frequencies and Enzyme Activity at GPI Locus of *Balanus amphitrite* (Cirripedia: Thoracica) in the Lagoon of Venice

V. Rossi[1], R. Antonietti[1], P. Bonilauri[1], Gi. Ferrari[2], Gr. Ferrari[1], G. Gentile[1], G. Magnaschi[1], C. Marchiani[1], and P. Menozzi[1]

ABSTRACT

From May 1997 to April 1998 we have studied the colonisation dynamics of 88 experimental panels, monitoring the allele and genotypic frequencies at GPI locus and GPI activity, of *Balanus amphitrite* in 3 sites from the lagoon of Venice. Two sampling stations are characterised by severe chemical (the Industrial Channel) and thermal pollution (the outflow of the cooling system of the Fusina electric power plant). A third station is located in the southern part of the lagoon (Chioggia) where chemical pollution is relatively low and thermal pollution absent. The mean number of successfully colonising animals is significantly different among sites and 80% of total collected individuals is from the power plant outflow. Very low variability at GPI locus is observed in all sites and on all dates; the frequency of the most common allele is higher than 90%. Most animals (91%) are homozygotes and significant deficiency of heterozygotes, with respect to the H-W equilibrium, is described in populations from the power plant outflow and from Chioggia after 160 and 310 days from panel immersion, respectively. The GPI activity is higher in homozygotes than in heterozygotes and is significantly different among stations. The low genetic variability at GPI locus cannot be linked to chemical and thermal pollution. Differences in enzymatic activity between genotypes and among sites indicate high fitness and plasticity of the homozygote organisms.

Introduction

Cosmopolitan species *Balanus amphitrite* is commonly associated with fouling and polluted environments and is considered a typical biomonitor to establish bioavailability of heavy metals in the marine habitat (Rainbow 1995). In North Adriatic Lagoons this is the most widespread barnacle species and Barbaro et al. (1978) described its potential use as an indicator of fluoride and several different heavy metals pollution. Previous work on populations from the lagoon of Venice suggested that selection by heavy metals acts in the Industrial Channel during the post-settlement phases and that significant differences in GPI activity exist among sites characterised by different level and type, chemical and thermal, of pollution, and between genotypes (Patarnello et al. 1991; Montero et al. 1994). The GPI is a typical marker used to investigate the genetic structure of marine organisms (Nevo et al. 1977; Fairbrother and Beaumont 1993; Hedgecock 1994; Holm and Bourget 1994; Viard et al 1994). Moreover, this locus is considered under selection in several species and codes for a soluble enzyme, the phospho-glucose isomerase, involved in glycolysis and gluconeogenesis pathways (Hummel and Patarnello 1994; Hu et al. 1993; da Silva and Solé-Cava 1994; Sloss et al. 1998). In *B. amphitrite* from Venice Lagoon, Montero et al. (1994) reported that mean activity of GPI is higher in homozygotes at this locus, the most frequent genotypic class, than in heterozygotes, and decreases significantly due to high concentration of Cu and Zn.

From May 1997 to April 1998 we have studied the colonisation dynamics of experimental panels, monitoring the allele and genotypic frequen-

[1] Dipartimento di Scienze Ambientali, Università di Parma, Parco Area delle Scienze 33/a, 43100 Parma, Italy
[2] ENEL, Direzione Costruzione, Laboratorio di Piacenza, Piacenza, Italy

F.M. Faranda, L. Guglielmo, G. Spezie (eds)
Mediterranean Ecosystems: Structures and Processes

cies at GPI locus and GPI activity of animals in the same three sites indicated by Patarnello et al. (1991) and Montero et al. (1994) for genetic and enzymatic analysis in spring 1989 and from May 1990 to June 1991. The first sampling station was located in the Industrial Channel where chemical pollution is very high due to industrial waste discharged directly for decades. The second one lies in front of the outflow of the electric power plant cooling system where temperatures are about 10°C higher than the rest of the lagoon. The third station is located in the southern part of the lagoon (Chioggia) where chemical pollution is relatively low (Brunetti et al. 1983; Donazzolo et al. 1984). The aim of our study is to continue the biomonitoring of such an area by the analysis of the genetic structure of *B. amphitrite* at GPI locus. Moreover we analyse the colonisation dynamics and evaluate the enzymatic activity of GPI in relationship to the genotype and the age of the animals.

Material and Methods

Colonisation dynamics was studied by counting barnacles that set 88 experimental fibrocement panels (24x34 cm) that were fixed to steel frames and suspended within 1 m of the surface in May 1997 approximately at the beginning of extended settlement season that ranges from late spring to early autumn. A mean number of 3 panels per site per date were removed after 20, 40, 80, 160, 230 and 310 days and transported to the laboratory in 40x30x22 cm and 20 l volume PVC boxes containing at least 10 l of sea water. In laboratory alive and dead (empty shells) specimens of *B. amphitrite* were identified and counted. For each date and site a sub sample of *B. amphitrite* was chosen, soft tissues were excised from the shells and frozen at –80°C for later genetic analysis and determinations of GPI enzyme activity and protein content. Starch gel electrophoresis was carried out to identify alleles at GPI locus (Harris and Hopkinson 1976). For GPI enzyme activity assay Sigma kits were used. The analyses were carried out at 30°C on single individuals scored for GPI genotype according to Montanini et al. (1998).

To compare colonisation success, mean number of specimens per panel and mean ratio between dead and alive organisms per panel in different stations and different dates, two way ANOVA tests (site, date), with homogeneity Levene's test, were performed. Panels colonised by less than 20 individuals were excluded from these analyses.

Allele and genotype frequencies were calculated and test for conformance to Hardy-Weinberg equilibrium with Levene correction for small sample size and coefficient for heterozygote deficiency or excess, D, were computed by Biosys-1 statistical package (Swofford and Selander 1981).

To evaluate differences in mean enzyme activity between genotypes (homozygotes and heterozygotes at GPI locus) among sites and sampling dates, one way ANOVA tests were computed. Sheffe F-tests were performed for pair comparisons.

Results

The mean number of colonising *B. amphitrite* is significantly different among sampling sites (Table 1) (Fig. 1). At the Industrial Channel the number of individuals per panel was less than 20 in all sampling dates. The highest values were detected at the power plant outflow where the species reaches a maximum density of 4718 ind/m^2 after 310 days from settlement.

Mortality of *B. amphitrite* was significantly related to the sampling date (Table 2) (Fig. 2). At the Industrial Channel the highest mortality (33%) was recorded after 80 days from panel immersion. At the power plant outflow, mortality increased significantly with the sampling date (r^2=0.809, P=0.0001) and presents the absolute maximum value (78%) after 310 days from panel immersion.

Considering samples with n>20, the frequency of the most common allele at GPI locus

Table 1. Two way ANOVA table for difference in mean number of colonising individuals per panel among sampling sites and sampling dates. Industrial Channel station was excluded from the analysis as the number of individuals per panel was <20 on all sampling dates

Source	df	Mean square	*P*
Site	1	52701.880	0.016
Date	5	11869.372	0.207
Sitexdate	4	16229.457	0.111
Error	15	7181.433	

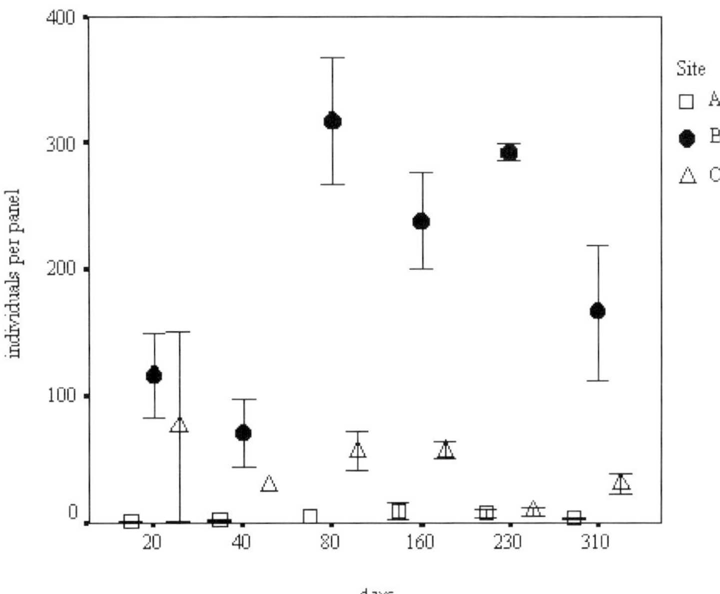

Fig. 1. Mean number of *B. amphitrite* per panel from different sampling sites, Industrial Channel (*A*), power plant outflow (*B*) and Chioggia (*C*) in different sampling dates after 20 (May 97), 40 (Jun 97), 80 (Aug 97), 160 (Nov 97), 230 (Jan 98) and 310 (Apr 98) days from panel immersion

Table 2. Two way ANOVA table for difference in mean dead organisms/alive organisms ratio per panel among sampling site and sampling dates. Industrial Channel station was excluded from the analysis as the number of individuals per panel was <20 in all sampling dates

Source	df	Mean square	*P*
Site	1	0.755	0.130
Date	5	0.930	0.038
Sitexdate	4	0.666	0.110
Error	15	0.294	

was very high, namely between 91 and 99%, at Chioggia in August and November, respectively (Table 3). Differences among stations and dates were not significant ($\varkappa^2$=43.1, df=32, P>0.05) while, with respect to the Hardy-Weinberg equilibrium, significant deficiency of heterozygotes were observed in populations from the power plant outflow in August 97 (D=–0.229, P=0.001) and in population from Chioggia in April 98 (D=–0.407, P=0.008). Since November 97 the frequency of the most common allele at the

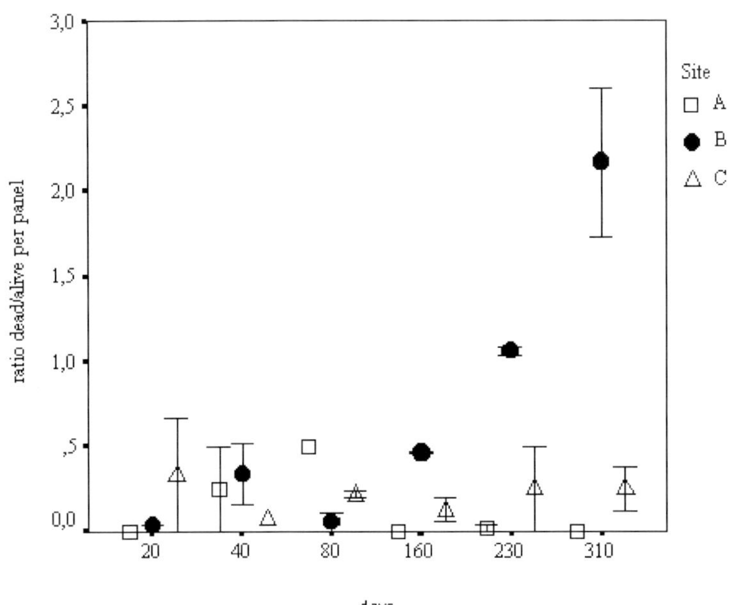

Fig. 2. Mean dead organisms/live organisms ratio per panel from different sampling sites, Industrial Channel (*A*), power plant outflow (*B*) and Chioggia (*C*) in different sampling dates after 20 (May 97), 40 (Jun 97), 80 (Aug 97), 160 (Nov 97), 230 (Jan 98) and 310 (Apr 98) days from panel immersion

Table 3. Gene frequencies for the most frequent allele at GPI locus in the 3 studied sites and on different sampling dates. Number of detectable individuals are reported *in brackets*

Date	Industrial Channel	Power plant outflow	Chioggia
Jun 97		0.889 (9)	0.977 (44)
Aug 97	1.000 (1)	0.927 (131)	0.905 (58)
Nov 97	0.920 (25)	0.962 (104)	0.990 (49)
Jan 98	0.933 (15)	0.963 (41)	1.000 (8)
Apr 98	0.833 (15)	0.988 (85)	0.938 (56)

Table 4. Mean overall GPI enzyme activity (U/mg protein) in homozygotes and heterozygotes for this locus

Genotype	*n*	Mean	SD
Homozygote	251	2.174	1.399
Heterozygote	19	1.184	0.866

Table 5. Mean GPI enzyme activity (U/mg protein) in homozygotes for this locus from different sampling stations

Site	*n*	Mean	SD
Industrial Channel	33	1.947	1.141
Power plan outflow	171	2.399	1.453
Chioggia	47	1.518	1.127

power plant outflow was constantly higher than 95%.

The GPI activity in homozygotes at this locus was higher than in heterozygotes (F=9.23, P<0.001) and was related to the sampling station and to the sampling date (Table 4). Particularly, considering only the homozygotes, the dominant genotype class, enzyme activity of GPI was different among stations (F=8.26, P<0.001) and was, on the whole, significantly higher in samples from the power plant outflow than from Chioggia (Sheffe F-test= 4.31) (Table 5) (Fig. 3). In general, GPI activity increases significantly until up to 230 days from panel immersion (January 1998). This relationship is well described in organisms from the power plant outflow where samples are more numerous (F=26.58, P<0.001) (Table 6).

Table 6. Mean GPI enzyme activity (U/mg protein) in homozygotes from the power plant outflow on different sampling dates

Date	*n*	Mean	SD
Aug 97	34	0.884	0.727
Nov 97	43	2.231	0.456
Jan 98	35	3.060	1.613
Apr 98	59	3.001	1.487

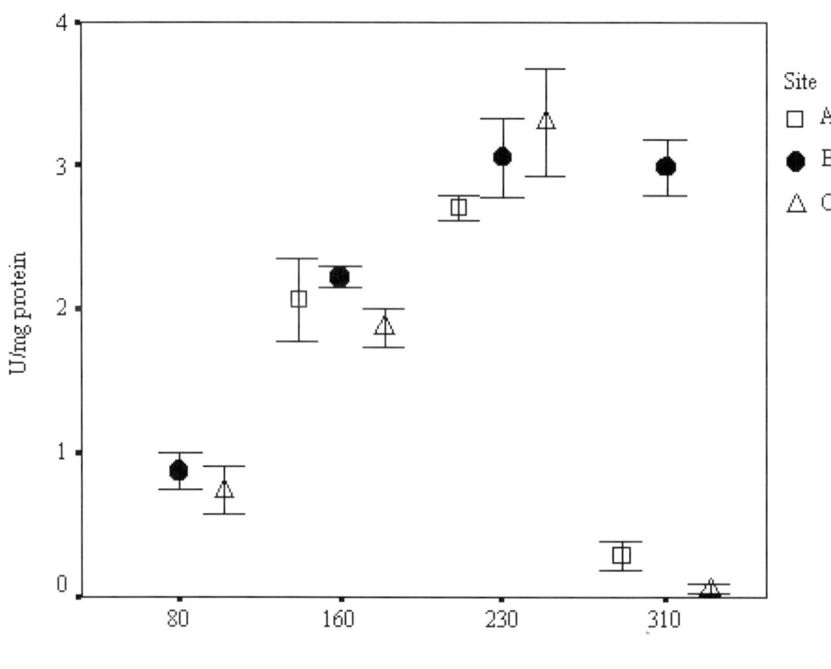

Fig. 3. Mean GPI enzyme activity of homozygotes for this locus from different sampling sites, Industrial Channel (*A*), power plant outflow (*B*) and Chioggia (*C*) in different sampling dates after 80 (Aug 97), 160 (Nov 97), 230 (Jan 98) and 310 (Apr 98) days from panel immersion

Discussion

The mean number of colonising *B. amphitrite* is significantly different among sites and this species, even not the most abundant one, is dominant at the power plant outflow. Among five barnacle species found, *B. eburneus, B. improvisus, B. perforatus, B. trigonus and B. amphitrite* (Rossi et al. in preparation), the last is considered the most pollution resistant (Relini and Relini Orsi 1970).

The lowest recruitment of barnacles, including *B. amphitrite,* observed at the Industrial Channel should indicate that, in this station a severe selection acts before or during larval settling. Indeed, several potential factors may contribute to differences in barnacle settlement and mortality among sites. According to O'Connor and Richardson (1998) variability in larval settling behaviour is a general and complex phenomenon. Settlement success of the larvae and variability in recruitment are influenced by various factors including delayed metamorphosis, larvae age, water flow, salinity, temperature, chemical and physical nature of the substratum's surface that is the presence of dissolved compounds or bacterial film (Judge et al. 1988; Pechenik et al. 1993; Anil and Kurian 1996; Satuito et al. 1996; Yamamoto et al. 1996; Bryan et al. 1996; O'Connor and Richardson 1998). The Industrial Channel population is probably subjected to high generalised mortality due, probably, to chemical pollution. Such an effect on an expected large cypris larvae population might explain the lack of significant departure from the Hardy-Weinberg equilibrium as previously described by Montero et al. (1994). Our preliminary data indicate a lower rostrocarenal mean length in *B. amphitrite* from this site than from the power plant outflow and from Chioggia, at least in 40-80 days from panel immersion. This result might reflect a low growth rate of organisms in this site but could support the hypothesis of high mortality during the precocious phase of benthic life, a very critical period (Gosselin and Qian 1996), coupled with continuous larval settlement. The low number of colonising *B. amphitrite* at Chioggia station as well as its dominance at the power plant outflow should be due to its ecological characteristics. At Chioggia station *B. amphitrite* might suffer interspecific competition, particularly with *B. eburneus* that is the most abundant species recorded in this site. On the other hand, *B. amphitrite* should outcompete and exclude *B. eburneus* from the power plant outflow. Indeed, even if, according to Scheltema and Williams (1982), neither larval survival and development of *B. eburneus* are affected by increase in temperature to 30 °C, this species is more sensitive to high temperatures and salinity than *B. amphitrite* whose settlement is not modified by thermal pollution (Relini 1980; Relini and Relini Orsi 1970).

We cannot relate the low genetic variability observed at GPI locus to the selective effect of both chemical and thermal pollution because no significant differences in allele frequencies among stations have been observed. Low polymorphism at this locus could be a genetic characteristic of the species that, indeed, is poorly investigated. For *B. amphitrite* populations from the inflowing and outflowing canals at the Haifa (Israel) electrical plant cooling system, Nevo et al. (1977) reported that values of estimated heterozygosity vary between 0.034 and 0.167. In fact, our data seem comparable as biased estimated heterozygosity at GPI locus is between 0.089, at the power plant outflow, and 0.184, at the Chioggia station. At the same time, we cannot exclude that some stabilising factor affects the genetic structure of the *B. amphitrite* population from the Lagoon of Venice. This should contribute to maintaining constant gene frequencies as shown by the result of the genetic monitoring carried out periodically over the last 10 years (Patarnello et al. 1991; Montero et al. 1994).

Our results confirm the data by Montero et al. (1994) indicating that the individuals of *B. amphitrite* homozygous at GPI locus present higher GPI activity than the heterozygotes. Moreover, we observed that homozygotes activity is higher in organisms from the power plant outflow than in ones from the other two sites. The highest temperature recorded at this station and the highest enzymatic activity observed in organisms from there might induce an increase in uptake and accumulation of pollutants (Barbaro et al. 1978; Montero et al. 1994) and may affect the survival of organisms. In fact, at the power plant outflow, total mortality is higher than in other sites and increases from August to January. As suggested by Montero et al. (1994) a significantly lower efficiency in heterozygotes might be coupled with differential fitness between genotype and produce the excess of homozygotes observed in August 1997 and then monomorphism, considering that, since this date, the frequency of the most common allele is higher than 95%. At the Chioggia station we can hypothesise a similar but very delayed situation due to the relatively low chemical pollution and to absence of thermal pollution. Barnacles are, indeed, able to uptake and accumulate heavy metals at a concentration factor of at least 10^3 (Barbaro et al. 1978). At the same time, they present mechanisms for the detoxification of the stored metals (Rainbow and White 1989). At this site, mortality never exceeds 20% but significant excess of homozygotes with respect to the Hardy-Weinberg equilibrium is observed only in April 98, after 310 days from panel immersion.

We suggest that, in general, environmental pollutants act as selective factors on recruitment of *B. amphitrite* at the Industrial Channel station. On the other hand thermal pollution seems important to account for dominance of *B. amphitrite* on experimental panels from the power plant outflow because its lower sensitivity to warmer temperature with respect to both *B. eburneus* and other congeners described in the Venice Lagoon (Nevo et al. 1977). Differences in enzymatic activity between genotypes and among sites seem to indicate the highest fitness and plasticity of the homozygote organisms. But, the relationship between enzymatic activity and sampling season might mask or twist the affect of other factors that could affect the population dynamics and the age-class structure. Moreover, competitive advantage of *B. amphitrite* and, in particular of homozygous individuals at GPI locus, might decrease in sites away from the highly thermally polluted site. The reason behind this situation is not clear but some cost, or adaptive trade-off, may result from possessing a more efficient detoxification mechanism eventually coupled with the highest enzymatic activity.

Acknowledgements. We would like to thank Prof. Battaglia, Dr. R. Cironi and Prof. T. Patarnello for aiding in the initial set up of the project and the staff of the Stazione Idrobiologica dell'Università di Padova (Chioggia) and of the Centrale ENEL di Fusina for use of facilities and assistance with the sampling work. This research was supported by ENEL (Ente Nazionale per l'Energia Elettrica) contract NO 352/22103 20.2.97 and by the research programme "Conservazione della biodiversità e gestione sostenibile dei biotopi salmastri delle coste italiane" (MURST 97).

References

Anil AC, Kurian J (1996) Influence of food concentration, temperature and salinity on the larval development of *Balanus amphitrite*. Mar Biol 127: 115-124

Barbaro A, Francescon A, Polo B, Bilio M (1978) *Balanus amphitrite* (Cirripedia: Thoracica) – a potential indicator of fluoride, copper, lead, chromium and mercury in North Adriatic Lagoons. Mar Biol 46: 247-257

Brunetti R, Marin M, Beghi L, Bressan M (1983) Study of pollution in the Venetian lagoon's lower basin during the period 1974-1981. Riv Idrobiol 22: 27-58.

Bryan PJ, Rittschof D, McClintock JB (1996) Bioactivity of echinoderm ethanolic body-wall extracts- an assessment of marine bacterial attachment and macroinvertebrate larval settlement. J Exp Mar Biol Ecol 196: 79-96

da Silva EP, Solé-Cava AM (1994) Genetic variation and population structure in the tropical marine bivalve *Anomalocardia brasiliana* (Gmelin) (Veneridae). In: Beaumont AR (ed) Genetics and evolution of aquatic organisms. Chapman and Hall, London, pp 159-168

Donazzolo R, Orio AA, Pavoni B, Perin G (1984) Heavy metals in sediments of the Venice Lagoon. Oceanol Acta 7: 25-31

Fairbrother JE, Beaumont AR (1993) Heterozygote deficiencies in a cohort of newly settled *Mytilus edulis* spat. J Mar Biol Assoc UK 73 647-653

Gosselin LA, Qian PY (1996) Early and post-settlement mortality of an intertidal barnacle: a critical period for survival. Mar Ecol Prog Ser 135: 69-75

Harris H, Hopkinson DA (1976) Handbook of enzyme electrophoresis in human genetics. North-Holland Publ Co, Amsterdam

Hedgecock D (1994) Does variance in reproductive success limit effective population size of marine organisms? In: Beaumont AR (ed) Genetics and evolution of aquatic organisms. Chapman and Hall, London, pp 122-134

Holm Er, Bourget E (1994) Balanoides in the Northwest Atlantic and Gulf of St-Lawrence. Mar Ecol Prog Ser 113: 247-256

Hu YP, Lutz RA, Vrijenhoek RC (1993) Overdominance in early-life stages of an american oyster strain. J Hered 84: 254-258

Hummel H, Patarnello T (1994) Genetics and pollution. In: Beaumont AR (ed) Genetics and evolution of aquatic organisms. Chapman and Hall, London, pp 425-434

Judge ML, Quinn JF, Wolin CL (1988) Variation in recruitment of *Balanus glandula* (Darwin 1854) along the central California coast. J Exp Mar Biol Ecol 119: 235-251

Montanini E, Antonietti R, Ferrari G, Marchiani C, Rossi V (1998) The effects of thermal pollution on gastropod populations in the Po River (Northern Italy). Verh Int Ver Limnol 26: 2103-2106

Montero C, Battaglia B, Stenico M, Campesan G, Patarnello T (1994) Effects of heavy metals and temperature in the genetic structure and GPI enzyme activity of the barnacle *Balanus amphitrite* Darwin (Cirripedia: Thoracica). In: Beaumont AR (ed) Genetics and evolution of aquatic organisms. Chapman and Hall, London, pp 447-439

Nevo E, Shimony T, Libni M (1977) Thermal selection of allozyme polymorphisms in barnacles. Nature 267: 699-701

O'Connor NJ, Richardson DL (1998) Attachment of barnacle (*Balanus amphitrite* Darwin) larvae: responses to bacterial films and extracellular materials. J Exp Mar Biol Ecol 226: 115-129

Patarnello T, Guinez R, Battaglia B (1991) Effects of pollution on heterozygosity in the barnacle *Balanus amphitrite*

(Cirripedia: Thoracica). Mar Ecol Prog Ser 70: 237-243

Pechenik JA, Rittschof D, Schmidt AR (1993) Influence of delayed metamorphosis on survival and growth of juvenile barnacles *Balanus amphitrite*. Mar Biol 115: 287-294

Rainbow PS (1995) Biomonitoring of heavy metal availability in the marine environment. Mar Pollut Bull 31: 183-192

Rainbow PS, White SL (1989) Comparative strategies of heavy metal accumulation by crustaceans: zinc, copper and cadmium in a decapod, an amphipod and a barnacle. Hydrobiologia 174: 245-262

Relini G (1980) Cirripedi toracici. In: Guida per il riconoscimento delle specie animali delle acque lagunari e costiere italiane. Aq/1/91, CNR, pp 116

Relini G, Relini Orsi L (1970) Fouling di zone inquinate. Osservazioni nel Porto di Genova. I. Cirripedi. Pubbl Staz Zool Napoli 38: 125-144

Satuito CG, Shimizu K, Natoyama K, Yamazaki M, Fusetani N (1996) Age-related settlement success by cyprids of the barnacle *Balanus amphitrite*, with special reference to consumption of cyprid storage protein. Mar Biol 127: 125-130

Scheltema RS, Williams IP (1982) Significance of temperature to larval survival and length of development in *Balanus eburneus* (Crustacea: Cirripedia). Mar Ecol Prog Ser 9: 43-49

Sloss BL, Romano MA, Anderson RV (1998) Pollution-tolerant allele in fingernail clams (*Musculium trasversum*). Arch Environ Contam Toxicol 35: 302-308

Swofford DL, Selander BR (1981) Biosys-1: a FORTRAN program for the comprehensive analysis of electrophoretic data in population genetics and systematic. J Hered 72: 282-283

Viard F, Delay B, Coustau C, Renaud F (1994) The *Mytilus-edulis Mytilus-galloprovincialis* complex. Mar Biol 119: 535-539

Yamamoto H, Tachibana A, Kawaii S, Matsumura K, Fusetani N (1996) Serotonin involvement in larval settlement of the barnacle *Balanus amphitrite*. J Exp Zool 275: 339-345

CHAPTER 43

Role of Macrodetritivores – Detritus Trophic Relationships in Phosphorus Dynamics at the Marine Water-sediment Interface: Laboratory Experiments Using ^{32}P

L. Rossi and G. Mancinelli

ABSTRACT

We investigated the combined effects on sediment ^{32}P uptake of bioturbation and particulate detritus processing by macrodetritivores in laboratory microcosms. Sediment and sediment + detritus microcosms were set using sediment and vascular plant detritus sampled in north-western Adriatic Sea. The effects of macrodetritivores at the water-sediment interface were tested by adding to microcosms individuals of the amphipod *Gammarus insensibilis* in different densities (541 and 1624 individuals m^{-2}, respectively). In defaunated controls, plant detritus caused a significant positive variation in sediment ^{32}P adsorption rates (ca. 40% increase). Under "sediment only" conditions, the presence of macrodetritivores stimulated both uptake and diffusion of the radiotracer in sediments. Furthermore, the analysis of water-amphipod concentration factors highlighted an effective inability of macrodetritivores to feed upon the finest organic particles constituting the bulk of sediment organic matter. Excluding any significant trophic interaction with the clayish matrix, amphipod bioturbation activity at the water-sediment interface resulted decisively in affecting ^{32}P dynamics. On the other hand, under "sediment + detritus" conditions, amphipod processing of plant detritus determined a significant, animal density-related increase of dissolved ^{32}P originally immobilised by detritus and/or sequestered by the microbial component.

Our experimental investigation highlighted a clear "short-circuiting" effect of the detritivores-detritus trophic interaction on benthic phosphorus dynamics, potentially significant also in natural environments. In this perspective, phosphorus regeneration in coastal systems, conventionally related to abiotic factors and microbial mineralization, might result strongly influenced also by macrodetritivores trophic exploitation of allochthonous detritus.

Introduction

Sediments can immobilise or release nutrients in aquatic systems, influencing the concentration of dissolved elements in overlying waters, and indirectly affecting both benthic and pelagic primary production. Even though only scattered information is available from comparative studies between freshwater and marine environments (*e.g.* Enell 1989), phosphorus uptake by sediment seems to represent in freshwater systems the primary process, whereas in salt marsh and estuarine environments regeneration from sediments prevails over immobilisation; P may thus behave as a conservative tracer of benthic decomposition (Caraco et al. 1990). In this perspective, in coastal benthic systems biotic factors directly taking part in decomposition processes (*e.g.* micro- and macrofauna) may play a crucial role in P dynamics.

Nutrient exchange at the sediment-water interface is deeply influenced by benthic organisms (Andersson et al. 1988; Clavero et al. 1994). Faunal trophic activity can accelerate the rate of organic matter decomposition, increasing phosphorus release from the sedimentary compartment (Holdren and Armstrong 1980; Gardner et al. 1981). On the other hand, burrowing and ventilation activity by macroinvertebrates may speed up oxygen diffusion into sediments and increase the thickness of the oxidised surface layer, enhancing sediment P binding capacity (Kamp-Nielsen et al. 1982).

Scarce efforts have been made to analyse the

Sezione Ecologia, Dip. di Genetica e Biologia Molecolare, Università di Roma I "La Sapienza", Via Lancisi 29, 00161 Roma, Italy

F.M. Faranda, L. Guglielmo, G. Spezie (eds)
Mediterranean Ecosystems: Structures and Processes

effects of epifaunal macrodetritivores on phosphorus dynamics in coastal marine systems in relation to both macrodetritus processing and superficial layer turbation. The trophic activity of this group of organisms has been demonstrated to affect plant detritus breakdown, microbial colonisation and decomposition rates (Fenchel 1970; Harrison 1977; Robertson and Mann 1980).

In our study, orthophosphate ^{32}P was used in simplified water-sediment systems to analyse both phosphorus distribution among compartments and trophic linkages between detritivores and other compartments. Two main questions were addressed: (1) What is the role of macrodetritus in phosphorus fluxes across the water-sediment interface? (2) Does the trophic interaction detritus-macrodetritivores cause significant variation in sediment ^{32}P adsorption and release?

Materials and Methods

Samples were collected from a coastal site in Northern Adriatic Sea (45°08.96' N, 12°23.21' E; depth: 26 m), located 4.28 nautical miles offshore. Previous investigations (*e.g.* Mancinelli et al. 1998) showed that the clay – silt sediments of the area were characterised by the occurrence of coarse plant detritus of allochthonous origins (fragments of the phanerogams *Zostera* sp. and *Cymodocea nodosa* Ascherson). The detritivorous amphipod *Gammarus insensibilis* Stock characterised the benthic epifauna all year round, reaching maximal densities in summer (ca. 1400 individuals m^{-2}).

In July 1995, sediment cores (5 replicates) were collected using a Reineck box-corer (sampling area: 170 cm^2). The sampling device permitted to retrieve undisturbed cores; on board, the oxidised layer (uppermost 5 cm) was cut and placed in plastic containers. After returning to the laboratory, sediment cores were wet-sieved (0.5 mm mesh size) to remove coarse particles and macrofauna; living *Gammarus insensibilis* specimens were simultaneously collected.

The freshly sieved sediment was homogenised and frozen at –80°C for 12 h. The procedure eliminated phytobenthos and meiofauna, without inactivation of the microbial component (Kristensen and Blackburn 1987). After complete defrosting at room temperature, sediment aliquots (30 ± 0.005 g wet weight, 19.66 ± 0.04 g dry weight) were transferred into 18 sterilised microcosms (Pyrex jars, 500 ml total volume, 73.9 cm^2 basal area) containing 220 ± 0.5 ml of pre-filtered (Corning® cellulose acetate filters, 0.45 μm pore size) sea water. Each microcosm was provided with an aeration system (flow was set under sediment resuspension level) and acclimated in a controlled temperature room for 60 h at 22°C.

Pre-weighted aliquots of *Cymodocea nodosa* fragments ("leaf packs": 200 ± 5 mg dry wt) were conditioned in an aerated suspension of fresh sediment for 72 h at 22 °C. At the end of the conditioning phase, *Cymodocea* leaf packs were carefully introduced in nine microcosms (1 leaf pack/microcosm; "sediment + detritus" condition); the remaining microcosms were left undisturbed ("sediment only" condition). After 12 h, three treatments were set for each condition: (1) "low density" animal treatment: 4 dimensionally similar *Gammarus insensibilis* specimens were added (corresponding to a spatial density of 541 individuals • m^{-2}); (2) "high density" animal treatment: 12 amphipods were added (1624 individuals • m^{-2}); (3) defaunated control. For each treatment three replicated microcosms were provided.

Immediately after amphipod introduction, 5 ml of a sterile solution containing 16.85 ± 0.37 mCi of orthophosphate ^{32}P were added to each microcosm. An identical amount of radiotracer was added to microcosms (3 replicates) containing 200 ml of sterile seawater and 10 amphipods to assess ^{32}P physical adsorption.

During the experiment, filtered (pre-weighted Sartorius® cellulose acetate filters, 0.2 μm pore size) water samples (50 μl, 3 replicas/microcosm) were taken on an exponential time scale until the 192th hour. Filters were oven dried (72 h, 60°C), and weighed.

The experiment was terminated after 192 h. Amphipods were collected and enumerated. They were consequently washed, oven-dried at 60°C for 72 h, weighed, and analysed for ^{32}P content. Detritus subsamples (4 replicas/microcosm) were taken from "sediment + detritus" microcosms and subjected to the same procedure; sediment cores (i.d. 0.5 cm) were collected (4 replicate/microcosm; coring operations in "sediment + detritus" microcosms resulted infeasible due to the large detritus particles interspersed in the sediment), frozen (48 h, –30°C), and cut into three layers (0-3, 3-6, 6-9 mm). For each layer, sediment was separated from interstitial water by squeezing under N_2 on pre-weighted filters (0.45 μm pore size). Filters were oven dried, weighed, and analysed for ^{32}P content. At the same time, interstitial water samples (50 μl, 4 subsamples/microcosm) were taken for radioactivity determination. Particulate detritus ("sediment + detritus" treatments) was separated from sediment by sieving (0.5 mm mesh), oven-dried, and weighed.

Filters, animal and detritus samples were weighed to the nearest 0.01 mg. Radioactivity was determined in both aqueous and solid samples after NCS digestion by liquid scintillation counting. Counts were corrected for colour quenching and radiotracer decay, and expressed as disintegration min^{-1} (1 DPM = 4.55 10^{-13} Curie).

In animal treatments, detritus processing rates were determined as the per cent loss of detritus dry weight during the experiment, and expressed as % h^{-1}; amphipods mortality was expressed as the loss of individuals per hour (*n* individuals h^{-1}).

The following parameters were used: (1) "activity density" (AD = DPM mg^{-1} dry weight; Whittaker 1961), as an index of ^{32}P concentration per unit dry weight; (2) "concentration factor" (CF) as an index of the intensity of ^{32}P transfer between source compartments (water, sediment, or detritus) and amphipods (CF = amphipods AD source AD^{-1}). The ^{32}P fluxes from water towards other compartments were estimated as the loss in the dissolved radiotracer pool per unit area of the water-sediment interface, and expressed as DPM cm^{-2} h^{-1}.

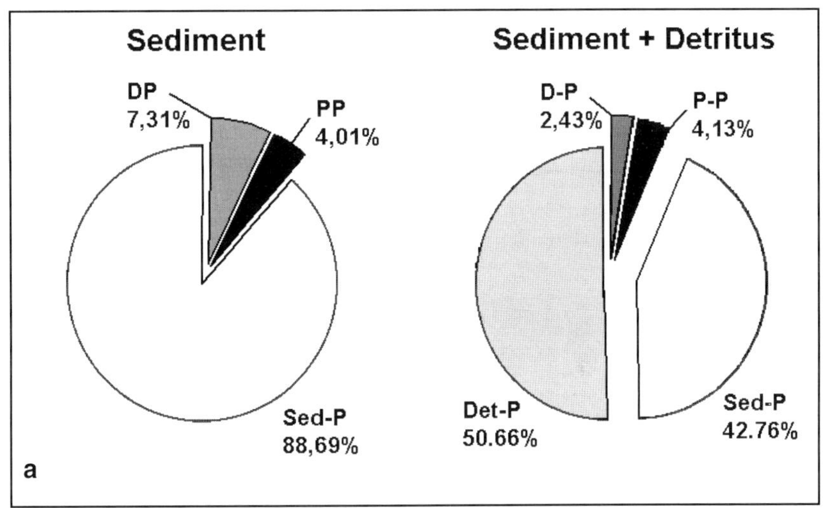

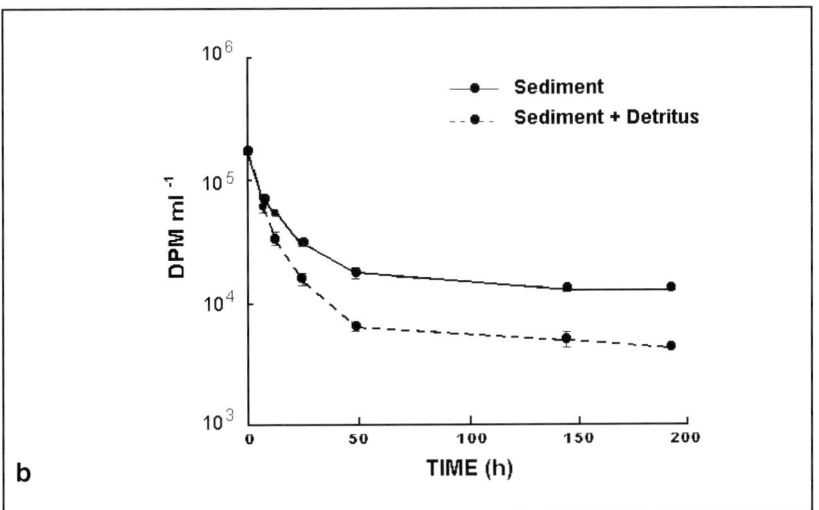

Fig. 1a,b. "Sediment only" and "sediment + detritus" conditions: **a** ^{32}P distribution among compartments; **b** temporal patterns of D-P concentration

Results and Discussion

In our study, macrodetritus represented the crucial factor determining both radiotracer final distribution among compartments and ^{32}P uptake patterns, whereas macrodetritivores activity contributed in different ways to ^{32}P dynamics in relation to plant detritus availability. In comparison to "only sediment" conditions, "sediment + detritus" defaunated systems showed a positive variation in the overall flux of dissolved ^{32}P versus other compartments (2531.51 ± 2.54 and 2655.33 ± 0.20 DPM cm^{-2} h^{-1} respectively; t-test unpaired samples: t = 48.61, P<0.001). Sediment demonstrated a significant "sink" effect when plant detritus was present (Fig. 1; interstitial ^{32}P pool was negligible and it is not reported). Furthermore, sediment concentration factor showed a significant increase (Table: 45.17 ± 6.34% increment; t-test unpaired samples: t = –7.13, P<0.05); in addition a positive variation (from 135.56 ± 0.01 to 426.09 ± 10.26) occurred in the total solid phase (sediment + detritus) concentration factor.

Plant detritus caused a net positive variation in radiotracer uptake during the early phases of the experiment (Fig. 1). Even though it determined a negligible variation in sediment dry weight (detritus: 0.2 g; sediment: 19.66 g) and bulk organic content (from 14.76 ± 0.56 to 15.52 ± 0.55 %; Mann-Whitney U-test, Z = –0.94, NS), at the end of the study more than 50% of the initial ^{32}P pool resulted sequestered by the detrital compartment. Two factors might have contributed to radiotracer uptake on detritus: physical adsorption and microorganisms. Physical adsorption is not likely to have occurred in our study, as it is mainly determined by humic compounds in highly refractory organic matter (Mesnage and Picot 1995). On the other hand, microorganisms are known to actively immobilise and store phosphorus, representing a significant fraction of the sedimentary nutrient pool (Hupfer and Uhlmann 1991; Gätcher et al. 1988; Gätcher and Meyer 1993). In our microcosms, detritus enrichment determined a general enhancement in the microbial activity (as indicated by the increase in sediment CF: Table) and represented a further, ideal substrate for growth and development.

Under "sediment-only" conditions, *G. insensibilis* activity caused an increase in ^{32}P flux from the aqueous compartment (Fig. 2; 1-WAY ANOVA: F=8.34, P<0.05), clearly unrelated to animal density. The ^{32}P final distribution among compartments (Fig. 2), sediment CFs (Table),

Table. ^{32}P concentration factors in different compartments. (*P-P*, particulate; *I-P*, interstitial water; *Sed-P*, sediment; *Det-P*, detritus; *An-P*, amphipods). ES *in brackets*

Sediment only condition

Treatment \ Compartment	P-P	I-P	Sed-P	An-P
Defaunated control	0.55 (0.21)	0.62 (3.10^{-5})	135.85 (0.01)	–
"Low density"	0.46 (0.01)	0.04 (2.10^{-2})	240.28 (2.57)	5617.16 (1847.50)
"High density"	0.41 (0.04)	0.03 (3.10^{-2})	225.62 (16.78)	3703.36 (410.32)

Sediment + detritus condition

Treatment \ Compartment	P-P	I-P	Sed-P	Det-P	An-P
Defaunated control	1.70 (0.82)	0.54 (0.01)	197.25 (8.58)	23328.90 (178.12)	—
"Low density"	0.75 (0.11)	0.05 (0.01)	284.85 (20.74)	5560.46 (303.62)	168463.17 (31921.48)
"High density"	1.19 (0.13)	0.03 (0.01)	238.04 (2.68)	2046.50 (319.52)	70024.37 (7719.42)

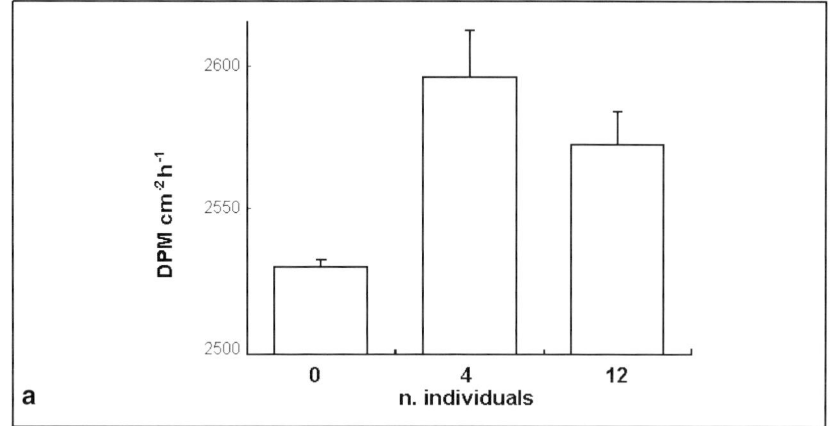

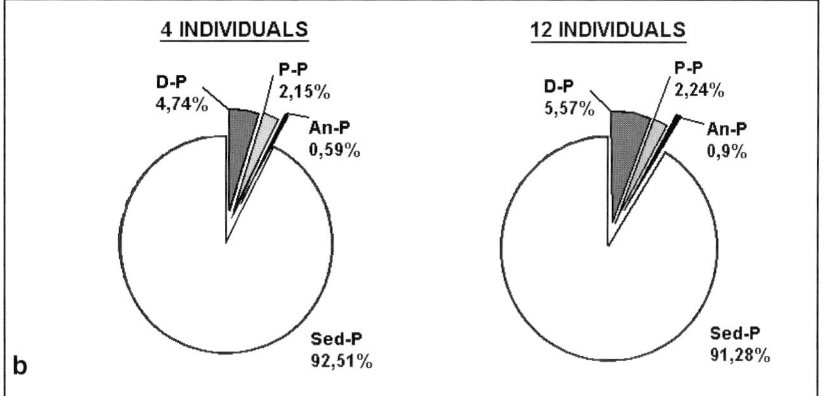

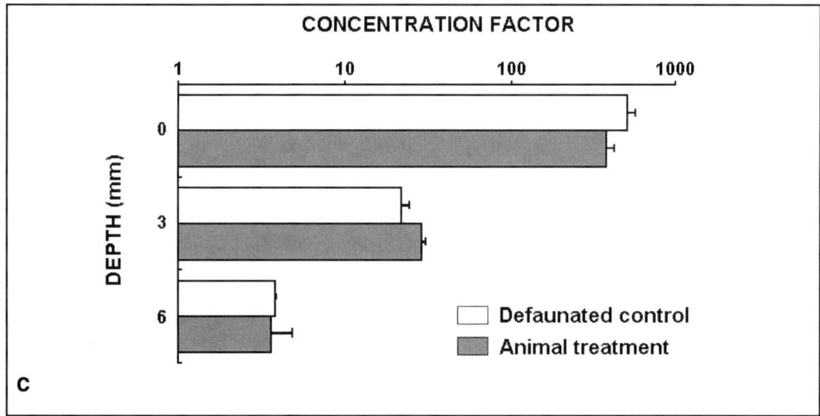

Fig. 2a-c. "Sediment only" condition, animal treatments: **a** variations in D-P fluxes; **b** ^{32}P distribution among compartments; **c** sediment concentration factors at different depths in the sedimentary column. Animal treatment cumulates both "low density" and "high density" treatments

and ADs (1691.87 ± 23.00, 1767.03 ± 10.03, 1744.41 ± 8.04 DPM/mg for 0, 4, and 12 individuals, respectively) indicated a significant increase of the sedimentary radiotracer pool. In Fig. 2, CFs of different layers in the sedimentary column are showed: *G. insensibilis* activity caused a decrease in surficial layer (0-3 mm) ^{32}P uptake and a contemporaneous increase in the layer 3-6 mm. Mechanical turbation of the water-sediment interface by amphipods may have determined a positive variation in O_2 penetration in the sedimentary column, and a consequent increase in both radiotracer diffusion rates and in phosphorus binding capacity of superficial sediments. Animal activity may strongly affect sediments redox potential and the oxidised surface layer thickness (Revsbech et al. 1980; Kamp-Nielsen et al. 1982; Rippey and Jewson 1982), determining in clayish sediments a positive variation in phosphate binding capacity (Froelich 1988;

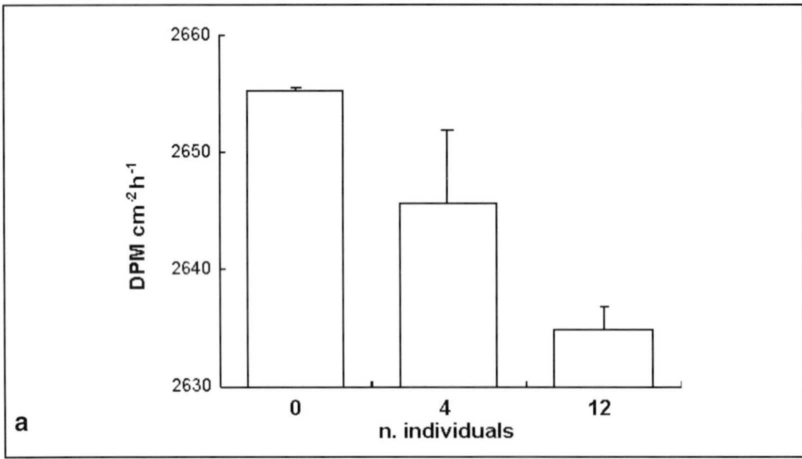

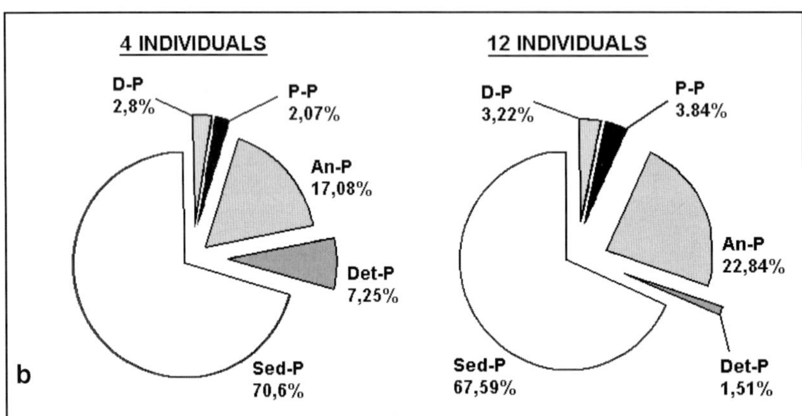

Fig. 3a,b. "Sediment + detritus" condition, animal treatments: **a** variations in D-P fluxes; **b** ^{32}P distribution among compartments

Goltermann 1995). On the other hand, the decrease in ^{32}P concentration factor observed for the most superficial layer may indicate a negative effect of amphipods on the development of a microbial mat and on bacterial P uptake (Alongi and Hanson 1985).

When detritus was available, macrodetritivores determined a completely different ^{32}P dynamics. In fact: (1) ^{32}P fluxes decreased both in "low density" and "high density" animal treatments (Fig. 3; 1-WAY ANOVA: F=10.42, P<0.05), reflecting an increase in dissolved ^{32}P pools (Fig. 3), significantly related to animal density (r=0.89, P<0.05) and detritus processing rates (0.24 ± 0.01 and 0.38 ± 0.02 % h^{-1}, "low density" and "high density" treatments, respectively; r=0.92, P<0.01); (2) significant negative variations occurred in detritus ^{32}P pools (Fig. 3), ADs (97329.62 ± 467.14, 26543.71 ± 836.36, 11242.12 ± 1620.67 DPM mg^{-1}, for defaunated control, "low density", and "high density" treatments, respectively), and CFs (Table); (3) the animal compartment sequestered a significant amount of radiotracer, ca. 27-fold higher than in the "sediment only" condition, where macrodetritivores showed a higher mortality rate (0.21 ± 0.04 versus 0.13 ± 0.03 individuals • h^{-1}) and a negligible final ^{32}P pool. Moreover, CFs showed non-significant differences with those determined in controls for physical adsorption (1-WAY ANOVA, F=1.09, NS).

Amphipods ingested and partially assimilated a significant fraction of the detrital radiotracer pool, transferring the nutrient along the food web and making it potentially available for higher trophic levels. At the same time, feeding and fragmentation of particulate detritus (Harrison 1977; Robertson and Mann 1980; Byren and Davies 1986), determined a quick release of ^{32}P originally present in the detrital pool, not counterbalanced by sediment uptake or by adsorption on the fine organic particles simultaneously produced by amphipods activity (*e.g.* as faecal pellets).

Our results indicated a significant "short-circuiting" effect on nutrient transfer processes from particulate to aqueous phase related to detritivores processing of macrodetritus, as already reported for freshwater benthic systems (*e.g.* Doremus and Clesceri 1982). In this perspective, regeneration of phosphorus sequestered into coarse allochthonous detritus in coastal benthic systems, conventionally related to abiotic factors and microbial mineralization processes, may result significantly amplified by macrodetritivores activity.

Acknowledgements. We wish to thank an anonymous reviewer for the valuable comments on the first draft of the manuscript. This research was supported by CNR -PRISMA1 and MURST funds.

References

Alongi R, Hanson RB (1985) Effect of detritus supply on trophic relationships within experimental benthic food webs. II. Microbial responses, fate and composition of decomposing detritus. J Exp Mar Biol Ecol 88: 167-182

Andersson G, Granéli W, Stenson J (1988) The influence of animals on phosphorus cycling in lake ecosystems. Hydrobiologia 170: 267-284

Byren BA, Davies BR (1986) The influence of invertebrates on the breakdown of *Potamogeton pectinatus* in a coastal marina (Zandvlei, South Africa). Hydrobiologia 137: 141-152

Caraco N, Cole J, Likens GE (1990) A comparison of phosphorus immobilization in sediments of freshwater and coastal marine systems. Biogeochemistry 9: 277-290

Clavero V, Niell FX, Fernandez JA (1994) A laboratory study to quantify the influence of *Nereis diversicolor* O.F. Muller in the exchange of phosphate between sediment and water. J Exp Biol Ecol 176: 257-267

Doremus C, Clesceri LS (1982) Microbial metabolism in surface sediments and its role in the immobilization of phosphorus in oligotrophic lake sediments. Hydrobiologia 91: 261-268

Enell, M, Fleischer S, Jansson M (1989) Phosphorus in sediments. Ambio 18: 137-138

Fenchel T (1970) Studies on the decomposition of the organic detritus derived from the turtle grass *Thalassia testudinum*. Limnol Oceanogr 15: 14-20

Froelich PN (1988) Kinetic control of dissolved phosphate in natural rivers and estuaries: a primer on the phosphate buffer mechanisms. Limnol Oceanogr 33: 649-668

Gardner WS, Nalepa TF, Quigley MA, Malczyk JM (1981) Release of phosphorus by certain benthic invertebrates. Can J Fish Aquat Sci 38: 978-981

Gätcher R, Meyer JS (1993) The role of micro-organisms in mobilization and fixation of phosphorus in sediments. Hydrobiologia 253: 103-121

Gätcher R, Meyer JS, Mares A (1988) Contribution of bacteria to release and fixation of phosphorus in lake sediments. Limnol Oceanogr 33: 1542-1558

Goltermann HL (1995) The role of ironhydroxide-phosphate-sulfide system in the phosphate exchange between sediments and overlying water. Hydrobiologia 297: 43-54

Harrison PG (1977) Decomposition of macrophyte detritus in seawater: effects of grazing by amphipods. Oikos 28: 165-169

Holdren GC, Armstrong DE (1980) Factors affecting phosphorus release from intact lake sediment cores. Environ Sci Technol 14: 79-87

Hupfer M, Uhlmann D (1991) Microbially mediated phosphorus exchange across the mud-water interface. Verh Int Ver Limnol 24: 2999-3003

Kamp-Nielsen LH, Mejer H, Sorensen SE (1982) Modelling the influence of bioturbation on the vertical distribution of sedimentary phosphorus in L. Esrom. Hydrobiologia 91: 197-206

Kristensen E, Blackburn TH (1987) The fate of organic carbon and nitrogen in experimental marine sediment systems: influence of bioturbation and anoxia. J Mar Res 45: 231-257

Mancinelli G, Fazi S, Rossi L (1998) Sediment structural properties mediating dominant feeding types patterns in soft-bottom macrobenthos of the Northern Adriatic Sea. Hydrobiologia 367: 211-222

Mesnage V, Picot B (1995) The distribution of phosphate in sediments and its relation with eutrophication of Mediterranean coastal lagoon. Hydrobiologia 297: 29-41

Revsbech NP, Sörensen J, Blackburn TH, Lomholt JP (1980) Distribution of oxygen in marine sediments measured with microelectrodes. Limnol Oceanogr 25: 403-411

Rippey B, Jewson DH (1982) The rates of sediment-water exchange of oxygen and sediment bioturbation in Lough Neagh, Northern Ireland. Hydrobiologia 92: 377-382

Robertson J, Mann KH (1980) The role of isopods and amphipods in the initial fragmentation of eelgrass detritus in Nova Scotia, Canada. Mar Biol 59: 63-69

Whittaker RH (1961) Experiments with radiophosphorus tracer in aquarium microcosms. Ecol Monogr 31: 137-186

CHAPTER 44

Structural and Trophic Variations in a Bathyal Community in the Ligurian Sea

S. Schiaparelli, M. Chiantore, R. Cattaneo-Vietti, F. Novelli, N. Drago, and G. Albertelli

ABSTRACT

The structure of a Mediterranean bathyal community living between 400 and 700 m depth in front of Portofino Promontory (Ligurian Sea) has been investigated, using sets of data collected in 1981 and 1996 respectively. Percent community composition, total density and biomass don't show any significant change, at least on the whole. Although total species number found in the two considered years is comparable, as well as diversity and evenness values, the specific composition shows an almost complete substitution. In particular, the disappearance of epifaunal species, mainly of large size, is outstanding. A real biocoenotical change did not occur, as evidenced by the qualitative (species dominance) and quantitative (numerical dominance) analyses of biocoenotical stocks, which show only a light increase in abundance of limicolous species and species linked to deep muds (VP) *versus* coastal terrigenous muds (VTC). Conversely, multivariate analysis evidences a clear clustering of 1996 and 1981 data due to presence/absence or different abundance of some key species: *Amphilepis norvegica, Spiophanes kroyeri rejssi* and *Amphiura filiformis* are present only in 1981, *Aricidea quadrilobata* and *Paradoneis lyra* appear in 1996. The analysis of community trophic structure highlights a significant increase of carnivorous species (mainly scavengers), mostly in terms of numerical dominance, opposed to a strong decrease of deposit-feeders.

The traditional use of descriptive tools, such as species number, total density and biomass lead to the classic hypothesis of general stability of bathyal communities. Conversely, changes in faunal composition of the bathyal benthic community of the Portofino Promontory were stressed only through the coupling of uni- and multi-variate approaches, suggesting that deep-sea benthic assemblages are not so stable as usually considered, probably partly because of small-scale changes in sedimentary regime (natural and anthropogenic) and heavy fishing trawling activities, both affecting the surface sediments, determining unfavourable conditions for the epifauna and favouring, instead, omnivorous and scavenger species.

Introduction

Deep-sea communities are considered to be characterised by a high degree of stability, due to the widespread uniformity of environmental parameters and food availability: increasing stability is correlated with increasing depth, as changes in the environment are decreased (Hessler and Sanders 1967; Rex 1981).

The *stability-time hypothesis* (Margalef, 1968; Sanders 1969; Slobodkin and Sanders, 1969) relates low diversity (= low degree of organisation) with severe and/or unpredictable hydrochemical environmental conditions. On the contrary, low variations of environmental parameters allow the stabilisation of biotic interactions which lead to high degree of specialisation. While there have been a great number of studies dealing with deep-sea benthic community structure from a static point of view, data on its persistence are still scarce (Bernstein and Meador 1979; Josefson 1981).

Lack of variation in the water column and in benthic environment below the thermocline have been a basic tenet of physico-chemical and ecological studies in the deep-sea. Whilst the

Dipartimento per lo Studio del Territorio e delle Sue Risorse, Università di Genova, 16132 Genova, Italy

F.M. Faranda, L. Guglielmo, G. Spezie (eds)
Mediterranean Ecosystems: Structures and Processes

concept of temporal stability is valid for much of the world oceans and over mesoscale terms, on a short basis periodic variability appears to be more common in certain areas of the deep-sea than originally supposed (Tyler 1988). Predictable and unpredictable processes may be responsible of variations in the deep-sea environment (Tyler 1988): physical features and seasonality in vertical flux are usually of a short time scale and predictable, while unpredictable events may be considered turbidity currents, sediment slides and benthic storms. The most considerable and predictable fraction of the vertical flux is constituted by the rain of small particles derived from surface primary and secondary production (organic and inorganic particles, macrophyte detritus, faecal pellets and large food falls of moults), while deadfall of carcasses and river run-off are unpredictable. Seasonal processes are mainly responsible for growth, spawning and recruitment of deep-sea benthic organisms (Gage and Tyler 1981; Tyler 1988), while unpredictable events can play a relevant role in determining small scale patchiness in mega-, macro- and meiobenthic organism distribution and relevant abundance of scavenger species (Dayton and Hessler 1972).

In order to evaluate deep-sea community structure and its stability, a comparison between present and past data sets was carried out off the Portofino Promontory (Eastern Ligurian Sea, Mediterranean Sea), comparing data recorded in 1981 and 1996. The investigated area (between 400 and 700 m depth) is located on a very steep continental slope, characterised by the presence of drainage lines, river run-off from the Entella river and strongly affected by anthropogenic activities (fishery and dumping).

The analysis of community structure was carried out taking into account biocoenotical and trophic features, evidencing analogies and discrepancies between recent and past data.

Materials and Methods

Samples were collected in April 1981 (Cattaneo and Albertelli 1983), on board of the H/V *L.F. Marsili* and in April 1996, on board of the H/V *Ammiraglio Magnaghi*, in an area off the Portofino Promontory around 44°14.2 N and 09°08.5 E (0.5 nautical mile radius; Fig. 1).

Sampling was performed using a Van Veen grab (0.1 m^2): 36 samples were collected in 1981, ranging in depth between 428 and 760 m, while 24 were collected in 1996, between 438 and 690 m depth.

Samples were sieved using a 1 mm mesh sieve, preserved in 10% formaline and sorted in the laboratory, using a stereomicroscope.

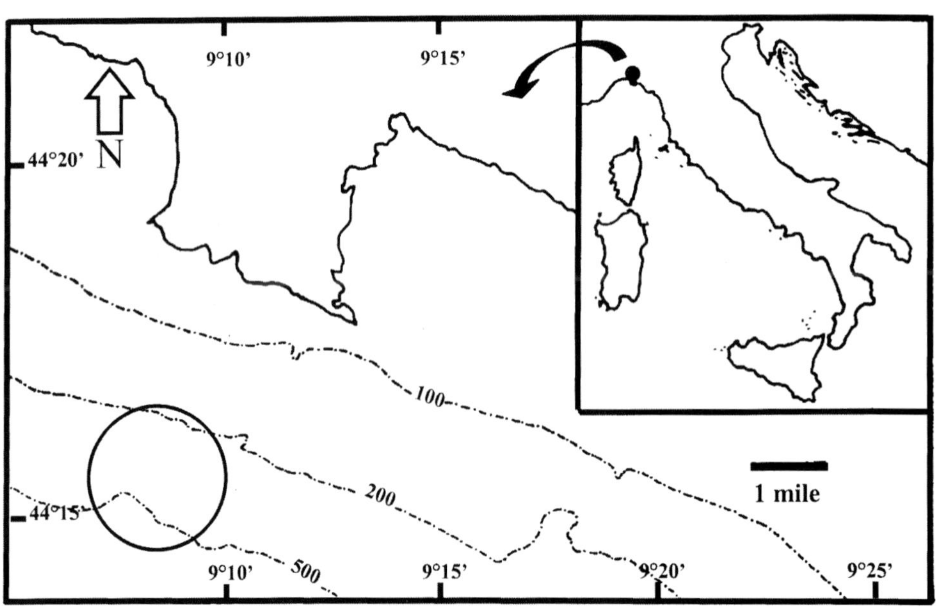

Fig. 1. Location of sampling area

Macrobenthic specimens were counted and identified at species level. Biomass values of different phyla were assessed in terms of wet weight (gWW).

The following Cabioch index (Cabioch et al. 1980) was applied:

$$Ci = (q-p) / (q+p)$$

where: p = frequency of species i in 1981 data set
q = frequency of species i in 1996 data set

The Cabioch index ranges between -1 (species disappearance) and +1 (species appearance).

Margalef species richness (Margalef 1958), Shannon-Weaver diversity (Shannon and Weaver 1963) and Pielou evenness (Pielou 1966) indexes were evaluated.

Table 1. Total density, biomass and species numbers in 1981 and 1996 samplings

		1981	1996
Total macrobenthic density (ind./m²)		126.4	101.6
Total macrobenthic biomass (gWW/m²)		6.6	6.3
Species number:	Polychaetes	46	38
	Crustaceans	16	22
	Molluscs	7	8
	Echinoderms	6	1
	Other taxa	6	6
	Total	81	75

Classification (cluster analysis) and ordination (MDS) of data were performed on square root transformed density data (1981 + 1996 data), grouping stations according to bathymetric levels (400-500 m, 500-600 m and > 600 m), using the Bray-Curtis similarity index (Bray-Curtis 1957). The contribution of species to groups dissimilarity was assessed through R-mode analysis. Classification, ordination and R-mode analysis were performed using the software PRIMER of the Plymouth Marine Laboratory.

Results

The comparison between present and past data sets has evidenced large discrepancies according to the different community descriptor parameters considered.

In fact, the macrobenthic community living off the Portofino Promontory in the last 15 years did not show any relevant variation in terms of density, biomass as well as total species number (Table 1; t-test, $P>5$ %).

Quite similar values of Margalef species richness, Shannon-Weaver diversity and Pielou evenness were also found comparing 1981 and 1996 data: 3.42 versus 3.46, 1.90 versus 2.07 and 0.98 versus 0.96, respectively.

Comparable results were also observed in terms of numerical dominance of the main different phyla (Fig. 2), except for the disappearance of ophiuroids, mainly *Amphilepis norvegica*, an exclusive species of Mediterranean deep muds

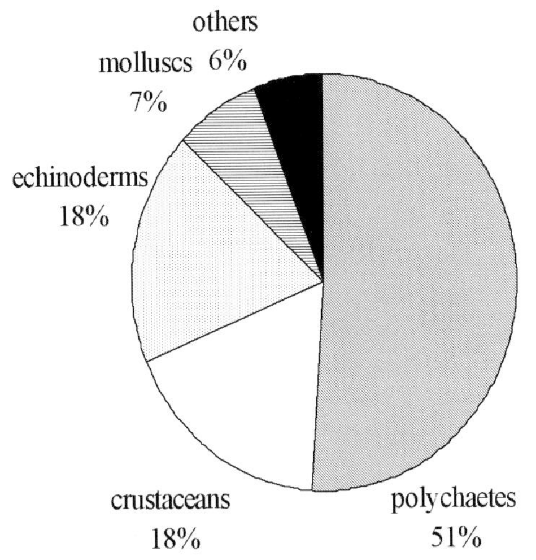

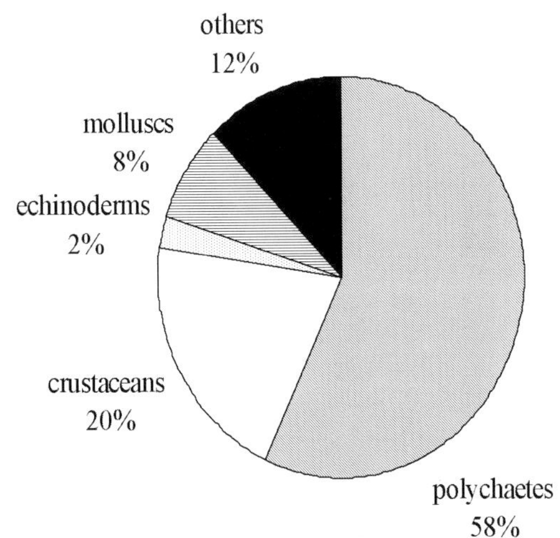

Fig. 2. Percent dominance of different taxa to total density in 1981 (*left*) and 1996 (*right*)

(VP: Peres and Picard 1964) (Carpine 1970; Picard 1972; Chardy et al. 1973).

Differently, significant variations were observed taking into account the contribution of the different phyla to total biomass (Fig. 3): in particular polychaetes showed a marked decrease in biomass values, dropping from 2.1 to 0.1 gWW/m^2.

Moreover, taking into account the species composition of the whole two data sets, large variations may be stressed. A total of 156 species was found on the whole, only 25 of which were recorded both in 1981 and 1996 cruises. These species showed quite different dominance values, as evidenced by the Cabioch index (Fig. 4).

In Table 2a,b the characterising species of the two data sets are reported with their numerical dominance values in respect of the whole community and of the phylum, to which they belong.

Trophic structure stressed an increase of carnivorous and scavenging species in 1996 (grouped together in Fig. 5) at despite of typical deposit-feeding species.

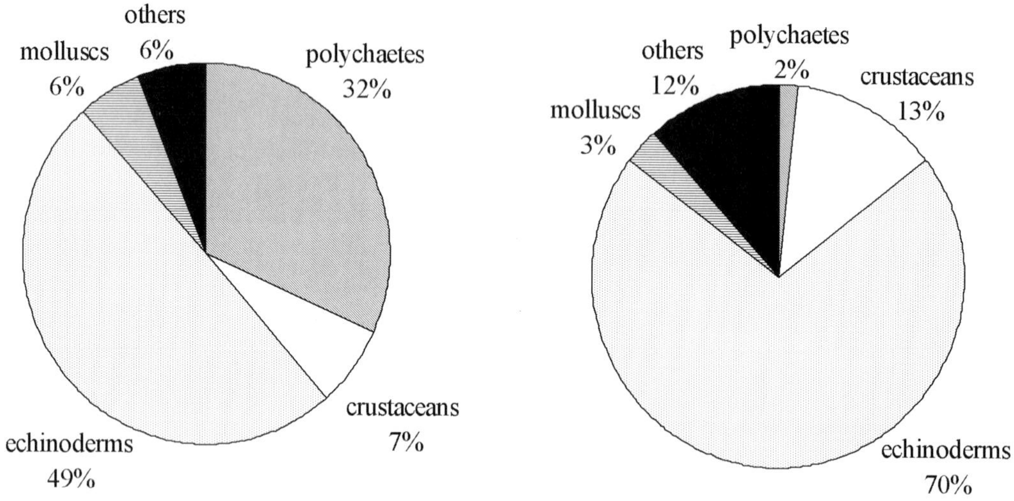

Fig. 3. Percent dominance of different taxa to total biomass in 1981 (*left*) and 1996 (*right*)

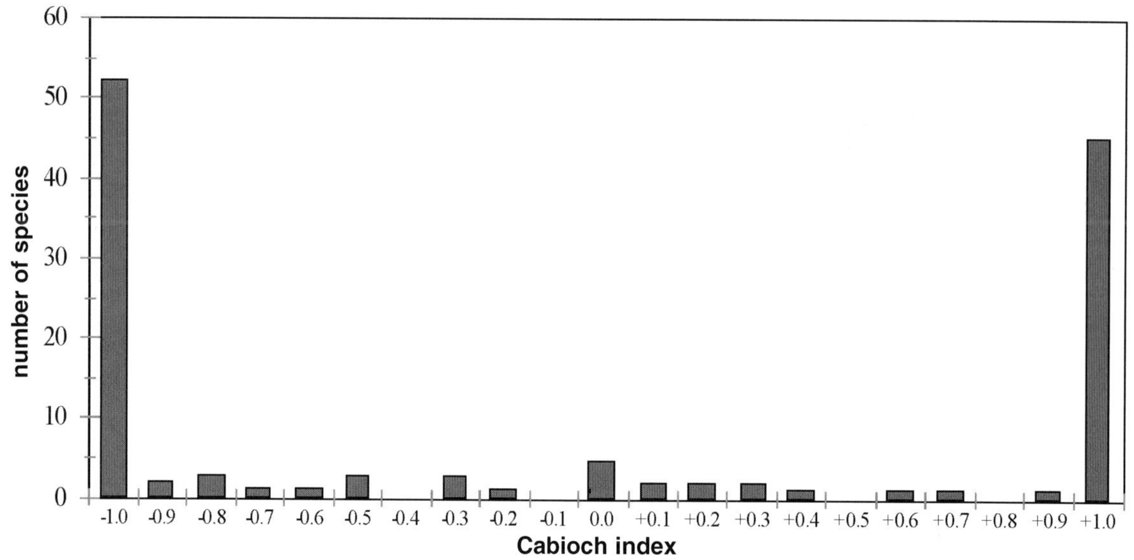

Fig. 4. Frequency distribution of Cabioch index

Table 2a,b. **a** 1981, and **b** 1996, characterising species: numerical dominance in respect of the phylum to which they belong and of the whole community

a	Phylum	1981	% on phylum	% on whole community
	Polychaetes	*Anobothrus gracilis*	12.9	6.6
		Melinna monoceroides	10.3	5.3
		Spiophanes kroyeri rejssi	10.3	5.3
	Crustaceans	*Calocaris macandreae*	27.2	4.8
		Eriopisa elongata	19.8	3.5
	Molluscs	*Abra longicallus*	42.4	3.1
		Lionucula tenuis aegeensis	18.2	1.3
		Thracia papyracea	15.2	1.1
	Echinoderms	*Amphilepis norvegica*	57.3	10.3
		Amphiura filiformis	24.4	4.4
	Other taxa	*Aspidosiphon mulleri*	53.8	3.1

b	Phylum	1996	% on phylum	% on whole community
	Polychaetes	*Aricidea quadrilobata*	11.5	3.5
		Onuphis lepta	11.5	3.5
		Prionospio cirrifera	10.8	3.3
	Crustaceans	*Anthelura fresi*	14.0	1.5
		Calocaris macandreae	12.0	1.3
		Eriopisa elongata	20.0	2.2
	Molluscs	*Abra longicallus*	36.8	1.5
		Dentalium agilis	15.8	0.7
	Echinoderms	*Molpadia musculus*	100	1.3
	Other taxa	*Onchnesoma steenstrupi*	20.0	1.3

The analysis of biocoenotical assemblages dominance evidenced only a slight increase of VP characteristic species, although remarkable absences may be stressed, such as the ophiuroid *Amphilepis norvegica*. Most remarkable differences were stressed about *habitus* of dominant species. In 1996 a decrease of tube-building (sediment stabilisers) species occurred, balanced by

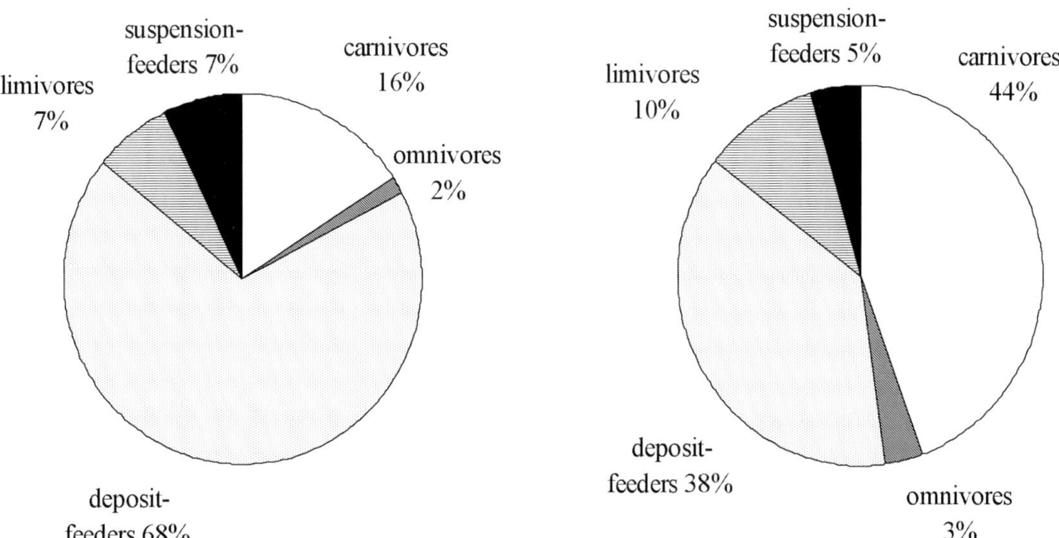

Fig. 5. Percent dominance of different trophic groups in 1981 (*left*) and 1996 (*right*)

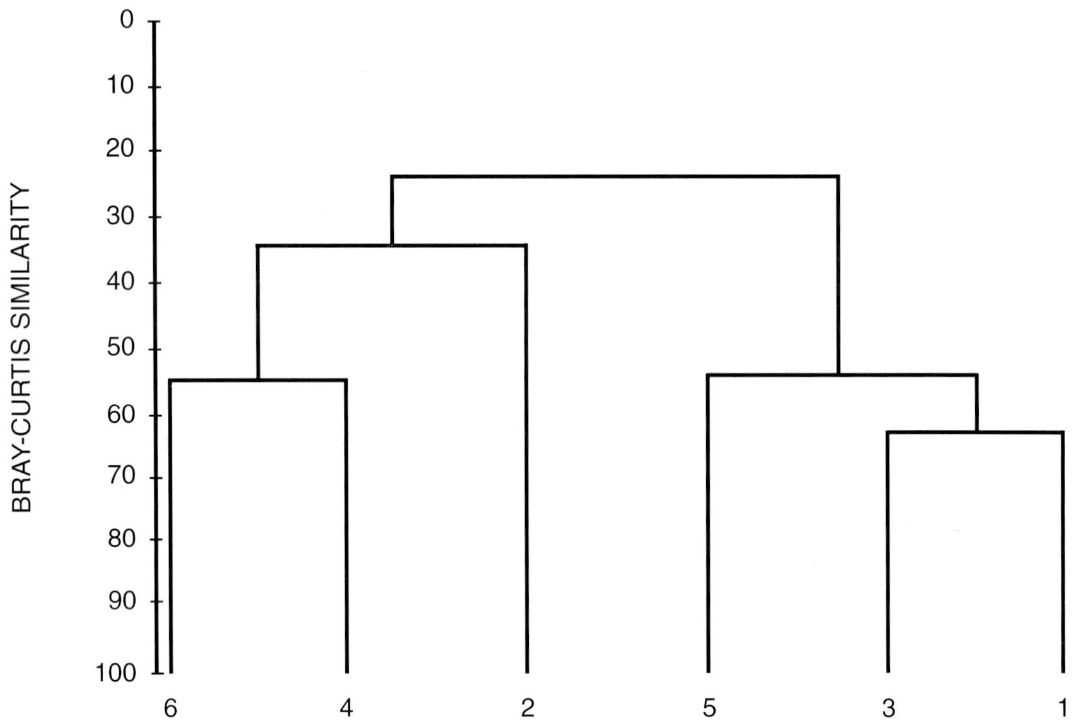

Fig. 6. Classification of density data. Stations grouped according to bathymetric levels [*1,* 400-500 m (1981); *2,* 400-500 m (1996); *3,* 500-600 m (1981); *4,* 500-600 m (1996); *5,* >600 m (1981); *6,* >600 m (1996)]

increase of vagile species (carnivores/scavengers) or sediment-destabilisers.

The differences in community structure were remarkably evidenced through classification (cluster analysis; Fig. 6) and ordination (MDS; Fig. 7) of data, taking into account samples grouped according to bathymetric gradient, showing a complete dichotomy between 1981 and 1996 data. From cluster as well as with *t*-test analyses, the uppermost area (400-500 m) evidenced larger differences between past and present data in terms of total density, biomass and numerical dominance (*t*-test, $P<0.05$), suggesting a more disturbed zone.

The comparison between single stations from the two data sets was also performed, but

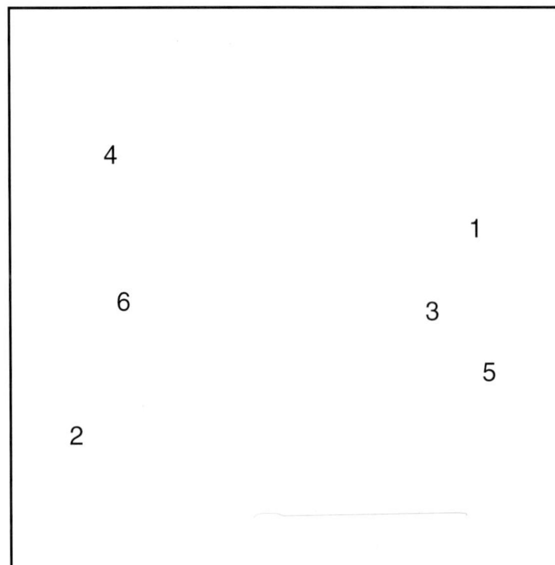

Fig. 7. Ordination of data (MDS). See legend in Fig. 6. 2-dimensional stress < 0.01, excellent representation

Table 3. Discriminating species, average abundance in the two data sets and percent contribution to dissimilarity between clusters

Species	1981	1996	%
Amphilepis norvegica	**1.10**	0.00	3.12
Aricidea quadrilobata	0.00	**0.56**	2.50
Paradoneis lyra	0.00	**0.56**	2.49
Spiophanes kroyeri rejssi	**0.61**	0.00	2.49
Amphiura filiformis	**0.51**	0.00	2.34
Prionospio steenstrupi	**0.44**	0.00	2.19
Anobothrus gracilis	**0.77**	0.11	2.09
Syllis cornuta	0.00	**0.35**	1.86
Prionospio cirrifera	0.05	**0.51**	1.86

results were highly contradictory, because of the organism paucity of each sampling station. The analysis of discriminating species through R-mode analysis is reported in Table 3.

Discussion and Conclusions

Comparing data regarding the composition of the bathyal macrobenthic assemblages living between 400 and 700 m depth off the Portofino Promontory (Ligurian Sea), relevant variations between 1981 and 1996 are evident.

No differences arose regarding biomass, density, species number and diversity, but a significant change occurred in the relationships among the trophic groups, with an increment of carnivores and scavengers in despite of deposit-feeders.

The lack of previous long-term data on the structure and dynamics of these deep-sea communities makes it difficult to assess whether these differences are really true or eventually only mirror their diversity due to biocoenoses overlapping (Cattaneo-Vietti 1991) or patchiness, as evidenced in other deep-sea communities (Hessler and Sanders, 1967; Jumars 1975).

This heterogeneity and structural complexity may be the natural result of different stochastic processes such as localised downward organic fluxes (from POM to large carcasses), cropper (*sensu* Dayton and Hessler 1972) and/or tube-builder presence (Gray 1974; Jumars 1975, 1976), allowing coexistence of various microsuccessional stages through contemporaneous disequilibria (Grassle and Sanders 1973).

In the case of real variations, questions arise about which were the main responsible events that affected the stability of benthic communities and influenced the dynamic nature of fine-grained sediments. Are they stochastic, biologically mediated or anthropogenic?

Major changes have been stressed in the uppermost investigated area (400-500 m), suggesting a littoral derived disturbance, probably linked to different and synergic factors. The study area is located close to the *Riviera di Levante* canyon, and it is affected by high density flowing down currents transporting unstable sediments sliding down the submarine slope (Fanucci et al. 1979; Corradi et al. 1980). Moreover these bottoms are influenced by the outflows of the Entella river (Piccazzo 1986) and by its floods.

Besides these natural influences, the area is affected by an intensive trawling activity (Massi et al. 1994), mainly linked to the red shrimp fishing (Relini and Orsi Relini 1987). It is reasonable to suppose that the continuous disturbance of the fishing gears causes a significant reworking of the sediment and difficulties for the bottom-stabilising organisms. Among these, *Funiculina quadrangolaris* and *Isidella elongata* for epifauna (Relini-Orsi 1973; Relini et al. 1986), and different tube-building polychaetes, for endofauna. The continuous fishing activity has caused quali- quantitative changes in the commercial catches (Relini-Orsi 1973; Relini et al. 1986) and an anomalous increase of large organic debris which favours the scavengers (Cattaneo-Vietti et al. 1993).

Besides this permanent impact, it is possible to take into account the authorised dumping of harbour muds that has occurred in the last decades which could have changed the superficial texture of the sediments up to 700 m depth.

A complex of factors favours in this area an environmental instability which influences the community structure, preventing the establishment of sediment-stabilising species (tube-builders, burrow dwellers), whose presence plays an important biological role controlling and maintaining diversity (Rhoads and Young 1970; Young and Rhoads 1971).

In the deep sea, these studies have a significant importance both in the evaluation of the long-term changes and/or of anthropogenic impact, since the only analysis of routine descriptive parameters (density, biomass, species richness and numerical dominance) are not exhaustive, as external factors could radically change the trophic pathways in these food-limited environments and consequently the species composition.

References

Bernstein BB, Meador JP (1979) Temporal persistence of biological patch structure in an abyssal benthic community. Mar Biol 51: 179-183

Bray JR, Curtis JT (1957) An ordination of the upland forest communities of Southern Wisconsin. Ecol Monogr 27: 325-349

Cabioch L, Dauvin JC, Mora Bermudez J, Rodriguez Babio C (1980) Effets de la marée noire de "Amoco Cadiz" sur le benthos sub-littoral de la Bretagne. Helgo Meeresunters 33: 192-208

Carpine C (1970) Écologie de l'étage bathyal dans la Méditerranée occidentale. Mem Inst Oceanogr Monaco 2

Cattaneo M, Albertelli G. (1983) Macrobenthos dei fondi batiali liguri. Atti V Congr AIOL: 251-260

Cattaneo-Vietti R (1991) Bathymetric distribution of soft-bottom opistobranchs along the Ligurian and Tuscany continental slope (Western Mediterranean). In: Proc. X Int Malacol Congr Tübingen, Klaus Meier-Brook, Unit Malacol, pp 327-334

Cattaneo-Vietti R, Burlando B, Senes L (1993) Life history and diet of *Pleurobranchea meckelii* (Opisthobranchia: Notaspidea). J Moll Stud 59: 309-313

Chardy P, Laubier L, Reyss D, Sibuet M (1973) Dragages profonds en mer Égée - données préliminaires, 1. Rapp Comm Int Mer Mediterr 22 (4): 107-108

Corradi N, Fanucci F, Firpo M, Piccazzo M, Traverso M (1980) L'olocene della piattaforma continentale ligure da Portofino alla Spezia. Ist Idrogr Mar FC 1099

Dayton PK, Hessler RR (1972) Role of biological disturbance maintaining diversity in the deep sea. Deep-Sea Res 19: 199-208

Fanucci F, Fierro G, Firpo M, Mirabile L, Piccazzo M (1979) La piattaforma continentale della Liguria appenninica. Conv Sci Nazl PF Oceanogr Fondi Mar, Roma 5-7 Marzo 1979, pp 1275-1289

Gage JD, Tyler PA (1981) Re-appraisal of age composition, growth and survivorship of the deep-sea brittle star *Ophiura ljungmani* from size structure in a sample time series from the Rockall Trough. Mar Biol 64: 163-172

Grassle JF, Sanders HL (1973) Life histories and the role of disturbance. Deep-Sea Res 20: 643-659

Gray JS (1974) Animal-sediment relationships. Oceanogr Mar Biol Annu Rev 12: 223-261

Hessler RR, Sanders HL (1967) Faunal diversity in the deep-sea. Deep-Sea Res 14: 65-78

Josefson AB (1981) Persistence and structure of two deep macrobenthic communities in the Skagerrak (West Coast of Sweden). J Exp Mar Biol Ecol 50: 63-97

Jumars PA (1975) Methods for measurement of community structure in deep-sea macrobenthos. Mar Biol 30: 245-252

Jumars PA (1976) Deep-sea species diversity: does it have a characteristic scale? J Mar Res 34: 217-246

Margalef R (1958) Temporal succession and spatial heterogeneity in phytoplankton. In: Buzzati-Traverso (ed) Perspectives in marine biology. Univ Calif Press, Berkeley, pp 323-347

Margalef R (1968) Perspectives in ecological theory. Univ Chicago Press, Chicago

Massi D, Fiorentino F, Zamboni A, Relini G (1994) Variazioni nella biomassa sbarcata e nella tipologia della pesca a strascico di Santa Margherita Ligure nell'ultimo quinquennio. Biol Mar Mediterr I (1): 325-326

Pérès JM, Picard J (1964) Nouveau manuel de bionomie benthique de la Mer Méditerranée. Rec Trav Stn Mar Endoume 31 (47): 1-137

Picard C (1972) Les peuplements de vase au large du Golfe de Fos. Tethys 3 (3): 569-618

Piccazzo M (1986) Caratteristiche geologiche e sedimentologiche della piattaforma continentale ligure ad Est di Genova. Quad Ist Geol Univ Genova 7 (3): 91-105

Pielou EC (1966) The measurement of diversity in different types of biological collections. J Theor Biol 13: 131-144

Relini G, Orsi Relini L (1987) The decline of the red shrimp stocks in the Gulf of Genoa. Inv Pesq 51 (Suppl 1): 245-260

Relini G, Peirano A, Tunesi L (1986) Osservazioni sulle comunità dei fondi strascicabili del Mar Ligure centro-orientale. Boll Mus Ist Biol Univ Genova 52 Suppl: 139-161

Relini-Orsi L (1973) I crostacei batiali del Golfo di Genova nelle osservazioni di Alessandro Brian e nelle condizioni attuali. In: Atti V Congr Nazl Soc Ital Biol Mar Salentina, pp 25-40

Rex MA (1981) Community structure in the deep-sea benthos. Annu Rev Ecol Syst 12: 331-353

Rhoads DC, Young DK (1970) The influence of deposit-feeding organisms on sediment stability and community trophic structure. J Mar Res 28: 150-178

Sanders HL (1969) Benthic marine diversity and the stability-time hypothesis. Brookhaven Symp Biol 22: 71-81

Shannon CE, Weaver W (1963) The mathematical theory of communication. Univ Illinois Press, Urbana, Illinois

Slobodkin LB, Sanders HL (1969) On the contribution of environmental predictability to species diversity. Brookhaven Symp Biol 22: 82-95

Tyler PA (1988) Seasonality in the deep sea. Oceanogr Mar Biol Annu Rev 26: 227-258

Young DK, Rhoads DC (1971) Animal-sediment relations in Cape Cod Bay, Massachusetts. I. A transect study. Mar Biol 11: 242-254

CHAPTER 45

Changes in Seaweed Biodiversity of the Gargano Coast (Adriatic Sea, Mediterranean Sea)

E. Cecere, G. Fanelli, A. Petrocelli, and O.D. Saracino

ABSTRACT

Benthic algal flora has been investigated along the Gargano coast (Adriatic Sea, Mediterranean Sea). The results were compared with those of previous researches carried out in the same area in 1960s in order to detect possible changes in species composition. Two hundred and thirty-four taxa have been identified. The literature data reported 162 taxa but only 82 of them were also found during the present study. In general, a decrease in "canopy-forming" species and an increase in "turf-forming" ones were observed. The decrease in Ulvaceae and the presence of sciaphilous taxa at low depths were also recorded. The present species composition might be related both to a reduction in light penetration and to an increase in sediment deposition.

Introduction

The present paper is a part of PRISMA 2, a multidisciplinary project aimed at the evaluation of the environmental conditions of the Adriatic Sea. Benthic algal flora composition has often been used as an indicator of environmental conditions (Levine 1984). In fact, algal assemblages can change in structure and composition in response to both occasional stress and long term modification of environmental conditions and record the interactions between biotic and abiotic factors. As the seaweed composition of the Gargano coast was previously studied in the sixties (Huvè et al. 1963; Rizzi Longo et al. 1967; Giaccone 1969; Giaccone and Bruni 1971; Bressan 1974; Giaccone 1978; Cinelli 1979), a comparison of the acquired floristic lists with the literature data was carried out in order to detect any possible differences.

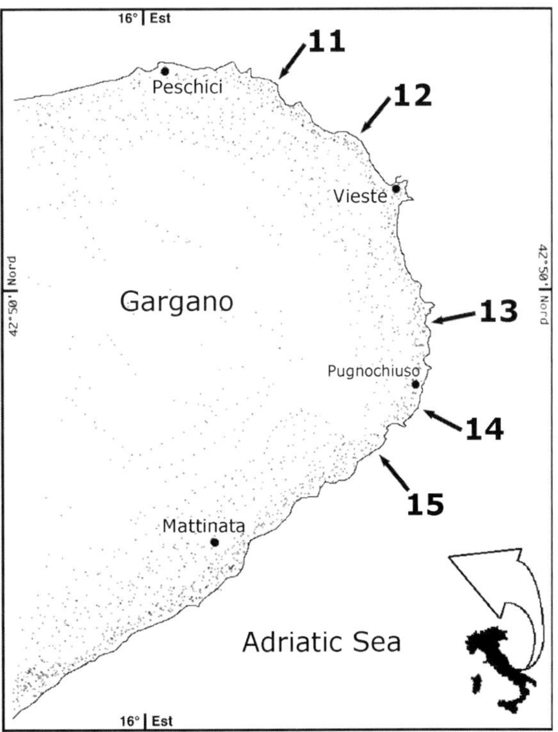

Fig. 1. Map of the area studied and location of the transects (*numbered arrows*). The *small picture* shows the localization of Gargano Promontory (Adriatic Sea)

Materials and Methods

The Gargano Promontory is located between the Middle and the Southern Adriatic (Fig. 1). Most of its coast is rocky, high, steep and wave-swept. Fifty samples were collected by SCUBA diving in

Istituto Sperimentale Talassografico, CNR, Via Roma 3, 74100 Taranto, Italy

F.M. Faranda, L. Guglielmo, G. Spezie (eds)
Mediterranean Ecosystems: Structures and Processes

May and October 1997 from rocky stations located along five transects perpendicular to the shore from the intertidal level up to 12 m. At greater depths the bottom was soft and without vegetation. Collected material was preserved in 4% formaldehyde in seawater. Voucher specimens are kept in the Herbarium of the Istituto Talassografico of Taranto.

The floristic list compiled using data from the literature was updated from both a taxonomic and a nomenclatural point of view to make it comparable to that of the present study. For each class, the proportion of each family was calculated as a part of the total list comprising the taxa present on both lists, those previously reported and not found in the present study and those newly recorded.

Cluster analysis and non-metric multidimensional scaling were performed on a data set of 14 floristic lists (presence/absence data): 4 from the literature (among the above mentioned papers, those of Giaccone and Bruni 1971, Bressan 1974, and Cinelli 1979, were not utilised because they only reported few species) and 10 from the five transects sampled during both the campaigns. Average linking was used for hierarchical clustering. Both analyses were based on the Bray-Curtis similarity index and were run on the PRIMER package (Clarke and Warwick 1994).

Results

During the present study 234 taxa of algae were identified at the specific and infraspecific level, 169 of which are Rhodophyceae, 33 are Fucophyceae and 32 are Chlorophyceae (data not shown; the floristic list is available on request). The literature data include 92 Rhodophyceae, 37 Fucophyceae and 33 Chlorophyceae, for a total of 162 taxa. From the comparison, it appears that the present flora is richer than that previously reported. A slight decrease in number was observed for Chlorophyceae and Fucophyceae, while the number of Rhodophyceae strongly increased. Only 82 taxa of the previous list were also found during the present study; therefore, 152 taxa are new records for the Gargano coast.

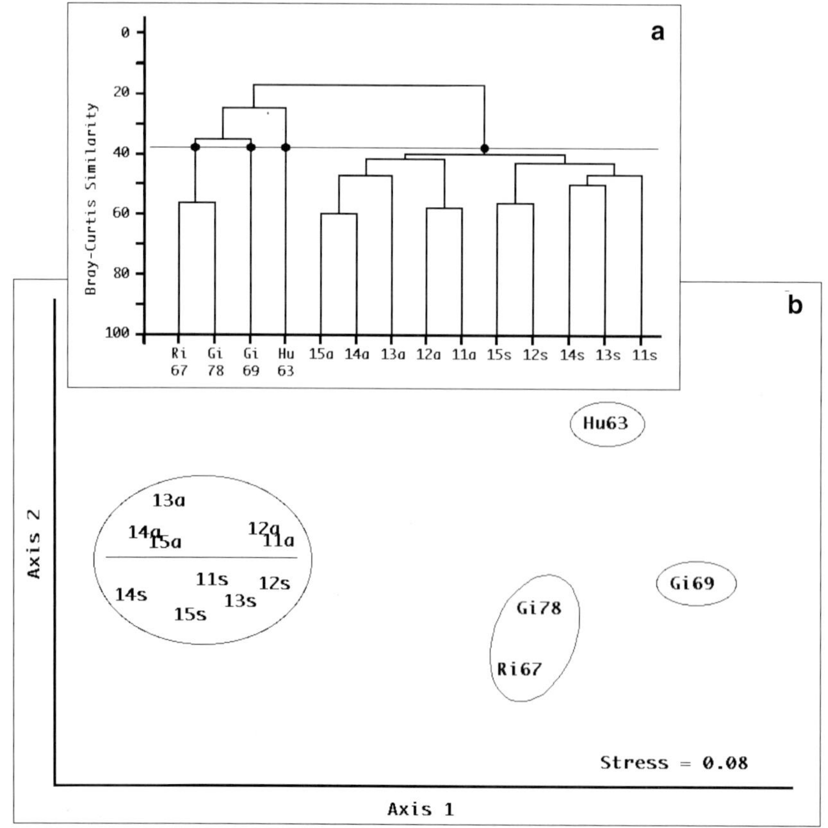

Fig. 2a,b. Comparison of floristic lists by: **a** cluster analysis; **b** non-metric multidimensional scaling [*s*, summer transects; *a*, autumn transects; *Hu63*, Huvé et al. (1963); *Ri67*, Rizzi Longo et al. (1967); *Gi69*, Giaccone (1969); *Gi78*, Giaccone (1978)]

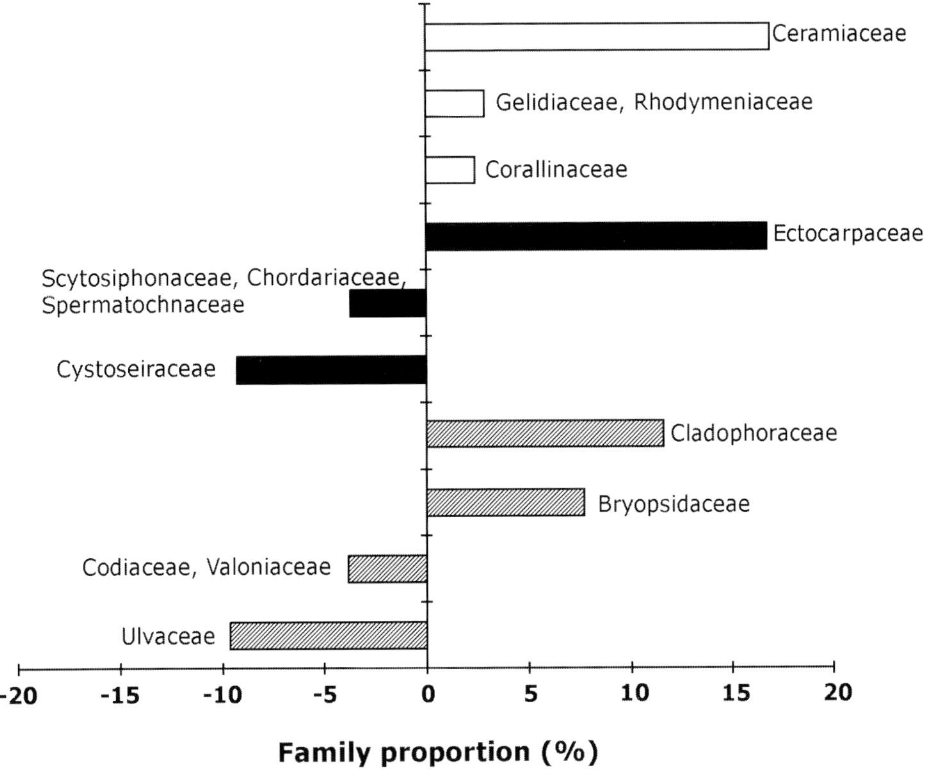

Fig. 3. Variation shown in the proportion of each family in Rhodophyceae (□), Fucophyceae (■), Chlorophyceae (▨). Only variations greater than 2% and lower than −2% are reported

Cluster analysis and non-metric multidimensional scaling (Fig. 2) showed that the compared floristic lists are clustered in two different groups: that of the present study and that of the 1960s papers. Spring and autumn transects showed little differences.

When species were grouped at the family level, it became evident that in the last three decades the proportion of Ceramiaceae increased considerably (16.8%). However, the increase in Rhodophyceae proportion was also due to other families (Fig. 3). Among the Fucophyceae, only the Ectocarpaceae showed a percentage increase (16.7%), while Cystoseiraceae decreased (9.3%) (Fig. 3) and many families disappeared (*i.e.* Sargassaceae, Arthrocladiaceae, Giraudiaceae, Ralfsiaceae, Chordariaceae, Scytosiphonaceae, Spermatochnaceae). In particular, these changes were due to the disappearance of *Cystoseira corniculata* (Turner) Zanardini, *C. schiffneri* Hamel f. *tenuiramosa* Ercegovic (Giaccone), *C. spinosa* Sauvageau v. *spinosa*, *Sargassum acinarium* (Linnaeus) Setchell and *S. hornschuchii* C. Agardh.

Among the Chlorophyceae, there was an increase in species number for Cladophoraceae (11.5%) and Bryopsidaceae (7.7%), while Ulvaceae noticeably decreased (9.6%) (Fig. 3). Many deep water species, such as *Aglaothamnion byssoides* (Arnott *ex* Harvey) Boudouresque *et* Perret-Boudouresque, *A. tripinnatum* (C. Agardh) Feldmann-Mazoyer and *Antithamnion tenuissimum* (Hauck) Schiffner were found in the upper sublittoral on flat rock.

Discussion

The comparison of the present floristic list with the literature data showed a decrease in those families comprising most of the canopy-forming species (Cystoseiraceae, Sargassaceae) and an increase in those including most of the turf-forming species. These latter comprise filamentous small algae which cover the substrate in thick, dense mats (Hay 1981; Airoldi et al. 1995) (mostly Ceramiaceae, Ectocarpaceae, Cladophoraceae), and those algae, such as Gelidiaceae, which showed high cover values (data not reported), whose thallus is a tightly clumped aggregate of upright and lateral branches (Hay 1981). Moreover, the upward migration of deep water species was also observed.

The detection of these changes would have not been possible without the availability of past data. This sort of comparison is not very frequent in the literature (but see Wilkinson and Tittley 1979; Cecere et al. 1991 and Nielsen 1991, for marine benthic flora), despite the importance of long-term studies advocated for many years (see references in Strayer et al. 1986, Likens 1989). However, even though the importance of time-series data is unquestionable, such observations are not sufficient to detect the causes of the changes and it is necessary to integrate the descriptive approach with both the experimental and the theoretical one (Tilman 1989). Therefore, time series should be used to generate hypotheses to be tested experimentally (Coull 1985): this is the only approach useful to reveal the reasons for changes in natural populations.

On the basis of collected data we advance some hypotheses to explain the observed changes. The most evident was the disappearance of some Fucales which are very sensitive and characteristic of the photophilic communities on hard substrata in unpolluted Mediterranean waters (Perrera and Giaccone 1986). Such a disappearance might be due to different factors, acting separately or in synergy: eutrophication, chemical pollution, reduction of light penetration, sedimentation, fishery activities, overgrazing by sea urchins. In the studied area there is no evidence of an increase in eutrophication, on the contrary in the last decades in the Southern Adriatic a nutrient lowering was observed (Vukadin et al. 1988; Zore-Armanda et al. 1991). Along the Gargano coast there are no industrial plants, even though the effects of chemical pollution can not be excluded; in fact, the general circulation of the Adriatic Sea is cyclonic, with a surface current flowing southwards along the Italian coast, which, especially in winter, consists of cold and fresh water coming from the Po river and other streams (La Violette and Gacic 1990). A reduction in light penetration occurred from the sixties to the eighties in both coastal and open waters of the Southern Adriatic Sea, as reported by Morovic and Domijan (1991) who recorded a decrease in the euphotic zone. In the Middle Adriatic Sea there is no evidence of increased sedimentation, even though Gargano coastal waters are presumably affected by sediment deposition and movement, as suggested by the large quantities of sediment observed both when sampling and washing the samples before examination. We can exclude the possibility that other human activities such as fishery or anchoring could affect the sampled area because of its proximity to the coast line. Finally, during both sampling campaigns the sea urchin density was rather low ($< 1\ m^{-2}$) in all the sampling stations leading to the exclusion of their grazing as playing an important role along the Gargano coast.

Therefore, the most probable hypothesis to explain the upward migration of deep water species and the disappearance of some species of *Cystoseira* and *Sargassum*, is the reduction in light penetration due to the high sedimentation rate. This is also in agreement with the increase in richness of turf-forming species. In fact, it is known from the literature that the turf morphology makes seaweeds more resistant to abrasion (Hay 1981), making turf-forming algae generally more abundant in habitats stressed by sediment (Littler et al. 1983; Airoldi et al. 1995) which is trapped in the algal matrix (Stewart 1983). Moreover, the change from canopy- to turf-dominated communities also occurred in *Cystoseira*-dominated rockpools as a result of the experimental removal of canopy species (Benedetti Cecchi and Cinelli 1992).

Furthermore, the decrease in Ulvaceae might be due to nutrient lowering as well as to the outcompetition by turf-forming species. In fact, algal turfs may preclude the settlement of other algal spores both by pre-empting the substrate by vegetative propagation (Sousa 1979), and by trapping the sediment which fills the space among axes (Sousa et al. 1981).

Similar changes in the composition of the benthic algal flora also occurred in the nearby Tremiti Islands, where both reduction in light penetration and sedimentation have also been supposed to be the most probable causes of the observed changes (Cormaci and Furnari 1999; Cormaci et al. this volume).

Conclusions

Over the last three decades a change in benthic algal flora composition of the Gargano coast has occurred, mainly consisting of: (1) a reduction in

Ulvaceae and the canopy-forming, photophilous species; (2) an increase in turf-forming species; (3) the upward migration of deep water species. The observed changes might be related mainly to the reduction in light penetration and to increased sedimentation rather than to eutrophication and other human activities, but these hypotheses need to be tested by means of appropriate experiments.

Acknowledgements. We are indebted to M. Cormaci and G. Furnari for both the nomenclatural and the systematic updating of the literature data. This work was supported by a grant from the Italian C.N.R. within the project PRISMA 2, sub-project 3. Two anonymous referees provided useful comments to improve the paper.

References

Airoldi L, Rindi F, Cinelli F (1995) Structure, seasonal dynamics and reproductive phenology of a filamentous turf assemblage on a sediment influenced, rocky subtidal shore. Bot Mar 38: 227-237

Benedetti Cecchi L, Cinelli F (1992) Canopy removal experiments in *Cystoseira*-dominated rockpools from the Western coast of the Mediterranean (Ligurian Sea). J Exp Mar Biol Ecol 155: 69-83

Bressan G (1974) Rodoficee calcaree dei mari italiani. Boll Soc Adriat Sci Trieste 59: 1-132

Cecere E, Cormaci M, Furnari G (1991) The marine algae of Mar Piccolo, Taranto (Southern Italy): a re-assessment. Bot Mar 34: 221-227

Cinelli F (1979) *Acetabularia acetabulum* (L.) Silva, *Acetabularia parvula* Solms-Laubach and *Dasycladus vermicularis* (Scopoli) Krasser (Chlorophyta, Dasycladaceae): ecology and distribution in the Mediterranean Sea. In: Bonotto S, Kefeli V, Puiseux-Dao S (eds) Developmental biology of *Acetabularia*. Elsevier-North Holland Biomedical Press, Amsterdam, pp 3-14

Clarke KR, Warwick RM (1994) Changes in marine communities: an approach to statistical analysis and interpretation. Nat Environ Res Counc, UK

Cormaci M, Furnari G (1999) Changes of the benthic algal flora of the Tremiti Islands (Southern Adriatic) Italy. In: Kain (Jones) JM, Brown MT, Lahaye M (eds), Proc 16th Int Seaweed Symp, Cebu City, Philippines, April 1998. Dev Hydrobiol 137, pp 190-194

Cormaci M, Furnari G, Alongi G, Catra M, Pizzuto F, Serio D (this volume) Spring marine vegetation on rocky substrata of the Tremiti Islands (Adriatic Sea, Italy). In: Faranda FM, Guglielmo L, Spezie G (eds), Structure and processes in the Mediterranean ecosystems, Atti 1st Conv Nazl Sci Mare, Ischia, Italy 11-14 Novembre 1998. Springer-Verlag, Berlin Heidelberg New York Tokyo

Coull BC (1985) The use of long-term biological data to generate testable hypotheses. Estuaries 8: 84-92

Giaccone G (1969) Raccolte di fitobenthos sulla banchina continentale italiana. G Bot Ital 103: 485-514

Giaccone G (1978) Revisione della flora marina del mare Adriatico. Annu WFF Parco Mar Miramare Trieste 6 (19): 5-118

Giaccone G, Bruni A (1971) Le Cistoseire delle coste italiane. I. Contrib Ann Univ Ferrara (N.S.) Sez IV Bot 4 (3): 45-70

Hay ME (1981) The functional morphology of turf-forming seaweeds: persistence in stressful marine habitats. Ecology 62 (3): 739-750

Huvé H, Huvé P, Picard J (1963) Aperçu préliminaire sur le benthos littoral de la côte rocheuse adriatique italienne. Rapp Comm Int Mer Mediterr 17: 93-102

La Violette PE, Gacic M (1990) Some circulation features in the Adriatic Sea – a satellite view. Rapp Comm Int Mer Mediterr 32 (1): 182

Levine HG (1984) The use of seaweeds for monitoring coastal waters. In: Shubert LE (ed) Algae as ecological indicators. Academic Press Inc, London, pp 189-210

Likens GE (1989) Long-term studies in ecology: approaches and alternatives. Springer-Verlag, Berlin Heidelberg New York Tokyo

Littler MM, Martz DR, Littler DS (1983) Effects of recurrent sand deposition on rocky intertidal organisms: importance of substrate heterogeneity in a fluctuating environment. Mar Ecol Prog Ser 11: 129-139

Morovic M, Domijan N (1991) Light attenuation changes in the middle and southern Adriatic. Acta Adriat 32 (2): 621-635

Nielsen R (1991) Vegetation of Tønnenberg Banke, a stone reef in the Northern Kattegat, Denmark. In: CNR (ed) Atti Eur Meet Mar Phytobenthos Stud Appl, Taranto, 17-20 Settembre 1990, Oebalia 17 Suppl 1: 199-211

Perrera G, Giaccone G (1986) Il mare costiero visto dal biologo. Stampatori Tipolitogr Assoc, Palermo, pp 152

Rizzi Longo L, Pignatti S, De Cristini P (1967) Contribuzione alla flora algologica del litorale garganico meridionale fra Manfredonia e Mattinata. G Bot Ital 101 (2): 131-132

Sousa WP (1979) Experimental investigation of disturbance and ecological succession in a rocky intertidal algal community. Ecol Monogr 49: 227-254

Sousa WP, Schroeter SC, Gaines SD (1981) Latitudinal variation in intertidal algal community structure: the influence of grazing and vegetative propagation. Oecologia 48: 297-307

Stewart JG (1983) Fluctuations in the quantity of sediment trapped among algal thalli on intertidal rock platforms in southern California. J Exp Mar Biol Ecol 73: 205-211

Strayer D, Glitzenstein JS, Jones CG, Kolasa J, Likens GE, McDonnell MJ, Parker GP, Pickett STA (1986) Long-term ecological studies: an illustrated account of their design, operation and importance to ecology. Occas Publ Inst Ecosyst Stud, 2. Mary Flagler Cary Arboretum, Millbrook, New York

Tilman D (1989) Ecological experimentation: strengths and conceptual problems. In: Likens GE (ed) Long-term studies in ecology: approaches and alternatives. Springer-Verlag Berlin Heidelberg New York Tokyo, pp 136-157

Vukadin I, Zvonaric T, Stojanosky L, Kuspilic L (1988) Hydrographic and chemical properties of Middle and South Adriatic Sea water. Rapp Comm int Mer Měditerr 31 (2): 46

Wilkinson M, Tittley I (1979) The marine algae of Elie, Scotland: a re-assessment. Bot Mar 22: 249-256

Zore-Armanda M, Bone M, Dadic V, Morovic M, Ratkovic D, Stojanoski L, Vukadin I (1991) Hydrographic properties of the Adriatic Sea in the period from 1971 through 1983. Acta Adriat 32 (1): 5-544

Meiofaunal Biodiversity on Hydrothermal Seepage off Panarea (Aeolian Islands, Tyrrhenian Sea)

M.A. Colangelo[1], F. Bertasi[2], P. Dall'Olio[3], and V.H. Ceccherelli[1]

ABSTRACT

Meiofauna biodiversity was investigated at two sites in the submerged crater off Panarea (Aeolian Islands), characterised by different degree of hydrothermal seepage. At the first site, a gasohydrothermal seepage of a carbon dioxide and sulphide (mean 1.96 vol% H_2S) was observed. The second site was characterised by a few centimetres deep layer of white colloidal sulphur deposits. At each site, two stations were sampled with high (H) and moderate (M) seepage respectively, while a third station (C) with no hydrotermal activity was selected as control. Meiobenthic organisms were counted, sorted by taxa and copepods classified to species level. A very low number of all meiobenthic organisms was observed in the two stations with colloidal sulphur deposits. Copepods resulted to be the dominant taxon in both sites. Multivariate analyses showed significant differences in the copepod community structure both between sites and among stations within sites. Diversity indexes displayed significantly lower values at the two control stations. The coarse grain size can be invoked to explain copepod dominance. But intermediate disturbance due to gas bubbling seems to play a role in promoting copepod biodiversity. However sites of high deposition of colloidal sulphur, by clogging interstitial spaces, seem to create a prohibitive environment for meiobenthos.

Introduction

Despite shallow water hydrothermal venting phenomena are widespread along marine coastal environments, their effects on benthic faunal composition remain far from being fully understood. Only a few studies reported informations on species assemblages and community structure of meiofauna in these particular ecosystems (Fricke et al. 1989; Dando et al. 1995; Kamenev et al. 1993; Thiermann et al. 1994, 1997). The island of Panarea belongs to the Aeolian archipelago, consisting of seven main islands lying on the Sicilian-Calabrian continental slope. East of the Panarea island there is a wide hydrothermal area ("caldera") in the centre of a submerged volcanic structure (Fig. 1). In an area of about 4 km^2 inside this caldera and up to a maximum depth of 30 m, there is a large number of underwater gas discharges which are distributed on exhala-

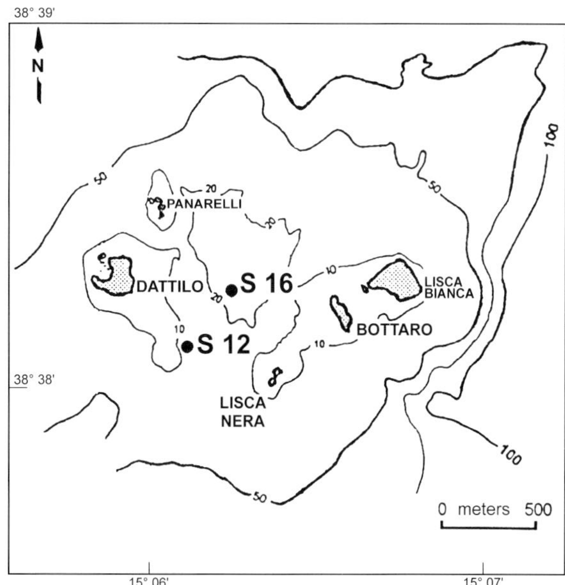

Fig. 1. Map of Panarea's caldera and sampling sites

[1] Dipartimento di Biologia Evoluzionale e Sperimentale, Università di Bologna, Via Selmi 3, 40126 Bologna, Italy
[2] Dipartimento di Biologia, Università di Ferrara, Via L. Borsari 46, 44100 Ferrara, Italy
[3] Corso di Laurea in Scienze Ambientali, Università di Bologna, Via Tombesi dall'Ova 55, 48100 Ravenna, Italy

F.M. Faranda, L. Guglielmo, G. Spezie (eds)
Mediterranean Ecosystems: Structures and Processes

tive fields each covering generally about a few hundred square metres (Italiano and Nuccio 1991). They are recognisable by columns of gaseous bubbles seeping from hard and sandy substrates. The mean composition of the gas mixture, measured from different seeps, is mainly 94.71 vol.% CO_2 and 1.96 vol.% H_2S (max 6.5 vol.%). Also lesser percentages of CH_4 and H_2 are present (Italiano and Nuccio 1991). Around exhalative fields with higher H_2S content, the interactions of the gases with marine sulphates and with dissolved oxygen in the sea-water give rise to patches of white, snow-like colloidal sulphur deposits (Italiano and Nuccio 1991; Gabbianelli et al. 1996). The present paper deals with investigations on meiobenthos from this hydrothermal area, with particular reference to the biodiversity of the Copepod assemblages.

Materials and Methods

Underwater investigations were carried out in two exhalative fields of the Panarea's caldera characterised by different kinds of seepage. In the study area two sites were marked as S12 and S16 (Fig. 1). At the two sites bottom water temperature, salinity and oxygen saturation were measured by a Hydrolab" CTD probe (Table 1) and meiobenthos was sampled by SCUBA divers in July 1995. Site S12 (38°38'11" N, 15°06'05" E) was located on a ground sloping down to 14 m, with sandy sediments interposed to sparse patches of *Posidonia oceanica*, and characterised by simple gaseous bubblings of different intensity and extent. At this site three stations were chosen at about 10-15 m away from each other and with different intensity of gas seepage: one station, S12H, showing a high seepage intensity, i.e. with a remarkable bubbling; another station, S12M, with medium intensity phenomena, i.e. characterised by moderate bubbling; and one control station, S12C, which showed no hydrothermal activities. Site S16 (38°38'15" N, 15°06'18" E) was located on a plateau at about 21 m depth and was characterised by bare sandy areas covered by a layer of few centimetres of snow-like colloidal material recognised as sulphur. Three stations were selected at about 5-10 m away from each other and characterised by different amounts of sulphur deposits: at station S16H the sand was completely covered by sulphur flakes; station S16M was characterised by a patchy cover; and station S16C, apparently devoid of sulphur, served as control. At each station 5 replicate sediment cores (4.5 cm^2 surface area) were taken for meiobenthos analysis and an additional one for grain size analysis (according to Buchanan 1984). The top 5 cm of each core were fixed in 8% buffered formaldehyde solution and stained with Rose Bengal. Owing to the very coarse sand (Table 1), it was possible to sort meiofaunal organisms directly from the entire samples. Major taxa were counted and copepods classified and counted to species levels.

Table 1. Values of temperature, salinity and oxygen saturation of the bottom water at the investigated sites and values of sediment mean grain size (±C.L.) at each station

Sites	T (°C)	S (PSS)	O_2 (%)	Stations	Mean diameter (mm)	±C.L. 95%
				S12H	0.76	0.39-1.49
S12	23.3	40.5	98	S12M	0.71	0.42-1.19
				S12C	0.85	0.48-1.51
				S16H	1.15	0.47-2.80
S16	19.5	41.2	91	S16M	1.21	0.61-2.41
				S16C	1.33	0.70-2.55

Clustering and non-metric multi-dimensional scaling (MDS) ordination with the Bray-Curtis similarity measure were performed on 4th root-transformed species densities to determine differences in copepod community structure among stations. Significance of the differences among community structures were tested by one-way Analysis of Similarity (ANOSIM). The following diversity indexes were calculated: number of species, species richness (Margalef's index), species diversity (Hill's N1) and evenness (Hill's N10). All analyses were performed using the PRIMER package developed at the Plymouth Marine Laboratory (Clarke and Warwick 1994).

The "jackknife" technique (Dixon 1993) was used in order to obtain a higher accuracy of the diversity index values. According to this technique the bias and the standard deviation of all the above statistics were estimated by recalculating the statistics on a subset of data obtained by pooling four by four the five replicates of each station. When required, one-way ANOVA was carried out both on log(x+1) transformed densities of taxa and on diversity index jackknifed values (after arcsin transformation for Hill's N10). When ANOVA's results were significant, the Student-Newman-Keuls (SNK) post-hoc test was

Table 2. Average densities ind.10 cm^{-2} (*SD*) of the main taxa at each station

	S12H		S12M		S12C		S16H		S16M		S16C	
Turbellaria	7.9	*4.6*	3.5	*3.7*	9.2	*9.8*	–		–		–	
Gnathostomulida	–		–		0.4	*1.0*	–		–		–	
Nematoda	58.5	*38.3*	147.8	*153.7*	10.1	*5.7*	1.8	*2.9*	1.8	*2.9*	4.4	*6.4*
Polychaeta	7.9	*7.7*	8.8	*7.8*	7.0	*7.2*	0.4	*1.0*	–		4.8	*10.8*
Gastropoda	0.4	*1.0*	0.4	*1.0*	–		–		–		–	
Ostracoda	7.5	*8.5*	0.9	*2.0*	0.9	*1.2*	–		–		–	
Copepoda	91.5	*47.5*	222.6	*214.0*	198.4	*97.4*	4.0	*1.8*	2.6	*1.0*	61.2	*38.2*
Nauplii	0.4	*1.0*	0.4	*1.0*	–		–		–		–	
Halacaridae	0.4	*1.0*	3.1	*3.3*	1.8	*1.8*	0.4	*1.0*	0.4	*1.0*	–	
Total	174.7	*81.3*	387.6	*375.1*	227.9	*97.5*	6.6	*4.9*	4.8	*4.8*	70.4	*29.4*

performed for significance of all pairwise comparisons between stations.

Results

The two sites showed marked differences in their sedimentary characteristics. Stations of the site S12 had fairly homogeneous sediments (Table 1) classified as "moderately sorted coarse sand". In contrast, all stations of the site S16 showed high-

Table 3. *P*-values from pairwise comparisons of total meiofauna, nematodes and copepods densities between stations of the two sites using SNK test after ANOVA (*F*-values are reported too). *Underlined* values indicate significant differences

Total meiofauna; $F_{(5,24)}$=36.08; P<0.0001

S12M	0.40				
S12C	0.49	0.55			
S16H	0.001	0.001	0.001		
S16M	0.001	0.001	0.001	0.42	
S16C	0.06	0.01	0.03	0.001	0.001
	S12H	S12M	S12C	S16H	S16M

Nematodes; $F_{(5,24)}$=17.85; P<0.0001

S12M	0.23				
S12C	0.001	0.001			
S16H	0.001	0.001	0.04		
S16M	0.001	0.001	0.02	1.00	
S16C	0.001	0.01	0.06	0.70	0.40
	S12H	S12M	S12C	S16H	S16M

Copepods; $F_{(5,24)}$=28.90; P<0.0001

S12M	0.15				
S12C	0.22	0.81			
S16H	0.001	0.001	0.001		
S16M	0.001	0.001	0.001	0.54	
S16C	0.37	0.06	0.07	0.001	0.001
	S12H	S12M	S12C	S16H	S16M

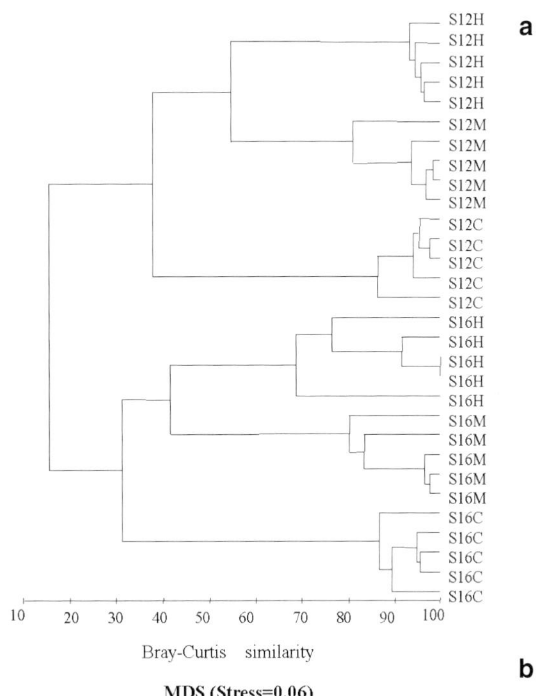

MDS (Stress=0.06)

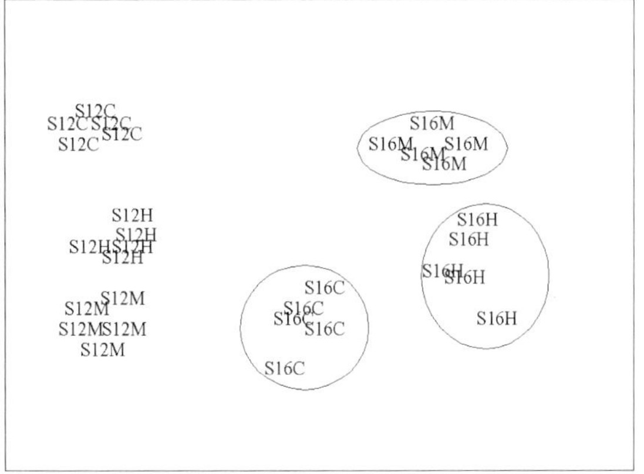

Fig. 2a,b. Multivariate analysis on copepod assemblages: (**a**) Dendrogram and (**b**) Two dimensional non-metric MDS plot of all replicate samples at each station

er values of mean grain size (Table 1) in the range of the "moderately sorted very coarse sand" (Buchanan 1984).

Densities of the major taxa were low at the two stations completely or partially covered by colloidal sulphur (S16H and S16M) (Table 2). Also at station S16C, apparently devoid of sulphur deposits, total meiofaunal density resulted to be significantly lower than those at S12M and S12C (Table 3). No significant differences in meiofaunal densities were observed between stations of the site S12 (Table 3). Copepods represented always the dominant taxon and, with the exception of S16H and S16M, their average densities did not differ significantly among the other stations (Table 3). At station S12C, nematode densities were significantly lower than those at stations with evident gas seepage, i.e. S12H and S12M (Table 3).

Clustering and MDS plot of the jackknife replicates of all stations (Fig. 2) showed a clear distinctness of the copepod community structures, both between sites and, within sites, among stations. These results were confirmed by the ANOSIM global result (R=0.998, P<0.001) as well as by the significance of all subsequent pairwise comparison tests between stations.

The results of the diversity indexes calculated on the jackknife replicates at each station are reported in Fig. 3, together with the results of the SNK test pairwise comparisons between stations (all ANOVAs gave highly significant F statistics). The results for the two stations with colloidal sulphur deposits (S16H and S16M) should be regarded as rather spurious being represented only by single figure specimens (Table 4). As for the other four stations, S12M, with a moderate gaseous seepage, showed the highest values of all

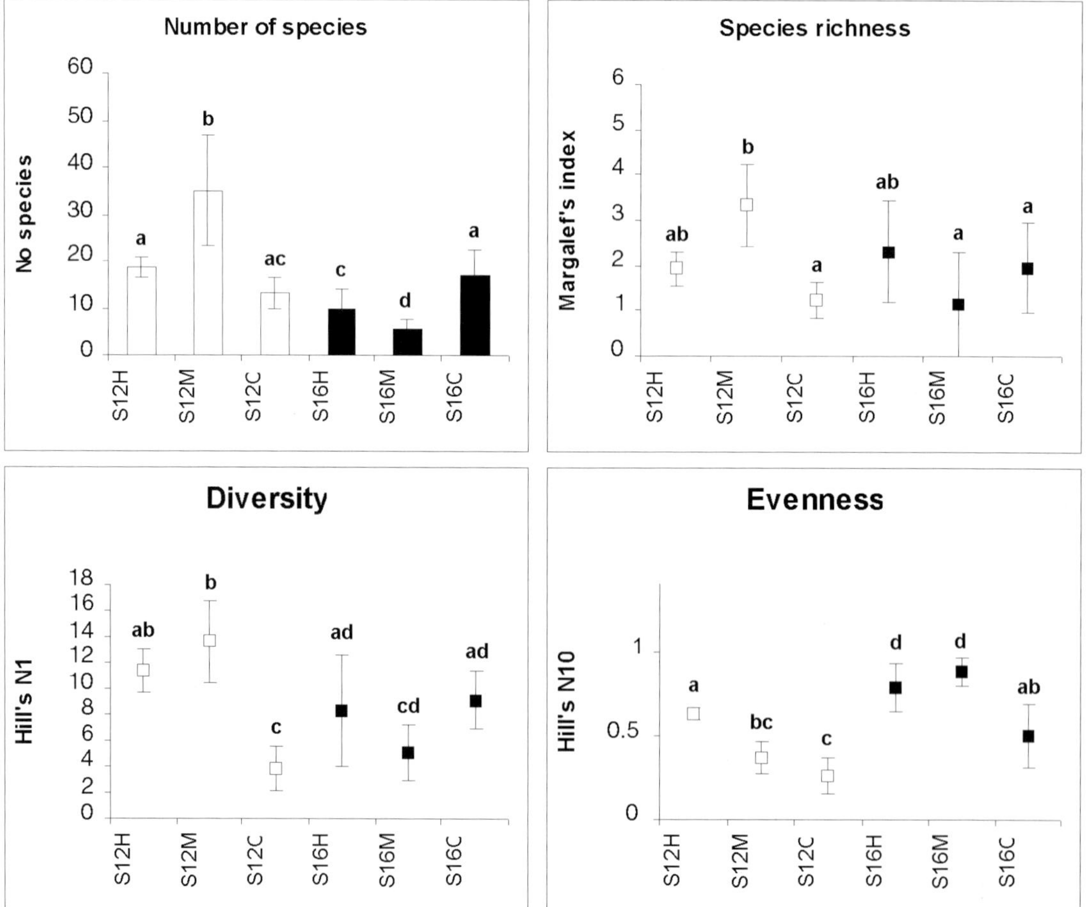

Fig. 3. Average jackknifed values (± SD) of diversity indexes for the copepod assemblages of each station (H, M, C, see text) at sites S12 and S16 respectively. Averages sharing the same letter are not significantly different after SNK test

Table 4. Average densities ind. 10 cm^{-2} (*SD*) of copepod species at each station

	S12H		S12M		S12C		S16H		S16M		S16	
CEctinosoma litorale	1.8	*2.9*	11.0	*10.0*	15.0	*24.5*	–		–		–	
Ectinosoma dentatum	–		–		–		0.9	*2.0*	–		0.4	*1.0*
Ectinosoma melaniceps	–		–		–		–		–		2.2	*3.8*
Halectinosoma sp.	–		0.4	*1.0*	–		–		–		–	
Pseudobradya sp.	–		1.3	*2.0*	–		–		–		–	
Tetanopsis mediterranea	–		0.4	*1.0*	–		–		–		–	
Halophytophilus fusiformis	–		–		–		0.4	*1.0*	–		–	
Hastigerella cfr. psammae	–		1.8	*3.9*	–		–		–		–	
Tisbe holothuriae	–		–		–		1.3	*1.2*	0.9	*1.2*	10.1	*12.8*
Tisbe sp.0.9	*2.0*	0.9	*1.2*	–		–		–		–		
Thalestridae sp.	–		–		–		–		–		0.9	*1.2*
Dactylopusia micronyx	–		1.3	2.0	–		–		–		–	
Dactylopusia tisboides	1.8	*1.8*	–		–		–		–		12.8	*18.8*
Amphiascus minutus	19.4	*12.1*	34.8	*35.1*	1.3	2.0	–		–		–	
Amphiascus sinuatus	3.5	4.6	–		–		–		–		–	
Amphiascus varians	–		–		–		–		–		9.2	*10.5*
Amphiascus sp.	–		–		0.4	*1.0*	–		–		–	
Amphiascopsis cinctus	–		2.6	*2.9*	–		–		–		–	
Pseudamphiascopsis attenuatus	2.6	*2.4*	3.5	*3.3*	–		–		–		2.2	*4.9*
Robertgurneya ilievecensis	7.5	*8.5*	–		0.4	*1.0*	–		–		–	
Robertgurneya sp.	0.9	*1.2*	0.4	*1.0*	–		–		–		–	
Amphiascoides neglectus	16.7	*19.5*	0.9	*2.0*	21.1	*22.7*	0.4	*1.0*	0.9	*1.2*	0.4	*1.0*
Ameira parvula	3.1	*4.3*	4.0	*2.9*	–		–		0.4	*1.0*	11.0	*11.8*
Ameira cfr longicaudata	2.6	*2.4*	45.8	*57.0*	–		0.4	*1.0*	–		0.4	*1.0*
Nitocra typica	–		0.9	*1.2*	–		–		–		–	
Ameiridae sp.	–		0.4	*1.0*	–		–		–		–	
Karllangia tertia	3.5	3.3	19.8	*23.5*	1.3	*3.0*	–		–		–	
Karllangia cfr arenicola	9.7	*4.0*	4.8	*5.5*	–		–		–		–	
Parevansula sp.	–		0.4	*1.0*	–		–		–		–	
Phyllopodopsyllus bradyi	–		1.3	*3.0*	–		–		–		–	
Stenocaropsis pristina	0.9	*2.0*	5.3	*9.4*	–		–		–		–	
Tryphoema cfr bocqueti	2.6	*2.9*	27.7	*62.0*	0.4	*1.0*	–		–		–	
Laophonte cornuta	–		36.1	*55.7*	–		–		–		–	
Heterolaophonte stroemi	–		1.3	*2.0*	–		0.4	*1.0*	–		2.6	*2.9*
Paralaophonte dieuzeidei	1.3	*2.0*	–		–		–		0.4	*1.0*	0.4	*1.0*
Asellopsis hispida	–		–		1.3	*2.0*	–		–		–	
Klieonychocamptus kliei	–		0.4	*1.0*	–		–		–		–	
Hemicyclopina sp.	–		1.3	*2.0*	86.2	*56.1*	–		–		–	
Cyclopina sp.	12.8	*13.5*	13.6	*9.0*	70.8	*49.0*	–		–		8.4	*10.9*
Total	91.5	*47.5*	222.6	*214.0*	198.4	*97.4*	4.0	*1.8*	2.6	*1.0*	61.2	*38.2*

the diversity indexes, except for evenness. In particular, the actual number of species resulted significantly higher than those found in the other stations. In the control station, S12C, of the bubble stream site, the significantly lowest value of diversity was recorded, as a consequence of lower values of both species number and evenness. Station S12H, with more intense seepage, and S16C, the control station in the sulphur site, generally showed intermediate values of the various indexes. But the evenness at station S12H was significantly higher than at the other stations of the same site.

Discussion

Studies carried out on meiobenthic communities of hydrothermal venting areas have focussed mainly on the dominating nematode community of muddy sediments (Fricke et al. 1989; Kamenev et al. 1993; Dando et al. 1995; Thiermann et al. 1997). These studies have generally showed higher densities of this taxon than those recorded in the present investigation and a reduction of diversity in areas close to seepage. On the contrary, in the exhalative fields of the Panarea's caldera, copepods resulted to be the dominant taxon and their diversity was high in areas with a moderate gas seepage or sulphur deposit. The primary influence of the sediment characteristic on the composition of meiobenthic assemblages cannot be disregarded (Wieser 1959). The total copepod densities are in the range of values recorded in sandy habitats (Hicks and Coull 1983) and seem to conform to the rule that copepods dominate or become numerically abundant as the particle size of the sediment increases (Coull 1985). The coarse to very coarse grain size of the sandy patches in Panarea's caldera, with their wide interstitial spaces, could explain the copepod dominance at this site. Nematode densities increased their relevance at stations characterised by bubble streams (S12H and S12M) (Table 2). Probably, in those areas, there is a sufficient production of sulphur bacteria (Acunto et al 1995) to support populations of some specialised nematode species (Giere 1992). Moreover, the drawdown of interstitial water to replace water displaced by the rising gas, would bring detrital material and phyoplankton into the sediment (Dando et al. 1995) with benefit to all meiobenthic organisms.

At site S16, which is probably located in a hydrothermal field with higher H_2S content (max 6.5 vol.%; Italiano and Nuccio 1991) in the gas mixture, colloidal sulphur precipitation occurred. The disturbance effects on the whole meiobenthic community appeared clearly evident and were likely due to the direct clogging of the interstitial spaces among the very coarse sand grains due to sulphur deposition. Such a kind of disturbance probably affected, to a minor extent, also meiobenthos at the control station (S16C, Table 2), which could have been affected by very fine colloidal sulphur particles, though not macroscopically evident.

At S12, characterised by a less relevant hydrothermal seepage (i.e. simple bubble stream) the three stations showed significantly different copepod assemblages (Table 4, Figs. 2, 3). At the control station (S12C), there were less species (10) and *Cyclopina sp.* and *Hemicyclopina sp.* were dominant. At station S12M, where a moderate bubbling occurred, the highest number of species was observed (27) but evenness was reduced due to the dominance of a few species (*Amphiascus minutus*, *Ameira longicaudata*, *Tryphoema cfr. bocqueti* and *Laophonte cornuta*). Finally at station S12H, with more intense gas emissions, an intermediate value of the number of species (17) with higher evenness was reported (Fig. 3).

On the whole, the overall situation of the species biodiversity, emerging by the present investigation on meiobenthic communities exposed to increasing hydrothermal seepage, could be interpreted in the light of "the intermediate disturbance" theory of diversity maintenance (Huston 1979). Where disturbance is minimal (e.g. at S12C) diversity is reduced because of competitive exclusion between species; with a slightly increased level or frequency of disturbance (e.g. at S12H and even more at S12M) competition decreases, resulting in an increased diversity. At higher or more frequent levels of disturbance (e.g. at site S16 and partly at S12H) species start to become eliminated by stress, so that diversity falls again.

References

Acunto S, Maltagliati F, Benedetti-Cecchi L, Lardicci C, Cinelli F, Cognetti G (1995) Osservazioni sui popolamenti bentonici di un'area interessata da fenomeni di vulcanismo secondario nei pressi dell'Isola di Panarea (Me). In: Faranda FM (ed) Caratterizzazione ambientale marina del sistema Eolie e dei bacini limitrofi di Cefalù e Gioia (EOCUMM94). G. Lang, Genova, pp 251-256

Buchanan JB (1984) Sediment analyses. In: Holme NA, McIntyre AD (eds) Methods for the study of marine benthos. (IPB Handbook, 16). Blackwell Scientific Publications, London, pp 41-65

Clarke KR, Warwick RM (1994) Changes in marine communities: an approach to statistical analysis and interpretation. Nat Environ Res Counc, UK

Coull BC (1985) Long-term variability of estuarine meiobenthos: an 11 year study. Mar Ecol Prog Ser 24: 205-218

Dando PR, Hughes JA, Thiermann F (1995) Preliminary observations on biological communities at shallow hydrothermal vents in the Aegean Sea. In: Parson LM, Walker CL, Dixon DR (eds) Hydrothermal vents and processes. Geol Soc, London, Spec Publ 87, pp 303-317

Dixon PM (1993) The Bootstrap and the Jackknife: describing the precision of ecological indices. In: Scheiner SM, Gurevitch J (eds) Design and analysis of ecological experiments. Chapman and Hall, New York, pp 290-318

Fricke H, Giere O, Stetter K, Alfredsson GA, Kristjansson JK, Stoffers P, Svavarsson J (1989) Hydrothermal vent communities at the shallow subpolar Mid-Atlantic ridge. Mar Biol 102: 425-429.

Gabbianelli G, Cortecci G, Capra A, Giacomelli L, Pompilio M, Rossi PML (1996) Lineamenti geo-vulcanologici ed ambientali dell'area craterica sottomarina di Dattilo-Lisca Bianca (Isola di Panarea, Arcipelago Eoliano). In: Faranda FM, Povero P (eds) Caratterizzazione ambientale marina del sistema Eolie e dei bacini limitrofi di Cefalù e Gioia (EOCUMM95). G. Lang, Genova, pp 475-484

Giere O (1992) Benthic life in sulfidic zones of the sea – ecological and structural adaptations to a toxic environment. Verh Dtsch Zool Ges 85: 77-93

Hicks GRF, Coull BC (1983) The ecology of marine meiobenthic harpacticoid copepods. Oceanogr Mar Biol Annu Rev 21: 67-175

Huston M (1979) A general hypothesis of species diversity. Am Nat 113: 81-101

Italiano F, Nuccio PM (1991) Geochemical investigations of submarine volcanic exhalations to the east of Panarea, Aeolian Islands, Italy. J Volcanol Geothermal Res 46: 125-141

Kamenev GM, Fadeev VI, Selin NI, Tarasov VG (1993) Composition and distribution of macro- and meiobenthos around sublittoral hydrothermal vents in the Bay of Plenty, New Zealand. NZ J Mar Freshwater Res 27: 407-418

Thiermann F, Windoffer R, Giere O (1994) Selected meiofauna around shallow water hydrothermal vents off Milos (Greece): ecological and ultrastructural aspects. Vie Milieu 44: 215-226

Thiermann F, Akoumianaki I, Hughes JA, Giere O (1997) Benthic fauna of a shallow-water gaseohydrothermal vent area in the Aegean Sea (Milos, Greece). Mar Biol 128: 149-159

Wieser W (1959) The effect of grain size on the distribution of small invertebrates inhabiting the beaches of Pudget Sound. Limnol Oceanogr 4: 181-194

CHAPTER 47

Change and Diversity: the Mediterranean Deep Corals from the Miocene to the Present

C. Corselli

ABSTRACT

During the last eighteen million years plate tectonics have deeply modified the physiography and the morphology of the Mediterranean Sea, with changes in basin oceanography, chemistry and biology. The presence and the depth of the western sill made it a semi-closed basin, heavily conditioned by the climate in its oceanographic features. In addition during the Pleistocene, the global climatic deterioration affected the circulation of the superficial and deep waters of the Mediterranean and the composition of its biota. The reconstruction of the chemical and physical characteristics of the deep waters of the Mediterranean sea, by the corals, shows the progressive isolation of the basin during the last 12 million years. Associated to this isolation a progressive decrease in the specific diversity in the deep coral fauna occurs, only partly arrested during the glacial phases.

Introduction

According to Picard (1985) a macro-ecosystem changes in terms of geologic time and a meso-ecosystem, instead, has a temporal length from some millennia to a few centuries. Among the climatic factors the temperature in particular influences the duration of the macro-ecosystems and in the deep sea ecosystems the temperatures of some oceans have not substantially changed in the last 30-40 millions of years. The stable temperature that has characterized the bathyal and abyssal depths, in the considered time, has certainly influenced the evolution of the organisms relatively inducing a high biological diversity with a low density. The autoecology of the still living coral species allows to reconstruct the past environments, while the specific diversity testifies to changes that have occurred in the physical and chemical parameters of a marine basin. Nevertheless the age of a biological species could be extremely variable from ten thousand to some million years and the biologists have not always taken into account the age of a still living species. Frequently they have preferred to institute a new species without verifying the possibility that the living organism corresponds to a fossil taxon, instituted for long time. An insufficient documentation (lack or loss of the original types) or few accurate morphological descriptions has often represented the alibi for such a choice. The paleontological record of some coral deep taxa documents a long temporal duration, that, in some cases, can also reach some million years. During the last eighteen million years plate tectonics has deeply modified the physiography and morphology of the Mediterranean Sea, with changes in basin oceanography, chemistry and biology. Here, as in the other semi-closed sea basins, the temperature of the bottom has undergone, in the same temporal interval, a series of fluctuations that have deeply influenced the bathyal and abyssal fauna and particularly the coral species. In this paper are illustrated three case histories of the deep coral assemblages of the Mediterranean Sea during the Late Miocene, Pliocene-Early Pleistocene, Late Pleistocene-Recent; such assemblages testify the dramatic changes undergone by the basin.

Dipartimento di Scienze Geologiche e Geotecnologie, Università degli Studi di Milano-Bicocca, Piazza della Scienza 4, 20126 Milano, Italy

F.M. Faranda, L. Guglielmo, G. Spezie (eds)
Mediterranean Ecosystems: Structures and Processes

Miocene (12-5 Ma B.P.)

At the beginning of the Miocene the Mediterranean was a wide corridor between two large oceans. The closing of the Eastern channels transformed it in a broad Atlantic gulf (Fig. 1a). Among the about ten species of corals recovered in the S.Agata Fossili Formation (Tortonian, Late Miocene) of northern Italy, *Deltocyathus italicus* (Michelotti 1838) is particularly representative of the depth and of the physical and chemical characteristics of the Mediterranean during the Late Miocene. *Deltocyathus conicus* Zibrowius 1980 (Fig. 2b) is a solitary coral actually present, beginning from 1100 metres depth, in the bathyal and abyssal areas of the eastern Atlantic ocean. Before the institution of this new taxon by Zibrowius (1980, 1991), the previous authors had ascribed specimens of *Deltocyathus* of the Atlantic and Indian oceans to the Michelotti's species (Cairns 1981). Cairns (1979) had attributed, by comparison, his samples of the western Atlantic to *Deltocyathus italicus* (Michelotti 1838). Figure 2a,b shows some specimens of the two taxa: the specimens of *Deltocyathus italicus* originate from the type-area of the S. Agata Fossili Formation of the Tortonian of northern Italy, while the specimens of *Deltocyathus conicus* are the original types of the species of the eastern Atlantic. Considering the alteration produced by diagenesis (from aragonite to calcite), the resemblance among the two species is remarkable. The authocthonous fossil assemblages recovered with *D. italicus* in the S. Agata Fossili Formation marls (Robba 1968), represent a paleo-community of the bathyal environment and paleoecological evidence supports a strong identity among the two taxa. However any taxon of the genus *Deltocyathus* has re-entered the Mediterranean after the Messinian Salinity Crisis, justifying a different oceanographic condition of the Mediterranean during the late Miocene. Figure 2c shows the distribution of the water masses along the two channels that in the Tortonian joined the Mediterranean Sea and the Atlantic ocean. According to Benson et al. (1991) the Betic Passage was deeper and this allowed an important entry of the Psychrosphere (water masses with temperature lower than 10°C) in the Mediterranean basin. The Psychrosphere penetrated well inside the basin, also occupying the oriental part of it, reaching the Ligure-Piedmontese Basin (area of the S. Agata Fossili Formation). The entry of Atlantic deep water with low temperature has thus allowed, in the Late Miocene, the occurrence of a coral deep fauna well diversified and similar to that currently present in the Atlantic ocean. In the Late Miocene (Messinian), the basin remained entirely isolated from the Atlantic Ocean, with a concomitant dramatic biological crisis. The whole marine fauna and flora disappeared from the basin, transformed in a deep endoreic depression, like a desert (Fig. 1b).

a

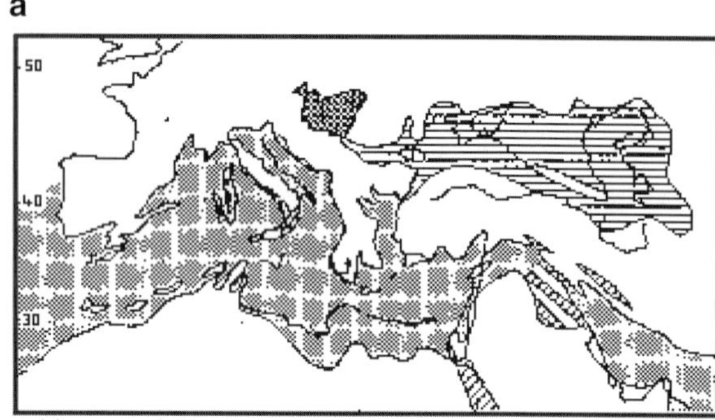

Upper Miocene

b

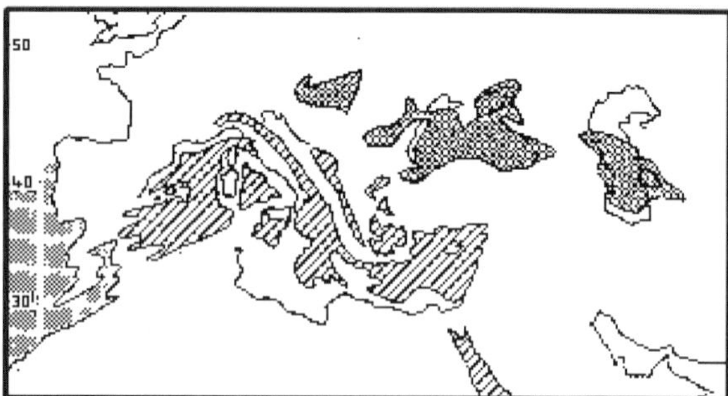

Messinian

c

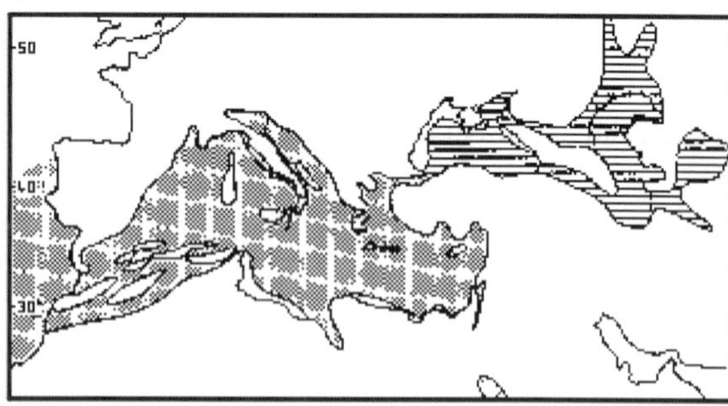

Lower Pliocene

Fig. 1a-c. Paleogeographic reconstructions of Mediterranean area: **a** Late Miocene (Tortonian); **b** Late Miocene (Messinian); **c** Pliocene. (*1*, Marine facies; *2*, Brackish facies; *3*, endemic fauna areas; *4*, Evaporitic facies; *5*, emerged areas). From Cita and Corselli 1993

Fig. 2a-c. a specimens of *Deltocyathus italicus* from the type-area of S.Agata Fossil Formation (Tortonian, Northern Italy); **b** specimens of *Deltocyathus conicus* Zibrowius (1980) as figured by the Author; **c** Paleogeographic reconstruction of the Betic and Rifian passages near the Tortonian-Messinian boundary (from Benson et al. 1991)

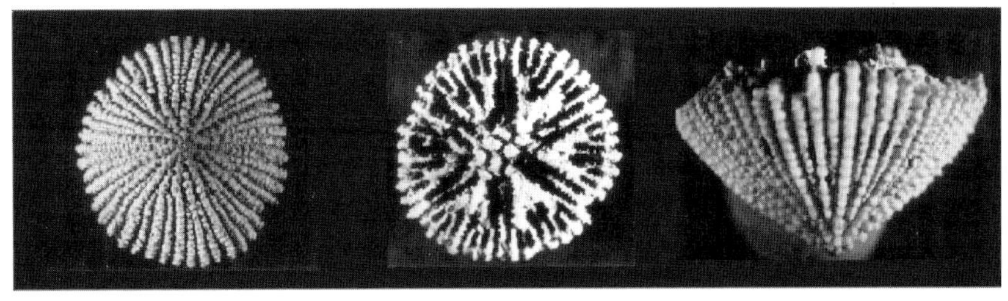

a *Deltocyathus italicus* (Michelotti 1838)
S. Agata Fossili Formation, Tortian

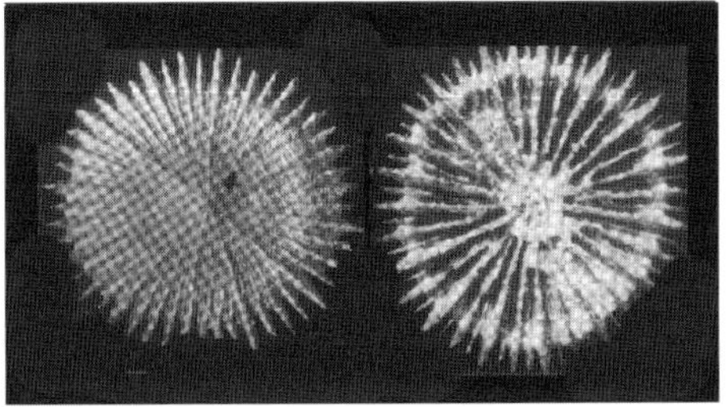

b *Deltocyathus conicus* Zibrowius (1980)
Gulf Ibero-Marocain -1473 m

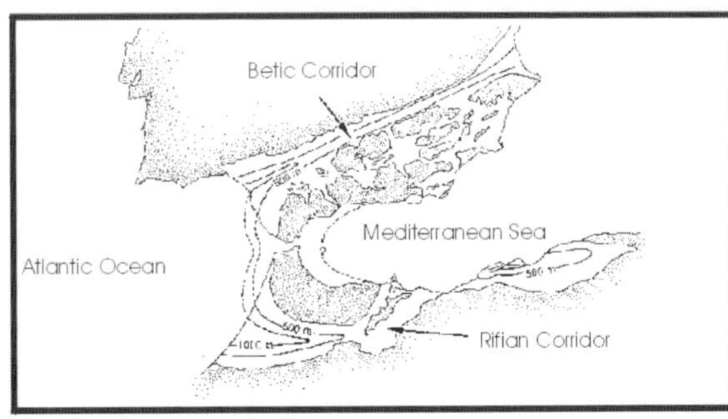

c Paleogeographic reconstruction of the Betic and Rifian passages near the Tortonian-Messinian boundary (Benson et al. 1991)

Pliocene-Early Pleistocene (5-1 Ma B.P.)

In the Early Pliocene the opening of the Straits of Gibraltar recreated the connections with the Atlantic ocean and a new fauna and flora of Atlantic origin colonized the Mediterranean Sea (Fig. 1c). However the basin was deeply different from the wide gulf of the late Miocene. The presence and the depth of the western sill made it a semi-closed basin , heavily conditioned by the climate in its oceanographic features. During the Pliocene and in the Early Pleistocene a Psycrosphere similar to that of the Miocene seems to characterize the whole basin. In fact deep coral similar to actual species of Atlantic cold waters have been found again in the sediments of northern Italy and of Rhodos (Montanaro 1931). But taxa such as *D. italicus* don't re-enter the basin anymore, probably documenting a less deep threshold, able to filter the entry of some deep faunas. As a whole the deep assemblages of corals remain still well diversified and with remarkable affinities with the Atlantic faunas. *Enallopsammia scillae* (Seguenza 1864) is a colonial coral, with unifacial calices, recovered in the clayey sediments of the Pliocene and the early Pleistocene of Southern Italy (Placella 1979). In 1980, Zibrowius pointed out the strong similarity among Seguenza's taxon and the specimens of *Enallopsammia_ampheliodes* (Alcock 1902), a species living to day in the oceans and characteristic of inclusive depth between 229 and 1683 metres (Zibrowius 1980, 1987). According to the French author the disap-

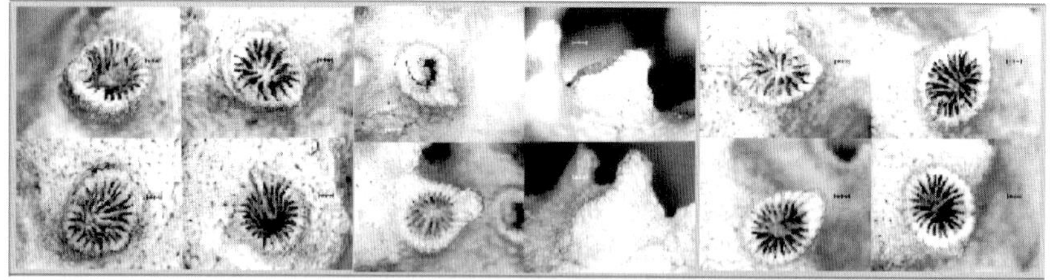

a Enallopsammia scillae (Seguenza 1864)
Plio-Pleistocele, Capo Milazzo (Sicily)

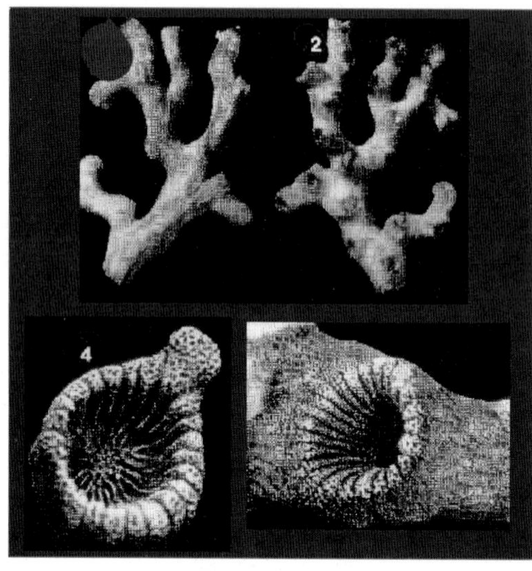

b Enallopsammia rostrata (Pourtalès 1878)
47°21'S 147°52'E, 910-915 m

Fig. 3. a. specimens of *Enallopsammia scillae* (Seguenza, 1864) from the Plio-Pleistocene sediments of Capo Milazzo (Messina, Sicily), type-area of Seguenza's species; **b.** specimens of Enallospammia rostrata (Pourtalés, 1878) from Cairns (1982)

pearance of the Seguenza's types didn't allow to compare the living specimens with the fossil ones and to set in synonymy the taxon of Alcock. Recently Cairns (1982), describing the species of Scleractinia from the Antarctic and Subantarctic regions, has pointed out a continuous intergradation between *E. rostrata* (Pourtalés 1878) and *E. ampheliodes* (Alcock 1902), setting this last in synonymy. In Fig. 3a are illustrated some specimens of the species of Seguenza, collected in one of the type localities of the Sicilian author; in Fig. 3b are represented some specimens of *E. rostrata* derived from the collections studied and shown by Cairns (1982). The specimens belong without uncertainty to the same species and *Enallopsammia rostrata* (Pourtalés 1878) is a synonym of *E. scillae* (Seguenza 1864). Zibrowius (1991) notes the similarity among living species and fossil specimens of Sicily and Calabry. The *E. scillae* is according to Cairns (1981) a widespread taxon of cold waters broadly distributed except for the eastern Pacific, to inclusive depth between 229 and 2165 metres. The scenario drawn by the Scleractinia species of the deep sediments deposited in the Pliocene and in the Early Pleistocene in the Mediterranean area, shows the deep waters of the basin still influenced by the presence of a Psycrosphere. This last is partly conditioned by the presence of a relatively deep threshold which prevents the entry of some taxa with a peculiar ecology. The deep coral fauna is still well diversified and similar to that of the Atlantic ocean.

Late Pleistocene-Recent (1-0,015 Ma B.P.)

During the Late Pleistocene, the global climatic deterioration affected the circulation of the superficial and deep waters of the Mediterranean and the composition of its biota. At the same time the tectonic activity along the Gibraltar Strait changed the depth of the threshold. The circulation of the basin assumed its present configuration, characterized by a homeotherm structure of the Mediterranean deep waters. The basin is filled by the Thermosphere (Benson et al. 1991); only some of the Atlantic deep species are able to live there. During the glacial and inter-glacial phases the temperature of the Thermosphere changes with oscillations of some degrees around the actual value of 13,5°C.

Such events are well documented by the deep corals which, at different times, lived along the continental slopes and in the abyssal plains of the basin. According to Delibrias and Taviani (1984), the C^{14} datings implemented on specimens of some coral fossil species of the Late Pleistocene, document an age of life between 30.000 and 15.000 years, during the phase of the glacial maximum. The ecology of *Madrepora oculata* Linné, 1758, *Caryophyllia sarsiae* Zibrowius, 1974 and *Desmophyllum cristagalli* Milne Edwards and Haime, 1848, species broadly diffused during the glacial phases, justifies the age attributed by the C^{14} datings. The optimal temperature of life of *Caryophyllia sarsiae* is around 10-11°C, and probably such values are also valid for the others two taxa (Di Geronimo 1979; Zibrowius 1980). A production of colder deep water, during the glacial phases, could have only lowered to such values the temperature of the Mediterranean Thermosphere, the thermohaline circulation of the basin being unchanged. When, during the warm climatic phases of the Late Pleistocene, the temperature of the Thermosphere rises, also the deep Scleractinia, more resistant, start to decline, reducing their geographic distribution to the few favourable zones. The decline of the Scleractinia is the same to that of other bathyal and abyssal taxa, making the bottom of the Mediterranean an area of low diversity and density.

Conclusion

The reconstruction of the chemical and physical characteristics of the deep waters of the Mediterranean sea, by the corals, shows the progressive isolation of the basin during the last 12 million years. Associated to this isolation is a progressive decrease in the specific diversity in the deep coral fauna, only in part arrested during the glacial phases, occurs (Fig. 4). The progressive increase of the temperature of bottom water in the basin during the last million years, is strictly connected to the low depth of the threshold between the Atlantic Ocean and the Mediterranean Sea. The presence of the threshold prevents the entry of the colder Atlantic waters and the faunas associated to them. During the Late Miocene and the Pliocene the climate of the area was decidedly warmer than the Recent, as is testified by the faunas and floras with a tropical affinity living in the shallow waters of the continental shelf. Nevertheless the ample and deep connection with the Atlantic Ocean allowed the entry of the Psycrosphere and deep oceanic faunas in the whole Mediterranean. The low

Atlantic Ocean Mediterranean Sea

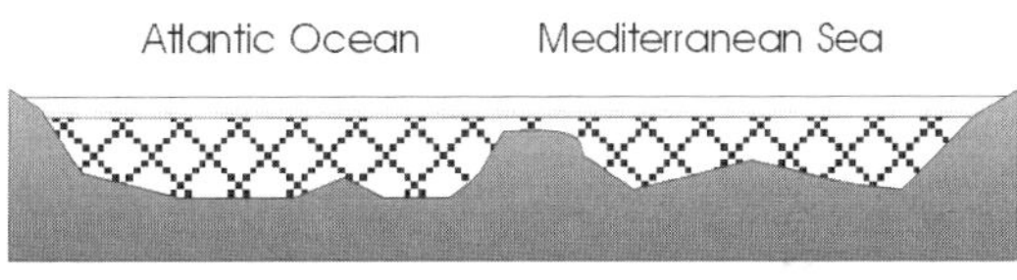

Tortonian: 8Ma

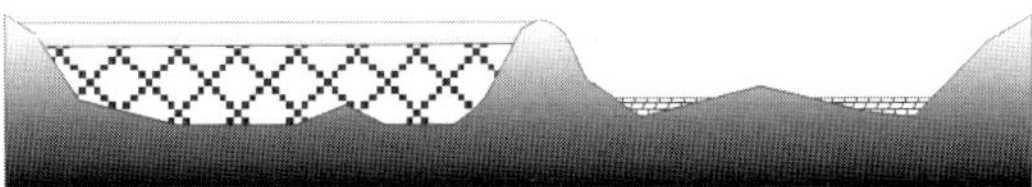

Messinian: Salinity crisis 5Ma

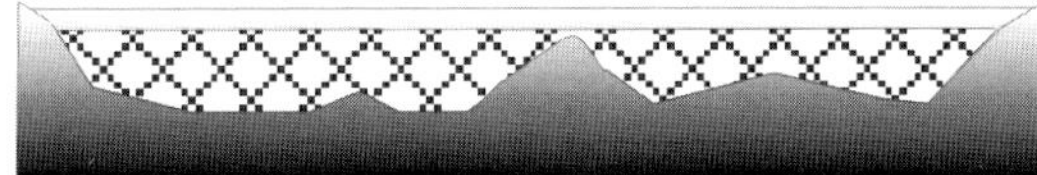

Pliocene - Early Pleistocene: 5-1Ma

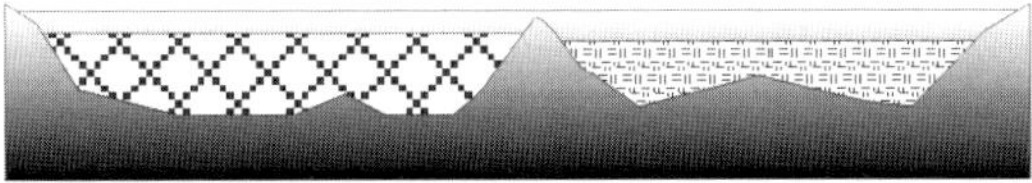

Late Pleistocene: glacial

Recent

Atlantic surface water
Psychrosphere
Thermosphere <13°C
Thermosphere >13°C
Evaporite/Lago Mare

Fig. 4. Hypothetical reconstruction of the late Neogene, Quaternary and Recent water masses distribution in the Atlantic Ocean and in the Mediterranean Sea diversity of the actual deep faunas of the basin is not tied to the increase of the temperature of the bottom water but to the effects of plate tectonics which have deeply modified the morphology of the boundary region between the Atlantic Ocean and the Mediterranean Sea and therefore the pattern of the shallow and deep currents (from an Atlantic broad gulf to a semi-enclosed basin).

References

Benson RH, Rakic-El Bied K, Bonaduce G (1991) An important reversal (influx) in the Rifian Corridor (Morocco) at the Tortonian-Messinian boundary: the end of Tethys ocean. Paleoceanography 6 (1): 164-192

Cairns SD (1979) The deep-water Scleractinia of the Caribbean sea and adjacent waters. In: Wagenaar Hummelinck P, Van der Steen LJ (eds) Studies on the fauna of Curaçao and other Caribbean islands, 180. pp 1-341

Cairns SD (1981) Marine flora and fauna of the northeastern United States: Scleractinia. NOAA Tech Rep NMFS Circ 438: 1-15

Cairns SD (1982) Antarctic and Subantarctic Scleractinia. biology of the Antarctic seas XI. Antarct Res Ser 34 (1): 1-74

Cita MB, Corselli C (1993) Messiniano: vent'anni dopo. Mem Soc Geol Ital 49: 145-164

Delibras G and Taviani M (1984) Dating the death of the Mediterranean deep-sea Scleractinian corals. Mar Geol 62/1-2: 175-180.

Di Geronimo (1979) Il Pleistocene in facies batiale di Valle Pallione. Boll Malacol 15 (5-6): 85-156

Montanaro E (1931) Coralli pliocenici dell'Emilia. Paleontogr Ital XXXI: 63-91

Picard JM (1985) Réflexions sur les ecosystèmes marins bentiques: hiérarchisation, dynamique spatio-temporelle. Téthys 11/3-4: 230-242

Placella B (1979) Nuove osservazioni sulla corallofauna delle argille pleistoceniche di Archi (Reggio Calabria). Boll Soc Nat Napoli 87: 1-31

Robba E (1968) Molluschi del Tortoniano-tipo (Piemonte). Riv Ital Paleontol Stratigr 74/2: 457-646

Seguenza G (1864) Disquisizioni paleontologiche intorno ai Corallarii fossili delle rocce terziarie del distretto di Messina. Mem R Accad Sci Torino 2/21: 399-560

Zibrowius H (1980) Les Scléractiniares de la Méditerranée et de l'Atlantique nord-oriental. Mem Inst Oceanogr 11

Zibrowius H (1987) Scléractiniaires et Polychètes Serpulidae des faunes bathyales actuelle et plio-pléistocène de Méditerranée. In: Barrier P, Di Geronimo I, Montenat C (eds) Le Détroit de Messine (Italie) évolution tectono-sédimentaire rècente (Pliocéne et Quaternaire) et environment actuel. Doc Trav IGAL 11: 255-257

Zibrowius H (1991) Les Scléractiniaires du Miocene au Pleistocéne de Sicile et de Calabre de Giuseppe Seguenza (1864, 1880) (Cnidaria, Anthozoa). Atti Accad Peloritana Pericolanti Cl Sc Mat Fis Nat LXVII 1: 79-135 (Suppl)

CHAPTER 48

Effect of Submerged Structures on the Diversity of Macrozoobenthos in the Northern Adriatic Sea

R. Crema[1], D. Prevedelli[1], and A. Castelli[2]

ABSTRACT

Offshore exploitation of oil and other hydrocarbons requires the installation of platforms or similar structures fixed to the sea bottom. Generally, such structures remain *in situ* for long time after the end of drilling and production activities. The macrozoobenthic communities of a sea area located in the Northern Adriatic Sea surrounding one of these platforms were studied on two occasions: the first prior to the installation of the platform (1985) and the second 8 years later (1993), after production and other activities had terminated. On both occasions, samplings were performed according to similar sampling grids extending about 3.5 km from the platform site. Comparing the community composition of the two periods, a remarkable decrease of surface deposit feeders was observed in the second period. Ordering techniques, applied to the entire set of community samples of both periods, pointed out not only the expected separation of the sampling points of the two periods but also a different pattern of dispersion. In fact, the samples of the first period were more dispersed in the model than those of the second period. The Bray-Curtis similarity values between samples exhibit a significantly greater similarity between pairs of samples of the second period. In spite of the fact that the diversity values of each individual sample are not different in the two sets, the different patterns of the increase of diversity, when plotted against the number of samples, further confirms the greater faunal uniformity among the samples of the second period. The results are discussed in light of sedimentological studies, performed in the same area, which highlight the role of the submerged structure in increasing the hydrodynamic energy near the bottom. This is consistent with the more recent findings in the literature on the dynamic variables that control the infauna distribution.

Introduction

The impact of oil and gas search and exploitation on the marine life derives mainly from the well drilling phase, which involves the installation of large platforms fixed to the bottom and discharge into the sea of cuttings extracted from the bottom and drilling muds. The fate and effect of these discharges have been extensively studied (Harper et al. 1981; Davies et al. 1984; Hartley 1984, 1996; Kingston 1987; Neff et al. 1989; Reiersen et al. 1989; Gray et al. 1990; Kroncke et al. 1992; Daan et al. 1994; Olsgard and Gray 1995) and depend largely on the specific mud formulation. The water-based muds, consisting of a suspension of inert weighting materials (mainly barite) in water, cause a non-selective reduction in the benthic macrofauna, whereas the oil-based muds, in which the inerts are suspended in oil or in a synthetic pseudo-oil base fluid, cause the occurrence of opportunistic species like in many other observed cases of organic pollution.

After the end of the drilling phase, the environmental impact of platform activities becomes less severe. During production, oil-field brines are frequently discharged in addition to the wastes of a small manned station that operates the platform. When the oil or gas reserve is exhausted, the platforms generally remain *in situ* for a long time, before their final decommissioning, in the absence of any relevant activity and any discharge to the sea.

In Italy search for and exploitation of offshore oil and gas reserves is particularly well

[1] Dipartimento di Biologia Animale, Università di Modena, Via dell'Università 4, 41100 Modena, Italy
[2] Dipartimento di Zoologia ed Antropologia Biologica, Viale Regina Margherita 15, 07100 Sassari, Italy

F.M. Faranda, L. Guglielmo, G. Spezie (eds)
Mediterranean Ecosystems: Structures and Processes

developed in the northern Adriatic Sea. As part of a research program aimed to investigate on the effects of this activity, the "history" of a platform was studied by means of several field surveys; these were carried out over a long period, beginning immediately prior to the installation of the platform and extending up to 8 years after, when all platform activity had terminated.

The platform is located off the city of Ravenna, on muddy bottoms of about 15 m depth. Data utilized in this paper pertain to surveys performed at the start and the end of the investigation, namely the baseline, pre-installation survey and the long-term final survey performed long after platform activity had ended.

The aim was to investigate, excluding the short-term physical or toxic effects of platform discharges, the effects of the mere presence in the site of the submerged structure on the surrounding biocenoses.

Materials and Methods

The first environmental investigation in the oilfield was performed in June 1985, prior to the installation of the platform. From August 1985 to October 1986, 9 wells were drilled. The platform produced about 4,000 tons of cuttings and 200 tons (wet material) of water-based muds. Drilling discharges have been completely stopped since late 1986. Eight years later (1993, late August) the final investigation was performed in the same seasonal period and site.

Samples were collected on the same sampling grid in 18 sampling points in the first investigation and in 30 sampling points in the second (Fig. 1) The overall grid size was about 3 km wide in WSW-ENE direction by 4.5 km long in the NNW-SSE direction. Sampling was performed with a hydraulic grab having a collecting surface of 0.07 m^2. Three replicates were collected at each

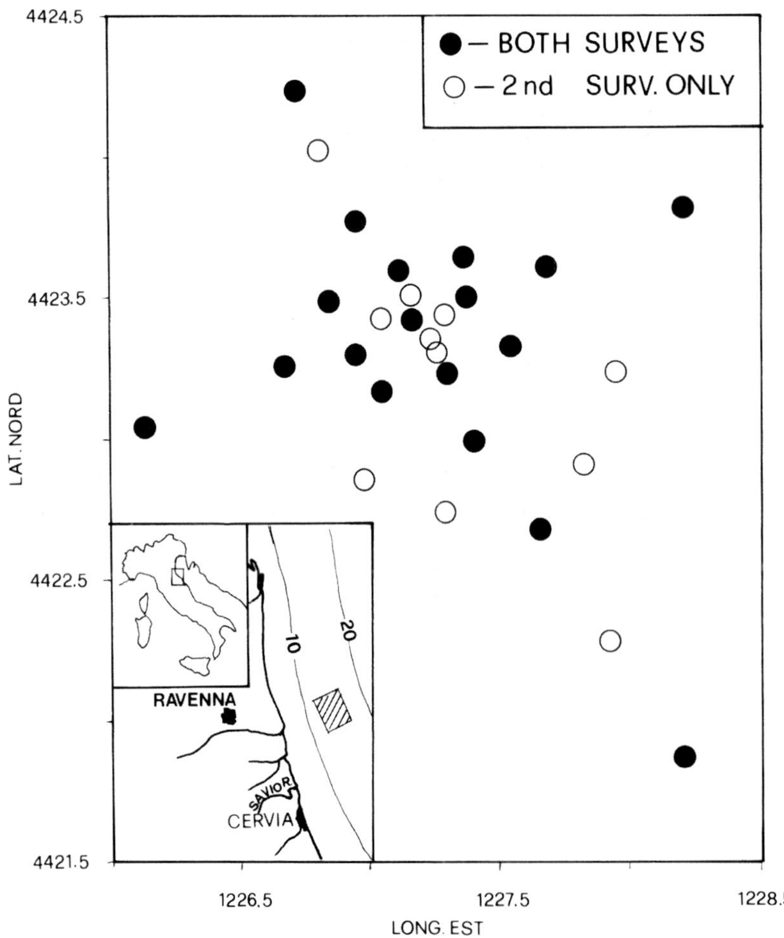

Fig. 1. Location and sampling grid

Table 1. List of collected species (•, present)

	Species	Survey 1st	Survey 2nd		Species	Survey 1^{sr}	Survey 2nd
1	*Ampharete acutifrons*	•	•	55	*Poecilochaetus fauchaldi*	•	•
2	*Ancistrosyllis groenlandica*	•	•	56	*Polycirrus* sp.		•
3	*Aphelochaeta marioni*	•		57	*Polydora ciliata*		•
4	*Aricidea assimilis*	•	•	58	*Polydora flava*	•	•
5	*Aricidea catherinae*	•	•	59	*Polydora* sp.		•
6	*Aricidea claudiae*	•	•	60	*Polymnia nesidensis*	•	
7	*Capitella capitata*		•	61	*Prionospio cirrifera*	•	•
8	*Capitomastus minimus*	•		62	*Prionospio malmgreni*	•	•
9	*Caulleriella caput-esocis*	•	•	63	*Prionospio multibranchiata*		•
10	*Chaetopteridae* n.c.		•	64	*Protodorvillea kefersteini*	•	
11	*Chaetozone setosa*	•	•	65	*Pseudoleiocapitella fauveli*	•	
12	*Chone longiseta*		•	66	*Pseudopolydora antennata*		•
13	*Cossura soyeri*		•	67	*Scolelepis tridentata*	•	•
14	*Diopatra neapolitana*	•	•	68	*Sigambra tentaculata*	•	•
15	*Euclymene collaris*	•		69	*Spio decoratus*		•
16	*Euclymene oerstedi*	•		70	*Spiochaetopterus costarum*	•	•
17	*Euclymene* sp.	•		71	*Spiophanes kroyeri reyssi*	•	
18	*Eulalia* sp.	•	•	72	*Sternaspis scutata*	•	•
19	*Eunice vittata*	•	•	73	*Streblospio shrubsolii*	•	
20	*Exogone dispar*	•		74	*Terebellidae* n.c.		•
21	*Glycera capitata*		•	75	*Terebellides stroemi*	•	
22	*Glycera rouxii*	•	•	76	*Abra alba*	•	•
23	*Gyptis rosea*	•	•	77	*Abra prismatica*		•
24	*Harmothoe johnstoni*	•		78	*Corbula gibba*	•	•
25	*Harmothoe lunulata*	•		79	*Dosinia lupinus*	•	
26	*Heteromastus filiformis*		•	80	*Hiatella arctica*		•
27	*Lanice conchilega*		•	81	*Laevicardium oblongum* juv.		•
28	*Laonice cirrata*	•	•	82	Loripinus fragilis	•	•
29	*Levinsenia gracilis*	•	•	83	*Modiolaria subpicta*		•
30	*Levinsenia oculata*	•	•	84	*Mysella bidentata*		•
31	*Litocorsa stremma*		•	85	*Nucula nucleus*	•	•
32	*Lumbrineris latreilli*	•	•	86	Phapia aurea	•	
33	*Magelona alleni*	•		87	*Phaxas adriaticus*		•
34	*Magelona rosea*	•	•	88	*Pitar rudis*	•	•
35	*Maldane sarsi*	•	•	89	*Plagiocardium papillosum*	•	
36	*Mediomastus capensis*	•	•	90	*Scapharca inaequivalvis*		•
37	*Melinna palmata*	•	•	91	*Spisula subtruncata*	•	•
38	*Micronephtys sphaerocirrata*	•	•	92	*Tellina distorta*	•	•
39	*Monticellina heterochaeta*	•	•	93	*Tellina nitida*	•	•
40	*Myriochele oculata*	•		94	*Tellina* sp.	•	
41	*Neanthes succinea*	•	•	95	*Timoclea ovata*		•
42	*Nephthys incisa*	•	•	96	*Dentalium inaequicostatum*	•	
43	*Notomastus aberans*		•	97	*Acteon tornatilis*	•	•
44	*Ophiodromus flexuosus*	•	•	98	*Aporrhais pespelecani*		•
45	*Owenia fusiformis*		•	99	*Cylichna cylindracea*	•	•
46	*Paradoneis lyra*	•	•	100	*Eulinia glabra*		•
47	*Pectinaria (Lagis) koreni*	•	•	101	*Haminoea hydatis*	•	•
48	*Pectinaria auricoma*		•	102	*Hyala vitrea*	•	
49	*Pherusa* sp.		•	103	*Nassarius mutabilis*	•	
50	*Phyllodoce* cfr. *madeirensis*		•	104	*Nassarius pygmaeus*		•
51	*Phyllodoce lineata*	•		105	*Nassarius reticulatus*	•	•
52	*Phyllodoce* sp.1		•	106	*Natica stercosmuscarum*	•	
53	*Phyllodoce* sp.2		•	107	*Polinices macilenta*	•	•
54	*Pilargis verrucosa*		•	108	*Turritella communis*	•	

sampling point. Sediments were processed through a sieve with mesh size of 0.5 mm. The retained material was preserved in 4% neutral formalin in seawater. The animals were extracted from the residual sediment and the polychaete and mollusc fraction, which comprises over 90% of the total macrofauna, was identified to species. A matrix was composed comprising the number (pooled over three replicates) of individuals of each species at each sampling occasion and site. A variety of structural and diversity indices were calculated, but only the Shannon-Wiener index will be discussed.

Three basic multivariate techniques were used: a Factorial Analysis of Correspondences (FAC) (Benzecrì 1982) and a Multidimensional Scaling (MDS) (Shepard 1962) as ordination techniques and a cluster analysis, based on the Bray-Curtis similarity coefficient, as a classification technique (Field et al. 1982).

In addition, in both the investigated periods, the Shannon diversity was plotted against the sample dimension. To do this, the two sample sets were first sorted in order of increasing diversity, the first two ordered samples pooled and the diversity recalculated on the pooled data. This procedure was repeated until all the samples were pooled together.

Results

A total of 108 polychaete and mollusc species, reported in Table 1, were found on the two sampling occasions. Of these, 48 were common to both periods, whereas the remaining 60 were exclusive to the first or the second period, evidencing a great qualitative difference between the communities of the two periods. The list and the densities of the 10 species, common to both periods, that exhibited the greatest variation in abundance between the periods, are reported in Table 2.

Ordering the entire set of samples by FAC generates on the first two axes (48.6% of total variance; $P<0.05$, Lebart's test) the model represented in Fig. 2. The model is clearly divided into two parts, both with regard to the species and the sample points. This is, evidently, explained by the variations undergone by the community in its specific composition, from the first to the second period. Moreover, the same model evidences a different pattern of dispersion of the samples of the two periods. In fact, the sample points of the first period locate close together in a small area on negative values of the first axis, whereas those of the second period are much more dispersed on positive values of the same axis.

Table 2. Mean density (individuals/sample) of the 10 common species that exhibited the greatest density variation across the 2 periods

Species	1st Survey Mean ± SD	2nd Survey Mean ± SD
Aricidea claudiae	33.66 ±18.99	15.53 ± 19.71
Laonice cirrata	4.66 ± 6.63	0.03 ± 0.18
Lumbrineris latreilli	13.61 ± 7.50	28.93 ± 18.16
Nephthys incisa	4.16 ± 5.59	35.93 ± 15.48
Paradoneis lyra	8.77 ± 5.57	0.53 ± 2.11
Pectinaria (Lagis) koreni	0.38 ± 0.50	23.56 ± 16.38
Polydora flava	15.61 ± 12.89	2.96 ± 13.99
Prionospio cirrifera	10.55 ± 8.08	6.20 ± 9.72
Prionospio malmgreni	45.38 ± 61.18	8.23 ± 13.71
Corbula gibba	210.22 ±153.96	362.33 ±150.64

Figure 3 reports the dendrogram of classification of the same samples. It evidences not only the expected separation of the samples belonging to the two periods but also a greater number of samples or sample groups with higher levels of similarity, in the second period.

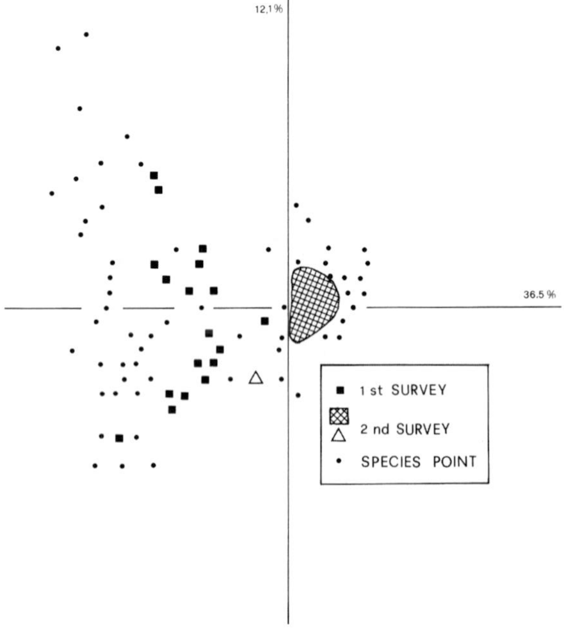

Fig. 2. FAC ordination model of the entire data set (I and II survey)

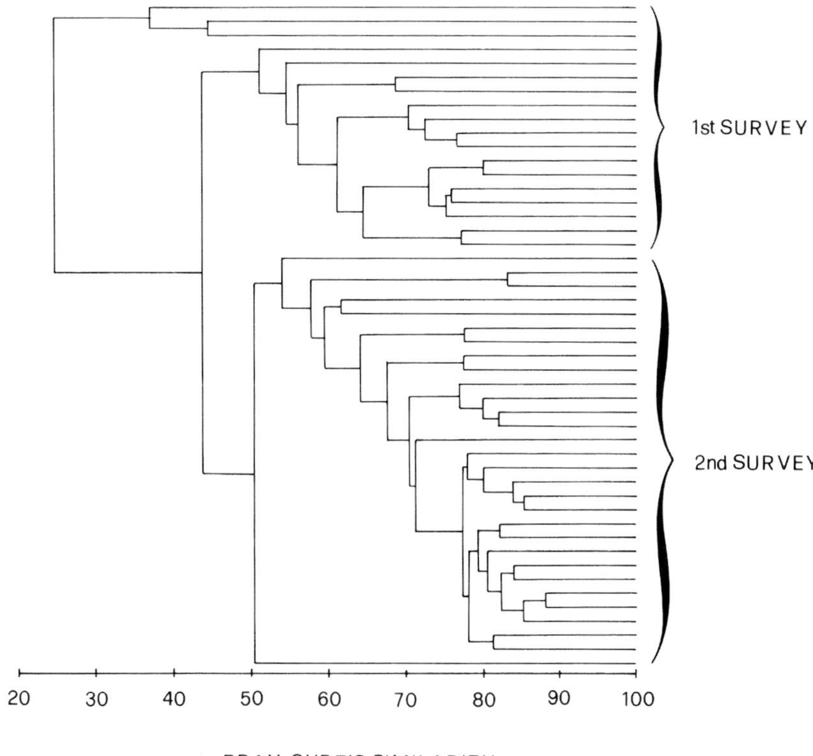

Fig. 3. Classification analisys based on Bray-Curtis similarity of the entire data set (I and II survey)

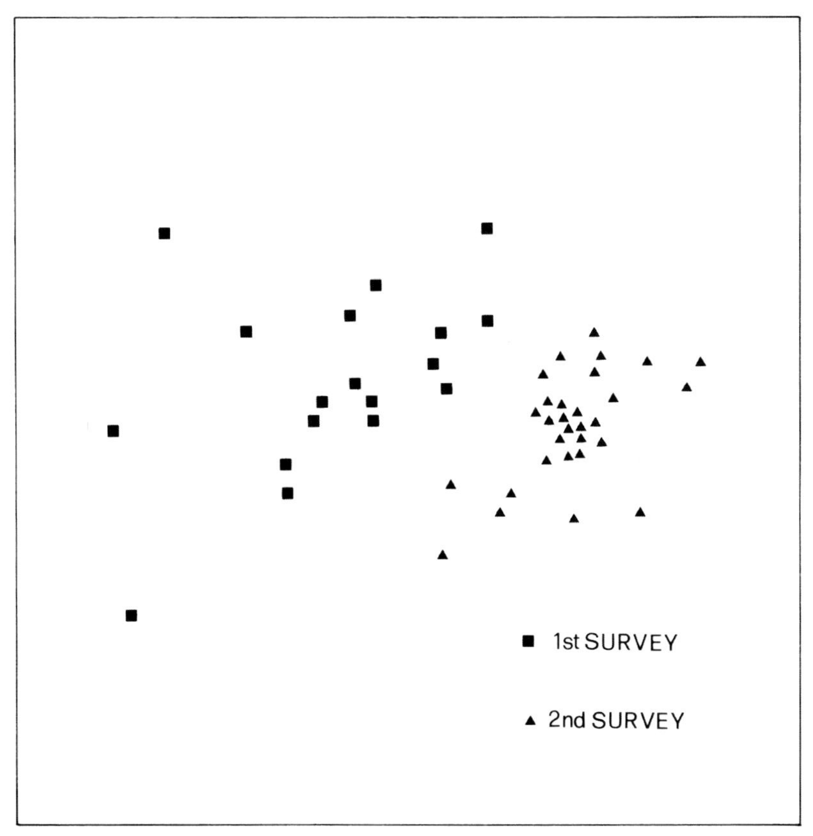

Fig. 4. MDS ordination (stress = 0.12) of the entire data set (I and II survey)

Table 3. Means and Student's t statistics of Shannon diversity and Bray-Curtis similarity in the two investigated periods

	Mean		t	df	*P*
	1st survey	2nd survey			
Shannon diversity	3.37 ± 0.68	3.29 ± 0.44	0.505	46	0.616
Bray-Curtis similarity	52.47 ±10.82	66.51 ±15.16	12.34	586	<0.001

The ordination model, performed by MDS from the same similarity matrix (Fig. 4), gives similar results, generating two separate sample-point groups of which those pertaining to the second period are much closer than those of the first period.

Results of an independent t-test for the differences between means of Shannon diversity of the single samples and of Bray-Curtis similarity between samples are given in Table 3. They point out that, whereas the diversity of the single samples is the same in the two periods, the level of similarity between samples is significantly greater in the second period.

Figure 5 shows the pattern of diversity increase, plotted against the number of samples. It evidences a greater rate of diversity increase in the first period, again indicating that communities from the samples of the first period tend to be more different from each other as compared to the more uniform situation of the second period.

Discussion

The benthic community of the investigated area exhibits a great variation between the two periods of study. In the first period, the community was dominated by typical muddy species that are common in the Northern Adriatic at about 15 m depth (Castelli et al. 1990; Crema et al. 1991), such as *Corbula gibba, Prionospio malmgreni, Aricidea claudiae* and *Lumbrineris latreilli*. The density of several detritivorous polychaetes that colonize the upper sediment layers such as, in addition to the above mentioned *P. malmgreni* and *A. claudiae, Prionospio cirrifera, Polydora flava, Aricidea assimilis* and *Paradoneis lyra* was particularly conspicuous.

A sharp reduction of many of these fine sediment and low energy bottom species occurs in the second period. Their reduction is balanced by the increase of some other species more suited to sustain high energy bottoms, such as *Nephtys incisa* (Crema et al. 1993; Castelli and

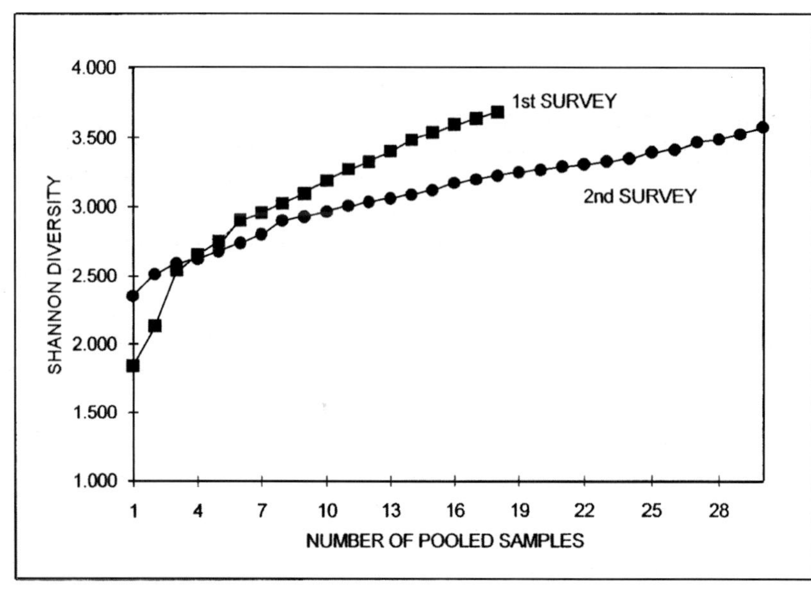

Fig. 5. Shannon diversity plotted against the number of pooled samples (see text)

Prevedelli 1994), whose density increases dramatically (from a mean of 4.2 ± 5.9 to 35.9 ± 15 individuals/sample, Table 3). Besides *N. incisa*, other tube dwelling polychaetes that can stabilise the sediment with their tubes, exhibit a similar trend. Among these are, *Pectinaria (Lagis) koreni* that exhibits a strong increase from the first to the second period and *Owenia fusiformis* that is only present in the second period.

Among molluscs, the bivalve *Corbula gibba*, reported as an indicator of sedimentary instability (Salen Picard 1981), dominates the communities of both the periods. Although *C. gibba* populations frequently exhibit great density fluctuations over the years, which are considered casual (Curini Galletti 1987), we can note that its density in the second period is about two-fold that in the first period. Moreover, the density of *Lumbrineris latreilli,* a polychaete associated with *C. gibba* as indicator of sedimentary instability, exhibits a similar trend.

Besides the above-mentioned variations in community composition and in species density, all the results of community analyses support the notion that, although the "point" diversity is not different in the two periods, the level of biotic variation, *i .e.* the amount of change of species compositions across the samples, is lower in the second as compared to the first period. This finding suggests that the amplitude of the zones in which most species succeed in establishing themselves, defined as realized zones (Pielou 1975) is increased from the first to the second period, causing a fall in the species turnover rate (β diversity *sensu* Whittaker 1972).

Our results can be related to the findings of two sedimentological studies performed on the same site and periods, aimed at detecting the impact of platform activities on the surrounding bottoms (Frascari and Bonvicini Pagliai 1986; Frascari et al. 1994).

Both studies emphasize the role exerted *per se* by the submerged structure that interacts with waves and currents thus enhancing their energy near the bottom. When circulation currents are active, erosion surfaces have been detected as far as 1 km from the platform, in their direction. Storms interfere with the submerged structure depending on their intensity and direction and cause erosion and transport to great distances, even outside the area of our sampling grid.

A classical generalization about infaunal invertebrate distribution is that a clear association between animals and specific sediment types exist. According to this generalization, infauna distribution should be determined by only a few parameters, such as grain size, organic content, sediment stability and microorganisms (Gray 1974; Rhoads 1974; Pérès 1982). In addition to these traditional factors, the leading role of hydrodynamic processes, such as the boundary layer flow and sediment transport regimen, have been increasingly emphasized (Riedl 1971; Snelgrove and Butman 1994). Arguing with the numerous studies that have correlated infaunal invertebrate distribution with single parameters of the sediment, Snelgrove and Butman (1994) conclude that there is little evidence that animal distribution is determined by any of the classical above mentioned factors. They propose a shift in focus towards understanding the relationship between organism distribution and the dynamic sedimentary and hydrodynamic environment. Given that the bottom sediment reflects the boundary-layer flow and sediment transport regimen, animal distribution should not be determined by some particular aspect of the sediment but by the same physical processes that created that particular sedimentary environment.

In this context, many recent studies in larval ecology are focused on the physical aspects of larval dispersion and settlement. Laboratory experiments and experimental field manipulations designed to evaluate the importance of bottom energy on larval distribution suggest that the hydrodynamics influence benthic larvae distribution (Eckman 1979, 1983; Kern and Taghon 1986; Savidge and Taghon 1988; Butman 1989). Moreover, benthic flow may also redistribute settled individuals. This may be an important way of dispersion for direct developers, and it has been observed in some benthic species, such as *Mya arenaria* and *Macoma baltica* (Emerson and Grant 1991; Gunther 1991).

The patterns we have described for the communities of the two periods are consistent with these views. In fact, concomitant with an increase in bottom hydrodynamic energy, we detected a sharp reduction of surface deposit feeders that seemingly suffer from the erosion of their specific micro-habitat (Castelli and Prevedelli 1994), and a new pattern of distribution evidencing an increase of the realized zones of most faunal invertebrates and a decrease of their spatial turnover rate.

Acknowledgements. The original data were collected during a monitoring programme promoted by AGIP and its permission to use this information is gratefully acknowledged.

References

Benzecrì JP (1982) L'analyse des données. Dunod, Paris

Butman CA (1989) Sediment-trap experiments on the importance of hydrodynamical processes in distributing settling invertebrate larvae in near bottom waters. J Exp Mar Biol Ecol 134: 37-88

Castelli A, Prevedelli D (1994) Effects of water movement disturbance on polychaete stratification in infralittoral muddy bottoms. In: Dauvin JC, Laubier L, Reish DJ (eds) Actes 4ème Conf Int Polychètés. Mem Mus Natl Hist Nat 162: 617-618

Castelli A, Becchi S, Crema R (1990) Distribuzione verticale della macrofauna di fondi mobili nel profilo del sedimento. Oebalia 16: 623-625

Crema R, Castelli A, Prevedelli D (1991) Long term eutrophication effects on macrofaunal communities in Northern Adriatic Sea. Mar Pollut Bull 22: 503-508

Crema R, Castelli A, Bonvicini Pagliai AM, Zunarelli Vandini R (1993) Natural and anthropogenic eutrophy of Northern Adriatic sea: its reflects on coastal biocenoses. Stud Materi Oceanol 64 Mar Pollut 3: 317-329

Curini Galletti M (1987) Recovery of a soft bottom community after extensive dredging. I. Mollusca. FAO Fish Rep 352 Suppl: 54-63

Daan R, Mulder M, Van Leeuwen A (1994) Differential sensitivity of macrozoobenthic species to discharges of oil-contamined drill cuttings in the North Sea. Neth J Sea Res 33: 113-127

Davies JM, Addy JM, Blackman R, Blanchard JR, Moore DC, Sommerville HJ, Whitehead A, Wilkinson T (1984) Environmental effects of oil based mud cuttings. Mar Pollut Bull 15: 363-370

Eckman JE (1979) Small-scale patterns and processes in a soft-substratum intertidal community. J Mar Res 37: 437-457

Eckman JE (1983) Hydrodynamic processes affecting benthic recruitment. Limnol Oceanogr 28: 241-257

Emerson CW, Grant J (1991) The control of soft-shell clam (*Mya arenaria*) recruitment on intertidal sandflats by bedload sediment transport. Limnol Oceanogr 36: 1288-1300

Field JG, Clarke KR, Warwick RM (1982) A pratical strategy for analysing multispecies distribution patterns. Mar Ecol Prog Ser 8: 37-52

Frascari F, Bonvicini Pagliai AM (1986) Studio delle alterazioni litologiche, geochimiche e biologiche dei fondali marini interessati dagli scarichi dei fluidi di perforazione della piattaforma "Antares" (AGIP). Alto Adriat AGIP Int Rep

Frascari F, Bergamini MC, Spagnoli F, Marcaccio M (1994) Analisi di impatto ambientale e socio economico dell'attività petrolifera nell'offshore nazionale e identificazione di metodologie di valutazione: indagine sui processi geochimico-sedimentologici. Ist Geol Mar Bologna Internal Rep

Gray JS (1974) Animal-sediment relationships. Oceanogr Mar Biol Annu Rev 12: 223-261

Gray JS, Clarke KR, Warwick RM, Hobbs G (1990) Detection of initial effects of pollution on marine benthos: an example from the Ekofish and Eldfield oilfield, North Sea. Mar Ecol Prog Ser 66: 285-299

Gunther CP (1991) Sediment of *Macoma balthica* on an intertidal sandflat in the Wadden Sea. Mar Ecol Prog Ser 76: 73-79

Harper DE Jr, Potts DL, Salzer RR, Case RJ, Jaschek RL, Walker CM (1981) Distribution and abundance of macrobenthic and meiobenthic organisms. In: Middleditch BS (ed) Environmental effect of offshore oil production – the Buccaneer gas and oil field study. Plenum Press, New York, pp 749-756

Hartley JP (1984) The benthic ecology of the Forties Oilfield (North Sea). J Exp Mar Biol Ecol 80: 161-195

Hartley JP (1996) Environmental monitoring of offshore oil and gas drilling discharges – a caution on the use of barium as a tracer. Mar Pollut Bull 32: 727-733

Kern JC, Taghon GL (1986) Can passive recruitment explain harpacticoid copepod distributions in relation to epibenthic structure? J Exp Mar Biol Ecol 101: 1-23

Kingston PF (1987) Field effects of platform discharges on benthic macrofauna. Philos Trans R Soc London B 316: 545-565

Kroncke I, Duineveld GCA, Raak S, Rachor E, Daan R (1992) Effects of a former discharge of drill cuttings on the macrofauna community. Mar Ecol Prog Ser 91: 277-287

Neff JM, Bothner MH, Maciolek NJ, Grassle JF (1989) Impacts of exploratory drilling for oil and gas on the benthic environment of Georges Bank. Mar Environ Res 27: 77-114

Olsgard F, Gray JS (1995) A comprehensive analysis of the effects of offshore oil and gas exploration and production on the benthic communities of the Norwegian continental shelf. Mar Ecol Prog Ser 122: 277-306

Pérès JM (1982) Major benthic assemblages. In: O Kinne (ed) Ocean Management, part I. (Marine ecology, V) John Wiley, Chichester, England, pp 373-522

Pielou EC (1975) Ecological diversity. Wiley and Sons, New York

Reiersen LO, Gray JS, Palmork KH, Lange R (1989) Monitoring in the vicinity of oil and gas platforms, results from the Norwegian sector of the North Sea and recommended methods for forthcoming surveillance. In: Engelhardt FR, Ray JP, Gillam AH (eds) Drilling wastes. Elsevier Applied Science, London, pp 91-117

Rhoads DC (1974) Organism-sediment relations on the muddy sea floor. Oceanogr Mar Biol Annu Rev 12: 263-300

Riedl R (1971)Water movement – animals. In: O Kinne (ed) Environmental factors, part 2. (Marine ecology, I) John Wiley, Chichester, England, pp1123-1156

Salen-Picard C (1981) Evolution d'un peuplement de la vase terrigene cotiere soumis a des rejets de dragages, dans le Golfe de Fos. Tethys 10: 83-88

Savidge WB, Taghon GL (1988) Passive and active components of colonization following two types of disturbance on intertidal sandflat. J Exp Mar Biol Ecol 115: 137-155

Shepard RN (1962) The analysis of proximities: multidimensional scaling with an unknown distance function. Psikometrika 27: 125-140

Snelgrove PVR, Butman CA (1994) Animal-sediment relationships revisited: cause versus effect. Oceanogr Mar Biol Annu Rev 32: 111-177

Whittaker RH (1972) Evolution and measurement of species diversity. Taxon 21: 213-251

Recent Changes in Biodiversity in the Ligurian Sea (NW Mediterranean): is there a Climatic Forcing?

C. Morri[1] and C.N. Bianchi[2]

ABSTRACT
A few case studies from the Ligurian Sea provide examples of how climate change has influence on marine biodiversity. This influence may be either direct (bryozoan population dynamics, coral growth, sea-grass shoot density) or indirect, mediated by biotic interactions (bivalves in a sandy bottom community) or marine currents (fish and rocky bottom communities, entering of southern species). The difficulty of distinguishing between anthropogenic and climatic actions on the marine biota emphasises the need to study marine biodiversity in the context of the physical setting and the relevant space and time scales.

Introduction

It has often been assumed, in the past, that marine communities were naturally stable, and that most changes (seasonal cycles apart) were probably caused by man (Lewis 1996). Only in the last decades the natural variability of marine communities has been fully recognised, and its relation with climate fluctuations hypothesised (Bianchi 1997).

Angel (1991) suggested that monitoring biogeographic boundaries would give an unambiguous signal of climate change. Evidence in that sense has been collected in recent years in different regions of the globe (Barry et al. 1995; Southward et al. 1995; Parker and Dixon 1998; Bianchi et al. 1998a).

The Ligurian Sea, situated in the north-eastern corner of the western Mediterranean, is one of the coldest areas of the Mediterranean Sea. Accordingly, the biota of the Ligurian Sea are characterised by a strong diminution of the subtropical elements and by a more marked presence of species of cold temperate waters (Rossi 1969). However, warm-water organisms have recently increased their occurrence in the Ligurian Sea (Bianchi and Morri 1993, 1994), providing a clue for climate change (Francour et al. 1994; Astraldi et al. 1995).

The aim of the present paper is to explore existing data on change in the Ligurian Sea biota in the search for further evidence for a climatic influence. First, the Ligurian Sea climate will be characterised in short and a climatic descriptor identified. The time trend of the climatic descriptor will be then compared with biological changes for different marine communities, studied especially in recent years.

Ligurian Sea Climate

The Ligurian Sea is tightly connected with the Gulf of Lion, which is subject to periodic intrusions of the northerly winds from the Rhone valley. They generate very energetic weather conditions, which are particularly severe in winter. The major large-scale hydrodynamic feature (Astraldi et al. 1994) is a well-defined cyclonic circulation active year round, which contributes to keeping the mean surface temperature lower than that of the adjacent Tyrrhenian Sea.

The greater cooling of the Ligurian Sea induces water and temperature losses, which in turn draw a seasonal flux of warmer Tyrrhenian Sea waters to restore the hydrological budget (Astraldi and Gasparini 1992). Analysing jointly current regimes and invertebrate settlement off

[1] Dipartimento Territorio e Risorse, Università di Genova, Corso Europa 26, 16132 Genova, Italy
[2] Centro Ricerche Ambiente Marino, ENEA Santa Teresa, 19100 La Spezia, Italy

F.M. Faranda, L. Guglielmo, G. Spezie (eds)
Mediterranean Ecosystems: Structures and Processes

Capraia Island (Fig. 1), Aliani and Meloni (1996, 1999) demonstrated that this Tyrrhenian current is the carrier of the dispersal stages of warm-water species into the Ligurian Sea. Notwithstanding seasonal variability, the flux is always directed northward and constitutes a sort of one-way bottleneck.

The different climatology of the two basins can be deduced from a number of hydrographic data, among which temperature is of primary importance. Many temperature measurements of Ligurian Sea waters are available since 1909 (Bruschi and Sgorbini 1986). Unfortunately, due to a temporal inhomogeneity in data sampling, it is not possible to derive long-term trends.

However, Astraldi et al. (1995) demonstrated that Ligurian Sea surface water temperatures are correlated to air temperature taken at the Meteorological Observatory of Genoa. Genoa air temperature in the last century shows important year-to-year fluctuations and two major warm periods: the first centred around the '40s, the second started in the mid '80s and continuing at present (Fig. 2). The temperature trend from the '50s onward mostly parallels the North Atlantic Oscillation, which is considered as the main determinant of the marine climate in the Northern Hemisphere (Mann and Lazier 1996). Vignudelli et al. (1999) showed the influence of the North Atlantic oscillation on the circulation of the western Mediterranean Sea.

The present increase in air temperature coincides with the observed warming in deep (Bethoux et al. 1990) and intermediate waters (Sparnocchia et al. 1994), and is also mirrored in the nearly stable presence of warm-water species in the Ligurian Sea. There is also evidence that at least some of them, such as the labrid fish *Thalassoma pavo*, reproduce in the Ligurian Sea, thus being independent from the larval supply by the Tyrrhenian Sea (Vacchi et al. 2000). In summary, it seems reasonable to use Genoa air temperature time series as a synthetic descriptor of Ligurian Sea climate.

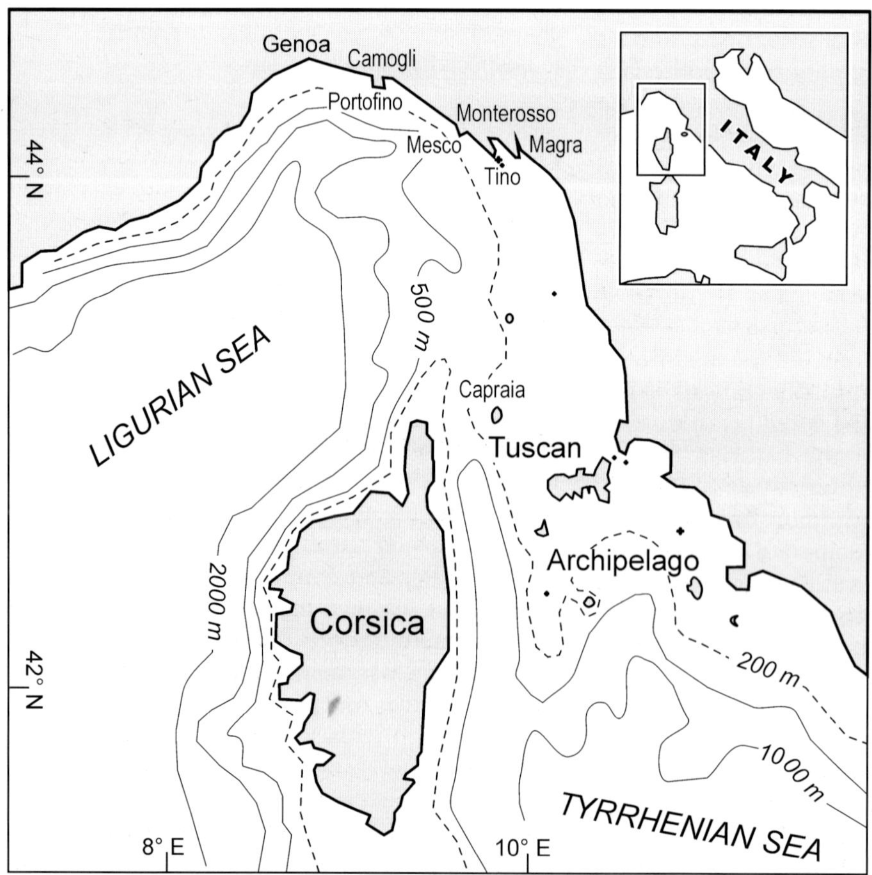

Fig. 1. Geographical setting of the Ligurian Sea, showing the boundary with the Tyrrhenian Sea (Tuscan Archipelago) and the localities mentioned in the text

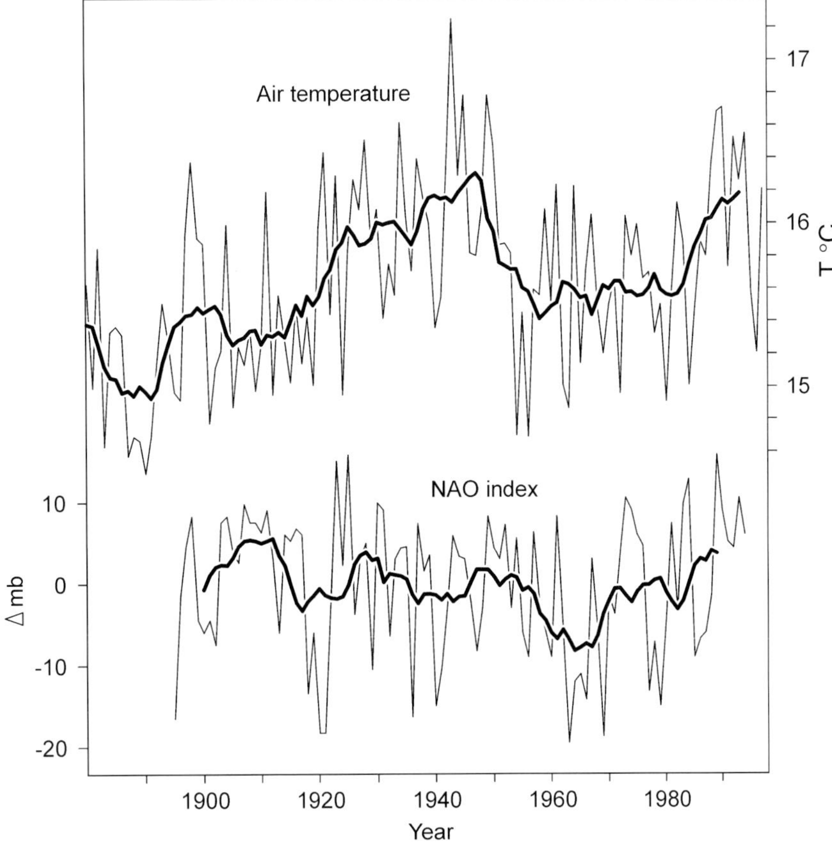

Fig. 2. Trend of air temperature (annual means, T °C) at Genoa (Meteorological Observatory), 1880 to 1997; and trend of the North Atlantic oscillation index (average pressure differences between the Azores and Iceland, Δ mb), 1895 to 1994 (adapted from Mann and Lazier, 1996). In both graphs the *thick line* is the 11-year running mean

Change in Marine Communities

Sandy Bottom Benthos

Soft-bottom benthic communities of the Ligurian Sea have been studied for many years (Albertelli and Cattaneo 1985). Seasonal and interannual changes in a bivalve-dominated community inhabiting shallow sand were recently described by Albertelli et al (1993, 1994a, 1994b).

A short time series is available from the sandy bottom in front of the Magra River estuary (Fig. 1), where the benthic community was monitored for 7 years, 1987 to 1993 (De Biasi 1994, 1997). Major changes in community composition and structure were observed in 1988 and 1991: to relate the observed changes to the activity of a new sewage treatment plant (started in 1988, completed in 1990-1991) was tempting (Bianchi 1997), but a sharp decrease in mean air temperature, soon followed by a new rise, was observed in the same period (Fig. 2). Species richness, expressed as Margalef's $D = (S\text{-}1)/\text{Ln}n$ (where S is the number of species and n the number of individuals), showed a trend nearly specular to that of air temperature. Dominance by the bivalve *Spisula subtruncata* coincided with the "warm" period 1988-1990, whereas another bivalve, *Chamelea gallina*, replaced *S. substruncata* when temperature started rising again after 1991 (Fig. 3).

Rocky Bottom Benthos

Time series on epibenthic communities living on Ligurian Sea rocky reefs are scarce. A permanent station was fixed on a subtidal rock at Tino Island (Fig. 1), where a bryozoan-dominated community has been studied since 1993. The first results illustrate the importance of episodic climatic events, which are unpredictable in timing and extent, for the population dynamics of the dominant organism, the bryozoan *Pentapora fascialis* (Cocito et al. 1998).

Peirano et al. (1998) suggested using sclerochronology to compare the growth of the coral

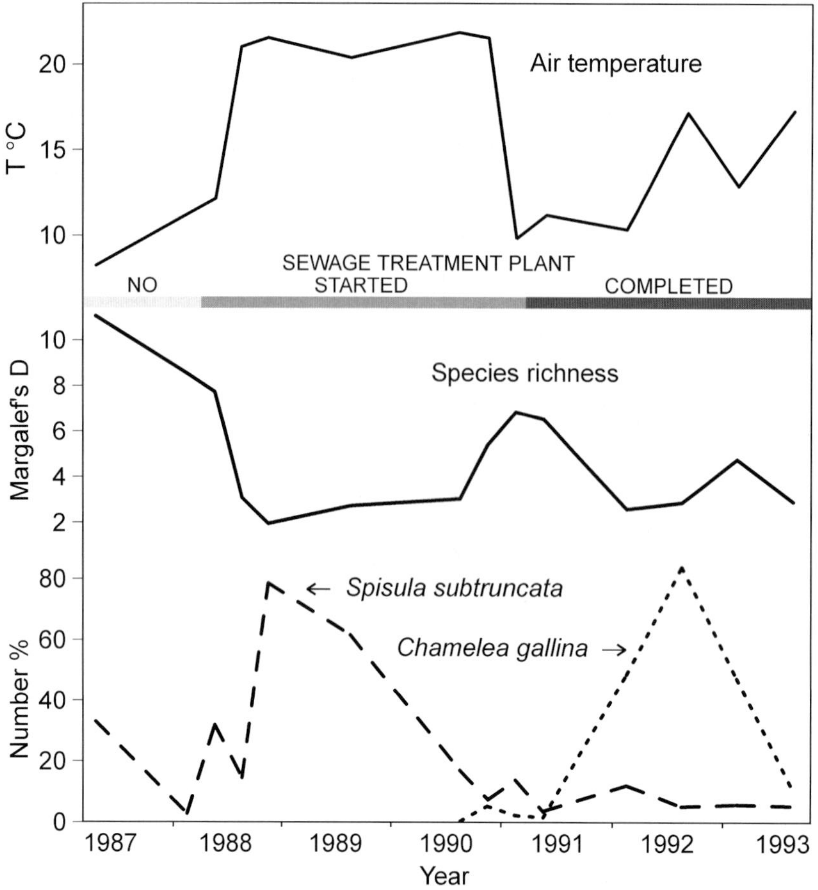

Fig. 3. Climate, anthropogenic, and biotic change in a soft bottom community, 1987 to 1993. *Top to bottom*: trend in air temperature at Genoa (each point represents the mean of the previous 3 months, *T* °C); operation of a big sewage plant in the lower estuary of the Magra River; trend in species richness (Margalef's D) of the community inhabiting the sandy bottom in front of the Magra River estuary; quantitative importance (number %) of the two dominant species in the same community

Cladocora caespitosa with climate fluctuations. Timing of growth-band deposition in several colonies from eastern Ligurian Sea proved to be correlated to monthly temperature and irradiance (Peirano et al. 1999a). Retrospective analysis, through X-radiography, of the oldest colonies showed that higher growth rates coincided with a warmer period in the 40s, and lower ones with a colder period in the 70s, but showed little correlation with temperature in other periods (Morri et al. 2000).

Another way to approach the study of long-term change in rocky bottom communities is to "re-visit" sites already studied in the past. In the Ligurian Sea, two pioneering studies on rocky subtidal communities were done by Tortonese (1958, 1961) and Rossi (1965) thanks to the co-operation of amateur divers.

Tortonese (1958, 1961) worked at Portofino (Fig. 1) in the '50s and reported on the unexpected occurrence of several warm-water species. The previous decade coincided with a warm period (Fig. 2), but these species were also found in other years in the Ligurian Sea (Bianchi and Morri 1994). We re-visited Tortonese's sites in 1991 and 1993 and concluded that no dramatic change had occurred in nearly half a century on these subtidal bottoms down to about 40 m depth (unpublished data). Nevertheless, we found as common a number of species, such as the algae *Pseudochlorodesmis furcellata* and *Zanardinia prototypus*, that Tortonese (1958, 1961) did not mention. Conversely, species reported as abundant in the '50s were unnoticeable in '90s. The most striking examples were the bivalve *Spondylus gaederopus* and the sponge *Calyx nicaeensis*. The virtual disappearance of the first might be due to a disease (Relini 1992) but we are unable to offer any explanations for the second: *C. nicaeensis* is a conspicuous species, easily to recognise underwater (see plate III B in Tortonese 1961), so it is difficult to think that it had escaped attention. Maybe it has been the object of excess collection as a curio, maybe its population collapsed for some other reason. However, the problem remains that Tortonese

(1958, 1961) used descriptions and samples taken by Duilio Marcante and the divers of the "Centro Subacqueo" of Genoa: the lack of precise, quantitative data prevents any conclusions.

A few years after Tortonese, in 1959 and 1962, Rossi (1965) worked at Punta Mesco (Fig. 1) with the help of the underwater photographer Gianni Roghi. Thanks to photography, she was able to provide (semi-)quantitative data of the sessile biota living on rocks 17 to 44 m deep. In 1990-1991, Peirano and Sassarini (1992) took photos in different stations of the same site, whereas in 1996 Peirano et al. (2000) were able to relocate both the 1959/1962 and 1990-1991 stations, taking new photographs of the epibenthic communities. Change was measured using the dissimilarity index 1-Sk, where Sk is the Kulczynski's similarity coefficient (Boudouresque 1971). Dissimilarity was significantly higher (1-way ANOVA, $P = 0.036$) between 1996 and 1959/1962 than between 1996 and 1990-1991 (Fig. 4). Peirano et al. (2000) related the bigger change from Rossi's time to increased water turbidity in the area. Based on the measurable raise of the compensation depth for seagrass (Torricelli and Peirano 1997), it can be calculated that water transparency (Secchi depth) passed from as much as 23 m to the present mean value of 11.4 m. Although human impact was presumed to be the main cause of this increased turbidity, the influence of climate fluctuations cannot be excluded, the '90s having been warmer than the late '50s (Fig. 4).

Sea-Grass Meadows

Posidonia oceanica is a seagrass species endemic to the Mediterranean Sea. Its meadows show at present alarming signs of degradation, especially

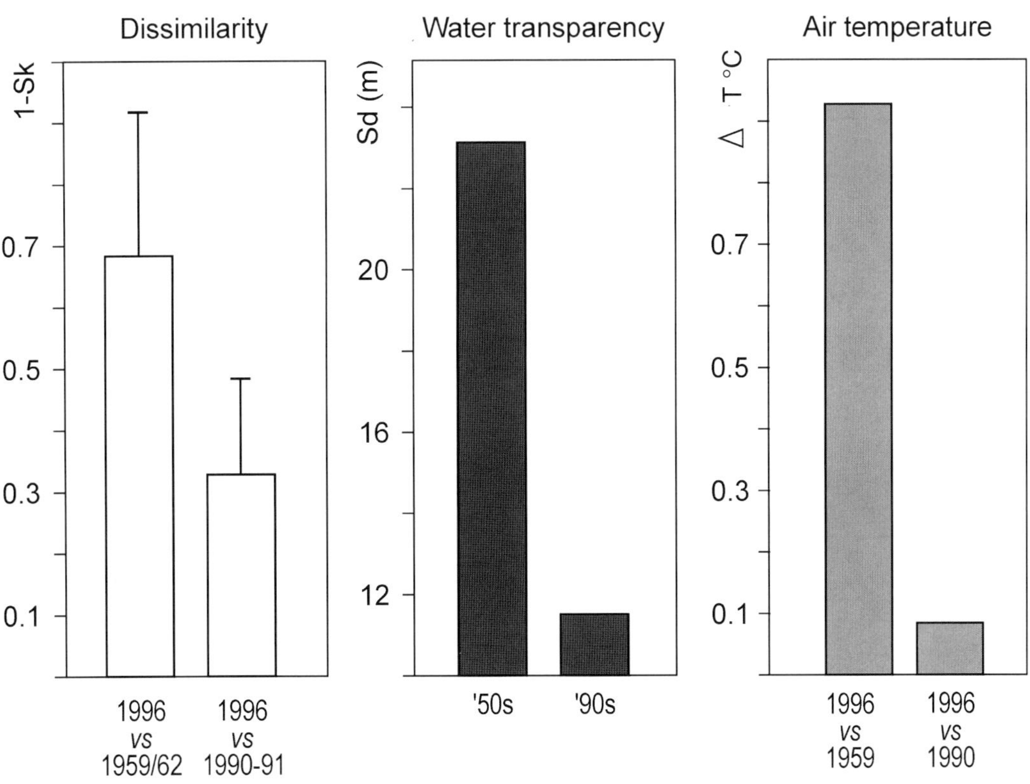

Fig. 4. *Left to right*: difference in subtidal rock epibenthic communities at Mesco Point, expressed as mean (± standard deviation) dissimilarity (1-Sk) between 1996 and 1959/62 and between 1996 and 1990-91; water transparency in the same area in the '50s and in the '90s; difference in air temperature at Genoa (Δ *T* °C) between 1996 and 1959 and between 1996 and 1990 (in all cases the means of the 5 previous years were used)

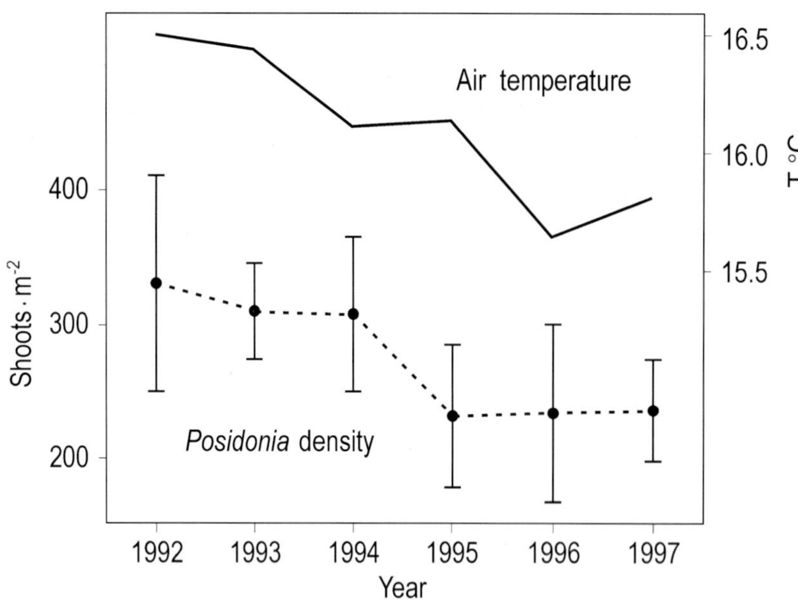

Fig. 5. Change in climate and in a *Posidonia oceanica* meadow, 1992 to 1997. *Top*: annual mean air temperature at Genoa (*T* °C). *Bottom*: annual mean (± standard deviation) of the density (shoots m^{-2}) at 10 m depth in the *Posidonia* meadow of Monterosso

in the northern parts of the Mediterranean. This degradation may be explained either by pollution, resulting in greater water turbidity and hence diminished plant vitality; or by the natural decline of the plant, which is believed to have had its climatic optimum around 6000-2750 years B.P. (Pérès 1984).

Regression of *Posidonia* meadows has been surely accelerating in recent decades: in the Ligurian Sea, for example, it can be reckoned that nearly 30% of their original surface area has been lost in the '60s, during the period of rapid urban and industrial development along the Ligurian coast (Bianchi and Peirano 1995; Peirano and Bianchi 1997). These years, however, also corresponded to a cold phase in the secular temperature trend (Fig. 2). In recent years, characterised by an apparent sea water warming (Bianchi 1997), flowering, fruit maturation and seedling occurrence have been more frequent (Piazzi et al. 1999).

A long-term study of the *Posidonia oceanica* meadow of Monterosso (Fig. 1) was started in 1991 and is still in progress. This meadow underwent heavy human aggression in the '60s and '70s (Góngora Gonzáles et al. 1996), but is now included in a Marine Protected Area which should ensure its protection. A number of parameters has been monitored and their variation with time correlated with climatic data. Whether there is such a correlation or not is unclear at present (Bianchi 1997; Peirano et al. 1999b). However, shoot density shows a pluriannual trend which parallels that of air temperature, thus suggesting an overriding influence of climate (Fig. 5).

Fish Communities

Fishery statistics often provide the longest time series of marine biological data available (Bianchi 1997). One of the best known series on Ligurian Sea fisheries was that of the so-called "tonnarella" of Camogli (Fig. 1). This tonnarella, described in detail by Cattaneo-Vietti (1985), was a fixed fishing net operating at the same location throughout spring and summer for many years. Balestra et al. (1976) reported catch data for the period 1950 to 1974. Although total catch was said to remain relatively constant over time (Boero 1996), its fluctuations grossly coincided with those of minimum air temperature. Species composition changed dramatically from a fish assemblage dominated by *Auxis rochei* in the '50s to one dominated by *Sarpa salpa* thereafter (Fig. 6). Balestra et al. (1976) explained the decreased abundance of the former as due to overfishing, and the increased abundance of the latter as a consequence of replacing nets made with hemp or nylon with nets made with coconut fibre, the latter being more easily fouled by algae which could attract the herbivorous fish *S. salpa*.

These *ad hoc* explanations may be convincing, but if we look at the trend of annual minimum air temperatures, we can see that most winters of *Auxis* years were less cold than those of *Sarpa* years. It might be not a coincidence that *A. rochei* is a warm-water species widely distributed in all tropical and subtropical seas of the globe, whereas *S. salpa* is a temperate water species, living in the Atlantic-Mediterranean region (Dubelius 1997).

Discussion

This short review of a few case studies shows that recent changes in Ligurian Sea biodiversity might well reveal a climatic forcing. According to Southward et al. (1995), climate change can influence marine communities by a combination of: (1) direct effect on the organisms (direct influence of temperature, causing changes in survival, reproductive success, dispersal pattern and

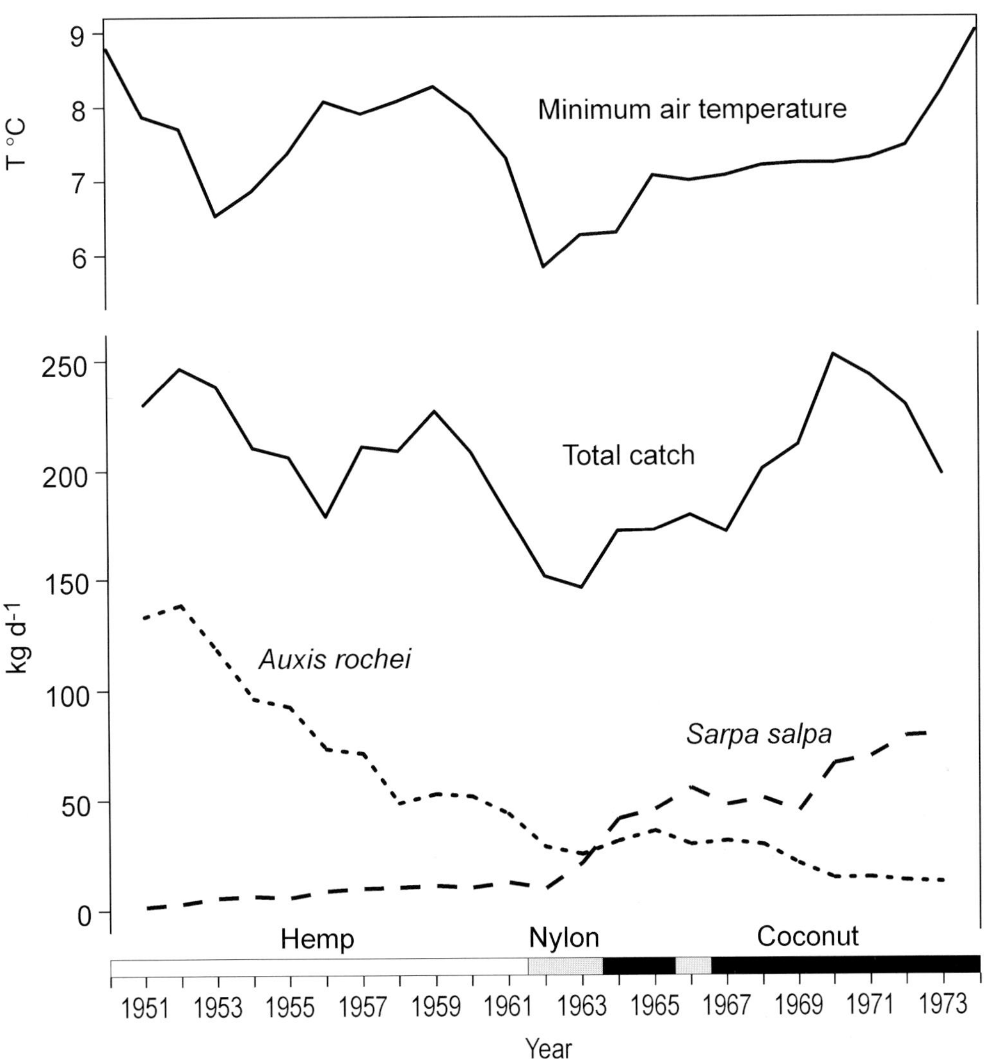

Fig. 6. Change in climate, fish catch and net fabric, 1950 to 1974. *Top to bottom*: trend in annual minimum air temperature at Genova (T °C); trends in total catch and in the catches of the two most important fish species (kg d^{-1}) at the "tonnarella" (= fixed fishing net) of Camogli; fabric of the nets used in the tonnarella in different years: both hemp and nylon fibres inhibited covering by biofouling, coconut fibre did not

behaviour); (2) effects mediated by biotic interactions (conferral of competitive advantage to one of a pair of overlapping species); (3) indirectly through ocean currents (changes in climate may alter the emphasis of water flow and the pattern of water circulation). Note that this meaning of "direct" and "indirect" effects is different from that commonly used in community ecology (Strauss 1991). The data from the Ligurian Sea apparently provided examples for all these mechanisms.

Population dynamics of *Pentapora fascialis*, growth of *Cladocora caespitosa*, and change in shoot density of *Posidonia oceanica* seem to stand for a direct effect of climate on the organisms, whereas the alternate dominance between *Spisula subtruncata* and *Chamelea gallina* in sandy bottom communities might correspond with an effect mediated by conferral of advantage to one of them in turn. Both species are filter-feeding bivalves, sharing the same habitat, and a competition between them can be suspected.

The same hypothesis is not tenable in the case of the alternation between *Auxis rochei* and *Sarpa salpa*. The adults of two species occupy very different niches in the marine ecosystems and, although we probably do not know enough about the ecology of their larval phases, a competition is unlikely. Rather, we can think of the influence of an altered pattern of water flow. Changes in subtidal rock communities might also be due to such collateral effects, whereas the influence of current variability upon the occurrence of southern species in the Ligurian Sea seems sufficiently proved (Astraldi et al. 1995; Aliani and Meloni 1999).

Supply side ecology (Roughgarden et al. 1987) emphasises the role of physical transport processes in structuring marine communities, and Barry and Dayton (1991) coined the expression "hydrodynamic biological oceanography" to underline that the variations in the distribution and abundance of marine organisms are strictly linked to the movement of water masses. Veron (1995) introduced the concepts of "surface circulation vicariance" to indicate that the fluctuations in surface currents have a major impact on shallow marine life.

The alternation observed within bivalves and fish is consistent with the idea that climatic fluctuations favour the coexistence of species potentially redundant (Blandin 1987), thus acting as a biodiversity pump on both evolutionary and ecological scales (Sarà 1985; Astraldi et al. 1995).

An aspect that turned out to be obvious in our analysis of these Ligurian Sea examples is that for most of them the observed change can be equally driven either by climate or by anthropogenic causes. Distinguishing between the two is often difficult (Bianchi 1997) and can also cause animated debates, as in the case of the alleged eutrophication of Skagerrak (Buhl-Mortensen 1996; Gray 1996). It is true that anthropogenic and climatic actions can combine their effects on the marine biota (Bourcier 1996). All this encourages caution when examining changes in marine ecosystems and emphasises the need of planning studies in the context of the physical setting and the relevant space and time scales (Levin 1992).

Finding an appropriate space and time scale was the main problem we met when designing the layout of this paper. After consulting both published data and our own archives, we clearly realised that most of the wealth of scientific research on the Ligurian Sea (Bianchi et al. 1987; Cattaneo-Vietti et al. 1988) concerns studies of local relevance and/or carried out for short periods. In part, this reflects the common Mediterranean research attitude, traditionally directed toward the identification and description of biocoenosis-types rather than the dynamic aspects (Bianchi 1997; Bianchi et al. 1998b). We are now getting fully aware that, contrary to the demand of most funding agencies, any small-scale and short-term approach to ecological monitoring and research will not enable us to understand ecosystem change, both climate or man-induced, but will likely result an unproductive investment of time and resources.

Acknowledgements. Change of marine ecosystems is the core topic of the national research project SINAPSI (Seasonal, INterannual and decAdal variability of the atmosPhere, oceanS and related marIne ecosystems). Research on *Cladocora caespitosa* was started within the project "Gli Cnidari del benthos marino: ecologia e biogeografia" of the University of Genoa (60% MURST). Studies on seagrass beds, epibenthic communities on rocky reefs and southern species occurrence in the Ligurian Sea fall within the scope of the activities on marine biodiversity at the Marine Environment Research Centre of La Spezia (5% MURST). We wish to thank S. Tucci and F. Guercio-Skara (University of Genoa) for kindly providing air temperature data from the Meteorological Observatory of Genoa.

References

Albertelli G, Cattaneo M (1985) Macrobenthos dei fondi molli del Mar Ligure. In: Cinelli F, Fabiano M (eds) Atti 6° Congr Assoc Ital Oceanol Limnol. AIOL, Genova, pp 87-97

Albertelli G, Bonomi A, Covazzi A, Della Croce N, Fraschetti S (1993) Macrobenthos y parámetros ambientales en fondos arenosos del Mar de Liguria, Italia. Publ Espec Inst Esp Oceanogr 11: 305-312

Albertelli G, Chiantore M, Covazzi A (1994a) Persistence and changes in suspension and deposit feeding species in a shallow soft-bottom macrobenthic community. Biol Mar Mediterr 1 (1): 277-278

Albertelli G, Chiantore M, Fraschetti S (1994b) A one-year study of a sandy bottom community: seasonal fluctuations compared to a previous sampling period. In: Albertelli G, Cattaneo-Vietti R, Piccazzo M (eds) Atti 10° Congr Assoc Ital Oceanol Limnol. AIOL, Genova, pp 529-542

Aliani S, Meloni R (1996) Relationship between transport of benthic species and hydrological characteristics: note on settlement in the Corsica Channel. In: Albertelli G, De Maio A, Piccazzo M (eds) Atti 11° Congr Assoc Ital Oceanol Limnol. AIOL, Genova, pp 259-269

Aliani S, Meloni R (1999) Dispersal strategies of benthic species and water current variability in the Corsican Channel (Western Mediterranean). Sci Mar 63 (2): 137-145

Angel MV (1991) Variations in time and space: is biogeography relevant to studies of long-time scale change? J Mar Biol Assoc UK 71: 191-206

Astraldi M, Gasparini GP (1992) The seasonal characteristics of the circulation in the North Mediterranean basin and their relationship with the atmospheric-climatic conditions. J Geophys Res 97 (C6): 9531-9540

Astraldi M, Gasparini GP, Sparnocchia S (1994) The seasonal and interannual variability in the Ligurian-Provençal Basin. In: La Violette PE (ed) Seasonal and interannual variability of the Western Mediterranean Sea. (Coastal and estuarine studies, 46) Am Geophys Union, Washington DC, pp 93-113

Astraldi M, Bianchi CN, Gasparini GP, Morri C (1995) Climatic fluctuations, current variability and marine species distribution: a case study in the Ligurian Sea (north-west Mediterranean). Oceanol Acta 18 (2): 139-149

Balestra V, Boero F, Carli A (1976) Andamento del pescato della tonnarella di Camogli dal 1950 al 1974: valutazioni bio-statistiche. Boll Pesca Piscic Idrobiol 31 (1-2): 105-115

Barry JP, Dayton PK (1991) Physical heterogeneity and the organization of marine communities. In: Kolasa J, Pickett STA (eds) Ecological heterogeneity. Ecological Studies, 86. Springer, Berlin Heidelberg New York Tokyo, pp 270-320

Barry JP, Baxter CH, Sagarin RD, Gilman SE (1995) Climate-related, long-term faunal changes in a California rocky intertidal community. Science 267: 672-675

Bethoux JP, Gentili B, Raunet J, Taillez D (1990) Warming trend in the western Mediterranean deep water. Nature 347: 660-662

Bianchi CN (1997) Climate change and biological response in the marine benthos. In: Piccazzo M (ed) Atti 12° Congr Assoc Ital Oceanol Limnol, vol 1. AIOL, Genova, pp 3-20

Bianchi CN, Morri C (1993) Range extension of warm-water species in the northern Mediterranean: evidence for climatic fluctuations? Porcupine Newsl 5 (7): 156-159

Bianchi CN, Morri C (1994) Southern species in the Ligurian Sea (northern Mediterranean): new records and a review. Boll Ist Mus Biol Univ Genova 58-59 (1992-1993): 181-197

Bianchi CN, Peirano A (1995) Atlante delle fanerogame marine della Liguria: *Posidonia oceanica* e *Cymodocea nodosa*. ENEA, Cen Ric Amb Mar, La Spezia

Bianchi CN, Morri C, Peirano A, Romeo G, Tunesi L (1987) Bibliografia ecotipologica sul Mar Ligure: elenco preliminare. ENEA, Roma

Bianchi CN, Morri C, Sartoni GF, Wirtz P (1998a) Sublittoral epibenthic communities around Funchal (Ilha da Madeira, NE Atlantic). Boll Mus Mun Funchal Suppl 5: 59-80

Bianchi CN, Boero F, Fonda Umani S, Morri C, Vacchi M (1998b) Successione e cambiamento negli ecosistemi marini. Biol Mar Mediterr 5 (1): 117-135

Blandin P (1987) Évolution des écosystèmes et spéciation: le rôle des cycles climatiques. Bull Ecol 18 (1): 59-61

Boero F (1996) Episodic events: their relevance to ecology and evolution. PSZN I: Mar Ecol 17 (1-3): 237-250

Boudouresque CF (1971) Méthodes d'étude qualitative et quantitative du benthos (en particulier du phytobenthos). Tethys 3 (1): 79-104

Bourcier M (1996) Long-term changes (1954 to 1982) in the benthic macrofauna under the combined effects of anthropogenic and climatic action (example of one Mediterranean bay). Oceanol Acta 19 (1): 67-78

Bruschi A, Sgorbini S (1986) Banche dati ambientali: idrologia del Mar Mediterraneo. Acqua Aria 6: 565-578

Buhl-Mortensen L (1996) Type-II statistical errors in environmental science and the precautionary principle. Mar Pollut Bull 32 (7): 528-531

Cattaneo-Vietti R (ed) (1985) La pesca in Liguria. Centro Studi Unioncamere Liguri, Genova

Cattaneo-Vietti R, Sirigu AP, Tommei A (1988) Sea of Liguria, 2nd edn. Res Cen Union Ligurian Chamber Commerce, Genoa

Cocito S, Sgorbini S, Bianchi CN (1998) Aspects of the biology of the bryozoan *Pentapora fascialis* in the north-western Mediterranean. Mar Biol 131 (1): 73-82

De Biasi AM (1994) Variabilità interannuale in una comunità sabulicola del Mar Ligure Orientale. In: Albertelli G, Cattaneo-Vietti R, Piccazzo M (eds) Atti 10° Congr Assoc Ital Oceanol Limnol. AIOL, Genova, pp 555-562

De Biasi AM (1997) Stability in a fine sand community: seven years of observations. In: Piccazzo M (ed) Atti 12° Congr Assoc Ital Oceanol Limnol. AIOL, Genova, pp 255-263

Deubelius H (1997) Mediterranean and Atlantic fish guide. IKAN, Frankfurt

Francour P, Boudouresque CF, Harmelin JG, Harmelin-Vivien ML, Quignard JP (1994) Are the Mediterranean waters becoming warmer? Information from biological indicators. Mar Pollut Bull 28 (9): 523-526

Góngora Gonzáles E, Immordino F, Peirano A, Stoppelli N (1996) Granulometric and geomorphologic features of the Bay of Monterosso al Mare (SP) and their relationship with the evolution of *Posidonia oceanica* meadow. In: Albertelli G, De Maio A, Piccazzo M (eds) Atti 11° Congr Assoc Ital Oceanol Limnol. AIOL, Genova, pp 395-404

Gray JS (1996) Environmental science and a precautionary approach revisited. Mar Pollut Bull 32 (7): 532-534

Levin SA (1992) The problem of pattern and scale in ecology. Ecology 73 (6): 1943-1967

Lewis J (1996) Coastal benthos and global warming: strategies and problems. Mar Pollut Bull 32 (10): 698-700

Mann KH, Lazier JRN (1996) Dynamics of marine ecosystems: biological-physical interactions in the oceans, 2nd edn. Blackwell Science, Cambridge, Mass

Morri C, Peirano A, Bianchi CN, Rodolfo-Metalpa R (2000) On-going studies on the Mediterranean coral *Cladocora caespitosa*. Reef Encounter 27: 22-25

Parker RO, Dixon RL (1998) Changes in a North Carolina reef fish community after 15 years of intense fishing - global warming implications. Trans Am Fish Soc 127 (6): 908-920

Peirano A, Bianchi CN (1997) Decline of the sea grass *Posidonia oceanica* in response to environmental disturbance: a simulation-like approach off Liguria (NW Mediterranean Sea). In: Hawkins LE, Hutchinson S, Jensen S, Williams AC, Sheader M (eds) Responses of marine organisms to their environment. Univ Southampton, UK, pp 87-95

Peirano A, Sassarini M (1992) Analisi delle caratteristiche distributive di alcune *facies* di substrato duro dei fondali delle Cinque Terre (Mar Ligure). Oebalia 17 Suppl: 523-528

Peirano A, Morri C, Mastronuzzi G, Bianchi CN (1998) The coral *Cladocora caespitosa* (Anthozoa, Scleractinia) as a bioherm builder in the Mediterranean Sea. Mem Descr Carta Geol Ital 52 (1994): 59-74

Peirano A, Morri C, Bianchi CN (1999a) Skeleton growth and density pattern of the zooxanthellate scleractinian *Cladocora caespitosa* (L) from the Ligurian Sea (NW Mediterranean). Mar Ecol Prog Ser 185: 195-201

Peirano A, Savini D, Bianchi CN, Farina G (1999b) Long-term monitoring on the *Posidonia oceanica* meadow of Monterosso al Mare (NW Mediterranean): first results. Atti Assoc Ital Oceanol Limnol 13 (1):129-135

Peirano A, Salvati E, Bianchi CN, Morri C (2000) Long-term change in the subtidal epibenthic assemblages of Punta Mesco (Ligurian Sea, Italy) as assessed through underwater photography. Porcupine Mar Nat Soc Newsl 5: 9-12

Pérès JM (1984) La régression des herbiers à *Posidonia oceanica*. In: Boudouresque CF, Jeudy de Grissac A, Olivier J (eds) Int Workshop *Posidonia oceanica* beds, vol 1. GIS Posidonie, Marseilles, pp 445-454

Piazzi L, Acunto S, Cinelli F (1999) *In situ* survival and development of *Posidonia oceanica* (L.) Delile seedlings. Aquat Bot 63: 103-112

Relini G (1992) Impoverishment and protection of Italian marine fauna. Boll Mus Ist Biol Univ Genova 56-57 (1990-1991): 29-52

Rossi L (1965) Il coralligeno di Punta Mesco (La Spezia). Ann Mus Civ Stn Natl Genova 75: 144-180

Rossi L (1969) Considerazioni zoogeografiche sul Bacino NW del Mediterraneo, con particolare riguardo al Mar Ligure. Arch Bot Biogeogr Ital Ser XLV-4ª 14 (4): 139-152

Roughgarden J, Gaines SD, Pacala SW (1987) Supply side ecology: the role of physical transport processes. In: Gee JHR, Giller PS (eds) Organization of communities: past and present. Blackwell, Oxford, pp 491-528

Sarà M (1985) Ecological factors and their biogeographic consequences in the Mediterranean ecosystems. In: Moraitou-Apostolopoulou M, Kiortsis V (eds) Mediterranean marine ecosystems. Plenum Press, New York, pp 1-17

Southward AJ, Hawkins SJ, Burrows MT (1995) Seventy years' observations of changes in distribution and abundance of zooplankton and intertidal organisms in the western English Channel in relation to rising sea temperature. J Therm Biol 20 (1/2): 127-155

Sparnocchia S, Manzella GMR, La Violette PE (1994) The interannual and seasonal variability of the MAW and LIW core properties in the Western Mediterranean Sea. Coast Estuarine Stud 46: 177-194

Strauss SY (1991) Indirect effects in community ecology: their definition, study and importance. Trends Ecol Evol 6 (7): 206-210

Torricelli L, Peirano A (1997) Produzione primaria fogliare della prateria di *Posidonia oceanica* (L) Delile di Monterosso al Mare (SP): variazioni lungo un gradiente batimetrico. Biol Mar Mediterr 4 (1): 81-85

Tortonese E (1958) Bionomia marina della regione costiera fra Punta della Chiappa e Portofino (Riviera ligure di levante). Arch Oceanogr Limnol 11 (2): 167-210

Tortonese E (1961) Nuovo contributo alla conoscenza del benthos della scogliera ligure. Arch Oceanogr Limnol 12 (2): 163-183

Vacchi M, Sara G, Morri C, Modena M, La Mesa G, Guidetti P, Bianchi CN (2000) Dynamics of marine populations and climate change: lessons from a Mediterranean fish. Porcupine Mar Nat Hist Soc Newsl 3: 13-17

Veron JEN (1995) Corals in space and time: the biogeography and evolution of the scleractinia. Univ New South Wales, Sidney

Vignudelli S, Gasparini GP, Astraldi M, Schiano ME (1999) A possible influence of the North Atlantic Oscillation on the circulation of the Western Mediterranean Sea. Geophys Res Lett 26 (5): 623-626

CHAPTER 50

Sexual Reproduction and Recruitment in *Posidonia Oceanica* (L.) Delile, a Genetic Diversity Study

L. Orsini[1], S. Acunto[2], L. Piazzi[2], and G. Procaccini[1]

ABSTRACT

The aim of our analysis was to provide evidence of the contribution that sexual reproduction has for increasing genetic diversity in *P. oceanica* meadows. The genetic diversity of three established populations and three adjacent seedling patches located along the Tuscany coast were surveyed. Seedling patches had higher values of heterozygosity and number of distinct genotypes but comparable number of alleles in respect to adult meadows, as ascertained using microsatellite marker. Assuming that seedlings are the result of sexual reproduction, high levels of inbreeding and self-fertilization can account for the slight increase of genetic variability. Our results also indicate that recruitment can occur from propagules originating outside the local population, limiting population genetic isolation at the given spatial scale.

Introduction

The genetic structure of seagrass populations is the result of sexual reproduction, clonal growth and gene flow among localities (Les 1988; Procaccini and Mazzella 1998). Sexual reproduction plays the role of introduction and inheritance of genetic variation, ensuring adequate genetic plasticity for survival and evolution of populations in changing environments (Les 1988).

Posidonia oceanica is an endemic Mediterranean seagrass, that forms widespread meadows along the coasts of the whole basin. The *P. oceanica* meadows are composed by clonal patches (Caye and Meinesz 1992; Procaccini and Mazzella 1998), the expansion of which would depend upon the past and/or recent history of successful sexual reproductive events and recruitment in a given locality. Nonetheless, sexual reproduction is sporadic and the overall low genetic diversity recently found in Western Mediterranean populations of such species (Procaccini et al. 1996; Procaccini and Mazzella 1996, 1998) suggests that clonal propagation is the main reproductive mode for maintaining and expanding established meadows. Climatic conditions are changing in the Mediterranean basin, and man-induced modifications of the coastline are determining habitat loss and meadow regression especially in highly urbanized areas. The capability of *P. oceanica* to adapt to new and changing environmental conditions through the adaptation of new and well fit genotypes is therefore important to ascertain.

Sexual reproduction is a sporadic event in *P. oceanica* meadows and few records report in situ germinating seeds along the coasts of Italy (e.g. Meinesz et al. 1993; Gambi et al. 1996; Piazzi et al. 1996). Posidonia is a monoecious genus with populations composed by a mosaic of clonal patches of different sizes. Because of that, fecundation might occur between male and female flowers belonging to the same shoot, to the same clone or to genetically similar individuals. The level of inbreeding, in fact, appears to be extremely high in natural populations (Procaccini and Mazzella 1998).

In this paper we wish to describe the levels of genetic diversity observed as a result of sexual reproduction. Microsatellite DNA markers were surveyed to establish population genetic variation in three well established populations and in three adjacent patches of germinating seeds (seedlings) located along the Tuscany coast. The mature meadows can be the result of either clon-

---

[1] Stazione Zoologica "A. Dohrn", Laboratorio di Ecologia del Benthos, 80077 Ischia (Napoli), Italy
[2] Dipartimento di Scienze dell'Uomo e dell'Ambiente, Università di Pisa, 56126 Pisa, Italy

F.M. Faranda, L. Guglielmo, G. Spezie (eds)
Mediterranean Ecosystems: Structures and Processes

al growth or sexual reproduction, while seedlings can only originate by sexual reproduction. By assessing gene flow among different localities we established the dispersal range of seedlings through the relative genetic contribution of the different adult meadows to the newly formed populations.

Materials and Methods

Study area and sampling strategy

Posidonia oceanica (L.) Delile individual shoots were sampled in three localities (Vada, Meloria and Gulf of Follonica) along the coasts of Tuscany (Italy), both in well established populations (Vada A, Meloria A and Follonica A) and in extensive patches of seedlings adjacent to the main meadow (Vada S, Meloria S and Follonica S). All seedling populations were distributed at 5-8 m depth, while the adult populations were sampled both at 7 and 15 m depth. Almost 40 shoots were collected for each population and DNA was extracted in CTAB buffer (Procaccini et al. 1996; Procaccini and Mazzella 1998).

Microsatellite analysis

Six polymorphic loci, one chloroplastic (*Poc-trn*) and five nuclear (*Poc*-5, *Poc*-26, *Poc*-35, *Poc*-42, *Poc*-45), previously identified from a *Posidonia oceanica* genomic library (Procaccini and Waycott 1998), were amplified in the 219 individual shoots sampled (for detailed technique see Procaccini and Waycott 1998).

Statistical analysis

Genetic parameters were calculated considering only one ramet of each genotype per population to avoid an artificial increase in *N* (Procaccini and Mazzella 1998). Weir and Cockerham's (1984) estimators of the level of inbreeding, *f* and *F*, were obtained using FSTAT (version 1.2, Goudet 1995) computer package. Standard deviations were calculated jackknifing over loci and over populations. Observed and expected heterozygosity levels, allele frequency, number of polymorphic loci per population were calculated using GENEPOP 1.2 (Raymond and Rousset 1995) computer package. The *Poc-trn* locus of the chloroplastic DNA was not included in analysis of the heterozygosity and related statistics, since the chloroplast is maternally inherited in angiosperms. Genotypic diversity in different populations was calculated by the G/N ratio, where G is the number of distinct genotypes and N is the total number of samples. In order to estimate levels of gene flow among populations, we calculated two statistical indices: ϑ (Weir and Cockerham 1984) and *Rho* (Goodman 1997). The significance of ϑ and Rho were assessed by a permutation test and over 1000 bootstrap replicates, respectively. The number of migrants per generation was then calculated for $\vartheta(Nm^{\vartheta})$ and for Rho (Nm^{Rho}). $\delta\mu^2$ was applied to calculate genetic distances (Goldstein et al. 1995). ϑ was calculated with FSTAT (Goudet 1995); *Rho* and $\delta\mu^2$ were calculated with RST Calc package (Goodman 1997). The UPGMA analysis was performed on $\delta\mu^2$ genetic distance.

Results

Five of the six microsatellite loci screened for this study were polymorphic in the six populations analysed. The monomorphic locus (*Poc*-5) was considered in the analysis to allow comparison with published data. Sixteen alleles were present in total (Table 1). The distribution of allele frequencies were similar in all populations, with high frequency alleles characterizing the electrophoretic profile of each locus. Five private alleles, three in seedling populations and two in adult populations, were present in total (Table 1). Private alleles, when present, were detected in a maximum of two individuals over the total number of samples screened for each population.

In the three localities, expected heterozygosity values were higher in the seedling populations (average value= 0.245) in respect to the mature ones (average value= 0.290) (Table 2). Values of *f* and *F*, calculated jackknifing over populations were positive and significantly different from zero, indicating high homozygosity both within the single populations and in the whole set of samples (Table 3). Among the three more polymorphic loci, *Poc*-35 gave the higher values of homozygosity ($f = 0.741$, SD= 0.118, F= 0.731, SD= 0.121).

Of the total 219 ramets sampled in the six populations, less than 50% had distinct geno-

Table 1. Allele frequencies for the six microsatellite loci analysed. For sequence and size of microsatellite regions see Procaccini and Waycott (1998). The monomorphic *Poc*-5 locus has been included in the analysis to allow comparison with published data (Procaccini and Mazzella 1998). Private alleles are in *italic*/**bold**

Locus and alleles	Meloria A	Meloria S	Vada A	Vada S	Follonica A	Follonica S
Poc-5						
.176	0.000	0.000	0.000	0.000	0.000	0.000
.173	1.000	1.000	1.000	1.000	1.000	1.000
Poc-26						
.289	0.000	0.000	0.000	***0.059***	0.000	0.000
.282	1.000	1.000	1.000	0.941	1.000	1.000
.275	0.000	0.000	0.000	0.000	0.000	0.000
Poc-35						
.206	0.000	0.000	0.000	0.000	0.000	***0.048***
.200	0.643	0.600	0.545	0.618	0.706	0.643
.197	0.036	0.000	0.000	0.000	0.000	0.048
.194	0.286	0.400	0.455	0.382	0.294	0.262
.191	0.000	0.000	0.000	0.000	0.000	0.000
.185	***0.036***	0.000	0.000	0.000	0.000	0.000
Poc-42						
.176	0.107	0.350	0.318	0.294	0.206	0.375
.170	0.893	0.650	0.682	0.706	0.794	0.625
Poc-45						
.168	0.679	0.450	0.091	0.265	0.471	0.525
.165	0.000	0.000	***0.045***	0.000	0.000	0.000
.144	0.321	0.550	0.864	0.735	0.529	0.475
Poc-trn						
.409	0.571	0.450	0.182	0.442	0.412	0.750
.401	0.000	***0.050***	0.000	0.000	0.000	0.000
.395	0.429	0.500	0.818	0.558	0.588	0.250

types, some of which were shared by more than one population. The G/N values for individual populations (expression of the percentage of new genotypes within a population) ranged from 0.275 in Vada A (11 genotypes) to 0.60 in Vada S (17 genotypes; Fig. 1). The number of unique genotypes varied in the six populations (Fig. 1), with the higher values in Follonica S.

The mean value of ϑ and Rho calculated for all loci were positive and significantly different from zero, showing the existence of low but significant genetic differentiation among the six populations (Table 3). The existence of gene flow was evidentiated by the two estimates of the mean number of migrants per generation (Nm^{ϑ} and Nm^{Rho}) that were higher than 1, considered

Table 2. Genetic variation averaged over six microsatellite loci in the six *Posidonia oceanica* populations analysed

Population	Sample size	# distinct genotypes	Mean number of allels per locus	Percentage of polymorphic loci	Expected heterozygosity*
Meloria A	36	14	2.0 ± 0.5	0.6	0.234 ± 0.110
Meloria G	40	20	1.6 ± 0.2	0.6	0.293 ± 0.120
Vada A	40	11	1.8 ± 0.4	0.6	0.246 ± 0.109
Vada G	28	17	1.8 ± 0.2	0.8	0.286 ± 0.096
Follonica A	39	17	1.6 ± 0.2	0.6	0.256 ± 0.108
Folloniga G	36	20	1.8 ± 0.4	0.6	0.294 ± 0.120

* for nuclear loci only

Table 3. Estimates of *f*, *F*, for nuclear loci and of ϑ, Rho, Nm^{ϑ} and Nm^{Rho} for all loci in the six *Posidonia oceanica* populations

f (std. dev.)	*F* (std. dev.)	θ (std. dev.)	*Rho* (SE)	Nm^{θ}	Nm^{Rho}
0.490 (0.154)	0.496 (0.129)	0.036 (0.031)	0.243 (0.004)	6.69	1.812

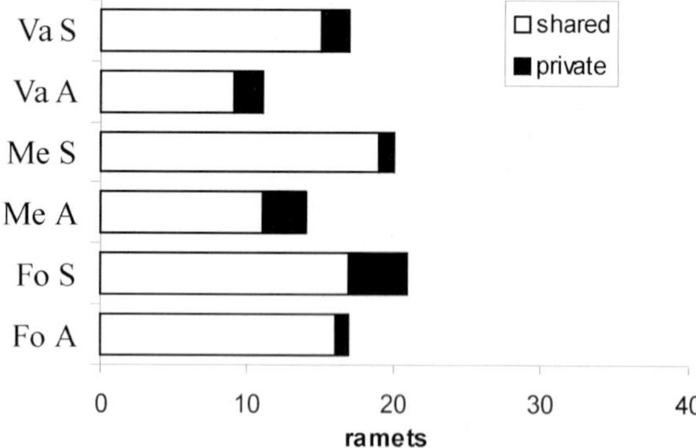

Fig. 1. Total number of genotypes in the six populations over the total number of sampled ramets. The number of genotypes is always slightly higher in seedlings (*S*) compared to adult populations (*A*)

to be the threshold value for population differentiation (Table 3). The UPGMA cluster of the $\delta\mu^2$ distance shows adult and seedling populations of Vada grouped in a separate branch and isolated from the other two localities. Adult and seedling populations of Follonica and Meloria clustered without accordance with the relative geographic position (Fig. 2).

Discussion

The goal of our paper was to investigate if sexual reproduction results in an increase of genetic diversity in natural populations of *P. oceanica*. Average values of the statistical parameters calculated over young and well established populations indicate that genetic variability is slightly higher in seedlings than in adult plants. Seedlings are the result of sexual reproduction and are supposed to be distinct from each other and from the parental plants. In spite of this, common genotypes are found in mature plants and in seedlings, suggesting that sexual reproduction occurs between genetically similar individuals or even clonemates. The values of observed heterozygosity are always lower than expected heterozygosity, suggesting high levels of inbreeding. The mean value of *f* indicates a 49% overall excess of homozygosity supporting the notion of inbreeding within population. We could hypothesize that *P. oceanica* meadows are structured by fragmented clonal patches of different size resulting by a few initial genotypes spreading within the meadow by asexual reproduction (Caye and Meinesz 1992; Procaccini and Mazzella 1998). To date, it is not clear if the possibility of self-fertilization exists in this species. A mixed mating model has been suggested for the congeneric *P. australis*, with considerable variation in outcrossing rates reflecting the low uniformity of the water pollination system (Waycott and Sampson 1997).

The water transport ensures also seedling dispersion. Our results suggest that recruitment can be allochtonous in *P. oceanica*. The cluster analysis, in fact, shows that seedlings do not always group with the adjacent mature populations. Only in Vada recruitment seems to be autocthonous and can be related to the local hydrodynamic regime that does not favour high rates of dispersion from the parental meadow.

In conclusion, the observation of slightly higher levels of genetic diversity in seedling patches in respect to mature meadows of *P. oceanica* indicates that sexual processes could offer a long term solution to maintain genetic

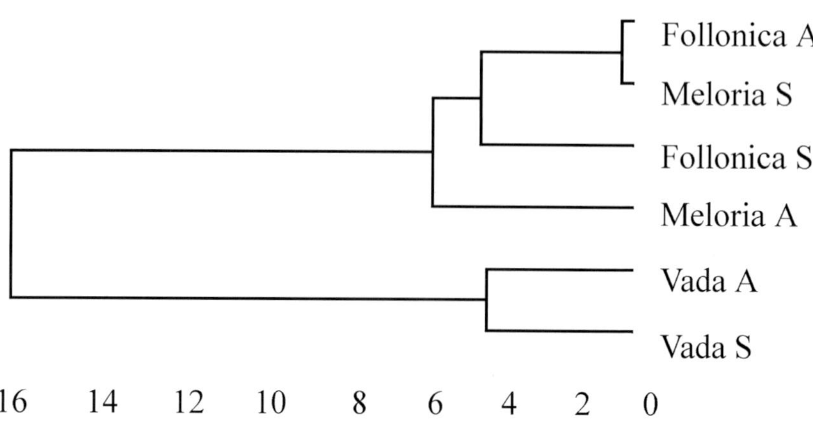

Fig. 2. UPGMA dendrogram of the six populations based on the $\delta\mu^2$ distance values (Goldstein et al. 1995)

diversity in this species. Moreover, allocthonous seedling recruitment in two of the three localities suggests that habitat fragmentation is not resulting in complete genetic isolation of populations. However, the genetic erosion observed in this important species, also related to high levels of inbreeding and possibly self-fertilization, can result in a low capacity to deal with the more rapid anthropogenic and climatic changes that are under way in the Mediterranean basin.

Acknowledgements. The research was partially supported by "C.N.R. Target Project on Biotechnology".

References

Caye G, Meinesz A (1992) Analyse des modalités de la multiplication végétative et de la reproduction sexuée de *Posidonia oceanica* et de ses conséquences sur la constitution génétique des herbiers. Rapp Comm Int Mer Méditerr 33:32

Gambi MC, Buia MC, Mazzella L (1996) Record of a diffuse germination of *Posidonia oceanica* (L.) Delile in the central Adriatic Sea (Croatia). Biol Mar Mediterr 3(1): 467-470

Goldstein DB, Ruiz Linares A, Cavalli-Sforza LL, Feldman MW (1995) Genetic absolute dating based on microsatellites and the origin of the modern humans. Proc Natl Acad Sci USA 92: 6723-6727

Goodman SJ (1997) RST CALC: a collection of computer programmes for calculating unbiased estimates of genetic differentiation and determining their significance for microsatellite data. Mol Ecol 6: 881-885

Goudet J (1995) Fstat version 1.2: a computer programme to calculate F-statistics. J Hered 86: 485-486

Les DH (1988) Breeding systems, population structure and evolution in hydrophilous angiosperms. Ann Missouri Bot Gard 75: 819-835

Meinesz A, Caye G, Locques F, Molenaar H (1993) Polymorphism and development of *Posidonia oceanica* transplanted from different parts of the Mediterranean into the National Park of Port-Cros. Bot Mar 36:209-216

Piazzi L, Acunto S, Balestri E, Cinelli F (1996) Osservazioni preliminari sulla germinazione di semi e sviluppo in situ di piantine di *Posidonia oceanica* (L.) Delile. Inf Bot Ital 28: 61-66

Procaccini G, Alberte RS, Mazzella L (1996) Genetic structure of the seagrass *Posidonia oceanica* in the Western Mediterranean: ecological implications. Mar Ecol Prog Ser 140: 153-160

Procaccini G, Mazzella L (1996) Genetic variability and reproduction in two Mediterranean seagrasses. In: Kuo J, Phillips RC, Walker DI, Kirkman H (eds) Seagrass biology. Proc Int Workshop, Sciences UWA, Nedlands, Western Australia, pp 85-92

Procaccini G, Mazzella L (1998) Population genetic structure and gene flow in the *seagrass Posidonia oceanica* (L.) assessed using a microsatellite analysis. Mar Ecol Prog Ser 169: 133-141

Procaccini G, Waycott M (1998) Microsatellite loci identified in the seagrass *Posidonia oceanica* (L) Delile. J Hered 89(6): 562-568

Raymond M, Rousset F (1995) GENEPOP (version 1.2): a population genetics software for exact tests and ecumenism. J Hered 86: 248-249

Waycott M, Sampson J (1997) The mating system of an hydrophilous angiosperm Posidonia australis (Posidoniaceae). Am J Bot 84(5): 621-625

Weir BS, Cockerham CC (1984) Estimating F-statistics for the analysis of population structure. Evolution 38(6): 1358-1360

CHAPTER 51

Biodiversity at Ecogenetic Level in Three Species of Beach Fleas

L. Zane, E. Angelini, N. Longo, S. Marcato, and P.M. Bisol

ABSTRACT

Patterns of fine-scale niche diversification were investigated in three co-generic species of littoral amphipods. Population samples of *Orchestia montagui*, *O. gammarella* and *O. mediterranea* were collected from the same sampling site, in the Venice Lagoon, and analysed by means of protein electrophoresis and RAPD techniques. The three species exhibited significant differences at the protein and the nuclear DNA level. Specific markers were identified that allow the correct species assignment of females, that are morphologically indistinguishable. Two species, *O. montagui*, that occurs under vegetal debris, and *O. mediterranea*, found on muddy infra-littoral substrates, were further investigated with regard to the resistance to high temperature and to glucose-phosphate isomerase (GPI) biochemical activity. The two species were found to have different resistance to high temperatures, with *O. montagui* showing the higher survival rate. The thermal stability of the common GPI genotype and the biochemical activity of this enzyme was higher in *O. montagui*. Our results strongly suggest that a relationship exists between habitat preference and metabolic properties that can explain the competition avoidance in these animals.

Introduction

The preservation of biological diversity has become an increasingly important issue in recent years. The number of endangered species is rapidly growing, due to several factors including habitat loss, introduction of allocthonous species, overexploitation, and pollution. Several reasons have been proposed to justify the preservation of biodiversity including the economic value of bioresources, ecosystem services, aesthetics and the right of living organisms to exist (reviewed in Frankham 1995; OECD 1996).

The World Conservation Union (IUCN) now recognizes the need to maintain biodiversity at three levels: genetic diversity, species diversity, and ecosystem diversity (World Conservation Monitoring Centre 1992). These indications result from the importance of preserving the different functional units together with the habitat in which they live.

Under this regard, the preservation of functionality of lagoons is of particular importance due to their high economical and ecological value. Among others, these habitats often sustain anthropic activities with high productivity like local fisheries and mollusc harvest industries, and are an important feeding ground for migratory birds.

Moreover, like other transition zones, lagoons are particularly important because they represent natural laboratories for the study of adaptive processes (Battaglia and Bisol 1988). Due to the strong temporal variation in environmental parameters such as salinity and temperature, organisms that inhabit these areas show peculiar specializations that enable them to tackle strong selective pressures.

The present study deals with the assessment of genetic and functional diversity in three talitrid species, *Orchestia montagui*, *O. gammarella* and *O. mediterranea* from a littoral lagoon area. Talitrids are, with isopods, the only crustaceans which were able to colonize a terrestrial habitat; this colonization is thought to date to the upper Cretaceous (70 million years ago), and was accompanied by an extensive radiation, which led talitrids to fill all the ecological niches they

Dipartimento di Biologia, Università degli Studi di Padova, Via Ugo Bassi 58/B, 35121 Padova, Italy

F.M. Faranda, L. Guglielmo, G. Spezie (eds)
Mediterranean Ecosystems: Structures and Processes

inhabit today. The remarkable adaptive success of these animals is testified by the 200 species described to date, with an overall estimate of about 1000 existent species (Bousfield 1983).

Special adaptations of terrestrial talitrids include acquisition of a waxy exoskeleton (Moore and Francis 1985), reduction in the gill area (Moore and Taylor 1984), acquisition of a highly specialized hemocyanin (Marsden 1984; Taylor and Spicer 1986) and the evolution of a refined osmoregulative system (Morrit 1988). Like other amphipods, the life cycle of talitrids is characterized by internal fecundation and formation of about twenty lecitotrophic eggs; eggs then undergo direct development in specialized pockets, and subadult forms are finally released. These characteristics are noteworthy because they enable reproduction to occur also in absence of water.

The whole of these adaptations however is not enough to qualify talitrids as fully terrestrial organisms, and they are, differently from Isopods, confined to high humidity habitats (Wildish 1988). Talitrids, in fact, are commonly found in sandy littoral habitats all around the world (Bousfield 1984). To cope with daily variation of environmental parameters, talitrids evolved a migratory behaviour, synchronized with tidal and nictemeral cycles, that enable them to locate the suitable patches of habitat on the shore (Pardi and Ercolini 1986).

Sandy littoral habitats are highly fragmented, and are ephemeral, because modification of coast line is a relatively fast process. To be a stable component of the fauna of these habitats means to possess a strong ability to recolonize suitable habitats and to disperse. Typically, at our latitudes, talitrid populations experience, during the spring-summer period, an exponential growth characterized by repeated cycles of reproduction, starting from the small number of individuals that survived to the previous winter season (Louis 1977). Their ability to recolonize a given area is enhanced from the fact that broods are kept by females in specialized pockets; thus rafting of ovigerous females provides an effective way to offspring dispersal (Wildish 1988).

Talitrid species are morphologically extremely conserved. Females of *Orchestia montagui*, *O. gammarella* and *O. mediterranea*, the three species under investigation, are morphologically indistinguishable. The study of these species is therefore difficult when they occur in sympatric conditions, such as in the Venice lagoon.

In the present study we use DNA and protein markers to develop a molecular tool for the classification of these species. In addition we aim at understand whether the lack of morphological differentiation is parallelled by low divergence at the genetic level. We also examine survival ability at high temperatures, in relation to peculiar adaptations at the biochemical level, to obtain a better characterization of the species on a functional basis. With this regard we investigate GPI biochemical activity, because this enzyme seems to be under strong selection in many organisms. Differential fitness associated with different GPI genotypes has been reported for pollution resistance in marine gastropods (Lavie and Nevo 1982), for resistance to low oxygen concentration in isopods (Shihab and Heath 1987), and for resistance to prolonged flight in *Colias* butterflies (Watt 1992). The latter study is of particular interest as superiority of GPI genotypes was temperature dependent. With regard to amphipods, in *Gammarus insensibilis*, following exposure to high temperature, a higher survival rate was reported for individuals heterozygous at the GPI locus, possibly due to a higher specific activity of the enzyme in heterozygous condition (Patarnello et al. 1989).

Materials and Methods

Population samples of *Orchestia montagui*, *O. gammarella* and *O. mediterranea* were collected in the Venice Lagoon (North-Eastern Italy). Classification of these species is based on the morphology of the sexually dimorphic second gnathopod (gn2), that is a diagnostic character in males only (Bellan Santini et al. 1993). Animals were kept alive until arrival in the laboratory, where they were either deep frozen (-40 °C) or used for survival experiments. For RAPD-PCR analysis, the frozen specimens were stored in individual vials containing ethanol.

Allozyme analysis of 12 different loci was performed on about 50 individuals per species (Table 1). For each specimen, a small piece of tail muscle was mechanically dissolved in 200 µl of extraction buffer (Tris-HCl 0.02 M, 0.25% Bromophenol blue, pH 8); one microlitre of the solution, clarified by centrifugation at 13,000 RPM, was then subjected to electrophoretic

Table 1. Locus name, EC number and acronyms of the protein loci studied

Enzyme	E.C. number	Acronym
Alkaline phosphatase	3.1.3.1	ALP
Arginine kinase	2.7.3.3	AK
Malic enzyme	1.1.1.40	ME
Esterase	3.1.1.1	ES
Glycerol-3-phosphate dehydrogenase	1.1.1.72	GPD
Glutamate oxaloacetate transaminase	2.6.1.1	GOT
Glucose phosphate isomerase	5.3.1.9	GPI
Glutamate pyruvate transaminase	2.6.1.2	GPT
Isocitrate dehydrogenase	1.1.1.42	ICD
Malate dehydrogenase	1.1.1.37	MDH
Mannose phosphate isomerase	5.3.1.8	MPI
Piruvate kinase	2.7.1.40	PK

migration on cellulose-acetate gel, except for ALPH, ES and GOT-2 that were resolved on starch gels. Hystochemical staining was conducted following Grunbaun (1981), with slight modifications (Biasiolo et al. 1989).

In order to estimate genetic variability, the percentage of polymorphic loci (*P*), the average number of alleles, and the frequency of the observed (h_{obs}) and expected (h_{exp}) heterozygotes were calculated. BIOSYS computer package (Swofford and Selander 1989) was used to test for deviation from Hardy-Weinberg equilibrium and to compute Nei's genetic distance between species (Nei 1978).

The RAPD-PCR analysis (Williams et al. 1990) was performed on three individuals for each species, with 20 random decamer primers (Kit. A, Operon Technologies). The DNA was extracted from tail muscle with a standard proteinase K-phenol-chloroform protocol (Sambrook et al. 1989). Following ethanol precipitation, DNA was resuspended in 100 µl of TE (Tris 10 mM, EDTA 1 mM, pH 8), incubated for 1 h at 37°C with 10 µg of RNAase A (Castiglione et al. 1993), and re-extracted as before. Finally, purified DNA was dissolved in water and quantified by "spot test" against a standard containing a known amount of DNA (Sambrook et al. 1989). One microliter of a diluted DNA solution containing 2-4 ng of DNA, was used in 25 µl of standard PCR reaction mix containing one single decamer primer (Williams et al. 1990). Amplification was performed on a Perkin Elmer Cetus Thermal Cycler, with an initial denaturation step of 3' at 94°C, followed by 40 cycles of 10" at 94°C, of 1' at 35°C, and 1' at 72°C. Amplification products were separated by size on 1.5% agarose gel in TBE 1X (Tris 0.1 M, boric acid 0.083 M, EDTA 0.001 M, pH 8), containing 0.5 µg/ml of ethidium bromide, visualized under UV light (at 302 nm of wavelength) and photographed with a Polaroid camera.

Survival rates were determined for 90 individuals of *Orchestia montagui* and 90 of *O. mediterranea*, after incubation in a thermostatic room with saturated humidity, for 30' at 45°C.

Measurement of GPI activity was carried out for animals that were homozygous for the most common allele in each species. Measures were performed by means of a spectrophotometer, using the *Phosphoexose isomerase* kit (Sigma). The GPI specific activity and residual activity after incubation at 47°C were estimated for animals kept at 6°C and saturated humidity, conditions for which survival in laboratory is maximized. Moreover, GPI activity at 20°C was estimated for the homozygous animals that were still alive after the survival experiment.

Results and Discussion

Allozyme analysis yielded reliable results when applied to the three species studied. No overall deviation from Hardy Weinberg expectations was observed. The percentage of polymorphic loci and the observed heterozygosity were slightly different in the three species (Table 2).

The majority of alleles were shared between at least two of the species studied (Table 3). However species showed differences in allelic frequencies at many loci, and three loci (namely GPT, MPI, and PK) bore species diagnostic alleles.

Values of Nei genetic distance between species indicate the presence of genetic separation, yielding values of distance ranging from 0.3 to 0.6 for all pairwise comparisons (Fig. 1). These values are higher than those reported for allozymic differentiation between species in a variety of animal *phyla* (Nei 1987). The importance of such a differentiation is strengthened if we consider that allopatric species of talitrids, belonging to the different genera *Talitrus* and *Talorchestia*, were previously found to have genetic distances in the range of 0.6-0.8 (De Matthaeis et al. 1994). Therefore allozyme data indicate that *Orchestia montagui*, *O. gammarella* and *O. mediterranea*, despite the extreme simi-

Table 2. Allelic frequencies at the different gene loci assayed

Locus	Alleles	*O. montagui*		*O. mediterranea*		*O. gammarella*	
ALP	A[a]	*38*[b]	0.171[c]	*52*	0.442	*47*	0.511
	B		0.829		0.558		0.489
AK	A	*44*	0.000	*56*	1.000	*54*	1.000
	B		1.000		0.000		0.000
ME	A	*52*	1.000	*63*	0.000	*54*	1.000
	B		0.000		0.008		0.000
	D		0.000		0.992		0.000
ES	A	*48*	0.479	*63*	0.071	*53*	0.462
	B		0.000		0.889		0.538
	C		0.521		0.040		0.000
GPD	A	*52*	1.000	*66*	1.000	*54*	1.000
GOT-1	A	*22*	0.000	*51*	1.000	*30*	1.000
	B		1.000		0.000		0.000
GOT-2	A	*48*	1.000	*63*	1.000	*54*	1.000
GPI	A	*52*	0.942	*66*	0.000	*54*	0.000
	C		0.029		0.000		0.000
	D		0.029		0.969		0.888
	F		0.000		0.023		0.009
	H		0.000		0.000		0.029
	I		0.000		0.008		0.074
GPT	B	*52*	0.000	*63*	1.000	*51*	0.000
	C		1.000		0.000		0.000
	D		0.000		0.000		0.889
	F		0.000		0.000		0.009
	H		0.000		0.000		0.028
	I		0.000		0.000		0.074
ICD-1	A	*50*	0.000	*65*	1.000	*54*	1.000
	C		1.000		0.000		0.000
ICD-2	B	*43*	1.000	*66*	1.000	*54*	1.000
MDH-1	A	*45*	1.000	*49*	1.000	*54*	1.000
MDH-2	A	*52*	0.029	*54*	0.000	*54*	0.000
	B		0.971		1.000		1.000
MPI	A	*52*	1.000	*63*	0.000	*53*	0.000
	C		0.000		0.833		0.000
	D		0.000		0.000		0.991
	F		0.000		0.167		0.000
	G		0.000		0.000		0.009
PK	A	*8*	0.000	*56*	0.009	*51*	0.000
	C		0.000		0.000		1.000
	D		0.000		0.991		0.000
	E		1.000		0.000		0.000

[a] A represent the fastest migrating allele, B the second slowest and so on for all talitrid species analysed in our laboratory

[b] Number of individuals analysed

[c] Allelic frequencies

Table 3. Genetic variability of the three species studied

	% P[a]	n. of alleles [b]	H_{obs} [c]	H_{exp} [d]
Orchestia gammarella	26.7	1.4	0.089	0.0082
Orchestia mediterranea	40.0	1.5	0.070	0.072
Orchestia montagui	23.5	1.3	0.036	0.056

[a] Percentage of polymorphic loci with 95% criterion
[b] Average number of alleles scored
[c] Observed heterozygosity
[d] Expected heterozygosity under the Hardy-Weinberg equilibrium hypothesis

larity in morphology and the sympatric occurrence, represent highly differentiated evolutionary units.

High genetic divergence between species was confirmed by RAPD-PCR analysis. In fact, RAPD patterns were highly species-specific (Fig. 2) for all the primers used. Not one of the about 130 RAPD markers amplified in each species was reliably scored in the other two. Moreover, some monomorphic bands were scorable with high confidence, due to the lack of any co-migrating

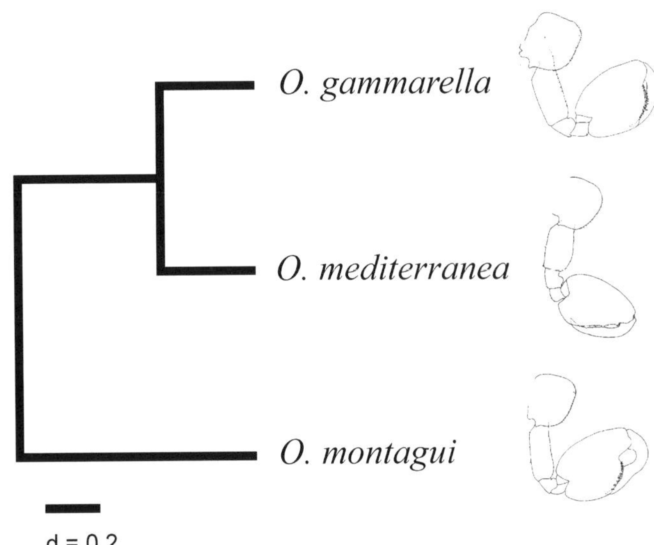

Fig. 1. Nei genetic distance between the three *Orchestia* species studied, and diagnostic Gnathod of the male

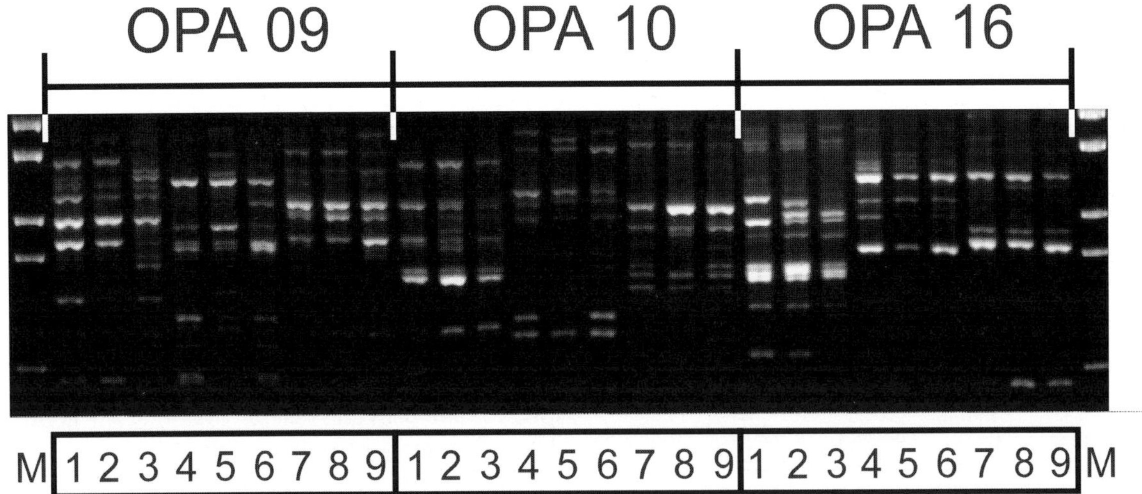

Fig. 2. Examples of RAPD phenotypes with three different decamer primers (OPA09 - OPA10 - OPA16) in *Orchestia montagui* (1,2,3), *O. mediterranea* (4,5,6) and *O. gammarella* (7,8,9). (*M*, Molecular marker)

band that could be a homologous gene, in the other two species.

Lack of shared bands between species provides compelling evidence against the existence of hybridization among the three species, thus indicating that *Orchestia montagui*, *O. gammarella* and *O. mediterranea* have evolved some mechanism of reproductive isolation. An orthodox explanation for this situation (high degree of genetic differentiation, no introgression and present day sympatric occurrence) is that the three species diverged in the past in allopatric conditions, and that the present day distribution is the result of secondary overlapping of species ranges of distribution.

Result of allozyme and RAPD analyses are interesting as they provide an useful tool to reliably classify females, which are otherwise indistinguishable from a morphological point of view. Protein electrophoresis of GPT, MPI or PK, or one PCR reaction with any one of the 20 decamer primers used in this study can correctly assign females to one of the three species. The RAPD analysis in particular allows to work with ethanol-preserved specimens, thus providing an useful field tool. The degree of confidence of a RAPD "molecular classification" is high in the populations analysed, due to the fact that each primer amplified several independent species-specific bands. The confidence can however be increased by selection of primers that produce an easily scorable pattern and that amplify the highest number of bands specific for the assignment of interest (Table 4). However, the reliability of our method needs to be confirmed by more extensive investigation on the degree of polymorphisms, studying other populations from different geographic areas.

We are then faced with the problem of understanding how can these species maintain their sympatric condition, that is to understand

Table 4. Selected set of primers that produced the better species-specific pattern

	Primers[a] (bands)[b]		
Orchestia gammarella	OPA09(3)	OPA10(3)	OPA16(3)
Orchestia mediterranea	OPA09(3)	OPA10(3)	OPA16(3)
Orchestia montagui	OPA09(2)	OPA10(4)	OPA16(4)

[a] Primer designation according to the Operon Technologies catalogue
[b] Number of easily scorable species specific bands

if there is some mechanism preventing competition, or if the distribution that we observe is the result of a dynamic interaction between species.

The three species seem to have slightly different habitat preferences: *O. montagui* occurs under vegetal debris, whereas *O. mediterranea* is found on muddy infra-littoral substrates. The *O. gammarella* instead occurs in a less defined habitat, in the driest supralittoral area. The question is then whether this distribution is the result of displacement between competing species, or it is due to habitat choice. Our sampling activity provides evidence of the fact that at least *O. mediterranea* and *O. montagui* cannot occupy the other species habitat: also if one of the two species were absent from one area, the other was found still confined to the "optimal" habitat of occurrence. This is specially noteworthy if we consider that talitrid species occur on infra-littoral substrates and regulate their position according to tidal cycles, undergoing extensive horizontal migrations (Pardi and Ercolini 1986).

Then the choice of the optimal habitat is performed actively, day by day, and with high specificity by these organisms.

We conducted further investigations on the two species that showed the strongest habitat preference, *O. montagui* and *O. mediterranea*, to test for differential resistance to high temperatures in a high humidity environment. This experimental setup was chosen on the basis of a preliminary survey of *O. mediterranea* that suggested a weak ability to resist dehydration, even in the range of temperatures normally experienced by these animals. Survival was in fact abolished after 65' incubation at 25°C and 45% r.h.; this result is somehow expected, because talitrids exoskeleton is known to have a substantial permeability to water (Morritt 1987), and indicates that *O. mediterranea* can survive only by actively avoiding dry habitats. We then focused our attention on the effect of temperature on survival, incubating the animals for 30' at 45°C with saturated humidity, and measuring the survival rate; this value of temperature is extremely high and it is not in the range normally experienced by the species, but provides useful information about the resistance to extreme temperatures.

Results are reported in Table 5. *Orchestia montagui* have a higher resistance to high temperatures (86.7%) than *O. mediterranea* (53.3%). This result indicates that the width of the niche

Table 5. Survival and GPI activity

		O. mediterranea	*O. montagui*
% survival[a]		53.3	86.7
GPI specific activity at 20°C [b]			
Control		1.52±0.31	2.12±0.21
Survived animals		1.71±0.27	2.76±0.40
GPI residual activity [c]			
After incubation at 47°C for:	10'	58	91
	15'	44	85
	20'	34	83

[a] Of animals that survived after 30' at 45°C
[b] Expressed as units/mg of total protein
[c] Expressed as percentage of activity respect to untreated enzyme

of each species is substantially different with respect to temperature and suggests that the two species are distributed on the patchy habitat according to their specific survival skills.

The GPI activity was different between species when measured in control animals, with *O. montagui* showing the highest specific activity. Successive analysis of surviving animals showed that incubation at high temperature determined a differential increase in enzyme activity.

Moreover, analysis of residual activity after heating at 47°C showed that *O. montagui* GPI have an higher thermal stability than that of *O. mediterranea*.

Data on GPI activity indicate that *O. montagui* possesses a GPI enzyme which is more stable and active at high temperatures. Metabolic processes are fastened by incubation at high temperature, and the cost to sustain this increment must then be lower in *O. montagui*, that lost less of its GPI, than in *O. mediterranea*. Regardless of whether GPI itself is involved in resistance at high temperatures or is simply an indicator of a better adaptation to thermal stresses, the differential survival in these two species could be explained by differential costs associated with the response to environmental factors.

Acknowledgements. We gratefully thank Prof. S. Ruffo for classification of the samples studied, and Prof. B. Battaglia for useful discussion on the topic. This manuscript was improved by comments and suggestions of L. Bargelloni. This paper is dedicated to the memory of Angelo Alfini. Without his help and his friendship this work would never have been possible.

References

Battaglia B, Bisol PM (1988) Environmental factors, genetic differentiation, and adaptive strategies in marine animals. In: Rotschild BJ (ed) Toward a theory on biological-physical interactions in the world ocean. Kluwer Academic Publishers, New York, pp 393-410

Bellan-Santini D, Karaman G, Krapp-Schickel G, Ledoyer M, Ruffo S (1993). The Amphipoda of the Mediterranean. Part 3. Gammaridea (Meliipidae to Talitridae), Ingolfiellidea, Caprellidea. In: Ruffo S (ed) The Amphipoda of the Mediterranean. Mem Inst Oceanogr Monaco 13, Principauté de Monaco, pp 742-753

Biasiolo A, Bisol PM, Battaglia B (1989) Indagine sui polimorfismi proteici in Anfipodi del genere *Orchestia*. I. *Orchestia mediterranea* A. Costa della laguna di Venezia. Rend Accad Naz Lincei 73: 299-305

Bousfield EL (1983). An updated phyletic classification and paleohistory of the Amphipoda. In: Schram FR (ed) Crustacean phylogeny. Balkema, Rotterdam, pp 257-277

Bousfield EL (1984) Recent advances in the systematics and biogeography of landhoppers (Amphipoda: Talitridae) of the Indo-Pacific Region. In: Radovsky FJ, Raven PH, Sohmer SH (eds) Biogeography of the tropical Pacific. Bishop Mus Spec Publ 72: 171-210

Castiglione S, Wang G, Damiani G, Bandi C, Bisoffi S, Sala F (1993) RAPD fingerprints for identification and for taxonomic studies of elite poplar (*Populus* spp.) clones. Theor Appl Gene 87: 54-59

De Matthaeis E, Cobolli M, Mattoccia M, Saccoccio P, Scapini F (1994) Genetic divergence between natural populations of Mediterranean sandhoppers (Crustacea, Amphipoda) In: Beaumont AR (ed) *Genetics and evolution of aquatic organisms*. Chapmann and Hall, London, pp 15-29

Frankham R (1995) Conservation genetics. Annu Rev Gene 29: 305-327

Grunbaun BW (1981) Handbook for forensic individualization of human blood and bloodstains. Sartorius GmbH, Göttingen, Germany

Lavie B, Nevo E (1982) Heavy metal selection of phosphoglucose isomerase allozymes in marine gastropods. Mar Biol 71: 17-22

Louis M (1977) Etude des populations de Talitridae des étangs littoraux Mediterranéens. I. Identification des cohortes, cycles et fécondité. Bull Ecol 8: 75-86

Marsden ID (1984) Effects of submersion on the oxygen consumption of the estuarine sandhopper *Transorchestia chiliensis* (Milne-Edwards 1840). J Exp Mar Biol Ecol 79: 263-276

Moore PG, Francis CH (1985) On the water relations and osmoregulation of the beach-hopper *Orchestia gammarellus* (Pallas) (Crustacea: Amphipoda). J Exp Mar Biol Ecol 94: 131-150

Moore PG, Taylor AC (1984) Gill area relationships in an ecological series of gammaridean amphipods (Crustacea). J Exp Mar Biol Ecol 74: 179-186

Morritt D (1987) Evaporative water loss under desiccation stress in semiterrestrial and terrestrial amphipods (Crustacea: Amphipoda: Talitridae). J Exp Mar Biol Ecol 111: 145-157

Morritt D (1988) Osmoregulation in littoral and terrestrial talitroidean amphipods (Crustacea) from Britain. J Exp Mar Biol Ecol 123: 77-94

Nei M (1978) Estimation of average heterozigosity and genetic distance from a small number of individuals. Genetics 89: 583-590

Nei M (1987) Molecular evolutionary genetics. Columbia University Press, New York

OECD (1996) Saving biological diversity. (Economic incentives) Paris

Pardi L, Ercolini A (1986) Zonal recovery mechanisms in talitrid crustaceans. Boll Zool 53: 139-160

Patarnello T, Bisol PM, Battaglia B (1989) Studies on differential fitness of GPI genotypes with regard to temperature in *Gammarus insensibilis* (Crustacea: Amphipoda). Mar Biol 102: 355-359

Sambrook J, Fritsch EF, Maniatis T (1989) Molecular cloning: a laboratory manual. Cold Spring Harbor Lab Press, Cold Spring Harbor, New York

Shihab AF, Heath DJ (1987) Components of fitness and the GPI polymorphism in the freshwater isopod *Asellus aquaticus* (L.). 2. Zygotic selection. Heredity 58: 289-295

Swofford DL, Selander RB (1989) BIOSYS-1: a computer programme for the analysis of allelic variation in population genetics and biochemical systematics. Ill Nat Hist Surv, Ill, USA

Taylor AC, Spicer JI (1986) Oxigen-transporting properties of the blood of two semi-terrestrial amphipods, *Orchestia gammarellus* (Pallas) and *O. mediterranea* (Costa). J Exp Mar Biol Ecol 97: 135-150

Watt WB (1992) Eggs, enzymes, and evolution-natural genetic variants change insect fecundity. Proc Natl Acad Sci USA 89: 10608-10612

Wildish DJ (1988) Ecology and natural history of aquatic Talitroidea. Can J Zool 66: 2340-2359

Williams JFK, Kubelik AR, Livak KJ, Rafalsky JA, Tingey SV (1990) DNA polymorphisms amplified by arbitrary primers are useful as genetic markers. Nucleic Acids Res 18: 6531-6535

World Conservation Monitoring Centre (1992) Global biodiversity: status of the Earth living resources. Chapman and Hall, London

Functional Diversity in the *Posidonia oceanica* Ecosystem: an Example with Polychaete Borers of the Scales

M.C. Gambi and G. Cafiero

ABSTRACT

Species composition and spatio-temporal distribution of polychaete borers in *Posidonia oceanica* (L.) Delile scales, were studied in 4 stations located in two meadows off the Island of Ischia (Gulf of Naples) with different typology and at different depths: Castello at 1 m and 3-5 m depth, Lacco Ameno at 3-5 m and 22 m depth. Five species, all belonging to the family Eunicidae, were found as borers of scales: *Lysidice collaris* (Grube) (59% of the specimens collected), *Lysidice ninetta* Audouin and Milne-Edwards (32%), *Nematonereis unicornis* Schmarda (8%), *Palola siciliensis* (Grube) and *Marphysa fallax* Marion and Bobretzky with only 2 specimens each. The spatio-temporal analysis showed the occurrence of polychaete borers (mainly the two species of *Lysidice*) at all stations and in all sampling months, except at the shallowest site (1 m) at the Castello meadow. The mean values of rhizomes affected by borers ranged from 19-21% at the shallow sites of both studied meadows (3-5 m) to 29% at the deep site (22 m) of the Lacco meadow. Traces of boring have been also frequently observed (37%) in the dead rhizomes, where specimens of 4 of the 5 recorded species were also found with a frequency of 23%. The mean number of individuals/m^2 was 66±22 and 84±25 for the shallow sites (Castello and Lacco, respectively), and 49±15 for the deep site at Lacco. The temporal analysis did not show any particular pattern in borer colonisation, but relatively homogeneous abundances in all studied months. Boring of living leaf bases (with detachment of the entire living shoot), and of the adjacent leaf tissues were sometime observed. The phenomenon of consumption of living tissues was very limited in the shallow meadows (Castello and Lacco 3-5 m, mean index of herbivory= 0.6% and 0.8% of the analysed rhizomes, respectively), while it was relatively more important in the deep site (Lacco 22 m, mean index of herbivory= 2.6%). The scales of *P. oceanica* represent an unique and preferential microhabitat for some boring polychaetes, whose colonisation resulted to be frequent, common to different meadows and depths and constant in time. The feeding and mechanical action of polychaete borers have strong ecological implications in enhancing scale fragmentation and decay by microbial activity, and in providing further microniches for other organisms (e.g. small polychaetes, sipunculids) that colonise the empty burrows. Last but not least the direct grazing on living plant tissues may have a direct impact on the system.

Introduction

The high animal biodiversity recorded in the *Posidonia oceanica* (L.) Delile ecosystem is related to the availability of microhabitats which are provided by the high spatial complexity of the plant and of the system (Mazzella et al. 1992). For the fauna associated to *P. oceanica* meadows it has been possible to recognize specific adaptations to microhabitats of the leaf and rhizome layers and of the "matte". The knowledge on the structural and functional diversity of each of these layers and microhabitats is different. While the vagile fauna of the leaf stratum is relatively well known, also from a trophic point of view (Gambi et al. 1992), the fauna inhabiting the rhizomes is far less studied (Somaschini et al. 1994), except for echinoderms and other large detritivores (Mazzella et al. 1992).

Within the compartment of the rhizomes,

Stazione Zoologica "A. Dohrn", Laboratorio di Ecologia del Benthos, Punta S. Pietro, 80077 Ischia (Napoli), Italy

F.M. Faranda, L. Guglielmo, G. Spezie (eds)
Mediterranean Ecosystems: Structures and Processes

the scales (remains of former leaf bases persisting along the rhizome) represent an important part of the detritic compartment of the *Posidonia* system (Romero et al. 1992), and a unique microhabitat, mainly colonized by microorganisms and fungi. Recently we have described some boring polychaetes and crustacean isopods, which burrow inside of *P. oceanica* scales (Gambi et al. 1997; Guidetti et al. 1997), and that represent the only Metazoans known to exploit this peculiar microhabitat. While isopod borers were previously observed in Australian seagrass systems (Brearley and Walker 1993, 1995, 1996), for polychaetes the boring habit has been reported so far only for the *P. oceanica* seagrass system (Guidetti et al. 1997; Gambi 2000). Polychaete and isopod borers may occur in the same bed and on some occasions in the same shoot, but they prefer scales of different age and structure. Polychaetes colonise older and aged scales, while isopods prefer younger scales (Guidetti et al. 1997). Finally, polychaetes occurred in most of the investigated *P. oceanica* beds, while isopod borers showed a more limited distribution (Gambi et al. 1997). However, our previous observations were spatially scattered along the Italian coasts (Gambi et al. 1997) and limited to short time of observation (Guidetti et al. 1997; Guidetti 2000).

The aims of this study were: verify species composition and distribution of scale boring polychaetes in *Posidonia* meadows with different typology and at different depths; verify the occurrence of these organisms in time; verify the occurrence of the boring species also in dead rhizomes, and confirm their preference for aged scale.

A preliminary evaluation of the feeding activity of the borer species, as well as of their role in scale fragmentation and of their ecological implications for the seagrass system has been reported in a different study (Gambi et al. 2000).

Study Sites – Material and Methods

The study has been conducted in two *P. oceanica* beds located on the northern side of the island of Ischia (Gulf of Naples): Castello Aragonese and Lacco Ameno (Cafiero 1998). The two meadows, where a previous survey revealed the occurrence of borer polychaetes (Gambi et al. 1997), were selected according to their different typology and shoot densities. The Castello is a relatively limited and sheltered meadow, ranging from 0.5 m to 6 m depth in front of the rocky promontory of Castello Aragonese. The *P. oceanica* meadow at Lacco is quite a large and continuous bed ranging from 1 m to about 30 m depth and has been intensively studied in the past for the meadow dynamics (Buia et al. 1992) as well as for the associated animal communities (Gambi et al. 1992).

Sampling of orthotropous rhizomes, from both living and dead shoots, was carried out by SCUBA diving in four stations at different depths: 1 m and between 3 and 5 m at the Castello meadow; between 3 and 5 m, and 22 m at the Lacco meadow. Stations were established on the basis of bottom topography, meadow features and shoot density. Samples were collected monthly, from October 1996 to October 1997, except in the deep station off Lacco (22 m) where rhizomes were collected on a seasonal basis. A total of 40 living rhizomes were collected in each station and at each month/season of sampling, except at Castello 1 m depth, where only 20 rhizomes were collected in order to limit the impact on this spatially limited area. However, due to the high shoot density of the bed in this zone (mean = 919 shoot/m^2), the number of analysed rhizomes is representative of the borer colonisation. To estimate the frequency and abundance of borers, various "indices" have been calculated: index of borers (IB= percentage of the number of rhizomes hosting borers over the total rhizomes analysed); index of traces (IT= % of the number of rhizomes with only empty traces of previous colonisation of borers – sinuous galleries on the scale mesophyll, holes and piercing grooves on the scale epidermis – over the total rhizomes analysed); index of colonisation (IC= IB + IT); index of herbivory (IH= % of the number of rhizomes with living tissues bored over the total rhizomes analysed) (Cafiero 1998; Gambi 2000).

Abundance of borers was calculated to the square metre multiplying the number of individuals found in each sample by the shoot density at that given station, and dividing by the number of rhizomes analysed in that sample.

In order to evaluate the microdistribution of the borers along the rhizomes, the scales of the rhizomes collected at all stations in July were separated into different groups according to their increasing age: the first scale; group 1= the group going from the second scale to that one with the lowest thickness; group 2= the group following the previous one up to the next scale with the lowest thickness; group 3= the remaining scales up to the end of the rhizomes. These groups may roughly represent different years, according to the lepidochronology technique (Pergent 1990). Borers found in the first scale and in each group have been checked.

Results and Discussion

A total of 1503 living rhizomes and 201 dead rhizomes were analysed. In total 326 specimens of boring polychaetes have been collected belonging to 5 species of Eunicidae: *Lysidice collaris* (Grube) (59% of the specimens sampled); *Lysidice ninetta* Audouin and Milne-Edwards (32%), *Nematonereis unicornis* Schmarda (8%); *Palola siciliensis* (Grube) (2 specimens); *Marphysa fallax* Marion and Bobretzky (2 specimens). While the two species of *Lysidice* and *N. unicornis* have been previously reported to bore on scales (Gambi et al. 1997; Guidetti et al. 1997), *P. siciliensis* and *M. fallax* are newly recorded as scale borers. Morphological description of these species can be found in George and Hartmann-Schroeder (1985), and in Martin (1987), while information on distribution and ecology can be found in Cantone (1993), as regards the Italian coasts.

Generally, in all the samples examined, in each affected shoot only a single polychaete borer was recorded. In a few cases two specimens were record in the same shoot, while only on a single occasion two specimens were observed inside of the same scale (Fig. 1a).

The spatio-temporal analysis of the index of borers (IB) showed the occurrence of polychaete borers (mainly the two species of *Lysidice*) at all stations and in all sampling months (Fig. 2a), except at the shallowest site (1 m) at the Castello meadow where polychaetes were occasional. The mean values of the IB ranged from 19-21% at the shallow sites of both studied meadows (3-5 m) to 29% at the deep site (22 m) of the Lacco meadow. Traces of previous colonisation of these organisms (index of traces, IT) were present in all stations and seasons, and contributed to the index of colonisation (IC) whose means ranged between 42 and 58%, indicating that the phenomenon of boring is relatively frequent.

The temporal analysis did not show any particular pattern in borer colonisation, but relatively homogeneous abundances in all months (Fig. 2b). The mean number of individuals/m^2 was 66±22 and 84±25 for the shallow sites (Castello and Lacco, respectively), and 49±15 for the deep site off Lacco. In the Castello meadow at

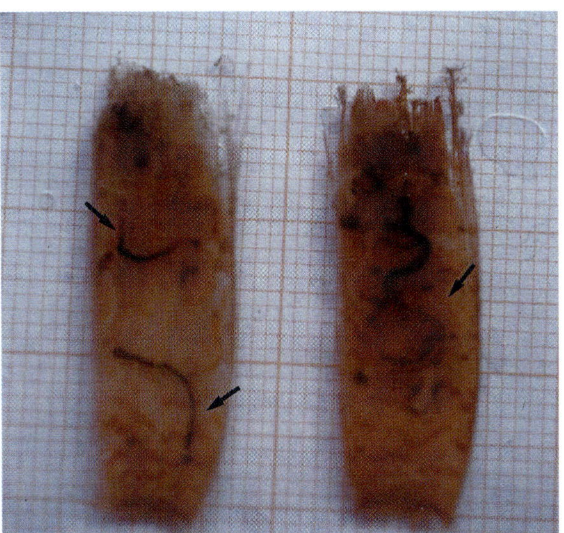
a

b

Fig. 1. a Two *Posidonia oceanica* scales with polychaete borers (*arrows*) inside; note that the *right* scales contain two specimens. **b** Living bases of *Posidonia* leaves with evident traces of grazing by polychaete borers (note the necrotic tissues around the grazed margins)

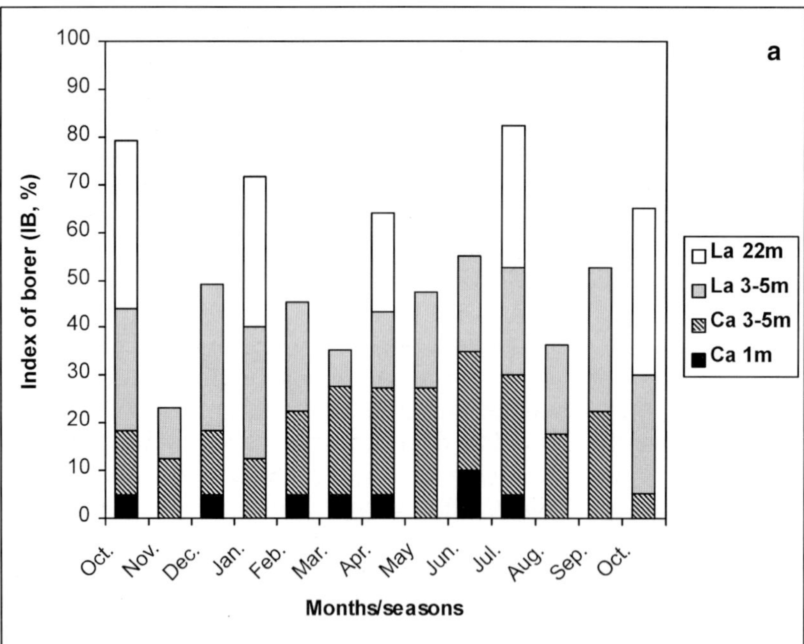

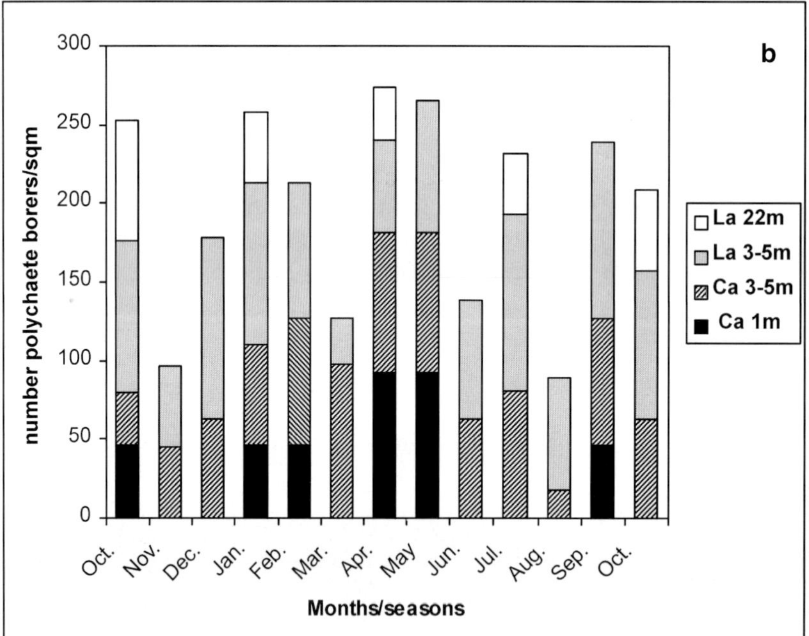

Fig. 2a,b. **a** Seasonal trend of the Index of borers (*IB*) in the studied stations. **b** Seasonal trend of the number of individuals/m^2 in the studied stations. (*La*, Lacco Ameno Meadow; *Ca*, Castello Meadow)

1 m depth, boring polychaetes, when present, were relatively abundant (e.g., in May and June) due to the high shoot density recorded at this site (mean = 919 shoots/m^2). At 3-5 m depth at Castello polychaete abundances ranged between 18 (in August) and 98 (in March) individuals/m^2, and showed a constant trend, increasing slightly from February to May. In the Lacco meadow at 3-5 m depth borer abundances were higher than at Castello at the same depth, and ranged between 29 (in March) and 115 (in December) individuals/m^2 (Fig. 2b). This may be due to the slightly higher shoot density recorded at Lacco (mean = 375 shoots/m^2) with respect to Castello (mean= 358 shoots/m^2) at the same depth (3-5 m). Finally at the deep stations off Lacco (22 m), despite the relatively high values of the IB (Fig. 2a), the borer abundances ranged from 34 (in April) to 69 (in October) individuals/m^2, probably related to the relatively low density of the meadow at this

depth (mean= 131 shoots/m^2). No other data on temporal colonisation of polychaete borers can be compared with these results. However, the temporal analysis pointed out a certain degree of variability in borer abundance and colonisation, and the opportunity to sample at least on a seasonal basis better to assess the incidence of the boring phenomenon in a specific *Posidonia* meadow.

Boring of living leaf bases and leaf meristem (with evident tissue necrosis and detachment of the entire living shoot), and of the adjacent leaf tissues were sometime observed (Fig. 1b). The phenomenon of consumption of the living tissues was very limited in the shallow stations (Castello and Lacco 3-5 m: mean index of herbivory= 0.6 and 0.8%, respectively), while it was relatively more important in the deep site (Lacco 22 m: mean index of herbivory= 2.6%) (Gambi et al. 2000). Direct herbivores of *P. oceanica* are quite rare (Mazzella et al. 1992), and in regard to polychaetes the only known species able to graze on living *Posidonia* tissue is the nereidid *Platynereis dumerilii* (Audouin and Milne-Edwards) (Gambi and Di Meglio 1996; Gambi et al. in press). However, the direct grazing of this nereidid was observed only under experimental laboratory conditions (Gambi and Di Meglio 1996). Thus, the two species of *Lysidice* borers of the scales, represent the first documented polychaetes that graze on *Posidonia* in natural conditions.

Traces of boring have also been frequently observed in the scales of dead rhizomes, where specimens of 4 of the 5 recorded species were also found (except *M. fallax*). Values of the Indices of borers (IB), of traces (IT), and of colonisation (IC) in the dead rhizomes were comparable to those (annual means) of the living ones (Fig. 3).

The distribution of the polychaete borers along the rhizomes (July samples) is shown in Fig. 4. The analysis revealed that polychaetes were occasional in the first scale (only one specimen of *L. ninetta* was found), while their numbers increased in the following groups, being maximum in the final group representing the most aged scales, especially for the species *L. collaris* and *N. unicornis* (Fig. 4). This pattern confirms a previous analysis performed on different *P. oceanica* meadows, where the first and the younger scales are generally selected by boring isopods (Guidetti et al. 1997).

On the whole these results demonstrate that

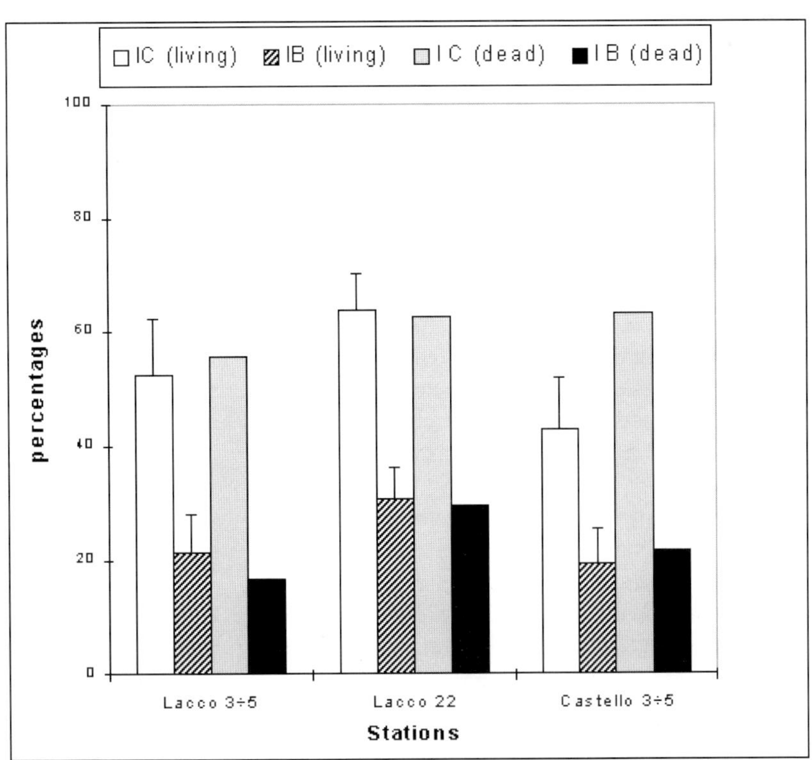

Fig. 3. Trends of the Index of borers (*IB*) and index of colonisation (*IC*) in dead and living rhizomes. Values of living rhizomes represent annual means, and their standard deviations, of the studied stations, while the values of the dead rhizome represent a single group of 201 rhizomes

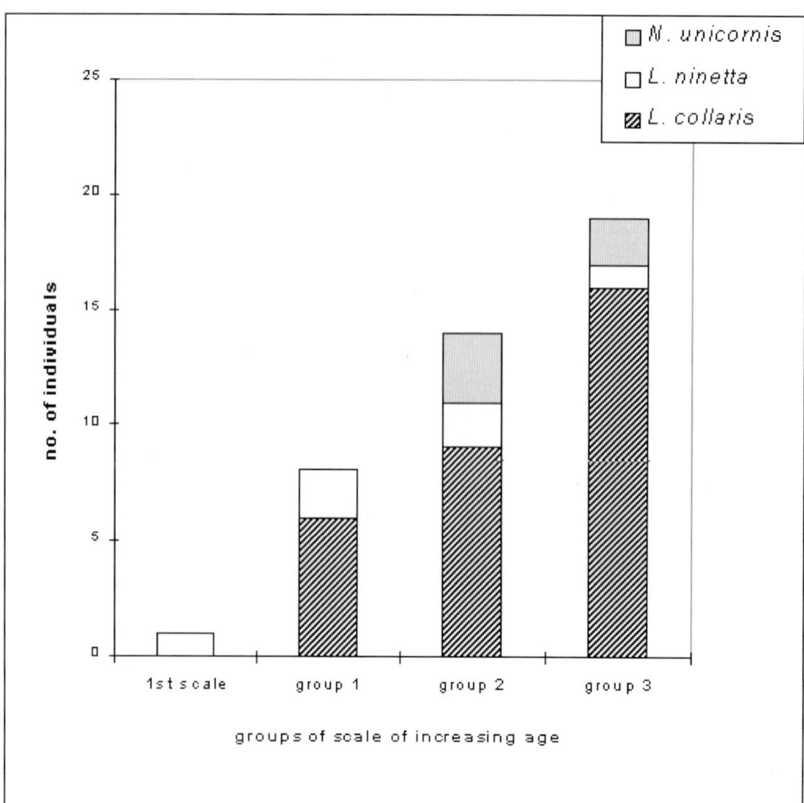

Fig. 4. Abundances of polychaete borers in groups of scales of different ages (rhizomes of all studied stations sampled in July)

Posidonia scales represent a unique and preferential microhabitat for some polychaetes, whose colonisation resulted to be frequent, common to different meadows and depths and relatively constant in time.

The role of boring polychaetes in the *Posidonia* system is multiple. The feeding and mechanical action of boring have strong ecological implications in enhancing scale fragmentation and decay by microbial activity. In fact, a consumption of single scale biomass, with means ranging from 10.5 to 11.6%, was estimated in a previous study (Gambi et al. 2000). The galleries formed by boring polychaetes provide often microniches for other organisms that colonize the empty burrows. Among them we have observed many other polychaetes, both sedentary (Terebellidae, Spionidae, Cirratulidae), and motile (Syllidae, Phyllodocidae, Nereididae, Glyceridae) forms, as well as sipunculids. Old galleries are often filled with fecal pellets or sediment.

A preliminary analysis of the gut content of the two species of *Lysidice* (Gambi et al. 2000) revealed the presence of high percentages of *Posidonia* tissues (both living and dead), as well as the occurrence of macroalgae and diatoms. This fact demonstrates that scale tissues, removed during boring, are effectively ingested and may in part be utilized as food by the worms. In the food-web of the *Posidonia* system, boring polychaetes represent a specialised group of plant detritivores, consuming a less exploited and "palatable" resource of the *Posidonia* system, but in some cases also a rare group of direct herbivores. In this respect, the grazing on living plant tissues, sometimes with necrosis phenomena and detachment of the living shoot, have a direct impact on the system; although the observations are still limited, this phenomenon seems more frequent in the deep stand, and may not be negligible for the meadow health in the long term.

Future studies will be focused on the evaluation of detritus consumption, and herbivore pressure at the system level, as well as on the ecophysiological adaptations of these organisms as regards to the digestion and assimilation of the highly refractory scale tissues.

Acknowledgements. We wish to thank Prof. R. Bargagli (University of Siena), tutor of the thesis of G. Cafiero, and Mr. M. Lorenti (Stazione Zoologica, Naples) for collaboration in sampling and data processing.

References

Brearley A, Walker DI (1993) Isopod borers in seagrass species from south-western Australia. In: Wells FE, Walker DI, Kirkman H, Lethbridge R (eds) The marine flora and fauna of Rottnest Island, Western Australia. Proc 5th Int Mar Biol Workshop, W Aust Mus, Perth, pp 415-428

Brearley A, Walker DI (1995) Isopod miners in the leaves of two Western Australian *Posidonia* species. Aquat Bot 52: 163-181

Brearley A, Walker DI (1996) Burrow structure and effects of burrowing isopods (Limnoriidae) in southwestern Australian *Posidonia* meadows. In: Kuo J, Phillips RC, Walker DI, Kirkman H (eds) Seagrass Biology. Proc Int Workshop, Rottnest Island, W Aust, 25-29 January 1996, Faculty of Science. The University of Western Australia, Nedlands, pp 261-268

Buia MC, Zupo V, Mazzella L (1992) Primary production and growth dynamics in *Posidonia oceanica*. PSZNI Mar Ecol 13: 2-16

Cafiero G (1998) Distribuzione spazio-temporale ed autoecologia di invertebrati perforatori di scaglie di *Posidonia oceanica* (L.) Delile (unpublished in Italian) Thesis, Univ Siena

Cantone G (1993) Censimento dei Policheti dei mari italiani: Eunicidae Berthold, 1927. Atti Soc Toscana Sci Nat Mem Ser B 100: 229-243

Gambi MC (2000) Polychaete borers of the Posidonia oceanica (L.) Delile scales: distribution patterns and ecological roles. Biol Mar Mediterr 7(2): 215-219

Gambi MC, Di Meglio MA (1996) Observations on feeding behaviour and food choice in the herbivore polychaete *Platynereis dumerilii* (Nereididae). In: Albertelli G, De Maio A, Piccazzo M (eds) Atti 11° Congr Assoc Ital Oceanol Limnol Sorrento, 24-26 ottobre 1994. Lang, Genova AIOL, pp 779-791

Gambi MC, Lorenti M, Bussotti S, Guidetti P (1997) Borers in *Posidonia oceanica* scales: taxonomical composition and occurrence. Biol Mar Mediterr 4: 384-387

Gambi MC, Lorenti M, Russo GF, Scipione MB, Zupo V (1992) Depth and seasonal distribution of some groups of the vagile fauna of *Posidonia oceanica* leaf-stratum: structural and trophic analyses. PSZNI Mar Ecol 13: 17-39

Gambi MC, Zupo V, Lorenti M (2000) Boring organisms of *Posidonia oceanica* scales: trophic role and ecological implications for the ecosystem. Biol Mar Mediterr 7(1): 253-261

Gambi MC, Zupo V, Buia MC, Mazzella L (2000) Feeding ecology of the polychaete *Platynereis dumerilii* (Audouin and Milne Edwards) (Nereididae) in the seagrass *Posidonia oceanica* system: role of the epiphytic flora. Ophelia (*in press*)

George JD, Hartmann-Schröder G (1985) Polychaetes: British Amphinomida, Spintherida and Eunicida. In: Kermack DM, Barnes RSK (eds), Linnean Soc London Estuarine Brackish Water Sci Assoc, Publ 32, EJ Brill and W Backhuiys, pp 221

Guidetti P (2000) Invertebrate borers in the Mediterranean sea grass *Posidonia oceanica*: biological impact and ecological implications. J Mar Biol Ass UK 80: 725-730

Guidetti P, Bussotti S, Gambi MC, Lorenti M (1997) Invertebrate borers in *Posidonia oceanica* scales: relationships between their distribution and lepidochronological parameters. Aquat Bot 58: 151-164

Martin D (1987) Anélidos Poliquetos asociados a las concreciones de algas calcàreas del litoral catalan. Misc Zool 11: 61-75

Mazzella L, Buia MC, Gambi MC, Lorenti M, Russo GF, Scipione MB, Zupo V (1992) Plant-animal trophic relationships in the *Posidonia oceanica* ecosystem of the Mediterranean Sea: a review. In: John DM, Hawkings SJ, Price JH (eds) Plant-animal interactions in the marine benthos. Syst Assoc Spec Vol 46, pp 165-187

Pergent G (1990) Lepidochronological analysis of the seagrass *Posidonia oceanica* (L.) Delile: a standardized approach. Aquat Bot 37: 39-54

Romero J, Pergent G, Pergent-Martini C, Mateo MA, Regnier C (1992) The detritic compartment in a *Posidonia oceanica* meadow: litter features, decomposition rates and mineral stocks. PSZNI Mar Ecol 13: 69-83

Somaschini A, Gravina F, Ardizzone GD (1994) Polychaete depth distribution in a *Posidonia oceanica* bed (rhizome and matte strata) and neighbouring soft and hard bottoms. PSZNI Mar Ecol 15: 133-151

Sediment Community Oxygen Consumption along a Shelf-slope Transect in the Western Gulf of Lions

P. Picon[1], A. Accornero[2], F. de Bovée[3], B. Charrière [1], and R. Buscail[1]

ABSTRACT

SCOC measurements were carried out on a seasonal basis during 1997 and 1998 along a transect from the shelf to the slope of the western Gulf of Lions (northwestern Mediterranean). Values ranged from 0.6 to 48.0 mmol O_2 m^{-2} d^{-1} and decreased with increasing depth. The overall mean was 11.2 ± 1.0 mmol O_2 m^{-2} d^{-1} for the shelf and 2.3 ± 1.9 mmol O_2 m^{-2} d^{-1} for the slope, with a significant difference between canyon axis (average 3.7 ± 2.1 mmol O_2 m^{-2} d^{-1}) and open slope sites (average 1.2 ± 0.8 mmol O_2 m^{-2} d^{-1}). The highest values were registered in all stations in late spring, when they were 1.5 to 4.2 times higher than in other seasons, except for the shelf deepest station, which showed a slight temporal variation. Temporal and spatial variations appeared to be related to the benthic biological activity and organic matter input. Our data confirm previous findings from other authors in the Gulf of Lions and highlight the importance of organic matter mineralization in this area in the frame of the processes occurring at the sediment-water interface.

Introduction

Despite the fact that they represent less than 20% of the world ocean area, continental margins have been demonstrated to play an important role in the marine biogeochemical carbon cycle (Walsh et al. 1981; Smith and MacKenzie 1987; Walsh 1991; Wollast 1991; Bauer and Druffel 1998). In this cycle the sediment-water interface can be considered as a key zone, because it represents the link between the carbon which is supplied from the water column, by downward particle fluxes, and the carbon which accumulates in marine deposits.

Once in the sediments, a part of this carbon undergoes deep burial and is thus substracted from the ocean-atmosphere system for periods ranging on geological time scales. Hence, the sediment-water interface constitutes a transitional zone between pelagic and sedimentary carbon pools. Once deposited at the seafloor, organic matter can follow different pathways, depending on its quality (e.g. labile molecules vs refractory compounds), the characteristics of the environment (e.g. temperature, reducing/oxidizing conditions) and the types and relative abundances of benthic organisms (heterotrophic bacteria, meio- and macrofauna). The extent and type of these transformations also depend on the rate at which organic matter is supplied, on the proportion of co-sedimenting inorganic material and on the presence of physical forcing factors (e.g. turbidity currents and resuspension events). Excellent review articles on the physical, chemical and biological processes which regulate the organic carbon cycling at the sediment-water interface can be found in Mantoura et al. (1991). In Fig. 1 fluxes of the different carbon forms across the sediment-water interface are schematized. The particulate organic carbon (POC) freshly deposited at the seafloor is subject to extensive degradation prior to burial: the largest portion of the metabolizable (i.e. not refractory) organic matter is mineralized by aerobic bacterial respiration in the oxic layer of surficial sediments, and thus partially returned to the water column (directly or via pore water) as dissolved inorganic carbon (DIC). Dissolved carbon can

[1] CEFREM, CNRS, Université de Perpignan, 52 avenue de Villeneuve, 66860 Perpignan cedex, France
[2] Istituto di Meteorologia ed Oceanografia, Istituto Universitario Navale, Via A. De Gasperi 5, 80133 Napoli, Italy
[3] Observatoire Océanologique de Banyuls-sur-mer, CNRS, Université de Paris VI, 66651 Banyuls-sur-mer cedex, France

F.M. Faranda, L. Guglielmo, G. Spezie (eds)
Mediterranean Ecosystems: Structures and Processes

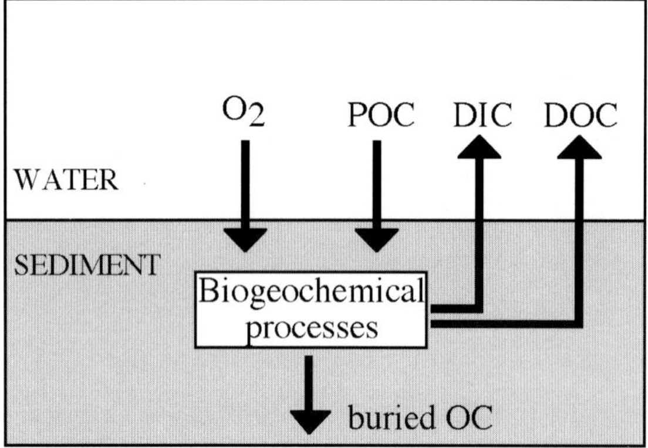

Fig. 1. Schematic representation of organic carbon pathways at the sediment-water interface

also diffuse from sediment to water in organic form (DOC), following the concentration gradient (Burdige and Homstead 1994).

The present work deals with the overall mineralization of organic matter at the sediment-water interface, by focusing on the sediment oxygen uptake rates measured by the use of benthic chambers. In the last decade the use of benthic chambers, together with the employment of other devices such as ROV (Remotely Operated Vehicles) and free respirometers, has greatly contributed to the assessment of material and energy fluxes through the benthic ecosystem, achieved by estimating the uptake of labelled substrata and/or the oxygen consumption.

The rate at which oxygen is taken up by sediments can also be defined as the sediment community oxygen consumption (SCOC), and is a parameter which exhibits a two-fold interest: from a biogeochemical point of view it allows to estimate the amount of organic carbon mineralized per unit of sediment surface (Canfield 1993; Henrichs 1992; Schulz et al. 1994), from a biological point of view it can give important information about the benthic community food energy demand (Deming and Baross 1993; Smith 1987).

Materials and Methods

This study was carried out as a part of the BBLL programme (which means "Banyuls Benthic Lander in the Gulf of Lions"), in the framework of the MATER and PNOC projects. The programme was focused upon studying the benthic biological activity and the chemical processes occurring at the sediment-water interface in the Gulf of Lions, in the northwestern Mediterranean. This area exhibits a narrow continental shelf, which continues as a steep slope incised by several submarine canyons (Got et al. 1985).

In the Gulf of Lions organic matter inputs come from different sources, including photosynthetic production, fluvial transport and sediment resuspension. The annual primary production in this area was reviewed by Lefevre et al. (1997): the shelf appears as the most productive zone, with values ranging from 236 to 389 mg C m^{-2} d^{-1}, while the slope is characterized by a lower productivity, between 213 and 290 mg C m^{-2} d^{-1}. The fluvial input is almost completely due to the Rhone river, which discharges 18 kT POC y^{-1} (Sempéré et al. 2000). Sediment resuspension and lateral advection have been proven to play a significant role in transferring organic particles to the slope of this area (Monaco et al. 1990; Monaco et al. 1999).

Five sampling sites (Fig. 2), located along a transect from the Bay of Banyuls to the Lacaze-Duthier Canyon, were selected. Station A (35 m depth) was the most inshore station, in the Bay of Banyuls, and Station B (87 m) was placed not far offshore, always within the limits of the continental shelf. Slope stations included Stn C (330 m) and D (785 m) on the open slopes of the Lacaze-Duthiers canyon, and Stn E (912 m) in the canyon axis. Sampling cruises were carried out on a seasonal basis during 1997 and 1998; values measured in June 1995 (EUROMARGE cruise) are also included in this study. In each cruise *in situ* processes were studied using a benthic lander and at the same time sediment samples were collected by a multicorer for the determination of sedimentary organic matter and benthic communities (microorganisms, meiofauna and macrofauna). The biological aspects of this study are illustrated in other works (de Bovée et al. 1998; Tholosan 1999; papers in preparations).

The SCOC was measured using an autonomous apparatus, the "Banyuls Benthic Lander", carrying four benthic chambers of cylindrical shape, each of which isolated a 177 cm^2 sediment surface area, overlaid by about 5 l of *in situ* bottom water. The confined water was kept gently mixed by a magnetic bar coupled to a motor located off the chamber, in order to prevent the

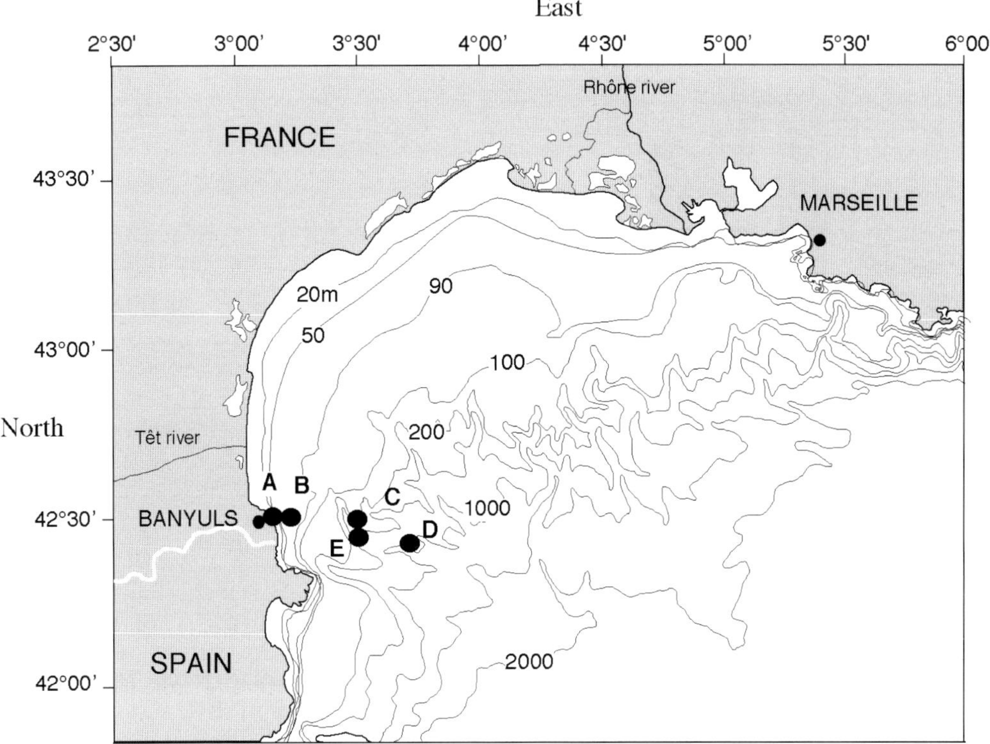

Fig. 2. Study area and locations of benthic chamber deployment

onset of concentration gradients. Further technical details about the lander can be found in Tenberg et al. (1995).

An oxygen probe continuously recorded the O_2 concentration in the overlying water inside the chamber. The O_2 flux was calculated from the slope of the linear decrease of its concentration in the water during the time of incubation. This latter ranged from 12 to 48 h, according to station location.

Results and Discussion

The SCOC values ranged from 0.6 to 48.0 mmol O_2 m^{-2} d^{-1}, and showed both spatial and seasonal variability (Fig. 3). The most coastal site (Stn A) was always characterized by the highest oxygen uptake rates, up to two orders of magnitude higher than at all other stations, independent of the sampling period. Values decreased with increasing depth, and a clear difference was found between Stn E, in the canyon axis, (average 3.7 ± 2.1 mmol O_2 m^{-2} d^{-1}) and Stn C and D, located on the Lacaze-Duthiers open slopes, (average of both stations: 1.2 ± 0.8 mmol O_2 m^{-2} d^{-1}). The overall mean was 11.2 ± 1.0 mmol O_2 m^{-2} d^{-1} for the shelf (Stn A and B, all sampling periods) and 2.3 ± 1.9 mmol O_2 m^{-2} d^{-1} for the slope (Stn C, D and E, all sampling periods). The highest values were found at each station in late spring, when SCOC values were 1.5 to 4.2 times higher than in February and August, except for Stn B, that showed a slight temporal variation.

Our results are consistent with previous measurements in the Gulf of Lions. Tahey et al. (1994), using on-board incubations, found average values of 6.8 ± 2.4 mmol O_2 m^{-2} d^{-1} on the shelf and 3.5 ± 1.3 mmol O_2 m^{-2} d^{-1} on the slope in November-December 1991. Helder (1989) calculated oxygen fluxes from pore water profiles in July 1987 and December 1988, resulting in a mean of 5.7 ± 2.6 mmol O_2 m^{-2} d^{-1} for the shelf and 3.3 ± 1.8 mmol O_2 m^{-2} d^{-1} for the slope. In his experiment summer values (average of all stations 5.0 ± 2.6 mmol O_2 m^{-2} d^{-1}) were always higher than winter ones (average of all stations 2.6 ± 0.6 mmol O_2 m^{-2} d^{-1}). In Table 1 oxygen consumption values from various regions of the world ocean (other than the Gulf of Lions) are displayed. Northwestern Mediterranean shows values comparable to other continental margins

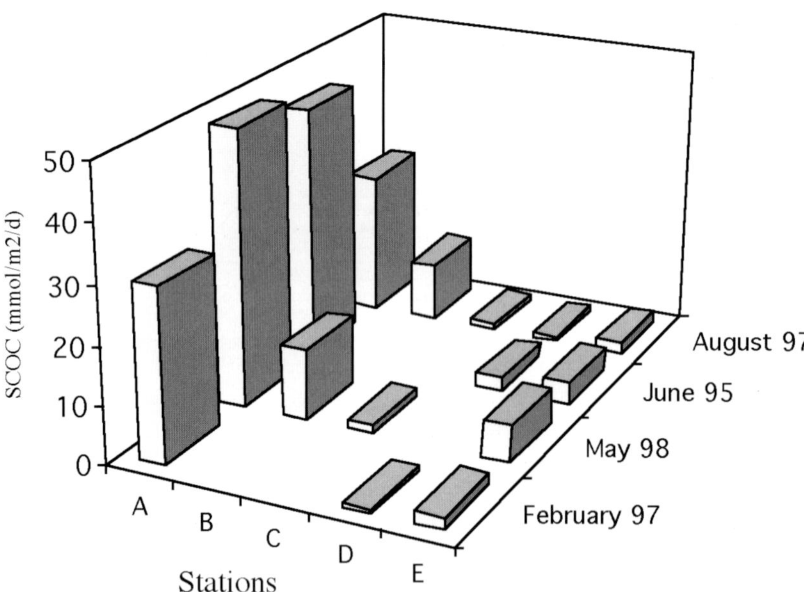

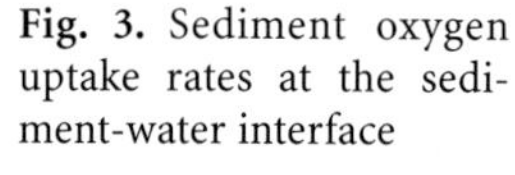
Fig. 3. Sediment oxygen uptake rates at the sediment-water interface

(only our Stn AS exhibits higher values). It is worth noticing that SCOC is relatively homogeneous in the different margin sites, despite the remarkable differences in environmental conditions, and exhibits the lowest values at the greatest depths.

Spatial and Temporal Variability

From the above results spatial and temporal trends may be pointed out: SCOC decreased with increasing depth, and increased from winter to late spring. Redox potential of surficial sediments was comprised between +200 and +300 mV at all stations and in all sampling periods; it decreased with depth in the sediment and, on average, became negative at 2 cm on the shelf and 7 and 5 cm on the canyon open slopes and axis respectively. It is known that in oxic sediments, which is our case, the rate at which organic matter is mineralized tends to balance the rate of its supply from the water column, which means the more and/or faster organic matter is supplied, the faster it is mineralized (Blackburn 1991; Henrichs 1992). It is obvious that in our case freshly produced organic matter is likely to reach the sea bottom faster in those sites with a shorter water column, i.e. in Stn A and B, which exhibited the highest SCOC values. Moreover, in spite of the lack of downward flux measurements on the shelf of the study area, we hypothesize the input of organic matter to the sediment-water interface is greater in shelf stations (A and B), as a consequence of the higher primary production and river input (Lefèvre et al. 1997). In the slope stations the highest values of the canyon axis (Stn E) as compared to the open slopes (Stn C and D) can be explained in the light of previous findings which indicated canyons as the main vectors of organic matter transport towards the continental slope, as well as preferential sites for organic carbon accumulation on the slope (Buscail and Germain 1997). As a matter of fact organic carbon content of surficial sediments (0-5 cm) was higher at the canyon axis (0.83%) than at open slope sites (0.69%) and exhibited values similar to the shelf stations (0.84%). Although we do not show data concerning the abundances of benthic organisms, we have found a significant correlation ($r = 0.97$, $n = 12$, $P < 0.001$) between our SCOC measurements and the spatial and seasonal trend of meiofaunal abundances (de Bovée et al. 1998). Literature data for this area also provide evidence that meiofaunal density (de Bovée et al. 1990), as well as bacterial activity (i.e. bacterial glutamate potential remineralization rate) (Tholosan 1999), decreases with increasing depth and increases in the warmer season as compared to winter. We thus suggest that the spatial and seasonal variability characterizing SCOC values may at least partially result from spatial and seasonal variability of the biological activity.

Carbon Mineralization versus Carbon Input

Data obtained from sediment trap measurements performed at the same sites as our slope stations resulted in an average annual input of particulate organic carbon to the benthic boundary layer in the slope region equal to 45 mg C m^{-2} d^{-1} (Heussner et al. 1996).

The average SCOC value we obtained for the slope stations is 2.3 mmol O_2 m^{-2} d^{-1}. Helder (1989) demonstrated that oxygen respiration is the main pathway through which organic carbon is mineralized in the Gulf of Lions sediments. We can thus use our SCOC values, expressed in mmol O_2 m^{-2} d^{-1} to calculate the average amount of organic carbon which is mineralized at the sediment-water interface (mg C m^{-2} d^{-1}), applying the appropriate respiratory quotient of 0.85, as suggested by Hargrave (1973). These calculations lead us to estimate that approximately 50% of the organic carbon deposited at the sediment surface by the downward flux from the overlying euphotic zone is mineralized, while the remaining 50% is available for accumulation in the sediment and eventually for long term storage in deep sediment layers. This result is very close to findings of other authors in the same area: Buscail et al. (1990), using ^{210}Pb dating, estimated that 49% of the organic carbon reaching the seafloor was mineralized prior to burial, in a station located in the Lacaze-Duthiers canyon axis. These high percentages are not surprising, considering the major role that continental margins play in the sedimentary organic carbon preservation (Hedges 1992; Reimers et al. 1992; Hedges and Keil 1995). Results from other continental margin sediments include preservation rates of 10-40% in the northwestern Atlantic (Anderson et al. 1994), 2-20% and about 50% in the central and south Californian margin respectively (Berelson et al. 1996). Conversely, in the pelagic environment, preservation rates are typically much lower (e.g. < 1%, Berelson et al. 1997). We hypothesize that the high preservation rate we calculated for our slope stations depends on the origin of the organic carbon which is deposited at the seafloor in this area. Monaco et al. (1990) and Buscail and Germain (1997) pointed out that most organic matter is not directly carried by downward fluxes from the overlying euphotic layer but is advected from the shelf. This material is conse-

Table 1. SCOC values measured in various regions of the world ocean (* Only mean values are given by the authors)

O_2 consumption (mmol m^{-2} d^{-1})	Location	Depth (m)	Method	Reference
1.40-7.61[a] (3.97 ± 2.64)[b] *n* = 8[c]	Middle Atlantic Bight (upper slope)	130-2010	Benthic chamber	Anderson et al. 1994
0.46-11.02 (2.47 ± 2.82) *n* = 15	Californian margin	95-3707	Benthic chamber	Berelson et al. 1996
1.70-3.40 (2.50 ± 0.49) *n* = 25	Bay of Biscay (NE Atlantic)	500-3000	Microelectrode	Etcheber et al. 1999
1.74-3.61 (2.61 ± 0.58) *n* = 14	Weddel Sea (Antarctica)	280-2514	Benthic chamber	Hulth et al. 1997
* (2.27 ± 0.39) *n* = 46	NE Atlantic continental shelf	2000	*In situ*	Lampitt et al. 1995
* (0.71 ± 0.12) *n* = 9	Porcupine abyssal plain (NE Atlantic)	4800	Incubations	

[a] Range of values
[b] Average ± SD
[c] Number of measurements

quently much more refractory and less easily metabolizable, and is thus more likely to be accumulated in the sediment.

Conclusions

Summarizing, although obtained by the use of different methods, our results confirm the range of SCOC values already registered by other authors in the Gulf of Lions (Helder 1989; Tahey et al. 1994). In this area the sediment oxygen uptake rates vary both temporally and spatially and seem to be related to biological parameters such as meiofaunal density and bacterial activity. On the slope, SCOC displays higher values in the canyon axis as compared to open slopes, which may possibly be explained on the basis of the higher rates of organic matter input. With respect to this input, we have calculated that, on average, 50% of the carbon that reaches the seafloor at the slope stations is mineralized, confirming thus the importance of mineralization in the frame of the processes occurring at the sediment-water interface. Future steps will include an effort to extend these calculations to other sites in the Gulf of Lions, with the aim of understanding the relative importance of the shelf and slope areas with respect to carbon transformations.

Acknowledgements. The authors are grateful to the Observatoire Océanologique de Banyuls sur Mer for logistical assistance and to the CIRMED for providing ship time. This research has been undertaken in the framework of the French "Programme National d'Océanographie Côtière" and of the Mediterranean Targeted Project (MTP) phase II-Mater. We acknowledge the support from the European Commission's Marine Science and Technology Programme (MAST III) under contract MAS3-CT96-0051.

References

Anderson RF, Rowe GT, Kemp PF, Trumbore S, Biscaye PE (1994) Carbon budget for the mid-slope depocenter of the Middle Atlantic Bight. Deep-Sea Res II 41: 669-703

Bauer JE, Druffel ERM (1998) Ocean margins as a significant source of organic matter to the deep open ocean. Nature 392: 482-485

Berelson WM, Anderson RF, Dymond J, DeMaster D, Hammond DE, Collier R, Honjo S, Leinen M, McManus J, Pope R, Smith C, Stephens M (1997) Biogenic budgets of particle rain, benthic remineralization and sediment accumulation in the equatorial Pacific. Deep-Sea Res II 44: 2251-2282

Berelson WM, McManus J, Coale KH, Johnson KS, Kilgore T, Burdige D, Pilskaln C (1996) Biogenic matter diagenesis on the sea floor: a comparison between two continental margin transects. J Mar Res 54: 731-762

Blackburn TH (1991) Accumulation and regeneration: processes at the benthic boundary layer. In: Mantoura RFC, Martin J-M, Wollast R (eds) Ocean margin processes in global change. John Wiley and Sons, New York, pp 181-195

Burdige DJ, Homstead J (1994) Fluxes of dissolved organic carbon from Chesapeake Bay sediments. Geochim Cosmochim Acta 58: 3407-3424

Buscail R, Germain C (1997) Present-day organic matter sedimentation on the NW Mediterranean margin: importance of off-shelf export. Limnol Oceanogr 42: 217-229

Buscail R, Pocklington R, Daumas R, Guidi LD (1990) Fluxes and budget of organic matter in the benthic boundary layer over the northwestern mediterranean margin. Cont Shelf Res 10: 1089-1122

Canfield DE (1993) Organic matter oxidation in marine sediments. In: Wollast R, Mackenzie FT, Chou L (eds) Interactions of C, N, P and S: biogeochemical cycles and global change. (NATO ASI Ser 14) Springer-Verlag, Berlin Heidelberg New York Tokyo, pp 333-363

de Bovée F, Guidi LD, Soyer J (1990) Quantitative distribution of deep-sea meiobenthos in the northwestern Mediterranean (Gulf of Lions). Cont Shelf Res 10: 1123-1145

de Bovée F, Picon P, Medernach L et al (1998) In situ studies of biogeochemical processes occuring at the deep-sea floor: first results obtained with the Banyuls Benthic Lander. Eur Geophys Soc, Katlenburg-Lindau, Germany. Annales Gephys 16: C731 (Suppl II)

Deming JW, Baross JA (1993) The early diagenesis of organic matter: bacterial activity. In: Engel MH, Macko SA (eds) Organic geochemistry, principles and applications. Plenum Press, New York, pp 119-144

Etcheber H, Relexans J-C, Beliard M, Weber O, Buscail R, Heussner S (1999) Distribution and quality of sedimentary organic matter on the Aquitanian margin (Bay of Biscay). Deep-Sea Res II 46: 2249-2288

Got H, Aloisi J-C, Monaco A (1985) Sedimentary processes in Mediterranean deltas and shelves. In: Stanley DJ, Wezel FC (eds) Geological evolution of the Mediterr basin. Springer-Verlag, Berlin Heidelberg New York Tokyo, pp 355-376

Hargrave BT (1973) Coupling carbon flow through some pelagic and benthic communities. J Fish Res Board Can 30: 1317-1326

Hedges JI (1992) Early diagenesis of organic matter in marine sediments: progress and perplexity. Mar Chem 39: 67-93

Hedges JI, Keil RG (1995) Sedimentary organic matter preservation: an assessment and speculative synthesis. Mar Chem 49: 81-115

Helder W (1989) Early diagenesis and sediment-water exchange in the Golfe du Lion. In: Martin JM, Barth H (eds) EROS 2000 (European River Ocean System) project. Water Pollut Res Rep, CEC Directorate-Gen Sci Res Dev, Brussels 13, pp 87-101

Henrichs SM (1992) Early diagenesis of organic matter in marine sediments: progress and perplexity. Mar Chem 39: 119-149

Heussner S, Calafat AM, Palanques A (1996) Quantitative and qualitative features of particle fluxes in the North-Balearic basin. In: Canals M, Casamor JL, Cacho I, Calafat AM, Monaco A (eds) EUROMARGE-NB Final Rep. MAST II Progr, vol II, pp 41-66

Hulth S, Tenberg A, Landen A, Hall POJ (1997) Mineralization and burial of organic carbon in sediments of the southern Weddell Sea (Antarctica). Deep-Sea Res 44: 955-981

Lampitt RS, Raine RCT, Billett DSM, Rice AL (1995) Material supply to the European continental slope: a budget based on benthic oxygen demand and organic supply. Deep-Sea Res 42: 1865-1880

Lefèvre D, Minas HJ, Minas M, Robinson C, Williams PJ Le B, Woodward EMS (1997) Review of gross community production, primary production, net community production and dark community respiration in the Gulf of Lions. Deep-Sea Res II 44: 801-832

Mantoura RFC, Martin J-M, Wollast R (eds) (1991) Ocean margin processes in global change. John Wiley and Sons, New York

Monaco A, Courp T, Heussner S, Carbonne J, Fowler SW, Deniaux B (1990) Seasonality and composition of particulate fluxes during ECOMARGE-I, western Gulf of Lions. Cont Shelf Res 9: 959-987

Monaco A, Durrieu de Madron X, Radakovitch O, Heussner S, Carbonne J (1999) Origin and variability of downward biogeochemical fluxes on the Rhone continental margin (NW Mediterranean). Deep-Sea Res 46: 1483-1511

Reimers CE, Jahnke RA, McCorkle DC (1992) Carbon fluxes and burial rates over the continental slope and rise off central California with implications for the global carbon cycle. Global Biogeochem Cycles 6: 199-224

Schulz DS, Dahmke A, Schinzel U, Wallmann K, Zabel M (1994) Early diagenesis processes, fluxes and reaction rates in sediments of the South Atlantic. Geochim Cosmochim Acta 58: 2041-2060

Sempéré R, Charrière B, Van Wambeke F, Cauwet G (2000) Carbon inputs of the Rhône River to the Mediterranean Sea: biogeochemical implications. Global Biogeochem Cycle 14: 669-681

Smith Jr KL (1987) Food energy supply and demand: a discrepancy between particulate organic carbon flux and sediment community oxygen consumption in the deep ocean. Limnol Oceanogr 32: 201-220

Smith S, MacKenzie F (1987) The ocean as a net heterotrophic system: implications for the carbon biogeochemical cycle. Global Biogeochem Cycle 1: 187-198

Tahey TM, Duineveld GCA, Berghuis EM, Helder W (1994) Relation between sediment-water fluxes of oxygen and silicate and faunal abundance at continental shelf, slope and deep-water stations in the northwest Mediterranean. Mar Ecol Prog Ser 104: 119-130

Tenberg A, de Bovée F, Hall P, Berelson W, Chadwick B, Ciceri G et al (1995) Benthic chamber and profiling lander in oceanography: a review of design, technical solutions and functioning. Prog Oceanogr 35: 253-294

Tholosan O (1999) Activités microbiennes dans les eaux et les sédiments profonds: rôle de la pression hydrostatique. Thèse Doctorat, Univ Méditerranée, Aix-Marseille 2

Walsh JJ (1991) Importance of continental margins in the marine biogeochemical cycling of carbon and nitrogen. Nature 350: 53-55

Walsh JJ, Rowe GT, Iverson RL, McRoy P (1981) Biological export of shelf carbon is a sink on the global CO_2 cycle. Nature 291: 196-201

Wollast R (1991) The coastal organic carbon cycle: fluxes, sources, and sinks. In: Mantoura RFC, Martin J-M, Wollast R (eds) Ocean margin processes in global change. John Wiley and Sons, New York, pp 365-381

Seasonal Influence of the Front Area on Sediment Deposition in the Northwestern Adriatic Sea

F. Alvisi, M. Frignani, and M. Ravaioli

ABSTRACT

In order to improve our knowledge of short time-scale variations of seasonal particle fluxes into sediments, 28 short sediment cores were collected along six transects in two areas influenced by the frontal system formed by the Po River waters moving into the Adriatic Sea. Radioisotopes ^{234}Th and ^{137}Cs were measured in surficial sediment sections and used as tools to evaluate the rates of sediment deposition, accumulation and mixing. Sedimentological and mineralogical parameters were also analysed.

The ^{137}Cs distribution points out that the area south of the Po Delta, characterised by muddy sediment scarcely colonised by macrofauna, is directly influenced by the input of fine material delivered by the Po River. Here the surficial ^{234}Th activities are lower in winter than in summer. This suggests a front influence that in the warm season couples a higher concentration of suspended matter with an enhanced capacity of interaction with particle-reactive species. The lithology of the area in front of the Marche Region is otherwise characterised by an alternance of silt and clay. Together with the presence of thanatocoenoses, this pattern reflects pulsed sedimentation with colonisation stages and sudden inputs of sediment that bury the benthic fauna. High values of ^{234}Th activities were found at the stations close to the front, where the amount of particulate material is high. Behind the front line, thorium activities are lower, due to both the shallow environment and the kind of particles. The maximum values of ^{234}Th activities, characteristic of the two offshore stations, reflect the high depth of the water column, the enhanced residence time of particles and the very fine grain size of the sedimentary material.

Introduction

Many key processes in the Adriatic system are influenced by the seasonal variation of the density structures along the water column. This is, in fact, completely mixed in winter and strongly stratified in summer (Artegiani et al. 1997a). This cycle determines the dynamics of biological and sedimentological processes influencing the vertical and horizontal transport of particles within the basin (Frascari et al. 1988).

To improve our knowledge of the causes of phenomena such as algal blooms and mucilage production in the Northwestern Adriatic Sea, it is very useful to study the biogeochemical cycles of nutrients at the frontal system. This is formed by the contact between the continental waters and those coming from the North Adriatic, and delimits a coastal area, directly influenced by the anthropogenic input, physically separated from the offshore oligotrophic area. The intensity of the front is variable as a function of the strength of the thermohaline gradients between the two water masses. Therefore, the density gradient acts as a barrier that influences the dispersal of the suspended matter and dissolved materials delivered by the Po River and by the minor rivers discharging along the Romagna coast.

This work has been carried out within the framework of the Italian Project Prisma 2, subproject "Biogeochemical cycles", which had the general aim to evaluate the fluxes of both dissolved and particulate matter across the front area, with respect to its seasonal location and intensity. In particular, our purpose was to contribute to the quantitative knowledge of the bio-

Istituto di Geologia Marina, CNR, Via P. Gobetti 101, 40129 Bologna, Italy

F.M. Faranda, L. Guglielmo, G. Spezie (eds)
Mediterranean Ecosystems: Structures and Processes

geochemical cycles through the estimate of short time-scale accumulation of sedimentary material and related chemicals on the sea floor, the effects of physical and biological mixing, and the influence of the presence of the front area on sediment-water interactions. Therefore, we studied a number of short sediment cores collected along land-sea transects crossing the front area and a few cores collected at the same stations in different seasons. Furthermore, we focused our attention on the surficial layer of the collected cores representing the short and very short time-scale variations by using ^{234}Th and ^{137}Cs to understand some of the mechanisms leading to the formation of sedimentary layers.

The water dynamics of the Adriatic system

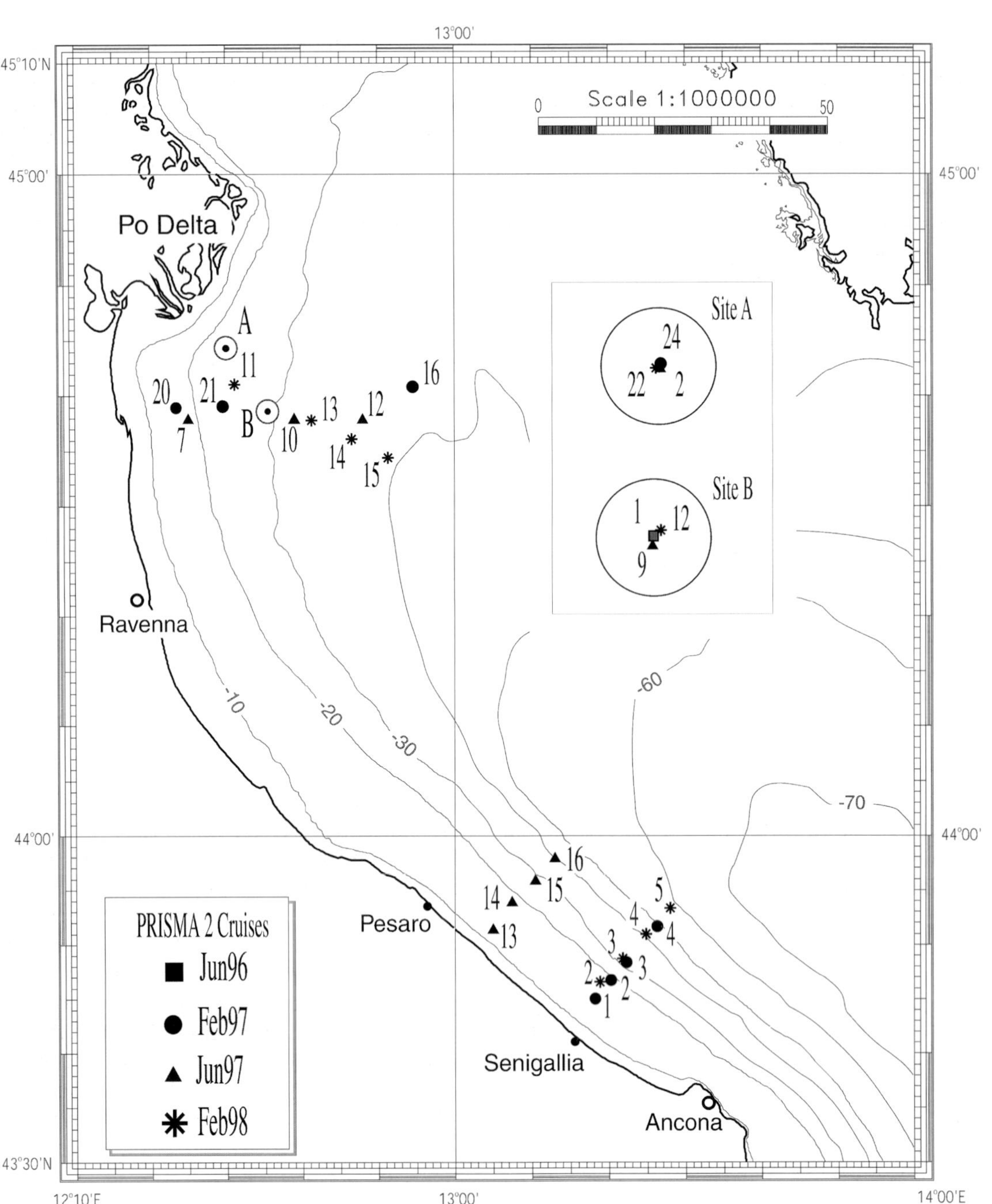

Fig. 1. Study areas with sampling locations and details of core locations in sites A and B

are dominated by a cyclonic gyre which causes an overall transport and deposition of sediment along the Italian coast. The hydrodynamic of the area fairly complicated and very dependent on thermohaline factors. The general circulation of the Northern Adriatic Sea has two different seasonal regimes (Malanotte Rizzoli and Bergamasco 1983; Artegiani et al. 1997a, b). In winter the currents are essentially of cyclonic type and force southward the Italian river plumes with a water column that is more or less homogeneous. In summer these currents are weaker and delimit small scale eddies, and the water column is frequently characterized by horizontal stratification (Franco et al. 1982; Orlic et al. 1992).

The location and the mesoscale seasonal evolution of the front structure were the object of collaboration between different subprojects of Prisma 2. Satellite-derived images were used together with field data collected during marine surveys preceding the sampling cruises, to locate the frontal system. The two most representative study areas were chosen in relation to the hydrological and biological peculiarities of the front system in the two seasons. The sampling grids were defined taking into account the strongest land-sea gradient of physico-chemical characteristics of the water body.

Figure 1 shows the study area and sampling locations. These were chosen in order to: (1) spatially follow the frontal area evolution; (2) reconstruct the time evolution of sedimentary input at fixed locations. The two areas forming the object of this study are one immediately south of the Po Delta and the other the coastal area in front of Marche Region. The northern area, located close to the delta, was chosen to evaluate direct and indirect effects of the frontal system produced by the Po River plume offshore the Romagna coast, and the southern one to study the frontal system formed between the shallow coastal waters of continental origin and the deeper water of the Marche coast.

Materials and Methods

Four cruises were carried out in June 96 (Jun96), February and June 97 (Feb97 and Jun97), and February 98 (Feb98) onboard of the R/V Urania. Sampling (Fig. 1) was developed along transects crossing the front line in two different seasons: June and February were taken as representative of a summer and a winter situation respectively. The sampling strategy was based on the collection of two/three cores both landward and seaward with respect to the front line. During all cruises, some other stations were sampled to compare different sedimentological environments and in some cases replicates were taken at the same site to evaluate the spatial variability. During the first cruise (June 96), all stations were located within the frontal system. The sediment cores (10-45 cm long) were collected using a multicorer (mod. Midi), equipped with 5.7 and 9.5 cm internal diameter Perspex tubes. All cores were radiographed and then cut longitudinally, photographed and the lithology was described. For the first 3 cm, 0.5 cm slices were sampled and dried at 60°C to quantify the water content and to calculate porosity and bulk dry density according to Berner (1971), assuming a sediment density of 2.5 g cm^{-3}. Dry sediment was analysed for grain size, and slightly desegregated for radiotracers, organic nitrogen and carbon determinations. Grain size analyses were performed by a pre-treatment with H_2O_2 and subsequent wet sieving at 60 microns to separate the muddy fraction. The sandy fraction was then isolated by dry sieving at 250 micron to separate the shell debris. The ^{234}Th and ^{137}Cs surficial activities were measured by gamma spectrometry (Frignani et al. 1991) and then calculated by using the Gammaplus programme. Total organic carbon and nitrogen were measured by means of a Fisons CHN Elemental Analyser after a 2N HCl treatment to eliminate the carbonate fraction.

Results and Discussion

In this paper, we present the results of radiochemical and organic matter analyses (Figs. 2, 4, 5 and 6) relative to the surface layers (0-0.5 cm). However, in the case of $^{234}Th_{xs}$, both surficial and total inventories (first 3 cm) are discussed. In the figures, the stations of each transect are shown along the horizontal axis according to water depth (i.e. distance from the coast). Front lines are represented in each diagram by vertical lines. In the northern area, during the cruise Feb97, the frontal system was located offshore, where no recent accumulation occurs and the sediment consists of relict sands, difficult to sample. Therefore, at that time, most cores were collected

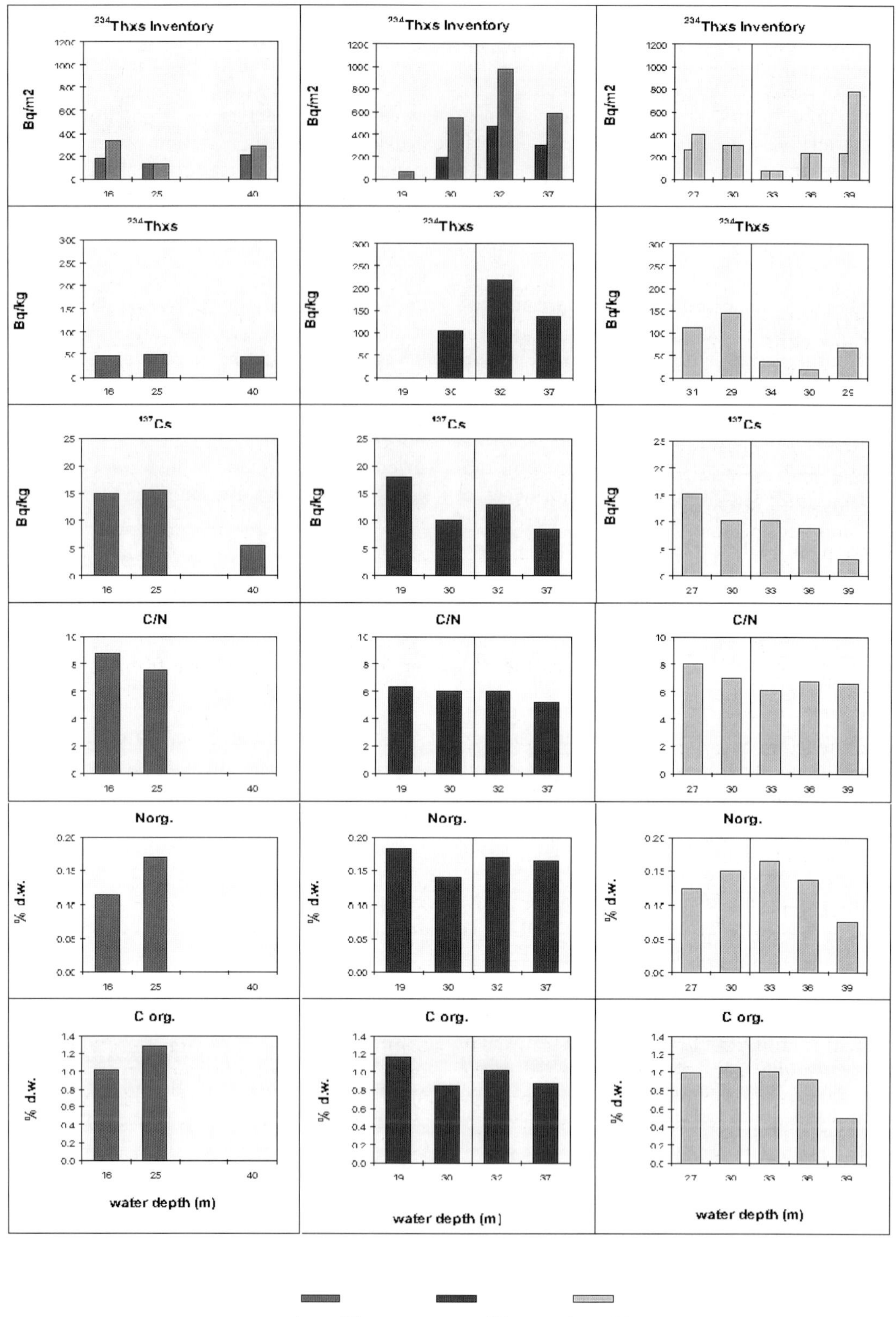

Fig. 2. Seasonal distribution of surficial sample values of the analysed parameters along the study transects of the northern area. From *left to right*, the transects of Feb97 (cores 20, 21 and 16), Jun97 (cores 7, 9, 10 and 12) and Feb98 (cores 11, 12, 13, 14 and 15). *Vertical lines* represent front locations. (*d.w.*, dried weight)

onshore, in the mud belt, far from the front. In the figures, they are compared to the stations sampled in Jun97 that were in different seasons and with different front influences.

Northern Area

The area south of the Po Delta is characterised by silty clay sediment mainly colonised by *Polychaetes* of different species and rarely by molluscs (*Turritella, Corbula*). Core stratigraphies show a different degree of bioturbation due to the activity of the invertebrates that tend to destroy the primary sedimentary structures represented by flat laminations. Some of the offshore stations show a different lithology with higher percentages of sand, up to 50%, with abundant shell debris and layers rich in mollusc remains. These sites are located in the transitional area between the pelitic mud belt and the offshore relict sands where no net deposition occurs.

The ^{137}Cs shows a pattern clearly not dependent on the frontal position (Fig. 2), with values decreasing from 15-18 to 5-8 Bq/kg, as the distance and the water depth from the delta increase. This behaviour confirms the predominant influence in this area of the ^{137}Cs supplied by the river with respect to the atmospheric, strictly "marine" delivery, as already recognised by previous Authors (Albertazzi et al. 1984, 1992; Frignani and Langone 1991). The surficial activity of excess ^{234}Th (^{234}Th$_{xs}$), over the background in equilibrium with ^{238}U, spans between 0 and 250 Bq/kg with the highest value close to the frontal system (summer 97, depth 32 m). In general, the activities are lower in winter than in summer, but they are greatly influenced by the distance from the front line. In fact, we can see in Fig. 2 that higher activities are recorded at the location immediately behind or on the front line both in summer and in winter. Furthermore, if we compare the surficial activities in the two winter seasons, we record values of around 150-170 Bq/kg in the presence of the frontal system (Feb98). These values decrease to one half (77 Bq/kg) with the frontal system, displaced further offshore. These results suggest a general strong front influence on the fast deposition of suspended matter inside the frontal system, which is also due to the shallow depth of the water column. The influence seems to be even higher in the warm season that couples a higher concentration of suspended matter with an enhanced surficial biological activity and a stratification of the water column that slows down the recycling of the particulate matter sinking to the bottom. The ^{234}Th$_{xs}$ profiles in the sediment cores show a deeper penetration during the warm season probably due to the enhanced biological activity on the sea floor linked to the higher input of particulate matter and warmer temperature on the bottom. Furthermore, if we look at the activity/depth profiles of the two radionuclides (Fig. 3), we can often observe an inverse correlation with relatively low ^{234}Th$_{xs}$ activity corresponding to high cesium activity and vice versa. The different origin and behaviour of the two radioisotopes in this area could then be used to recognize in the sedimentary record periods of sediment deposition with a predominant riverine versus marine sources. In this area, in fact, ^{234}Th is derived by the direct decay of ^{238}U dissolved in the water column. On the contrary, even if small amounts are directly derived from atmospheric fallout, ^{137}Cs is mainly delivered to the marine system adsorbed onto the river particulate matter. These particles derive from soil erosion of the drainage basins, where ^{137}Cs has been widely deposited during the Chernobyl accident in 1986 (Frignani and Langone 1991; Albertazzi et al. 1992).

The distribution of nutrients in surficial sediments, organic C and N, do not show a clear signal linked to the frontal system. However, some patterns can be highlighted (Fig. 2). Organic nitrogen shows generally low values, between 0.12 and 0.18% dry weight. The N content is similar in the two seasons but it shows higher percentages in shallower water in summer and lower in winter, and it seems slightly depleted just behind the front. The organic carbon has the same behaviour with concentrations between 0.9-1.3% dry weight. The C/N ratio shows values always lower than 10 with higher values in wintertime, especially close to the coast. This behaviour suggests a more important contribution of organic riverborne material to the sediment during wintertime.

Sites A and B

Two sites were sampled in different seasons to understand the time evolution of the sedimentation at fixed locations. Site A is located offshore

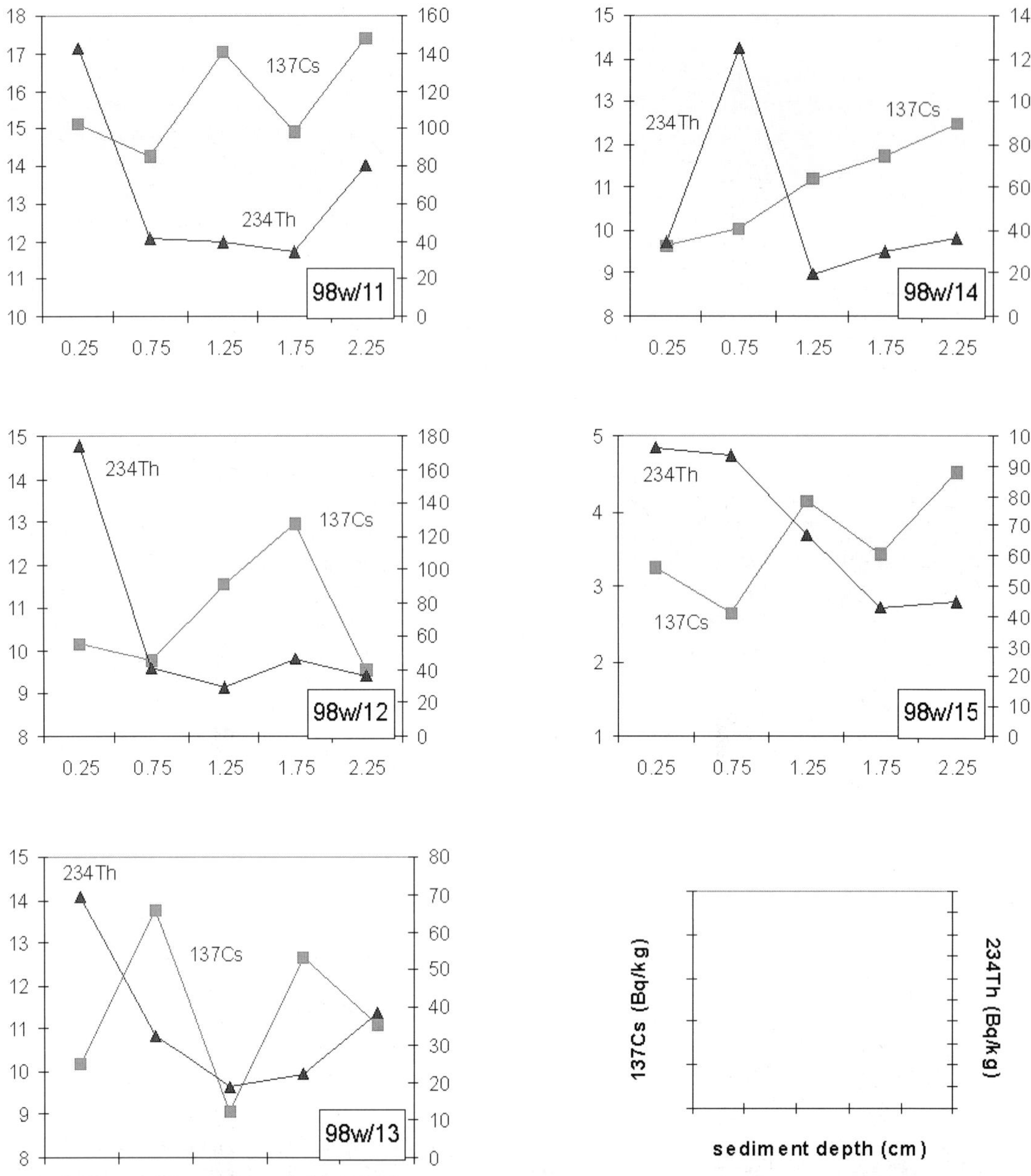

Fig. 3. Activity/depth profiles of ^{234}Th and ^{137}Cs along the transect of the northern area during winter 1998 (cruise Feb98)

the Po di Goro mouth, at about 24 m water depth, and site B is further southeast, at 29 m depth (Fig. 1).

The surficial sediments at site A show two winter situations and a summer one (Fig. 4). The ^{137}Cs and nutrients show a clear decreasing pattern during the period considered. The $^{234}Th_{xs}$ shows a higher surficial activity (165 vs 63-102 Bq/kg) in summer with respect to winter season, whereas its inventories increase. During the winter 1997, the location of the frontal system was more offshore in comparison to the other two sampled seasons. The behaviour of ^{137}Cs and the contemporary increase of thorium inventories suggest a decrease of the river discharge with an increase of the relative importance of marine particulate matter during the study period. The C/N ratio seems to remain fairly constant with a slight decrease in summer 1997 suggesting a possible seasonal influence on the delivery of organ-

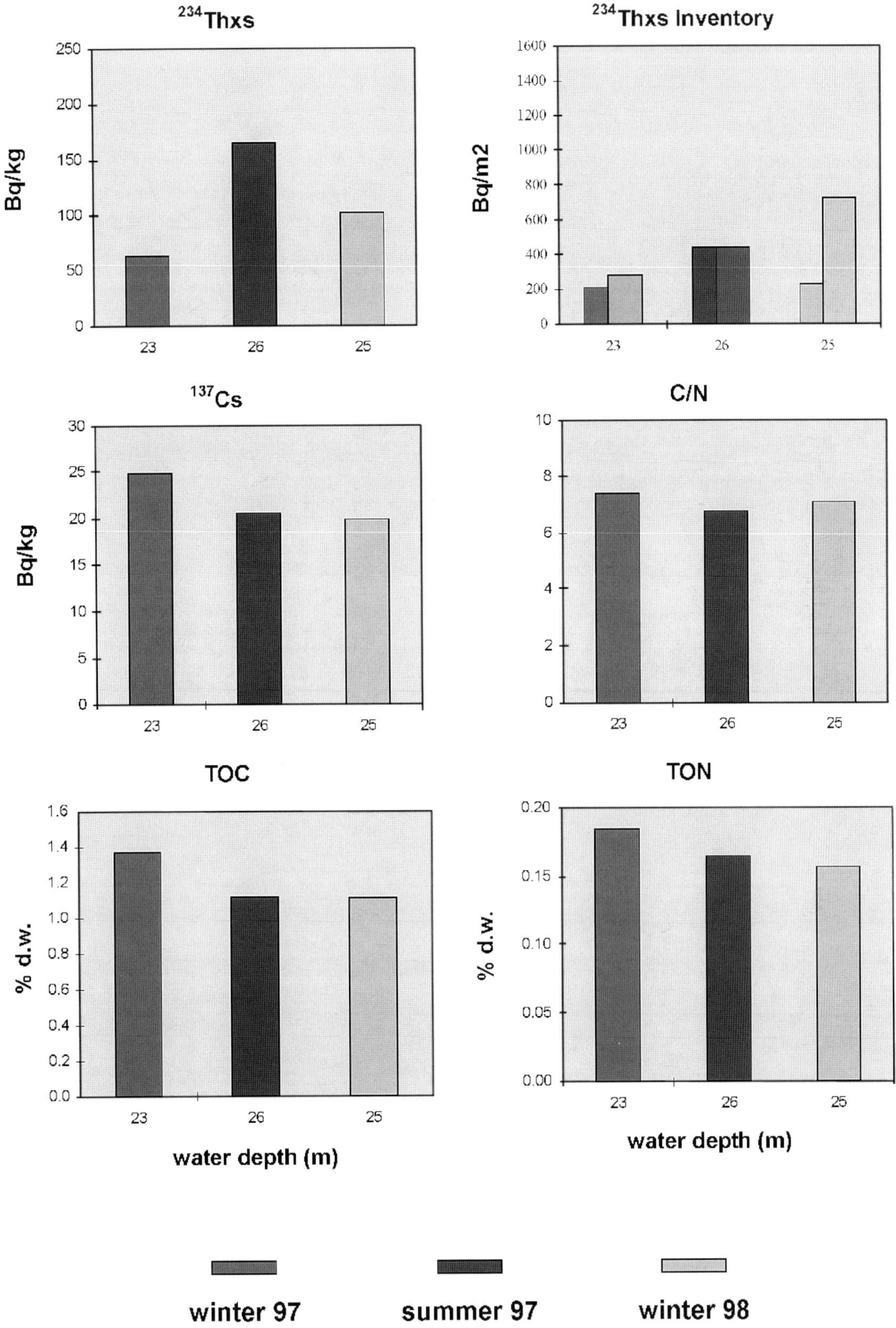

Fig. 4. Seasonal distribution of surficial sample values of the analysed parameters at site A. From *left to right* in each diagram, cores Feb97-24, Jun97-2 and Feb98-22. (*d.w.*, dried weight)

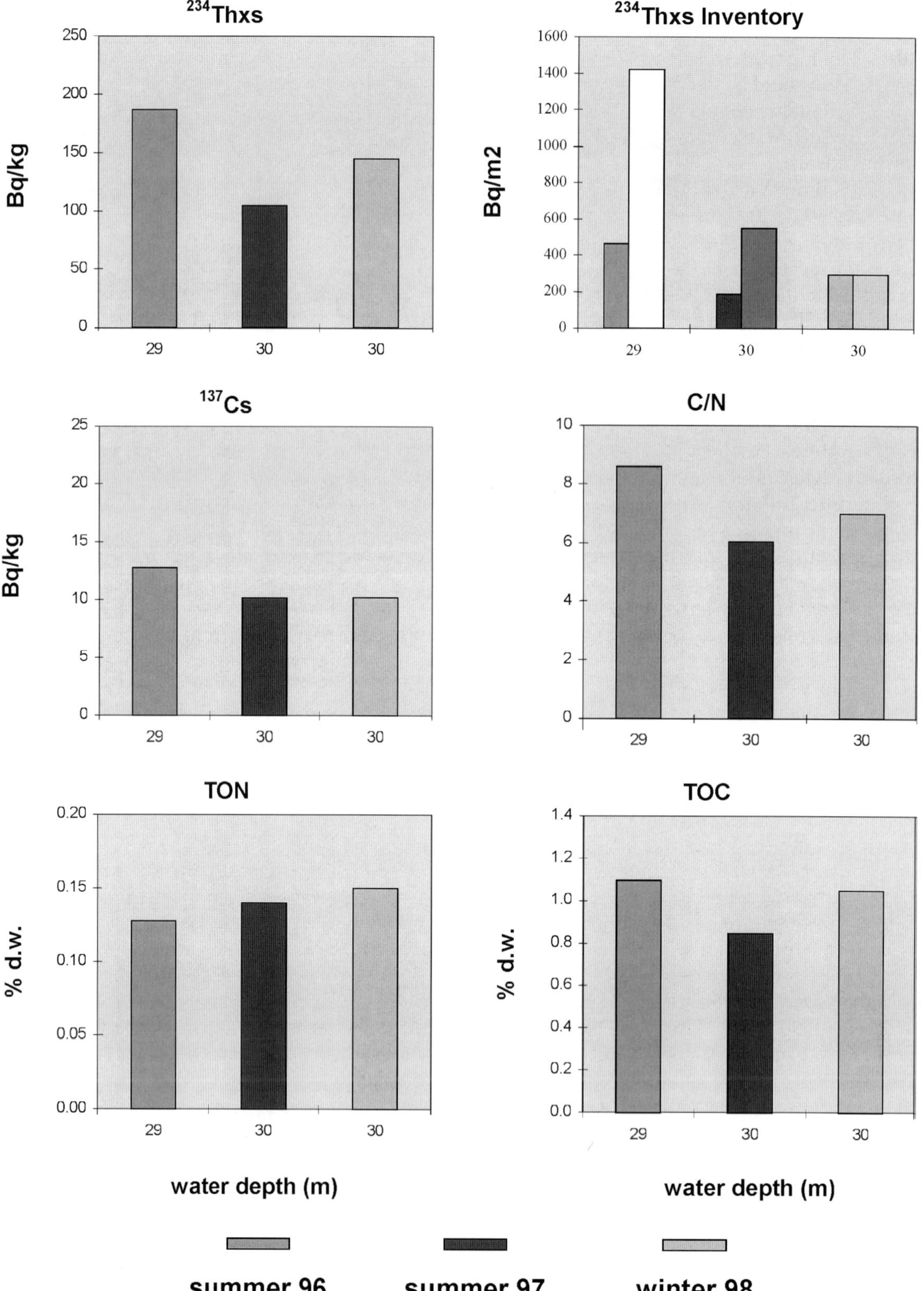

Fig. 5. Seasonal distribution of surficial sample values of the analysed parameters at site B. From *left to right*, cores Jun96-1, Jun97-9, and Feb98-12. (*d.w.*, dried weight)

ic matter to the sediment instead of a general diminishing trend as shown by cesium, organic C and N. The high activity of thorium during the warm season could be due to the closest proximity of the frontal system to this location and/or to the quality of the scavenging material to which thorium adsorbs.

Site B shows two summer situations and a winter one (Fig. 5). The ^{137}Cs seasonal surfical activities follow the same trend as at site A, but with the expected lower surficial values (10-13 Bq/kg) due to the increased distance from the Po Delta. The $^{234}Th_{xs}$ shows different surficial activities (105-187 Bq/kg) in the three seasons, with a minimum in summer 1997. This behaviour is opposite to that of site A. But for that location we do not have the summer 1996 situation. The sediment depth reached by the base of the thorium profile is higher in both summers (1.5 cm) than in winter (0.5 cm), thus suggesting that the mixing mechanisms are more intense during the warm season. Excess thorium inventories and activities show a strong decrease during the period considered and in particular between the two summer seasons. These results suggest that an important deposition of particulate matter occurred in summer 1996 that was much more productive than summer 1997. This difference was probably favoured by a higher river runoff as suggested by the discharge curve of the Po River recorded at Pontelagoscuro (M.G. Tomasino, personal communication) that supplied more new organic and inorganic matter to the frontal system (Tomasino 1996). The higher surficial values of ^{137}Cs, organic carbon and C/N ratio (Fig. 5) support this hypothesis. The opposite trend, shown by the organic nitrogen, with percentages slightly increasing from 0.128 to 0.150, confirms the higher river contribution of the organic matter during 1996. The $^{234}Th_{xs}$ values recorded in summer 1997 suggest a lower input of particles to the sea floor with respect to 1996 values, probably as a consequence of low river discharge, reduced productivity and biological activity in the overlying water.

Southern Area

The lithology of the sediment cores in front of the Marche Region is characterised by an alternance of silt and clay. Together with the presence of thanatocoenoses, this pattern reflects pulsed sedimentation with colonisation stages and sudden inputs of sediment that bury the benthic fauna. The shallower stations show a clear influence of the sandy input due to wave resuspension of the very close sandy shores during storm events, as observed during the winter cruises. The storm events can be recognized in the sedimentary record until a water depth of 18-20 m. They are then progressively mixed up by bioturbation with finer sediment in the sedimentary column.

Unfortunately, we had some analytical problems with two stations of the summer transect concerning thorium quantification. Thus, the comparison between the different seasons can not be very reliable. However, some considerations about thorium behaviour are still possible. High activities were found at the stations close to the front (Fig. 6), where the amount of particulate material is high. Behind the front line, thorium activities are lower, due to both the shallow environment and the sediment grain size. The maximum values of $^{234}Th_{xs}$ activities, characteristic of the two distal stations, reflect the high depth of the water column, the enhanced residence time of particles and the very fine grain size of the sedimentary material. Furthermore, the high inventories in deep water sediments, recorded especially in winter, suggest a possible shift of sediment offshore, due to the enhanced winter water dynamics (Artegiani et al. 1997a, b), leading to phenomena of resuspension and redeposition on the sea floor. The higher summer inventories of thorium underneath the front system seem to confirm the presence of lower water dynamic on the bottom, favoured also by water column stratification, that permits the vertical deposition of the suspended material. In this area, the ^{137}Cs distribution in surficial samples shows a better correlation with the frontal system (Fig. 6), probably related to the fine particle dynamics and to the lack of an important river input. Nevertheless, we can observe a decrease of the activities during the study period that suggests a decreasing riverine discharge. The nutrient behaviour seems to be mainly influenced by the lithology of the sea floor with lower values close to the coast and higher values corresponding to the muddy wedge. However, it could be also an effect of the depletion of nutrients behind the front line due to the enhanced benthic activity induced by the frontal system itself. Nutrient concentrations in the three seasons

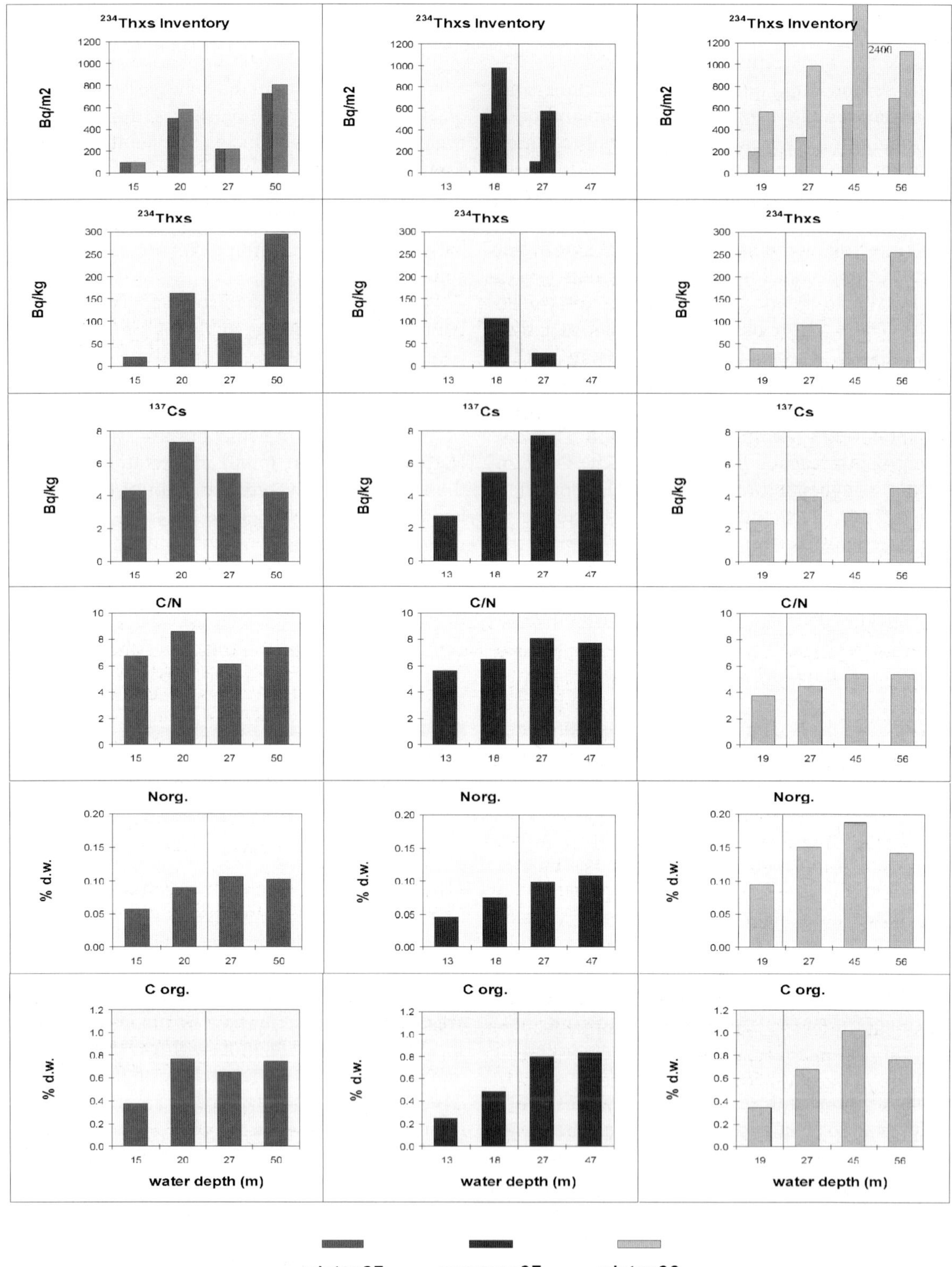

Fig. 6. Seasonal distribution of surficial sample values of the analysed parameters along the study transects of the southern area. From *left to right*, the transects relative to Feb97 (cores 1, 2, 3 and 4), Jun97 (cores 13, 14, 15 and 16) and Feb98 (cores 2, 3, 4 and 5). *Vertical lines* represent front locations. (*d.w.*, dried weight)

show some interesting features. Organic nitrogen displays the same pattern but with increasing values in winter and the lowest values during summer 1997. Organic carbon shows a shift of the highest value toward deeper waters during the same period. The C/N ratio shows a decrease in the last season suggesting an increased importance of marine organic matter. The recognized trend of ^{137}Cs and nutrients seems to point out a general decrease of the influence of riverborne material as seen in the northern area, and an increased input of marine particulate matter deposition onto the sea floor. The general lower values of nutrients, and in particular of nitrogen, during summer 1997, account for a low productive summer, with a possible depletion of nitrogen in the sediment due to the benthic activity or to a deposition of particulate matter already nitrogen-depleted.

Conclusions

The frontal system seems to exert a strong influence on the very short time-scale deposition of particulate matter in the northern area, as pointed out by the ^{234}Th behaviour. The influence is enhanced by the large spatial shift of the frontal system over this shallow area. ^{234}Th and ^{137}Cs can help to characterise the particulate matter produced by marine processes from that introduced by riverborne discharges, which are characterised by relatively higher contents of radiocaesium. In the southern area, the shifts of the frontal system are less wide, influencing in a similar way the deposition during the whole year. Nevertheless, the land-sea slope gradient and the presence of seasonal currents parallel to the coast at different water depths influence the deposition and resuspension of the fresh sedimentary material with a net offshore transport.

Both areas show a decreasing input of riverborne material to the sea floor during the study period, suggested by decreasing ^{137}Cs surficial activities and C/N values. The lowering of river input of particulate matter seems to affect the activity of organisms at the frontal system with a consequent decrease of fresh sedimentary material reaching the sea floor. In fact, it seems that summer 1997 was less productive than 1996, as suggested by the low excess thorium activities and inventory at site B and the low nutrient concentrations in the bottom sediment of the southern area.

Acknowledgements. Funds were provided by the National Italian Project "Prisma 2". We wish to thank Captains E. Gentile and V. Lubrano, their officers and crew for the assistance during the four cruises Prisma 2 subproject 2 onboard the R/V Urania, E. Manini, E. Vio, R. Danovaro and the research group of Ancona University for the fundamental collaboration in the fieldwork, T. Bernardi, A. Filippi, R. Lorenzelli and L. Tositti for some ^{234}Th analyses, G. Marozzi for core photos and X-ray scans, A. Magagnoli and M. Mengoli for their assistance and technical support with sedimentological analyses, G. Rovatti for the CHN analyses, G. Zini for the drawings.
This is paper N. 1219 from the Istituto di Geologia Marina, CNR, Bologna.

References

Albertazzi S, Alboni M, Frignani M, Langone L, Ravaioli M, Tesini E (1992) Chernobyl derived radiocesium in marine sediments near the Po River Delta. Rapp Comm Int Mer Mediterr 33: 337

Albertazzi S, Bopp RF, Frignani M, Hieke Merlin O, Menegazzo Vitturi L et al (1984) Cs-137 as a tracer for processes of marine sedimentation in the vicinity of the Po River Delta (Northern Adriatic Sea). Mem Soc Geol Ital 27: 447-459

Artegiani A, Bregant D, Paschini E, Pinardi N, Raicich F, Russo A (1997a) The Adriatic Sea general circulation. I. Air-sea interactions and water mass structure. J Phys Ocean 27: 1492-1514

Artegiani A, Bregant D, Paschini E, Pinardi N, Raicich F, Russo A (1997b) The Adriatic Sea general circulation. II. Baroclinic circulation structure. J Phys Ocean 27: 1515-1532

Berner RA (1971). Principles of chemical sedimentology. McGraw-Hill, New York

Franco P, Jeftic L, Malanotte Rizzoli P, Michelato A, Orlic M (1982) Descriptive model of the Northern Adriatic. Oceanol Acta 5: 379-389

Frascari F, Frignani M, Guerzoni S, Ravaioli M (1988) Sediments and pollution in the Northern Adriatic Sea. In: Living in a chemical world. Ann New Acad Sci Spec Publ 534: 1000-1020

Frignani M, Langone L (1991) Accumulation rates and ^{137}Cs distribution in sediments off the Po River delta and the Emilia-Romagna coast (northwestern Adriatic Sea, Italy). Cont Shelf Res 11: 525-542

Frignani M, Langone L, Ravaioli M, Sticchi A (1991) Cronologia di sedimenti marini – analisi di radionuclidi naturali ed artificiali mediante spettrometria gamma. Tech Rep 24, CNR-IGM, Bologna, Italy

Malanotte Rizzoli P, Bergamasco A (1983) The dynamics of the coastal region of the Northern Adriatic Sea. J Phys Ocean 13: 1105-1130

Orlic M, Gacic M, La Violette PE (1992) The currents and circulation of the Adriatic Sea. Oceanol Acta 15: 109-124

Tomasino MG (1996) Is it feasible to predict "slime blooms" or "mucilage" in the northern Adriatic Sea? Ecol Model 84: 189-198

CHAPTER 55

The Link between Continental and Marine Geochemistry as shown by Mercury and Arsenic Anomalies in Sediments from Southern Tuscany and Tyrrhenian Sea

M. Dall'Aglio[1], O. Ferretti[2], F. Manfredi Frattarelli[2], and I. Niccolai[2]

ABSTRACT

Presenting and discussing new data stressing the link between marine and continental trace element geochemistry is the aim of this paper.

1. Among the maps from geochemical exploration surveys, applied to environmental issues, it is worth mentioning the one that shows the distribution of Hg in the stream sediments from Tuscany, where 1,000 samples were examined from an area of 20,000 km^2 (1966). Fifteen years later huge Hg anomalies were picked up in marine sediments from Tyrrhenian sea and 20 years later Hg contents in local fisheries were frequently found to be higher than the allowed Maximum Admissible Concentration (MAC) values.
2. Data on the anomalous As contents, in both continental and marine environments, are presented. In particular, the new maps of As distribution in the stream sediments of southern Tuscany, as well as in marine sediments from the Piombino channel, show huge and wide anomalies. These last data are impressive, because the average value of all the 57 samples is as high as 66 mg/kg (vs an average value for the Earth's crust of about 2 mg/kg) and the samples collected around Elba Island have an average value as high as 252 mg/kg, with peaks of over 400 mg/kg. The isoconcentration curves clearly identify the origin of the dispersion aureole on the eastern coast of the Island. With reference to As distribution, two very extensive and distinct aureoles can be distinguished: one originating from the mining sector of eastern Elba Island (high arsenopyrite content in sulfide deposits) and the other located in mainland Tuscany and represented by multiple dispersion sources.

The need for urgently initiating research in two related complementary areas to get more information on the As aureoles and on their effects on the biota is also discussed.

Introduction

Studying geochemical anomalies of toxic trace elements in dispersion media related to the biosphere, taking into account continental and marine environments as a whole, is the best way to evaluate the environmental and health effects caused by geochemical anomalies.

The abundance and the bioavailability of chemical elements have, in fact, always had a strong influence on the survival and development of the biosphere, affecting either the different ecosystems in general or individual living species in particular.

The distribution of essential, critical and toxic elements corresponds, to a first approximation, to standards of uniformity within the different geochemical spheres. The anomalous distribution of these elements, especially within geochemical spheres which mostly interact with the biosphere, may have direct effects on the health of animals, in general, and of man, in particular. The scientific community – moreover – has just recently realized that the appearance of life on and the subsequent biological evolution of our planet are closely related to and dependent on the geochemical evolution that occurred at the Earth's surface in geological times (Dall'Aglio 1995).

[1] Dipartimento di Scienze della Terra, Università "La Sapienza", Roma, Italy
[2] Centro Ricerche Ambiete Marino, ENEA St. Teresa, Pozzuolo di Lerici 316, 19100 La Spezia, Italy

F.M. Faranda, L. Guglielmo, G. Spezie (eds)
Mediterranean Ecosystems: Structures and Processes

Natural processes may bring about highly anomalous values of these elements at the Earth's surface. Among these processes the most effective ones are those connected with hydrothermal circulation in high heat flow areas. Hydrothermal circulation supports geothermal fields at present and supported minerogenetic processes in the past. There is a logical link between these two resources, which are related to each other, in space and time, by geochemical, tectonic, and structural features.

In the last years, the assessment of natural risks has undergone a scientific revolution. Among these risks, the geochemical one ranks among the most severe. Italy is one of the countries in the world most exposed to these risks. Past and present natural processes, mainly due to hydrothermal circulation, generate intense and widespread anomalies in trace elements of toxicological concern at the Earth surface. (Dall'Aglio and Marinelli 1994).

The geochemical risk may be defined as the health hazard originating by huge anomalies of toxic elements at the Earth's surface interacting with the biosphere. (Dall'Aglio 1996).

One of the most serious issues that mankind has to face now, and especially in the near future, concerns the anthropic change of the environment and, in particular, in the concentration of elements in geochemical cycles and different

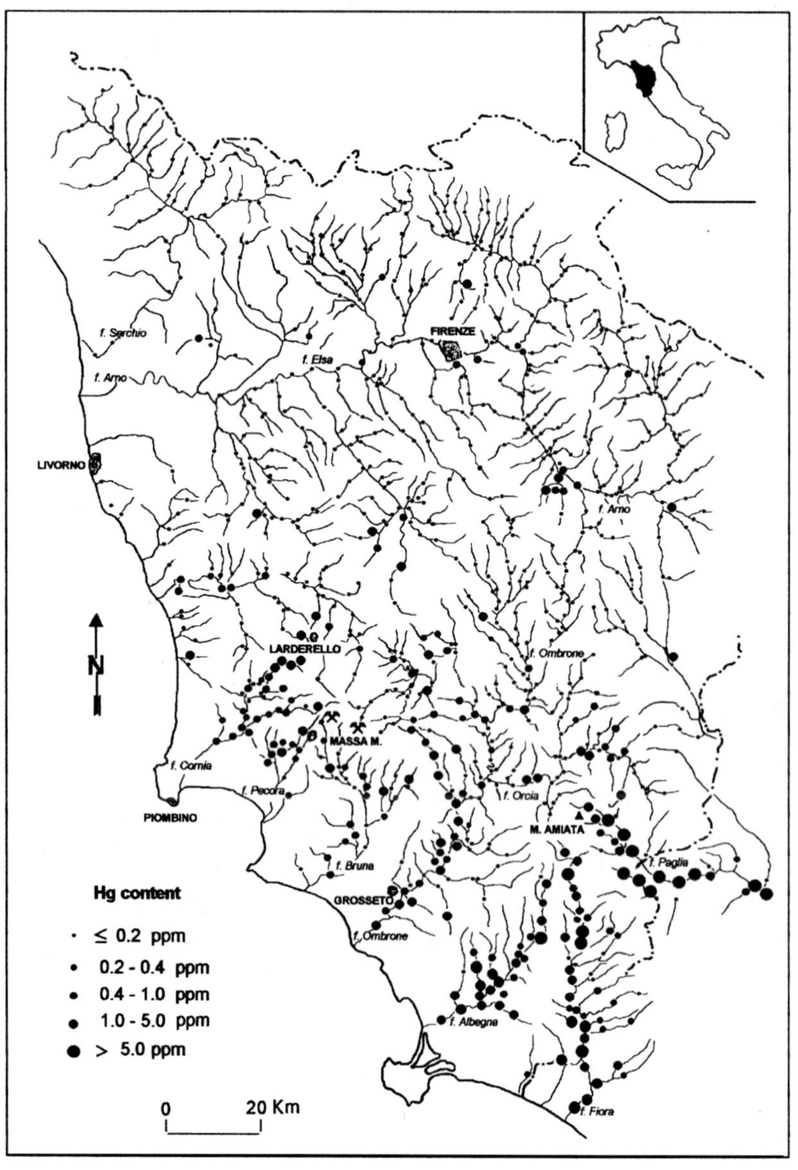

Fig. 1. Hg distribution in stream sediments from Tuscany region

environments. Estimations of anthropic changes, and the necessary regulatory framework to cut down emissions and releases, must be based on reliable data on the capacities of natural reservoirs, as well as on the fluctuations in the content of chemical elements that may occur even under undisturbed conditions.

This paper covers the environmental and health effects that marked geochemical anomalies may entail, in general; particular attention will be paid to new examples dealing with Hg and As distribution in stream and marine sediments from the Tuscany region, both on land and in the Tyrrhenian Sea in front of this region. Tuscany is in fact, one of the areas in the world that is richest in thermal manifestations and in hydrothermal metal deposits, and it is thus supposed to have significant anomalies of all trace elements that are selectively mobilized and deposited under hydrothermal conditions.

Mercury Anomalies in Stream Sediments, Marine Sediments and Marine Biota from Tuscany Region

Among old maps from geochemical exploration surveys, applied to environmental issues, it is worth mentioning the one, which shows the distribution of Hg in the stream sediments from

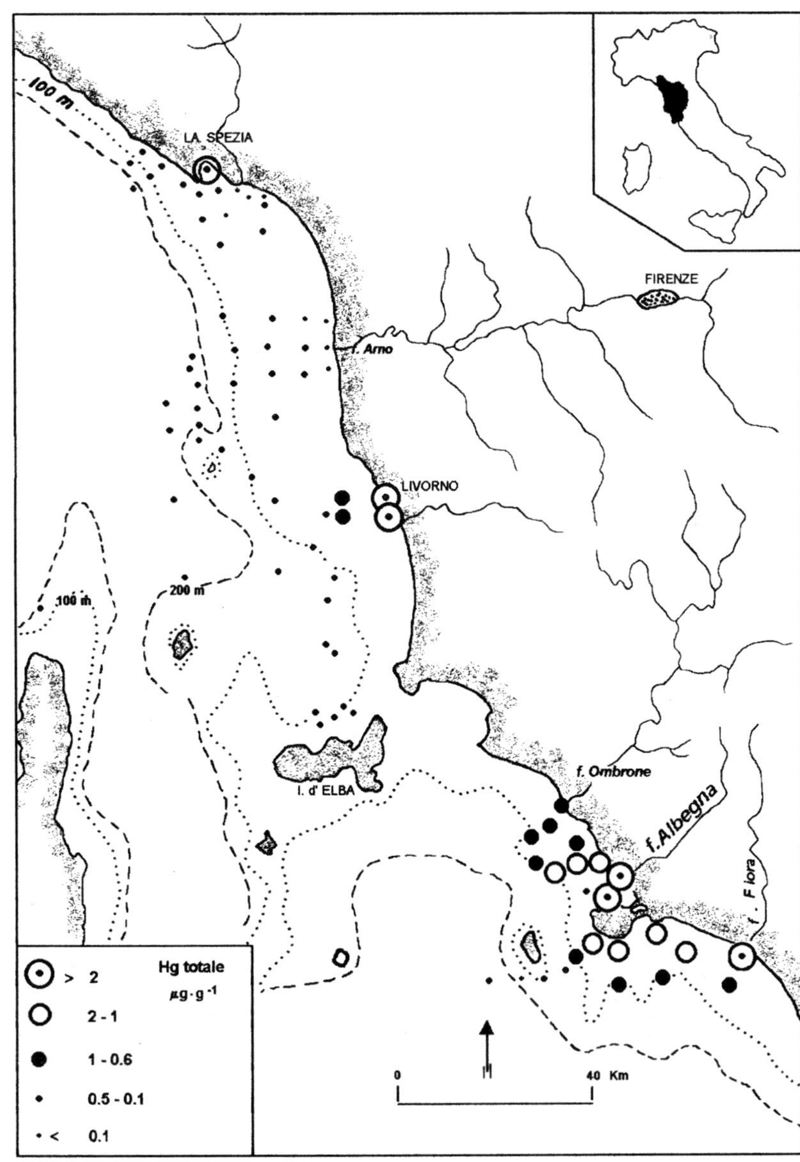

Fig. 2. Mercury distribution in marine sediments from the Tyrrhenian Sea. Notice the low content found in samples north of Elba Island and the huge anomalies in correspondence to Ombrone, Albegna and Fiora rivers (from Baldi et al. 1979)

Tuscany, where 1,000 samples were examined from an area of 20,000 km^2 (Dall'Aglio et al. 1966a). Figure 1 reports the Hg distribution in stream sediments. This geochemical map resulted to be very significant either for mineral exploration (new Hg mineralizations were found in the vicinity of some anomalies) or for environmental purposes.

Fifteen years later huge Hg anomalies were picked up in marine sediments from the Tyrrhenian Sea. Figure 2 shows the map of mercury distribution: notice the high values of the anomalies induced in marine sediments by the input from Ombrone, Albegna and Fiora rivers, whereas no significant Hg anomalies are found north of Elba Island (Baldi et al. 1979).

Twenty years later Hg content in local fisheries was frequently found to be higher than the allowed MAC values (Bernhard and Buffoni 1982). The diagram in Fig. 3 shows the differences between the Hg content in fisheries (sardines) from the Mediterranean Sea and from the Atlantic sea. Inside the Mediterranean Sea we have other Hg contributions (for instance from Spain) apart those from the Tuscany geochemical province.

More recent publications, as well as controls on fisheries, indicated that frequently Hg MAC values are higher than the allowed MAC values, particularly in the area in the vicinity of the huge anomalies in marine sediments in front of the rivers coming from the Hg mineralized areas of continental Tuscany.

Arsenic Anomalies in Stream and Marine Sediments from Tuscany

As Distribution in the Stream Sediments from Southern Tuscany

As a result of research jointly conducted by RIMIN S.p.A. (ENI Group) and the Earth Science Department, Rome1 University "La Sapienza", new data were collected on the distribution of trace elements of mining and environmental interest in Latium (Dall'Aglio 1997; Dall'Aglio et al. 1992). Subsequently, similar data obtained in

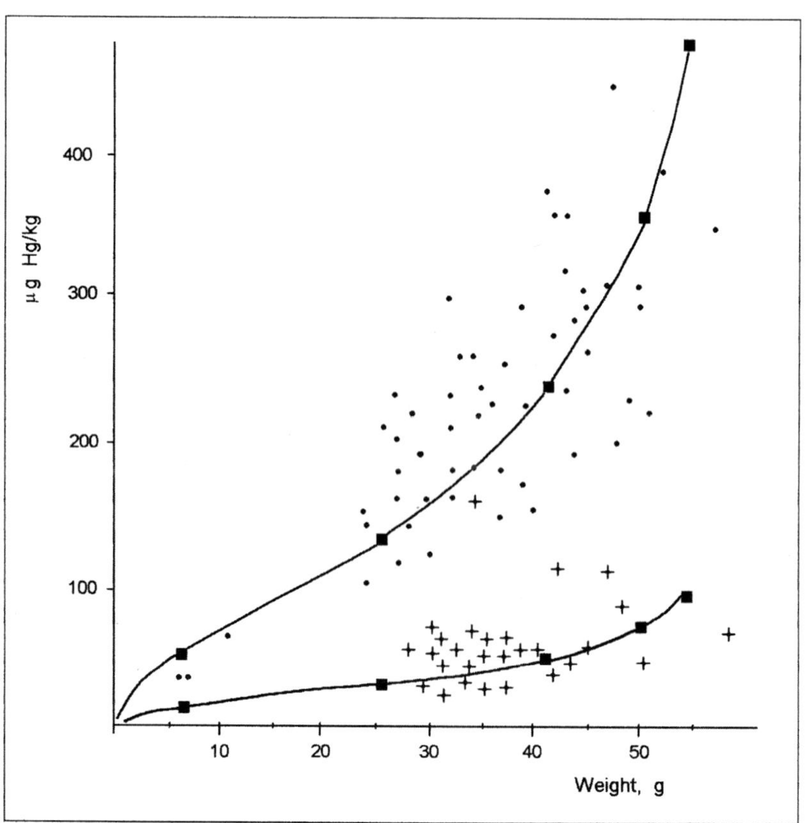

Fig. 3. Differences between the Hg content in fisheries (sardines) from the Mediterranean sea (•) and from the Atlantic sea (+) (from Bernhard and Buffoni 1982)

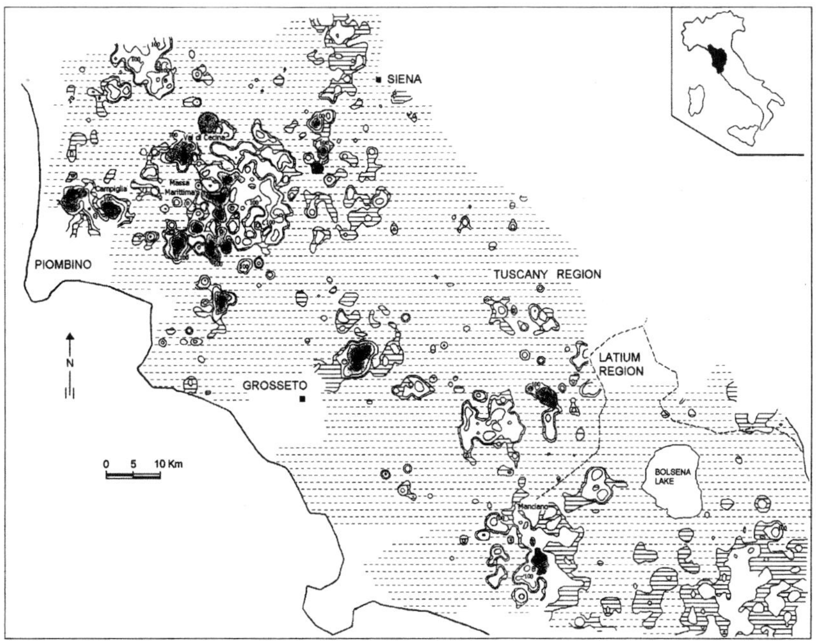

Fig. 4. As distribution in stream sediments from southern Tuscany. *Darkest areas* point out As content > 200 mg/kg

Table. General data related to the geochemical map of As distribution in the stream sediments from southern Tuscany shown in Figure 4

Studied area	6600 km^2
Sample type	Stream sediments
No. of samples	20,191
Sampling density	3-4 samples/km^2
Analised elements	Si, Al, Fe, Ca, Mg, Ti, Mn, P, K, Ag, As, B, Ba, Be, Cd, Co, Cu, Hg, Li, Nb, Ni, Pb, Sb, Sn, Sr, V, W, Zn, Zr

other Italian regions were processed and sorted. Here, a new map of the distribution of As in the stream sediments of southern Tuscany is presented. Figure 4 shows the map of As isoconcentration curves. The pertinent basic data are exhibited in the Table.

The mixed-sulfide mineralizations of the Campiglia area display major anomalies in the content of As, which reaches approx. 10,000 mg/kg in the Massa area, in the Tuscan ridge and in the epithermal halos of Val di Cecina, Amiata and Manciano.

As Distribution in the Marine Sediments of the Piombino Channel

Leoni et al. (1991) reported the results of surveys on marine sediments from the Tyrrhenian sea dealing with the distribution of some trace elements; As was analysed too and no significant anomalies were detected.

Recently, new data have been obtained on the distribution of As in the marine sediments of the Piombino Channel (Dall'Aglio et al. 1996b). Fifty-seven sediment samples were collected in the sea around Elba Island, namely in the area connecting mainland Italy with the island. Analyses were carried out starting from the minus 250 mm fraction. Flameless AAS analyses were carried out of the total content of some trace elements of environmental relevance, but here only the results concerning the distribution of As will be discussed.

Figure 5 shows the map of As isoconcentration curves in the area under review, as well as elementary statistical data on the different populations that were investigated.

These data are impressive, because the average value of all the 57 samples is as high as 66 mg/kg (vs an average value for the Earth's crust of about 2 mg/kg) and the samples collected around Elba Island have an average value as high as 252 mg/kg, with peaks of over 400 mg/kg. The isoconcentration curves clearly identify the origin of the dispersion aureole on the eastern coast of the Island.

It is worth making a few points on the overall findings from the two campaigns (stream and marine sediments) as they provide a relatively

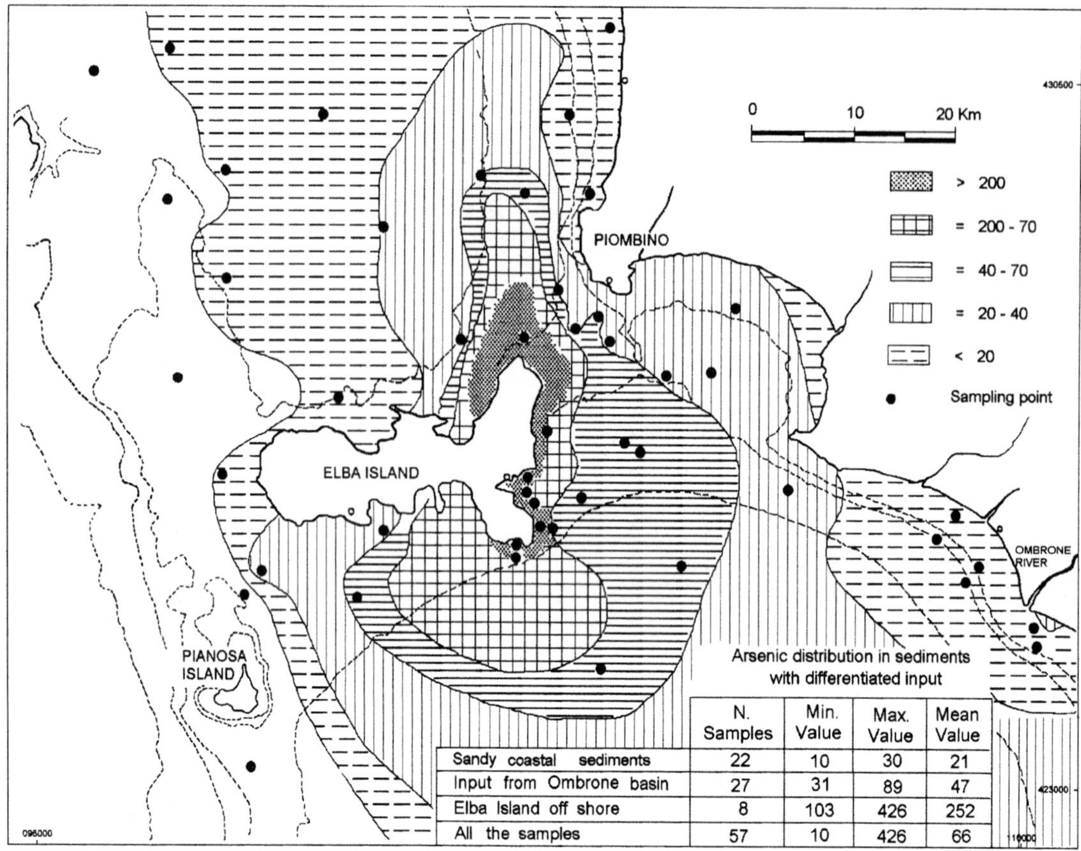

Fig. 5. Isoconcentration curves of total As (mg/kg d.w.) in marine sediments around Elba Island

complete picture of the situation, in spite of the lack of exhaustive data. Two very extensive and distinct aureoles can be distinguished: one originating from the mining sector of the eastern Elba Island (high arsenopyrite content in sulfide deposits) and the other located in mainland Tuscany and represented by multiple dispersion sources. Conversely, the scale of both marine and continental aureoles is completely unexpected.

Conclusions

The data presented stress the interrelations existing between continental and marine sediments with respect to geochemical anomalies of Hg and As in the case of the Tuscany region and the Tyrrhenian Sea. As distribution, in particular, showed unexpected aureoles both in continental and marine sediments. These observations point to the need for urgently initiating research in these two complementary areas, with a view to correctly assessing the related geochemical risk. In effect, it is necessary and urgent: (1) to determine the distribution of As in the various surface dispersing media of both Elba Island and of the suspected mainland areas as accurately as possible; (2) to gain greater insight into the bioavailability of As, its critical pathways and possible transmission along trophic chains. There is a poor availability of data especially on the marine environment. Furthermore, the EU might introduce acceptability limits for terrains and sediments having such high As concentrations as those reported here.

Shareable and sustainable development can only rely on the responsible assessment and maximization of the value of natural resources and on mitigation of natural risks, among which the geochemical risk is the one with the greatest number of potential consequences in our country. The issues related to this risk were discussed by Dall'Aglio (1996, 1997) and Dall'Aglio and Marinelli (1994). "Prevention is better than cure" is the leitmotiv of a recent volume published by

UNESCO (1995), which proposes a programme of worldwide geochemical research for collection of new data on the distribution of trace elements in all surface dispersing media, so as correctly to assess and preserve natural resources by responsibly taking into account the geochemical risk. Scientific communities and individual governments should rapidly acquire all the relevant data, with the necessary level of detail.

Italy fails to have reliable and comprehensive statistics on the characteristics of natural resources and namely on natural waters as well as on surficial dispersion media such as soil, stream and marine sediments (ISTAT 1996).

The data presented strongly demand for implementing research activities, with a fully interdisciplinary approach, in order to collect all the necessary information for the best use of natural resources and to avoid health injury due to anomalies of highly toxic elements in dispersion media interacting with the biosphere. Among these activities we can recall the following:

1. Gathering full information on the distribution, speciation and bioavailability of the highly toxic elements in surficial dispersion media.
2. Reducing the lack of information related to some basic processes dealing with the interrelations between the substratum features and living species; as a matter of fact no information is available on the bioavailability of As in marine environment.
3. Implementing toxicological research on long term effect on animal health of low dose response to highly toxic trace elements.
4. Initiating systematic epidemiological research on the possible health effects in areas of huge geochemical anomalies of toxic elements (accomplishing this point strongly demands for gathering the necessary information under 1).
5. Carefully examining the impact of geochemical anomalies on ecosystems.

Scientific conclusions and their impact on land and water use, society and finance underline the urgency of addressing and solving these problems. A time-bomb has been activated: the longer the time elapsing without taking the necessary actions to minimize and prevent damage, the heavier its effects. All governmental institutions involved, such as technical services, basin authorities, ministries (Public Works, Environment, Health, etc.) and regional authorities should urgently take the above research and development initiatives and make up for lost time.

References

Baldi F, Bargagli R, Renzoni A (1979) The distribution of mercury in the surficial sediments of the northern Tyrrhenian Sea. Mar Pollut Bull 10: 301-303

Bernhard M, Buffoni G (1982) Mercury in the Mediterranean, an overview. In: Anagnostopoulos A. (Ed.) Proc. Intern. Conf. Environ. Pollution. Univ. Thessaloniki, Greece. 458-484

Dall'Aglio M (1995) Cicli biogeochimici in condizioni indisturbate ed alterate. In: Pignatti A (ed) Ecologia Vegetale, UTET SPA, pp 259-273

Dall'Aglio M (1996) Geochemical risk assessment: a tool in evaluating and minimizing natural risks. Plinius 16: 88-91

Dall'Aglio M (1997) Aree italiane con elevato rischio geochimico come illustrato da nuove carte geochimiche. In: Atti VIII Congr Naz Soc Ital Ecol. Soc Ital Ecol Atti 18, pp 381-384

Dall'Aglio M, Da Roit R, Orlandi C, Tonani F (1996a) Prospezione geochimica del mercurio: distribuzione del mercurio nelle alluvioni della Toscana. Ind Min XVII: 391-398

Dall'Aglio M, Ferretti O, Manfredi-Frattarelli F, Niccolai I (1996b) Distribution of arsenic in marine sediments from Piombino Channel as depicted by new geochemical maps. Plinius 16: 91-92

Dall'Aglio M, Marinelli G (1994) Geochemical provinces, minerogenetic processes and environment. In: Proc Workshop Minerogenesi dell'Appennino SGI-Simp. Firenze, September 1992. Mem Soc Geol Ital 48, pp 593-603

Dall'Aglio M, Paoloni A, Stea B (1992) Distribuzione di As, Hg e Sn nei sedimenti fluviali del Lazio ottenuta da campagne di prospezione geochimica: implicazioni ambientali. In: Atti Conv Nazi Soc Ital Ecol, Cosenza, Ottobre 1990, pp 815-825

ISTAT (1996) "Statistiche ambientali"

Leoni L, Sartori F, Damiani V, Ferretti O, Viel M (1991) Trace element distributions in surficial sediments of the northern Tyrrhenian sea: contribution to heavy metal pollution assessment. Environ Geol Water Sci 17: 103-116

UNESCO (1995) A global geochemical database for environmental and resource management. Earth Sci 19

W H O (1993) Guidelines for drinking water quality, 2nd Edn. 1. Recommendation. Geneva

Environmental Constraints on Pathways of Organic Detritus in a Semi-enclosed Marine System (W-Mediterranean)

M. Fabiano[1], G. Sarà[2], A. Mazzola[2], and A. Pusceddu[3]

ABSTRACT

In order to assess seasonal and spatial changes in water-sediment interaction processes in a semi-enclosed marine system of Western Sicily (Marsala lagoon; W-Mediterranean), the biochemical composition of suspended and sediment organic matter was studied, during a one-year sampling period. The observed dynamic balance of resuspension vs. sedimentation processes and the macroalgal and vascular plant coverage appear major factors in affecting both amounts and biochemical composition of suspended and sedimentary organic matter and allowed us to identify two different sub-systems. The northern area, characterised by frequent wind-induced sediment resuspension events and by a scant vegetation, displayed higher amounts of high quality (i.e. protein) sedimentary organic matter and higher amounts of particulate organic matter. By contrast, the southern area, characterised by less wind exposure and a deeper water column, displayed lower particulate organic matter concentrations and increasing sedimentary organic matter concentrations. The oligotrophy of the system (phytoplankton biomass < 0.1 µg Chl-a l^{-1}) and the relatively low concentrations of photosynthetic pigments in the sediments (on annual average 3.1 µg g^{-1}) suggest that this system is largely dominated by organic material of detrital and/or heterotrophic origin. The overall picture of the study area suggests that the Marsala lagoon is a "pulsing" environment characterised by no clear seasonal change of the primary production processes and by the accumulation of large amounts of organic detritus in the sediments. Such uncoupling between amounts and nutritional value of suspended and sediment organic detritus suggests that this particular environment behaves probably as a detrital "trap".

Introduction

Coastal lagoons are controlled by the dynamic balance between external water inputs (marine and fresh waters) and wind and wave auxiliary energy inputs. Lagoons of semi-arid regions, as those facing the Tyrrhenian Sea, show little tidal ranges and scant or absent continental inflows. Thus, in these environments, wind become the main physical factor in conditioning biological functions both in the water column and sediments (Millet and Cecchi 1992; Pusceddu 1999).

A large number of studies dealing with ecological features of shallow coastal areas are available in the literature. Most deal with the role of unpredictable fluctuations of physical variables (Painchaud et al. 1990; Navarro et al. 1993; Navarro and Thompson 1995; Chassany de Casabianca et al. 1995; Clavier et al. 1995; Savenkoff et al. 1995; Pusceddu et al. 1996). Other papers also showed that spatial and temporal changes in quality and quantity of organic detritus, rather than changes in physical variables could affect both benthic and plankton (Pusceddu et al. 1997a; Scilipoti et al. 1997).

In lagoon environments, primary production from phytoplankton, macrophytes and salt marsh plants generally exceeds consumption by herbivores (Newell 1982). Thus, a large amount of primary organic matter might be directly avai-

[1] Dipartimento per lo Studio del Territorio e delle Sue Risorse, Università di Genova, Viale Benedetto XV, 16100 Genova, Italy
[2] Dipartimento di Biologia Animale, Università di Palermo, Via Archirafi 18, 90123 Palermo, Italy
[3] Istituto di Scienze Marine, Università di Ancona, Via Brecce Bianche, 60131 Ancona, Italy

F.M. Faranda, L. Guglielmo, G. Spezie (eds)
Mediterranean Ecosystems: Structures and Processes

lable to consumers but, in most cases, it needs to be fragmented and processed through decomposer pathways to become easily digestible (Hansen et al. 1992). This gap leads to a well-documented lack of coupling between primary and secondary production processes (Carrada and Fresi 1988). Nevertheless, only few papers dealing with this matter are detailed up to the biochemical scale of observation (Sarà et al. 1995; Pusceddu et al. 1996, 1997a, 1997b; Pusceddu 1999; Pusceddu et al. 1999).

In the present paper, we describe how some physical factors (e.g. wind and algal coverage) might influence spatial distribution and temporal patterns of quality (i.e. biochemical composition) and quantity of suspended and sediment organic matter in a semi-enclosed marine system of Western Sicily (Marsala lagoon, W-Mediterranean). The aim of this paper is to highlight the role of organic detritus as a high-definition descriptor of the trophic state and functioning of Mediterranean coastal lagoons.

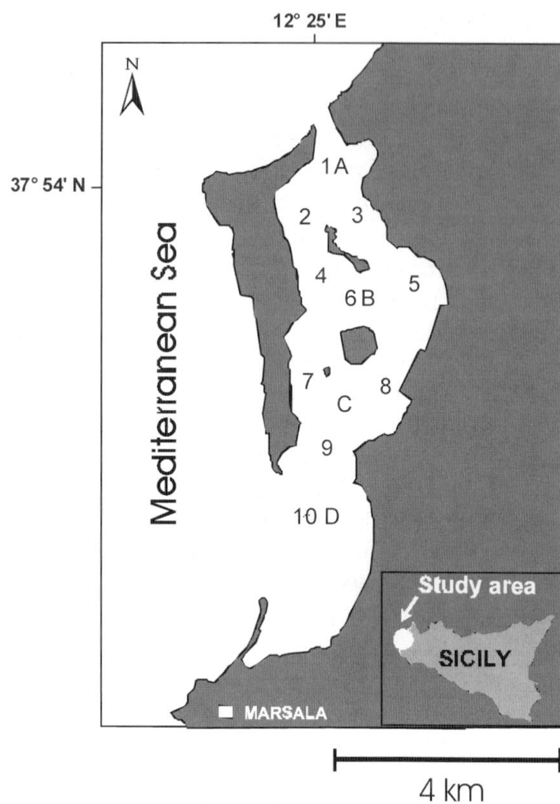

Fig. 1. The study area and sampling stations. Arabic numbers and letters indicate particulate matter and sediment sampling stations, respectively

Materials and Methods

Study area and Sampling

The study was carried out in the Marsala lagoon, characterised by two main communication channels with the open sea (Fig. 1). The basin is very shallow, depth ranging from 2 m along eastern shore of the platform to 0.50 m in the western area. Basin depth increases gradually to about 2.5 m in the southernmost area, close to the open sea. The northern channel is 450 m wide and characterised by occasional turbulent inputs of seawater. In this area sediments are characterised by scant (low frequency) algal coverage. The southern mouth, 1450 m wide, is open to a continuous sea water inflow and is characterised by internal tides. The hydrography of the central area is greatly influenced by two little islands, which act as mechanical obstacles to the water circulation and generate turbulent currents. Southernmost to this area a luxuriant *Posidonia oceanica* reef is present, influencing water circulation and silting and, in particular, determining a *de facto* separation of the basin in two different sub-systems. In fact, water exchanges with the open sea are ensured by currents with mean speed that in the southern mouth (annual average, 4.92 ± 1.54 cm s^{-1}) are about two-fold higher than in the northern area (2.34 ± 0.87 cm s^{-1}) (Mazzola and Sarà 1995; Campolmi 1998). No continental inputs are present.

Samples of superficial water (0.5 m) were collected monthly from January to December 1994, using 10-l Niskin bottles, at 10 stations located along a north-south transect. A series of sediment samples was collected monthly by scuba diving at four stations. Three replicate cores per station and period were collected randomly from three quadrats (20 cm in length, 400 cm^2 surface area) belonging to a larger quadrat (1 m^2). Sediment cores were obtained by inserting PVC tubes (internal diameter, 4.7 cm). Samples were brought back to the laboratory and processed within 4 h. The top 1-cm layer of each core, used for analyses of carbohydrates, proteins, lipids and photosynthetic pigments, was immediately frozen at −20 °C and stored until analysis. Temperature and salinity were measured in situ monthly using a Hydrolab multiprobe.

Chemical Analyses

To analyse total suspended matter (TSM), suspended pigments, particulate carbohydrate, protein and lipid concentrations, sub-aliquots (100-1000 ml) of the samples were filtered under gentle vacuum onto Whatman GF/F glass-fibre filters (0.8-μm nominal pore size). To determine TSM, Whatman GF/F filters (0.45 μm nominal pore size) were weighed after desiccation (60 °C, 24 h, Strickland and Parson 1972) using a Mettler M3 balance (accuracy ± 1 μg). Particulate chlorophyll-a (P-Chl-a) analysis was carried out according to Lorenzen and Jeffrey (1980). Pigments were extracted with 90% acetone. Phaeopigments (Phaeo) were determined after acidification with 0.1 N HCl. Particulate organic carbon (POC) and nitrogen (PON) were determined with a Perkin-Elmer CHN Elemental Analyser (Mod. 2400), using acetanilide at 925 °C as a standard, the inorganic carbon having been removed using HCl (Hickel 1984; Iseki et al. 1987). Particulate carbohydrates (P-CHO) were determined according to Dubois et al. (1956). d (+) Glucose was used as the standard. Particulate proteins (P-PRT) were determined according to Hartree (1972), using bovine serum albumin (BSA) as the standard. Particulate lipids (P-LIP) were extracted according to Bligh and Dyer (1959) and analysis was performed by carbonisation according to Marsh and Weinstein (1966). Tripalmitine was used as the standard. The biopolymeric fraction of particulate organic carbon (P-BPC) was defined as the sum of carbohydrate, protein and lipid carbon (Fabiano and Pusceddu, 1998).

Analyses of sediment chlorophyll-a (S-Chl-a) and phaeopigments (S-Phaeo) were carried out according to Lorenzen and Jeffrey (1980). Pigments were extracted with 90% acetone (24 h in the dark at 4 °C). After centrifugation, the supernatant was used to determine the functional Chl-a and acidified with 0.1 N HCl to estimate the amount of phaeopigments (Plante-Cuny 1974). Sedimentary chloroplastic pigment equivalents (S-CPE) were calculated as sum of chlorophyll-a and phaeopigment concentrations. Microphytobenthic carbon was calculated by converting S-Chl-a concentrations to carbon content using a conversion factor of 40 (de Jonge 1980).

After sonication in deionized water, total sediment lipids (S-LIP) were extracted by direct elution with chloroform and methanol according to Bligh and Dyer (1959) and Marsh and Weinstein (1966). Data are reported in tripalmitine equivalents. Sedimentary protein (S-PRT) analyses of three replicates were conducted following extraction with NaOH (0.5 M, 4 h) and were determined according to Hartree (1972) as modified by Rice (1982), to compensate for phenol interference and expressed as BSA equivalents. Sedimentary carbohydrates (CHO) of three replicate samples were analysed according to Gerchacov and Hatcher (1972) and expressed as glucose equivalents. The assay, which reacts with reducing saccharides and is widely utilised in the literature, is based on the same principle as the widely utilised method of Dubois et al. (1956), but it is specifically adapted for carbohydrate determination in sediments.

For each biochemical analysis, blanks were performed using sediments that had been precombusted at 450 °C for 2 h. Sedimentary lipid, carbohydrate and protein were converted into carbon equivalents using 0.75, 0.40 and 0.49 g C g^{-1} conversion factors, respectively (Fabiano and Pusceddu 1998). The sedimentary biopolymeric organic carbon (S-BPC) was defined as the sum of carbohydrate, protein and lipid carbon (Fabiano et al. 1995).

Statistical Analyses

Temporal and spatial fluctuations were investigated by means of analysis of variance (ANOVA; Underwood, 1997) with time (month) and space (station) as sources of variation. When a significant difference ($p < 0.05$) for the main effect was observed, means were analysed by a Tukey's multiple comparison test to determine the differences between sampling stations and months. Data were transformed, only when necessary, to meet the assumptions of the parametric statistics.

Results

Physical Measurements, Total Suspended Matter and Suspended Pigments

Temperature and salinity showed a clear seasonal trend (Fig. 2). The minimum temperature was observed in January (12.38 ± 0.53 °C), while the maximum was measured in August (27.65 ±

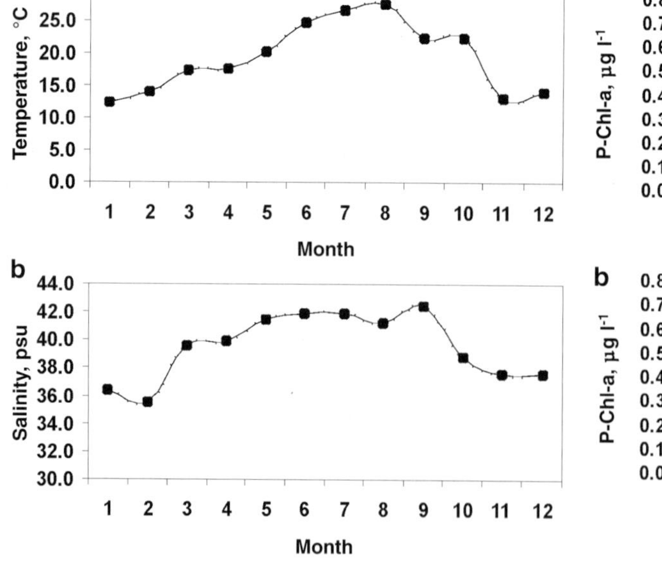

Fig. 2a,b. Temperature (**a**) and salinity (**b**) seasonal patterns in the Marsala lagoon. *psu*, practical salinity units

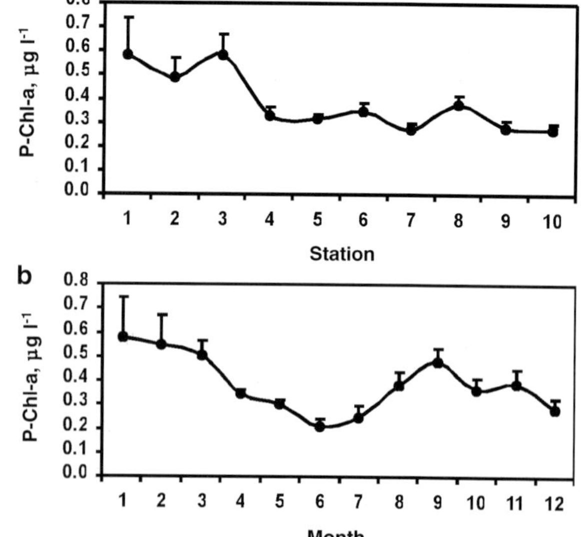

Fig. 4a,b. Spatial and temporal changes in suspended chlorophyll-a (S-Chl-a) concentrations in the Marsala lagoon. *Error bars* indicate standard deviations among stations (spatial) and months (temporal)

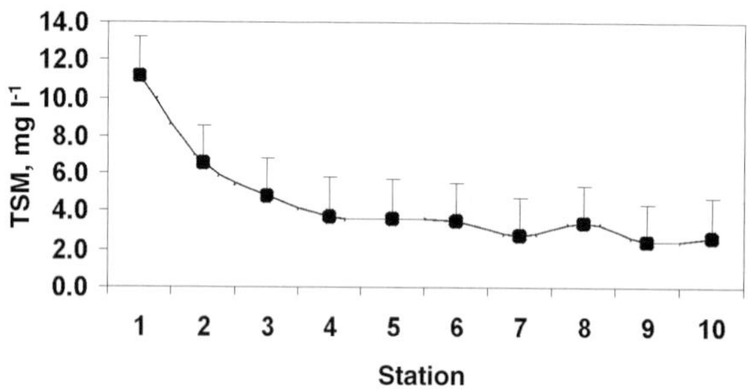

Fig. 3. Spatial changes in total suspended matter (TSM) concentrations in the Marsala lagoon. *Error bars* indicate standard deviations among months

0.45 °C). The minimum value of salinity was recorded in January (36.4 ± 0.53), while the maximum was measured in September (42.50 ± 1.58). Total suspended matter (TSM) concentrations were fairly high, ranging from 1.6 mg l^{-1} to 11.2 mg l^{-1} and displayed a significant decrease along the north-south transect (Fig. 3). TSM was dominated by its inorganic fraction (annual average, 2.9 mg l^{-1}, accounting for approximately 60%, range 1-90%). Chlorophyll-a concentration (annual average 0.4-µg l^{-1}) ranged from 0.02 µg l^{-1} to 2.0 µg l^{-1}. Phaeopigment concentration (annual average, 0.13 µg l^{-1}) ranged from 0.01 to 0.67 µg l^{-1} and represented, on annual average, about 25% of the total chloroplastic pigment concentration. The concentration of chloroplastic pigment equivalents (CPE) showed significant ($p < 0.05$) spatial and temporal changes (Fig. 4).

Elemental and Biochemical Composition of Particulate Organic Matter

Particulate organic carbon and nitrogen concentrations ranged from 32.6 µg C l^{-1} to 1288.3 µg C l^{-1} and 5.3 µg N l^{-1} to 130.7 µg N l^{-1}, respectively. Generally, POC and PON concentrations showed a decreasing pattern from the northern to the southern stations (Fig. 5a,b), with the exception of station 5, where concentrations reached an annual average of 306.6 µg C l^{-1} and 37.1 µg N l^{-1}. POC and PON concentrations (Fig. 5c,d) showed three accumulation periods ($p > 0.05$) in January-February, May-June and September-October.

Particulate protein concentrations (annual average 390 µg l^{-1}) ranged from 34 to 2237 µg l^{-1}. Particulate proteins showed significant temporal and spatial changes ($p < 0.05$) (Fig. 6a,b). Peaks appeared in January-February (average, 574 µg l^{-1}) and in July (451 µg l^{-1}). The highest concentrations were observed in the northern area (maximum, 610 µg l^{-1} at station 1), with concentrations decreasing continuously from the northern to

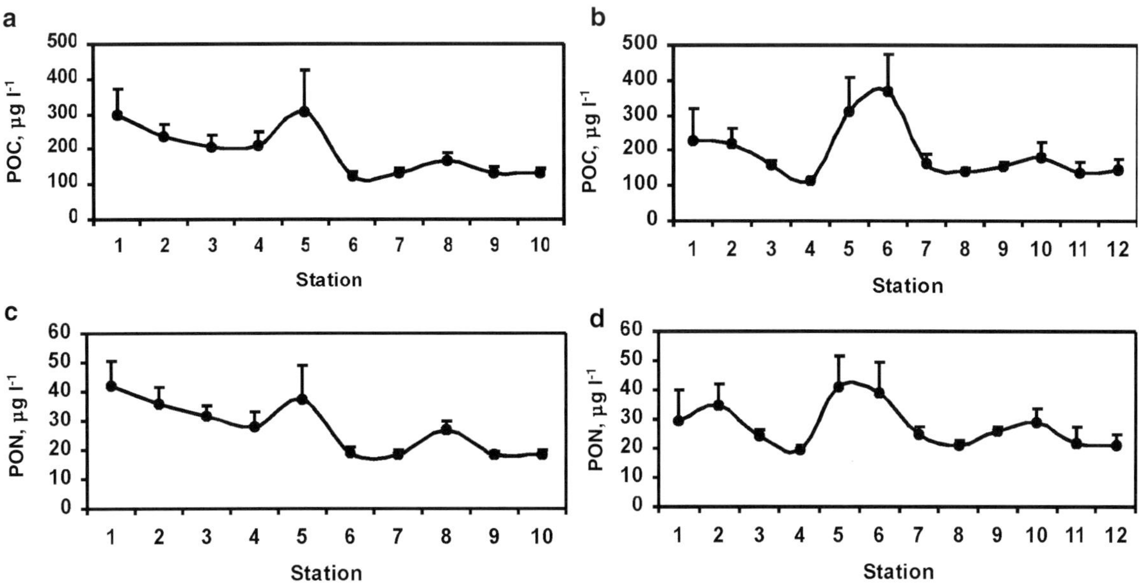

Fig. 5a-d. Spatial and temporal changes in POC (**a,b**) and PON (**c,d**) concentrations in the Marsala lagoon. *Standard error bars* indicate standard deviations among stations (spatial) and months (temporal).

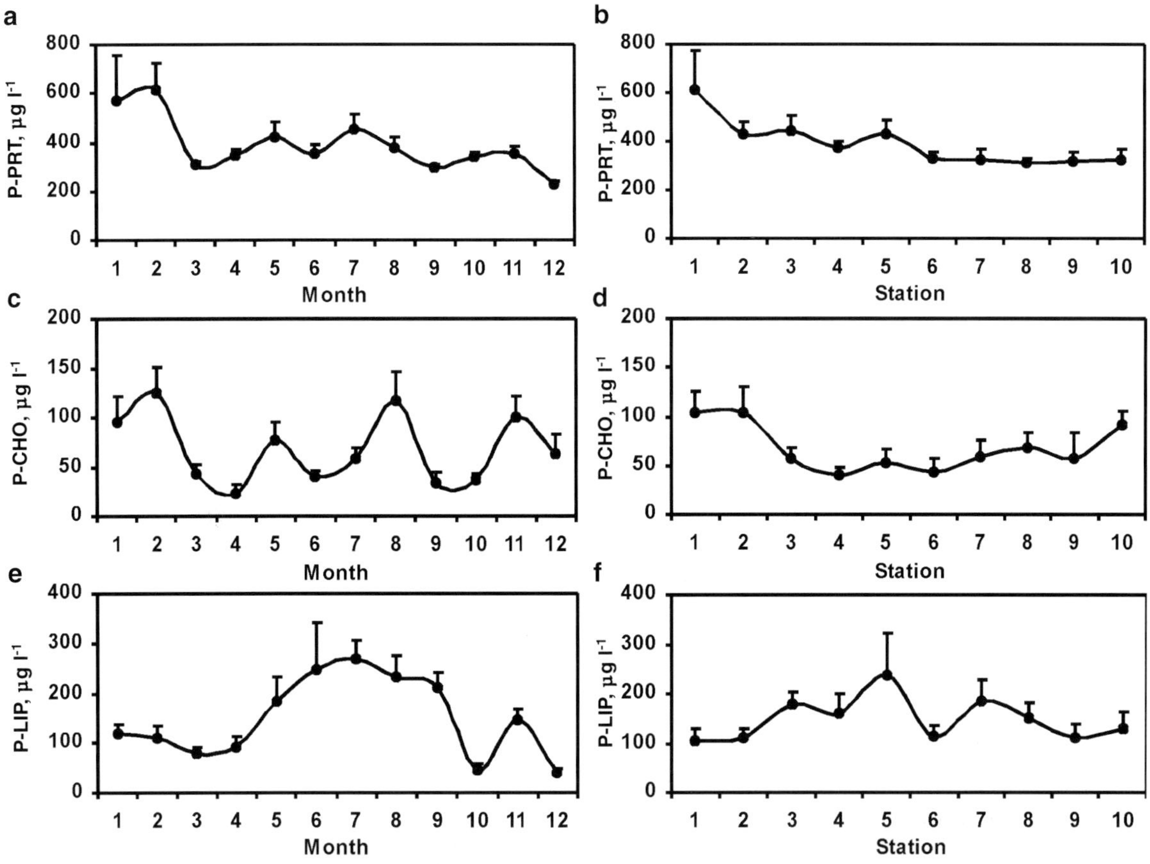

Fig. 6a-f. Spatial and temporal changes in suspended protein (**a, b**), carbohydrate (**c, d**) and lipid (**e, f**) concentrations in the Marsala lagoon. *Standard error bars* indicate standard deviations among stations (spatial) and months (temporal)

the southern area (mean value between stations 8 and 10, 318 μg l^{-1}). Particulate carbohydrate concentrations ranged from undetectable to 503 μg l^{-1}. Carbohydrate concentrations fluctuated widely and displayed significant seasonal changes (Fig. 6c). No significant spatial changes were observed (Fig. 6d), although concentrations decreased from station 1 to station 4 (annual average 104 μg l^{-1} and 41 μg l^{-1} respectively), then increased up to station 10 (91 μg l^{-1}). Particulate lipid concentrations (annual average 184 μg l^{-1}) ranged from 7 to 1441 μg l^{-1}. Particulate lipids did not display significant spatial changes, while significant seasonal changes were observed (Fig. 6e,f). A significant peak occurred in July (270 μg l^{-1}). Particulate proteins were the most abundant biochemical compound, accounting for an annual average of approximately 66% of biopolymeric carbon, followed by lipids (23.0%) and carbohydrates (11.0%). Biopolymeric organic carbon concentrations (data not shown; annual average, 355 μg C l^{-1}) ranged from 100 μg C l^{-1} to 1381 μg C l^{-1}. P-BPC concentrations did not show significant spatial changes, while significant seasonal changes were observed, with maximum values in February, July and November (average 446, 535 and 339.1 μgC l^{-1}, respectively).

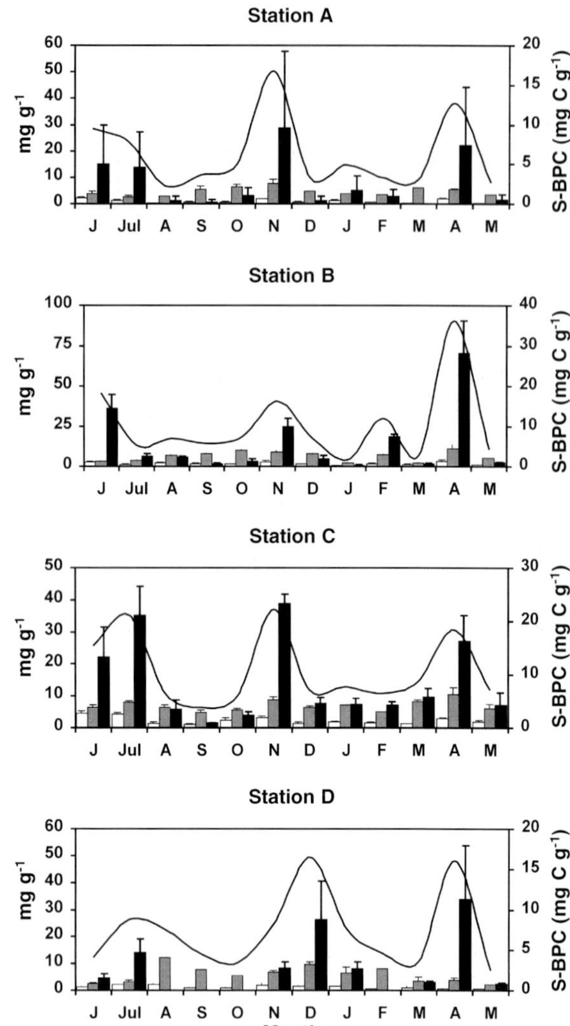

Fig. 8. Temporal changes in sedimentary organic matter at the four investigated stations in the Marsala lagoon. Values are given for sedimentary lipid (□) protein (▒), carbohydrate (■) and biopolymeric organic carbon (S-BPC, *curve*). *Left ordinate*, protein, carbohydrate, lipid; *right-ordinate*, S-BPC. *Error bars* indicate standard deviations among replicates ($n=3$)

Biochemical Composition of Sediment Organic Matter

Sedimentary chlorophyll-a concentrations in the study area were, on annual average, 3.1 μg g^{-1}. ANOVA results did not show any significant seasonal or spatial changes. However, September-October and February-March appeared to be the accumulation periods (Fig. 7). Highest concentration of chloroplastic pigments (CPE) was measured in station B (43.7 μg g^{-1}, February). S-Chl-a carbon accounted for a negligible fraction of S-BPC (on average, 2.2%). Phaeopigments concentrations were always higher than active chlorophyll-a and ranged from 0.6 to 38.3 μg g^{-1}. Sedimentary carbohydrate, lipid and S-BPC temporal patterns were characterised by a strong seasonality, while sedimentary protein concentrations did not vary significantly among seasons (Fig. 8). The four stations did not show significant differences for the protein and car-

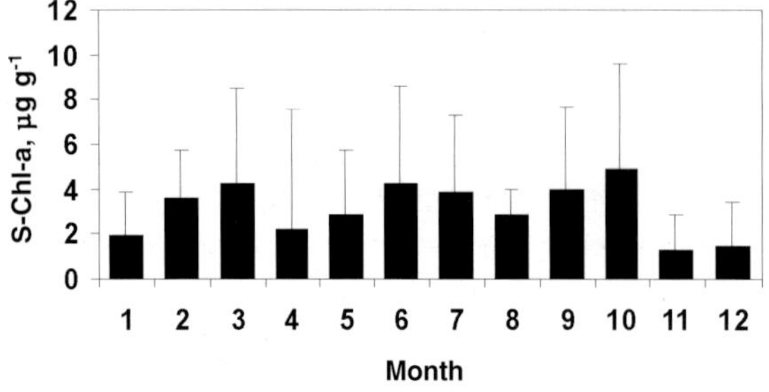

Fig. 7. Temporal changes in chlorophyll-a concentrations in the sediments of the Marsala lagoon. *Error bars* indicate standard deviations among stations

bohydrate content of sediments, while their contribution to the S-BPC varied seasonally.

Sedimentary carbohydrate concentrations ranged between 0.8 mg g^{-1} (station B, January) and 70.5 mg g^{-1} (station B, April). Soluble carbohydrates accounted on annual average for 8.3% (range 0.3-26.7%) of the total carbohydrate pool. Protein concentrations ranged between 2.2 mg g^{-1} (station B, January) and 12.1 mg g^{-1} (station D, August). Lipid concentrations ranged between 0.3 mg g^{-1} (station A, March) and 4.5 mg g^{-1} (station C, June). The biopolymeric fraction of sedimentary organic carbon (S-BPC) varied seasonally. Carbohydrates were the dominant class among labile organic compounds (annual average, 51.2%), followed by proteins (39.0%) and lipids (9.8%). The carbohydrate contribution to the sedimentary BPC was higher in stations 3 and 4 (52.5% and 59.6%, respectively), whereas proteins increased their relative importance (36.9% and 31.1%) with respect to stations A and B (44.9% and 43.1%, respectively). On annual basis, lipid contribution to sedimentary BPC ranged between 8.4% (station 1) and 10.7% (station C).

Discussion

Origin and Biochemical Composition of Suspended and Sediment Organic Matter

Little difference in phytoplankton biomass (in terms of chlorophyll-a concentrations) was found when compared to the results of other studies previously carried out in the same area (Magazzù 1982; Giani et al. 1995) or in the adjacent open sea (Carrada et al. 1996). The study area can be defined oligotrophic, as P-Chl-a concentrations had an annual average of <1.0 µg l^{-1}. However, photosynthetic rates have been reported to be quite high (2 mg C m^{-3} h^{-1}), with high specific phytoplankton growth rates ($\mu = 0.93 \pm 0.55$; Magazzù 1982). These results may indicate the presence of rapid nutrient recycling and are supported by the high biomass values and activity of heterotrophic bacterial assemblages found by Genovese (1969). Throughout the study period, P-Chl-a concentrations were consistently dominant (on average, 75% of CPE) with respect to phaeopigments, indicating the presence of photosynthetically active phytoplankton. Nevertheless, phytoplankton biomass accounted on average for a very low fraction of the total suspended matter pool (annual average approximately 0.1%). Assuming an average P-Chl-a carbon content of 54 µg C µg^{-1} (Nival et al. 1972), with respect to total particulate organic carbon, chloroplastic pigments accounted for an annual average of approximately 13%. Thus, we may assume that most of the suspended organic matter in the study site (annual average approximately 87%) was of detrital (non-living) and/or heterotrophic origin. Particulate proteins were the dominant biochemical compound, accounting for approximately 66% of the biopolymeric fraction of particulate organic matter, followed by lipids (23.0%) and carbohydrates (11.0%). Dominance of particulate proteins over carbohydrates is generally characteristic of highly productive areas. Indeed, proteins have been reported to dominate in coastal lagoons (Pusceddu et al. 1996), estuarine lagoons (Navarro et al. 1993; Sorokin et al., 1996) and mariculture impacted areas (Mazzola et al. 2000). Particulate protein concentrations in the study area were higher than those encountered in open sea waters, comparable to those encountered in an estuarine lagoon in southern Chile (Navarro et al., 1993) but about two-fold lower than those found in a true Mediterranean lagoon (Pusceddu et al. 1996). Since chlorophyll-a represented a very limited fraction of the particulate organic matter throughout the year, we might expect most of the particulate protein not to be associated with phytoplankton. POC:PON ratio values were relatively low in the whole basin, ranging from 4.9 to 10.9 (annual average, 6.9). In contrast, as the POC:PON ratio for phytoplankton is reported to range from approximately 6 (e.g. green algae) to approximately 11 (e.g. Peridineans) (Valiela 1984), we could expect a large fraction of POC to be accounted by phytoplankton. However, as discussed above, phytoplankton biomass (in terms of chlorophyll-a) accounted for only 13% of the POC. Thus, other ON sources have to be present. To this regard a possible explanation could be found in the large biomass of bacterioplankton reported for the study area (Genovese 1969) that might provide freshly generated nitrogen, as the POC:PON ratio for bacteria ranges from 3.5 to 5.0 (Lee and Fuhrman 1987). However, only data on bacterial abundance and biomass will clarify and confirm our contention.

Sediment chlorophyll-a and CPE concentrations in the sediments of the study area, when compared to other enclosed marine basins or Mediterranean coastal lagoons, were quite low indicating that this system, although extremely rich in organic detritus, receives low inputs of primary organic matter deriving from microalgae. Moreover, the contribution of primary organic carbon to the sedimentary BPC, calculated assuming the conversion factor of 40 µg C µg S-Chl-a^{-1} (de Jonge 1980), was quite low (on average, 2.2%), being 30-times lower than values encountered in the oligotrophic Ligurian Sea (Fabiano et al. 1995). Although, the C to Chl-a ratio can vary from about 10 to 100, the contribution of primary organic carbon to the sedimentary biopolymeric fraction of organic matter (OM) in the present study remains still low (0.6% and 5.6%, using 10 and 100 as conversion factors, respectively). Thus, we can conclude that most of the sedimentary OM in the Marsala lagoon originates from sources different from microphytobenthos or deposited phytoplankton.

In the sediments of the Marsala lagoon, the three main biochemical classes of organic compounds showed the highest concentrations ever reported in the literature. The biochemical composition of the sedimentary organic detritus showed the dominance of carbohydrates (on average, 51.2%) followed by proteins (39.0%) and lipids (9.8%). The dominance of carbohydrates over proteins is characteristic of highly oligotrophic or detrital (such as the deep-sea) environments (Danovaro et al. 1993). Organic nitrogen (i.e. protein) is widely considered to be the major limiting factor for deposit-feeders. Protein concentrations in the sediments of the Marsala lagoon were very high, even when compared with those encountered in extremely highly productive areas such as the coastal sediments of the Baltic Sea (Meyer-Reil 1983). However, protein accounted only for less than 40% of the sedimentary biopolymeric carbon indicating that, although large amounts of detrital organic matter were present during the study period in the sediments of the Marsala lagoon, this detritus was relatively of low nutritional quality. This appeared even more clear when looking at the protein to carbohydrate (PRT:CHO) values. Low values of the PRT:CHO ratio suggest the presence of aged organic matter (Danovaro et al. 1993). No significant seasonal or spatial changes were found in PRT:CHO ratio values in the study area. Moreover, for most of the study period, the ratio remained lower than 1.0. This result confirms that the sediments of the Marsala lagoon, for almost the entire year of investigation, were characterised by the presence of large amounts of aged and/or non-living organic matter.

Environmental Constraints on Quality and Quantity of Organic Matter

Temperature and salinity reported for the study area are similar to those of other typical Mediterranean coastal environments (Chassany de Casabianca 1979; Hamon et al. 1979; Zaouali and Baeten 1983; Friligos 1989; Pusceddu and Fabiano 1996). Although these variables showed significant seasonal changes, they have been reported as having only a secondary role in affecting seston dynamics in the Marsala lagoon. Indeed, Pusceddu et al. (1997a,b) showed that the southern area of the study site is characterised by constant sea vivification and by the presence of freshly generated detritus. In contrast, they found that the northern area is hardly influenced by the open sea and is characterised by the presence of particulate organic matter deriving mostly from wind-wave induced resuspension. The Marsala lagoon is characterised also by a clear southward positive gradient of algal coverage (Calvo et al. 1996). In the northern sector, highly turbulent water dynamics affect significantly the possibility for algae to colonise these sediments. Here, large amounts of total suspended matter have been measured and related to the frequent and persistent sediment resuspension. The large amount of mineral seston in the northern area (annual average about 90% of TSM concentrations at stations 1-3), and the large dilution of the biopolymeric fraction of particulate organic matter (in terms of POM:TSM ratio, approximately 17%), could be further indirect signals of sediment resuspension. Accordingly, the lowest amounts of sediment organic matter (in terms of S-BPC), characterised by the co-dominance of proteins and carbohydrates, were found in the northern area (station A).

Moving southward and approaching the central area, *Cymodocea nodosa* becomes dominant, gradually substituted by a *Posidonia oceanica* reef. In this area, concentrations of total suspended matter and of all its biochemical components, with the exception of particulate proteins, gradually decrease, as a result of the increase in sedimentation effects. Indeed, moving from the northern area (stations 1-3) to the mid-basin (stations 4-6), total suspended matter concentrations decrease approximately two-fold. Such a decrease was also evident for the POC concentrations, which decreased approximately 2.3-times moving from station 1 and approaching station 7. Here, the presence of the *Posidonia oceanica* reef acts as a mechanical barrier to the lateral drifting of seston, giving rise to an increase in the sinking velocities of particles (Pusceddu et al. 1997a). To this regard, other studies showed that seagrass meadows strongly affect the sinking velocity of particles (Pirc and Wollemweber 1988). In addition, in this area, corresponding to stations B and C, S-BPC concentrations in the sediments increase up to 1.7-times with respect to concentrations at station 1. Moving from station A to stations B and C, the contribution of carbohydrates to S-BPC increases, although not significantly, from 47% to about 52% against a decrease of the protein contribution from 45% to 37%. The carbohydrate content of sediments in this sector (stations B and C) is probably caused by the release of *Posidonia* fragments rich in structural carbohydrates (Danovaro 1996). Since the gross biochemical composition of *Posidonia oceanica* is highly refractory (Lawrence et al. 1989), a large fraction of organic matter accumulated in the sediments of the southern area of the study site may be not directly available for benthic consumers. Kenworthy and Thayer (1984) showed by in situ experiments, that 50-60% of the organic carbon of the seagrass leaves was lost within 170 days. Thus, although *Posidonia* reefs provide an important input of organic detritus to the sediments, only a small fraction (i.e. proteins, soluble carbohydrates and free amino acids; Pirc and Wollemweber 1988) might be potentially available to benthic consumers.

We are aware that further direct investigation of particle sedimentation and sediment OM diagenesis rates is needed to clarify the depicted picture of the Marsala lagoon. However, results of the present study suggest that quality, quantity and spatial distribution of suspended and sediment organic matter in the study area are drastically controlled by the dynamic balance and spatial repartition of sediment resuspension and sedimentation processes, to a larger extent than seasonal changes in temperature and salinity do.

Concluding Remarks on the Food Availability of Organic Matter

Results of this study indicate that both quality and quantity of particulate and sediment organic matter in the Marsala lagoon are controlled by a wide and complex array of variables. Although the amount of primary organic matter is quite low, as the low particulate and sediment chlorophyll-a concentrations suggest, relatively large amounts of organic detritus are present throughout the year in the whole basin.

The biochemical composition of POM (Particulate Organic Matter) revealed the dominance of large amounts of readily available compounds (e.g. proteins). Nevertheless, this pool of suspended organic matter is highly diluted in a dominant inorganic fraction, as the POM:TSM ratio values suggest (annual average, 19%; range, 2.1-49.5%). The POM:TSM ratio, used as a food index (Navarro et al. 1993), is quite low if compared with that found in a Mediterranean shallow coastal lagoon (average 50%; range 30-100%, Pusceddu et al. 1996) and in an estuarine lagoon (30-100%, Navarro et al. 1993). Thus, in terms of food availability, we may conclude that a suspension feeder in the Marsala lagoon would filter a sea water volume about 3-times higher than in other coastal lagoons to reach the same amount of particulate food. This could be considered a partial confirmation in the scarce filter feeding mollusks community inhabiting the sediments of the Marsala lagoon (Chemello and Riggio, personal communication).

As in other oligotrophic environments (in terms of phytopigment concentrations), in the Marsala lagoon we might expect a large fraction of the autochthonous organic matter production to be rapidly canalised throughout planktonic and sedimentary microbial loops. However, as indicated by the value of the LOM (Labile Organic Matter): TOM ratio (on average, about 14%, Pusceddu et al. 1999), most of the sediment OM is unaccounted by labile organic compounds. This uncoupling between the large amounts of sediment OM and its low nutritional value suggests that this particular environment probably behaves as a detrital "trap" in which the large gross primary production of macroalgae and seagrasses, not directly available to benthic consumers, tends to accumulate. However, further studies (now in progress) will have to deal with dynamics and secondary production of bacterial assemblages to focus on the fate of the large detrital pool in the Marsala lagoon.

Acknowledgements. We thank an anonymous referee for precious suggestions and useful comments. We are indebted to Prof. Masala Tagliasacchi (University of Cagliari) for kindly providing laboratory facilities. We thank Dr. Monica Armeni and Dr. Elena Manini (University of Ancona) for help in chemical analyses. The Ministero Politiche Agricole and Ministero Università Ricerca Scientifica Tecnologica of the Italian government partially supported this work.

References

Bligh EG, Dyer WJ (1959) A rapid method for total lipid extraction and purification. Can J Biochem Physiol 37: 911-917

Calvo S, Ciraolo G, La Loggia G, Malthus TJ, Savona E, Tomasello A (1996) Monitoring *Posidonia oceanica* meadows in the Mediterranean Sea by means of airborne remote sensing techniques. Intern Airborne Rem Sensing Conf Exibition, June 1996, San Francisco, USA 3: 659-668

Campolmi M (1998) Zooplankton community in the Stagnone di Marsala (West MED). PhD Dissertation, University of Messina, Italy, 205 pp

Carrada G, Fresi E (1988) Le lagune salmastre costiere. Alcune riflessioni sui problemi e sui metodi. In: GC Carrada, Cicogna F, Fresi E (eds) Coastal lagoons: research and managment. CLEM, Massa Lubrense, Napoli, pp 35-56

Carrada GC, Mangoni O, Sgrosso S (1996) Distribuzione spaziale di clorofilla a e di feopigmenti in diverse frazioni dimensionale del fitoplancton. In: FM Faranda, P Povero (eds) Caratterizazzione ambientale del sistema Eolia e dei bacini limitrofi di Cefalù e Gioia (EOCUMM95). CoNISMa Data Rep, pp 197-216

Chassany de Casabianca ML (1979) Les phénomès de décomposition et de transformation en milie lagunaire et le C, N, et C/N particulaire. Comm Intern Explor Sci Mer Méditerranée, Etang Salés et Lagunes 25/26(3): 109-112

Chassany de Casabianca ML, Boonne ML, Semroud R (1995) Principal component analysis of relationship between physicochemical variables in a Mediterranean lagoon. CR Hebd Seances Acad Sci (III), Paris 310: 397-403

Clavier J, Chardy P, Chevillon C (1995) Sedimentation of particulate matter in the south-west lagoon of New Caledonia: Spatial and temporal pattern. Estuar Coastal Shelf Sci 40: 281-294

Danovaro R (1996) Detritus-bacteria-meiofauna interactions in a seagrass bed (*Posidonia oceanica*) of the NW Mediterranean. Mar Biol 127: 1-13

Danovaro R, Fabiano M, Della Croce N (1993) Labile organic matter and microbial biomasses in deep-sea sediments (Eastern Mediterranean Sea). Deep Sea Res 40: 953-965

de Jonge VE (1980) Fluctuations in the organic carbon to chlorophyll a ratios for estuarine benthic diatom populations. Mar Ecol Progr Ser 2: 345-353

Dubois M, Gilles KA, Hamilton JK, Rebers, PA, Smith F (1956) Colorimetric method for determination of sugars and related substances. Anal Chem 28: 350-356

Fabiano M, Danovaro R, Fraschetti S (1995) A three-year time series of elemental and biochemical composition of organic matter in subtidal sediments of the Ligurian Sea (northwestern Mediterranean). Cont Shelf Res 15: 1453-1469

Fabiano M, Pusceddu A (1998) Total and hydrolizable particulate organic matter (carbohydrates, proteins and lipids) at a coastal station in Terra Nova Bay (Ross Sea, Antarctica). Polar Biol 19: 125-132

Friligos N (1989) Nutrients status in a eutrophic Mediterranean lagoon. Vie Milieu 39(2): 63-69

Genovese S (1969) Données écologique sur le "Stagnone" de Marsala (Sicile occidentale). Rapp Comm Int Mer Médit 19: 823-826.

Gerchacov SM, Hatcher PG (1972) Improved technique for analysis of carbohydrates in sediment. Limnol Oceanogr 17: 938-943

Giani M, Mecozzi M, Gesumundo C, Ciuffa G, Sunseri G, Andaloro F (1995) Physico-chemical features and nutrients distribution in the Marsala lagoon (Italy). Rapp Comm Int Mer Médit 34: 81

Hamon PY, Tournier H, Arnaud P (1979) Cycles annuels del quelques paramètres physico-chimiques de l'étang de Thau. Commission Internationale pour l'Exploration Scientifique de la Mer Méditerranée, Etang Salés et Lagunes, 25/26(3): 11-12

Hansen JA, Klump DW, Alongi DM, Dayton PK, Riddle MJ (1992) Detrital pathways in a coral reef lagoon. II. Detritus deposition, benthic microbial biomass and production. Mar Biol 113: 363-372

Hartree EF (1972) Determination of proteins: a modification of the Lowry method that gives linear photometric response. Anal Biochem 48: 422-427

Hickel W (1984) Seston in the Wadden sea of sylt (German Bight, North Sea). Proc 4th Int Wadden Sea Symp 10: 113-131

Iseki K, McDonald RW, Carmack E (1987) Distribution of particulate matter in the south-eastern Beaufort Sea in late summer. Polar Biol 1: 35-46

Kenworthy WJ, Thayer GW (1984) Production and decomposition of the roots and rhizomes of seagrasses *Zostera marina* and *Thalassia testudinum* in temperate and subtropical marine ecosystems. Bull Mar Sci 35: 364-379

Lawrence JM, Boudouresque Ch-F, Maggiore F (1989) Proximate costituents, biomass and energy in *Posidonia oceanica* (Potamogetonaceae). Publ Staz Zool Napoli (I Mar Ecol) 10: 263-270

Lee S, Fuhrman JA (1987) Relationships between biovolume and biomass of naturally derived marine bacterioplankton. Appl Environ Microbiol 53: 1298-1303

Lorenzen CJ, Jeffrey SW (1980) Determination of chlorophyll and phaeopigments spectrophotometric equations. Limnol Oceanogr 12: 343-346

Magazzù G (1982) La crescita fitoplanctonica in alcuni ambienti lagunari del Mar Mediterraneo. Naturalista Siciliano 6(2): 337-359

Marsh JB, Weinstein WJ (1966) A simple charring method for determination of lipids. J Lipid Res 7: 574-576

Mazzola A, Mirto S, Danovaro R, Fabiano M (2000) Fish farming effects on benthic community structure in coastal sediments: analysis of meiofaunal resilience. ICES Marine science (*in press*)

Mazzola A, Sarà G (1995) Caratteristiche idrologiche di una laguna costiera mediterranea (Stagnone di Marsala - Sicilia Occidentale): ipotesi di un modello qualitativo di circolazione lagunare. Naturalista Siciliano 19(3-4): 229-277

Meyer-Reil LA (1983) Benthic response to sedimentation events in the Western Kiel Bight. II Analysis of benthic bacterial populations. Mar Biol 77: 247-256

Millet B, Cecchi P (1992) Wind-induced hydrodinamic control of the phytoplankton biomass in a lagoon ecosystem. Limnol Oceanogr 37(1): 140-146

Navarro JM, Clasing E, Urrutia G, Asencio G, Stead R, Herrera C (1993) Biochemical composition and nutritive value of suspended particulate matter over a tidal flat of southern Chile. Estuar Coastal Shelf Sci 37: 59-73

Navarro JM, Thompson RJ (1995) Seasonal fluctuaction in the size spectra, biochemical composition and nutritive value of the seston available to suspension-feeding bivalves in a cold ocean environment. Mar Ecol Progr Ser 125: 95-106

Newell RC (1982) The energetic of detritus utilization in coastal lagoon and in nearshore waters. Oceanol Acta Proceedings International Symposium on coastal lagoons, SCOR/IABO/UNESCO, Bordeaux, France, September 1981, pp 347-355

Nival P, Charra R, Malara G, Boucher D (1972) La matière organique particulaire de la Méditerranée occidentale en mars 1970. Mission Mediprod du "Jean Charchot". Annales Istitute Oceanographique, Paris 48: 141-156

Painchaud J, Lefaivre D, Tremblay GH, Therriault J-C (1990) Analysis of the distribution of suspended particulate matter, bacteria, chlorophyll a and PO_4 in the upper St. Lawrence Estuary, using a two dimensional box model. In: Michaelis W (ed) Coastal and Estuarine Studies, vol. 36, Springer, Berlin Heidelberg New York, pp 59-65

Pirc H, Wollemweber B (1988) Seasonal changes in nitrogen, free amino acids and C/N ratio in Medieterranean seagrasses. PSZNI Mar Ecol 9: 167-179

Plante-Cuny MR (1974) Evaluation par spectrophotométrie des teneurs en chlorophyl-a fonctionelle er en phaeopigments des substrates meubles marins. ORSTOM Nosy-Bé, 45 pp

Pusceddu A (1999) Organic detritus in coastal lagoon systems: role and pathways. Proceedings of the XIII Italian Association Oceanography and Limnology Symposium, 13(1): 55-64

Pusceddu A, Fabiano M (1996) Short-term changes in the biochemical composition of particulate organic matter in a Mediterranean lagoon (Southern Sardinia): the role of wind and rain. Proceedings of the XI Italian Association Oceanography and Limnology Symposium, pp 571-581

Pusceddu A, Sarà G, Armeni A, Mazzola A, Fabiano M (1999) Seasonal and spatial changes in sediment organic matter composition of a semi-enclosed marine system (W-Mediterranean Sea). Hydrobiologia 397: 59-70

Pusceddu A, Sarà G, Manini E., Puccia E (1997a) Short-term changes in the biochemical composition of particulate organic matter in a mediterranean shallow sound (Western Sicily). Proceedings of the XII Italian Association Oceanography and Limnology Symposium, pp 299-310

Pusceddu A, Sarà G, Mazzola A, Fabiano M (1997b) Relationships between suspended and sediment organic matter in a semi-enclosed marine system: the Stagnone di Marsala (western Sicily). Water Air Soil Poll 99: 343-352

Pusceddu A, Serra E, Sanna O, Fabiano M (1996) Seasonal fluctuations in the nutritional value of particulate organic matter in a lagoon. Chem Ecol 13: 21-37

Rice DL (1982) The detritus nitrogen problem: new observations and perspectives from organic geochemistry. Mar Ecol Progr Ser 9: 153-162

Sarà G, Pusceddu A, Mazzola A, Fabiano M (1995) Variazioni della composizione biochimica del materiale organico particellato nello Stagnone di Marsala (Sicilia occidentale): osservazioni preliminari. Biol Mar Medit 2: 127-129

Savenkoff C, Chanut JP, Vézina AF, Gratton Y (1995) Distribution of biological activity in the St. Lawrence Estuary as determined by multivariate analysis. Estuar Coastal Mar Sci 40: 647-664

Scilipoti D (1997) Fish community in the Stagnone of Marsala: distribution and resource repartition as a function of different habitat complexity degrees. PhD Dissertation, University of Messina, Italy, 256 pp

Sorokin Y, Sorokin P, Giovanardi O, Dalla Venezia L (1996) Study of the ecosystem of the lagoon of Venice with emphasis on anthropogenic impact. Mar Ecol Prog Ser 141: 247-261

Strickland JDH, Parsons T (1972). A practical handbook of seawater analysis. Bull Fish Board Can 167: 310

Underwood AJ (1997) Experiments in ecology. Their logical and interpretation using analysis of variance. Cambridge University, Cambridge, p 504

Valiela I (1984) Marine ecological processes. Springer, Berlin Heidelberg New York, p 546

Zaouali J, Baeten S (1983) Impact de l'eutrophisations dans la lagune de Tunis (partie nord). 2ème partie: analyse de corrispondance. Rapp Comm Int Mer Médit 28(6): 327-332

Distribution and Biochemical Composition of Suspended and Sedimentary Organic Matter in the Northern Adriatic

M. Fabiano[1], C. Misic[1], E. Manini[2], S. Chiatti[2], P. Povero[1], and R. Danovaro[2]

ABSTRACT

In order to study the benthic-pelagic coupling in the northern Adriatic Sea, sediment and water samples were collected between July 1996 and March 1997 in two zones of the northern Adriatic, differently influenced by river outflow. In the northern area the concentrations of organic matter decreased from coast to open sea [particulate organic carbon concentrations in the coastal stations were on average 327.0 (from 124.4 to 963.0) μgC l^{-1} during summer and 374.0 (from 88.6 to 1196.9) μgC l^{-1} in early spring], thus increasing the local sedimentary organic carbon concentrations [in the coastal area on average 2860.2 (from 2734.1 to 2986.2) μgC g^{-1} during summer, with a decrease in early spring to 868.4 (from 550.2 to 1186.5) μgC g^{-1}]. By contrast, the open-sea area showed generally much lower organic matter concentrations (2 to 6 times lower than at coastal stations in water and sediments, respectively). In the southern area the absence of almost continuous and strong allochthonous inputs brought from the river, coupled with the presence of well defined frontal areas, resulted in increased sedimentary organic matter concentrations in the intermediate stations, that also displayed clear seasonal changes [2314.5 (2102.8-2526.2) μgC g^{-1} during summer, only 991.5 (856.4-1126.5) μgC g^{-1} during early spring]. In the water column, high values of suspended organic matter in the coastal stations were generally observed [on average 93.5 (31.2-252.5) μgC g^{-1} in summer and 242.3 (from 56.1 to 1358.0) μgC l^{-1} in early spring]. In both areas, proteins were the main biochemical class of organic compounds. The higher values of PRT:CHO ratios in the sediment as compared to the water column (7.8±1.7 vs 2.2±0.6, respectively) suggest the presence of a protein accumulation in the sediments. At the same time, the lower values of the RNA:DNA ratios observed in the water column of the northern area [summer values of 1.87 (0.01-12.79) and early spring values of 1.04 (0.01-4.62)] as compared to the southern area [summer values of 1.93 (0.05-18.10) and early spring values of 2.69 (0.07-34.91)] suggest the presence of a larger fraction of detrital DNA likely to be provided by the river discharge.

Although the relations between organic matter distribution and characterisation pointed out a clear pelagic-benthic coupling, in the northern Adriatic energy and materials transfer might be affected by spatially and temporally confined functional uncoupling, changing the balance between supply and exploitation of organic matter, thus forcing the systems to different trophic patterns.

Introduction

Oceanographic properties within the northern Adriatic Sea exhibit high vertical and horizontal gradients, largely due to the freshwater discharge of the Po river (Gilmartin et al. 1990; Revelante and Gilmartin 1992), which are exacerbated by the shallow depths and by the strong seasonal stratification of the water column. Thus, the resulting effects on water and sedimentary systems are usually pronounced in the western side of the northern Adriatic (Franco et al. 1982; Gilmartin et al. 1990), and generally decrease moving southward, where local terrigenous

[1] Dipartimento per lo Studio del Territorio e delle Sue Risorse, Università di Genova, C.so Europa 26, 16132 Genova, Italy
[2] Istituto di Scienze Marine, Università di Ancona, Via Brecce Bianche, 60131 Ancona, Italy

F.M. Faranda, L. Guglielmo, G. Spezie (eds)
Mediterranean Ecosystems: Structures and Processes

inputs may achieve higher importance. Po waters, especially during stratification periods, are involved in a very complex circulation, with the presence of several gyres (Cerovecki et al. 1990), which affect the distribution of the inputs of allochthonous suspended matter, their exploitation by the autochthonous organisms and the distribution of organic and inorganic matter in the sediments. In this contest, the organic fraction of the suspended matter is destined to exhibit large and rapid temporal and spatial variations in response to physical and biological factors. In the coastal environment, the deposition of large amounts of suspended organic matter on the sediments, although not continuous, allows a tight coupling between the two compartments (i.e., loss system *sensu* Peinert et al. 1989). Thus, the concentration and distribution of organic matter in the sediments may be considered as a record of the biological phenomena occurring in the overlying waters (Graf 1992).

The analysis of the composition of suspended and sedimentary organic matter can be used to evaluate the quantity and quality of the materials potentially available to consumers, discriminating between the organic compounds which are rapidly mineralised and which are characterised by lower degradation rates (Fabiano et al. 1993, 1995). Moreover, the relationship between the different components of organic matter as well as nucleic acids may give clues on the metabolic state of the organic matter (Fabiano et al. 1993; Danovaro et al. 1993).

The aim of this paper was to study the temporal and spatial variations of the main biochemical components of suspended and sedimentary organic material in two areas of the northern Adriatic, during two different seasonal periods (summer and early spring), focusing on the occurrence and on the intensity of the pelagic-benthic coupling.

Materials and Methods

Sediment and Water Sampling

Sediment and water samples were collected between June-August 1996 and February-March 1997 in two areas of the northern Adriatic, differently influenced by river outflow: the first next to the Po delta and the second located between Rimini and Ancona (Fig. 1). Sampling was defined in order to cover the coastal area, largely affected by river waters, and the open-sea zone, scarcely influenced by freshwater.

In each sampling area, the water sampling was performed on 4 to 5 transects, constituted by 3 to 5 stations each, in order entirely to cover the fresh and seawater mixing areas with a grid of about 12-22 stations, depending on the mixing area extension. The sampling was carried out four times during the study period. Samples, collected by means of Niskin bottles equipped on a GO rosette sampler, were immediately prefiltered on 200 μm net and then filtered on Whatman GF/F filters. Sampling depths (generally from 2 to 4) were selected in order to gather information on the surface layer, on the layer above the halocline and on the water layer next to the bottom. All samples were frozen at –20°C until analysis.

The sediment sampling, performed four times for each area during the study period, was carried out on grids of 6 stations, which included 2 coastal, 2 central and 2 open-sea stations. Undisturbed sediment cores were collected using a multiple corer (Mod. Midi, $n = 4$ with 5.7 and 9.5 cm inner diameter). For the analysis of the biochemical composition of the sedimentary organic matter, 2-3 corers were collected at each station. The top 6 cm of sediment cores were stored at –20°C until analysis.

Biochemical Composition of Sedimentary and Particulate Organic Matter

Protein analysis was conducted according to the Hartree method (1972). The absorbance was evaluated at 650 nm. Bovine albumin solutions were used as standard. Carbohydrates were analysed according to Dubois et al. (1956). The absorbance was measured at 490 nm; D (+) glucose solutions were used as standard. Lipids were extracted according to the Bligh and Dyer method (1959) and measured following Marsh and Weinstein (1966). The absorbance was evaluated at 375 nm. Tripalmitine solutions were used as standard. Nucleic acids (DNA and RNA) in sedimentary matter were determined according to Zachleder (1984), modified by Danovaro (1996). The absorbance of total nucleic acids was measured at 260 nm. For particulate matter analysis the Berdalet and Dortch (1991) method

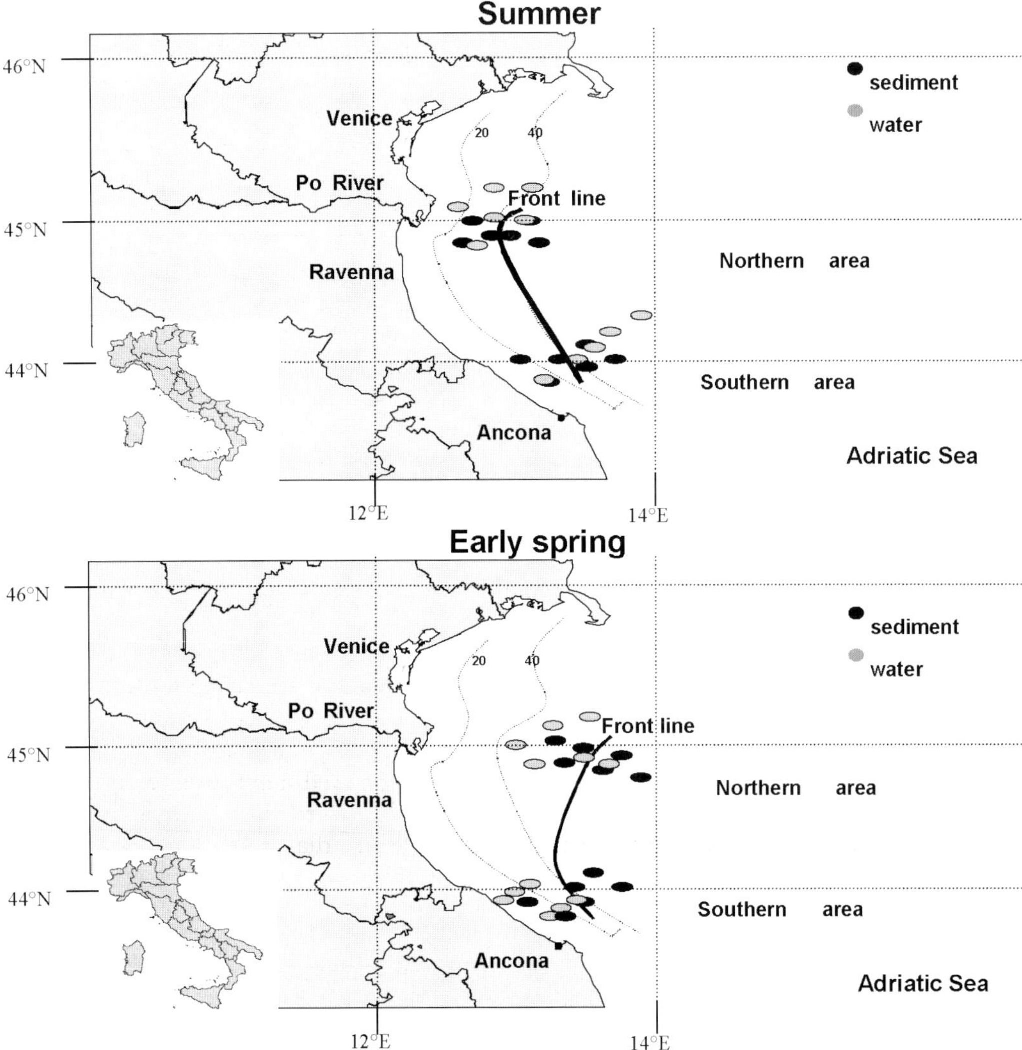

Fig. 1. Study areas and location of the sampling stations

was followed. Particulate (C-POM) and sedimentary (C-OM) organic matter was expressed as the sum of carbohydrates, proteins and lipids converted to carbon content using: 0.4, 0.49 and 0.75 µgC µg^{-1} conversion factor, respectively.

Results and Discussion

The high and persistent allochthonous inputs coming from the Po river outflow are known to influence strongly the northern Adriatic Sea, mainly its north-western section (Gillmartin et al. 1990; Revelante and Gillmartin 1992) and, to a minor extent, the southern area.

In our northern sampling area organic matter concentrations decreased, both in the water column and in the sediments, from the coast to the open sea (Fig. 2). The river outflow carries in the marine waters high quantities of suspended organic matter [in the coastal stations on average 327.0 (124.4-963.0) µgC l^{-1} during summer and 374.0 (88.6-1196.9) µgC l^{-1} in early spring] which generally settle rapidly on the seabed, thus increasing sedimentary organic matter carbon concentrations [in the coastal area on average 2860.2 (2734.1-2986.2) µgC g^{-1} during summer, with a decrease in early spring to 868.4 (550.2–1186.5) µgC g^{-1}]. By contrast, the offshore area showed lower organic matter concentrations (Fig. 2) both in the water column [organic carbon values of 178.7 (90.9-337.4) µgC l^{-1} in

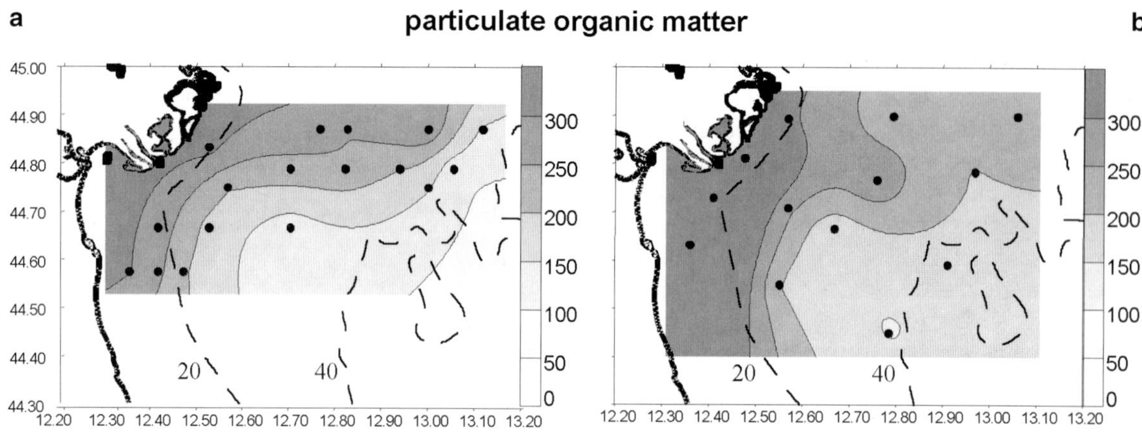

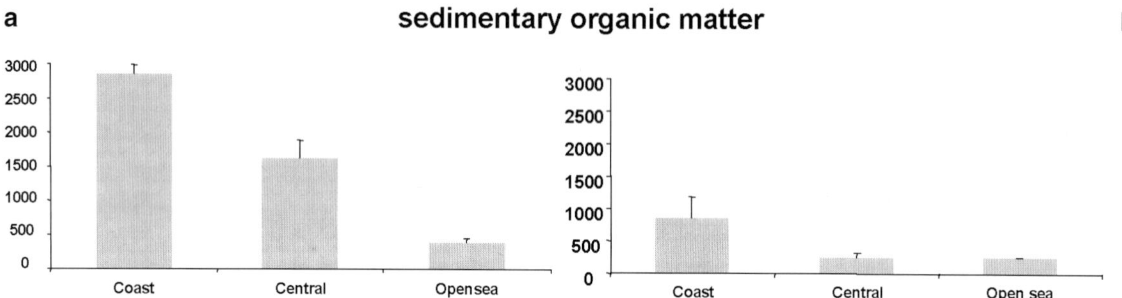

Fig. 2a,b. Particulate (C-POM, µgC l^{-1}) and sedimentary (C-OM, µgC g^{-1}) organic matter distributions in the northern area: **a** summer; **b** early spring

summer and 190.3 (98.8-529.2) µgC l^{-1} in early spring] and in the sediments [on average 390.3 (325.4-455.3) µgC g^{-1} in summer and 258.2 (255.5-261.0) µgC g^{-1} in early spring]. The organic matter distributions suggest the presence of a strong benthic-pelagic coupling. Moreover, the effects of an enhanced production due to the frontal area are probably masked by the presence of large inputs of allochthonous organic matter.

In the southern area, a particulate organic matter concentration decrease from coast to open sea in the water column was recorded during the early spring period, when local river outflow is higher due to the increase of rain fall. The highest C-POM concentrations of the water column and a remarkable seasonal variability were observed in coastal stations [on average 93.5 (31.2-252.5) µgC l^{-1} in summer and 242.3 (56.1-1358.0) µgC l^{-1} in early spring], whereas in open-sea stations lower values and narrower seasonal changes were observed [on average 56.9 (35.1-80.2) µgC l^{-1} in summer and 91.4 (40.2-144.5) µgC l^{-1} in early spring] (Fig. 3). In the sediments, it was possible to observe an accumulation of organic matter in the central stations during summer, probably due to deposition events deriving from a previous, temporary frontal area that occurred in the water column. Such as increase was strongly seasonal and variable. During summer the central stations show the higher values [mean value of intermediate stations 2314.5 (2102.8-2526.2) µgC g^{-1}, while in coastal and open-sea stations the average value was 1229.8 (1226.0-1233.5) and 838.9 (706.4-971.4) µgC g^{-1} respectively]. In early spring, instead, the organic matter distribution showed an increasing concentration gradient from inshore to offshore stations, pointing out the absence of remarkable differences between the intermediate stations [991.5 (856.4-1126.5) µgC g^{-1}] and all other stations [mean value of 688.1 (604.3–772.0) and 1271.2 (1266.7-1275.8) µgC g^{-1} respectively for coastal and offshore areas]. In conclusion, the lack of significant and continuous allochthonous inputs played an important role in the distribution of both suspended and sedimentary organic matter, thus weakening the strength of the benthic-pelagic coupling and increasing the influence of temporary events as the assessment of frontal, highly productive areas.

In the Northern area proteins are the main biochemical class of organic compounds (Table and Fig. 4), and their concentration is slightly higher in summer [60.8% (42.6-71.5%) in water and 73.0% (68.1-79.8%) in sediments]. In the water column, the diminished protein contribution during spring (52.5%, 32.4-70.4%) is largely due to the increase of the lipid fraction [17.5% (7.8-30.5%) in summer and 26.2% (5.9-50.7%) in early spring]. By contrast, in the sediments the biochemical composition of organic matter showed limited seasonal changes (percentage values of proteins during early spring 66.9%, from 58.2 to 77.3%) (Table and Fig. 4).

In the southern area, the percentage of proteins in the water column, despite their dominance (Fig. 5), is more variable than in the northern area [59.8% (39.6-86.3%) in summer and 38.6% (19.5-61.4%) in early spring]. Lower variability but higher protein contents were observed in the sediments [proteins amounting to 78.1% (74.2-82.0%) in summer and 75.7 % (71.7–80.5%) in early spring] (Fig. 5).

The higher values of the PRT:CHO ratio in the sediment as compared to the water column (on average 7.8±1.7 vs 2.2±0.6, respectively, Table), might be related to an increase of the bacterial secondary production induced by the enhanced C-POM rain. The accumulation of the protein fraction in the sediments is a typical feature of the loss systems and of highly productive environments (Fabiano et al. 1997). Moreover, in the coastal waters of the northern Adriatic, where the input of allochthonous material provided by the Po river is higher, the chemical link between proteins and carbohydrates of vegetal origin may produce complex macromolecular compounds which can contribute to the *gelbstoff* production (Ferrari and Mingazzini 1995). Keil and Kirchman (1993) observed that protein-sugar complexes have lower degradation rates than

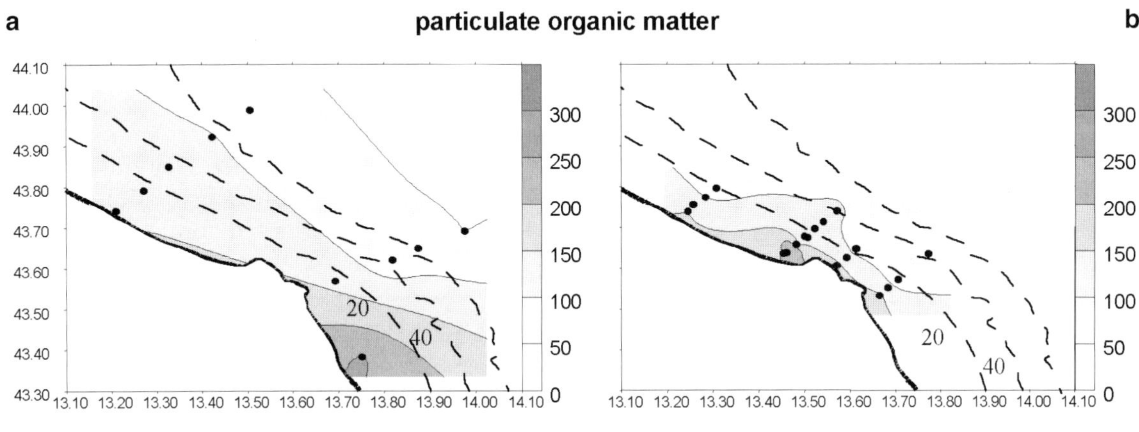

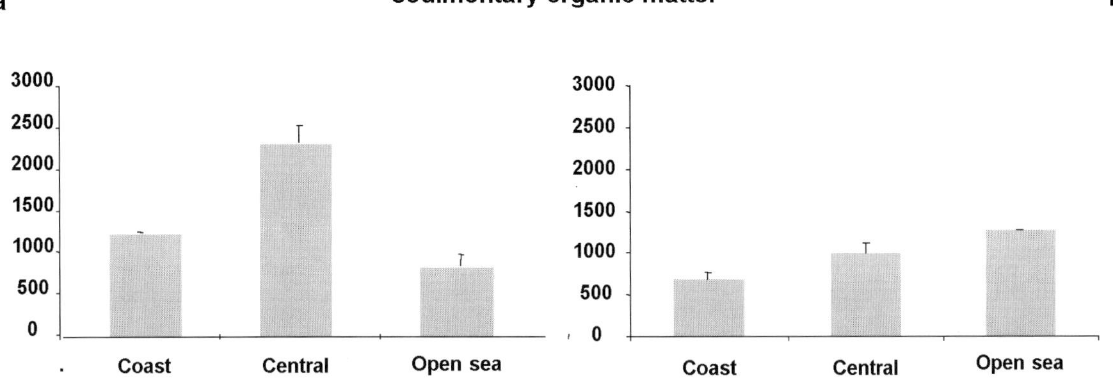

Fig. 3a,b. Particulate (C-POM, µgC l^{-1}) and sedimentary (C-OM, µgC g^{-1}) organic matter distributions in the southern area: **a** summer; **b** early spring

Table. Concentrations of the biochemical parameters (average, minimum and maximum values) in the sampling areas during the study period for water column and sediment samples. (proteins – *PRT*, carbohydrates – *CHO*, lipids – *LIP*, nucleic acids – *DNA* and *RNA*, protein/carbohydrate ratio – *PRT/CHO* and *RNA/DNA* ratio)

Season	Area	Type		PRT ($\mu g\ l^{-1}$)	CHO ($\mu g\ l^{-1}$)	LIP ($\mu g\ l^{-1}$)	DNA ($\mu g\ l^{-1}$)	RNA ($\mu g\ l^{-1}$)	PRT/CHO	RNA/DNA
Summer	Northern	Water	avg	309.0	141.1	63.0	2.9	3.4	2.4	1.9
			min	113.8	46.9	14.5	0.2	0.0	1.0	0.0
			max	894.9	845.5	286.3	9.5	17.3	5.0	12.8
	Southern	Water	avg	99.3	53.5	16.0	3.6	4.9	2.9	1.9
			min	41.9	4.1	2.6	0.7	0.1	0.0	0.1
			max	281.1	154.8	70.7	11.8	17.8	13.0	18.1
Early spring	Northern	Water	avg	315.6	155.3	108.1	2.5	1.2	2.2	1.0
			min	73.6	41.4	20.6	0.3	0.0	0.5	0.0
			max	1102.9	897.7	502.8	10.3	17.2	4.9	4.6
	Southern	Water	avg	132.1	137.0	67.8	3.7	7.0	1.2	2.7
			min	23.3	19.8	15.9	0.3	0.0	0.4	0.1
			max	801.1	1854.6	298.1	12.5	37.0	2.9	34.9
Season	**Area**	**Type**		**PRT ($\mu g\ g^{-1}$)**	**CHO ($\mu g\ g^{-1}$)**	**LIP ($\mu g\ g^{-1}$)**	**DNA ($\mu g\ g^{-1}$)**	**RNA ($\mu g\ g^{-1}$)**	**PRT/CHO**	**RNA/DNA**
Summer	Northern	Sediment	avg	2273.1	250.2	546.7	73.4	21.3	8.3	0.3
			min	425.2	65.9	120.9	33.6	2.6	4.1	0.1
			max	4151.6	418.4	1072.7	128.6	61.5	11.0	0.5
	Southern	Sediment	avg	2204.4	295.2	350.5	37.2	11.2	8.5	0.6
			min	1046.9	153.8	151.1	22.2	0.8	5.2	0.0
			max	3722.2	645.7	622.8	76.4	36.1	12.7	1.6
Early spring	Northern	Sediment	avg	609.7	110.2	157.4	17.9	14.9	5.1	1.5
			min	216.8	70.1	74.9	8.1	9.5	3.1	0.4
			max	1600.4	242.6	407.1	47.7	19.8	8.6	2.6
	Southern	Sediment	avg	1404.4	154.0	311.8	15.3	43.7	9.5	4.0
			min	944.3	86.0	143.0	7.4	13.4	7.4	0.7
			max	1777.8	230.4	445.2	26.3	84.1	11.0	6.7

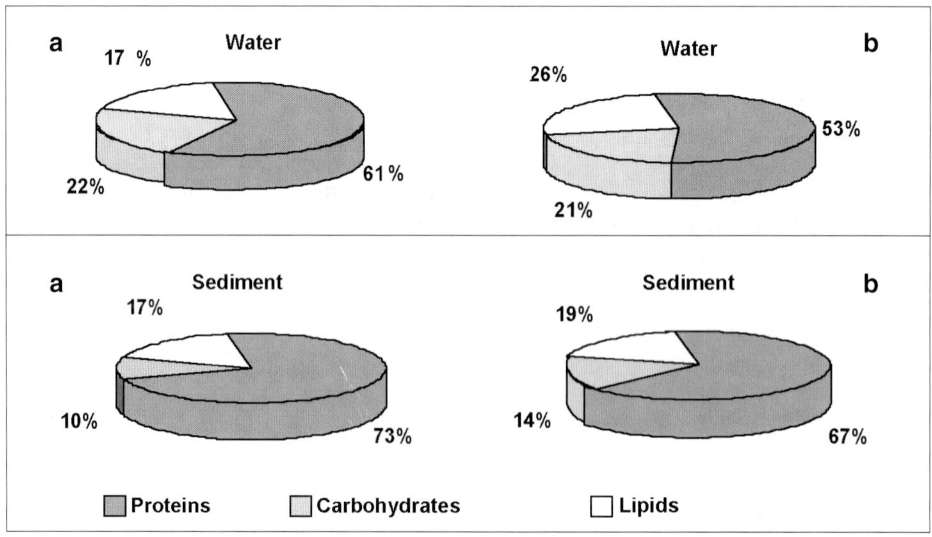

Fig. 4a,b. Biochemical composition of organic matter in water and sediment in the northern area: **a** summer; **b** early spring

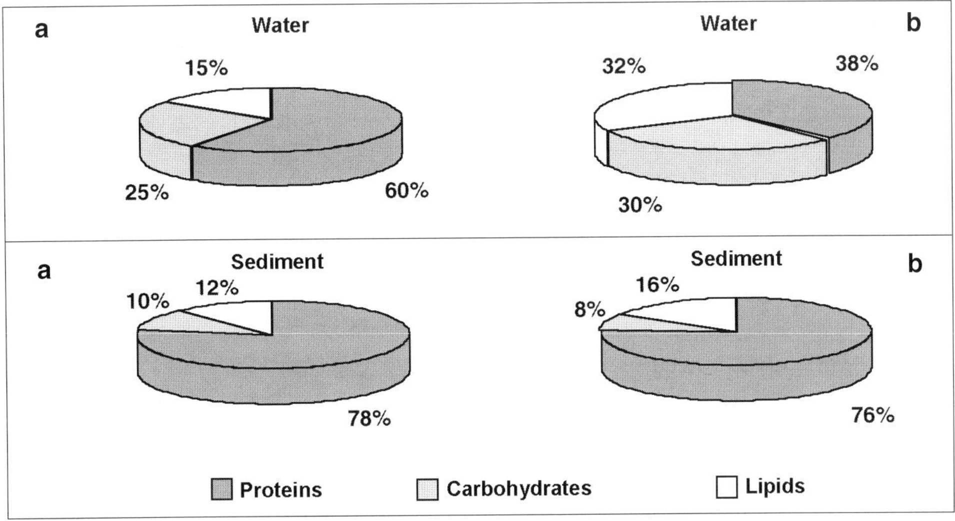

Fig. 5. Biochemical composition of organic matter in water and sediment in the southern area: **a** summer; **b** early spring

proteins alone, and thus determine the presence of proteins highly refractory to degradation which may have a large effect on early diagenetic processes.

The RNA:DNA ratio has been recently proposed as a metabolic activity index of the microbial communities associated to suspended and sedimentary organic matter (Danovaro et al. 1993; Fabiano et al. 1993). In the water column, the lower values of the RNA:DNA ratios in the northern area [summer values of 1.87 (0.01-12.79) and values of 1.04 (0.01-4.62) in early spring, Table] as compared to the southern area [summer values of 1.93 (0.05-18.10) and early spring values of 2.69 (0.07-34.91)] are probably due to the large detrital DNA river input in the northern area (which amounts up to 10 µg l^{-1}, while often the RNA concentrations are under the detection limits, Table). This leads to functional differences between water and sediments. In fact, during summer water-column stratification, the RNA:DNA ratios in water column are higher than in the sediments [values of 0.29 (0.07-0.49) and 0.61 (0.01-1.63) for northern and southern areas, respectively, Table], probably due to refractory organic matter accumulation on the bottom, which may slow and, sometimes, inhibit the aerobic metabolic activity of the sediment (Graf 1992; Paerl et al. 1998). By contrast, during winter and early spring water mixing, the presence of fresh organic matter inputs induces a clear increase of the metabolic activities in the sediments and therefore an increased RNA:DNA ratio in the sediment [1.48 (0.38-2.61) and 4.02 (0.65-6.70) for the northern and southern area respectively], as compared to the water column (Table).

Conclusions

The analysis of the organic matter distribution and composition in the two studied areas suggests the presence of temporal and spatial changes in the strength and efficiency of the benthic-pelagic coupling. Sedimentation events taking only a few days after spring phytoplankton blooms might produce short time scale couplings leading to a quite immediate sediment response (Graf et al. 1982). This applies particularly in shallow waters where large amounts of almost undegraded organic matter reach the sediments (Graf 1992). The absence of continuous organic matter inputs, observed in the southern area, may weaken the benthic-pelagic coupling (Wassman 1990) and even create uncoupling conditions which are evident from the analysis of organic matter distribution and accumulation in water and sediment of the southern area. Moreover, water column and sediments are subjected to different temporal scales, so that post-bloom conditions characterised by low suspended organic matter concentrations might be opposed to the finding of large organic matter accumulation in the sediments, thus suggesting an apparent uncoupling. The continuous inputs of organic materials from Po river is the main force sustaining a permanent coupling in the northern area, but the organic matter amounts reaching the sediments might be so high that benthic microbial activities may be induced to a self feed-back control by temporary ipoxic conditions. We hypothesise that in these conditions of clear pelagic-benthic coupling, organic matter transfer might be reflected by spatially and temporally confined functional uncoupling. Therefore, seasonal variations in the chemical and biological features of the water – sediment interactions may affect the balance between supply and exploitation of organic matter, thus forcing the systems to different trophic patterns.

References

Berdalet E, Dortch D (1991) New double-staining technique for RNA and DNA measurement in marine phytoplankton. Mar Ecol Prog Ser 73: 295-305

Bligh EG, Dyer W (1959) A rapid method for total lipid extraction and purification. Can J Biochem Physiol 37: 911-917

Cerovecki I, Pasaric Z, Zuzmic M, Orlic M (1990) Observations of currents and temperature on the Adriatic shelf in summer. Rapp P V Reun Comm Int Explor Sci Mer Méditerr, 32: 163

Danovaro R (1996) Detritus-bacteria-meiofauna interactions in a seagrass bed (*Posidonia oceanica*) of the NW Mediterranean. Mar Biol 127: 1-13

Danovaro R, Fabiano M, Della Croce N (1993) Labile organic matter and microbial biomasses in deep-sea sediments (eastern Mediterranean Sea). Deep-Sea Res 40, 953-965

Dubois M, Gilles K, Hamilton JK, Rebers PA, Smith F (1956) Colorimetric method for determination of sugars and related substances. Anal Chem 28: 350-356

Fabiano M, Chiantore M, Povero P, Cattaneo-Vietti R, Pusceddu A, Misic C, Albertelli G (1997) Short-term variations in particulate matter flux in Terra Nova Bay, Ross Sea. Antarct Sci 9, 2: 143-149

Fabiano M, Danovaro R, Fraschetti S (1995) A three-year time series of elemental and biochemical composition of organic matter in subtidal sandy sediments of the Ligurian Sea (northwestern Mediterranean). Cont Shelf Res 15: 1453-1469

Fabiano M, Povero P, Danovaro R (1993) Distribution and composition of particulate organic matter in the Ross Sea (Antarctica). Polar Biol 3: 525-533

Ferrari GM, Mingazzini M (1995) Synchronous fluorescence spectra of dissolved organic matter (DOM) of algal origin in marine coastal waters. Mar Ecol Prog Ser 125: 305-315

Franco P, Jeftic LJ, Malanotte-Rizzoli P, Michelato A, Orlic M (1982) Descriptive model of the northern Adriatic. Oceanol Acta 5: 379-389

Gilmartin M, Degobbis D, Revelante N, Smodlaka N (1990) The mechanism controlling plant nutrient concentrations in the northern Adriatic Sea. Int Rev Gessameite Hydrobiol 75: 425-455

Graf G (1992) Benthic-pelagic coupling: a benthic view. Oceanogr Mar Biol Annu Rev 30: 149-190

Graf G, Bengtsson W, Diesner U, Theede H (1982) Benthic response to sedimentation of a spring phytoplankton bloom: process and budget. Mar Biol Prog Ser 67: 201-208

Hartree EF (1972) Determination of proteins: a modification of the Lowry method that gives a linear photometric response. Anal Biochem 48: 422-427

Keil RG, Kirchman DL (1993) Dissolved combined aminoacids in marine waters: chemical form and utilization by heterotrophic bacteria. Limnol Oceanogr 38: 1256-1270

Marsh JB, Weinstein WJ (1966) A simple charring method for determination of lipids. J Lipid Res 7: 574-576

Paerl HW, Pinckney JL, Fear JM, Peierls BL (1998) Ecosystem responses to internal and watershed organic matter loading: consequences for hypoxia in the eutrophying Neuse River Estuary, North Carolina, USA. Mar Ecol Prog Ser 166: 17-25

Peinert R, Bodungen B von, Smetacek VS (1989) Food web structure and loss rate. In: Rep Dahlem-workshop. Willey Interscience Publ 44, Chichester, UK, pp 35-48

Revelante N, Gilmartin M (1992) The lateral advection of particulate organic matter from the Po Delta region during summer stratification, and its implications for the northern Adriatic. Estuarine Coast Shelf Sci 35: 191-212

Wassman P (1990) Relationship between primary production in the boreal coastal zone of the North Atlantic. Limnol Oceanogr 35: 464-471

Zachleder V (1984) Optimization of nucleic acids assay in green and blue-green algae: extraction procedures and the light-activated diphenilamine reaction for DNA. Arch Hydrobiol 67: 313-328

Dissolved Organic Carbon Seasonality in the Northern Adriatic and its Possible Involvement in the Aggregate Formation

A. Puddu, R. Pagnotta, L. Patrolecco, M. Pettine, and A. Zoppini

ABSTRACT
Dissolved organic carbon (DOC) concentrations in northern Adriatic coastal waters show seasonal variations with increasing values from winter (75-95 µM in February) to spring-summer periods (120-190 µM in June 1996 and 85-123 µM in June 1997). Riverine inputs of refractory organic matter and abiotic transformations of in situ produced biogenic matter contribute to the accumulation of DOC. Phosphorus limitation of bacteria, whose productivity varied widely with time and space (0.01 to 7 µg C l^{-1} h^{-1}), probably concurs to this seasonal DOC behaviour making microbial community unable to contrast the enrichment in DOC concentrations. Possible involvement of DOC accumulation in the gel production in the Adriatic sea are discussed.

Introduction

Dissolved organic carbon is by far the major carbon pool in oceans, outweighing any other marine source of organic carbon by at least a factor of 10 (Kepkay 1994). Furthermore, despite the fact that algal biomass in oceans is only 0.2% of that of terrestrial plants, annual phytoplankton photosynthesis is roughly equivalent to that of plants since algae have duplication rates on the order of days compared to months to years for plants (Chisholm 1992). Biogenic carbon which is produced in the photic layer of oceans is transferred to deep and bottom waters by the biological pump consisting of sinking particles (Thingstad and Rassoulzadegan 1995), and by diffusion and downwelling of DOC which may seasonally accumulate in surface waters (Carlson et al. 1994). While this organic matter descends to deep and bottom waters it is slowly degraded to CO_2, making deep oceans a massive oversaturated reservoir of CO_2. The average C age of DOC in deep waters is between 4,000 and 6,000 years (Druffel et al. 1992), thus suggesting a strong refractive character of this pool. However, even in surface waters only a fraction of the freshly produced biogenic carbon may be biodegradable. The magnitude of this fraction ranges from 10 to 50% of the DOC with 19% found as a mean in a recent review (Søndergaard and Middelboe 1995). For this reason algal blooms are generally followed by partial accumulation of the DOC produced, and the quantity of accumulated carbon may increase significantly in the case that phytoplankton blooms deplete nutrients necessary for bacterial activity. Thingstad et al. (1997) have summarized surface and deep water concentrations of DOC in different areas, highlighting that concentrations differed in many cases between surface and deep waters due to a fraction of slowly degradable DOC. The fate of carbon in oceans has a large influence on the trophic food chain and affects the exchange of carbon at seawater-atmosphere interface. Seasonal variations in the production/respiration ratio produce an annual cycle in the concentrations of CO_2 and O_2 in the atmosphere (Sherr and Sherr 1996). Seawater/atmosphere exchange processes have strongly stimulated scientists interest in studying limiting factors in the so-called high nutrient, low chlorophyll (HNLC) central equatorial Pacific. Research efforts have been addressed to a possible limitation by iron (Martin 1992; Landry et al. 1997) and silicon (Dugdale and Wilkerson 1998).

The North Adriatic sea is a small basin and its influence on a planetary scale is limited. However, this basin is one of the most productive

Istituto di Ricerca sulle Acque, CNR, Via Reno 1, 00198 Roma, Italy

F.M. Faranda, L. Guglielmo, G. Spezie (eds)
Mediterranean Ecosystems: Structures and Processes

regions over all the Mediterranean sea and its coastal waters are widely exploited for tourism. In recent years this basin has been affected by massive occurrences of mucilaginous aggregates which have heavily affected fishing activity and tourism and produced serious damage to the ecosystem (Degobbis et al. 1995). Then, it is a priority task to go deep into processes responsible for these phenomena. The hypotheses proposed to explain the origin of mucilage aggregates took into account a high number of variables, namely: trophic conditions of the northern Adriatic, high N/P ratios, strong stratification during the spring-summer period, limited grazing control of algal growth by zooplankton, suitable hydrodynamic conditions (calm sea, slow currents and wind speeds), changes in the community structure with increased dominance of diatoms, modifications in environmental conditions induced by P-substitutes in detergency products, the presence of high C/P and slow-to-degrade organic matter due to a faster cycle of P compared to C, mucus excretion from bacterial capsules (Degobbis et al. in press; Pettine et al. 1995). All these factors may indeed cooperate in the formation of organo-mineral particles, of various extents, which are difficult to quantify. However, in order that aggregates become very large and acquire a mucilaginous property particular conditions are needed. Particles have either to become entrapped in a polysaccharide matrix, generated by plankton and bacteria, or to undergo surface modifications leading to the production of sticking material which favours coagulation-flocculation processes.

This paper presents data on dissolved organic carbon, colloidal organic carbon and bacterial activity in the context of the PRISMA-2 research project, subproject "Biogeochemical cycles", whose goal was to evaluate biological and chemical phenomena affecting C cycle along the western coast of the northern Adriatic. Measurements were performed in frontal regions defined by sharp gradients between freshwater inputs (mainly due to the Po discharge) and the waters of the interior basin. Temporal and spatial variability of data and possible relationships among these variables are discussed. Results of laboratory DOC degradation runs on samples, spiked with inorganic and organic nutrients, are also discussed to highlight possible limitations of bacterial activity which would favour DOC accumulation in the field.

Material and Methods

Water samples were collected with Niskin bottles on a rosette with a Sea-Bird CTD system in two frontal regions in the northern Adriatic basin (Fig. 1) during four cruises on the R/V Urania (June 1996, February and June 1997 and February 1998). In every area and in each cruise, sampling was conducted in a grid of stations along 2-4 transects across the front at 4-6 depths. A bench salinometer Autosal Guildline was used to measure the bottle salinities. The sampling generally took 6 days per grid to be completed. Samples were chosen in order to be representative of different water masses in the frontal region. The total number of measurements was 401, 232 and 30, respectively for DOC, BCP and COC.

For the analysis of dissolved organic carbon, water samples were filtered aboard, immediately after collection, through precombusted (4 hours at 480°C) Whatman GF/F glass fibre filters (0.7 µM nominal pore size) and transferred directly in duplicate into 25 mL high density polyethylene (HDPE) bottles, previously treated with HNO_3 1.2 M at 50°C for 1 h. The HDPE bottles were quick-frozen in an aluminium block at −20°C. In the laboratory, filtered samples were thawed, acidified to pH 2 with ultrapure HCl and purged with N_2 for about 10 minutes to remove inorganic carbon. The DOC was assayed by high temperature catalytic oxidation (HTCO) using a Shimadzu TOC-5000 A analyser and, on a limited number of samples, it was fractionated in dissolved (< 1 kDa) and colloidal (> 1 kDa) organic carbon (Pettine et al. 1999).

Bacterial production was measured on natural samples by the incorporation of 3H-thymidine in all cruises and 3H-leucine only in '97 and '98 cruises. Triplicate 1.7 mL aliquots and one killed control (TCA 5%) were spiked with 20 nM (final concentration) radiotracer and incubated for 1 h, from 10 to 11 am local time, at in situ temperature. The extraction, with 5% ice-cold TCA for thymidine and room temperature 5% TCA for leucine and subsequent washing with 5% TCA and 80% ethanol, was performed by the microcentrifugation method, according to Smith and Azam (1992). To estimate the bacterial carbon production from rates of thymidine incorporated (BCPTdR), the most widely used conversion factor of 2 x 10^{18} cells per mole thymidine incorporated (Bell 1993) and a bacteria-to-carbon conversion factor of 20 fgC $cell^{-1}$ were used

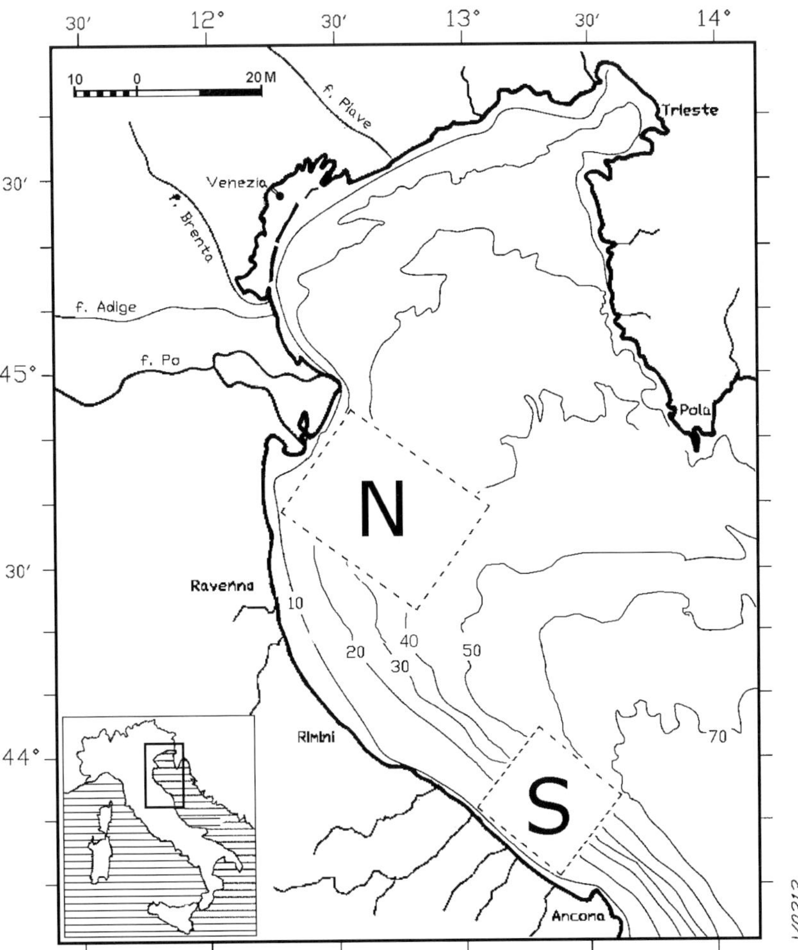

Fig. 1. Location of sampling areas

(Lee and Furhman 1987). Rates of leucine incorporation were transformed in C units (BCPLeu) using the conversion factor of 3.1 kg C produced per mole leucine incorporated (Kirchman 1993). Measurements with the two methods gave similar results with an average BCPTdR/BCPLeu ratio of 1.2±0.6. In the discussion of our results we will refer only to BCPTdR, hereafter simply referred as BCP, for which a complete set of data for all the cruises is available.

The effect of nutrients on bacterial growth was investigated using two samples, from surface and bottom waters, respectively, collected during the four PRISMA-2 cruises and in a following cruise in June 1998. Immediately after collection, samples were filtered through 0.4 µM and 1 µM polycarbonate membranes (Nuclepore), to obtain respectively dilution water and the predator-free inoculum. Filtered water was kept in the dark at 4°C until utilization within 24 h. Batch cultures, prepared by diluting 10 fold the inoculum in dilution water, were carried out in 100 mL BOD-type bottles in the dark and at in situ temperature. To evaluate the effect of inorganic nutrients on bacterial growth, parallel treatments in duplicate were spiked with nitrogen (5 µM NO_3^- + 5 µM NH_4^+) and phosphate (1 µM PO_4^{3-}) 2 and 20 fold greater than their ambient water concentrations, respectively, in most cases. To test the response of heterotrophic bacteria to the addition of a labile organic C substrate, in each experiment a further parallel treatment was spiked with a mixture of inorganic nutrients and glucose in concentrations close to the natural DOC content. In the two last experiments (February and June 1998) N and P were added separately in the molar ratio C:N:P=80:10:1. In June 1998 experiment, glucose was added alone and in combination with inorganic nutrients. In order to measure the time course of changes in the cultures, aliquots were withdrawn at 12-24 hour intervals and monitored for 96 hours.

Bacterial numbers were counted with acridine orange staining (Hobbie et al. 1977) and bacterial production by 3H-leucine incorporation. All materials in contact with the samples were washed with 1 N HCl and ultrapure water (Millipore Milli-Q) prior to use. All the glassware was heated before use (500°C). Specific growth rates (μ d^{-1}) were estimated from the increase in bacterial numbers during the exponential growth phase.

Results

Field Measurements

Variability ranges of DOC and BCPTdR together with mean values and their respective 95% confidence intervals resulting from all the data in the upper and bottom layers are summarized in the Table for the two investigated frontal regions in each cruise. The overall set of DOC values ranged from 53 to 281 μM while in winter cruises it was 53-124 μM. The significance of differences in the median values among different data groups was analysed by using the Mann-Whitney Rank Sum Test. Concentrations of DOC in the upper and bottom layers were reasonably constant in winter surveys in the two areas. The differences between layers were small, but significant ($P<0.001$). In the south, the data set of DOC in February 1997 was limited to only 4 surface values and the resulting average value is not representative of the area; then, it may not be compared with the other averages derived from much more numerous data sets. The bottom average in winter periods is substantially the same in both years and areas, 74±2 μM considering all data together, and can be considered as a threshold value for this coastal ecosystem. Contrary to winter results, concentrations of DOC in summer surveys varied in the two investigated years. Mean value differences were significant between layers in the northern region in both summer cruises ($P<0.001$) and in the south only in June 1996 ($P<0.05$). The average values of corresponding layers in the two summer cruises differed both in the north ($P<0.001$) and in the south ($P<0.01$), while differences between areas were significant only for surface waters in June 1996 ($P<0.001$).

Values of BCP from all the cruises in the two frontal regions varied in the range 0.01–7.1 μg C l^{-1} h^{-1} (Table). The variability in winter periods was much more limited with values in the range 0.01-1.6 μg C l^{-1} h^{-1}. In the north, the difference between surface averages resulting from consecutive winter cruises was not significant, while bottom averages were significantly different

Table. Summary of dissolved organic carbon (DOC) concentrations and bacterial carbon production rates (BCP) in the two sampled areas in different cruises, in upper (≤5 m) and bottom layers, *C.I.* is the 95% confidence interval for the mean

		DOC (μM)		BCP (μgC l^{-1} h^{-1})	
		North	South	North	South
June '96	Min-max	74-281	88-219	0.03-1.26	0.02-0.92
	n	66	37	30	31
	Upper layer mean±C.I.	191±16	153±15	0.60±0.28	0.38±0.11
	Bottom layer mean±C.I.	137±13	123±13	0.10±0.02	0.10±0.11
February '97	Min-max	53-123	84-153	0.03-0.67	0.05-1.61
	n	76	4	30	24
	Upper layer mean±C.I.	95±8	111	0.23±0.08	0.62±0.38
	Bottom layer mean±C.I.	74±3		0.21±0.10	0.16±0.09
June '97	Min-max	66-162	71-142	0.09-7.09	0.08-0.62
	n	81	23	30	30
	Upper layer mean±C.I.	123±8	100±27	2.45±1.8	0.36±0.39
	Bottom layer mean±C.I.	81±9	84±11	0.15±0.05	0.11±0.04
February '98	Min-max	65-124	58-115	0.02-1.19	0.01-0.45
	n	63	51	30	27
	Upper layer mean±C.I.	95±10	83±7	0.37±0.34	0.18±0.10
	Bottom layer mean±C.I.	75±5	73±4	0.03±0.01	0.01±0.004

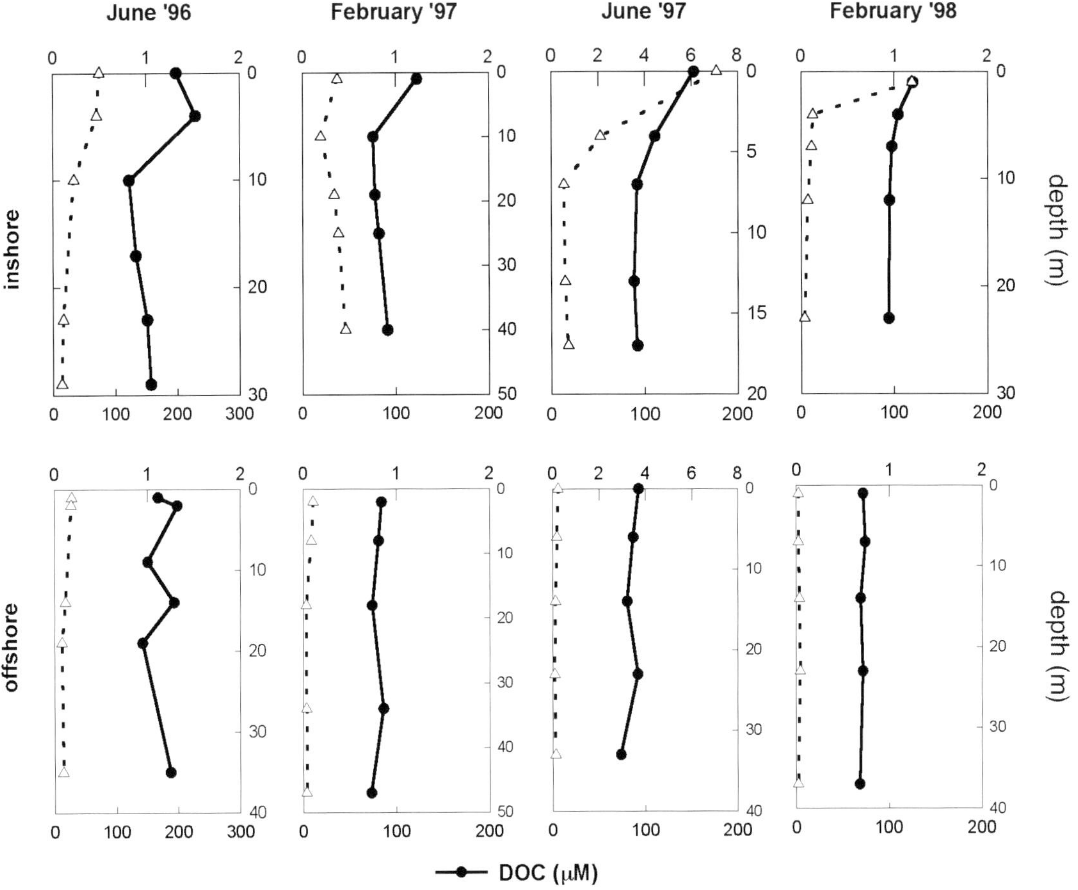

Fig. 2. Vertical distribution of DOC and BCP in the four cruises, on stations representative of inshore and offshore waters

($P<0.001$); differences between layers were significant in February 1998 ($P<0.05$). In the south, surface averages in February 1997 and 1998 were significantly different ($P<0.05$) from the corresponding bottom values and differences between layers in the same cruise were significant in 1997 ($P<0.01$). In summer cruises there was always a significant difference between surface and bottom layers ($P<0.01$), with a higher bacterial productivity at surface compared to deep waters. The significance of differences between averages in consecutive years was limited to surface northern waters ($P<0.05$). Differences between regions were only poorly significant although values were generally higher in the north with the exception of February 1997.

Figure 2 shows typical DOC and BCP profiles with decreasing concentrations from surface to deep waters. Gradients of DOC concentrations and heterotrophic activities are generally more marked in summer than in winter cruises and in inshore compared to offshore stations. In some sporadic cases, depending on local conditions, profiles could show deviations from the typical distribution. An interesting exception is given in February 1997 in the north by heterotrophic activities in four coastal and intermediate stations characterized by a strong vertical salinity gradient. In these cases thymidine and leucine incorporation rates in bottom waters were similar or higher than those measured in the upper layer. Figure 3 shows mean values of DOC and BCP with related standard deviations in the internal, intermediate and external sides of the northern frontal region. From this figure it results that DOC concentrations decrease from coastal to offshore stations in all the cruises except June 1996, when DOC levels showed a reasonably uniform distribution. The BCP averages were reasonably constant in February 1997, while they decreased in the offshore direction in the other cruises with marked changes in June 1997.

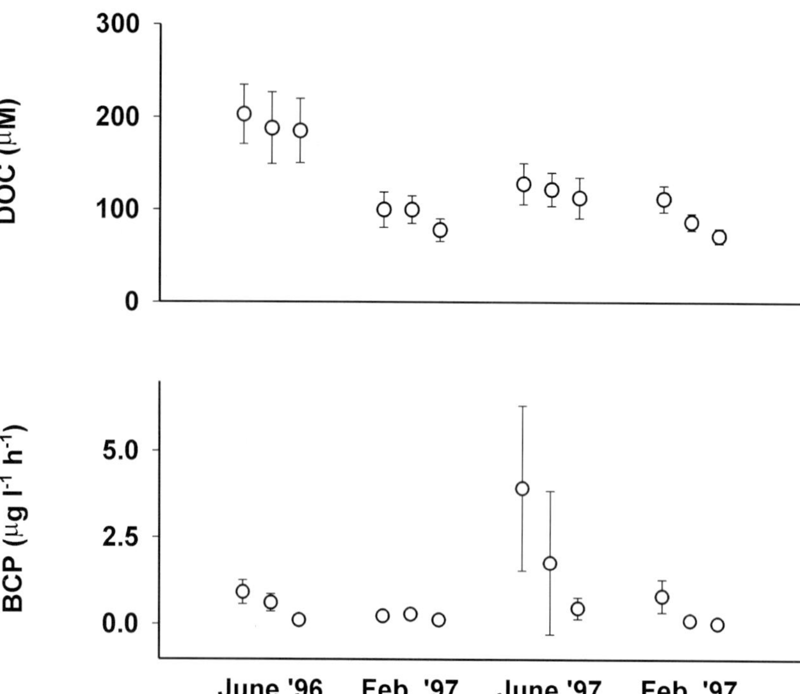

Fig. 3. Mean values of DOC, BCP and salinity in the northern area and in the upper column layer (≤5 m), in the coastal (*left*), intermediate (*centre*) and offshore (*right*) stations respectively in each cruise

During our investigations, 30 samples, overall, were taken in the two areas at various depths and processed with the ultrafiltration system (12 in June 1996 and 6 in each of other cruises). Since we did not find any significant difference between regions and as a function of depth, results are discussed all together. Concentrations of the overall COC ranged from 49 to 98 µM and from 44 to 80 µM in summer and winter periods, respectively. Average values calculated from all the data were 67.5 µM and 39.5 µM for COC and truly dissolved organic components (DOC < 1 kDa), respectively. Overall COC concentrations were made up of 52 to 74% of total DOC with an average of 63% in summer and 54 to 83% with an average of 66% in winter, hence in both periods a major portion of the whole DOC pool was in the colloidal fraction.

Laboratory Experiments

Growth rates of predator-free bacterial populations, collected in February and June 1997 in surface (ul) and bottom (bl) samples are shown in Fig. 4. The quality of the natural substrate and a high water temperature gave higher growths in June (0.4 and 0.8 µ d^{-1} in the surface and bottom samples respectively) than in February (<0.1 µ d^{-1} at both depths). However, bottom samples in both February and June did not show any significant effect of N and P additions, while the combined additions of glucose and inorganic nutrients produced a strong increase in the bacterial growth, with values of 0.2 and 1.7 µ d^{-1}, respectively, in the two months. Contrary to bottom samples, surface samples showed a clear positive effect of N and P additions (0.5 and 0.9 µ d^{-1} in February and in June) with respect to the control (0.05 and 0.4 µ d^{-1}, respectively) and the addition of glucose produced a further stimulation (0.7 and 1.7 µ d^{-1} respectively). In this first series of experiments nitrogen and phosphorus salts were added together, then we could not conclude whether both these nutrients or one of them were responsible for the effect produced.

Figure 5 shows time course of changes in bacterial production, measured as leucine incorporation, in surface samples collected in February and June 1998 after separate additions of N and P. Bacterial abundances showed a similar behaviour. The P addition enhanced in both cases the bacterial activity with respect to the control; increasing factors were 2.0 and 2.3 in February and June, respectively. The addition of N did not produce any significant effect. The combined additions of inorganic N and P and

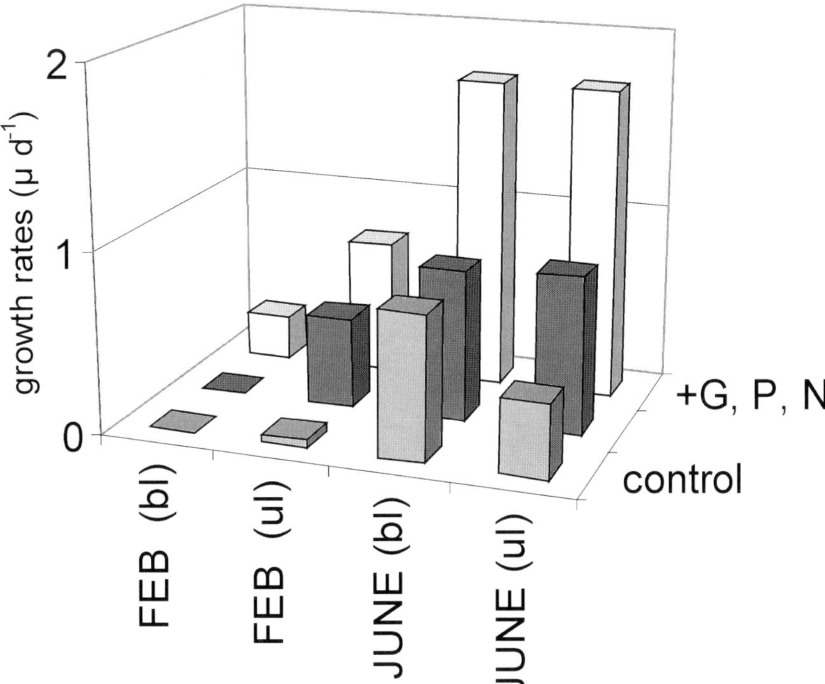

Fig. 4. Growth rates (μ day^{-1}) in batch cultures of natural predator-free bacterial assemblage collected in February and June 1997, in upper (*ul*) and bottom layers (*bl*), amended with phosphate (*P*), ammonia plus nitrate (*N*) and organic carbon as glucose (*G*)

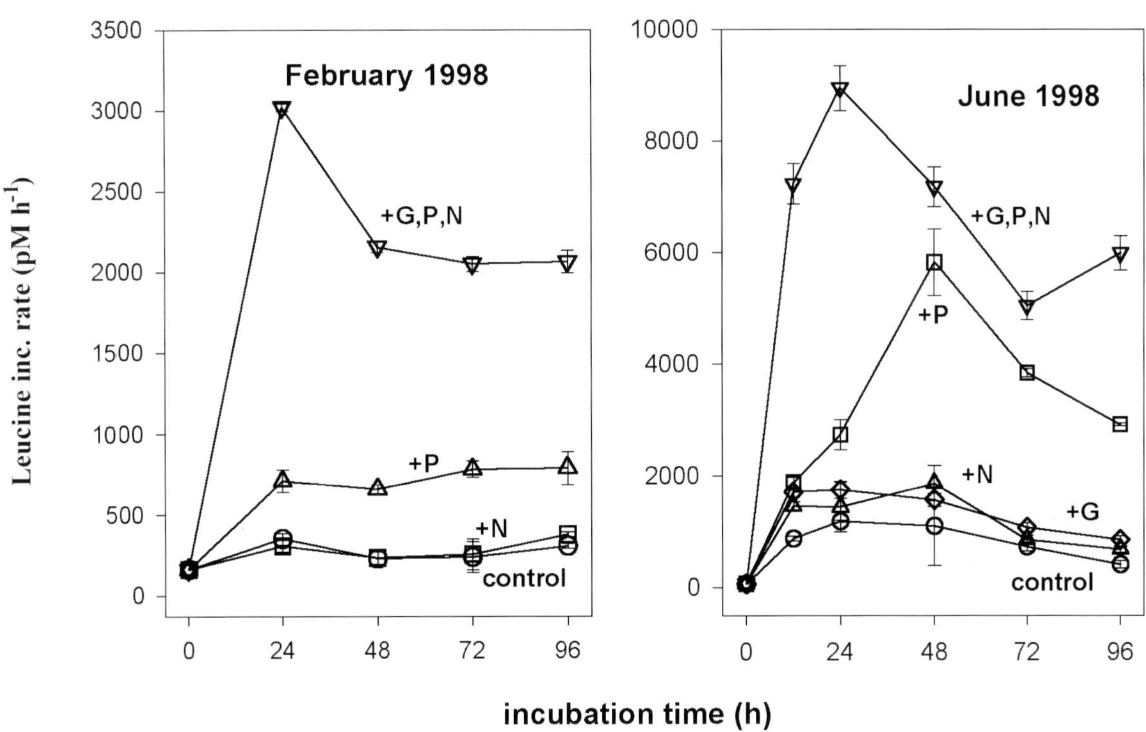

Fig. 5. Bacterial production rates, as leucine incorporation, in natural predator-free bacterial assemblage collected in 1998 in the surface layer, amended with phosphate (*P*), ammonia plus nitrate (*N*) and organic carbon as glucose (*G*)

glucose produced in both cases maximum values in leucine incorporation rates with increasing factors with respect to the control of 8.6 and 7.6. It is worth noting that the only addition of glucose did not produce any significant stimulation in the June 1998 sample.

Discussion

The DOC concentrations in northern Adriatic waters show seasonal changes with the lowest values in winter (74±2 μM for bottom samples) comparable to DOC levels of surface oceanic waters (70-80 μM; Guo et al. 1995). Our surface DOC values in winter (92±5 μM) correspond to maxima values measured over a 12 month period by Copin-Montegut and Avril (1993) in the north western Mediterranean basin between 0 and 50 m depths (92-100 μM). Deep Mediterranean waters (150-2000 m depth) showed in this study values in the range 50 to 58 μM. Our summer DOC concentrations are significantly higher than winter values, although the increasing factors in the two investigated years are different. Average surface values in June 1996 and 1997 are a factor of 2.0 and 1.3 higher than those recorded in the respective following February surveys. Thus, degradation processes and water mass exchanges between the northern and middle Adriatic are able to keep DOC concentrations at low levels despite the significant external (50 10^4 tons year^{-1}; Pettine et al. 1999) and internal (170 10^4 t year^{-1} , Degobbis and Gilmartin 1990; or 320 10^4 t year^{-1} estimated from recent PRISMA-1 results, not yet published) organic matter loads which affect this basin. It may be noted that the summer of 1997 was again characterized by massive mucilage occurrences. These did not reach the extension recorded in 1988, 1989 and 1991 years, but were markedly more intensive compared to those of 1996 summer (Degobbis et al. in press). The presence of mucilaginous materials may scavenge dissolved materials favouring their transfer to the particulate phase. This might have contributed to the marked differences in concentrations between the two investigated summers. That DOC concentrations show seasonal changes also results from previous findings. In the Gulf of Trieste DOC was found to vary in the ranges 102-297 and 113-176 μM in September 1994 and June 1995, respectively (Fonda Umani et al. 1997), while in the same gulf Faganeli and Herndl (1991) found distinct seasonal variations ranging from 66-154 μM during the autumn and winter to up to 500 μM during the summer in the Istrian coast. Finally, Vojvodić and Cosović (1996) found that DOC concentrations varied approximately from 100-140 to 130-200 μM in winter and summer, respectively, over the period 1984-1993.

Average bacterial carbon production rates represent, according to values reported by Ducklow and Carlson (1992) for the euphotic zones of major marine habitats, a variety of situations from estuaries (1.8-3.3 μg C l^{-1} h^{-1}) to the ocean (0.1-0.2 μg C l^{-1} h^{-1}). The high range of bacterial production rates measured in this area, 0.2-177 pM thymidine incorporated h^{-1}, has a parallel only in the estuarine environment of the Chesapeake Bay (0.5-500 pM TdR h^{-1}, Ducklow and Shiah 1993). This wide variability reflects relatively quick adaptations of microorganisms to environmental changes in the water column produced by biological processes and external forcing. Bacterioplankton reacts primarily to variations in the dissolved organic substrate composition. The BCP variations observed are consistent with increased DOC concentrations in surface waters, which are more marked in summer than winter, and with a high bioavailability of fresh biogenic organic matter resulting from exudation or sloppy feeding compared to aged products dominant in bottom waters. Previous findings have shown a lack of synchrony in the cycles of activity which characterizes heterotrophic bacteria and autotrophic phytoplankton in this area in different seasonal periods (Puddu et al. 1998). Diel variations in the bacterioplankton activity also reflect short-term variations in enzymatic activities and diel periodicity in bacterivorous flagellates which influence bacterial biomass (Riemann and Søndergaard 1984; Karner and Rassoulzadegan 1995).

As shown in Fig. 6, where the behaviour of bacterial productions are reported as a function of DOC, BCP variations show in all cruises a tendency to increase with increasing DOC concentrations, but the slope of the linear regression in June 1996 (0.098, $P<0.01$) is a factor of about 0.5 compared to those in June 1997 (0.202, $P<0.001$) and in winter (0.188, $P<0.001$). Changes in the BCP versus DOC slopes reflects differences in the quality of available organic substrates and in the nutrient optimal ratios which are required for

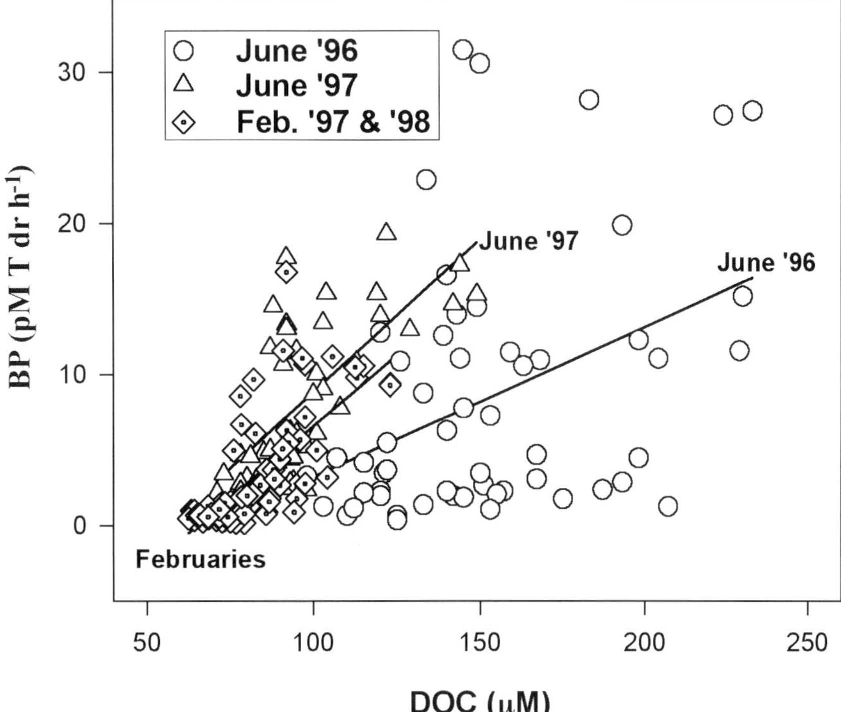

Fig. 6. Behaviour of bacterial production (*BP*), as ^{3}H-thymidine incorporation rates, as a function of DOC concentrations separately in the two summer periods and in two Februarys together

bacterial growth. The post bloom situation which has been suggested by Lipizer et al. (1997) in June 1996, leading to aged organic substrates and low P concentrations, may probably be responsible for the lower slope found in that period.

Biodegradation runs give an insight into factors controlling the heterotrophic activity in our system. In deep waters where the phytoplankton activity is reduced, the bacterioplankton is not in competition for nutrients and the substrate bioavailability of the carbonaceous substrate is the dominant controlling factor. In the February 1997 enrichment experiments, the lack of bacterial growth suggests that the natural organic substrate is not degraded, even in the presence of inorganic nutrients. While in this same sample the addition of a bio-labile compound such as glucose stimulate bacteria to grow. Contrary to February 1997, in the June 1997 bottom sample we observed a high bacterial growth rate, probably due to a higher bioavailability of the organic substrate derived from phytoplankton activity in the photic layer; however, the addition of inorganic nutrients did not produce any significant increase in this case either, probably because release processes from sediments made nutrients not limiting in that case. Runs with separate additions of nutrients highlight that orthophosphate strongly limits bacterial growth in our system while inorganic nitrogen did not exert any effect, neither in winter nor in summer samples. The not significant increase in leucine incorporation rates, measured in the June 1998 sample spiked with glucose, highlights a scarce ability of bacteria to consume also organic carbonaceous substrates, even if as highly labile as glucose, suggesting a strong limitation by inorganic nutrients during summer periods. Although nutrient limitation is not easy to prove unambiguously, our findings support a P- rather than N-limited environment in the northern Adriatic sea. Such conclusions are also in accordance with previous observations on phytoplankton P-limitation in this area (Zoppini et al. 1995) and with a general microbial P limitation in the Mediterranean (Zohary and Robarts 1998; Thingstadt et al. 1998). A phosphorus deficiency limiting primary and secondary production in northern Adriatic coastal waters is also consistent with the very high phosphatase activities measured in samples from this area (R. Zaccone, personal communication).

Inorganic P concentrations determined in this project by the research group of the Ist. Talassografico Sperimentale of Trieste (Catalano

G., personal communication) extend information on P availability to natural samples. Mean orthophosphate concentrations in the northern area in June 1996 (0.02±0.003 µM PO_4^{3-}) were found to be significantly lower than in June 1997 (0.04±0.006 µM PO_4^{3-}). The very low inorganic nutrient concentrations observed in June 1996, possibly due to a "post bloom" situation (Lipizer et al. 1997), may explain the increase in DOC values with a scarce uptake of DOC by bacteria in this period. Unusual high bottom bacterial productions measured in the north in February 1997, correspond to inorganic P concentrations > 0.10 µM PO_4^{3-}. These levels have been very seldom measured during the project, strengthening the important role of low phosphorus concentrations in controlling secondary production in our system.

Compared to data for other marine environments, where COC concentrations gave percentage contributions to DOC of 10 to 64% (Benner et al. 1992; Guo et al. 1994, 1995; Martin and Dai 1995), our results on colloidal organic matter show a shift in the molecular weight distribution toward the large size fraction. Since COC contributions to DOC were found to increase in riverine waters compared to seawaters (76% in the Amazon river, Benner and Hedges 1993), this shift is consistent with an important role of freshwater inputs. The strong correlations of both DOC and BCP values as a function of salinity along with estimates of a bacterial carbon demand higher than primary production in low productivity periods (Puddu et al. 1998) also strongly support a key role of allochthonous inputs in controlling the trophic dynamics in this basin.

Setting up of strong vertical stratification and eddy circulation, which characterizes the northern Adriatic in the summer period, increases freshwater residence time and favours the accumulation of low degradable organic matter directly discharged by rivers. However, the accumulation of organic matter in the refractory pool may partially be counteracted by photoxidation processes which tend to increase the liability of aged DOM (Wetzel et al. 1995). In spring-summer periods the in situ production of DOC, mainly due to phytoplankton activity, is also strongly enhanced and often phosphorus becomes exhausted after microalgae blooms. Under these conditions, the fraction of biogenic carbon which is not immediately degraded may undergo, with ageing, abiotic processes which are strongly driven by interactions with terrestrial-derived DOC. The enrichment in organic polymers and particles gives rise to high colloidal components and organo-mineral aggregates. Under these conditions, the probability that organic macromolecules spontaneously assemble to form microgels and that secondary annealing of assembled gels occurs is very high, according to findings by Chin et al. (1998). Furthermore, the shift we have found in the molecular weight distribution toward the large size fraction favours DOM assembling, since the probability of spontaneous assembly of organic networks increases with the second power of the polymer length (Chin et al. 1998). Then, seasonal increase in DOC concentrations along with its high colloidal organic fraction and high molecular weight are necessary conditions which may lead to the formation of mucillaginous aggregates.

Conclusions

Labile to refractory conversions of aged organic matter and a restricted bacterial consumption of DOC are two interrelated factors which concur, along with refractory matter directly discharged by rivers and escaping photoxidation processes, to the observed seasonal DOC enrichment in the North Adriatic. Grazing or viral infections may also contribute in limiting bacterial activity, but information on this matter is very scarce. While we have environmental conditions (high COC levels and high molecular weights) which are apparently favourable to the formation of mucilaginous aggregates in our system, their occurrence shows wide interannual variations (in size and weight) which are difficult to explain only on a chemico-physical basis. However, mucilaginous aggregates are not simply physico-structural components which add to the liquid and particulate matrices; they behave as preferential microhabitats due to their strong scavenging properties which make them rich in nutrients compared to surrounding water. High primary and secondary productions have been measured inside these aggregates in previous studies (Muller et al. 1994) and biological processes may profoundly influence the fate of aggregates. A high residence time in the water column and low wind-induced mixing are required in order that these aggregates reach as large sizes as those with which we have been faced in recent years (Riebesell 1992).

Acknowledgements. This study is part of the PRISMA Project, founded by the Italian Ministry for University and Scientific and Technological Research. Thanks are due to the crew of the R/V Urania and to Mr. F. Bacciu, of this Institute, for the collection of samples and analysis of DOC.

References

Bell RT (1993) Estimating production of heterotrophic bacterioplankton via incorporation of tritiated thymidine. In: Kemp PF, et al (eds) Handbook of methods in aquatic microbial ecology. Lewis Publishers, Boca Raton, pp 495-503

Benner R, Hedges JI (1993) A test of accuracy of fresh water DOC measurements by high-temperature catalytic oxidation and UV-promoted persulfate oxidation. Mar Chem 41: 161-165

Benner R, Pakulski JD, McCarthy M, Hedges JI, Hatcher G (1992) Bulk chemical characteristics of dissolved organic matter in the oceans. Sciences 255: 1561-1564

Carlson CA, Ducklow HW, Michaels AF (1994) Annual flux of dissolved organic carbon from the euphotic zone in the northwestern Sargasso Sea. Nature 371: 405-408

Chin WC, Orellana MV, Verdugo P (1998) Spontaneous assembly of marine dissolved organic matter into polymer gels. Nature 391: 568-572

Chisholm S (1992) What limits phytoplankton growth? Oceanus 35(3): 36-46

Copin-Montegut G, Avril B (1993) Vertical distribution and temporal variation of dissolved organic carbon in the North-Western Mediterranean Sea. Deep-Sea Res 40(10): 1963-1972

Degobbis D, Gilmartin M (1990) Nitrogen, phosphorus and biogenic silicon budgets for the northern Adriatic sea. Oceanol Acta 13: 31-45

Degobbis D, Fonda-Umani S, Franco P, Malej A, Precali R, Smodlaka N (1995) Changes in the northern Adriatic ecosystem and the hypertrophic appearance of gelatinous aggregates. Sci Total Environ 165: 43-58

Degobbis D, Malej A, Fonda Umani S (2000) The mucilage phenomenon in the northern Adriatic – a critical review of the present scientific hypothesis. Proc Int Workshop State Art New Sci Hypotheses Phenomenon Mucilages Adriatic Sea, 24-25 March 1997, Trieste. Ann Ist Super Sanità (in press)

Druffel ER, Williams PM, Bauer JE, Ertel R (1992) Cycling of dissolved and particulate organic matter in the open ocean. J Geophys Res 97: 639-659

Ducklow HW, Carlson CA (1992) Oceanic bacterial production. In: Marshall KC (ed) Advances in microbial ecology. Plenum Press, New York, pp 113-181

Ducklow HW, Shiah FK (1993) Bacterial production in estuaries. In: Ford TE (ed) Aquatic microbiology: an ecological approach. Blackwell, Oxford, pp 261-287

Dugdale RC, Wilkerson FP (1998) Silicate regulation of new production in the equatorial Pacific upwelling. Nature, 391: 270-273

Faganeli J, Herndl G (1991) Dissolved organic matter in the waters of the Gulf of Trieste (Northern Adriatic). Thalassia Jugosl 23: 51-63

Fonda Umani S, Cauwet G, Cok S, Martecchini E, Predonzani S (1997) Chemical and biological seasonal patterns of the Gulf of Trieste: the example of early and late summer. In: Cauwet G (ed) PALOMA – production and accumulation of labile organic matter in adriatic (contract n. EV5V-CT94-0420). EEC Environ Climate Progr Final Rep 2

Guo L, Coleman CH, Santschi PH (1994) The distribution of colloidal and dissolved organic carbon in the Gulf of Mexico. Mar Chem 45: 105-119

Guo L, Santschi PH, Warnken KW (1995) Dynamics of dissolved organic carbon (DOC) in oceanic environments. Limnol Oceanogr 40(8): 1392-1403

Hobbie JE, Daley RJ, Jasper S (1977) Use of Nuclepore filters for counting bacteria by fluorescence microscopy. Appl Environ Microbiol 33(5): 1225-1228

Karner M, Rassoulzadegan F (1995) Extracellular enzyme activity: indications for high short-term variability in a coastal marine ecosystem. Microb Ecol 30: 143-156

Kepkay PE (1994) Particle aggregation and the biological reactivity of colloids. Mar Ecol Prog Ser 109: 293-304

Kirchman DL (1993) Leucine incorporation as a measure of biomass production byheterotrophic bacteria. In: Kemp PF, et al (eds) Handbook of methods in aquatic microbial ecology. Lewis Publishers, Boca Raton, pp 509-512

Landry MR, Barber RT, Bidigare RB, Chai F, Coale K, Dam HG et al (1997) Iron and grazing constraints on primary production in the central equatorial Pacific. Limnol Ocanogr 42(3): 405-418

Lee S, Fuhrman JA (1987) Relationships between biovolume and biomass of naturally derived marine bacterioplankton. Appl Environ Microbiol 53(6): 1298-1303

Lipizer M, Cozzi S, Catalano G (1997) Accumulation of DON and DOP and "new" production in the northern Adriatic Sea. In: Dehairs F,Elskens M, Goeyens L (eds) Integrated marine system analysis. In: Proc 2nd Network Meet Brussels, 29-31 May 1997, pp 85-97

Martin JH (1992) Iron as a limiting factor in oceanic productivity. In: Falkowski P, Woodhead AD (eds) Primary productivity and biogeochemical cycles in the sea. Plenum, New York, pp 123-127

Martin JM, Dai MH (1995) Significance of colloids in the biogeochemical cycling of organic carbon and trace metals in the Venice Lagoon (Italy). Limnol Oceanogr 40(1): 119-131

Muller-Niklas G, Schuster S, Kaltenböck E, Herndl G (1994) Organic content and bacterial metabolism in amorphous aggregations of the northern Adriatic sea. Limnol Oceanogr 30(1): 58-68

Pettine M, Pagnotta R, Liberatori A (1995) Composition of mucilaginous macroaggregates and hypotheses for their formation. Ann Chim 85: 431-441

Pettine M, Patrolecco L, Manganelli M, Capri S, Farrace MG (1999) Seasonal variations of dissolved organic matter in the northern Adriatic Sea. Mar Chem 64: 153-169

Puddu A, La Ferla R, Allegra A, Bacci C, Lopez M, Oliva F, Pierotti C (1998) Seasonal and spatial distribution of bacterial production and biomass along a salinity gradient (Northern Adriatic Sea). Hydrobiol 363: 271-282

Riebesell U (1992) The formation of large marine snow and its sustained residence in surface waters. Limnol Oceanogr 37(1): 63-76

Riemann B, Søndergaard M (1984) Measurements of diel rates of bacterial secondary production in aquatic environments. Appl Environ Microbiol 47(4): 632-638

Sherr EB, Sherr BF (1996) Temporal offset in oceanic production and respiration processes implied by seasonal changes in atmospheric oxygen: the role of heterotrophic microbes. Aquat Microb Ecol 11:91-100

Smith DC, Azam F (1992) A simple, economical method for measuring bacterial protein synthesis rates in seawater using 3H-leucine. Mar Microb Food Webs, 6(2): 107-114

Søndergaard M, Middelboe M (1995) A cross-system analysis of labile dissolved organic carbon. Mar Ecol Prog Ser 118: 283-294

Thingstad TF, Hagstrom A, Rassoulzadegan F (1997) Accumulation of degradable DOC in surface waters: is it caused by a malfunctioning microbial loop? Limnol Oceanogr 42(2): 398-404

Thingstad TF, Rassoulzadegan F (1995) Nutrient limitations, microbial food webs, and 'biological C-pumps': suggested interactions in a P-limited Mediterranean. Mar Ecol Prog Ser 117: 299-306

Thingstad TF, Zweifel UL, Rassoulzadegan F (1998) P limitation of heterotrophic bacteria and phytoplankton in the northwest Mediterranean. Limnol Oceanogr 43(1): 88-94

Vojvodić V, Cosović B (1996) Fractionation of surface active substances on the XAD-8 resin: Adriatic Sea samples and phytoplankton culture media. Mar Chem 54: 119-133

Wetzel RG, Hatcher PG, Bianchi TS (1995) Natural photolysis by ultraviolet irradiance of recalcitrant dissolved organic matter to simple substrates for rapid bacterial metabolism. Limnol Oceanogr 40(80): 1369-1380

Zohary T, Robarts RD (1998) Experimental study of microbial P limitation in the eastern Mediterranean. Limnol Oceanogr 43(3): 387-395

Zoppini A, Pettine M, Totti C, Puddu A, Artegiani A, Pagnotta R (1995) Nutrients, standing crop and primary production in western coastal waters of the Adriatic Sea. Estuarine Coast Shelf Sci 41:493-513

CHAPTER 59

Evolution of the Trophic Conditions and Dystrophic Outbreaks in the Sacca di Goro Lagoon (Northern Adriatic Sea)

P. Viaroli, R. Azzoni, M. Bartoli, G. Giordani, and L. Tajé

ABSTRACT

Dystrophy and vulnerability have been analysed in the Sacca di Goro lagoon (Po river Delta, Italy), where long term research was started in 1989 with respect to the abnormal growth and decomposition of the seaweed *Ulva rigida*. The research addresses: (1) trends in hydrochemical variables; (2) biomass, growth rates and the elemental composition of the dominant macroalgae; (3) nutrient uptake and retention within the macroalgal biomass; (4) oxygen metabolism; (5) decomposition processes and their effects on benthic fluxes of nitrogen, phosphorus and sulphide; (6) iron buffering of the phosphorus and sulphur cycles.

In the last decade, annual patterns have been observed with the abnormal spring growth of the seaweed *Ulva rigida*, which is usually followed by a sudden collapse and a prolonged oxygen deficiency in early summer. Nitrates are the most important nitrogen source for the macroalgae. Soluble reactive phosphorus concentrations attain significant peaks only during summer anoxia. In the worst years (1989-92 and 1997), *Ulva* growth commences in early Spring at temperatures above 10°C. The biomass increases at high rates (0.10-0.25 d^{-1}) and reaches the highest standing crop (10^3 g dw m^{-2}) and the maximum spreading (10 km^2) at the end of May. From mid June, the macroalgal biomass starts to decompose causing anoxia and sulphide production. The biomass accumulation results in the temporary nitrogen retention within the organic pool leading to a prolonged nitrogen deficiency in the water column which allows only *Ulva* to grow. The impact of the dystrophy is related to the sedimentary cycles of sulphur and iron. Microbiologically reducible iron seems to buffer against free sulphide, precipitating it as insoluble iron monosulphide and pyrite. However, a considerable production of sulphide may occur in the water column associated with the decomposition of the thick floating macroalgal mats, where its concentration is independent of the potential iron buffering capacity of the sediment. In conclusion, the vulnerability (or buffer capacity) of such coastal systems seems to depend on the extent of the macroalgal blooms, amplitude of dissolved oxygen fluctuations at different time scales (from days to weeks), and sedimentary iron availability.

Introduction

In coastal areas, the development of large and fast growing primary producers plays a key role in oxygen production and consumption, and regulation of the entire benthic metabolism (Borum 1996; Castel et al. 1996). Usually, fast growing macroalgae exhibit nutrient uptake and storage capacities greatly in excess of their physiological needs. There is also experimental evidence of a strict relationship between growth and decomposition rates: fast growing plants tend to have high nutrient concentrations and low fibre content and therefore also decompose more rapidly because their high N/C and P/C ratios stimulate microbial growth. Fast macroalgal growth followed by decomposition of the accumulated biomass is also considered to be one of the most important factors in the occurrence of dystrophic crises. Hence, growth patterns and distribution of macroalgae may explain the changes which occur in other processes within coastal ecosystems (for detailed reviews see Sand-Jensen and Borum 1991; Borum 1996; Flechter 1996; Valiela et al. 1997).

Dipartimento di Scienze Ambientali, Università degli Studi di Parma, Parco Area delle Scienze 33/A, 43100 Parma, Italy

F.M. Faranda, L. Guglielmo, G. Spezie (eds)
Mediterranean Ecosystems: Structures and Processes

This study has focused on the dystrophic Sacca di Goro, the southern-most embayment of the Po river delta, Northern Italy. Disturbance and vulnerability of these ecosystems have been considered particularly with respect to: (1) trends in hydrochemical variables; (2) the species composition, biomass, growth rates and the elemental composition of the dominant macrophytes; (3) nutrient uptake and nutrient retention by the plants and the seasonality of these processes; (4) oxygen production and consumption rates over the growth season; (5) decomposition processes of the plant biomasses and their influence on benthic fluxes of nitrogen, phosphorus and sulphide. In this paper, we present trends in *Ulva* biomass, dissolved oxygen, nitrates and soluble reactive phosphorus concentrations in the water column measured at station 17 (Fig. 1) at approximately monthly intervals from March 1989 to December 1997. Factors affecting macroalgal growth are also analysed considering three critical phases of the *Ulva* life cycle in 1997, a typical "worst year" scenario, due to the large extent of macroalgal spreading. Finally, we discuss the role of biogeochemical sedimentary processes by comparing summer (August 1997) and winter (January 1998) data.

The Sacca di Goro is a shallow-water embayment of the Po River Delta (44° 47' –44° 50' N and 12° 15' –12°20' E) (Fig. 1). The bottom is flat and the sediment is composed of typical alluvial muds with high clay and silt contents in the northern and central zones. Sand is more abundant near the southern shore-line, whilst sandy muds occur in the eastern area. Numerical models have demonstrated a clear zonation of the lagoon with the low energy eastern area separated from two higher energy zones including both the western area influenced by freshwater inflow from the Po di Volano and the central area influenced by the sea (O'Kane et al. 1992). Moreover, the eastern zone is very shallow (maximum depth 1 m) and accounts for one half of the entire area and for one fourth of the total water volume ($40 \cdot 10^6$ m^3), while the central area is on average 1.5 deep. The lagoon is subject to anthropogenic eutrophication, which causes extensive growth of the seaweed *Ulva rigida* C. Ag. (Chlorophyta: Ulvales) in the sheltered eastern area and phytoplankton blooms in the deeper central zone. Macroalgal growth is also responsible for summer anoxia and dystrophy, which usually occur in the eastern area (Viaroli et al. 1992; 1995).

Materials and Methods

Water Quality

Temperature, salinity and oxygen were determined throughout the water column using an automatic multiparametric probe (IDRONAUT

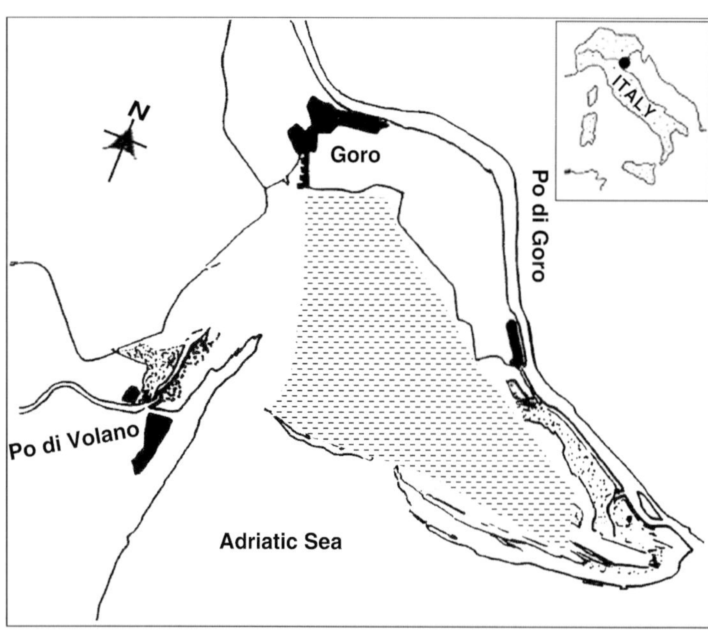

Fig. 1. Map of the Sacca di Goro lagoon, with the sampling station (17) indicated. The *shaded area* represents the maximum spreading of *Ulva* on 27 May 1997

OCEAN SEVEN, MILAN, ITALY). Calibration tests were also performed monthly with the iodometric method of Winkler (APHA 1975). Soluble reactive phosphorus (Valderrama 1977) and nitrate (APHA 1975) were determined on water subsamples obtained by mixing samples collected from both surface and near-bottom layers.

Macroalgal Biomass, Elemental Composition, Growth Rates and Oxygen Metabolism

For each sampling date triplicate macroalgal samples were collected quantitatively with a benthic grab from an area of 2 to 10 m^2 of sediment surface. We computed the areal biomass from the dry weight and surface area sampled. In 1989 biomass values were not determined. In 1990 we estimated mean biomass values using the data reported by Piccoli and Godini (1994). In the laboratory *Ulva* thalli were cleaned to remove epiphytes, rinsed with tap water to remove salt, and oven-dried at 70°C to determine the dry weight (DW). Triplicate wet subsamples (approx. 100 mg) were extracted overnight with 90% acetone and chlorophyll-a was determined according to Jensen (1978). Dried subsamples ranging from 20 to 50 mg were analysed for carbon and nitrogen content by elemental analysis (PERKIN ELMER 4200 CHNS).

In 1994 (Viaroli et al. 1996) and 1997 net growth rates (NGR) of *Ulva* were estimated by means of incubations of *Ulva* thalli inside five cylindrical enclosures positioned 20 cm above the sediment surface. Enclosures (Ø=20 cm, volume = 10.5 litres) were built using a plastic fencing net (mesh = 8 mm) to allow free water exchange. Here we present data of the experiments which were performed from 23rd January to 6th February, 8th to 17th April, and 27th May to 10th June 1997. At the beginning of each experiment, each cage was filled with 20 g of wet *Ulva* thalli. At the end of each incubation, the wet and dry weight (DW) was determined. The NGR were estimated considering an exponential growth pattern (DeBoer et al. 1978). Since the mesh is large enough to allow grazing and losses by thallus fragmentation, we consider the measured rates as the net biomass increase resulting from the equilibrium between growth and losses.

Along with the NGR measurements, photosynthetic rates were estimated by means of P/I curves, recorded in the laboratory at selected temperatures. Macroalgal thalli were cut into small disks (Ø 25 mm). Four cylindrical beakers were then filled with 100 ml of filtered lagoon water with five disks added. Beakers were sealed with floating lids, incubated under stirring at increasing light intensities (I = 25, 50, 100, 200, 400 and 800 $\mu Em^{-2}s^{-1}$) for 30 minutes each and in the dark for 3 hours. Dissolved oxygen (Winkler's method, APHA 1975) and temperature were determined at the beginning and at the end of each incubation interval. Net oxygen production was calculated as $mgO_2g^{-1}h^{-1}$. Light was supplied by a halogen cold white lamp (LAMPITALIA R7S 500 W, ITALY) regulated with a rheostat. Light intensities were recorded with a quantum radiometric detector with cosine correction (HD 9021 RAD/PAR DELTA OHM, PADOVA, ITALY).

Sedimentary Iron, Phosphorus and Reduced Sulphur

In the Sacca di Goro, benthic fluxes and features of the surficial sediment have been analysed regularly since 1994 (Bartoli et al. 1996; Giordani et al. 1996, 1997). Here we consider as an example, two samplings conducted on 26th August 1997, during the summer hypoxia, and on 27th January 1998. On each date, five sediment cores (i.d. 5 cm, length 30 cm) were collected at station 17 with a hand corer. Cores were kept in an aerated water bath until they were sliced (within 1 day). The upper part of each core was sliced into three sections (0-2, 2-5 and 5-10 cm), here we consider only the 0-2 cm section.

Iron pools were extracted immediately with 0.5M HCl and analysed by the ferrozine method (Lovley and Phillips 1987) according to the procedures reported by Giordani et al. (1997).

The pore water PO_4, the loosely bound PO_4 (exchangeable-PO_4), the PO_4 that is bound to Fe and Al (Fe~PO_4) and the calcium bound PO_4 (Ca~PO_4) were determined according to Ruttemberg (1992).

Reduced sulphur pools, namely acid volatile sulphide (AVS) and chromium reducible sulphur (CRS), were determined by means of two sequential distillations under anoxic conditions (Fossing and Jørgensen 1989). Free sulphide was measured by the methylene blue method (Cline 1969).

Sulphate reduction rates (SRR) were determined separately in five sediment cores following the procedures of Fossing and Jørgensen (1989). Details dealing with incubation and extraction techniques and analytical methods are reported by Giordani et al. (1997).

Results

A Decadal trend of Macroalgal Growth and Water Quality

Lagoon trophic status and water quality depend mostly on the growth and decomposition of *Ulva*. In the sheltered zone of the lagoon (approximately 10 km^2) abnormal macroalgal growths have been frequently observed in spring, with biomass peaks towards the end of May (Fig. 2a). Nitrate (Fig. 2b) is the most important nitrogen source for *Ulva* (Viaroli et al. 1993, 1996). Nitrate availability in the water mass is usually controlled by the equilibrium between macroalgal assimilation and riverine discharge. Nitrification and denitrification pathways seem to be less important as shown by the comparison of rates measured in denuded and vegetated sediments (Bartoli et al. in this volume). Dissolved oxygen reflected the life cycle of *Ulva* (Fig. 3a). In spring, macroalgal growth leads to oxygen supersaturation throughout the water column. From June onwards, hypoxia and anoxia frequently occur mostly in the bottom water. The sudden crash of macroalgal productivity with oxygen deficiency in the water column and dystrophy usually occurs in June-July. Soluble reactive phosphorus concentrations attain significant peaks during the summer anoxia (Fig. 3b), which coincide with the sudden collapse and decomposition of the macroalgal biomasses.

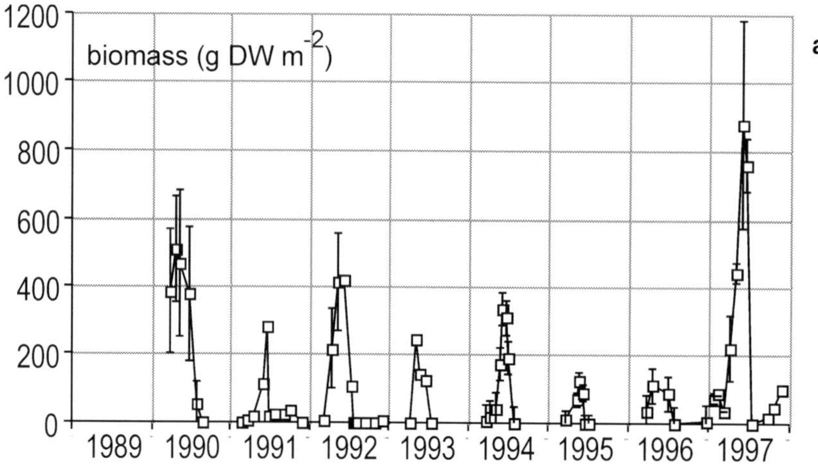

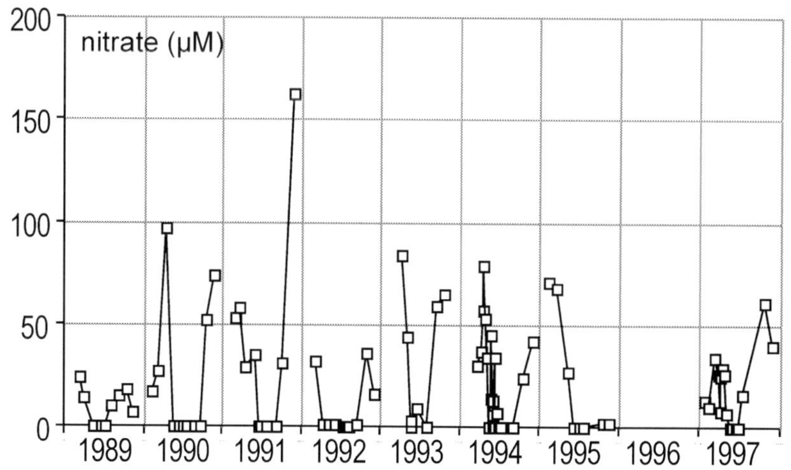

Fig. 2a,b. Annual trends of: **a** *Ulva* biomasses; **b** dissolved nitrate in the water column at station 17 in the sheltered zone of the Sacca di Goro lagoon

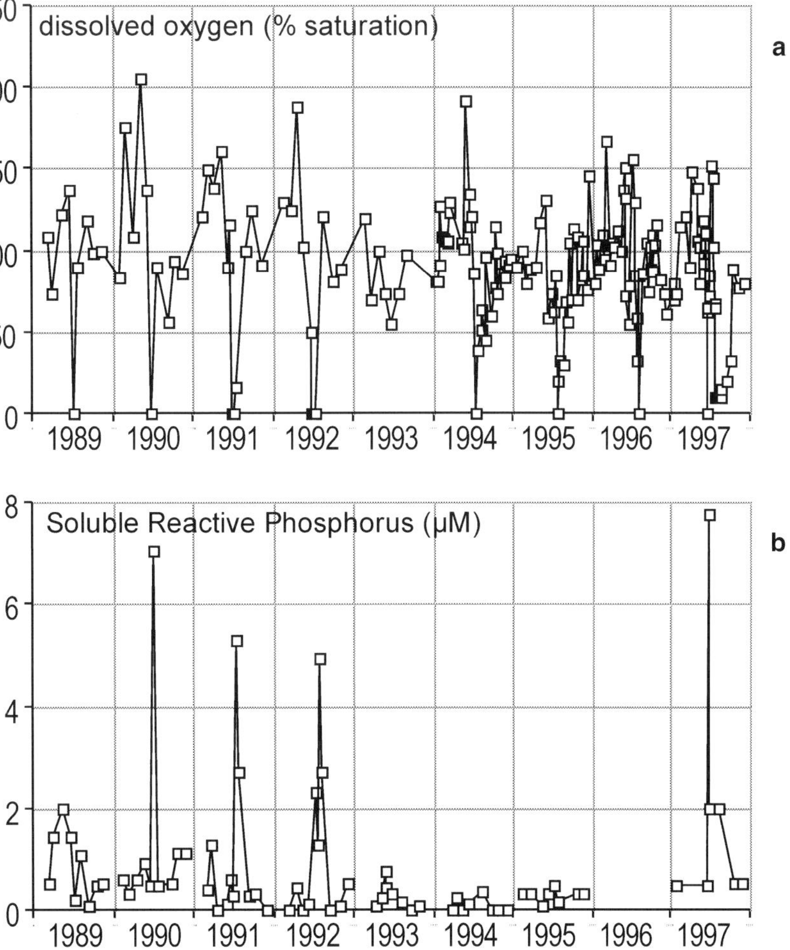

Fig. 3a,b. Annual trends of: **a** dissolved oxygen saturation; **b** soluble reactive phosphorus (SRP) concentrations in the water column at station 17 in the sheltered zone of the Sacca di Goro lagoon

Conditions and Factors Related with Ulva Growth and Biomass Decomposition

In 1997, a typical "worst year" scenario, a bloom of *Ulva* persisted through the winter period in the eastern part of the lagoon, with biomasses of up to 100 gDWm^{-2}. Active macroalgal growth recommenced in March when the water temperature was 10-12°C and light conditions became favourable. Subsequently, the biomass increased rapidly and reached the maximum spreading at the end of May (Table 1). On the 27th May, the biomass peak at station 17 was 878±302 gDWm^{-2}, whilst at the edge of the macroalgal beds up to 300 gDWm^{-2} were attained. The standing biomass peak was followed by a sudden collapse of macroalgal mats, which started to decompose causing anoxia and sulphide release, mostly at station 17 (Fig. 4). In the sheltered area of the lagoon, *Ulva* growth recommenced in mid September and persisted through the Winter at approximately 100 gDWm^{-2}.

The seasonal pattern of *Ulva* biomass correlated with their internal total nitrogen reserves and C:N ratio (Table 1). Thallus nitrogen content showed a peak of 4.89±0.21% DW in January and decreased during the Spring growth phase attaining a minimum of 2.14±0.47% DW on the 27th May. The ability of *Ulva* to store nitrogen may have relevant effects not only upon the macroalgal growth, but also on nutrient retention and recycling within the lagoon ecosystem. For example, on the 27th May at station 17 a standing biomass of 878±302 gDWm^{-2} was equivalent to a net retention of 19.6±10.7 gNm^{-2}, whilst at the edge of this macroalgal bed the accumulated nitrogen attained 6.4 gNm^{-2}. From the 1st January to 27th May, the estimated nitrogen load discharged into the lagoon by the Po di Volano canal and some minor tributaries was approximately 18 gNm^{-2} (P. Viaroli, unpublished

Table 1. *Ulva* biomass (gDWm^{-2}), and total carbon (% DW), total nitrogen (% DW) and atomic C:N ratios in macroalgal thalli at station 17 in the Sacca di Goro in 1997. Mean values±standard deviation of three replicates are reported

Date	Biomass	Carbon	Nitrogen	C:N
23rd January	74 ± 17	33.8 ± 1.47	4.89 ± 0.21	8.07 ± 0.12
17th April	219 ± 96	28.9 ± 0.18	3.34 ± 0.02	10.09 ± 0.68
27th May	878 ± 302	30.1 ± 0.59	2.14 ± 0.47	16.91 ± 3.56

Table 2. Net growth rates (NGR) of *Ulva rigida* incubated within suspended cages located at a site close to station 17 in the Sacca di Goro. Mean values±standard deviation of five replicates as well as water temperature ranges are reported

Date	Temperature range (°C)	NGR (d^{-1})
23.01-06.02	5-8	0.029±0.003
08.04-17.04	12-16	0.143±0.040
27.05-10.06	19-27	0.048±0.024

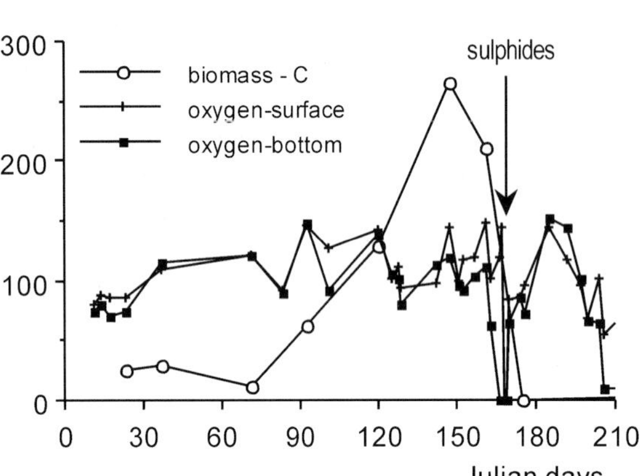

Fig. 4. Typical seasonal pattern of a dystrophic crisis in the sheltered zone of the Sacca di Goro lagoon (January-July 1997). The biomass is represented as carbon (g Cm^{-2}), the dissolved oxygen concentration is given as % saturation

Table 3. Chlorophyll-*a* (mg gDW^{-1}) and some photosynthetic parameters inferred from the P/I curves of *Ulva rigida* in the Sacca di Goro in 1997. [P_{max}, maximum net productivity (mgO_2 gDW^{-1} h^{-1}); *Isat*, light intensity (μEm^{-2}s–1) at which the onset of photosynthesis saturation occurs; *DR*, dark respiration (mgO_2 gDW^{-1} h^{-1})]. Mean values±standard deviation of three replicates are reported

Date	Chlorophyll-*a*	P_{max}	I_{sat}	DR
23rd January	5.86±0.51	10.75±0.90	64±15	0.34±0.09
17th April	8.10±1.19	21.57±0.88	108±14	0.40±0.04
27th May	4.51±0.51	17.01±0.87	125±19	0.60±0.09

data), thus during this period most of the N-load was assimilated into the *Ulva* biomass. A more detailed description of nitrogen cycling in the Sacca di Goro is reported by Bartoli et al. (this volume).

Net growth rates (NGR) attained maximum values (0.143±0.040 d^{-1}) in April, whereas minimum rates were coincident with the lowest temperatures (6°C) in January and with the highest standing biomass in May (Table 2). The maximum productivity, dark respiration rates and saturating light intensities inferred from the P/I curves followed a clear seasonal trend (Table 3). At 6°C saturation was reached at 64±15 μEm^{-2}s^{-1}, while at 18-20°C photosynthetic saturation was attained at 125±19 μEm^{-2}s^{-1} equivalent to approximately 15% of the maximum daily photon flux. The seasonal trend of P_{max} was also related to net growth rates and the thallus chlorophyll content.

Sediment Features and Dystrophy

In 1997, during the macroalgal collapse, anoxia led to sulphide production in the water column

Table 4. Sulphate reduction rates (*SRR*, nmol cm^{-3}h^{-1}), acid volatile sulphide (*AVS*, μmol cm^{-3}), chromium reducible sulphur (*CRS*, μmol cm^{-3}), reduced iron [*Fe(II)*, (μmol cm^{-3})] and total iron [*Fe(II)+Fe(III)*, (μmol cm^{-3})] measured in the surficial sediment of the Sacca di Goro in two representative periods of the annual cycle. Mean values±standard deviation of five replicates are reported

Date	SRR	AVS	CRS	Fe(II)	Fe(II)+Fe(III)
26th August 1997	17.1±11.7	11.1±7.1	23.9±12.2	85.3±11.5	82.8±13.6
27th January 1998	1.6±0.8	3.4±1.2	32.9± 6.9	69.9± 5.6	75.2± 7.4

Table 5. Speciation of inorganic phosphate in the surficial sediment of the Sacca di Goro in two representative periods of the annual cycle. (*PW-PO$_4$*, porewater phosphate; *Exch-PO$_4$*, exchageable phosphate; *Fe~PO$_4$*, iron bound phosphate; *Ca~PO$_4$*, calcium bound phosphate). *Units*: nmol cm^{-3}. Mean values±standard deviation of five replicates are reported

Date	PW-PO$_4$	Exch-PO$_4$	Fe~PO$_4$	Ca~PO$_4$
26th August 1997	16.7±12.1	427±189	41±30	7.911±1.410
27th January 1998	2.3± 1.0	320± 37	25±13	15.028±4.800

probably due to the direct decomposition of macroalgal biomasses (Fig. 4). A high SRR activity was measured in August and caused the complete reduction of the microbiologically available iron pools (Table 4). Nevertheless, in August free sulphides were not detected either in the sediment or in the water column. This could be explained by an increase of iron monosulphide as demonstrated by the high AVS concentration. The SRR was ten times lower in January coinciding with the lowest temperatures. The SRR decrease was also accompanied by the decrease of AVS and the increase of CRS, with a partial recovery of the oxidised iron pool. The iron-monosulphide system seemed to control the mobility of the available phosphate in the pore water, although most of the inorganic phosphate is calcium bound (Table 5).

Discussion

In polluted coastal systems ephemeral macroalgae may predominate and undergo abnormal growth and sudden crashes, which are usually unpredictable. The causes of these blooms have been analysed mostly considering the ecophysiological characteristics of the bloom-forming species (for an updated review see Valiela et al. 1997; Raffaelli et al. 1998). Amongst these, many chlorophytes have been found to integrate short term environmental variability, by virtue of either light or nutrient saturation characteristics. Thus, at a relatively low growth rate they can control fluctuating and extreme environments, gaining a competitive advantages.

The decade of monitoring in the Sacca di Goro demonstrates that trophic status and water quality are controlled by *Ulva* growth and collapse. Usually, macroalgal growth starts in early spring when the temperature increases above 10°C. Thereafter, the biomass increases rapidly, reaching the highest standing crop and the maximum extent of spreading at the end of May. When the biomasses exceed 200-300 gDWm^{-2}, decomposition processes take place, and cause anoxia and sulphide accumulation in the water column (Viaroli et al. 1995, 1996). In the worst years (1989-1992) and 1997 anoxia lasted from days to weeks with dramatic consequences mostly in the sheltered part of the lagoon.

These outbreaks are likely multifactorial and include both bottom-up controls and internal biochemical and physiological regulators (Lapointe 1997). A general precept is that the relative abundance of carbon, nitrogen and phosphorus controls the macroalgal productivity. Hence, the deficit of one of these nutrients may act as a switch determining the onset of the bloom collapse. The optimum C:N:P ratio for many *Ulva* species was assumed to be 336:35:1 (Atkinson and Smith 1983). In this concept, C:N>9.6 and N:P<35 would be an indication of N-limitation. For the *Ulva* in the Sacca di Goro the C:N ratio values demonstrated increasing N-limitation until the biomass peaks were attained (Viaroli et al. 1992, 1993, 1996). In 1997, the highest C:N ratio (16.91±3.56) coincided with the lowest nitrogen content in the macroalgal tissues (2.14±0.47 %DW). This value was less than 2.4 %DW (Fujita et al. 1989) and close to 2.0%DW (Lavery and McComb 1991) and 2.17%DW (Pedersen and Borum 1994), values which were considered to be the critical levels for the continued growth of *U. rigida* and *U. lactuca*, respectively.

The strict relationship between *Ulva* growth and nitrogen uptake may have relevant effects not only upon the macroalga itself, but also on nutrient retention and recycling within the lagoon ecosystem. In 1997, during the spring growth phase, *Ulva* exhibited a potential nitrogen uptake up to 600 mgNm^{-2}d^{-1}. This caused a prolonged nitrogen deficiency in the water column which allowed only *Ulva* to grow, outcompeting other primary producers as well as denitrification processes (Bartoli et al. this volume). Before the macroalgal collapse, the overall *Ulva* standing biomass accounted for 174 (133-221) tons of nitrogen. This nitrogen bulk was stored from late March to early June, which corresponded to an external total nitrogen loading of 540 tons of which 175 tons was comprised of dissolved inorganic N (P. Viaroli, unpublished data). Therefore, most of the inorganic load was retained in the macroalgal biomasses. However, nitrogen accumulation within the *Ulva* was only a temporary N-sink, due to its short life cycle. In mid June, the macroalgal community shifted from active production to rapid decomposition, leading to a rapid nutrient loss and recycling.

The bloom collapse also had a significant impact on dissolved oxygen, phosphorus and sulphide concentrations (Viaroli et al 1992; 1993; Giordani et al. 1996; 1997). In 1997 we observed an extended dystrophic crisis in mid-June and several dystrophic episodes in July and August. Moreover, a persistent hypoxia lasted until mid September. Several studies have demonstrated that the iron-iron monosulphide-pyrite system buffers against free sulphides (Howarth and Stewart 1992; Smolders et al. 1995). However, simple measurements of these variables are not sufficient to estimate the true buffering capacity of the sedimentary processes (Giordani et al. 1996; 1997). Additionally, the decomposition of the accumulated biomasses may occur directly in the anoxic bottom water, where the sulphide concentrations are independent of the potential buffering capacity provided by sedimentary iron pools. The buffering capacity of these ecosystems appears to depend mostly on the extent of macroalgal blooms, and seems to be influenced by the timing and rates of biomass growth and decomposition. Therefore, the abnormal growth of ephemeral macroalgae may lead to a high level of intrasystem disturbance and instability, which results in oxygen deficiencies that frequently degenerate into dystrophic crisis.

Acknowledgements. This research has been jointly supported by the EU projects ROBUST (contract ENV4-CT96-0218) and NICE (contract MAS3-CT96-0048) of the ELOISE network of the Commission of the European Communities. We also gratefully acknowledge S. Bencivelli, G. Grigatti and T. Zappata (Consortium for the Sacca di Goro Management) for field assistance and measurements.

References

APHA, AWWA, WPCF (1975) Standard methods for the examination of water and wastewaters. 14th edn. APHA, Washington

Atkinson MJ, Smith SV(1983) C:N:P ratios of benthic marine plants. Limnol Oceanogr 28: 568-574

Bartoli M, Cattadori M, Giordani G, Viaroli P (1996) Benthic oxygen respiration, ammonium and phosphorus regeneration in surficial sediments of the Sacca di Goro (Northern Italy) and two French coastal lagoons: a comparative study. Hydrobiologia 329: 143-159

Bartoli M, Castaldelli G, Nizzoli D, Gatti LG, Viaroli P (2000) Benthic fluxes of oxygen, ammonium and nitrate and coupled and uncoupled denitrification rates in two eutrophic coastal lagoons with different primary producer communities. (this volume)

Borum J (1996) Shallow waters and land/sea boundaries. In: Jørgensen BB, Richardson K (eds) Eutrophication in coastal marine ecosystems: coastal and estuarine studies. Am Geophys Union 52: 179-203

Castel J, Caumette P, Herbert R (1996) Eutrophication gradients in coastal lagoons as exemplified by the bassin d'Arcachon and the Étang du Prévost. Hydrobiologia 329: ix-xxviii

Cline JD (1969) Spectrophotometric determination of hydrogen sulphide in natural waters. Limnol Oceanogr 14: 454-459

DeBoer JA, Guigli HJ, Israel TL, D'Elia CF (1978) Nutritional studies of two red algae. I. Growth rate as a function of nitrogen source and concentration. J Phycol 14: 261-266

Flecht RL (1996) The ocurrence of "green tides" – a review. In: Schramm W, Nienhuis PH (eds) Marine benthic vegetation: recent changes and the effect of eutrophication. (Ecological studies, 123), Springer, Berlin Heidelberg New York Tokyo, pp 7-43

Fossing H, Jorgensen BB (1989) Measurement of bacterial sulphate reduction in sediment: evaluation of a single-step chromium reduction method. Biogeochemistry 8: 205-222

Fujita RM, Wheeler PA, Zedler JB (1989) Assessment of macroalgal nitrogen limitation in a seasonal upwelling region. Mar Ecol Prog Ser 53: 293-303

Giordani G, Cattadori M, Bartoli M, Viaroli P (1996) Sulphide release from anoxic sediments in relation to iron availability and organic matter recalcitrance and its effects on inorganic phosphorus recycling. Hydrobiologia 329: 211-222

Giordani G, Azzoni R, Bartoli M, Viaroli P (1997) Seasonal variations of sulphate reduction rates, sulphur pools and iron availability in the sediment of a dystrophic lagoon (Sacca di Goro, Italy). Water Air Soil Pollut 99: 363-371

Howarth RW, Stewart JWB (1992) The interactions of sulphur with other element cycles in ecosystems. In: Howart RW, Stewart JWB, Ivanov MU (eds) Sulphur cycling on the continents: wetlands, terrestrial ecosystems and associated water bodies: SCOPE 33. J Wiley and Sons, New York, pp 67-84.

Jensen A (1978) Chlorophylls and carotenoids. In: Hellebust JA, Craigie JS (eds) Handbook of phycological methods. Physiological and biochemical methods. Cambridge Univ Press, Cambridge, pp 59-70.

Lapointe BE (1997) Nutrient thresholds for bottom-up control of macroalgal blooms on coral reefs in Jamaica and southeast Florida. Limnol Oceanogr 42: 1119-1131

Lavery PS, McComb AJ (1991) The nutritional eco-physiology of *Chaetomorpha linul* and *Ulva rigida* in Peel Inlet, Western Australia. Estuarine Coast Shelf Sci 33: 1-22

Lovley DR, Phillips EJP (1987) Rapid assay for reducible ferric iron in aquatic sediments. Appl Environ Microbiol 53: 1536-1540

O'Kane JP, Suppo M, Todini E, Turner J (1992) Physical intervention in the lagoon of Sacca di Goro: an examination using a 3-D numerical model. Sci Total Environ 1992, pp 489-510 (Suppl)

Pedersen MF, Borum J (1997) Nutrient control of estuarine macroalgae: growth strategy and the balance between nitrogen requirements and uptake. Mar Ecol Prog Ser 161: 155-163

Piccoli F, Godini E (1994) Ricerche qualitative e quantitative sulla vegetazione della Sacca di Goro: anni 1989-90. In Bencivelli S, Castaldi N, Finessi D (eds) Sacca di Goro: studio integrato sull'ecologia. Franco Angeli, Milano. I. pp 227-243

Raffaelli DG, Raven JA, Poole LJ (1998) Ecological impact of green macroalagal blooms. Oceanogr Mar Biol Annu Rev 36: 97-125

Ruttemberg KC (1992) Development of a sequential extraction method for different forms of phosphorus in marine sediments. Limnol Oceanogr 37: 1460-1482

Sand-Jensen K, Borum J (1991) Interactions among phytoplankton, periphyton, and macrophytes in temperate freshwaters and estuaries. Aqat Bot 41: 137-175

Smolders A, Nijboer RC, Roelofs JGM (1995) Prevention of sulphide accumulation and phosphate mobilization by the addition of iron (II) chloride to a reduced sediment: an enclosure experiment. Fresh Water Biol 34: 559-568

Valderrama JC (1977) Methods used by the Hydrographic Department of National Board of Fisheries, Sweden. In: Grasshoff K (ed) Report of the Baltic Intercalibration Workshop. Annex Interim Comm Protection Environ Baltic Sea, pp 14-34

Valiela I, McLelland J, Hauxwell J, Behr PJ, Hersh D, Foreman K (1997) Macroalgal blooms in shallow estuaries: controls and ecophysiological and ecosystem consequences. Limnol Oceanogr 42: 1105-1118

Viaroli P, Pugnetti A, Ferrari I (1992) *Ulva rigida* growth and decomposition processes and related effects on nitrogen and phosphorus cycles in a coastal lagoon (Sacca di Goro, Po River Delta). In: Colombo G, Ferrari I, Ceccherelli VU, Rossi R (eds) Marine eutrophication and population dynamics. Olsen and Olsen, Fredensborg, pp 77-84.

Viaroli P, Naldi M, Christian R, Fumagalli I (1993) The role of macroalgae and detritus in the nutrient cycles in a shallow-water dystrophic lagoon. Verh Int Ver Limnol 25: 1048-1051

Viaroli P, Bartoli M, Bondavalli C, Naldi M (1995) Oxygen fluxes and dystrophy in a coastal lagoon colonized by *Ulva rigida* (Sacca di Goro, Po River Delta, Northern Italy). Fresenius Environ Bull 4: 381-386

Viaroli P, Naldi M, Bondavalli C, Bencivelli S (1996) Growth of the seaweed *Ulva rigida* C. Agardh in relation to biomass densities, internal nutrient pools and external nutrient supply in the Sacca di Goro lagoon (Northern Italy). Hydrobiologia 329: 93-103

CHAPTER 60

Microbial Loop Structure along Trophic Gradients in the Adriatic Sea

P. Del Negro[1], G. Civitarese[2], P. Ramani[1], and S. Fonda Umani[1]

ABSTRACT

Seasonal observations (May 1995-February 1996) carried out along the whole Adriatic basin have significantly contributed to a better understanding of the distribution of bacterial and heterotrophic nanoplankton communities in different trophic areas. The average density values along the water column ranged from 1×10^5 to 1.6×10^6 cell mL^{-1} and from 1.2×10^2 to 1.9×10^3 cell mL^{-1} for bacteria and nanoheterotrophs respectively.

The relationships between bacterial and heterotrophic nanoplankton abundance changed along the trophic gradient suggesting that bacteria should be regulated more strongly by predation in eutrophic systems.

Introduction

Bacteria are the less variable component of the plankton in terms of both total density and biomass (Cole et al. 1993). Bacterial growth and abundance may be tightly regulated within a relatively narrow range by a combination of resource limitation (bottom-up control) and grazing (top-down control). There has been an antagonism between the two theories as to which is the principal regulating force. Bottom-up control of bacterial community is supported by the results of Billen et al. (1990), Simon et al. (1992), Weinbauer and Höfle (1998) that analysed several marine and freshwater systems characterized by very different trophic conditions using various approaches. Pace and Cole (1994) found no evidence in experimental studies that protozoa effectively regulate bacterial abundance. On the contrary, Berninger et al. (1991) and Sanders et al. (1992) described a high relationship between bacterial and heterotrophic nanoflagellate (HNF) abundance, and suggested that significant predatory control of bacteria occurs. However, it has also been shown that bacteria and HNFs are not strongly coupled across systems, and, consequently, HNFs do not always control bacterial abundance (Gasol and Vaqué 1993) probably because of predatory control exerted on HNFs by larger zooplankton (Gasol 1994). Evidence has shown the potentially important role for cladocera, bacterivorous ciliates, mixotrophic algae in removing bacteria from aquatic ecosystems (Sanders et al. 1992). By eliminating a fraction of bacteria through cell lysis, also viruses are likely to be significant agents in the control of bacteria (Fuhrman 1999).

Much of the evidence used in support of bottom-up or top-down hypothesis in marine waters has been obtained from estuarine or coastal systems and surface waters, and there are still few data for the open sea, especially throughout the water column (Dufour and Torréton 1996).

The aim of this study was to examine the processes regulating bacterial abundance in the whole water column along four transects of the Adriatic Sea presenting different trophic status.

Study Area

The Adriatic Sea covers about six degrees of latitude, from 39.5 to 45.5°N, and about eight degrees of longitude, from 12 to 20°E. Its NW-SE oriented longitudinal axis is about 800 km long, while its width varies from 100 to 200 kilometres.

Usually, the Adriatic Sea is divided into three

[1] Laboratorio di Biologia Marina, Via A. Piccard 54, 34010 Trieste, Italy
[2] Istituto Sperimentale Talassografico, CNR, Trieste, Italy

F.M. Faranda, L. Guglielmo, G. Spezie (eds)
Mediterranean Ecosystems: Structures and Processes

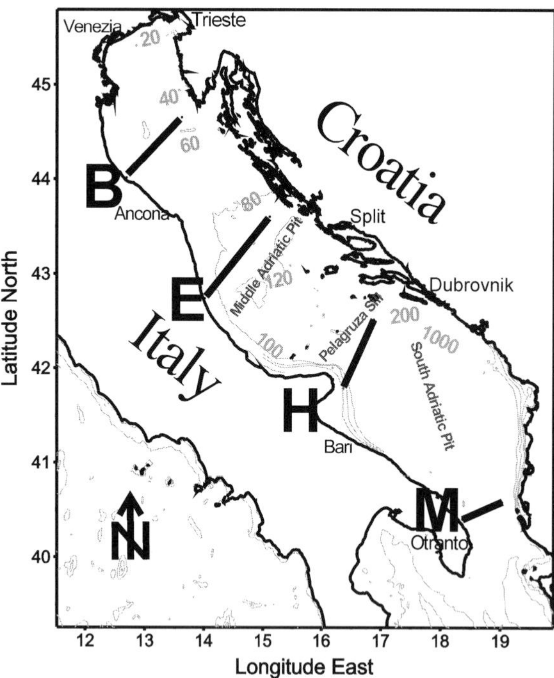

Fig. 1. Bathymetric map of the Adriatic Sea. Sampling transects are illustrated

distinct sub-basins on the basis of their oceanographycal dynamics and bathymetric features (Fig. 1). The Northern Adriatic is one of the most eutrophic areas of the Mediterranean due to its shallow depth (average 35 m) and the high discharge rate (about 2900 m^3s^{-1}) of freshwater coming from one of the most industrialized areas in Europe (Smodlaka et al. 1999). In the Middle Adriatic the bathymetry continues to slope down to the 170 m deep of Pelagosa sill that represents a discontinuity area separating the relatively shallow and productive northern shelf from the Southern Adriatic. This is the deepest portion of the Adriatic (maximum depth about 1200 m) and its characteristics are mainly determined by the dynamic of the exchanges with the Ionian Sea through the Strait of Otranto.

The Northern Adriatic shelf area is a source of two water masses, a surface one, relatively fresh, due to the influence of the riverine inflow, and the other one cold and dense, formed in winter at the bottom. Their southward spreading along the Italian coast constitute one of the main component of the general circulation of the Adriatic Sea. Their path is the route along which the physical, chemical and biological signatures of the productive northern basin are advectively exported to the rest of the basin.

Materials and Methods

In the framework of Prisma-1 project (sub-project: Descriptive and Dynamical Oceanography of the Adriatic Sea) some stations located along four inshore-offshore transects in the Adriatic Sea were investigated during four seasonal cruises (May, August and November 1995 and February 1996). Transect B was located in the northern shallow zone, transect E in the Middle Adriatic, transect H over the steep continental slope of the southern basin and transect M at the southern opening to the Ionian Sea (Fig. 1).

Seawater samples were collected at selected depths (6 along transect B, 8 along transcct E, 10 along transect H and 12 along transect M) using a Seabird Electronics SBE 9/11 plus coupled with a Carousel water sampler SBE 32 carrying 24 bottles of 12 L each.

Enumeration of bacteria and autotrophic picoplankton was carried out at ten stations along transect B, eight along transect E and six both in H and M transects while enumeration of heterotrophic nanoplankton was carried out at 3 stations for each transect.

Subsamples (10 mL) for enumeration of bacteria were fixed with prefiltered formaline (final concentration 2%), stained directly with 4',6 diamidino-2 phenylindole (DAPI; final concentration 1 μg mL^{-1}) for 5 min and then a 3 mL subsample was filtered onto black 0.2 μm pore-size polycarbonate Nuclepore filters (Porter and Feig 1980).

Subsamples (100 mL) for heterotrophic nanoplankton (HN) enumeration were preserved with prefiltered 10% glutaraldehyde prepared in natural seawater to obtain a final preservative concentration of 1%. Aliquots of 20-50 mL were taken and filtered onto black 0.6 μm pore size polycarbonate Nuclepore filters. The HN were stained with DAPI (final concentration 1μg mL^{-1}) for 5 min thereafter with primulin according to Martinussen and Thingstad (1991).

All filters were examined under Leitz Dialux 20EB epifluorescence microscopy equipped with HBO 50W mercury vapour lamp at 1000 x final magnification. Over 400 bacterial cells were counted under UV excitation (330 to 380 nm with a 420 nm barrier filter) in at least 20 fields. Picophytoplankton were enumerated using blue excitation (450-490 nm with a 510 barrier filter) by observing at least 20 fields. According to Murphy and Haugen (1985) picocyanobacteria fluoresce bright yellow while other picophytoplankton appear deep red. The HN emitting yellow fluorescence under blue excitation (410 to 485 nm with 515 nm barrier filter) and blue fluorescence under UV excitation were counted for at least 20 fields; over 100 cells per filter were counted.

Results

Bacterial abundance data used in this study are summarized in Table 1. The average density values along the water column ranged from 1 x 10^5 to 1.6 x10^6 cell mL^{-1} over the four transects. There were significant differences among coastal and off-shore stations along each of the transects (Kruskal Wallis test, $P<0.05$).

Seasonal patterns are shown in Fig. 2. During spring the bacterial density (median value of all the determinations) was similar to north, central and south Adriatic areas, while the lowest abundance in the Otranto Strait significantly differed (Kruskal Wallis test; $P<0.05$). In summer bacteria reached the highest densities along all the transects showing the same distribution pattern observed in spring. During autumn bacterial densities decreased in the whole basin and particularly in the Northern Adriatic. In winter increasing abundances were observed in north and central basin while they remained constant in the most southern areas.

Picophytoplankton abundance ranged between 2.8 x 10^3 to 5.6 x 10^4 cell mL^{-1}. Both

Table 1. Bacterial abundance (106 cell mL[1]): each datum represents the mean value and standard deviation (*SD*) of bacterial densities along the water column

Station	Spring		Summer		Autumn		Winter	
	Mean	SD	Mean	SD	Mean	SD	Mean	SD
B01	0.780	0.107	1.614	1.096	0.326	0.086	0.467	0.132
B02	–	–	1.332	0.632	0.432	0.114	0.613	0.183
B03	0.495	0.169	0.546	0.157	0.207	0.060	0.502	0.086
B05	0.246	0.032	0.389	0.181	0.203	0.038	0.656	0.647
B07	0.241	0.057	0.492	0.214	0.232	0.091	0.607	0.250
B08	–	–	0.922	0.365	0.164	0.076	0.692	0.287
B09	0.252	0.081	0.913	0.239	0.291	0.083	0.632	0.276
B10	–	–	0.512	0.074	0.258	0.127	0.816	0.320
B11	0.157	0.089	0.451	0.259	0.120	0.061	0.662	0.081
B12	0.362	0.060	0.785	0.123	0.105	0.027	0.477	0.140
E02	0.325	0.066	1.091	0.353	0.279	0.175	1.019	0.123
E03	0.254	0.081	0.714	0.187	0.233	0.170	0.477	0.139
E05	0.288	0.095	0.850	0.200	0.244	0.093	0.530	0.116
E06	0.281	0.150	0.681	0.140	0.366	0.105	0.223	0.051
E08	0.275	0.054	0.432	0.180	0.561	0.143	0.538	0.063
E09	0.400	0.172	0.781	0.163	0.520	0.243	0.490	0.147
E11	0.279	0.064	0.774	0.306	0.244	0.063	0.448	0.127
E12	0.402	0.155	0.666	0.202	0.696	0.183	0.551	0.127
H02	0.227	0.121	0.827	0.216	0.614	0.228	0.462	0.167
H04	0.226	0.106	0.902	0.352	0.464	0.173	0.492	0.187
H06	0.295	0.060	0.646	0.226	0.435	0.148	0.176	0.070
H08	0.333	0.053	0.335	0.132	0.343	0.148	0.110	0.325
H10	0.354	0.113	0.642	0.311	0.383	0.137	0.325	0.187
H12	0.300	0.094	0.590	0.233	0.263	0.138	0.326	0.079
M01	0.488	0.170	1.017	0.226	0.794	0.348	0.154	0.041
M03	0.313	0.042	0.739	0.202	0.743	0.263	0.159	0.106
M05	0.195	0.079	0.584	0.245	0.563	0.369	0.297	0.122
M07	0.114	0.084	0.442	0.215	0.411	0.232	0.104	0.048
M09	0.137	0.077	0.396	0.271	0.420	0.233	0.239	0.148
M11	0.169	0.104	0.390	0.232	0.220	0.144	0.229	0.130

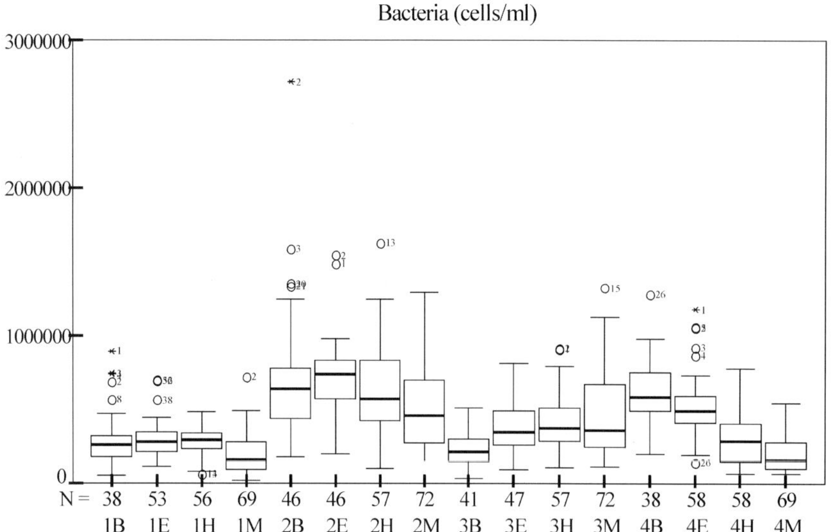

Fig. 2. Box and whisker plots showing the range and the median bacterial abundance along each transect in four different seasons (*1*, Spring; *2*, Summer; *3*, Autumn; *4*, Winter)

Table 2. Picophytoplankton abundance (10^4 cell mL^{-1}): each datum represents the mean value and standard deviation (*SD*) of picophytoplankton densities along the water column

Station	Spring		Summer		Autumn		Winter	
	Mean	SD	Mean	SD	Mean	SD	Mean	SD
B01	2.207	0.369	14.580	7.627	8.488	4.806	0.633	0.171
B02	-	-	5.816	5.830	16.129	10.060	0.797	0.191
B03	5.710	2.816	3.219	1.065	5.543	1.653	1.104	0.465
B05	1.710	1.032	3.148	2.135	1.628	1.320	0.958	0.220
B07	1.784	1.392	3.534	2.383	1.336	0.760	1.051	0.693
B08	-	-	4.098	2.153	1.992	0.846	1.338	0.568
B09	1.654	0.829	5.460	3.163	2.932	1.288	1.427	0.601
B10	-	-	2.780	0.693	2.525	1.342	1.371	0.690
B11	1.671	1.250	2.922	1.324	1.105	0.717	1.700	0.169
B12	2.214	1.090	4.223	1.786	2.071	0.973	0.897	0.419
E02	0.899	0.289	3.926	1.653	5.479	1.929	0.772	0.119
E03	1.305	1.129	2.264	1.367	1.168	0.598	0.589	0.299
E05	0.918	0.699	2.260	1.325	1.690	1.024	0.691	0.193
E06	0.740	0.564	1.927	1.278	1.125	0.905	0.553	0.287
E08	10.915	0.615	1.596	1.004	1.609	1.220	0.732	0.196
E09	2.100	2.046	2.271	1.786	1.735	1.494	0.606	0.273
E11	1.483	1.214	2.464	2.110	1.866	1.360	0.686	0.204
E12	1.177	0.735	2.571	1.829	2.455	1.956	1.036	0.229
H02	1.317	0.748	3.235	2.201	4.041	2.878	0.562	0.393
H04	0.566	0.237	2.341	2.104	1.143	1.027	1.169	0.840
H06	2.135	3.033	0.850	0.732	1.178	1.123	1.033	0.556
H08	2.553	1.991	0.917	1.023	0.837	1.117	0.626	0.377
H10	2.940	1.570	1.544	1.611	1.056	1.183	1.128	0.698
H12	2.281	2.028	0.977	1.183	0.840	1.122	0.866	0.410
M01	1.912	0.934	4.779	0.855	5.538	1.734	0.520	0.109
M03	2.161	1.052	3.470	1.871	3.017	2.337	0.452	0.230
M05	1.215	1.174	1.239	1.896	1.780	2.548	0.414	0.227
M07	0.526	0.913	0.851	1.505	0.947	1.422	0.284	0.315
M09	0.389	0.539	0.711	1.246	0.920	1.466	0.565	0.446
M11	0.791	0.679	1.153	1.662	0.500	0.802	0.703	0.731

eukaryotic and prokaryotic organisms have been found and phycoerytrin-rich cyanobacteria represented the main group (Table 2).

Heterotrophic nanoplankton abundances ranged from 1 x 10^2 to 1.9 x 10^3 cell mL^{-1} and were mostly due to small flagellates (3-5 μm size) (Table 3). Seasonal patterns are shown in Fig. 3. The highest values were generally reached in the northernmost part of the basin and only in winter the nanoplankton abundance was highest along the transect of the central basin and of the Otranto Strait. Comparing bacteria and heterotrophic nanoplankton seasonal pattern a specular trend was observed. In spring, low bacterial densities corresponded to the highest nanoplankton abundances that decreased during summer when bacteria increased.

Chlorophyll *a* concentrations data were derived from Gacic et al. (1999).

In order to examine the influence of trophic conditions on bacterial population the correlation between bacterial abundance and chlorophyll supply was calculated along each transect. The use of a regression fit between chlorophyll and bacteria is a common index of dependency of bacterial biomass on phytoplanktonic biomass in marine waters (Bird and Kalff 1984; Dufour and Torréton 1996). It is well known that phytoplankton production either directly or indirectly supplies dissolved organic matter

Table 3. Heterotrophic nanoplankton abundance (10^3 cell ml^{-1}): each datum represents the mean value and standard deviation (SD) of heterotrophic nanoplankton densities along the water column

Station	Spring		Summer		Autumn		Winter	
	Mean	SD	Mean	SD	Mean	SD	Mean	SD
B03	1.006	0.629	0.440	0.078	0.400	0.177	0.353	0.106
B07	1.067	0.381	0.648	0.250	0.571	0.398	0.212	0.065
B11	0.854	0.381	0.340	0.145	0.441	0.217	0.220	0.038
E02	1.901	0.774	0.535	0.205	0.490	0.103	0.526	0.042
E08	0.480	0.149	0.315	0.067	0.277	0.128	0.513	0.159
E11	0.577	0.332	0.276	0.069	0.187	0.079	0.506	0.157
H02	0.620	0.342	0.475	0.188	0.460	0.169	0.330	0.126
H06	1.002	0.434	0.357	0.166	0.175	0.052	0.176	0.038
H10	0.403	0.228	0.288	0.101	0.375	0.217	0.262	0.078
M01	0.401	0.122	0.210	0.089	0.355	0.077	1.180	0.406
M05	0.172	0.158	0.305	0.258	0.346	0.225	0.437	0.180
M11	0.199	0.149	0.162	0.045	0.124	0.070	0.281	0.179

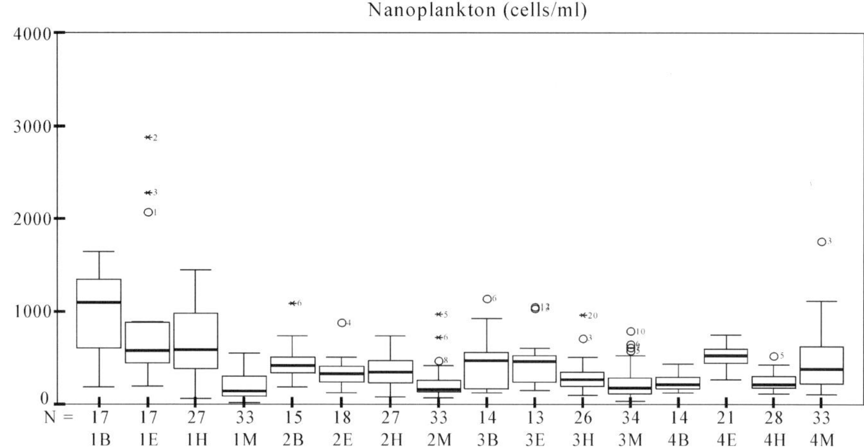

Fig. 3. Box and whisker plots showing the range and the median heterotrophic nanoplankton abundance along each transect in four different seasons (*1*, Spring; *2*, Summer; *3*, Autumn; *4*, Winter)

(DOM) to marine systems (via extracellular exudation, sloppy feeding by grazers, and particle decomposition), and bacteria represent the primary consumers of DOM (Alldredge et al. 1993; Furhman 1999).

To evaluate the predator-prey dependence, the correlation between bacterial and HN abundance was calculated. It has been demonstrated, in fact, that bacterial biomass is positively correlated with HN biomass across a wide range of marine and freshwater environments (Berninger et al. 1991; Sanders et al. 1992). The bacteria/HN abundance ratio was used as an index of grazing pressure (Krstulovic et al. 1998). The results obtained are shown in Table 4. The ratio between bacterial and HN abundance ranged from 582 to 3706.

Along transect B (Northern Adriatic Sea) bacteria were significantly correlated with chlorophyll *a* during the productive periods, while in winter bacterial abundance was inversely correlated with predator abundance and bacteria/HN ratio reached the highest values. In spring and summer the ratio was lower than 1000. Sanders et al. (1992) suggested that relationship between bacteria and HN was constant at about 1000:1 across a wide variety of ecosystems. Along transect E bacterial density was never correlated with chlorophyll *a* but was positively correlated with HN abundance in spring and summer and inversely in autumn. The bacteria/HN ratio was generally higher than in transect B. In the south basin the correlation between bacteria and chlorophyll *a* was significant in summer, autumn and winter. In spring the bacterial density was inversely correlated with HN density while in summer and winter the correlation was positive. Along the Otranto Strait chlorophyll *a* influenced bacterial density in spring, summer and autumn. In summer and autumn variation of bacterial abundance could be explained also by the variability of HN abundance, in fact the densities were significantly correlated and the ratios were very high. In winter the bacterial/HN abundance ratio reached a minimum value due to a high abundance of HN

Table 4. Ratio between bacterial and heterotrophic nanoplankton abundances (*B/HN*), correlation coefficient (*r*) between bacterial abundance and chlorophyll a cocentration and heterotrophic nanoplankton density (data are reported as mean ± standard deviation of *n* determination

	B/HN			Chl a			Autumn		
	Mean	SD	*n*	*r*	*n*	*p*	*r*	*n*	*p*
Spring	582	826	18	0.841	39	0.01	−0.022	18	–
Summer	1370	735	16	0.580	47	0.01	0.242	16	–
Autumn	813	1079	15	0.528	42	0.01	−0.187	15	–
Winter	3023	1559	15	−0.093	39	–	−0.577	15	0.01
Transect E									
Spring	707	531	18	−0.131	48	–	0.576	18	0.01
Summer	2998	1448	19	0.148	47	–	0.464	19	0.05
Autumn	1718	1167	14	0.187	48	–	−0.537	14	0.05
Winter	1312	549	22	0.235	59	–	0.086	22	–
Transect H									
Spring	1037	1510	28	0.062	56	–	−0.391	28	0.05
Summer	2264	1118	28	0.581	57	0.01	0.392	28	0.05
Autumn	1845	1253	27	0.556	54	0.01	0.107	27	–
Winter	1370	712	29	0.365	59	0.01	0.376	29	0.05
Transect M									
Spring	1953	2040	34	0.594	61	0.01	0	34	–
Summer	3706	2020	34	0.227	60	0.05	0.609	34	0.01
Autumn	2309	1394	35	0.701	59	0.01	0.769	35	0.01
Winter	887	916	34	0.024	52	–	−0.041	34	–

Discussion and Conclusion

This study offers, for the first time, a picture of bacterial and HN abundance in the whole Adriatic Sea. Previous information on bacterial distribution in the Adriatic Sea is very scarce and representative only of specific situations. Turk et al. (1990) and Del Negro et al. (1995) reported data on Gulf of Trieste, Fuks et al. (1991) and Puddu et al. (1998) described seasonal and spatial distribution of bacterial biomass along the northern Adriatic Sea while Krstulovic et al. (1998) evaluated bacterial densities in the central basin. Few data were available about southern basins (Bregant et al. 1992; La Ferla et al. 1993; Bregant et al. 1994). The HN distribution was studied only in coastal areas of the northern basins (Turk et al. 1990; Solic and Krstulovic 1995).

Regarding the current knowledge of the regulatory controls of bacterial development there are only few studies along the Adriatic Sea (Turk et al. 1990; Solic and Krstulovic 1995; Krstulovic et al. 1998) evidencing a strong top-down control during the warmer period of the year particularly in eutrophic areas.

From our results, bacterial abundance appeared significantly correlated with chlorophyll *a* in the northern basin in spring, summer and autumn suggesting an important role exerted by resources derived from photosynthetic production as reported for autotrophic systems (Bird and Kalff 1984). The phytoplankton release of dissolved organic carbon is considered as one of the primary sources for bacterial growth. In winter even if the availability of organic resource decreases due to photolimitation affecting primary producers (Harding et al. 1999), bacterial abundance was not controlled by nanoflagellate grazing. The bacterial/HN abundance ratio reached the maximum value due to HN abundance minimum. The temperature, in fact, seems to affect protozoan community (Solic and Krstulovic 1994).

In the central basin the variability of bacterial community could be explained by predator-prey dependence confirming the observation of Krstulovic et al. (1998) for the Kastela Bay. In spring and summer, in fact, a positive correlation with HN abundance was observed.

In the southern Adriatic basin HN influenced the bacterial community in summer and winter. In summer, autumn and winter also the organic matter availability, deriving from phytoplankton, played an important role in the regulation of bacterial densities.

In the Strait of Otranto a positive correlation between bacteria and HN were detected in summer and autumn while the chlorophyll *a* concentration was correlated with bacterial density in spring, summer and autumn. A correlation between bacterial and HN abundance was not always established along the Adriatic basin; in fact this has not always been noted in field data (Gasol and Vaqué 1993). A lack of correlation can be caused by different factors like utilization of alternative preys by HN (Solic and Krstulovic 1995) as photosynthetic picoplankton and particularly cyanobacteria (Caron et al. 1991). The abundance ratio was in accordance with 2-3 orders of magnitude as a general difference between bacterial and HN abundance (Fenchel 1986; Sanders et al. 1992). In all the basins the ratio showed the same trend: values increased from spring to summer and then progressively decreased. Only in the northern Adriatic the bacterial/HN abundance ratio reached the maximum value in winter.

In conclusion we can hypothesize that in the Northern Adriatic Sea the microbial loop structure was clearly influenced by phytoplankton as producer of DOM as prey for HN. In the central and southern basins as well as in the Otranto Strait HN abundance influenced bacterial community development in spring-summer, summer-winter and summer-autumn periods respectively, according to Solic and Krstulovic (1995). In the southern basins the bacterial community seemed also to be influenced by organic substrate availability. In these areas the lack of nutrients (Gacic et al. 1999) could favour a trophic web consisting of small producer cells as picophytoplankton, which is more efficient in the utilization of the resources in oligotrophic systems (Thingstad and Sackshaug 1990) and grazed efficiently by HN as alternative prey (Caron et al. 1991).

The general increase in the bacterial/HN abundance ratio from eutrophic to oligotrophic situations is consistent with the hypothesis of Sanders et al. (1992) that bacteria should be more strongly regulated by predation in eutrophic systems. On the other hand, the approach used in this study did not consider factors of top-down control, mainly viral lysis and grazing pressure exerted by ciliates, tintinnids and mesozooplanktonic organisms.

Acknowledgements. This research was supported by the Italian Ministry of University, Scientific Research and Technology, the National Council of Research and the Central Institute for Applied Marine Research within the project PRISMA 1 (Programma di RIcerca e Sperimentazione per il Mare Adriatico). We thank the crew members of the R. V. Urania for their technical assistance on board. We also thank V.Tirelli for her helpful comments. Finally, we acknowledge two anonymous reviewers for their constructive criticisms on the previous version of the manuscript.

References

Alldredge AL, Passow U, Logan B (1993) The abundance and significance of a class of large, transparent organic particles in the ocean. Deep - Sea Res I 40: 1131-1140

Berninger UG, Finlay BJ, Kuuppo-Leinikki P (1991) Protozoan control of bacterial abundances in freshwater. Limnol Ocenogr 36: 139-147

Billen G, Servais P, Becquevort S (1990) Dynamics of bacterioplankton in oligotrophic and eutrophic aquatic environments: bottom-up or top-down control? Hydrobiologia 207: 37-42

Bird DE, Kalff J (1984) Empirical relationships between bacterial abundance and chlorophyll concentration in fresh and marine waters. Can J Fish Aquat Sci 41: 1015-1023

Bregant D, Allegra A, Azzaro F, Civitarese G, Crisafi E, La Ferla R, Luchetta A, Rabitti S (1992) Condizioni idrologiche nell'Adriatico Meridionale. Aprile 1990. Atti IX Congr AIOL, Nov 1990, pp 25-33

Bregant D, Azzaro F, Civitarese G, Crisafi E, La Ferla R, Leonardi M, Luchetta A, Polimeni R, Raicich F (1994) Condizioni idrologiche nell'Adriatico Meridionale. Novembre 1991. Atti X Congr AIOL, Nov 1992, pp 37-46

Caron DA, Lim EL, Miceli G, Waterbury JB, Valois FW (1991) Grazing and utilization of chrococcoid cyanobacteria and heterotrophic bacteria by protozoa in laboratory cultures and a coastal plankton community. Mar Ecol Prog Ser 76: 205-217

Cole JJ, Pace ML, Caraco NF, Seinhart GS (1993) Bacterial biomass and cell size distributions in lakes: more and larger cells in anoxic waters. Limnol Oceanogr 38: 1627-1632

Del Negro P, Ramani P, Martecchini E, Celio M (1995) Distribuzione annuale del picoplancton in una stazione costiera del Golfo di Trieste. Atti XI Congr AIOL, Oct 1994, pp 747-756

Dufour PH, Torréton J-P (1996) Bottom-up and top-down control of bacterioplankton from eutrophic to oligotrophic sites in the tropical northeastern Atlantic Ocean. Deep-Sea Res 43: 1305-1320

Fenchel T (1986). Ecology of heterotrophic microflagellates. Adv Microb Ecol 9: 57-95

Fuhrman JA (1999) Marine viruses and their biogeochemical and ecological effects. Nature 399: 541-548

Fuks D, Devescovi M, Precali R, Krstulovic N, Solic M (1991) Bacterial abundance and activity in the highly stratified estuary of the Krka River. Mar Chem 32: 333-346

Gacic M, Civitarese G, Ursella L (1999) Spatial and seasonal variability of water and biogeochemical fluxes in the Adriatic Sea. In: Malanotte-Rizzoli P, Eremeev VN (eds) The Eastern Mediterranean as a laboratory basin for assessment of contrasting ecosystems. Kluwer Acad Publ, Dordrecht, pp 335-357

Gasol JM (1994) A framework for the assessment of top-down versus bottom-up control of heterotrophic nanoflagellate abundance. Mar Ecol Prog Ser 113: 291-300

Gasol JM, Vaqué D (1993) Lack of coupling between heterotrophic nanoflagellates and bacteria: a general phenomenon across aquatic systems? Limnol Oceanogr 38: 657-665

Harding LW, Degobbis D, Precali R (1999) Production and fate of phytoplankton: annual cycles and interannual variability. In: Malone TC, Malej A, Harding LW, Smodlaka N, Turner RE (eds) Ecosystems at the land-sea margin: drainage basin to coastal sea. Coast Estuar Stud Am Geophys Union, Washington DC, 55, pp 131-172

Krstulovic N, Solic M, Marasovic I (1998) Regulation of bacterial abundance along the trophic gradient in the central Adriatic. Rapp Comm Int Mer Mediterr 35: 360-361

La Ferla R, Monticelli LS, Giacobbe MG, Crisafi E (1993) Microbial aspects of the Southern Adriatic Sea: present knowledge and future prospects. In: Workshop Ric Mar Mediterr 2000 Oceano. Roma, Oct 1993

Martinussen I, Thingstad TF (1991) A simple double staining technique for simultaneous quantification of auto- and heterotrophic nano- and picoplankton. Mar Microb Food Webs 5: 5-11

Murphy LS, Haugen EM (1985) The distribution and abundance of phototrophic ultraplankton in the North Atlantic. Limnol Oceanogr 30: 47-58

Pace ML, Cole JJ (1994) Comparative and experimental approaches to top-down and bottom-up regulation of bacteria. Microb Ecol 28: 181-193

Porter KG, Feig YS (1980) The use of DAPI to identifying and counting aquatic microflora. Limnol Oceanogr 25: 943-948

Puddu A, La Ferla R, Allegra A, Bacci C, Lopez M, Oliva F, Pierotti C (1998) Seasonal and spatial distribution of bacterial production and biomass along a salinity gradient (Northern Adriatic Sea). Hydrobiologia 369: 271-282

Sanders RW, Caron DA, Berninger U-G (1992) Relationships between bacteria and heterotrophic nanoplankton in marine and freshwaters: an inter-ecosystem comparison. Mar Ecol Prog Ser 86: 1-14

Simon M, Cho BC, Azam F (1992) Significance of bacterial biomass in lakes and the ocean: comparison to phytoplankton biomass and biogeochemical implications. Mar Ecol Prog Ser 86: 103-110

Smodlaka N, Malone TC, Malej A, Harding LH (1999) Introduction. In: Malone TC, Malej A, Harding LW, Smodlaka N, Turner RE (eds) Ecosystems at the Land-Sea Margin: drainage basin to coastal sea. Coast Estuarine Stud Am Geophysi Union, Washington DC, 55: 1-6

Solic M, Krstulovic N (1994) Role of predation in controlling bacterial and heterotrophic nanoflagellate standing stocks in the coastal Adriatic sea: seasonal patterns. Mar Ecol Prog Ser 114: 219-235

Solic M, Krstulovic N (1995) Bacterial carbon flux through the microbial loop in Kastela Bay (Adriatic Sea). Ophelia 41: 345-360

Thingstad TF, Sackshaug E (1990) Control of phytoplankton growth in nutrient recycling ecosystems: theory and terminology. Mar Ecol Prog Ser 63: 261-272

Turk V, Lipej L, Malej A (1990) Heterotrophic plankton dynamic in the stratified water-column in the Gulf of Trieste. Rapp Comm Int Mer Mediterr 32: 216

Weinbauer MG, Höfle MG (1998) Significance of viral lysis and flagellate grazing as factors controlling bacterioplankton production in a eutrophic lake. Appl Environ Microbiol 64: 431-438

Enzymatic Activities and Carbon Flux through the Microbial Compartment in the Adriatic Sea

R. La Ferla, R. Zaccone, G. Caruso, and M. Azzaro

ABSTRACT

Carbon flux through the microbial community by the determination of biomass, heterotrophic bacteria, aminopeptidase and respiratory activities has been studied in two areas of the Adriatic sea with different trophic characteristics during four oceanographic surveys, carried out in June 96,97 and February 97,98. In front of the Po delta (area A), the average rates of the carbon released by aminopeptidase activity ranged from 4.9 to 9.9 µg C l^{-1} h^{-1} and near the Ancona coast (area B) from 3.1 to 7.6µg C l^{-1} h^{-1}, whereas the microbial respiration as metabolic carbon production (CO_2) ranged from 0.19 to 2.29 and from 0.24 to 1.40µg C l^{-1} h^{-1} in the two areas, respectively.

In area A, both the microbial activities showed a seasonal trend with higher values in summer than in winter. In area B, respiration increased during years whilst aminopeptidase activity decreased; bacterioplankton abundance increased on average in the second year of the research (June 97 and February 98). However, all these differences were not statistically significant. The biovolume resulted always significantly higher during summer.

Relationships between bacterial carbon production and organic carbon content were discussed. Bacterial growth efficiency ranged from 17 to 38% and from 13 to 44% in areas A and B, respectively. The relationship between hydrolysis of peptides and bacterial carbon production indicated a good coupling of processes during June 97. Various ratios between bacterial activities were reckoned with the aim to define the role of bacteria in the biogeochemical flux of C in the Northern Adriatic Sea. The C transfer from the biotic versus the abiotic compartment seems to flow better in summer, particularly in area A.

Introduction

The Northern Adriatic Sea represents a semi-enclosed basin in which strong river runoff occurs causing important organic matter load and nutrient enrichment. Moreover, the peculiar hydrological characteristics of the environment consist of shallow depths, occurrence of frontal systems, strong wind mediated winter mixing and summer stratification (Artegiani et al. 1997a, b). As a consequence, this peculiar environment often suffers distrophic crisis and occurrence of mucilagenous phenomena, particularly in summer, with dramatic implications on the ecology of the whole Adriatic Sea.

In aquatic ecosystems, the role of bacterial community in the biogeochemical fluxes is very important. In fact, the C flux is mainly ruled by bacterial activity that, by means of extracellular hydrolytic enzymes, transforms organic macromolecules into soluble products so that the resulting monomers can be widely available for incorporation into bacterial biomass. If hydrolysis and uptake are coupled, monomers liberated by hydrolysis will be taken up. So enzymatic hydrolysis plays a key role in the initial step of the carbon flux in aquatic ecosystems (Azam et al. 1993). The fate of such biomass could be the export to higher trophic levels or the sinking through the water column towards the bottom. The following metabolic oxidation of biomass by respiration liberates carbon dioxide newly available for autotrophic production (Heip et al. 1995; Del Giorgio et al. 1997).

Istituto Sperimentale Talassografico, CNR, Spianata S. Raineri, 98122 Messina, Italy

F.M. Faranda, L. Guglielmo, G. Spezie (eds)
Mediterranean Ecosystems: Structures and Processes

In the framework of the Prisma 2 research programme, subproject: biogeochemical cycles, bacterioplankton abundance, heterotrophic bacteria, aminopeptidase and respiratory activities have been studied in the Northern Adriatic Sea. The aim of this paper was two-fold: to describe the changes of microbial biomass and activities and to quantify the C fluxes through the microbial community during a two-year study period. Previous research had shown that the hydrological features of the Northern Adriatic Sea do not have a strong influence on the bacterial abundance but affect both the bacterial composition and activities with consequences on the trophic web. In fact, near the coast in low salinity and high nutrient waters, the bacterial production was very high. Moreover, in productive periods, the bacterial carbon demand was supported by phytoplankton production while, in less productive periods, it was sustained by external carbon sources (Puddu et al. 1998). In conclusion, bacteria can utilise dissolved organic carbon of autochthonous origin, derived from primary production, or of allochthonous origin from river inputs (Ducklow and Carlson 1992).

Materials and Methods

During 1996-1998, four oceanographic cruises were performed in June (96; 97) and February (97; 98) on board the R/V Urania in two areas of the Adriatic Sea: area A, located in the Northern basin, in front of the Po river delta and area B, in the central basin facing the city of Ancona. A total number of 120 (in the area A) and 116 (in the area B) sea water samples were collected at different depths at the stations indicated in Fig. 1.

Analyses of extracellular enzymatic activity (leucine aminopeptidase, AMP), as indicator of peptide hydrolysis, were performed immediately after sampling according to the method proposed by Hoppe (1983). The stock solution (5mM) of Leu-MCA substrate (L-leucine -4-methyl-coumarinylamid hydrochloride, Sigma) was added to triplicate water samples yielding final concentrations of 10, 25, 50, 100, 200 µM. Increase of fluorescence was measured immediately and after 3 h of incubation at *in situ* temperature. Boiled samples were used as a blank. A calibration curve of MCA standard (Sigma) was performed in prefiltered and autoclaved sea water;

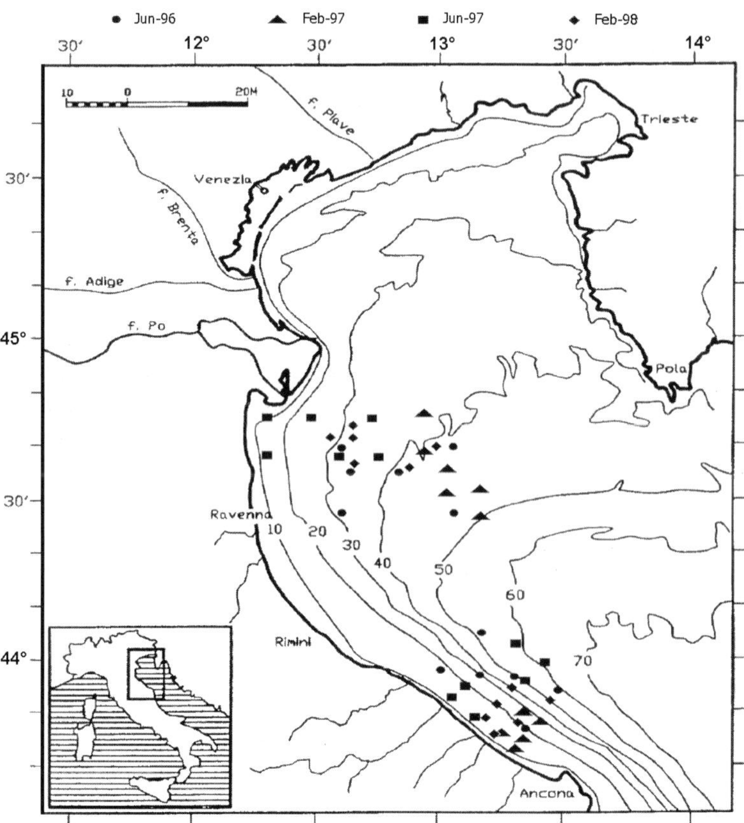

Fig. 1. Location of the sampling sites

fluorescence was measured with a Hitachi spectrofluorometer at 380 and 440 nm excitation and emission wavelength, respectively. Velocity of hydrolysis (v, µg C l^{-1} h^{-1}) for each concentration of substrate was calculated, in terms of C content, according to Hoppe (1993). The Vmax (maximum velocity of hydrolysis) and Km (Michaelis constant) were obtained by plotting 1/[S] versus 1/v using the Michaelis-Menten equation rearranged in a linear form: Lineweaver-Burk transformation (Michal 1983).

The respiration rates and the consequent metabolic production of CO_2 were detected measuring the electron transport system activity (ETS). The method is based on the conversion of tetrazolium salt in formazan (Packard 1971) and details of the procedure are quoted by La Ferla et al. (1996). Results are reported as Vmax and converted into C by using a respiratory quotient (R.Q.) of 1 (Vosjan and Nieuwland 1987). This choice was done to better compare respiratory rates with peptidase activity data.

Bacterial carbon biomass (BB) was calculated using a factor of 20 fgC/cell (Ducklow and Carlson 1992) on estimates of bacterial abundance determined by DAPI staining method (Porter and Feig 1980). In area B, bacterioplankton cell volumes and morphotypes were determined by the microphotographic technique, using AO staining, according to La Ferla et al. (1998).

Heterotrophic viable bacteria counts (MA) were determined on Marine Agar 2216 (DIFCO), by spread plate and incubation at 22°C for 7 days (Zaccone et al. 1998).

Respiration, aminopeptidase and bacterial biomass data have been statistically analysed by ANOVA to test the data means for differences and by Pearson correlation coefficient to measure the associations between pairs of variables.

In this paper, the concept of "potential activity" is taken into account because the rates are expressed as Vmax.

Results

Ranges and mean values with standard deviation of the aminopeptidase activity (AMP), respiratory carbon production (R), bacterial biomass (BB) and viable heterotrophic bacteria (MA) together with salinity and temperature, determined in the two areas, are reported in Table 1. In the area A, AMP increased in summer (1.5 times with respect to winter), reaching maximum in June 96 (mean 9.97±10.72 µg C l^{-1} h^{-1}) and decreased in winter (minimum in February 98, mean 4.86±8.60 µg C l^{-1} h^{-1}). R also showed seasonal changes since, in June 96, it was 4.5 fold higher than February 97 and in June 97, 6 fold higher than February 98. Maximum occurred in June 97 (mean 2.29±3.20 µg C l^{-1} h^{-1}) and minimum in February 97 (mean 0.19±0.11 µg C l^{-1} h^{-1}). Bacterial biomass (BB) derived from DAPI direct counts showed increasing values during the whole research, from June 96 to February 98 (BB mean range 9.22–43.90 µg C l^{-1}). The MA ranged from 1.3 to 0.4×10^3 CFU/ml in February 97 and 98, respectively.

In area B, all the studied parameters showed lower values with respect to area A, without any relationships with seasons. Interannual differences were observed, with decreasing averaged values of AMP from June 96 to June 97 (2.4 fold) and from February 97 to February 98 (1.6 fold) (Table 1). Increases in R were observed from June 96 to June 97 (3.8 times) and from February 97 to February 98 (1.8 times); also BB increased (1.2 times both in summer and winter). The MA showed higher values in winter and lower in summer (6 and 2.3 fold, respectively).

The statistical analysis with ANOVA did not reveal any significant differences for the studied parameters between areas, seasons and years.

Mean biovolume changed between warm and cold periods as reported in Table 2, with larger cell volumes in summer (0.155 and 0.138 µm^3 in June 96 and 97, respectively) and smaller in winter (0.048 and 0.021 µm^3, in February 97 and 98, respectively). Significant differences were registered by ANOVA test comparing summer to winter, and winter months to each other but no difference resulted between the summer months. Bacterial morphotypes also changed in different seasons with the predominance of bacilli and vibrios, together with spirilli, in summer while, during winter, cocci prevailed.

Since most microbial activity was confined to the upper 10 m (Zaccone et al. 1999), Fig. 2 reports the integrated values of AMP and R calculated in this depth range. In area A, AMP values generally showed a negative gradient from near-shore to offshore; this decrease was not observed for R. Overall, the AMP potentially hydrolysed 108.9 mg C m^{-2} h^{-1} in warm months and 54.6 mg C m^{-2} h^{-1} in cold months. The

Table 1. Ranges, mean values and standard deviations (*SD*) of aminopeptidase activity (*AMP*), respiratory activity (*R*), bacterial biomass (*BB*) and heterotrophic bacteria (*MA*) together with temperature and salinity in areas A and B

		Area A				Area B			
		n	Range	Mean	SD	*n*	Range	Mean	SD
AMP (µgC/l/h)	Jun '96	30	0.85-44.25	9.97	10.72	32	0.14-27.27	7.60	9.06
	Feb '97	30	0.24-37.59	6.66	8.85	23	0.03-28.82	6.00	9.01
	Jun '97	30	0.61-59.45	8.29	14.59	30	0.49-8.02	3.09	2.64
	Feb '98	30	0.51-31.08	4.86	8.60	27	0.35-11.16	3.66	4.09
R (µgC/l/h)	Jun '96	30	0.13-3.82	0.99	0.84	32	0.09-1.01	0.37	0.25
	Feb '97	30	0.03-0.44	0.19	0.11	23	0.06-0.63	0.24	0.15
	Jun '97	30	0.19-17.40	2.29	3.20	30	0.20-4.03	1.40	1.03
	Feb '98	30	0.01-1.33	0.38	0.28	27	0.08-1.28	0.45	0.32
BB (µgC/l)	Jun '96	30	2.93-21.30	9.22	5.22	32	1.04-15.50	7.97	3.04
	Feb '97	30	6.83-37.40	14.80	6.05	23	2.20-41.20	7.72	8.55
	Jun '97	28	9.76-37.60	22.30	7.69	30	3.00-18.40	9.32	4.16
	Feb '98	30	23.80-279	43.90	45.60	27	4.60-15.50	9.74	2.69
MA (CFU/ml)	Jun '96	30	60-11500	1200	2048	32	100-3132	820	673
	Feb '97	30	45-4100	1329	1433	23	40-26500	5098	7785
	Jun '97	30	40-2717	874	687	30	100-1045	461	224
	Feb '98	30	55-1190	387	284	27	40-4950	1061	1190
T°C	Jun '96	30	9.71-20.57	17.54	5.58	36	9.79-21.55	18.28	3.49
	Feb '97	30	9.33-11.61	10.45	0.60	23	7.80-11.72	10.64	1.13
	Jun '97	30	12.08-24.67	17.54	4.32	28	12.59-21.87	17.76	3.59
	Feb '98	25	8.41-10.99	9.84	0.97	26	7.05-11.99	10.14	1.86
S	Jun '96	30	34.15-38.10	36.74	1.49	36	35.18-38.22	36.70	1.03
	Feb '97	30	31.56-37.94	36.65	1.59	23	32.76-37.94	37.04	1.47
	Jun '97	30	23.07-37.94	35.47	2.89	30	35.91-38.01	37.19	0.62
	Feb '98	30	31.81-38.33	37.05	1.77	27	33.69-38.55	37.13	1.82

Table 2. Bacterioplankton cell volumes, biovolumes and morphotypes in the different seasons

	June '96	Feb '97	June '97	Feb '98
Cell volumes (µm^3)	0.155	0.048	0.138	0.021
Biovolume/ml	0.165	0.070	0.234	0.091
Morphotypes (%)				
Cocci	37.54	60.07	25.91	79.38
Bacilli	46.84	35.91	47.87	15.98
Vibrios	9.97	3.02	21.04	4.12
Spririllae	5.65	1.01	5.18	0.52

organic carbon oxidised by respiration was equivalent to 20.0 mg C m^{-2} h^{-1} in warm periods and decreased 7-fold in cold periods (2.9 mg C m^{-2} h^{-1}).

In area B, the distribution of AMP showed again this trend but, in June 96, high activities were also observed at the bottom, probably due to down-welled fresh water input (salinity at bottom 35.2-35.7). In area B, R showed the coast-offshore decreasing gradient only in June 97. The amount of organic carbon potentially hydrolysed by AMP in both the seasons resulted similar (40.2 and 42.8 mg C m^{-2} h^{-1}, in warm and cold periods, respectively). Conversely, the organic carbon oxidised decreased about 3-fold from 9.7 mg C m^{-2} h^{-1} in warm months to 3.5 mg C m^{-2} h^{-1} in cold months.

Pearson correlation coefficients computed among the different examined parameters in the whole water column are reported in Table 3. Generally AMP and R were significantly correlated, with the only exception of February 97 in area A. The BB, MA, AMP and R showed instead different correlations with other parameters in the two study areas. In area A, AMP and MA showed most correlation both with hydrological and metabolic parameters, as well as with the organic C pool, as opposed to BB, which was seldom related with other parameters. The R showed a good level of correlation with all other parameters in June 97 and February 98.

In area B, R, AMP and MA were nearly always significantly correlated with each other and with a considerable number of examined parameters during the whole study period. Conversely to that observed for area A, in area B, BB was highly correlated with a large number of parameters.

In both areas A and B, AMP and R resulted negatively correlated with salinity while, if present, correlations with temperature were negative in cold periods and positive in warm periods.

Discussion

No data on respiratory and aminopeptidase activities are available yet in the Adriatic Sea. Only Karner et al. (1992) measured aminopeptidase activity in surface samples in spring (range 11.8-123 μg C l^{-1} h^{-1}) and summer (21.4-9.6 μg C l^{-1} h^{-1}), along a trophic gradient in the eastern coast of the Northern Adriatic basin; these values are in accordance with our data in the western part of the Northern Adriatic Sea. In the superficial layers of the Mediterranean Sea various authors have studied representative specific situations. In the Northern Adriatic basin, aminopeptidase activity reached the highest levels (7.45 and 5.09 μg C l^{-1} h^{-1}, area A and B, respectively) with respect to the up-welling area of the Straits of Messina (6.36 μg C l^{-1} h^{-1}; G. Caruso, personal communication) and the Liguro-Provencal basin (1.44 μg C l^{-1} h^{-1}; Karner and Rassoulzadegan 1995). The lowest values were determined in the Sicily Channel near the Egadi Islands (0.03 μg C l^{-1} h^{-1}, Caruso et al. 1998).

Respiration also reached the highest values in the Northern Adriatic Sea. Respiratory activity in the surface layers of the Mediterranean generally decreased from west to east, from 0.94 μg C l^{-1} h^{-1} in the Western Mediterranean Sea (Martinez et al. 1990) to 0.19 μg C l^{-1} h^{-1} in the Levantine basin (Azzaro 1997) as well as from north to south, from 0.96 and 0.62 μg C l^{-1} h^{-1} in the Northern Adriatic Sea (areas A and B, respectively) to 0.28 μg C l^{-1} h^{-1} in the Otranto Straits (La Ferla et al.

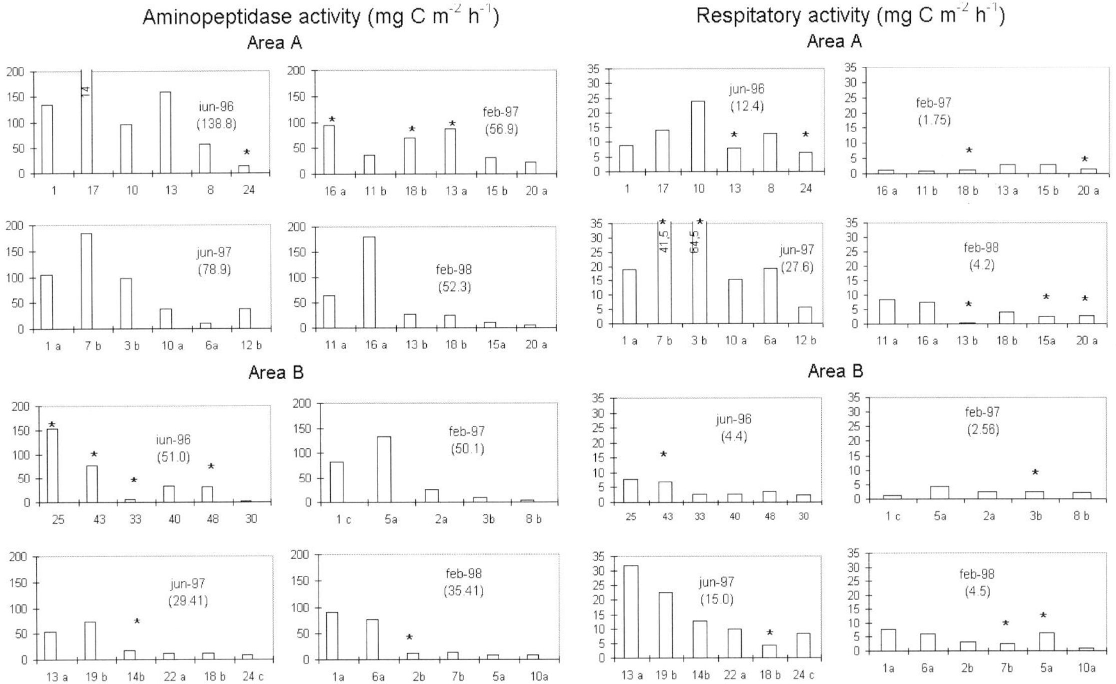

Fig. 2. Distribution of AMP and R integrated values in the layer 0-10 m; stations from coast to offshore on x-axis and mean integrated values in brackets are reported; in the stations indicated with an asterisk, there is a tendency to increasing R and AMP towards the bottom

Table 3. Pearson correlation coefficients of R, AMP, BB and MA vs. other parameters (*ND*, not determined)

	Area A				Area B			
	June '96	**Feb '97**	**June '97**	**Feb '98**	**June '96**	**Feb '97**	**June '97**	**Feb '98**
R *versus*:								
T	0.44*		0.47**	−0.52**		−0.5		−0.74**
S	−0.58*		−0.65**	−0.69**	−0.62**		−0.78**	−0.61**
DOC	0.39*			0.66**			0.69**	0.78**
POC			0.74**	ND		0.61**	0.86**	ND
BB					0.35*	0.54**	0.72**	0.50**
MA			0.61*		0.79**	0.60**		0.61**
CHL			0.83**	0.67**	0.49**	0.50*	0.79**	0.78**
AMP	0.39*		0.55**	0.67**	0.37*	0.62**	0.80**	0.79**
BCP			0.53**	0.73**	0.77**	0.54**		0.79**
AMP *versus*:								
T	0.49**		0.61**	−0.65**		−0.90		−0.93**
S	−0.75**	−0.42*	−0.94**	−0.95**	−0.48**	−0.85**	−0.56**	−0.89**
DOC	0.47*	0.81**	0.67**	0.73**			0.56**	0.87**
POC		0.55**	0.94**	ND		0.93**	0.91**	ND
BB						0.57**	0.76**	0.70**
MA			0.63**	0.49**	0.48**	0.90**		0.71**
CHL	0.49**	0.68**	0.83**	0.87**	0.51**	0.71**	0.84**	0.99**
R	0.39*		0.55**	0.67**	0.37*	0.62**	0.80**	0.79**
BCP	0.75**	0.47**	0.99**	0.98**	0.51**	0.98**		0.98**
BB *versus*:								
T				−0.53**		−0.56**	−0.57**	−0.76**
S				−0.71**		−0.46*		−0.68**
DOC								0.74**
POC				ND			0.73**	ND
MA						0.56**		0.53**
CHL						0.79**	0.62**	0.72**
R					0.35*	0.54**	0.72**	0.50**
AMP						0.57**	0.76**	0.70**
BCP		0.42*			0.41*	0.52*		0.67**
MA *versus*:								
T		−0.55**		−0.44*	0.49**	0.74**		0.64**
S	0.50**	−0.60**	−0.66**	−0.46*	0.77**	0.77**		0.55**
DOC		0.61**	0.51**				0.56**	
POC	0.52**	0.52**	0.71**					
BB						0.56**		0.53**
CHL								
R			0.61**		0.79**	0.60**		0.61**
AMP			0.63**	0.49**	0.48**	0.90**		0.71**
BCP	0.49**	0.42*	0.61**	0.45*	0.75**	0.88**		0.77**

* $P < 0.05$; ** $P < 0.01$

1999). All respiratory rates were reckoned using a R.Q. 1 better to compare them to our data.

Bacterial abundance was higher in area A than in area B and, on the whole, was lower than those expected from an estuarine area (Heip et al. 1995), but confirmed previous data reported for the Northern Adriatic Sea by Puddu et al. (1998). Estimates of biovolumes, instead, showed the occurrence of smaller or larger bacterial cells which alternated in cold and warm periods. Previous investigations in the Adriatic Sea had shown no seasonal differences in the bacterial abundance, whereas greater variations had been observed in the bacterial structure - cell size and genera composition - and activity (Zaccone and Caruso 1996; La Ferla et al. 1998; Puddu et al. 1998; Zaccone et al. 1998). The succession of different microbial communities may reasonably be associated to different trophic conditions since smaller size bacteria normally inhabit low nutrient waters while larger size bacteria live in enriched environments (Fuhrmann et al. 1980; Holligan et al. 1984). Moreover, larger bacteria generally live attached to the particulate organic matter while small free living bacteria depend on dissolved organic matter (Unanue et al. 1992). The viable fraction of heterotrophic bacteria on the total bacteria, contributed in a significant way to microbial activity; this population, ranging from 0.02 to 0.31% in area A and from 0.02 to 8% in area B, surely constituted a fraction of active bacteria. However, in marine environments, another fraction of total bacteria - the viable but not cultivable cells - can display some biological activities such as enzyme production, even though it has lost the ability to replicate itself on media (Roszack and Colwell 1987).

Our observations showed the highest microbial activities in the upper water layers, within 10 m depth and sediment trap studies revealed that less than 10% of the particulate organic matter, deriving from primary production, reached the bottom (S. Rabitti, personal communication). As a consequence, we hypothesised that the organic matter deriving from phytoplankton, inducting enzyme production, constituted an available source of carbon for the bacteria metabolism (Middelboe et al. 1995). The correlations with POC, DOC and Chl supported this hypothesis and the viable fraction of heterotrophic bacteria contributed to decomposition of organic compounds by their enzymatic spectrum (Zaccone et al. 1998). However, also the fresh water supply assumed a predominant role influencing the microbial activities and distribution as underlined by the numerous negative correlations with salinity. In conclusion, the course of biological processes is affected both by coupling between production and mineralisation and by salinity gradients (Naush et al. 1998; Puddu et al. 1998). The analysis of AMP and R integrated data in the first 10 m has underlined that twice the quantity of organic carbon is on average hydrolysed and oxidised in area A with respect to area B. In fact the integrated average AMP value calculated was 81.7 e 41.5 mg C m^{-2} h^{-1} for areas A and B, respectively; the integrated average value of respiratory activity was 11.5 and 6.6 mg C m^{-2} h^{-1} in areas A and B, respectively.

With the aim of estimating the microbial processes coupling, the AMP/BCP ratio was calculated on the averaged values (Fig. 3). In area A, the ratio (>45) indicated a higher production of enzymes (AMP) with respect to incorporation rates (BCP) in June 96, Feb. 97, Feb. 98. In area B, in Feb. 97, the ratio was reduced (<30) indicating a higher coupling of processes. In June 97, in both the areas, the ratio further decreased (<20) and the two processes were more balanced and the C released by enzymatic hydrolysis could be available for new biomass production. Substrate hydrolysis may exceed substrate uptake mainly for particle associated bacteria (Hoppe et al. 1988; Karner and Herndl 1992; Smith et al. 1992, 1995; Middelboe et al. 1995; Unanue et al. 1998).

The quantification of carbon flux into bacteria closely depends on the accuracy of the estimates of the BCP/R ratio which indicates the balance existing between production of new bacterial biomass and consumption by means of oxidative metabolism. Our estimates ranged from 0.2 to 0.6 in area A and from 0.15 to 0.78 in area B (Fig. 3). In Feb. 97, in both the areas, the ratio reached the maximum values suggesting that most of the respiratory activity may be sustained by bacterial production. Zweifel et al. (1995) determined a BCP/R ratio of 0.36 from sea water cultures of bacteria collected from the Northern Baltic Sea; Amon and Benner (1998) determined higher ratios of 0.6-1.7, taking into account BCP calculated from changes in bacterial cell abundance and 2.5-3.7 from bacterial leucine incorporation in samples collected in the Mississippi river plume.

Estimates of bacterial growth efficiency (BGE), as the percentage ratio between BCP and

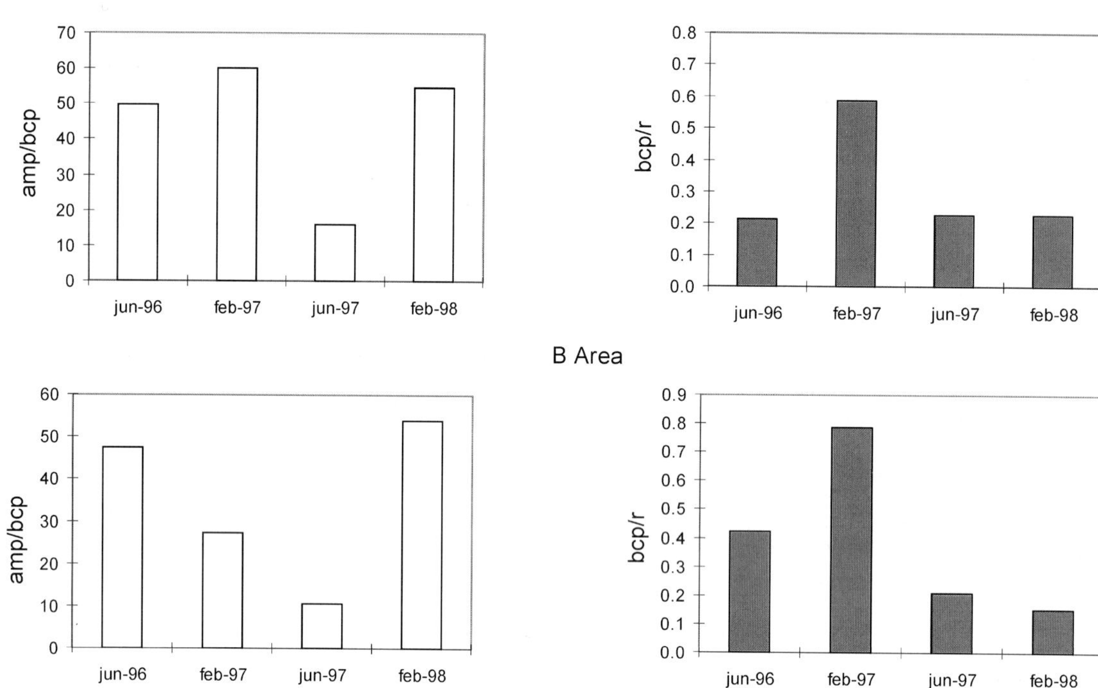

Fig. 3. Histograms showing the AMP/BCP and BCP/R ratios calculated in the different areas and seasons

Table 4. Bacterial growth efficiency (*BGE*) determined in the different areas

	Area A (%)	Area B (%)
Jun '96	17	29
Feb '97	38	44
Jun '97	18	17
Feb '98	18	13

Table 5. Respiratory Turnover Rates (*RTR*) *per* day in the layers between 0 and 10 m

	RTR (%)	
	Area A	Area B
Jun '96	1.47	0.50
Feb '97	0.34	0.32
Jun '97	4.15	2.93
Feb '98	0.91*	1.26*

* *RTR* calculated using DOC values only

the sum of BCP and R, quoted by Del Giorgio et al. (1997), ranged from less than 10 to 25% in the fresh water and marine ecosystems and from 5 to 60% in estuaries. In agreement with this, our estimates varied from 17 to 38% in area A and from 13 to 44% in area B. In both areas the higher values were recorded in February 97 (Table 4). Such an increase may be linked to changes in both the rate of supply and the nutritional quality of organic carbon substrates available for bacterial biomass. The percentage ratio between organic matter content (DOC+POC) and respiration in the layers between 0 and 10 m (Table 5) showed that 0.24-3.60% of organic matter is respired daily in area A and 0.24-2.64% in area B, with highest turnover rates in June 97. In conclusion the C transfer from the biotic to the abiotic compartment through the microbial community seems to flow better in the summer period, when all microbial processes seem to converge towards mineralization.

Acknowledgements. The Authors wish to thank M.Pettine and A.Puddu (IRSA, Roma), S.Rabitti (IBM, Venezia), P. Del Negro (LBM, Trieste), who kindly provided data (DOC, BCP, POC and bacterial abundance in area A, respectively) supporting this research. We are also very grateful to F. Azzaro, L.Monticelli and F. Soraci (Istituto Talassografico, CNR, Messina) and C. Walker (Laboratorio Biologia Marina, Trieste) who participated in sampling cruises. This work was carried out within the frame of the PRISMA 2 programme, financially supported by MURST.

References

Amon RMW, Benner R (1998) Seasonal patterns of bacterial abundance and production in the Mississippi River plume and their importance for the fate of enhanced primary production. Microb Ecol 35: 289-300

Artegiani A, Bregant D, Paschini E, Pinardi N, Raicich F, Russo A (1997a) The Adriatic Sea general circulation. I. Air-sea interactions and water mass structure. J Phys Oceanogr 27: 1497-1514

Artegiani A, Bregant D, Paschini E, Pinardi N, Raicich F, Russo A (1997b) The Adriatic Sea general circulation. II. Baroclinic circulation structure. J Phys Oceanogr 27: 1515-1532

Azam F, Smith DC, Steward GF, Hagstrom A (1993) Bacteria-organic matter coupling and significance for carbon cycling. Microb Ecol 28: 167-179

Azzaro M (1997) Attività respiratoria (ETSa) e biomassa microbica (ATP) nel Mediterraneo Orientale. Tesi laurea, Univ Stud Messina

Caruso G, Leonardi M, Azzaro M, Zaccone R (1998) Microbial activities and organic matter turnover in an oligotrophic environment (Egadi Islands). Biol Mar Mediterr 5: 27-34

Del Giorgio PA, Cole JJ, Cimbleris A (1997) Respiration rates in bacteria exceed phytoplankton production systems. Nature 385: 148-151

Ducklow HW, Carlson CA (1992) Oceanic bacterial production. In: Marshall KC (ed) Advances in Microbial Ecology, vol 12. Plenum Press, New York, pp 113-181

Fuhrman JA, Ammerman JW, Azam F (1980) Bacterioplankton in the coastal euphotic zone: distribution, activity and possible relationships with phytoplankton. Mar Biol 60: 201-207

Heip CHR, Goosen NK, Herman PMJ, Kromkamp J, Middelburg JJ, Soetaert K (1995) Production and consumption of biological particles in temperate tidal estuaries. Oceanogr Mar Biol Annu Rev 33: 1-149

Holligan PM, Harris RP, Newell RC, Harbour DS, Head RN, Linley EAS, Lucas MI, Tranter PRG, Weekly CM (1984) Vertical distribution and partitioning of organic carbon in mixed, frontal and stratified waters of the English Channel. Mar Ecol Prog Ser 14: 111-127

Hoppe HG (1983) Significance of exoenzymatic activities in the ecology of brackish water: measurements by means of methylumbelliferyl-substrates. Mar Ecol Prog Ser 11: 299-308

Hoppe HG (1993) Use of fluorogenic model substrates for extracellular enzyme activity (EEA) measurement of bacteria. In: Kemp PF, Sherr BF, Sherr EB, Cole JJ (eds) Handbook of methods in aquatic microbial ecology. Lewis Publishers, Boca Raton, pp 423-431

Hoppe HG, Kim S-J, Gocke K (1988) Microbial decomposition in aquatic environments: combined process of extracellular enzyme activity and substrate uptake. Appl Environ Microbiol 54: 784-790

Karner M, Herndl GJ (1992) Extracellular enzymatic activity and secondary production in free-living and marine snow associated bacteria. Mar Biol 113: 341-347

Karner M, Kuks D, Herndl GJ (1992) bacterial activity along a trophic gradient. Microb Ecol 24: 243-257

Karner M, Rassoulzadegan F (1995) Extracellular enzyme activity: indications for high short-term variability in a coastal marine ecosystem. Microb Ecol 30: 143-156

La Ferla R, Azzaro M, Chiodo G (1996) Microbial respiratory activity in the euphotic zone of the Mediterranean Sea. Microbiologica 19: 243-250

La Ferla R, Azzaro M, Chiodo G (1999) Microbial respiratory activity and CO_2 production rates in the Otranto Straits (Mediterranean Sea). Aquat Ecol 1:1-9

La Ferla R, Bacci C, Chiodo G, Parrino S, Zoppini A (1998) Stime di abbondanza e biovolume batterioplanctonico nell'Adriatico Settentrionale: confronto tra AO e DAPI. Biol Mar Mediterr 5(1): 750-753

Martinez R, Arnone RA, Velasquez Z (1990) Chlorophyll a and respiratory electron transport activity in microplankton from surface waters of the Western Mediterranean. J Geophys Res 95: 1515-1522

Michal G (1983) Determination of Michaelis constants and inhibitor constants. In: Bergmeyer HU (ed) Methods of enzymatic analysis, 3rd ed, vol 1. Verlag Chemie, Weinheim, pp 86-104

Middelboe M, Sondergaard M, Letarte Y, Borch NH (1995) Attached and free-living bacteria: production and polymer hydrolysis during a diatom bloom. Microb Ecol 29: 231-248

Nausch M, Pollehne F, Kersten E (1998) Extracellular enzyme activities in relation to hydrodynamics in the pomeranian bight (Southern Baltic Sea). Microb Ecol 36: 251-258

Packard TT (1971) The measurement of respiratory electron transport activity in marine phytoplankton. J Mar Res 29: 235-244

Porter KG, Feig YS (1980) The use of DAPI for identifying and counting aquatic microflora. Limnol Oceanogr 25: 943-948

Puddu A, La Ferla R, Allegra A, Bacci C, Lopez M, Oliva F, Pierotti C (1998) Seasonal and spatial distribution of bacterial production and biomass along a salinity gradient (Northern Adriatic Sea). Hydrobiologia 363: 271-282

Roszack DB, Colwell RR (1987) Metabolic activity of bacterial cells enumerated by direct viable count. Appl Environ Microbiol 53: 2889-2893

Smith DC, Simon M, Alldredge AL, Azam F (1992) Intense hydrolytic enzyme activity on marine aggregates and implications for rapid particle dissolution. Nature 359: 139-142

Smith DC, Steward GF, Long RA, Azam F (1995) Bacterial mediation of carbon fluxes during a diatom bloom in a mesocosm. Deep-Sea Res 42: 75-97

Unanue M, Ayo B, Azùa I, Barcina I, Iriberri J (1992) Temporal variability of attached and free living bacteria in coastal waters. Microb Ecol 23: 27-39.

Unanue M, Azùa I, Arrieta JM, Labirua-Iturburu A, Egea L, Iriberri J (1998) Bacterial colonization and ectoenzymatic activity in phytoplankton-derived model particles: cleavage of peptides and uptake of amino acids. Microb Ecol 35:136-146

Vosjan JH, Nieuwland G (1987) Microbial biomass and respiratory activity in the surface waters of the east Banda Sea and Northwest Arafura Sea (Indonesia) at the time of the Southeast monsoon. Limnol Oceanogr 32(3): 767-775

Zaccone R, Caruso G (1996) Stima dell'attività proteasica batterica in Adriatico Settentrionale mediante substrati fluorogenici. SiTE Atti 17: 87-90

Zaccone R, Caruso G, Calì C, Scarfò R (1998) Primi dati sulla caratterizzazione microbiologica delle acque dell'Adriatico Settentrionale . In: Atti XII Congr AIOL vol II, pp 487-497

Zaccone R, La Ferla R, Caruso G, Azzaro M (1999) Attività proteasica e respiratoria microbica in due aree dell'Adriatico Settentrionale. In: Atti XIII Congr AIOL, Portonovo Ancona 28-30 Settembre 1998, 13(1): 33-44

Zweifel UL, Wikner J, Hagstrom A (1995) Dynamic of dissolved organic carbon in a coastal ecosystem. Limnol Oceanogr 40(2): 299-305

Subject Index

Genera and Species Index